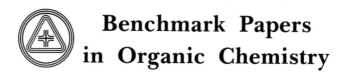 **Benchmark Papers in Organic Chemistry**

Series Editor: Calvin A. VanderWerf
University of Florida

Published Volumes and Volumes in Preparation

SYNTHESIS OF LIFE
    Charles C. Price
SOLID-PHASE SYNTHESIS
    E. C. Blossey and D. C. Neckers
ORGANO-PHOSPHORUS STEREOCHEMISTRY
    William McEwen and K. D. Berlin
MECHANISMS OF ELECTRON TRANSFER PROCESSES
    Raymond E. Dessy and Gary D. Howard
CARBON-13 MAGNETIC RESONANCE SPECTROSCOPY
    Gary E. Maciel
PHOTOCHEMISTRY
    D. C. Neckers
ORGANO-METALLIC CHEMISTRY
    Marvin Rausch and J. J. Zuckerman
FLUORINE CHEMISTRY
    Paul Tarrant, Jean M. Shreeve, and Gene Stump
STEREOSELECTIVE REDUCTIONS
    Michael P. Doyle and C. Thomas West
SINGLET MOLECULAR OXYGEN
    A. P. Schapp
CONFORMATION ANALYSIS
    Norman L. Allinger
ELECTRON PARAMAGNETIC RESONANCE
    Gareth R. Eaton
CHEMISTRY OF VISION
    Edwin W. Abrahamson and Stanford E. Ostroy

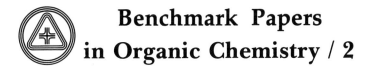

Benchmark Papers in Organic Chemistry / 2

A BENCHMARK® Books Series

SOLID-PHASE SYNTHESIS

Edited by
E. C. BLOSSEY
Rollins College

and

D. C. NECKERS
Bowling Green State University

Dowden, Hutchinson & Ross, Inc.
Stroudsburg, Pennsylvania

Distributed by
HALSTED PRESS *A Division of John Wiley & Sons, Inc.*

Copyright © 1975 by **Dowden, Hutchinson & Ross, Inc.**
Benchmark Papers in Organic Chemistry, Volume 2
Library of Congress Catalog Card Number: 75-1424
ISBN: 0-470-08346-8

All rights reserved. No part of this book covered by the copyrights hereon may be reproduced or transmitted in any form or by any means—graphic, electronic, or mechanical, including photocopying, recording, taping or information storage and retrieval systems—without written permission of the publisher.

77 76 75   1 2 3 4 5

Manufactured in the United States of America.

LIBRARY OF CONGRESS CATALOGING IN PUBLICATION DATA

Blossey, Erich C   comp.
   Solid phase synthesis.

   (Benchmark papers in organic chemistry ; v. 2)
   Bibliography: p.
   Includes indexes.
   1. Chemistry, Organic—Synthesis—Addresses, essays, lectures. 2. Solid state chemistry—Addresses, essays, lectures. I. Neckers, Douglas C., joint comp. II. Title.
QD262.B634   547'.2   75-1424
ISBN 0-470-08346-8

Exclusive Distributor: **Halsted Press**
A Division of John Wiley & Sons, Inc.

Permissions

The following papers have been reprinted with the permission of the authors and the copyright holders.

AMERICAN CHEMICAL SOCIETY
  Analytical Chemistry
    Instrument for Automated Synthesis of Peptides
  Biochemistry
    Reactions of Nucleosides on Polymer Supports: Synthesis of Thymidylylthymidylylthymidine
    Synthesis of Angiotensins by the Solid-Phase Method
  Chemical and Engineering News
    R. Bruce Merrifield—Designer of Protein-Making Machine
  The Journal of Organic Chemistry
    Solvation of the Polymer Matrix: Source of Truncated and Deletion Sequences in Solid Phase Synthesis
  The Journal of the American Chemical Society
    Activation of Homogeneous Catalysts by Polymer Attachment
    p-Alkoxybenzyl Alcohol Resin and *p*-Alkoxybenzyloxycarbonylhydrazide Resin for Solid Phase Synthesis of Protected Peptide Fragments
    Azide Solid Phase Peptide Synthesis
    Carboxyl-Catalyzed Intramolecular Aminolysis: A Side Reaction in Solid-Phase Peptide Synthesis
    Catalytic Reduction of Olefins with Polymer-Supported Rhodium(I) Catalyst
    Cyclization via Solid Phase Synthesis: Unidirectional Dieckmann Products from Solid Phase and Benzyl Triethylcarbinyl Pimelates
    The Directed Mixed Ester Condensation of Two Acids Bound to a Common Polymer Backbone
    Failure Sequences in the Solid Phase Synthesis of Polypeptides
    Investigation of a Synthetic Catalytic System Exhibiting Substrate Selectivity and Competitive Inhibition
    Oligonucleotide Syntheses on Insoluble Polymer Supports: I. Stepwise Synthesis of Trithymidine Diphosphate
    Oligonucleotide Synthesis on a Polymer Support
    Oligosaccharide Synthesis on a Light-Sensitive Solid Support: I. The Polymer and Synthesis of Isomaltose (6-*O*-α-D-Glucopyranosyl-D-glucose)
    Polymer-Based Sensitizers for Photooxidations
    Polymer-Protected Reagents: Polystyrene–Aluminum Chloride
    Popcorn Polymer as a Support in Multistep Syntheses
    Preparation of a *t*-Alkyloxycarbonylhydrazide Resin and Its Application to Solid Phase Peptide Synthesis
    Reactive Species Mutually Isolated on Insoluble Polymeric Carriers: I. The Directed Monoacylation of Esters
    Regulation of Rate of Reaction of Polyuridylic Acid Derivative by Use of Suppressor and Antisuppressor Molecules
    Resin-Bound Transition Metal Complexes
    Retension of Configuration in the Solid Phase Synthesis of Peptides
    Selective Catalysis Involving Reversible Association of a Synthetic Polymeric Catalyst and Substrate
    Selectivity in Solvolyses Catalyzed by Poly-(4-vinylpyridine)
    Solid Phase Synthesis: I. The Synthesis of a Tetrapeptide
    Solid Phase Synthesis of Oligosaccharides: I. Preparation of the Solid Support. Poly [*p*-(1-propen-3-ol-l-yl)styrene]
    Solid-Phase Synthesis of Oligosaccharides: II. Steric Control by C-6 Substituents in Glucoside Synthesis

Stepwise Synthesis of Oligodeoxyribonucleotides on an Insoluble Polymer Support
Studies of Polynucleotides: LXXII. Deoxyribooligonucleotide Synthesis on a Polymer Support
Synthesis of a Hydrophobic Potassium Binding Peptide
Synthesis of a Polypeptide with Lysozyme Activity
A Synthesis of Cyclic Peptides Utilizing High Molecular Weight Carriers
Use of Polymers as Chemical Reagents: I. Preparation of Peptides
Use of Polymers as Chemical Reagents: II. Synthesis of Bradykinin

THE AMERICAN SOCIETY OF BIOLOGICAL CHEMISTS, INC.—*The Journal of Biological Chemistry*
Acyl Carrier Protein
Synthesis of Ribonuclease A

THE CHEMICAL SOCIETY, LONDON
Journal of the Chemical Society
Synthesis of Oligosaccharides on Polymer Supports: I. 6-O-(p-Vinylbenzoyl) Derivatives of Glucopyranose and Their Copolymers with Styrene
Journal of the Chemical Society, Chemical Communications
The Effect of Ring Size on Threading Reactions of Macrocycles
Oligonucleotide Synthesis on Polar Polymer Supports: The Use of a Polypeptide Support
Removal of Protected Peptides from an *ortho*-Nitrobenzyl Resin by Photolysis
Wittig Resins: The Preparation and Application of Insoluble Polymeric Phosphoranes

ELSEVIER PUBLISHING CO.—*Carbohydrate Research*
Solid-Phase Synthesis of Oligosaccharides: III. Preparation of Some Derivatives of Di- and Tri-Saccharides via a Simple Alcoholysis Reaction

INTRA-SCIENCE RESEARCH FOUNDATION—*Intra-Science Chemical Reports*
The Solid Phase Synthesis of Ribonuclease A

MARCEL-DEKKER, INC.—*Journal of Macromolecular Science Chemistry*
Solid Phase Synthesis: Evidence for and Quantification of Intraresin Reactions

MICROFORMS INTERNATIONAL MARKETING CORPORATION—*Tetrahedron Letters*
Hydrogenation, Hydrosilylation and Hydroformylation of Olefins Catalyzed by Polymer-Supported Rhodium Complexes
A New Solid Support for Polypeptide Synthesis
A New Support for Polypeptide Synthesis in Columns
Organic Syntheses with Functionalized Polymers: I. Preparation of Polymeric Substrates and Alkylation of Esters
Organic Syntheses with Functionalized Polymers: II. Wittig Reaction with Polystyryl-*p*-Diphenylphosphoranes
Synthesis of Cyclic Peptides on Dual Function Supports
Unusual Brominations with N-Bromopolymaleimide

PERGAMON PRESS—*Tetrahedron Letters*
Insoluble Resins in Organic Synthesis: I. Preparation and Reactions of Polymeric Anhydrides
Oligonucleotide Synthesis on a Polymer Support Soluble in Water and Pyridine
Polymer Protected Reagents: II. Esterifications with ⓟ-AlCl$_3$
Polymeric Reagents: I. Synthesis of an Insoluble Polymeric Carbodiimide
Polymeric Reagents: II. Preparation of Ketones and Aldehydes Utilizing an Insoluble Carbodiimide
Polymers as Chemical Reagents: The Use of Poly(3,5 Diethylstyrene)—Sulfonyl Chloride for the Synthesis of Internucleotide Bonds
Solid-Phase Synthesis of Oligosaccharides: II. Synthesis of 2-Acetamido-6-O-(2-Acetamido-2-Deoxy-β-D-Glucopyranosyl)-2-Deoxy-D-glucose

SWISS CHEMICAL SOCIETY—*Helvetica Chimica Acta*
The Preparation of Merrifield-Resins Through Total Esterification with Cesium Salts

Series Editor's Preface

"Benchmark Papers in Organic Chemistry" is a series of single-topic volumes that contain direct reproductions of outstanding papers on basic and timely topics. The purpose of the series is to bring together those landmark papers which have advanced our knowledge to its present state of development in the major subject areas of intense contemporary interest and significance. Each volume in the series contains a collection of key papers—those whose contemporary impact, historical significance, and scientific elegance place them in the breakthrough category—chosen by a recognized expert from the worldwide primary literature. In addition to the reprinted papers, which are published in their original form, readers will find an authoritative introduction to the subject which covers its history, present state, and future prospects, as well as specific commentaries that place each of the carefully chosen papers in context and perspective. Subject and author-citation indexes are also included to add to the convenience of each volume.

Ever since 1828, when Friedrich Wöhler first prepared urea from ammonium cyanate, the literature in organic chemistry has been proliferating at an exponential rate. Published in many languages and in a wide variety of journals and reports, much of this literature is not readily available to, or easily assessed by, students, research workers, and interested laymen. Now each Benchmark volume provides the reader an expertly chosen and skillfully annotated collection of the classic seminal papers in a special field which, taken together, delineate the present state of the art and constitute a platform for future research.

Few developments in organic chemistry over the past twenty years promise to be as practically important or theoretically intriguing as that of solid-phase synthesis. Since its invention just over a decade ago, it has proved to be a remarkably useful synthetic tool in the armamentarium of the organic chemist. Solid-phase synthesis has been shown to be the method of choice for one type of reaction after another, and sound experimental and theoretical groundwork to extend the method on a broad basis has now been clearly established.

To bring together for the first time a penetrating survey of the brilliant breakthrough research which provides a springboard for dramatic new developments, we have chosen two experts who are themselves at the forefront of research in the field. Blossey and Neckers have, in *Solid-Phase Synthesis,* made a critical selection and review of the key papers in this revolutionary new field with a view toward pointing the way to future possibilities of uncharted dimensions.

Every organic chemist, and many chemists in other branches of the science, will wish to be thoroughly familiar with the history and potential of this significant and fertile area of contemporary chemistry. Whether your interest is in the experimental or the theoretical aspects—or both—, Blossey and Neckers' *Solid-Phase Synthesis* provides you the ready and direct means for doing just that, between the covers of one carefully edited comprehensive and compact volume. You will find it richly informative and eminently useful.

C. A. VanderWerf

Contents

Permissions	v
Series Editor's Preface	vii
Contents by Author	xvii

Introduction — 1

1 HENAHAN, J. F.: R. Bruce Merrifield—Designer of Protein-Making Machine — 4
Chem. Eng. News, 22–26 (Aug. 2, 1971)

I. HISTORICAL PERSPECTIVES

2 MERRIFIELD, R. B.: Solid Phase Synthesis: I. The Synthesis of a Tetrapeptide — 11
J. Amer. Chem. Soc., **85**, 2149–2154 (July 20, 1963)

3 LETSINGER, R. L., and M. J. KORNET: Popcorn Polymer as a Support in Multistep Syntheses — 17
J. Amer. Chem. Soc., **85**, 3045–3046 (Oct. 5, 1963)

II. POLYPEPTIDE SYNTHESIS

Editors' Comments on Papers 4 Through 21 — 20

4 GUTTE, B., and R. B. MERRIFIELD: Synthesis of Ribonuclease A — 24
J. Biol. Chem., **246**(6), 1922–1941 (1971)

5 MERRIFIELD, R. B.: The Solid Phase Synthesis of Ribonuclease A — 44
Intra-Sci. Chem. Rept., **5**(3), 184–198 (1971)

6 MARSHALL, G. R., and R. B. MERRIFIELD: Synthesis of Angiotensins by the Solid-Phase Method — 59
Biochemistry, **4**(11), 2394–2401 (1965)

7 FRIDKIN, M., A. PATCHORNIK, and E. KATCHALSKI: A Synthesis of Cyclic Peptides Utilizing High Molecular Weight Carriers — 67
J. Amer. Chem. Soc., **87**(20), 4646–4648 (1965)

8   FRIDKIN, M., A. PATCHORNIK, and E. KATCHALSKI: Use of Polymers as Chemical Reagents: I. Preparation of Peptides   69
    J. Amer. Chem. Soc., **88**(13), 3164–3165 (1966)

9   FRIDKIN, M., A. PATCHORNIK, and E. KATCHALSKI: Use of Polymers as Chemical Reagents: II. Synthesis of Bradykinin   71
    J. Amer. Chem. Soc., **90**(11), 2953–2957 (1968)

10 FLANIGAN, E., and G. R. MARSHALL: Synthesis of Cyclic Peptides on Dual Function Supports   76
    Tetrahedron Letters, No. 27, 2403–2406 (1970)

11 GISIN, B. F., and R. B. MERRIFIELD: Synthesis of a Hydrophobic Potassium Binding Peptide   80
    J. Amer. Chem. Soc., **94**(17), 6165–6170 (1972)

12 WANG, S., and R. B. MERRIFIELD: Preparation of a t-Alkyloxycarbonylhydrazide Resin and Its Application to Solid Phase Peptide Synthesis   86
    J. Amer. Chem. Soc., **91**(23), 6488–6491 (1969)

13 WANG, S.: p-Alkoxybenzyl Alcohol Resin and p-Alkoxybenzyloxycarbonylhydrazide Resin for Solid Phase Synthesis of Protected Peptide Fragments   90
    J. Amer. Chem. Soc., **95**(4), 1328–1333 (1973)

14 FELIX, A. M., and R. B. MERRIFIELD: Azide Solid Phase Peptide Synthesis   96
    J. Amer. Chem. Soc., **92**(5), 1385–1391 (1970)

15 GISIN, B. F.: The Preparation of Merrifield-Resins Through Total Esterification with Cesium Salts   103
    Helv. Chim. Acta, **56**(5), 1476–1482 (1973)

16 PARR, W., and K. GROHMANN: A New Solid Support for Polypeptide Synthesis   110
    Tetrahedron Letters, No. 28, 2633–2636 (1971)

17 BAYER, E., I. SEBESTIAN, G. JUNG, and I. HALÁSZ: A New Support for Polypeptide Synthesis in Columns   114
    Tetrahedron Letters, No. 51, 4503–4505 (1970)

18 RICH, D. H., and S. K. GURWARA: Removal of Protected Peptides from an *ortho*-Nitrobenzyl Resin by Photolysis   117
    J. Chem. Soc. Chem. Commun., 610–611 (1973)

19 BAYER, E., E. GIL-AV, W. A. KÖNIG, S. NAKAPARKSIN, J. ORO, and W. PARR: Retention of Configuration in the Solid Phase Synthesis of Peptides   118
    J. Amer. Chem. Soc., **92**(6), 1738–1740 (1970)

20 MERRIFIELD, R. B., J. M. STEWART, and N. JERNBERG: Instrument for Automated Synthesis of Peptides   121
    Anal. Chem., **38**(13), 1905–1914 (1966)

21	HANCOCK, W. S., D. J. PRESCOTT, G. R. MARSHALL, and P. R. VAGELOS: Acyl Carrier Protein *J. Biol. Chem.*, **247**(19), 6224–6233 (1972)	131

III. PROBLEMS WITH THE SOLID-PHASE METHOD

Editors' Comments on Papers 22 Through 25 — 142

22	BAYER, E., H. ECKSTEIN, K. HÅGELE, W. A. KÖNIG, W. BRÜNING, H. HAGENMAIER, and W. PARR: Failure Sequences in the Solid Phase Synthesis of Polypeptides *J. Amer. Chem. Soc.*, **92**(6), 1735–1738 (1970)	143
23	HANCOCK, W. S., D. J. PRESCOTT, P. R. VAGELOS, and G. R. MARSHALL: Solvation of the Polymer Matrix: Source of Truncated and Deletion Sequences in Solid Phase Synthesis *J. Org. Chem.*, **38**(4), 774–781 (1973)	147
24	GISIN, B. F., and R. B. MERRIFIELD: Carboxyl-Catalyzed Intramolecular Aminolysis: A Side Reaction in Solid-Phase Peptide Synthesis *J. Amer. Chem. Soc.*, **94**(9), 3102–3106 (1972)	155
25	SHARP, J. J., A. B. ROBINSON, and M. D. KAMEN: Synthesis of a Polypeptide with Lysozyme Activity *J. Amer. Chem. Soc.*, **95**(18), 6097–6108 (1973)	160

IV. SOLID-PHASE SYNTHESIS OF NUCLEOTIDES

Editors' Comments on Papers 26 Through 33

26	LETSINGER, R. L., and V. MAHADEVAN: Oligonucleotide Synthesis on a Polymer Support *J. Amer. Chem. Soc.*, **87**(15), 3526–3527 (1965)	176
27	LETSINGER, R. L., and V. MAHADEVAN: Stepwise Synthesis of Oligodeoxyribonucleotides on an Insoluble Polymer Support *J. Amer. Chem. Soc.*, **88**(22), 5319–5324 (1966)	178
28	LETSINGER, R. L., M. H. CARUTHERS, and D. M. JERINA: Reactions of Nucleosides on Polymer Supports: Synthesis of Thymidylylthymidylylthymidine *Biochemistry*, **6**(5), 1379–1388 (1967)	184
29	HAYATSU, H., and H. G. KHORANA: Studies on Polynucleotides: LXXII. Deoxyribooligonucleotide Synthesis on a Polymer Support *J. Amer. Chem. Soc.*, **89**(15), 3880–3887 (1967)	194

30	**MELBY, L. R., and D. R. STROBACH:** Oligonucleotide Syntheses on Insoluble Polymer Supports: I. Stepwise Synthesis of Trithymidine Diphosphate *J. Amer. Chem. Soc.*, **89**(2), 450–453 (1967)	202
31	**RUBINSTEIN, M., and A. PATCHORNICK:** Polymers as Chemical Reagents: The Use of Poly(3, 5 Diethylsytrene)—Sulfonyl Chloride for the Synthesis of Internucleotide Bonds *Tetrahedron Letters*, No. 28, 2881–2884 (1972)	206
32	**SELIGER, H., and G. AUMANN:** Oligonucleotide Synthesis on a Polymer Support Soluble in Water and Pyridine *Tetrahedron Letters*, No. 31, 2911–2914 (1973)	210
33	**CHAPMAN, T. M., and D. G. KLEID:** Oligonucleotide Synthesis on Polar Polymer Supports: The Use of a Polypeptide Support *J. Chem. Soc. Chem. Commun.*, 193–194 (1973)	214

V. THE SOLID-PHASE SYNTHESIS OF SACCHARIDES

Editors' Comments on Papers 34 Through 39		216
34	**FRÉCHET, J. M., and C. SCHUERCH:** Solid-Phase Synthesis of Oligosaccharides: I. Preparation of the Solid Support. Poly [p-(1-propen-3-ol-1-yl)styrene] *J. Amer. Chem. Soc.*, **93**(2), 492–496 (1971)	217
35	**FRÉCHET, J. M., and C. SCHUERCH:** Solid-Phase Synthesis of Oligosaccharides: II. Steric Control by C-6 Substituents in Glucoside Synthesis *J. Amer. Chem. Soc.*, **94**(2), 604–609 (1972)	222
36	**FRÉCHET, J. M., and C. SCHUERCH:** Solid-Phase Synthesis of Oligosaccharides: III. Preparation of Some Derivatives of Di- and Tri-Saccharides via a Simple Alcoholysis Reaction *Carbohydrate Res.*, **22**, 399–412 (1972)	228
37	**GUTHRIE, R. D., A. D. JENKINS, and J. STEHLICEK:** Synthesis of Oligosaccharides on Polymer Supports: I. 6-O-(p-Vinylbenzoyl) Derivatives of Glucopyranose and Their Copolymers with Styrene *J. Chem. Soc., Ser. C*, 2690–2696 (1971)	242
38	**EXCOFFIER, G., D. GAGNAIRE, J. P. UTILLE, and M. VIGNON:** Solid-Phase Synthesis of Oligosaccharides: II. Synthesis of 2-Acetamido-6-O-(2-Acetamido-2-Deoxy-β-D-Glucopyranosyl)-2-Deoxy-D-glucose *Tetrahedron Letters*, No. 50, 5065–5068 (1972)	249
39	**ZEHAVI, U., and A. PATCHORNIK:** Oligosaccharide Synthesis on a Light-Sensitive Solid Support: I. The Polymer and Synthesis of Isomaltose (6-O-α-D-Glucopyranosyl-D-glucose) *J. Amer. Chem. Soc.*, **95**(17), 5673–5677 (1973)	253

VI. SYNTHETIC APPLICATIONS OF THE SOLID-STATE METHOD

Editors' Comments on Papers 40 Through 53   260

40   CAMPS, F., J. CASTELLS, M. J. FERRANDO, and J. FONT: Organic Syntheses with Functionalized Polymers: I. Preparation of Polymeric Substrates and Alkylation of Esters   263
Tetrahedron Letters, No. 20, 1713–1714 (1971)

41   CAMPS, F., J. CASTELLS, J. FONT, and F. VELA: Organic Syntheses with Functionalized Polymers: II. Wittig Reaction with Polystyryl-p-Diphenylphosphoranes   265
Tetrahedron Letters, No. 20, 1715–1716 (1971)

42   McKINLEY, S. V., and J. W. RAKSHYS: Wittig Resins: The Preparation and Application of Insoluble Polymeric Phosphoranes   267
J. Chem. Soc. Chem. Commun., 134–135 (1972)

43   WEINSHENKER, N. M., and C. M. SHEN: Polymeric Reagents: I. Synthesis of an Insoluble Polymeric Carbodiimide   269
Tetrahedron Letters, No. 32, 3281–3284 (1972)

44   WEINSHENKER, N. M., and C. M. SHEN: Polymeric Reagents: II. Preparation of Ketones and Aldehydes Utilizing an Insoluble Carbodiimide   273
Tetrahedron Letters, No. 32, 3285–3288 (1972)

45   SHAMBHU, M. B., and G. A. DIGENIS: Insoluble Resins in Organic Synthesis: I. Preparation and Reactions of Polymeric Anhydrides   277
Tetrahedron Letters, No. 18, 1627–1629 (1973)

46   YAROSLAVSKY, C., A. PATCHORNIK, and E. KATCHALSKI: Unusual Brominations with N-Bromopolymaleimide   280
Tetrahedron Letters, No. 42, 3629–3632 (1970)

47   GRUBBS, R. H., and L. C. KROLL: Catalytic Reduction of Olefins with Polymer-Supported Rhodium (I) Catalyst   284
J. Amer. Chem. Soc., **93**, 3062 (1971)

48   GRUBBS, R. H., C. GIBBONS, L. C. KROLL, W. D. BONDS, Jr., and C. H. BRUBAKER, Jr.: Activation of Homogeneous Catalysts by Polymer Attachment   285
J. Amer. Chem. Soc., **95**(7), 2373–2375 (1973)

49   COLLMAN, J. P., L. S. HEDEGUS, M. P. COOKE, J. R. NORTON, G. DOLCETTI, and D. N. MARQUARDT: Resin-Bound Transition Metal Complexes   287
J. Amer. Chem. Soc., **94**(5), 1789–1790 (1972)

50   ČAPKA, M., P. SVOBODA, M. ČERNÝ, and J. HETFLEJŠ: Hydrogenation, Hydrosilylation and Hydroformylation of Olefins Catalyzed by Polymer-Supported Rhodium Complexes   289
Tetrahedron Letters, No. 50, 4787–4790 (1971)

51 NECKERS, D. C., D. A. KOOISTRA, and G. W. GREEN: Polymer-Protected Reagents: Polystyrene–Aluminum Chloride 293
J. Amer. Chem. Soc., **94**(26), 9284–9285 (1972)

52 BLOSSEY, E. C., L. M. TURNER, and D. C. NECKERS: Polymer Protected Reagents: II. Esterifications with Ⓟ-AlCl$_3$ 295
Tetrahedron Letters, No. 21, 1823–1826 (1973)

53 BLOSSEY, E. C., D. C. NECKERS, A. L. THAYER, and A. P. SCHAAP: Polymer-Based Sensitizers for Photooxidations 299
J. Amer. Chem. Soc., **95**(17), 5820–5822 (1973)

VII. SYNTHETIC APPLICATIONS OF POLYMER SUPPORTS IN SPECIFIC REACTIONS

Editors' Comments on Papers 54 Through 62 302

54 PATCHORNIK, A., and M. A. KRAUS: Reactive Species Mutually Isolated on Insoluble Polymeric Carriers: I. The Directed Monoacylation of Esters 305
J. Amer. Chem. Soc., **92**(26), 7587–7589 (1970)

55 KRAUS, M. A., and A. PATCHORNIK: The Directed Mixed Ester Condensation of Two Acids Bound to a Common Polymer Backbone 307
J. Amer. Chem. Soc., **93**(26), 7325–7327 (1971)

56 CROWLEY, J. I., and H. RAPOPORT: Cyclization via Solid Phase Synthesis: Unidirectional Dieckmann Products from Solid Phase and Benzyl Triethylcarbinyl Pimelates 309
J. Amer. Chem. Soc., **92**(21), 6363–6365 (1970)

57 CROWLEY, J. I., T. B. HARVEY, III, and H. RAPOPORT: Solid Phase Synthesis: Evidence for and Quantification of Intraresin Reactions 311
J. Macromol. Sci. Chem., **A7**(5), 1117–1126 (1973)

58 HARRISON, I. T.: The Effect of Ring Size on Threading Reactions of Macrocycles 321
J. Chem. Soc. Chem. Commun., 231–232 (1972)

59 LETSINGER, R. L., and T. J. SAVERSIDE: Selectivity in Solvolyses Catalyzed by Poly-(4-vinylpyridine) 323
J. Amer. Chem. Soc., **84**, 3122–3127 (Aug. 20, 1962)

60 LETSINGER, R. L., and I. S. KLAUS: Investigation of a Synthetic Catalytic System Exhibiting Substrate Selectivity and Competitive Inhibition 329
J. Amer. Chem. Soc., **87**(15), 3380–3386 (1965)

61 LETSINGER, R. L., and I. S. KLAUS: Selective Catalysis Involving Reversible Association of a Synthetic Polymeric Catalyst and Substrate 336
J. Amer. Chem. Soc., **86**, 3884–3885 (Sept. 20, 1964)

62 **LETSINGER, R. L., and T. E. WAGNER:** Regulation of Rate of
Reaction of Polyuridylic Acid Derivative by Use of
Suppressor and Antisuppressor Molecules   338
J. Amer. Chem. Soc., **88**(9), 2062–2063 (1966)

Bibliography   341
Author Citation Index   343
Subject Index   357

Contents by Author

Aumann, G., 210
Bayer, E., 114, 118, 143
Blossey, E. C., 295, 299
Bonds, W. D., Jr., 285
Brubaker, C. H., Jr., 285
Brüning, W., 143
Camps, F., 263, 265
Čapka, M., 289
Caruthers, M. H., 184
Castells, J., 263, 265
Černý, M., 289
Chapman, T. M., 214
Collman, J. P., 287
Cooke, M. P., 287
Crowley, J. I., 309, 311
Digenis, G. A., 277
Dolcetti, G., 287
Eckstein, H., 143
Excoffier, G., 249
Felix, A. M., 96
Ferrando, M. J., 263
Flanigan, E., 76
Font, J., 263, 265
Fréchet, J. M., 217, 222, 228
Fridkin, M., 67, 69, 71
Gagnaire, D., 249
Gibbons, C., 285
Gil-Av, E., 118
Gisin, B. F., 80, 103, 155
Green, G. W., 293
Grohmann, K., 110
Grubbs, R. H., 284, 285
Gurwara, S. K., 117

Guthrie, R. D., 242
Gutte, B., 24
Hägele, K., 143
Hagenmaier, H., 143
Halász, I., 114
Hancock, W. S., 131, 147
Harrison, I. T., 321
Harvey, T. B., III, 311
Hayatsu, H., 194
Hedegus, L. S., 287
Henahan, J. F., 4
Hetflejš, J., 289
Jenkins, A. D., 242
Jerina, D. M., 184
Jernberg, N., 121
Jung, G., 114
Kamen, M. D., 160
Katchalski, E., 67, 69, 71, 280
Khorana, H. G., 194
Klaus, I. S., 329, 336
Kleid, D. G., 214
König, W. A., 118, 143
Kooistra, D. A., 293
Kornet, M. J., 17
Kraus, M. A., 305, 307
Kroll, L. C., 284, 285
Letsinger, R. L., 17, 176, 178, 184, 323, 329, 336, 338
McKinley, S. V., 267
Mahadevan, V., 176, 178
Marquardt, D. N., 287
Marshall, G. R., 59, 76, 131, 147
Melby, L. R., 202

Merrifield, R. B., 11, 24, 44, 59, 80, 86, 96, 121, 155
Nakaparksin, S., 118
Neckers, D. C., 293, 295, 299
Norton, J. R., 287
Oró, J., 118
Parr, W., 110, 143
Patchornik, A., 67, 69, 71, 206, 253, 280, 305, 307
Prescott, D. J., 131, 147
Rakshys, J. W., 267
Rapoport, H., 309, 311
Rich, D. H., 117
Robinson, A. B., 160
Rubenstein, M., 206
Saverside, T. J., 323
Schaap, A. P., 299
Schuerch, C., 217, 222, 228
Sebestian, I., 114

Seliger, H., 210
Shambhu, M. B., 277
Sharp, J. J., 160
Shen, C. M., 269, 273
Stehlicek, J., 242
Stewart, J. M., 121
Strobach, D. R., 202
Svoboda, P., 289
Thayer, A. L., 299
Turner, L. M., 295
Utille, J. P., 249
Vagelos, P. R., 131, 147
Vela, F., 265
Vignon, M., 249
Wagner, T. E., 338
Wang, S., 86, 90
Weinshenker, N. M., 269, 273
Yaroslavsky, C., 280
Zehavi, U., 253

Introduction

The purpose of this collection of reprints is to show how useful the attaching of reactive residues to insoluble polymer supports can be in a wide variety of processes. Polymer supports have been employed chiefly in the synthesis of polypeptides and proteins. But they are applicable also to the synthesis of carbohydrates and their derivatives and of polynucleotides. They are useful in organic reactions like the Wittig reaction and the Dieckmann condensation and have been tried successfully in certain photochemical processes.

Virtually all the procedures described owe their existence to, and are patterned after, Merrifield's published work on solid-phase polypeptide synthesis, and the reader interested in experimental details is referred to Merrifield's fine papers.

Introduction to the Solid-Phase Technique

Attaching amino acids to a cross-linked polystyrene support for the purpose of facilitating the isolation of products is known as the Merrifield solid-phase method for polypeptide synthesis. The technique was invented by two American chemists operating independently. The first report was that of R. Bruce Merrifield in 1962 at the Federation of American Societies for Experimental Biology; a printed report followed in 1963 in the *Journal of the American Chemical Society* (Paper 2). Almost simultaneously, Robert Letsinger at Northwestern University reported the synthesis of a dipeptide on a "popcorn" polymer support. The difference in the two methods was that whereas Merrifield's group attached the polymer to the amino acid residue as the carboxylate ester, Letsinger's group attached easily swollen "popcorn" polymer to the amino acid via the amino group and as an amide.

The technique is fundamentally simple. In the Merrifield procedure, an N-blocked amino acid is attached to a polystyrene bead by means of the chloromethylated polymer. The amino acid–polymer attachment is as a benzylic carboxylate ester. Attaching the amino acid to the polymer immobilizes the amino acid and isolates it from other molecules like itself on the polymer backbone.

Introduction

The polymer isolated amino acid is then coupled, by standard techniques, to a second amino acid, then a third, and so on. The coupling reactions can be completed as often as possible until the polypeptide of desired composition is built up.

After the completion of the polypeptide synthesis on the polymer, the peptide can be removed, virtually intact, by a series of now standard cleavage reactions. This entire sequence of reactions is shown by means of the following equations:

Step 1: Forming the polymer–amino acid bond:

$$\text{(P)}-CH_2Cl + HOOC-\underset{R}{\overset{H}{C}}-NH-\boxed{\text{Blocking Group}} \xrightarrow{\text{base}}$$

$$\text{(P)}-CH_2OOC-\underset{R}{\overset{H}{C}}-NH-\boxed{\text{Blocking Group}}$$

Step 2: Unblocking the amine group:

$$\text{(P)}-CH_2OOC-\underset{R}{\overset{H}{C}}-NH-\boxed{\text{Blocking Group}} \xrightarrow{\text{acid}} \text{(P)}-CH_2OOC-\underset{R}{\overset{H}{C}}-NH_2$$

Step 3: Coupling the amino acid to a second amino acid (step 3 may be repeated *n* times):

$$\text{(P)}-CH_2OOC-\underset{R}{\overset{H}{C}}-NH_2 + HOOC-\underset{R'}{\overset{H}{C}}-NH-\boxed{\text{Blocking Group}} \xrightarrow[\text{agent}]{\text{coupling}}$$

$$\text{(P)}-CH_2OOC-\underset{R}{\overset{H}{C}}-NH\overset{O}{\overset{\|}{C}}-\underset{R'}{\overset{H}{C}}-NH-\boxed{\text{Blocking Group}}$$

$$\text{(P)}-CH_2OOC\underset{R}{\overset{H}{C}}-NH\left(-\overset{O}{\overset{\|}{C}}-\underset{R'}{\overset{H}{C}}-NH-\right)_n\boxed{\text{Blocking Group}}$$

Step 4: The final unblocking step:

$$\text{(P)}-CH_2OOC-\underset{R}{\underset{|}{\overset{H}{\overset{|}{C}}}}-NH\left(-\underset{R''}{\underset{|}{\overset{O}{\overset{\|}{C}}}}-\underset{}{\overset{H}{\overset{|}{C}}}-NH\right)_n-\underset{R'}{\underset{|}{\overset{O}{\overset{\|}{C}}}}-\overset{H}{\overset{|}{C}}-NH_2$$

Step 5: Removing the synthetic polypeptide from the polymer support:

$$\text{(P)}-CH_2OOC-\underset{R}{\underset{|}{\overset{H}{\overset{|}{C}}}}-NH\left(-\underset{R'}{\underset{|}{\overset{O}{\overset{\|}{C}}}}-\overset{H}{\overset{|}{C}}-NH\right)_n-\underset{R'}{\underset{|}{\overset{O}{\overset{\|}{C}}}}-\overset{H}{\overset{|}{C}}-NH_2 \xrightarrow{HX}$$

$$\text{(P)}-CH_2X + HOOC-\underset{R}{\underset{|}{\overset{H}{\overset{|}{C}}}}-NH\left(-\underset{R'}{\underset{|}{\overset{O}{\overset{\|}{C}}}}-\overset{H}{\overset{|}{C}}-NH\right)_n-\underset{R'}{\underset{|}{\overset{O}{\overset{\|}{C}}}}-\underset{R'}{\overset{H}{\overset{|}{R}}}-NH_2 \cdot HX$$

The advantages of the solid-phase method in polypeptide synthesis are obvious. Since peptide preparations require several successive, repetitive operations, attaching the substrate to a polymer support that is insoluble in the solvent in which the reaction is being carried out greatly facilitates the separation of products after the reaction is completed. A less obvious advantage, but still an important one, is that the growing peptide chains can be virtually isolated from one another on the polymer support, thereby eliminating bimolecular interactions between peptide chains.

Every synthetic chemist can visualize ways in which immobile polymer supports might be useful in synthetic operations. For example, polymer supports might be used, and in fact have been used, to hold and immobilize synthetic intermediates and catalysts. Polymer supports might also be employed to fix the relationships of groups with respect to one another so that their interactions in bimolecular processes might be studied.

The collected papers, many in communication form, demonstrate the viability of the technique, not only in polypeptide synthesis but in other operations as well. Each paper has been chosen because it represents a unique aspect of the uses of polymeric supports.

Obviously a great many of the papers cited originated in Merrifield's laboratory, and we would not have proceeded on this undertaking without his blessing. Paper 1 provides a biographical sketh of Merrifield's life and work. We hope that the readers of this volume will find it useful and will apply the solid-state method to their own research area where it appears convenient.

Copyright © 1971 by the American Chemical Society

Reprinted from *Chem. Eng. News*, 22–26 (Aug. 2, 1971)

The Chemical Innovators 15

John F. Henahan, *senior editor*

R. Bruce Merrifield
Designer of protein-making machine

On May 26, 1969, Bruce Merrifield recorded in his notebook: "There is a need for a rapid, quantitative, automatic method for the synthesis of long-chain peptides." Five years later he satisfied that need, and chemists, biologists, physicians, several instrument makers, and probably society at large are in his debt.

Dr. Merrifield, now professor of biochemistry at Rockefeller University, New York City, has made it possible for a chemist to make molecules in the space of a few weeks or months that might previously have taken much of his career. And, because enzymes, hormones, antibiotic peptides, and their analogs can be synthesized so much faster than before, obstacles to the solution of many important medical and physiological problems are now beginning to crumble.

The Rockefeller biochemist is modest about his accomplishments, but refreshingly direct in relating the history of how they came about. He tells his story with the same kind of self-effacing simplicity that appears to permeate his entire life style. Outside the lab, he is content to work in his basement shop or tend the garden and magnolia trees surrounding his old, three-story frame dwelling in Cresskill, N.J. In recent years, he has traveled considerably in Europe and Japan, telling other chemists about how his automated protein setup works. He appreciates the high points of travel, but tries to keep them in perspective with more fundamental considerations. "I liked the Japanese food very much," he confesses. "But you know after a while, I really began to miss the potatoes."

The 50-year-old biochemist was born in Fort Worth, Tex., but spent most of his early life in California. His father was a furniture salesman, and in the depression years the Merrifield family followed what little money there was up and down the California coast. "I must have attended about 40 different schools," Dr. Merrifield recounts.

John F. Henahan

The nearest that he ever came to calling any place home was Montebello, Calif., where he lived for five years and graduated from high school. It was there that an essay he wrote on his plans for a career was apparently persuasive enough that he decided to take up chemistry at Pasadena Junior College. He went there primarily because it was nearby and inexpensive, he says.

"At that time, the thought of applying to all those big name colleges never occurred to me. In fact, most young people did what I did in those years—went to the most convenient college available."

To get his B.A. in chemistry, Dr. Merrifield went to another nearby school, the University of California, Los Angeles. While there he earned extra money by synthesizing DOPA in a small amino acid manufacturing laboratory run by UCLA biochemistry professor Max Dunn. After graduation he worked for a company that made vitamin supplements for animal feeds and Dr. Merrifield's job was to clean out animal cages, conduct growth experiments, assay for vitamin D, and carry out routine chemical analyses. "If it had been a better job," he admits frankly, "I would probably have stayed there. Instead, I decided to go to grad school at UCLA."

Dr. Merrifield toyed with the idea of getting a master's degree, but a fellowship from Anheuser Busch came through that allowed him to think in terms of a Ph.D.

Effective bioassays. Appropriately enough, the beer company fellowship involved Dr. Merrifield in developing analytical techniques for purines and pyrimidines in yeast samples. At that time, the vital role of these important nucleic acid components in determining the hereditary characteristics and evolution of every living thing was only barely suspected. However, Dr. Merrifield succeeded in devising effective bioassays for both important substances, based on the extent to which they were required for the growth of certain bacteria. However, Dr. Erwin Chargaff's innovative use of chromatography was easier and more quantitative than Dr. Merrifield's method and quickly overshadowed it. "That's the trouble with any method," Dr. Merrifield says philosophically. "It's timely for a while and then. . . ."

After four and a half years at UCLA, Bruce Merrifield got his Ph.D. in 1949. He graduated on Sunday, got married on Monday, and left for his new job at what was then known as Rockefeller Institute on Tuesday. His wife, Elizabeth, had been studying zoology at UCLA at the time of their marriage. (They now have six children, ranging from six to 19 years old.)

At Rockefeller, Dr. Merrifield entered the biochemistry laboratory, then under the direction of the late Dr. D. W. Woolley. For a while he worked on the isolation and characterization of peptides that appeared to act as bacterial growth factors, returned to his earlier interest in nucleic acid chemistry for a while, and then concentrated on protein chemistry, one of the chief concerns of the Woolley laboratory. While working on the nucleic acids, Dr. Merrifield was given the desiccator of Dr. P. A. Levene, one of the foremost pioneers in the field. "The desiccator had a 'PAL' scratched on it," he recollects. "That was a really exciting thing for a postdoc like me. And I still have it."

By 1953, Dr. Merrifield was deeply involved in separating and determining the sequence of protein growth factors. Later he synthesized and made structural analogs of some of the peptides to compare their biological activity with variations in their chemical structure. He also synthesized a number of larger polypeptides (20 to 40 amino acids of random sequence) that seemed to act like catalytic enzymes in the hydrolysis of certain nitrophenyl esters. The project soon became bogged down because it took so long to synthesize the peptides—several months in the case of the simpler pentapeptides. It would have taken even longer to synthesize larger molecules with well-defined amino acid sequences. At that point (1959) Dr. Merrifield told himself there must be a better and quicker way to make polypeptides.

Better way. One of the most frustrating aspects of conventional peptide synthesis was the fact that the product had to be removed from the reaction vessel and purified each time a new amino acid was added to the growing peptide chain. The result was a case of decreasing returns, in which more of the desired product was lost with each purification step. Dr. Merrifield reasoned that yields

Photos: Bill Schropp

5

might be increased if that first amino acid could be anchored to a solid support of some type to act as a starting point for the addition of new amino acids. Then, the anchored amino acid would stay in the reactor while unneeded reagents and solvent were removed by filtration and thorough washing. His original idea was supported by the observation that peptides or amino acids could be held firmly to the particles of a chromatographic column, but others weren't.

Dr. Merrifield mulled the idea over in his mind for a couple of weeks, committed it to his notebook that May day in 1959, and presented his thoughts to Dr. Woolley as they shared an elevator ride to an upper floor at Rockefeller. "He didn't say much when I suggested the idea to him; just got off the elevator and that was that," Dr. Merrifield remembers. "Next day he came into my office and said, 'Do you know that idea you were talking about yesterday? Maybe you ought to try it.'"

Dr. Woolley's encouragement was enough to get Dr. Merrifield moving on a problem which he hoped to solve in two or three months. Three years later, he was still working on it without a single publication to dignify his labors. "You just can't get away with a dry spell like that at just any old place," he observes. "If I had been at Harvard or Berkeley, I think people would have begun to look at me a little funny."

One of the biggest problems in that three-year period was finding a good support that would hold the amino acid chain as it was being formed and then allow it to be released chemically without breaking down the peptide again. First he tried powdered cellulose because it was used extensively in column chromatography to separate proteins and appeared to be chemically compatible. Working with simple amino acids such as alanine and leucine, his scheme was to first block the amine end of the amino acid then esterify the carboxyl end with the hydroxyl groups of the cellulose powder. He succeeded in building up a two-amino-acid dipeptide that way, but everything turned to tar when he tried to remove it from the support with strong acid. For the next several months ("I was *sure* it would work," he says) Dr. Merrifield tried other polymers, such as methacrylate resins and finally hit the jackpot with tiny beads of polystyrene copolymerized with about 2% divinylbenzene.

Right handle. Before he could tie the first amino acid to the polymeric beads, Dr. Merrifield first had to equip the benzene rings of the polymer with chemically reactive "handles." That took a bit of doing, but he eventually found that if he chloromethylated the polymer with chloromethyl ether, using a stannic chloride catalyst, about 10 to 15% of the styrene's benzene rings were converted to the more reactive benzyl chloride derivative (**A**).

He then protected the amine end of the first amino acid (alanine) with a benzyloxycarbonyl group, made the triethylamine salt of the carboxyl end of the amino acid, then reacted that with the chloromethyl anchor (**B**).

Before adding another link to the peptide chain, Dr. Merrifield filtered solvents and unreacted reagents from the amino acid (now anchored to the styrene beads) and washed the beads thoroughly. He removed the Cbzo protecting group with dilute hydrogen bromide and then linked a "protected" molecule of leucine to the alanine with any one of several possible coupling agents. (One of the most effective was dicyclohexylcarbodiimide.) Then, he proceeded as before, removing the protecting group from the second amino acid, filtering off solvent and reagents, and washing thoroughly. The end result was a dipeptide consisting of alanine and leucine, which could be easily removed from the polymer support with anhydrous HBr in a solution of trifluoro acetic acid (**C** and **D**).

Since then many chemical aspects of the process have changed and been improved, but that first little dipeptide meant a lot to Dr. Merrifield. He realized that it might not impress other biochemists who were already synthesizing molecules as long as insulin (51 amino acids) by the traditional methods. But he was now very confident that his new method would eventually do the same thing faster, better, and easier.

The tetrapeptide. His next important success in peptide chain building was a tetrapeptide; leucine, alanine, glycine, and valine tied together in that order. The technique he used to make the tetrapeptide was generally similar to his earlier work, except that he had now introduced two different amino acids. In 1962, he reported his results at a meeting of the Federation of American Societies for Experimental Biology in Atlantic City. He told his audience that he got nearly a 100% yield in the coupling steps, that it took about four hours to add each amino acid and two or three weeks to purify and characterize his final product. Chemists who were not committed to the older techniques were very intrigued with the solid-phase method, but the traditionalists were openly dubious. One chemist questioned the completeness of his analytical data.

"He was somewhat skeptical and pointed out that I would have to characterize the product more completely if I expected to convince anybody," Dr. Merrifield says. "By the time we published in *JACS* we had met all his objections. Although it seemed that he was giving me a real hard time at the FASEB meeting, he was right as it turned out. His questions were certainly good ones and should have been asked."

With the synthesis of the model tetrapeptide safely tucked away in the literature, Dr. Merrifield decided that

John F. Henahan

it was time he used his method to make something of biological interest. He chose bradykinin, a peptide containing nine amino acids. Its chemical structure was well known, it had been synthesized by the older methods, and it was very easy to assay. In various organisms, including man, it lowers blood pressure, causes pain, and contracts smooth muscle.

Aside from its length and biological activity, the nonapeptide was a challenge because three of its amino acids (arginine, serine, and proline) had chemical side groups that might make it very difficult to adapt them to the Merrifield chain-building technique. Although it was about a year before Dr. Merrifield worked out all the methodological wrinkles, he was eventually able to make bradykinin by the solid-phase method in less than a week. For a better idea of how the new method compares to traditional techniques, it took Rockefeller's Dr. John M. Stewart 12 months to make three bradykinin analogs by the older technique. With the solid-phase method, he found he could make about 50 in the same period.

Automated proteins. As his method proved itself over and over again, Dr. Merrifield settled down to following the recommendation he wrote in his notebook several years earlier. At that time he predicted that if all aspects of the solid-phase method would become routine and uniform, there was no reason why a machine could not do the same things; add reagents to a reactor, stir them up for a while, remove unwanted solvents and reagents, and carry out all the other mechanical and chemical aspects of peptide synthesis.

"Originally, the idea was sort of helter-skelter," he says, "until John Stewart, who was also a ham radio man, cooperated on the electrical end of things. I functioned as plumber, and between the two of us we put something together in the basement at home."

The three chief elements of the automated system put together by the "plumber" and "electrician" were the *Reagent and solvent containers* for amino acids, deblocking and coupling agents, acids for cleaving the peptide from the support, various solvents, and the like. These fed into a *rotary valve system* developed by Nils Jernberg of the Rockefeller staff. ("The valves had to be precision made or they leaked like a sieve," Dr. Merrifield points out.) Fabricated from two Teflon disks mounted between steel plates, each valve contains 12 orifices that open and close according to a predetermined program.

"Originally, the idea was sort of helter-skelter until John Stewart, who was also a radio ham, cooperated on the electrical end of things"

Thirdly, *a rotating drum* that triggers as many as 30 different microswitches, one at a time. The triggering sequence, controlled by timers, is first mapped out on paper by the chemist who establishes the order in which solvents and reagents should be added, how long they stay in the reactor, when they should be removed, and which amino acid will be next added to the peptide chain then growing on the polystyrene beads. The chemist then duplicates the paper program on the drum by hammering small nylon pins into its surface. These trip the microswitches as the drum rotates.

When Dr. Merrifield and Dr. Stewart finally developed a working model in 1965, they found that their automated synthesizer could add six new amino acids in a period of 24 hours; each amino acid required nearly 90 steps on the programer drum. Since then the machine has been used both inside and outside Rockefeller University to make bradykinin; the hormones angiotensin and oxytocin; the antibiotic peptide valinomycin; insulin; a decapeptide from tobacco mosaic virus; the enzyme ribonuclease; and several other peptides.

Distinguished achievements. Each new achievement was distinguished in its own right. For example, valinomycin, with 12 residues, has a rather peculiar circular structure composed of alternating amino acids and hydroxy acids. The insulin molecule, on the other hand, contains two peptide chains, 21 and 30 amino acids long, respectively, connected by disulfide linkages. By traditional methods, its synthesis takes at least several months. Dr. Merrifield and Dr. A. Marglin (a young M.D. who took on the synthesis as a Ph.D. project) did it in about 20 days. Dr. Marglin, now at Tufts medical school, is using the technique to make insulin analogs and fragments to study fatty acid synthesis and carbohydrate metabolism.

Perhaps the most monumental accomplishment for the automated protein maker was the successful synthesis of the enzyme ribonuclease—a convoluted array of 124 amino acids—by Dr. Merrifield and Dr. Bernd Gutte, assistant professor. Their work capped a 30-year period in which ribonuclease had become, in effect, the Rockefeller enzyme. It was first isolated by Prof. René Dubós in 1938 and crystallized by Dr. Moses Kunitz in 1940. Its chemical structure was first determined by Dr. C. H. Hirs, William H. Stein, and Stanford Moore. Finally it was synthesized by Dr. Merrifield and Dr. Gutte in 1969. The synthesis required 369 chemical reactions and 11,391 steps in their machine.

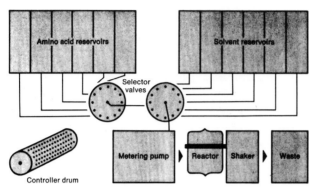

Dr. Bernd Gutte, left, and Dr. Merrifield made ribonuclease with the synthesizer

In the same issue of *JACS* as the Rockefeller chemists announced the synthesis of RNAseA, a 26-man team from Merck, Sharp and Dohme Laboratories in Rahway, N.J., reported that they had done a similar thing, but by a different synthetic process. Headed by Dr. Ralph Hirschmann and Dr. Robert G. Denkewalter, the Merck team used their own carboxyanhydride method to build up a series of small peptides that were then combined to give a synthesis of the part of the enzyme known as S-protein. When it was mixed with the other part, S-peptide, the enzyme formed.

Based on their work with smaller peptides, Dr. Merrifield and Dr. Gutte calculated that it should take about 21 days to put the RNAse molecule together. When they actually got working, it took about six weeks to make the enzyme and another four months to purify, characterize, and assay it. The original partly purified product had about 13% activity, but in later syntheses, the two Rockefeller researchers increased that figure to 78% by employing a battery of refined purification techniques.

Dr. Merrifield now suspects, however, that the original product may have been more pure than the assay indicated, because the enzyme may have contained a small amount of inhibitor that suppressed its activity.

Worldwide use. Since the RNAse synthesis was reported in 1969, researchers all over the world have adapted the Merrifield synthesis—automated or otherwise—to make molecules of profound biological interest, including human growth hormone and ferredoxin. Others are using the technique to study the effect of man-made antigenic peptides on the production of specific antibodies; still others hope that synthetic analogs of certain hormones might be used to inihbit the activity of hormones that sometimes have adverse physiological effects. An angiotensin analog, for example, might help lower blood pressure, since the hormone itself has been linked to hypertension. In addition, the ability to make enzymes or enzyme fragments efficiently and quickly might help pin down the nature of the actual site of the enzyme's catalytic activity, or perhaps lead to the synthesis of artificial enzymes.

Dr. Merrifield is reluctant to speculate on the medical ramifications his work may have, but its long-range value was recognized in 1969 when he received the Albert Lasker Award for Basic Medical Research.

A different kind of recognition is obvious from the fact that a number of manufacturers, including Beckman Instruments and Schwarz BioResearch in the United States, have begun to market their own sophisticated versions of the machine that was born in a basement.

For his part, Dr. Merrifield is happy to see that the automated synthesis has taken hold, even though he will probably receive no great financial rewards from its success. He has a patent on the automated process, which is now in the public domain, since it was developed with the help of funds from the National Institutes of Health. But even then, he only filed for the patent when an NIH official told him: "On the basis of your work, I have decided that you should get a patent."

Aware that many more chemists may soon be turning to the new protein-making machines to help them get better results quicker, Bruce Merrifield can't help feeling a twinge of apprehension. He knows what went into developing the solid-phase protein-synthesis technique and its automated descendant and that it took a lot of hard, careful work. He hopes that when other chemists build their own protein chains with the aid of a machine, they will realize that too.

"I can foresee where a research director might read that a new machine that is coming on the market is able to make all these interesting things. However, unless he has a good protein chemist to work it, there might be a constant feedback of poor results. You still can't just push a button to get out a new product."

I
Historical Perspectives

Copyright © 1963 by the American Chemical Society

Reprinted from *J. Amer. Chem. Soc.*, **85**, 2149–2154 (July 20, 1963)

Solid Phase Peptide Synthesis. I. The Synthesis of a Tetrapeptide[1]

By R. B. Merrifield

Received January 31, 1963

A new approach to the chemical synthesis of polypeptides was investigated. It involved the stepwise addition of protected amino acids to a growing peptide chain which was bound by a covalent bond to a solid resin particle. This provided a procedure whereby reagents and by-products were removed by filtration, and the recrystallization of intermediates was eliminated. The advantages of the new method were speed and simplicity of operation. The feasibility of the idea was demonstrated by the synthesis of the model tetrapeptide L-leucyl-L-alanylglycyl-L-valine. The peptide was identical with a sample prepared by the standard p-nitrophenyl ester procedure.

The classical approach to peptide synthesis has yielded impressive successes in recent years in the preparation of several biologically active peptides.[2] With the development of new reagents and techniques the synthesis of most small peptides has been placed within easy reach.[3] However, these procedures are not ideally suited to the synthesis of long chain polypeptides because the technical difficulties with solubility and purification become formidable as the number of amino acid residues increases. A new approach to peptide synthesis has been investigated in an effort to overcome some of these difficulties. The present report deals with the basic idea behind the new method and with a demonstration of its feasibility through the synthesis of a simple model tetrapeptide.

The general concept underlying the new method is outlined in Fig. 1. It depends on the attachment of the first amino acid of the chain to a solid polymer by a covalent bond, the addition of the succeeding amino acids one at a time in a stepwise manner until the desired sequence is assembled, and finally the removal of the peptide from the solid support. The reason for this approach is that when the growing peptide chain is

(1) (a) Supported in part by Grant A 1260 from the U. S. Public Health Service. (b) An abstract of this work was presented at the 46th Annual Meeting of the Federation of American Societies for Experimental Biology, April, 1962; R. B. Merrifield, *Fed. Proc.*, **21**, 412 (1962).

(2) (a) V. du Vigneaud, C. Ressler, J. M. Swan, C. W. Roberts, P. G. Katsoyannis and S. Gordon, *J. Am. Chem. Soc.*, **75**, 4879 (1953); (b) R. B. Merrifield and D. W. Woolley, *ibid.*, **78**, 4646 (1956); (c) H. Schwarz, M. Bumpus and I. H. Page, *ibid.*, **79**, 5697 (1957); (d) R. A. Boissonnas, S. Guttmann and P. A. Jaquenoud, *Helv. Chim. Acta*, **43**, 1349 (1960); (e) K. Hofmann, H. Yajima, N. Yanaihara, T. Liu and S. Lande, *J. Am. Chem. Soc.*, **83**, 487 (1961); (f) C. H. Li, J. Meienhofer, E. Schnabel, D. Chung, T. Lo and J. Ramachandran, *ibid.*, **83**, 4449 (1961); (g) H. Kappeler and R. Schwyzer, *Helv. Chim. Acta*, **44**, 1136 (1961).

(3) See J. P. Greenstein and M. Winitz, "Chemistry of the Amino Acids," Vol. 2, John Wiley and Sons, Inc., New York, N. Y., 1961.

firmly attached to a completely insoluble solid particle it is in a convenient form to be filtered and washed free of reagents and by-products. Thus the intermediate peptides are purified, not by the usual recrystallization procedures, but by dissolving away the impurities. This greatly simplifies the manipulations and shortens the time required for the synthesis of the peptides. It is hoped that such a method will lend itself to automation and provide a route to the synthesis of some of the higher molecular weight polypeptides which have not been accessible by conventional procedures.

The Polymer.—The first requirement was for a suitable polymer. It had to be insoluble in all of the solvents which were used and have a stable physical form which permitted ready filtration. It also had to contain a functional group to which the first protected amino acid could be firmly linked by a covalent bond. Many polymers and modes of attachment were investigated. Among the polymers were cellulose, polyvinyl alcohol, polymethacrylate and sulfonated polystyrene. The one which worked best was a chloromethylated copolymer of styrene and divinylbenzene. The resin, in the form of 200–400 mesh beads, possessed a porous gel structure which allowed ready penetration of reagents, especially in the presence of swelling solvents. Although diffusion and steric hindrance were no doubt important factors, they were not serious enough to prevent the desired reactions from proceeding to completion. The reaction rates were slower than corresponding ones in solution, but conditions were found which permitted all of the reactions to occur at useful rates in spite of the fact that the growing peptide chain was in the completely insoluble solid phase at all times. It was for this reason that the term solid phase peptide synthesis was introduced to describe the new method.

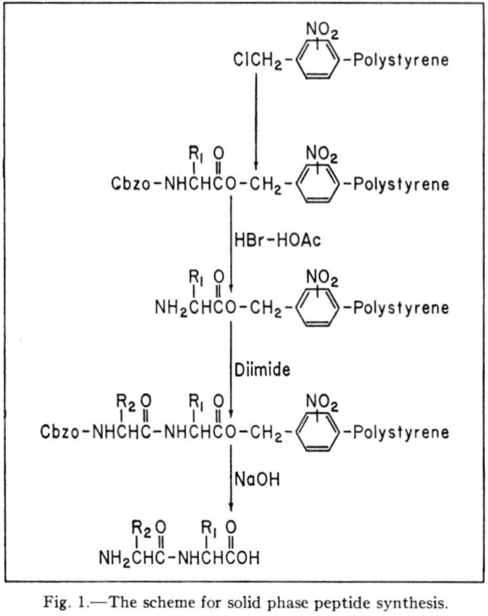

Fig. 1.—The scheme for solid phase peptide synthesis.

Attachment of the First Amino Acid to the Polymer.—To provide a point of attachment for the peptide the polystyrene resin was partially chloromethylated.[4] For reasons to be discussed later the product was then nitrated or brominated. The resulting substituted chloromethyl polystyrene was treated with the triethylammonium salt of the first protected amino acid in the proposed peptide chain to give a substituted benzyl ester linkage. This was the stable covalent bond which held the growing peptide chain in the solid phase on the supporting resin. The reaction was shown to go to completion with carbobenzoxyglycine at refluxing temperatures in dry solvents such as ethyl acetate, benzene or dioxane giving carbobenzoxyglycyl polymer.[5] It was also demonstrated to proceed with carbobenzoxy-L-valine and carbobenzoxynitro-L-arginine. Racemization of the C-terminal amino acid was not observed during the reaction. The quantity of amino acid attached to the resin was purposely limited to approximately 0.5 mmole per gram of substituted polymer. To avoid undesired alkylations in subsequent steps the unreacted chloromethyl groups were then esterified with an excess of triethylammonium acetate. Complete reaction was indicated by the fact that the product was nearly free of halogen.

Cleavage of the Amino-Protecting Group.—The protecting group which was used throughout the syntheses to be reported was the carbobenzoxy group. It was selected because it could be removed readily and completely by hydrogen bromide in glacial acetic acid.[6] Since the latter reagent also attacked the benzyl ester at a slow rate the loss of peptide from the resin was a serious problem. The difficulty was largely overcome by nitration or bromination of the polymer after the

(4) K. W. Pepper, H. M. Paisley and M. A. Young, *J. Chem. Soc.*, 4097 (1953).
(5) The abbreviated name carbobenzoxyglycyl polymer was used to designate the compound in which carbobenzoxyglycine was bound in ester linkage to the aromatic rings of the styrene–divinylbenzene copolymer through hydroxymethyl side chains. The number of substituents per molecule of polymer varied and the exact location on the ring or distribution among the rings was not known. The other polymer derivatives were named in an analogous manner.
(6) D. Ben-Ishai and A. Berger, *J. Org. Chem.*, **17**, 1564 (1952).

chloromethylation step. Table I shows that while the carbobenzoxy group cleaved rapidly from the unsubstituted carbobenzoxy-L-valyl polymer even in 10% HBr–acetic acid there was also considerable loss of ester. After nitration the rate of removal of carbobenzoxy was decreased, but the loss of ester was reduced to a very small level. With 30% HBr the carbobenzoxy group was removed in 2 to 4 hr., while the ester cleavage remained at a low level for at least 6 hr. In fact, the peptides could not be removed completely from the nitrated resin even with prolonged HBr treatment. The carbobenzoxy-L-valyl bromopolymer had properties intermediate between the nitrated and unsubstituted derivatives with respect to the stability of the carbobenzoxy and ester bonds. The carbobenzoxy group could be removed from it with dilute (10%) HBr and the ester could be cleaved with concentrated (30%) HBr.

TABLE I
CLEAVAGE OF CARBOBENZOXY AND ESTER GROUPS BY HBr–ACETIC ACID

	Extent of cleavage, %							
	Cbz-val-polymer		Cbz-val-nitropolymer				Cbz-val-bromopolymer	
Time, min.	10% HBr		10% HBr		30% HBr		10% HBr	30% HBr
	Cbz	Ester	Cbz	Ester	Cbz	Ester	Cbz	Ester	Ester
5	52	6					8	0.5	8
10	73	8	6		14		17	1.1	13
20	90	12	12		28		32	2.2	21
30	100	15	19		43	1.2			28
60	100	18	38	1.2	67		77	6	46
90			57	2.0	90	3.8			52
120			67	2.5	100	4.3	87	10	58
180			83		99		91	14	
240			94	3.2	108	5.4			69
300			99	3.2	98	4.8	100	18	

The Peptide-Forming Step.—After removal of the N-terminal protecting group by HBr the hydrobromide was neutralized with excess triethylamine and the free base was coupled with the next protected amino acid. Although the *p*-nitrophenyl ester method[7] at first appeared to be ideal for this step it finally proved to be unsatisfactory after many experiments. The yields were not high enough even with elevated temperatures (80°) and there was evidence for partial racemization (Table II). The N,N'-dicyclohexylcarbodiimide method,[8] on the other hand, proved to be very satisfactory. The reaction went in virtually quantitative yield in about 30 minutes at room temperature. The relatively insoluble by-product dicyclohexylurea and the rearrangement product carbobenzoxyaminoacyldicyclohexylurea, which was formed in appreciable quantities, were both easily removed by thorough washing. In conventional syntheses this is not always the case, especially when the peptide derivatives have become large and relatively insoluble themselves. The choice of solvent for the condensation was very important. Of the several solvents which were examined dimethylformamide was the best, methylene chloride was satisfactory, but dioxane, benzene, ethanol, pyridine and water gave very low yields and were not useful. The effectiveness of the solvent depended partly on its ability to swell the resin and partly on other factors. For example, the two effective solvents (dimethylformamide and methylene chloride) had high dielectric constants and also swelled the resin whereas benzene, which swelled the resin but had a low dielectric constant, was ineffective.

The coupling reaction was shown to occur between glycylvalyl polymer and the carbobenzoxy derivatives

(7) M. Bodanszky, *Nature*, **175**, 685 (1955).
(8) J. C. Sheehan and G. P. Hess, *J. Am. Chem. Soc.*, **77**, 1067 (1955).

TABLE II
AMINO ACID ANALYSES OF L-LEUCYL-L-ALANYLGLYCYL-L-VALINE

Polymer subst.	Coupling method	Cleavage method	Ratio[a] enzyme hydrolysate			Ratio[a] acid hydrolysate
			Leu	Ala	Gly	Val
NO$_2$	Active ester	NaOH	0.83	0.73	0.61	0.58
NO$_2$	Diimide	NaOH	1.00	1.05	0.97	..
Br	Diimide	HBr	1.06	1.00	1.02	1.01

[a] The concentration of each amino acid in a leucine aminopeptidase digest divided by the corresponding value in a 6 N HCl hydrolysate.

of leucine, isoleucine, valine, alanine, glycine, phenylalanine, O-benzyltyrosine, proline, serine, threonine, methionine, S-benzylcysteine, im-benzylhistidine, nitroarginine, γ-methyl glutamate and asparagine. Lysine and tryptophan derivatives were not studied.

Since it was very important that no unreacted amine remain after the peptide-forming reaction,[9] several precautions were taken. First, an excess of carbobenzoxyamino acid and of diimide was used and each condensation step was repeated with fresh reagents. Finally, after the coupling reactions were completed any trace of unreacted amine was acetylated with a large excess of acetic anhydride and triethylamine. This reduced the free amine to less than 0.1% of that originally present. The acetamido bond was completely resistant to 30% HBr–acetic acid for 18 hours at 25°. Consequently, the danger of re-exposure of these amino groups during subsequent steps was eliminated.

The steps described to this point completed one cycle, i.e., the peptide chain was lengthened by one amino acid residue. Further cycles were carried out in the same way by alternately deprotecting and coupling with the appropriate carbobenzoxyamino acid. The completed protected peptides were finally decarbobenzoxylated by HBr and the free peptides were liberated from the polymer by saponification, or, in the case of the brominated resin, by more vigorous HBr treatment. The liberated peptides were desalted and purified by ion-exchange chromatography, countercurrent distribution or other suitable procedures.

In order to establish the optimal conditions for the various reactions and to test the feasibility of the method as outlined, the simple tetrapeptide L-leucyl-L-alanylglycyl-L-valine was synthesized. Carbobenzoxy-L-leucyl-L-alanylglycyl-L-valyl polymer was made by the stepwise procedures described, using first the nitrophenyl ester method and later the more suitable diimide reaction. The latter was used on both nitrated and brominated polymer. The best procedure was with the nitropolymer using dicyclohexylcarbodiimide in dimethylformamide for the coupling reaction and saponification for the liberation of the free peptide.

Isolation and Purification of the Peptides.—The crude peptide product liberated from the resin was chromatographed on a Dowex 50-X4 column using 0.1 M pH 4.0 pyridine acetate buffer. The eluate was analyzed by the quantitative ninhydrin reaction of Moore and Stein[10] and by dry weight after evaporation of the volatile buffer from pooled fractions. Figure 2 shows the chromatogram resulting from the separation of the peptides prepared on 10 g. of polymer. A total of 332 mg. of tetrapeptide was isolated which accounted for 80% of the free amino acids and peptides

(9) If any amine remained unacylated after reaction with the carbobenzoxyamino acid it would be liable to react with the subsequent carbobenzoxyamino acids. In this way a peptide chain lacking one or more of the internal amino acid residues would begin to grow. A family of closely related peptides, difficult to separate, would thus result.
(10) S. Moore and W. H. Stein, J. Biol. Chem., 211, 907 (1954).

present in the crude product. In addition, an acetylated peptide fraction weighing 57 mg. passed through the column almost unretarded. It contained acetylvaline, acetylglycylvaline and acetylalanylglycylvaline in approximately equal amounts. It was concluded that the coupling reactions were not quite quantitative, but that the unreacted amine was available to acetic anhydride and could be prevented from further reactions in this way. The presence of small amounts of acetylated derivatives was not a serious difficulty, however, since these could be separated readily from the desired tetrapeptide. On the other hand, the presence of small amounts of the intermediate dipeptides (20 mg.) and tripeptides (33 mg.) was more serious and required the chromatographic separation for their removal from the tetrapeptide. Their presence indicated that either the acetylation step, or more likely, the cleavage of the carbobenzoxy group was incomplete. Thus the appearance of leucylalanylvaline and leucylvaline must mean that some carbobenzoxyvaline remained after the first HBr–acetic acid treatment and was then deprotected by the second or third treatments. This could, of course, become serious if the synthesis of large polypeptides were contemplated. The presence of two peaks (at 1.25 l. and 1.34 l.) both containing leucine, alanine, glycine and valine also deserves some comment. The main component (92.5%) was shown by carboxypeptidase and leucine aminopeptidase digestions to be the desired all-L-peptide and the minor peak (7.5%) was found to be L-leucyl-L-alanylglycyl-D-valine. The explanation is clear when it was subsequently found that the sample of commercial carbobenzoxy-L-valine from which the peptides in this particular run were made contained 12% of carbobenzoxy-D-valine.

The purity and identity of the tetrapeptide was established in the following ways: (1) Rechromatography on Dowex 50 gave a single homogeneous peak. (2) Countercurrent distribution[11] for 100 transfers gave a single symmetrical peak which agreed closely with the calculated theoretical curve. (3) Paper chromatography gave a single spot in three solvent systems.[12] (4) The ratios of amino acids determined by quantitative amino acid analysis[13] of an acid hydrolysate were leucine 1.00, alanine 1.00, glycine 0.99, valine 1.00. (5) Optical purity was demonstrated by digestion with two enzymes. Leucine aminopeptidase completely digested the peptide (Table II). This showed the optical purity of L-leucine and L-alanine. Carboxypeptidase A completely liberated L-valine, demonstrating the optical purity of this amino acid. (6) The final proof was a comparison with the same tetrapeptide synthesized by conventional methods. The two peptides were identical by chromatography, optical rotation and infrared spectra.

For the conventional synthesis the stepwise p-nitrophenyl ester method of Bodanszky[7] was used. L-Valine methyl ester was coupled with carbobenzoxyglycine p-nitrophenyl ester and the product hydrogenated to give glycyl-L-valine methyl ester which was condensed with carbobenzoxy-L-alanine p-nitrophenyl ester. The protected tripeptide was hydrogenated and coupled with carbobenzoxy-L-leucine p-nitrophenyl ester to give analytically pure carbobenzoxy-L-leucyl-

(11) L. C. Craig, J. D. Gregory and G. T. Barry, Cold Spring Harbor Symposia on Quant. Biol., 14, 24 (1949).
(12) The sprays used should have detected any impurities of free or protected amino acids or peptides. The first was the ninhydrin spray, made by dissolving 2 ml. of acetic acid, 3 ml. of pyridine and 200 mg. of ninhydrin in 100 ml. of acetone. The second was the hypochlorite–KI spray for amides; see C. G. Greig and D. H. Leaback, Nature, 188, 310 (1960), and R. H. Mazur, B. W. Ellis and P. S. Cammarata, J. Biol. Chem., 237, 1619 (1962).
(13) S. Moore, D. H. Spackman and W. H. Stein, Anal. Chem., 30, 1185 (1958).

L-alanylglycyl-L-valine methyl ester. Carbobenzoxy-L-leucyl-L-alanylglycyl-L-valine was also obtained analytically pure after saponification of the methyl ester. Catalytic hydrogenation over palladium-on-carbon gave the free tetrapeptide L-leucyl-L-alanylglycyl-L-valine.

These experiments demonstrate the feasibility of solid phase peptide synthesis. For the new method to be of real value it must, of course, be applied to the synthesis of much larger peptides, preferably to those with biological activity. It must also be automated if large polypeptides of predetermined sequence are to be made. Encouraging experiments along these lines are now under way.

Experimental

The Polymer.—Copolymers of styrene and divinylbenzene of varying degrees of cross linking were obtained from the Dow Chemical Co.[14] The preparations were in the form of 200–400 mesh beads. The degree of cross linking determined the extent of swelling, the effective pore size and the mechanical stability of the beads, and these properties in turn determined the suitability of the polymers for peptide synthesis. The 1% cross-linked beads were somewhat too fragile and became disrupted during the peptide synthesis to an extent which made filtration difficult. On the other hand, the 8 and 16% cross-linked beads were too rigid to permit easy penetration of reagents, causing slower and less complete reactions. The 2% cross-linked resin was most useful and was employed for all the reactions reported here. The commercial product was washed thoroughly with M NaOH, M HCl, H_2O, dimethylformamide and methanol and dried under vacuum at 100°. The resulting material consisted of nearly white spheres ranging from approximately 20 to 80 μ in diameter. No attempt was made to obtain particles of more uniform size.

Chloromethylpolymer.[4]—Copolystyrene–2% divinylbenzene (50 g.) was swelled by stirring at 25° for 1 hr. in 300 ml. of chloroform and then cooled to 0°. A cold solution of 7.5 ml. of anhydrous $SnCl_4$ in 50 ml. of chloromethyl methyl ether was added and stirred for 30 min. at 0°. The mixture was filtered and washed with 1 l. of 3:1 dioxane–water, and then with 1 l. of 3:1 dioxane–2 N HCl. The cream-colored beads were washed further with a solution which changed gradually from water to pure dioxane and then progressively to 100% methanol. Abrupt changes of solvent composition were avoided. The product was dried under vacuum at 100°; yield 55.0 g. The product contained 1.89 mmoles of Cl/g., from which it could be calculated that approximately 22% of the aromatic rings of the polymer were chloromethylated. Conditions which would give this degree of substitution were purposely selected in order to minimize cross linking and to avoid starting peptide chains at the more difficultly accessible sites.

Nitrochloromethylpolymer.—A 50-g. sample of the dry chloromethylpolymer (1.89 mmole of Cl/g.) was added slowly with stirring to 500 ml. of fuming nitric acid (90% HNO_3, sp. gr. 1.5) which had been cooled to 0°. The mixture was stirred 1 hr. at 0° and then poured into crushed ice. The light tan beads were filtered and washed as described above with water, dioxane and methanol and dried; yield 69.9 g. The nitrogen content was 6.38 mmoles/g. by Dumas analysis. This was equivalent to 1.03 nitro groups per aromatic ring.

Bromochloromethylpolymer.—An 8.00-g. sample of chloromethylpolymer (4.92 mmoles of Cl/g.) was suspended in 25 ml. of carbon tetrachloride containing 100 mg. of iodine. A solution of 5 ml. of bromine in 10 ml. of CCl_4 was added and the mixture was stirred for 3 days in the dark at 25°. The deep red suspension was filtered and the polymer was washed with CCl_4, dioxane, water, sodium bicarbonate, water and methanol and dried; yield 12.5 g. The extent of substitution was 4.56 mmoles of Br/g., which was equivalent to 0.97 bromine atom per aromatic ring.

The Esterification Step. Carbobenzoxy-L-valyl Nitropolymer.—A solution of 15.5 g. (61.7 mmoles) of carbobenzoxy-L-valine and 8.65 ml. (61.7 mmoles) of triethylamine in 350 ml. of ethyl acetate was added to 40.0 g. of chloromethylnitropolymer and the mixture was stirred under reflux. After 48 hr., a solution of −Cbz { 12 ml. of acetic acid and 28 ml. of triethylamine in 100 ml. of ethyl acetate was added and the stirring and heating continued for another 4 hr. The resin was filtered and washed thoroughly with ethyl acetate, ethanol, water and methanol by gradual change of solvent composition under vacuum; yield 46.9 g. A 30-mg. sample of the resulting esterified resin was hydrolyzed by refluxing for 24 hr. in a mixture of 5 ml. of acetic acid and 5 ml. of 6 N HCl. Valine in the filtrate was determined

(14) Obtained from the Technical Service and Development Dept. of the Dow Chemical Co., Midland, Mich.

by the ninhydrin method and found to be 0.56 mmole/g. A chlorine determination (Parr fusion) indicated that only 0.08 mmole of Cl/g. remained. When carbobenzoxynitro-L-arginine was used in the corresponding reaction, ethyl acetate was replaced by dioxane as solvent because of the increased solubility of the triethylammonium salt.

The Reaction Vessel.—A reaction vessel was designed to make possible the process of automatic synthesis. The remaining reactions to be described were conducted within this vessel. It consisted of a 45 × 125 mm. glass cylinder, sealed at one end and fitted with a 40-mm. medium porosity fritted disk filter at the other end. Immediately beyond the disk the tubing was constricted sharply and sealed to a 1.5-mm. bore stopcock through which solvents could be removed under suction. A side arm, fitted with a drying tube, was used to introduce reagents and to remove solid samples for analysis. The apparatus was attached to a mechanical rocker which rotated the vessel 90° between the vertical and horizontal positions and provided gentle but efficient mixing of solvents and polymer. At the end of each reaction the vessel was stopped in the vertical position with the fritted disk at the bottom so that opening the stopcock allowed the solvents to be removed by suction.

The Deprotection Step. L-Valyl Nitropolymer.—Carbobenzoxy-L-valyl nitropolymer (10 g., 5.6 mmoles of valine) was introduced into the reaction vessel and 30 ml. of 30% HBr in acetic acid was added. After shaking for 5 hr. at 25°, the suspension was diluted with 50 ml. of acetic acid, filtered and washed with acetic acid, ethanol and dimethylformamide (DMF). The resulting hydrobromide was then neutralized by shaking for 10 min. in 30 ml. of DMF containing 3 ml. of triethylamine. The solvent containing the triethylammonium bromide was removed by suction and the polymer was washed with DMF. The filtrate contained 5.4 mmoles of halogen (Volhard titration), indicating a quantitative removal of the carbobenzoxy group.

The time required for the deprotection step was determined in two ways. In the first method 100-mg. samples of carbobenzoxy-L-valyl polymers (either nitrated, brominated or unsubstituted) were treated with 1 ml. of 10% or 30% HBr–HOAc. At intervals the samples were filtered and washed with acetic acid, ethanol and water. The resulting valyl polymer hydrobromides were then analyzed for halogen by Volhard titration. A measure of the rate of ester cleavage during the reaction was obtained by analyzing the filtrates for valine by the ninhydrin reaction.[10] These data are summarized in Table I. Completion of the reaction with 30% HBr–HOAc on carbobenzoxy-L-valyl nitropolymer required about 2 hr. and resulted in a loss of 4.3% of valine. The second determination of the rate of the reaction depended on coupling the free α-amino derivatives with another carbobenzoxyamino acid according to the procedures to be described below. The amount of the new amino acid thus added to the peptide was determined by hydrolysis and quantitative measurement of the amino acids. For this purpose the tripeptide derivative carbobenzoxy-L-alanylglycyl-L-valyl nitropolymer was used and carbobenzoxy-L-leucine was coupled to it. The leucine in the resulting tetrapeptide derivative was found to be 0.05 mmole/g. for the 0.5-hr. deprotection time. For times of 2, 4 and 6 hr., the leucine values were 0.14, 0.31 and 0.29 mmole/g. These results showed that removal of the carbobenzoxy group was maximal under these conditions within 4 hr. In order to ensure a complete reaction at this step a standard time of 5 hr. was adopted.

The Peptide-Forming Step. Carbobenzoxyglycyl-L-valyl Nitropolymer. A. The Diimide Method.—Carbobenzoxy-L-valyl nitropolymer (10 g., 5.6 mmoles of valine) was deprotected and neutralized as described above. Immediately thereafter a solution containing 2.34 g. (11.2 mmoles) of carbobenzoxyglycine in 20 ml. of dimethylformamide was added and shaken for 10 min. to allow for penetration of the reactant into the solid. A solution of 2.31 g. (11.2 mmoles) of dicyclohexylcarbodiimide in 4.6 ml. of DMF was then added and the suspension was shaken for 18 hr. at 25°. The polymer beads were sucked dry, washed with DMF, and an additional portion (1.17 g.) of carbobenzoxyglycine in 20 ml. of DMF was added. After 10 min., 1.16 g. of dicyclohexylcarbodiimide was added, the mixture was shaken for 2 hr. and then filtered and washed. A solution of 3 ml. of acetic anhydride and 1 ml. of triethylamine in 20 ml. of DMF was then added. After 2 hr., the resin was filtered and washed four times each with DMF, ethanol and acetic acid to remove excess reagents and by-products. Although dicyclohexylurea is a relatively insoluble compound, it was readily removed in this manner (solubility about 4, 10 and 17 mg./ml. respectively, in DMF, ethanol and acetic acid). A hydrolysate of a dried sample of the product showed glycine 0.53 mmole/g. and valine 0.53 mmole/g.

The reaction time allowed for the peptide-forming step could probably be reduced to much less than the 18 hr. employed here, although the exact reaction rate will depend on the particular amino acid derivative, the length of the peptide chain and perhaps other factors. To determine the approximate rate of the coupling reaction, glycyl-L-valyl nitropolymer was condensed as just described with a twofold excess of carbobenzoxy-L-alanine and dicyclohexylcarbodiimide in DMF. Portions of the suspension were removed at intervals, filtered, washed and dried. Weighed samples were hydrolyzed in 6 N HCl–acetic acid and analyzed quantitatively for amino acids by column chromatography.[13] Alanine was found to be 0.1, 0.2 and 0.4 mmole/g. after coupling times of 5, 10 and 30 min., and remained at this level up to 18 hr.

B. The Nitrophenyl Ester Method.—A solution of 3.70 g. (11.2 mmoles) of carbobenzoxyglycine p-nitrophenyl ester in 30 ml. of benzene was added to 10 g. of L-valyl nitropolymer hydrobromide (5.6 mmoles of valine) and stirred 1 hr. at 25° to allow swelling of the resin and penetration of the reagent. The swelling step was particularly important in order to get a satisfactory yield in this reaction. A solution of 5 ml. of triethylamine in 15 ml. of benzene was added and the mixture was stirred for 18 hr. at 80°. The mixture was cooled, 3 ml. of acetic anhydride was added, and the mixture was stirred for 2 hr. at 25°. The product was then filtered, washed with benzene, ethanol, water, and methanol and dried. A hydrolysate showed glycine 0.29 mmole/g., and valine 0.43 mmole/g.

The effectiveness of the acetylation step in blocking free α-amino groups was demonstrated in the following way. Samples (100 mg.) of L-valyl nitropolymer were stirred with acetic anhydride (0.3 ml.) and triethylamine (0.1 ml.) in DMF (2 ml.) for various lengths of time at 25°. The products were filtered, washed, and then saponified in ethanol (1 ml.) and 2 N NaOH (0.3 ml.) at 25° for 2 hr. Free valine was determined in the filtrate by the ninhydrin reaction.[10] The unacetylated control contained 0.72 mmole/g., while the 1.5-hr. and 24-hr. acetylation samples each contained only 0.0029 mmole/g.

The stability of the acetylated product to HBr–HOAc was determined by treating samples (100 mg.) with 30% HBr–HOAc (0.4 ml.) at 25° for 1 to 24 hr. The samples were then saponified as before and analyzed for free valine. In this experiment the untreated control showed 0.0015 mmole/g. and the 24-hr. sample 0.0015 mmole/g.

Carbobenzoxy-L-leucyl-L-alanylglycyl-L-valyl Nitropolymer.—The peptide chains were extended in a manner exactly analogous to the general procedures just described for the dipeptides. Carbobenzoxy-glycyl-L-valyl nitropolymer (from part A above), still in the reaction vessel, was treated with HBr–HOAc, neutralized with triethylamine and coupled with carbobenzoxy-L-alanine to give the tripeptide derivative. Without further manipulations another cycle was carried out using carbobenzoxy-L-leucine to give the desired tetrapeptide derivative. The same protected tetrapeptide polymer was also made by the nitrophenyl ester method.

The Peptide Liberation Step. L-Leucyl-L-alanylglycyl-L-valine.—The carbobenzoxy-L-leucyl-L-alanylglycyl-L-valyl nitropolymer described above was decarbobenzoxylated with HBr–HOAc in the usual way. It was then saponified by shaking for 1 hr. at 25° in a solution of 40.5 ml. of ethanol and 4.5 ml. of 2 N aqueous NaOH. The filtrate, withdrawn through the fritted disk of the reaction vessel by suction, was neutralized at once with HCl and the saponification step was repeated on the resin with more ethanolic sodium hydroxide for another hour. The resin was washed with ethanol and water and the neutralized filtrates were combined. Quantitative ninhydrin analysis of the filtrate gave a yield of 1.98 mmoles (leucine equivalents) and analysis of an acid hydrolysate gave 6.92 mmoles. The ratio of 3.49 instead of the 4.0 expected for a tetrapeptide indicated that some amino acids or shorter chain peptides were also present. Since the ninhydrin value of an acid hydrolysate of the carbobenzoxy tetrapeptide polymer before saponification was 12.8 mmoles, the yield in the saponification step was 54%.

The tetrapeptide was purified and isolated by ion-exchange chromatography on a 2 × 120 cm. column of Dowex 50-X4 (200–400 mesh) resin using 0.1 M pH 4 pyridine acetate buffer. It was eluted at 1 ml./min. and collected in 5.4-ml. fractions. Aliquots (0.2 ml.) of the fractions were analyzed by the ninhydrin method. Tubes comprising the peaks were pooled and their contents were weighed after evaporation of the buffer. The fractions were identified by paper chromatography of their acid hydrolysates. The results are summarized in Fig. 2. The acetylated peptides emerged after only slight retardation as a group of four poorly resolved fractions. They were thus easily and completely separated from the desired tetrapeptide which emerged at 1.25 l. The portion of the fraction between 1.12 and 1.29 l. was combined, evaporated to dryness, dissolved in 20 ml. of ethanol, filtered and precipitated with 100 ml. of dry ether. The product was centrifuged, washed, and dried under vacuum at 60°. The yield was 265 mg. of a cream-colored, hygroscopic powder. Rechromatography on the same column gave a single sharp peak at 1.20 l. which indicated that L-leucyl-L-alanylglycyl-D-valine had been completely removed.

A 200-mg. sample of the tetrapeptide was countercurrented for 100 transfers in a solvent (10 ml. each phase) made from 2000 ml. of redistilled sec-butyl alcohol, 22.2 ml. of glacial acetic acid

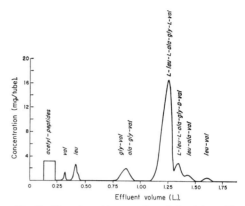

Fig. 2.—Chromatographic separation of the tetrapeptide L-leucyl-L-alanylglycyl-L-valine on a 2 × 120 cm. column of Dowex 50-X4 resin. Elution was with 0.1 M pyridine acetate buffer, pH 4.0. The concentration was calculated from the ninhydrin analysis of individual tubes and dry weight of pooled fractions.

and 1977 ml. of doubly distilled water. A single symmetrical peak resulted which was located by ninhydrin analysis on 0.2-ml. aliquots of both phases. The distribution coefficient was 0.32 and was constant between tubes 16 and 33. This fraction was combined and evaporated to dryness; yield 195 mg. The peptide was dissolved in 2 ml. of ethanol and crystallized by the addition of 5 ml. of dioxane; yield 76 mg., $[\alpha]^{21}\text{D}$ +18.0° (c 2, ethanol); R_f 0.71 (propanol–H$_2$O, 2:1), 0.73 (sec-BuOH–formic acid),[15] 0.49 (sec-BuOH–ammonia)[15]; amino acid ratios[13]: leucine 1.00, alanine 1.00, glycine 0.99, valine 1.00.

Anal.[16] Calcd. for C$_{16}$H$_{30}$N$_4$O$_5$: C, 53.6; H, 8.4; N, 15.6. Found: C, 52.6; H, 8.2; N, 15.2.

Optical Purity of the Peptides.—The possibility of racemization of the leucine and alanine residues was examined with use of leucine aminopeptidase (LAP) by a method similar to that described by Hofmann.[17] The peptide was incubated with Worthington LAP (1 mg./20 mg. of peptide) for 2 days at 37°. The digest was analyzed on a quantitative amino acid column and a comparison of the ratios of amino acids found after enzymatic hydrolysis to those found after acid hydrolysis was used as a measure of racemization (Table II). Within the experimental error of about 5% it was found that the diimide method caused no racemization in the synthesis of the tetrapeptide. On the other hand, the nitrophenyl ester procedure gave a product containing approximately 17% of D-leucine and 12% of D-alanine. Because the tetrapeptide isolated from the diimide route showed no evidence of racemized residues and because 92.5% of the total tetrapeptides in the reaction mixture was isolated as pure product, it was clear that racemization was not a problem with synthesis by the diimide method.

The configuration of valine was established by carboxypeptidase digestion using Worthington carboxypeptidase A.[18] In this case, hydrolysis was followed by paper chromatography in sec-butyl alcohol–ammonia.[15] The spot for tetrapeptide (R_f 0.49) progressively decreased while spots for valine (R_f 0.29) and leucyl-alanylglycine (R_f 0.33) increased. After 3 hr., the reaction was essentially complete and the total ninhydrin value had doubled, indicating that the glycylvaline bond was completely digested and therefore that the valine had not been racemized. The second tetrapeptide isolated from this run (peak at 1.34 l.) which represented 7.5% of the total was resistant to carboxypeptidase. After 20 hr. only unchanged tetrapeptide (R_f 0.49) was present. It therefore contained D-valine. This was presumed to have originated from the commercial carbobenzoxyvaline used as starting material in this run since the latter was subsequently shown to be 12% racemized.

(15) W. Hausmann, *J. Am. Chem. Soc.*, **74**, 3181 (1952).

(16) The several samples of L-leucyl-L-alanylglycyl-L-valine prepared both by the new method and by the conventional procedure have, in every case, given analytical values for C, H and N which were a few per cent lower than theory. The data for the different preparations, however, have always agreed well with one another. The values given were for samples dried for 18 hr. at 100° under vacuum over P$_2$O$_5$. Other conditions were tried but did not change the results.

(17) K. Hofmann and H. Yajima, *J. Am. Chem. Soc.*, **83**, 2289 (1961).

(18) H. Neurath, E. Elkins and S. Kaufman, *J. Biol. Chem.*, **170**, 221 (1947).

Conventional Synthesis of the Tetrapeptide. A. Carbobenzoxyglycyl-L-valine Methyl Ester.—To 7.13 g. (0.0425 mole) of L-valine methyl ester hydrochloride in 30 ml. of methylene chloride at 0° was added 12 ml. (0.085 mole) of triethylamine, followed by a solution of 14 g. (0.042 mole) of carbobenzoxyglycine p-nitrophenyl ester in 50 ml. of methylene chloride. After standing 24 hr. at room temperature the clear yellow solution was washed with N ammonium hydroxide until colorless, then with N hydrochloric acid and water. The dried product was an oil weighing 10.3 g. (75%). Thin layer chromatography on silicic acid in chloroform–acetone (10:1) gave one spot, R_f 0.75, by the hypochlorite–KI spray for amides.[12] The oil was used directly for the next step.

B. Carbobenzoxy-L-leucyl-L-alanylglycyl-L-valine Methyl Ester.—Carbobenzoxy-glycyl-L-valine methyl ester (5.1 g., 15.8 mmoles) was hydrogenated for 2 hr. at 25° in 50 ml. of methanol containing 15.8 ml. of N HCl and 500 mg. of 5% palladium-on-carbon. The product was dissolved in 20 ml. of methylene chloride, cooled to 0° and 4.4 ml. of triethylamine was added. Carbobenzoxy-L-alanine p-nitrophenyl ester (5.44 g., 15.8 mmoles) in 20 ml. of methylene chloride was added and the mixture was stirred for 3 days at 25°. The solution was washed with NH$_4$OH, HCl and water and dried. The solution of carbobenzoxy-L-alanylglycyl-L-valine methyl ester was evaporated to dryness and the oil (3.43 g., 55%) was dissolved in methanol for hydrogenation. The product was then coupled with 2.88 g. (7.65 mmoles) of carbobenzoxy-L-leucine p-nitrophenyl ester in a manner similar to that described for the previous step. The yield of crude carbobenzoxy-L-leucyl-L-alanylglycyl-L-valine methyl ester was 2.80 g. (64% from the protected tripeptide), m.p. 155–157°. For analysis the product was recrystallized twice from ethyl acetate–petroleum ether; m.p. 160–161°, $[\alpha]^{21}$D −31.8° (c 2, ethanol).

Anal. Calcd. for C$_{25}$H$_{38}$N$_4$O$_7$: C, 59.25; H, 7.57; N, 11.07. Found: C, 59.21; H, 7.26; N, 11.08.

C. Carbobenzoxy-L-leucyl-L-alanylglycyl-L-valine.—Carbobenzoxy-L-leucyl-L-alanylglycyl-L-valine methyl ester (662 mg., 1.3 mmoles) was dissolved in 2.5 ml. of ethanol and 0.28 ml. of 5 N NaOH was added. After 60 min. at 25° the solution was diluted with water and extracted with ethyl acetate. The aqueous phase was acidified with HCl and the product extracted into ethyl acetate. Evaporation of the dried solution gave 520 mg. (81%), m.p. 194°. Two recrystallizations from ethanol–water raised the m.p. to 199–200°, $[\alpha]^{21}$D −23.8° (c 2, ethanol).

Anal. Calcd. for C$_{24}$H$_{36}$N$_4$O$_7$: C, 58.52; H, 7.37; N, 11.37. Found: C, 58.55; H, 7.00; N, 11.30.

D. L-Leucyl-L-alanylglycyl-L-valine.—Carbobenzoxy-L-leucyl-L-alanylglycyl-L-valine (493 mg., 1.0 mmole) was dissolved in 50 ml. of ethanol and hydrogenated for 24 hr. in the presence of 50 mg. of 5% palladium-on-carbon. The mixture was filtered and evaporated to dryness. The residue was redissolved in 2 ml. of ethanol and filtered again and the tetrapeptide was then precipitated with dry ether; yield 343 mg. (96%), $[\alpha]^{21}$D +17.5° (c 2, ethanol); R_f 0.71 (propanol–H$_2$O), 0.73 (sec-butyl alcohol–formic acid),[14] 0.49 (sec-butyl alcohol–ammonia[15]); amino acid ratios[13]: leucine 0.98, alanine 1.00, glycine 1.00, valine 1.01.

Anal.[16] Calcd. for C$_{16}$H$_{30}$N$_4$O$_5$: C, 53.6; H, 8.4; N, 15.6. Found: C, 52.7; H, 8.1; N, 14.9.

Acknowledgments.—The author wishes to express his sincere appreciation to Dr. D. W. Woolley for his interest and advice during this work. The author also is grateful to Dr. John Stewart for many helpful suggestions and discussions and to Miss Angela Corigliano and Miss Gretchen Grove for their technical assistance.

Copyright © 1963 by the American Chemical Society

Reprinted from *J. Amer. Chem. Soc.*, **85**, 3045–3046 (Oct. 5, 1963)

Popcorn Polymer as a Support in Multistep Syntheses[1]

ROBERT L. LETSINGER and MILTON J. KORNET

Sir:

The recent publication by Merrifield[2] describing the synthesis of a tetrapeptide on polymer beads prompts us to report at this time experiments carried out independently which demonstrate the feasibility of using a modified "popcorn" polymer as a supporting matrix in repetitive-step syntheses.

In principle it appears that the manipulations involved in the stepwise synthesis of polypeptides and polynucleotides may be materially reduced, thereby rendering possible the construction of more complex substances, if the polymer is held to a solid support throughout the synthetic sequence. To function properly the support should (a) be insoluble in the solvents and inert to the reagents employed, (b) contain a functional group to which the initial monomer unit may be joined and from which the final product may be removed, and (c) possess a structure that will permit diffusion of reagents into the reactive sites and the product out into solution.

We selected styrene–divinylbenzene "popcorn" polymer[3] for study since insoluble polymers of this type may be obtained which have a very low degree of cross linking. Diffusion problems should therefore be less serious than with conventional resins, which must have a relatively high degree of cross linking to be suitable as a support. The experiments herein described were carried out with polymer 99.5% in styrene and 0.5% in divinylbenzene.[4] More recently, insoluble, low swelling polymer of excellent quality has been obtained which contains only 0.1% divinylbenzene.[5] Although unnecessary insofar as the chemical reactions are concerned, it was found convenient to cut the low density popcorn polymer into small particles with a Waring blendor. In this form the polymer can be packed to give a rapid draining column and may be separated from liquid suspensions readily by filtration.

Carboxyl groups (0.33 mequiv./g.) were introduced by treating the polymer, suspended in nitrobenzene, with diphenylcarbamyl chloride and aluminum chloride and hydrolyzing the resulting amide with a mixture of sulfuric acid, acetic acid, and water. In these and subsequent steps the polymer product was recovered quantitatively simply by filtration. Transformations at the side chains were followed by changes in the infrared absorption bands arising from the carbonyl groups. Thus, the diphenylcarboxamido polymer exhibited a strong band at 6.02 μ whereas the carboxy-polymer absorbed at 5.90 μ. Chemically, the functional groups joined to the polymer behaved normally. Carboxy polymer was converted quantitatively to a hydroxymethyl polymer by reduction with lithium aluminum hydride in ether, to the acid chloride (λ 5.58 μ) by treatment with thionyl chloride, and to a methyl ester (λ 5.90 μ) on reaction with diazomethane in ether. Reaction of the acid chloride with p-phenylenediamine and ethylenediamine afforded amides (λ 6.00 μ).

As a test of the synthetic applicability of the support, the hydroxymethyl derivative was used in the preparation of leucylglycine. Phosgene in benzene converted the hydroxymethyl polymer to the chloroformyl derivative (λ 5.62 μ), which with L-leucine ethyl ester hydrochloride in dimethylformamide in the presence of triethylamine afforded Ⓟ-leucine ethyl ester[6] (λ 5.78 μ). Alkaline hydrolysis at room temperature yielded Ⓟ-leucine (λ 5.80 μ; 0.25 mequiv. of acid/g. of polymer by titration). On successive treatment of the leucine derivative with (a) isobutyl chlorocarbonate and triethylamine in toluene and (b) glycine benzyl ester p-toluenesulfonate and triethylamine in dimethylformamide, Ⓟ-leucylglycine benzyl ester was formed. The dipeptide was cleaved from the polymer with 15% hydrobromic acid in acetic acid and was precipitated as the hydrobromide by addition of ether to the solution of cleavage products. Neutralization in methanol solution with Amberlite CG-400 (OH⁻) yielded leucylglycine. For characterization of the products from the popcorn polymer, the hydrobromide salt was chromatographed on paper (descending) with 1-butanol–ethanol–water (5:1:4), which gave two faint spots corresponding to glycine hydrobromide (R_f 0.25) and leucine hydrobromide (R_f 0.71) and a major spot corresponding to leucyl glycine hydrobromide (R_f 0.61). Elution of the last, hydrolysis with acid, and paper chromatography of the products revealed leucine and glycine as the constituents of the dipeptide. As further confirmation, the mixture of hydrobromide salts from the ether precipitation was analyzed[7] on a Beckman/Spinco Model 120 B amino acid analyzer according to the method of Moore, *et al.*[8] Leucylglycine, glycine, and leucine were found in the relative amounts 89.2, 7.2, and 3.6 mole %.

Experiments aimed at utilization of polymer supports in the synthesis of relatively large peptides and of oligonucleotides are in progress.

---

(1) This research was supported by the National Science Foundation, grant G25069, and by the Division of General Medical Sciences of the National Institutes of Health, grant GM 10265-01.
(2) R. B. Merrifield, *J. Am. Chem. Soc.*, **85**, 2149 (1963).
(3) For discussion of the chemistry of popcorn polymers see J. L. Amos, K. E. Coulter, and F. M. Tennant in "Styrene," edited by R. H. Boundy and R. F. Boyer, Reinhold Publishing Corp., New York, N. Y., 1952, p. 729; E. H. Immergut, *Makromol. Chem.*, **10**, 93 (1953); R. L. Letsinger and S. B. Hamilton, *J. Am. Chem. Soc.*, **81**, 3009 (1959).
(4) Some of the polymerization reactions were studied in these Laboratories by Dr. Merlin Guinard.
(5) Merrifield (ref. 2) used resin beads that were 2% in divinylbenzene. His attempts to use 1% cross-linked resin were unsuccessful due to fracturing of the beads, whereas with 8 and 10% cross-linked beads the reaction rates were too slow to be practical.
(6) For simplicity, the insoluble blocking group for the amino function is designated here by the symbol Ⓟ-.
(7) The analysis was performed by K. A. Thompson.
(8) S. Moore, D. H. Spackman, and W. H. Stein, *Anal. Chem.*, **30**, 1185 (1958); D. H. Spackman, W. H. Stein, and S. Moore, *ibid.*, **30**, 1190 (1958).

DEPARTMENT OF CHEMISTRY
NORTHWESTERN UNIVERSITY
EVANSTON, ILLINOIS

RECEIVED AUGUST 19, 1963

II

Polypeptide Synthesis

Editors' Comments on Papers 4 Through 21

4 **Gutte and Merrifield:** *Synthesis of Ribonuclease A*

5 **Merrifield:** *The Solid Phase Synthesis of Ribonuclease A*

6 **Marshall and Merrifield:** *Synthesis of Angiotensins by the Solid-Phase Method*

7 **Fridkin, Patchornik, and Katchalski:** *A Synthesis of Cyclic Peptides Utilizing High Molecular Weight Carriers*

8 **Fridkin, Patchornik, and Katchalski:** *Use of Polymers as Chemical Reagents: I. Preparation of Peptides*

9 **Fridkin, Patchornik, and Katchalski:** *Use of Polymers as Chemical Reagents: II. Synthesis of Bradykinin*

10 **Flanigan and Marshall:** *Synthesis of Cyclic Peptides on Dual Function Supports*

11 **Gisin and Merrifield:** *Synthesis of a Hydrophobic Potassium Binding Peptide*

12 **Wang and Merrifield:** *Preparation of a t-Alkyloxycarbonylhydrazide Resin and Its Application to Solid Phase Peptide Synthesis*

13 **Wang:** *p-Alkoxybenzyl Alcohol Resin and p-Alkoxybenzyloxycarbonylhydrazide Resin for Solid Phase Synthesis of Protected Peptide Fragments*

14 **Felix and Merrifield:** *Azide Solid Phase Peptide Synthesis*

15 **Gisin:** *The Preparation of Merrifield-Resins Through Total Esterification with Cesium Salts*

16 **Parr and Grohmann:** *A New Solid Support for Polypeptide Synthesis*

17 **Bayer, Sebestian, Jung, and Halász:** *A New Support for Polypeptide Synthesis in Columns*

18 **Rich and Gurwara:** *Removal of Protected Peptides from an ortho-Nitrobenzyl Resin by Photolysis*

19 **Bayer, Gil-Av, König, Nakaparksin, Oró, and Parr:** *Retension of Configuration in the Solid Phase Synthesis of Peptides*

20 **Merrifield, Stewart, and Jernberg:** *Instrument for Automated Synthesis of Peptides*

21 **Hancock, Prescott, Marshall, and Vagelos:** *Acyl Carrier Protein*

By far the most successful application of solid-phase synthesis (SPS) has been in the area of oligopeptide chemistry. It is here that the apparent need has existed for new techniques to overcome the many problems of polypeptide synthesis. Among other advantages, the most important has been the ease of purifying the desired unit from various side products attendant in this type of chemistry. The use of polypeptide–resin combination allows the easy separation of the desired amino acid unit from side reaction products since the former would be attached to an insoluble resin. The following papers were selected to demonstrate the various facets involved in solid-phase synthesis.

The total power of the SPS method is probably best depicted by Paper 4, of Gutte and Merrifield, and the accompanying report of Merrifield (Paper 5). Comparison of the yield and scale of the solid-phase synthesis to the classical method of Hirschmann et al. (1969) shows a dramatic difference. The SPS method produced milligram quantities in a relatively short period of time, whereas solution methods resulted in the isolation of only micrograms of the enzyme.

One of the earliest syntheses of a moderately large peptide by means of SPS is reported by Marshall and Merrifield in Paper 6. This work is interesting, in that direct coupling of the chloromethylated polystyrene and t-BOC phenylalanine was used. Analysis showed that only 15 percent of the chlorine site had been displaced by the amino acid. Later changes in the procedure increased the yield considerably for this type of coupling step.

A different type of polymer and general technique is found in Papers 7, 8, and 9 by Patchornik and coworkers. The first of the papers deals with a soluble acetylated poly-4-hydroxy-3-nitrostyrene, which gave comparable yields of cyclic peptide as a similar insoluble polymer. Papers 8 and 9 describe a unique technique whereby an activated ester attached by a phenolic linkage to the resin is displaced by a free amino acid (or amino acid ester). The resulting N-acyl amino acid is then released into solution. Thus the technique eliminates the usually difficult cleavage step. Also, the insoluble active ester can be used in large excess without contaminating the reaction mixture and yet achieve high yields on the coupling step. The major drawback is the use of a different amino acid ester polymer for each step.

Flanigan and Marshall (Paper 10) use a 4-(methylthio)phenyl group attached to polystyrene as a means of linking amino acids to the polymer. The interesting aspect is that the thio group is oxidized to a sulfone, making the cyclization of the peptide extremely facile. Obviously this method cannot be used with sulfur-containing amino acids because of the oxidation step involved.

Another example of the synthesis of a cyclic peptide is reported by Gisin and Merrifield in Paper 11. Some difficulty was encountered with the coupling of proline to the resin. Paper 12 describes the work of Wang and Merrifield in yet a different approach to attachment of the peptide chain to the resin. Here the main advantage is in the cleavage of the completed peptide from the polymer resin. In this case the hydrolyzed peptide is protected as a hydrazide derivative. Under these conditions the protected side chains of the peptide remain intact, allowing for possible further transformations. Yet another modification of the Merrifield resin, described by Wang in Paper 13, uses a p-methoxybenzyl ester of p-methoxybenzyloxycarbonylhydrazide as the linkage to the resin. The very large and long group attached to the polystyrene backbone may be responsible for its apparent success. Through the use of this type of attachment, even the first amino acid unit is quite remote from the polystyrene. The method is also attractive in that it gives high cleavage yields of both the blocking group and the completed peptide chain. The azide group on a polymer has also been used as the means of attachment of free amino groups (Paper 14). Felix and Merrifield demonstrate that this method has the advantages found in the use of a *t*-butyloxycarbonyl group (*t*-BOC) as an N-blocking group. The authors do not suggest that this method would replace the more conventional methods of solid-state synthesis.

One problem mentioned earlier is that the coupling steps must be of high yield, a problem of both solution and solid-phase synthesis involving many steps. Gisin demonstrates in Paper 15 that use of cesium salts of N-protected amino acids directly with chloromethylated polystyrene results in unusually high yields of coupling products. In comparison to the use of triethyl amine as the condensing base, the cesium salt method yields 14 times more substitution. Parr and Grohmann addressed the problems associated with chloromethylated polystyrene beads in Paper 16. They used a silanized glass bead in which *p*-bromomethylphenyl siloxy groups were bound in a monomolecular layer to the glass. Considerably shorter reaction times were noted in comparison to chloromethylated polystyrene. One apparent difficulty with this method is found in the cleavage of the completed polypeptide, which achieves only 50 percent conversion to the free peptide. Paper 17 of Bayer et al. reports a novel method for peptide formation employing columns. Here, as in the previous paper, a monomolecular layer is bound to silica and then placed in a chromotographic column. Increased reactivity is also noted, owing to the absence of diffusion through a normal bead.

Several of the preceding papers have touched on the problem of cleavage of the completed peptide from the resin. A recent communication of Rich and Gurwara (Paper 18) reports that cleavage can be accomplished by a photolysis reaction. The resin used was an *o*-nitrobenzyl type of polystyrene. One of the first ideas to suggest itself concerning the use of a photochemical cleavage process is the possibility of photolytic-induced reactions of tryptophan and tyrosine. The authors demonstrate that both amino acids can be recovered in fairly good yield. Similar questions can be raised with regard to the highly acidic conditions necessary for conventional cleavage reactions and the effect upon asymmetric centers. Paper 19 by Bayer et al. is addressed to this question. The authors also investigated other reaction conditions found in SPS and the effect upon optical activity. The solid-phase method is quite complicated and yet has a feature of redundancy once the plan of attack has been worked out. A tremendous advantage of the method would be to automate it. Paper 20, describing such an instrument for automation by Merrifield, Stewart, and Jernberg, appeared early in the literature of SPS. Since then a number of commercial instruments for automated peptide synthesis have appeared on the market.

Finally, the "state of the art" in SPS of polypeptides is probably best represented by the synthesis [Hancock et al. (Paper 21)] of a protein possessing acyl carrier protein activity. Here a 74-amino acid residue protein is prepared not only by the solid-phase method, but is accomplished on an automated basis (see Paper 20 for a description of an automated system). A number of problems associated with the formation of a large polypeptide are identified and successfully solved.

In addition to the selected papers found in this section, an excellent information source on practical problems of polypeptide synthesis is the book of J. M. Stewart and J. D. Young, *Solid Phase Peptide Synthesis* (1969).

References

Hirschmann, R., R. F. Nutt, D. F. Veber, R. A. Vitali, S. L. Varga, T. A. Jacob, F. W. Holly, and R. G. Denkewalter (1969). "Studies on the Total Synthesis of an Enzyme. V. The Preparation of Enzymatically Active Material," *J. Amer. Chem. Soc.*, **91,** 507.

Stewart, J. M., and J. D. Young (1969). *Solid Phase Peptide Synthesis*. W. H. Freeman and Company, San Francisco.

Copyright © 1971 by the American Society of Biological Chemists, Inc.

Reprinted from *J. Biol. Chem.*, **246**(6), 1922–1941 (1971)

The Synthesis of Ribonuclease A*

(Received for publication, September 4, 1970)

BERND GUTTE AND R. B. MERRIFIELD

From The Rockefeller University, New York, New York 10021

SUMMARY

A protected linear polypeptide of 124 amino acid residues with the sequence of bovine pancreatic ribonuclease A was synthesized by the solid phase method. The polypeptide was removed from the solid support and purified, and the four disulfide bonds were closed by air oxidation of the reduced form. The synthetic enzyme was fractionated by gel filtration and ion exchange chromatography, and the material that eluted at the same position as natural reduced-reoxidized RNase A was isolated. The product was further purified by incubation with trypsin and removal of the enzymically degraded components. A fractional precipitation with $(NH_4)_2 \cdot SO_4$ gave a purified synthetic RNase A with a specific activity of 78%.

The synthetic RNase A was indistinguishable from natural bovine pancreatic RNase A by gel filtration on Sephadex G-75, by chromatography on carboxymethylcellulose, and by electrophoresis. Amino acid analyses, peptide maps of tryptic digests, and the Michaelis constant agreed well with those of the natural enzyme. The synthetic enzyme showed the high substrate specificity to be expected of RNase A. It was highly active against yeast RNA and 2′,3′-cyclic cytidine phosphate and was completely inactive toward DNA, 2′,3′-cyclic guanosine phosphate, and the dinucleotides 5′-(3′-guanylyl)-cytidylic acid and 5′-(3′-adenylyl)-adenylic acid.

During the synthesis, samples were removed after 99 and 104 residues had been coupled and the corresponding polypeptides, des-(21–25)S-protein and S-protein, were isolated. They were then reduced and reoxidized in the presence of natural or synthetic S-peptide to give RNase S and des-(21–25)RNase S. The specific activities of the two products were approximately equal, thus indicating that the five NH_2-terminal residues of S-protein are not required for the protein to oxidize and fold in the presence of S-peptide to give an active enzyme.

The chemical synthesis of bovine pancreatic ribonuclease A was undertaken in an effort to provide a new approach to studies on the relation of structure to function in enzymes. The accessibility of synthetic enzymes, and of derivatives of them, can lead to information about binding sites, catalytic sites, cross-linking, and folding that is not readily attainable by other

* This work was supported in part by Grant A-1260 from the United States Public Health Service.

methods. Ribonuclease A was selected because it was a relatively small, stable, crystalline protein which had been studied in great detail in many laboratories (1–3). Both its linear amino acid sequence (4–8) and its crystallographic structure (9, 10) were known. Furthermore, it had been established (11–13) that the denatured, reduced chain could be reoxidized and refolded into a structure possessing full enzymic activity. Thus, the chemical synthesis of the linear sequence of ribonuclease could be expected to lead to a total synthesis of the enzyme.

The solid phase method (14, 15) was chosen for the synthesis because of the advantages in yield, speed, simplicity, and manpower requirements. The synthetic enzyme possessed high specific activity and closely resembled the natural enzyme. A preliminary account of our synthesis of ribonuclease A was published last year (16) and simultaneously a synthesis of ribonuclease S was reported by Hirschmann *et al.* (17). The details of our original synthesis together with some further findings are described here.

MATERIALS AND METHODS

Reagents

Boc-amino acids[1] were purchased from Schwarz BioResearch and Cyclo Chemical Company. Their thin layer chromatographic behavior, melting points, and optical rotations were checked prior to use. The 1% cross-linked styrene-divinylbenzene resin (100 to 200 mesh beads) was a gift from the Dow Chemical Company. Natural RNase A (RAF, phosphate-free) for comparative studies and yeast RNA were obtained from Worthington. Natural S-peptide and natural S-protein were a gift from Dr. Erhard Gross, National Institutes of Health, Bethesda, Maryland. All chemicals used were analytical reagents. Urea was deionized over AG 501-X8 (D) mixed bed resin. The optical densities of column effluents were recorded on an ultraviolet analyzer (model UA-2, ISCO, Inc.) at 254 nm, then the appropriate fractions were read at 280 nm in the Beckman spectrophotometer. When indicated, the protein concentration of a solution was determined by the ninhydrin reaction following an alkaline hydrolysis (18). Free peptides were hydrolyzed in 6 N HCl in sealed, evacuated tubes for 24 hours at 110°. Peptide resins (10 mg) were suspended in a mixture of 2 ml of 12 N HCl, 1 ml of acetic acid, and 1 ml of phenol,[2] and

[1] The nomenclature and symbols follow the Tentative Rules of the IUPAC-IUB Commission on Biochemical Nomenclature (*J. Biol. Chem.*, **241**, 2491 (1966) and **242**, 555 (1967)).
[2] First used by Dr. M. Zaoral during his tenure as a Visiting Investigator in this laboratory.

```
                Z  OBz1Bz1       Z      OBz1NO2     O   OBz1Bz1 Bz1 Bz1 Bz1           Bz1 Bz1 Bz1       Bz1 Bz1
      Boc-Lys-Glu-Thr-Ala-Ala-Ala-Lys-Phe-Glu-Arg-Gln-His-Met-Asp-Ser-Ser-Thr-Ser-Ala-Ala-Ser-Ser-Ser-Asn-Tyr-Cys-Asn-Gln
        1                           10                           20                                           Met→O
                                                                                                           30 Met→O
                                                                                                              Lys-Z
             Bz1 Bz1                   OBz1        Bz1 OBz1           Bz1             Z   Bz1 NO2 OBz1Z  Bz1  Ser-Bz1
        60 Gln-Ser-Cys-Val-Ala-Gln-Val-Asp-Ala-Leu-Ser-Glu-His-Val-Phe-Thr-Asn-Val-Pro-Lys-Cys-Arg-Asp-Lys-Thr-Leu-Asn-Arg-NO2
                                              50                              40
        Z-Lys
        Asn
        Val  Z          Bz1      Bz1 Bz1        Bz1 Bz1 Bz1 Bz1 O    Bz1    Bz1 OBz1Bz1 NO2 OBz1Bz1       Bz1 Bz1 Z  Bz1
        Ala
        Bz1-Cys-Lys-Asn-Gly-Gln-Thr-Asn-Cys-Tyr-Gln-Ser-Tyr-Ser-Thr-Met-Ser-Ile-Thr-Asp-Cys-Arg-Glu-Thr-Gly-Ser-Ser-Lys-Tyr
                            70                           80                                          90     Pro
                                                                                                            Asn
                                                                                                            Cys-Bz1
            Bz1      OBz1            Bz1        OBz1Bz1                        Z          Bz1 Bz1 Z         Ala
        Resin-Val-Ser-Ala-Asp-Phe-His-Val-Pro-Val-Tyr-Pro-Asn-Gly-Glu-Cys-Ala-Val-Ile-Ile-His-Lys-Asn-Ala-Gln-Thr-Thr-Lys-Tyr-Bz1
                      120                                110                                      100
```

FIG. 1. Fully protected synthetic ribonuclease A-resin

the tubes were cooled, evacuated, and sealed. They were heated at 110° for 24 hours, and then cooled before opening. The resin was removed by filtration and washing, and the filtrate was extracted with chloroform. It was evaporated to dryness and made to volume for amino acid analysis.

To avoid problems of determining dry weights of proteins all quantities of RNase samples and standards were determined from amino acid analyses. The mean of the molar ratios of all of the accurately measurable amino acids in an acid hydrolysate was used to calculate the concentration of the protein.

Yields for solid phase peptide synthesis were calculated from the amount of the first amino acid attached to the resin. These values were based on the limiting reactant rather than on the excesses of amino acid reagents that were used in the synthesis but, if desired, can be expressed in that way from the recorded excesses.

Ribonuclease A Assays

When yeast RNA was the substrate RNase A activities were determined either by the Kunitz method (19) from the initial velocities of the hypochromic shift at 300 nm or by spectrophotometric measurement at 260 nm of the acid-soluble oligonucleotides of the digestion mixtures (20, 21). The unknown samples were compared with natural RNase A standard solutions whose concentrations were 2, 1.5, and 1 μg per ml of assay mixture for the Kunitz method, and 0.0525, 0.0350, and 0.0175 μg per ml of assay mixture for the assay measuring the acid-soluble oligonucleotides produced by the action of RNase A.

Specific activity is expressed as the enzymic activity of the sample divided by the activity of an equimolar amount of pure native RNase A times 100.

N^α-*t*-*Butyloxycarbonyl*-L-*histidine*—N^α-*t*-Butyloxycarbonyl-N^{im}-benzyl-L-histidine (2.00 g) was dissolved in 75 ml of methanol containing 5 ml of water, 3 g of catalyst (5% palladium on barium sulfate, Engelhard Industries) were added, and hydrogenolysis was carried out at 50 p.s.i. for 24 hours. The mixture was filtered, and after partial evaporation and addition of water the filtrate was lyophilized. The resulting white, amorphous powder was dried in a vacuum at 80° over KOH pellets. Yield: 1.25 g (84%).

$$C_{11}H_{17}N_3O_4$$

Calculated: N 16.46, C 51.76, H 6.71
Found: N 16.24, C 51.54, H 6.68

R_F values from thin layer chromatography on silica gel (Solvent: 1-butanol-acetic acid-water-pyridine, 30:6:24:20); N^α-Boc-N^{im}-benzylhistidine, 0.84; N^α-Boc-histidine, 0.68.

EXPERIMENTAL PROCEDURE AND RESULTS

Solid Phase Synthesis of Protected Linear 124-Amino Acid Residue Polypeptide Chain of Ribonuclease A

The general procedures of the automated solid phase method were followed (22, 23), but with several changes in detail. The synthesis began with the attachment of the carboxyl-terminal amino acid, valine, to the resin support and continued with the stepwise addition of protected amino acids until the 124-amino acid chain was assembled on the resin (Fig. 1). The peptide was then cleaved from the resin, deprotected, purified, and isolated as described below.

Polystyrene-1%-divinylbenzene beads (100 to 200 mesh) were characterized by swelling experiments in two ways. Microscopic examination showed the average diameter of the dry beads to be 160 ± 20 μ (range 100 to 200 μ). After swelling in dry dimethylformamide the average diameter was 230 ± 16 μ, and in CH_2Cl_2 it was 290 ± 15 μ. The volume of 1.00 g of packed dry beads was 1.70 ml. After swelling in dimethylformamide the volume was 5.20 ml, and in CH_2Cl_2 it was 8.60 ml.

The styrene-1%-divinylbenzene beads (50 g) were thoroughly washed (14) and then were chloromethylated in 295 ml of redistilled chloromethylmethyl ether at 0° by the dropwise addition of 5 ml of stannic chloride dissolved in 50 ml of chloromethylmethyl ether, with stirring for 30 min, and were worked up as described previously (14). A 250-mg sample was quaternized by heating in 2.5 ml of pyridine for 1 hour at 100°, and the suspension was diluted with 25 ml of 50% acetic acid. The chloride was displaced by the addition of 5 ml of concentrated nitric acid and was determined quantitatively by a Volhard titration. The chloromethyl resin contained 1.19 mmole of chloride per g.

The chloromethyl resin (10 g) was esterified by refluxing in 75 ml of ethyl acetate with 868 mg (4.0 mmoles) of Boc-L-valine and 0.50 ml of triethylamine (3.6 mmoles) for 50 hours. The resin was filtered and washed with ethyl acetate, ethanol, water, and ethanol and dried in a vacuum at 25°. A sample of the Boc-valine-resin was deprotected in 50% (v/v) TFA[3]-CH_2Cl_2 and hydrolyzed in HCl-acetic acid-phenol, and valine was determined on the amino acid analyzer to be 0.21 mmole per g.

[3] The abbreviation used is: TFA, trifluoroacetic acid.

Boc-valine-resin (2.0 g, 0.42 mmole of valine) was placed in a small (40 ml total volume) reaction vessel of the automated peptide synthesizer similar to the one described before (22), but which had been modified with a stopper of improved design that eliminated the escape of solvents into the air outlet line. The instrument was programmed to perform the remainder of the synthesis automatically with the following reagents and conditions. All amino acids were protected on the α-amino position with the Boc group and the following side chain blocking groups were used: Asp(OBzl), Glu(OBzl), Cys(Bzl), Ser(Bzl), Thr(Bzl), Tyr(Bzl), Lys(Z), Arg(NO$_2$), Met(O). The imidazole side chain of histidine was unprotected. Boc groups were removed at each cycle of the synthesis by treatment with 25 ml of 50% (v/v) TFA-CH$_2$Cl$_2$ for 21 min. To avoid dilution of the TFA by the previous CH$_2$Cl$_2$ wash a 2-min preliminary wash with 50% TFA-CH$_2$Cl$_2$ was introduced. The 50% TFA-CH$_2$Cl$_2$ was found to be a milder reagent than the usual 4 N HCl in dioxane reagent with respect to the stability of the side chain blocking groups. Nonetheless, the ε-benzyloxycarbonyl group of lysine and the benzyl esters of aspartic acid and glutamic acid are not entirely stable and, as will be seen, this reagent also caused a loss of peptide chain from the resin. Recently the hexapeptide Phe-Leu-Ile-Val-Gly-Ala was synthesized by the solid phase method using 20% (v/v) TFA-CH$_2$Cl$_2$ as the deblocking reagent. A reaction time of 21 min was still found to be sufficient for complete deprotection even of the sterically hindered amino acid residues in this peptide. Following each deprotection step the free amino group was liberated by treatment with 25 ml of 10% triethylamine in chloroform.

The couplings were usually mediated with N,N'-dicyclohexylcarbodiimide (24) in CH$_2$Cl$_2$. Boc-amino acid (1.26 mmole, 3-fold excess) in 7 ml of CH$_2$Cl$_2$ was added to the peptide-resin, followed by a 5-ml rinse with CH$_2$Cl$_2$, and after 10 min, dicyclohexylcarbodiimide (1.26 mmole) in 7 ml of CH$_2$Cl$_2$ was pumped in, followed by another 5-ml rinse. Because of their low solubility in CH$_2$Cl$_2$, Boc-Arg(NO$_2$) and Boc-His were first dissolved in 4.8 ml of dimethylformamide and then diluted with 2.2 ml of CH$_2$Cl$_2$ before adding to the peptide-resin. In the case of Boc-His, 5-fold excesses of amino acid and dicyclohexylcarbodiimide in CH$_2$Cl$_2$ were used. The reaction time for all dicyclohexylcarbodiimide couplings was 5 hours. Boc-asparagine and Boc-glutamine were coupled as p-nitrophenyl esters in dimethylformamide (25), employing a 4-fold excess of reagent and a reaction time of 10 hours.

One cycle of the synthesis (dicyclohexylcarbodiimide coupling) consisted of: CH$_2$Cl$_2$, twice for 2 min; 50% TFA-CH$_2$Cl$_2$, 2 min; 50% TFA-CH$_2$Cl$_2$, 21 min; CH$_2$Cl$_2$, five times for 2 min; CHCl$_3$, three times for 2 min; 10% Et$_3$N-CHCl$_3$, 7 min; CHCl$_3$, three times for 2 min; CH$_2$Cl$_2$, three times for 2 min; Boc-amino acid, 7 min; dicyclohexylcarbodiimide, 300 min; CH$_2$Cl$_2$, twice for 2 min; EtOH, three times for 2 min. For an active ester cycle the CH$_2$Cl$_2$ wash before the Boc-amino acid p-nitrophenyl ester addition was replaced by a dimethylformamide wash, and the dicyclohexylcarbodiimide step was omitted.

The course of the synthesis was followed by removing 10-mg samples at intervals of 6 to 10 residues during the run and determining amino acid ratios on acid hydrolysates of the peptide-resin. In addition to the small analytical samples, larger samples were removed after the peptide chain reached 36, 75, 99, and 104 residues so that peptides representing partial sequences of ribonuclease could be isolated. A total of 1.70 g of samples, corresponding to 47% of the original 2 g of resin, was removed. The final weight of the fully protected 124-residue RNase-resin was 1.59 g. The product contained 4 NO$_2$-, 49 Bzl-, 10 Z-, and 4 sulfoxide groups and had a calculated molecular weight of 19,791. Amino acid analyses showed this to contain 0.0212 mmole of protein per g of protected RNase-resin, from which it could be calculated that the product contained 664 mg of protected peptide plus 926 mg of polystyrene, or 0.0364 mmole of peptide per g of polystyrene. Therefore, the retention of peptide chain on the resin was 17%.

After the primary sequence of RNase A had been assembled it was necessary to cleave the polypeptide from the solid support, to remove protecting groups, and to purify the resulting protein. Of the several experiments that were performed, under a variety of conditions, three runs (A, B, and C) will be described here.

Cleavage of Ribonuclease A from Resin

Run A—Protected RNase-resin (200 mg) was placed in a Daiflon (polytrifluorochloroethylene) reaction vessel of an HF cleavage apparatus (Toho Company, Japan), dried under high vacuum, and treated according to the general procedures of Sakakibara *et al.* (26) and Lenard and Robinson (27). TFA (2 ml) was added, followed by 2 ml of anisole as a trap for benzyl$^+$ and NO$_2^+$ ions, and the system was evacuated again. HF was collected in a storage vessel containing CoF$_3$, and 10 ml were redistilled into the reaction vessel which had been cooled to $-78°$. The temperature was raised to 0° and this temperature was maintained for 60 min, after which it was allowed to rise slowly to 20° over a period of 30 min. The HF and TFA were evaporated with a water pump, and the final traces of HF and most of the anisole were removed with an oil pump protected with a Dry Ice trap. Residual anisole and its derivatives were extracted with ether. The cleaved, deprotected peptide was dissolved in 15 ml of TFA and filtered from the resin. The filtrate and TFA washes were evaporated to dryness, and the solid protein residue was treated with 10 ml of 5% sodium bicarbonate in order to reverse any N → O acyl shift which might have occurred during the HF step (28). After 2 hours at pH 7.5, the solution was dialyzed in the cold against distilled water and was then lyophilized.

Amino acid analysis showed 1.77 μmoles of peptide, indicating a 41% yield in the cleavage step. The over-all yield was 7% calculated from the amount of COOH-terminal valine originally esterified to the resin. This experiment was repeated several times, giving a total of 5.69 μmoles of crude cleaved peptide.

Run B—A 203-mg sample of protected RNase-resin was treated with HF as before, but without added trifluoroacetic acid. The work-up was the same as described for Run A. Amino acid analysis showed 3.03 μmoles of peptide, indicating a 70% yield in this cleavage step. The over-all yield was 12%.

Run C—A 340-mg sample of the 124-residue polypeptide-resin was cleaved from the solid support and was partially deblocked by passing a stream of hydrogen bromide gas through a suspension of the peptide-resin in a mixture[4] of 10 ml of trifluoroacetic acid and 10 ml of CH$_2$Cl$_2$ (29) containing 2 ml of anisole. After 90 min at 25° the cleaved synthetic material was filtered, the resin was washed with 15 ml of TFA, and the combined filtrates were evaporated to dryness. The product was treated with 20 ml of 0.05 M NH$_4$HCO$_3$ buffer at pH 7.2. Lyophilization of this

---

[4] A. Marglin, unpublished.

solution gave the partially protected 124-residue polypeptide. It was dried in a vacuum at 40° for 5 hours. In order to remove the S-benzyl blocking groups from the 8 cysteinyl residues the synthetic product was treated with 10 ml of anhydrous HF in the presence of 1 ml of anisole, as described for Run A, and the NH_4HCO_3 treatment was repeated. Amino acid analysis of the hydrolyzed product showed 2.96 μmoles, indicating a cleavage yield of 41%. The over-all yield was 7%.

Purification of Synthetic Ribonuclease A

A. Ion Exchange Chromatography and Gel Filtration of S-Sulfonate of RNase A Cleaved by HF in Presence of Trifluoroacetic Acid

Oxidative Sulfitolysis (30, 31) of Liberated Polypeptide—The crude synthetic product (3 μmoles) was dissolved in 5 ml of 0.2 M phosphate buffer, 8 M in urea, pH 7.5. Then 150 mg of sodium sulfite (100-fold excess per disulfide bond) and 150 mg of sodium tetrathionate were added alternately in small portions during a period of 15 min. The reaction was allowed to proceed at room temperature for 15 hours. After a 5-hour dialysis to remove salts and urea, the solution was freeze-dried, yielding 40 mg of a white, fluffy material.

Paper electrophoresis was performed with a small sample of the crude $RNase(S-SO_3^-)_8$ in 2.4 M formic acid, 4 M in urea, pH 2.25, for 120 min at 1000 volts (Fig. 2). It showed a major Pauly-positive spot which moved toward the cathode with the same mobility (R_{His} 0.28) as that of the S-sulfonate of natural RNase A, together with two minor components of neutral and negative charge. Since RNase A has 19 positive charges at pH 2.25 and $RNase(S-SO_3^-)_8$ has 11, they were both expected to move toward the cathode. In order to remove the negatively charged impurity the $RNase(S-SO_3^-)_8$ was dissolved in 1 ml of 2.4 M formic acid, 4 M in urea, and applied on a Dowex 1-X2 column (2.1 × 15 cm) which was eluted with the same solvent. The anionic fraction was retained, the neutral and the cationic material emerged from the column as a single sharp peak. The protein-containing fractions were pooled (8.5 ml). Most of the urea was removed through a 2-hour dialysis. After lyophilization, 2 ml of water were added, the solution was adjusted to pH 5.2, and was then submitted to gel filtration on a Sephadex G-50 column (2.1 × 37 cm) in 0.1 M ammonium acetate, 5 M in urea, pH 5.2. Two peaks were obtained. The faster moving product eluted at the same volume as the S-sulfonate of natural RNase A. Both fractions were dialyzed 5 hours and lyophilized. The yields (calculated from amino acid analyses) were: Peak I (purified synthetic $RNase(S-SO_3^-)_8$), 21.5 mg (50%); Peak II (neutral fraction from Fig. 2), 6 mg (14%). The synthetic $RNase(S-SO_3^-)_8$ now moved as a single (but elongated) spot on paper electrophoresis and had the same R_{His} as the corresponding natural derivative (Fig. 2). The amino acid analysis is shown in Table I. The over-all yield from Boc-valine-resin was 3.5%.

B. Sephadex G-75 Gel Filtration of Ribonuclease A Cleaved by HF in Absence of Trifluoroacetic Acid

A 37-mg (2.7 μmoles) sample of the crude polypeptide was dissolved in 2 ml of 0.05 M ammonium bicarbonate buffer. A small amount of insoluble material was removed by centrifugation. Then the clear solution was fractionated on a Sephadex G-75 column (2.1 × 45.5 cm) by elution with 0.05 M ammonium bicarbonate buffer. The synthetic product was separated into

FIG. 2. Paper electrophoresis of natural and synthetic $RNase-(S-SO_3^-)_8$ and natural and synthetic RNase A in 2.4 M formic acid, 4 M in urea, pH 2.25, 1000 volts, 120 min. The *spots* were detected by the Pauly spray. Crude $RNase(S-SO_3^-)_8$ was purified by ion exchange chromatography on Dowex 1-X2 and gel filtration on Sephadex G-50. The synthetic RNase A used for the electrophoresis was a sample of the IRC-50-fractionated material from Run A, Peak I. R_{His} of natural and purified synthetic $RNase-(S-SO_3^-)_8$ was 0.28, that of natural and synthetic RNase A was 0.58.

three peaks. After lyophilization of the solutions the components were obtained in the following yields: Peak I ($V_E = 34$ to 47 ml), 5.5 mg (15%); Peak II ($V_E = 76$ to 92 ml), 16.5 mg (45%); Peak III ($V_E = 93$ to 140 ml), 11 mg (30%). Peak I was sharp and appeared at the void volume of the column. It consisted of interchain disulfide-linked aggregates of RNase A, which were derived by oxidation from the freshly cleaved mixture of reduced polypeptides during incubation with 5% sodium bicarbonate solution and during the following lyophilization. These steps did not provide the necessary conditions (11–13) for an exclusive formation of intramolecular disulfide bonds. Peak II, emerging from the column between 76 ml and 92 ml, agreed well with natural RNase A with respect to elution volume (V_E of natural RNase A, 76 to 90 ml) and electrophoretic mobility. The amino acid analysis of the material from Peak II, Run B, is given in Table I. The specific activity of this preparation was 5% when assayed against RNA. Peak III comprised a mixture of peptides with a lower molecular weight than that of the natural enzyme. These RNase A fragments were presumably obtained by incomplete coupling reactions during the assembly of the amino acid sequence of RNase A.

The protein aggregates under Peak I were reduced and then carefully reoxidized in dilute solution (0.02 mg per ml) under the conditions described below for $RNase(S-SO_3^-)_8$. The re-formed product then separated on Sephadex G-75 into two main components whose yields were 3.0 mg and 1.6 mg, respectively. Their positions agreed with those of Peaks II and III of the first fractionation on this column. This is explained by the conversion of the intermolecular disulfide bridges to intramolecular links through reduction and reoxidation under the proper conditions. Figures for this run are not shown because the patterns closely resembled those obtained for Run C.

C. Sephadex G-75 Gel Filtration of Ribonuclease A Cleaved in HBr and Deprotected with HF

A 1.46-μmole sample of the RNase A, derived from HBr and HF treatment, was dissolved in 1 ml of 0.05 M NH_4HCO_3 buffer. This solution was applied to a Sephadex G-75 column (2.1 × 45.5 cm) and was eluted with the same buffer. The elution diagram obtained by measuring the optical density of the fractions at 280 nm showed three peaks (Fig. 3A). The contents of the three peaks were isolated through lyophilization with the following yields: Peak I, 5.7 mg (29%); Peak II, 7.5 mg (38%);

TABLE I

Amino acid analyses[a]

Amino acid	Natural RNase A				Synthetic RNase A							Natural S-protein		Synthetic S-protein, Sephadex G-75 Peak II	Synthetic des-(21-25) S-protein, Sephadex G-75 Peak II	Synthetic S-peptide
					Run A		Run B			Run C						
								Trypsin-treated								
	Expected[b]	Found	HF-treated	HF-TFA-treated	RNase(SSO$_3^-$)$_8$	IRC-50 Peak I	Sephadex G-75 Peak II	Sephadex G-50 Peak T-I	Sephadex G-50 Peak T-II	Sephadex G-75 Peak II		Expected	Found			
Lys	10	10.0	9.7	9.2	10.3	9.1	9.2	9.9	8.3	10.0		8	8.0	7.0	7.9	2.2
His	4	3.8	3.9	3.5	3.3	3.3	3.3	3.7	2.6	3.6		3	2.7	2.1	2.4	0.9
Arg	4	4.1	4.0	3.9	4.6	4.2	3.7	4.4	5.9	3.8		3	3.0	3.1	3.1	1.0
Asp	15	15.7	15.6	16.2	15.9	16.0	16.2	14.9	14.7	15.4		14	14.3	14.6	12.4 (13)[c]	1.0
Thr	10	9.8	10.2	10.4	10.5		10.4	9.6	8.2	9.2		8	7.4	6.2	6.2	2.2
Ser	15	14.0	15.0	15.5	15.3	15.5	14.0	15.0	11.6	15.1		12	11.2	11.4	8.6 (9)[c]	2.9
Glu	12	12.1	12.2	12.8	12.4	12.1	12.1	11.8	10.8	12.5		9	9.7	9.1	9.5	3.0
Pro	4	3.8	3.9	3.6	3.9	3.9	4.3	3.5	2.1	4.3		4	4.0	4.2	4.1	
Gly	3	3.1	3.2	3.4	3.1	3.3	3.4	3.6	4.1	3.1		3	3.2	3.2	3.1	
Ala	12	12.0	12.0	12.0	12.0	12.0	12.0	12.0	12.0	12.0		7	7.0	7.0	7.0	5.2
Cys	8	7.8	7.6[d]	7.4[d]	6.7[d]	7.2[d]	7.3[d]	[e]	[e]	7.4[d]		8	7.8	7.6[d]	[e]	
Val	9	8.6	8.5	8.8	9.4	8.9	9.6	9.3	5.9	9.8		9	8.7	9.7	10.6	
Met	4	4.1	3.7	3.4	3.9	3.8	3.7	4.1	3.2	3.4		3	2.8	2.7	2.4	0.9
Ile	3	2.3	2.1 (2.7)[f]	2.2	2.1	2.1 (2.9)[f]	2.7	2.8	1.7	2.3		3	2.0	2.1	2.4	
Leu	2	2.2	2.1	2.3	2.9	2.6	2.6	2.3	2.6	2.3		2	2.1	2.8	2.6	
Tyr	6	5.6	5.7	5.9	5.1	5.2	5.1	5.4	3.8	5.0		6	5.6	5.1	4.3 (5)[c]	
Phe	3	3.3	3.0	3.2	3.2	3.5	3.4	3.2	3.2	2.7		2	2.2	2.5	2.8	1.0

[a] Values are expressed as moles per mole of RNase A with alanine set at 12.0. All values are uncorrected.
[b] Based on the published amino acid sequence of bovine pancreatic ribonuclease A.
[c] Expected.
[d] Determined on performic acid-oxidized sample.
[e] Not determined.
[f] Hydrolyzed 48 hours.

Peak III, 3.8 mg (19%). The elution volume and electrophoretic mobility of Peak II agreed exactly with those of the natural enzyme. The amino acid analysis of the material from Peak II, Run C, is shown in Table I. The specific activity of this preparation was 9%. The protein from Peak I was reduced and reoxidized as in Run B. The product was separated on Sephadex G-75 into two main peaks (Fig. 3B) which corresponded in position to Peaks II and III of Fig. 3A.

Reduction and Reoxidation of Purified Synthetic RNase(S-SO$_3^-$)$_8$

The general procedures that had been developed for the reduction and reoxidation of the disulfide bonds of RNase A (11-13) were used for the RNase(S-SO$_3^-$)$_8$. A 5-mg (0.35 μmole) sample of purified synthetic RNase(S-SO$_3^-$)$_8$ from Run A was dissolved in 1 ml of 8 M urea. Then 10 μl of β-mercaptoethanol were added and the mixture was flushed with nitrogen for 15 min. After adjusting to pH 8.2 with 5% methylamine, the reduction was allowed to proceed for 20 hours at room temperature. At the end of this time, 0.3 ml of glacial acetic acid was added. Reduced synthetic RNase A was separated from urea and excess mercaptoethanol on a Sephadex G-25 column (2.1 × 36 cm) in 0.1 M acetic acid. The protein-containing fractions were combined, the volume was brought to 250 ml with water (final protein concentration, 0.02 mg per ml), and 3 g of Tris (0.1 M) were added slowly with stirring. The pH was adjusted to 8.3 at the glass electrode with 4 N HCl. The protein was allowed to air-oxidize in a 400-ml beaker for 24 hours at room temperature to form the four disulfide bonds. An aliquot of the solution was diluted and assayed for RNase A activity. The crude preparation had a specific activity of 1.1 to 1.7%. In order to isolate the synthetic enzyme the reoxidation mixture was acidified to pH 4 with 4 N HCl and freeze-dried. The synthetic protein was separated from salts on the Sephadex G-25 column in 0.1 M acetic acid. Lyophilization of the protein fraction gave the salt-free synthetic enzyme.

Fractionation of Synthetic Enzyme on IRC-50

Further purification of the synthetic RNase A preparations was achieved through cation exchange chromatography on IRC-50 (32). The elution pattern obtained depended strongly on the previous work-up conditions for the protein.

Run A—When the protein chain was removed from the resin with anhydrous HF in the presence of trifluoroacetic acid, converted to the S-sulfonate, reduced with mercaptoethanol, and finally air-oxidized to form the four disulfide bridges of RNase A, the fractionation on an IRC-50 column (1 × 25 cm) with 0.2 M phosphate buffer, pH 6.47 (33), gave five peaks by ultraviolet analysis. In a typical experiment, 4.5 mg (0.33 μmole) of synthetic enzyme in 0.25 ml of the phosphate buffer were applied to the column and eluted in the same buffer. The protein contents of the five fractions (Fig. 4A) were determined by quantitative amino acid analysis. Aliquots of each of the pooled fractions

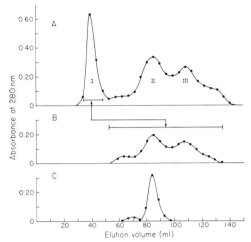

FIG. 3. Gel filtration of the cleaved, crude synthetic product and natural RNase A on Sephadex G-75. The column (2.1 × 45.5 cm) was eluted with 0.05 M ammonium bicarbonate buffer. *A*, 20 mg of crude synthetic RNase A (Run C) in 1 ml of 0.05 M ammonium bicarbonate buffer. *Peak I*, 5.7 mg; *Peak II*, 7.5 mg; *Peak III*, 3.8 mg. *B*, the synthetic material under Peak I of Chromatogram *A* was reduced and reoxidized in dilute solution (0.02 mg per ml), and then rechromatographed on the Sephadex G-75 column. *C*, 5 mg of natural RNase A in 0.25 ml of the eluent.

were assayed for enzymic activity against yeast RNA as substrate. The protein from Peak I had a specific activity of approximately 13% by the Kunitz assay and also by the Anfinsen assay (Fig. 5). The protein from Peak II showed 9.7% activity, which was partially due to the incomplete separation of Peak I from Peak II. The other three components were inactive. The fractions were lyophilized and phosphate was removed by gel filtration on Sephadex G-25 in 0.1 M acetic acid.

Yields: Peak I, 1.8 mg (40%); Peak II, 1.0 mg (22%); Peak III, 0.6 mg (13%); Peak IV, trace; Peak V, trace.

A typical amino acid analysis of material from the IRC-50 Peak I, Run A is shown in Table I.

On paper electrophoresis at pH 2.25 this synthetic RNase A was indistinguishable from the native enzyme (R_{His} 0.58, Fig. 2).

Run B—When the polypeptide was removed from the resin with anhydrous HF in the absence of trifluoroacetic acid, purified by gel filtration on Sephadex G-75, reduced and reoxidized, and then fractionated (5 mg) on an IRC-50 column (1 × 25 cm) with 0.2 M phosphate buffer, pH 6.47, only three peaks were observed (Fig. 4*B*). The fractions from each of the three peaks were combined, lyophilized, and desalted on Sephadex G-25 in 0.1 M acetic acid. Yields: Peak I, 3 mg (60%); Peak II, 0.45 mg (9%); Peak III, 0.3 mg (6%). RNase A activity was found only in Peak I. The isolated protein from Peak I had a specific activity of 13%.

Run C—When the polypeptide was cleaved and fully deprotected with HBr and HF in a two-step reaction, partially purified on Sephadex G-75, reduced, reoxidized, and then submitted to cation exchange chromatography on IRC-50, only one peak was obtained. The synthetic enzyme (5 mg) in 0.25 ml of 0.2 M

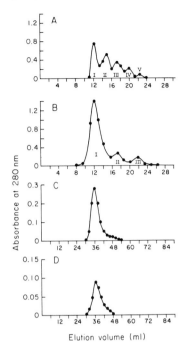

FIG. 4. Ion exchange chromatography on IRC-50 of synthetic and natural reduced and reoxidized RNase A. *Samples A and B* were eluted from a column (1 × 25 cm) with 0.2 M phosphate buffer, pH 6.47. *A*, 4.5 mg of synthetic RNase A (HF-TFA cleaved, converted to the *S*-sulfonate, then reduced and reoxidized) in 0.25 ml of 0.2 M phosphate buffer. *Peak I*, 1.8 mg; *Peak II*, 1.0 mg; *Peak III*, 0.6 mg; Peaks IV and V, not determined. *B*, 5.0 mg of synthetic RNase A (HF cleaved, fractionated on Sephadex G-75, reduced, and reoxidized) in 0.25 ml of 0.2 M phosphate buffer. *Peak I*, 3 mg; *Peak II*, 0.45 mg; *Peak III*, 0.3 mg. *Samples C and D* were eluted from a column (1.4 × 47.5 cm) with 0.2 M phosphate buffer, pH 6.47. *C*, 5.0 mg of synthetic RNase A (HBr-HF-treated, fractionated on Sephadex G-75, reduced, and reoxidized) in 0.25 ml of the eluent. *D*, 2.0 mg of reduced and reoxidized natural RNase A in 0.1 ml of 0.2 M phosphate buffer.

phosphate buffer, pH 6.47, was applied on an IRC-50 column (1.4 × 47.5 cm) and eluted in the same buffer (Fig. 4*C*). Aliquots of the eluted fractions were removed for alkaline hydrolysis followed by the ninhydrin reaction to estimate the protein concentration. The specific activity, measured as before with RNA as substrate, was 16%. The fractions eluted between 32 ml and 42 ml (Peak I) were pooled, lyophilized, and desalted on Sephadex G-25 in 0.1 M acetic acid. Yield: 3.6 mg (72%).

Fractionation of Synthetic Enzyme on CM-cellulose

A column (1.1 × 23.5 cm) of CM-cellulose was poured and equilibrated with a pH 6.0 sodium phosphate buffer, 0.01 M in Na^+ (11). The samples were applied in the 0.01 M buffer and washed with a total of 60 ml of the same buffer. A gradient was then started with pH 7.5 sodium phosphate buffer, 0.1 M in Na^+. The eluate was monitored at 254 nm and with the nin-

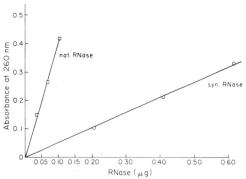

FIG. 5. RNase A assay using yeast RNA as substrate. The assay was performed according to the procedure of Anfinsen *et al.* (20) as modified by Egami, Takahashi, and Uchida (21); 3-mg samples of RNA in 1 ml of 0.05 M Tris-HCl buffer, pH 7.5, were incubated with different amounts of natural and synthetic RNase A (IRC-50 Peak I, Run A) for 15 min at 37°. The optical densities at 260 nm of the acid-soluble fractions of the digestion mixtures were plotted against the RNase A concentrations of the samples. From the ratio of the slopes of the two curves (0.53 for the synthetic, 4.0 for the natural enzyme) a specific RNase A activity of 13% was calculated.

hydrin reaction (Fig. 6). Natural RNase A, natural reduced-reoxidized RNase A, and synthetic RNase A (IRC-50 Peak I, Run B) all eluted at 50 ± 1 ml.

Further Purification of Synthetic Ribonuclease A by Treatment with Trypsin and by Ammonium Sulfate Fractionation

Natural RNase A was mixed with various amounts of trypsin in order to determine the limit of its resistance to tryptic hydrolysis. A typical experiment was carried out as follows. In 4 ml of 0.2 M ammonium bicarbonate buffer, 1 mg of trypsin (treated with L-1-toluenesulfonylamido-2-phenylethyl chloromethyl ketone, 265 units per mg, 91% protein, Worthington) was dissolved. To 0.5 ml of this solution, 2.5 mg of natural RNase A were added. The pH of the mixture was 7.7. The ratio of enzyme to substrate was 1:20. Incubation time was 13 hours at room temperature. An aliquot of the sample was then diluted to an RNase A concentration of 4 μg per ml. The RNase A activity of 1 ml of this solution was compared with that of the same amount of untreated RNase A by the Kunitz method. The two solutions were indistinguishable in this enzyme assay and also behaved the same on a Sephadex G-50 column. However, a decrease in activity and a small additional peak were observed when larger amounts of trypsin (1:4 or 1:2) were used. The enzymic activities of the various digestion mixtures of the natural RNase A are given in Table II. When RNase(S-SO$_3^-$)$_8$ was treated with trypsin (ratio, 1:20) it was readily digested. The incubation mixture was submitted to gel filtration on Sephadex G-50 in 0.1 M acetic acid and the material emerged between 15 and 27.5 ml. The controls of native RNase A and trypsin-treated RNase A both eluted between 9.5 and 17.5 ml.

A 1-mg sample of the synthetic RNase A from Sephadex G-75 Peak II, Run B, with 8% specific activity was dissolved in 0.25 ml of 0.2 M ammonium bicarbonate buffer containing 0.0625 mg of trypsin. The enzyme to substrate ratio was 1:16. The

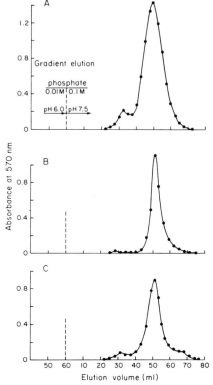

FIG. 6. Ion exchange chromatography of synthetic, reduced-reoxidized natural, and untreated natural RNase A on CM-cellulose. After the samples had been applied, the column (1.1 × 23.5 cm) was eluted with 60 ml of 0.01 M phosphate buffer, pH 6.0. Then gradient elution toward 0.1 M phosphate buffer, pH 7.5, was started. *A*, 5.0 mg of untreated natural RNase A; *B*, 1.9 mg of reduced-reoxidized natural RNase A; *C*, 1.9 mg of synthetic RNase A (HBr-HF-treated, fractionated on Sephadex G-75, reduced, and reoxidized), then 0.5-ml (*Chromatogram A*) and 1.0-ml aliquots (*Chromatograms B* and *C*) of each fraction of the eluate from the column were submitted to alkaline hydrolysis. The hydrolysates were reacted with ninhydrin and the optical densities at 570 nm were measured spectrophotometrically.

incubation was carried out for 12 hours at pH 7.7 at room temperature. Then 0.01 ml of the mixture (0.04 mg of synthetic RNase A) was removed for an enzyme assay. The specific activity was found to be about 2% by the Kunitz method. The rest of the digestion mixture was acidified with glacial acetic acid and fractionated on a column (1.1 × 23.5 cm) of Sephadex G-50 in 0.1 M acetic acid. A freshly packed Sephadex column was used for the gel filtration of the synthetic RNase A preparations in order to avoid any possible contamination by natural RNase A.

Two fractions were obtained. The first one, T-I, eluted between 10.5 ml and 17.5 ml, the second one, T-II, eluted between 17.5 ml and 30.5 ml. Immediately thereafter a 2-mg sample of untreated, natural RNase A was chromatographed on the same

TABLE II
Stability of natural RNase A toward trypsin

Ratio of trypsin to RNase A[a] (w/w)	Recovery of RNase A activity
	%
1:200	100
1:40	100
1:20	100
1:4	80
1:2	69

[a] Samples of natural RNase A (5 mg per ml) were incubated with trypsin in 0.2 M ammonium bicarbonate buffer at pH 7.7 for 13 hours at room temperature.

TABLE III
Increase of specific activity and total activity of synthetic RNase A upon purification by tryptic digestion and fractional precipitation

Purification stage	Weight of synthetic RNase	Specific activity	RNase units[a]
	mg	%	
Sephadex G-75 fractionated, reduced-reoxidized RNase A	1.0	8	0.08
Crude tryptic digest	1.0	2	0.02
Fractionation of tryptic digest on Sephadex G-50			
Peak I	0.69	61	0.42
Peak II	0.23	0	0.00
Fractional precipitation of Peak I			
Supernatant	0.41	78	0.32
Precipitate	0.21	16	0.03

[a] Expressed as milligrams of pure RNase A derived from 1.0 mg of the Sephadex G-75 fraction (Run B, Peak II).

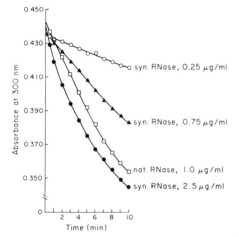

FIG. 7. Enzymic activity of purified synthetic RNase A. The synthetic RNase A that had been purified by trypsin treatment and $(NH_4)_2SO_4$ fractionation was assayed by the Kunitz method with yeast RNA as substrate. Solutions containing 2 mg of RNA in 2 ml of 0.1 M sodium acetate buffer, pH 5, were mixed with 4 μg of natural RNase A in 2 ml of water as the control and 10 μg, 3 μg, and 1 μg of synthetic RNase A in 2 ml of water, respectively. The reactions were followed spectrophotometrically at 300 nm for 10 min. The specific activities of the synthetic RNase A samples were calculated from the initial slopes of the curves.

column. A single peak emerged between 10.5 ml and 18.0 ml. Recovery of the natural enzyme from the Sephadex G-50 column was quantitative.

Fractions T-I and T-II of the tryptic digest of synthetic enzyme were lyophilized. Their yields, according to amino acid analyses (Table I), were 0.69 mg and 0.23 mg, respectively. Fraction T-II was of low molecular weight and had no RNase A activity. The amino acid analysis of T-II was distinctly different from that of pure RNase A or of Fraction T-I. Fraction T-I was dissolved in 1 ml of water and 0.1 ml of this solution was used for determination of RNase A activity. In the Kunitz assay this trypsin-treated synthetic RNase A had a specific activity of 61%.

The remaining 0.9 ml (0.62 mg) of Fraction T-I was lyophilized again and then dissolved in 0.1 ml of water. To this was added 0.16 ml of a saturated ammonium sulfate solution (pH 4.6), and the mixture was placed in the cold room. After 2 days, some amorphous precipitate had formed. It was centrifuged. An aliquot of the supernatant solution was assayed again by the Kunitz method. The activity against RNA of 0.25, 0.75, and 2.5 μg per ml of synthetic RNase A was compared with that of 1.0 μg per ml of natural enzyme. In this assay (Fig. 7) the specific activity of the purified synthetic material at the three enzyme levels was found to be 80, 75, and 80%; average specific activity, 78%. This value was confirmed by the Anfinsen assay.

The amorphous precipitate was dissolved in 0.5 ml of 0.05 M ammonium bicarbonate buffer. From quantitative amino acid analysis of an acid hydrolysate the yield of the amorphous fraction was found to be 0.21 mg. It showed a specific activity of 16%. Table III summarizes the activity data on the trypsin- and ammonium sulfate-treated synthetic RNase A preparations.

Preparation of Peptide Maps from Natural and Synthetic Ribonuclease A

Samples of natural and synthetic RNase A were oxidized with performic acid (33). The oxidized derivatives were then susceptible to tryptic digestion. Trypsin, 1 mg, was dissolved in 40 ml of 0.2 M ammonium bicarbonate buffer, pH 7.75, and 0.5 ml of this solution (containing 0.0125 mg of trypsin) was then added to 2.5 mg of oxidized RNase A. After an incubation time of 22 hours at room temperature the reaction was stopped by adjusting to pH 2. When peptide maps were not prepared immediately the samples were stored frozen.

Usually 0.2 ml of the digest (1 mg) was applied on Whatman No. 3MM chromatography paper. The mixture was resolved by high voltage electrophoresis at 2250 volts in a pH 3.5 buffer containing 2.2 ml of pyridine per liter and 22 ml of acetic acid per liter, in a LT-48A tank (Savant Instrument, Inc.) in the first dimension and descending paper chromatography in phenol-1-butanol-acetic acid-water (3:3:2:4) in the second dimension. The air-dried chromatograms were developed with the ninhydrin-cadmium acetate spray (34). In each experiment lysine was run along the edge of the paper in each direction as a control. The electrophoretic and chromatographic mobility of the tryptic peptides was compared to that of lysine ($R_{Lys} = 1$). The peptide maps thus obtained from natural and synthetic RNase A

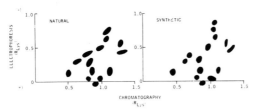

FIG. 8. Peptide maps of tryptic digests of oxidized natural (*left*) and synthetic RNase A (*right*). Samples (2.5 mg each) of performic acid-oxidized natural or synthetic RNase A (IRC-50 Peak I, Run A) were dissolved in 0.5 ml of 0.2 M ammonium bicarbonate buffer, pH 7.75, containing 0.0125 mg of trypsin. Incubation time was 22 hours at 25°. Aliquots (0.2 ml) of the digests (1 mg) were applied on Whatman No. 3MM chromatography paper. The tryptic peptides were resolved by high voltage electrophoresis at 2250 volts in pyridine-acetic acid, pH 3.5, and descending paper chromatography in phenol-1-butanol-acetic acid-water (3:3:2:4). The chromatograms were sprayed with the ninhydrin-cadmium acetate reagent. The positions of the tryptic peptides were compared to that of lysine ($R_{Lys} = 1$).

are shown in Fig. 8. The peptide maps were prepared two more times each and were found to be reproducible.

The relative positions of the 14 expected (18) ninhydrin-positive spots on the two chromatograms agreed quite well, but some of the corresponding peptide pairs gave spots of different shape and size. In addition, there was one spot on the peptide map of the synthetic material near the position of lysine which could not be seen in the natural RNase control. It was not identified, however, and its origin is not known.

Enzymic Digestion of Synthetic Ribonuclease A with Papain and Aminopeptidase M

In order to establish its optical purity the synthetic enzyme (IRC-50 Peak I, Run A) was treated with papain followed by aminopeptidase M.[5] After natural RNase A was shown to be digested by this procedure, 0.14 mg of synthetic material (10 mμmoles) was dissolved in 95 μl of 0.05 M ammonium acetate buffer, pH 5.3, then 5 μl of a 3% solution of mercaptoethanol and a suspension of 10 μg of papain (15 units per mg, Worthington) in 1 μl of 0.05 M sodium acetate, pH 4.5, were added. The mixture was incubated at 37° for 2 hours, at which time the papain was inactivated with 2 drops of glacial acetic acid and the solution was lyophilized.

The partially digested material was dissolved in 80 μl of 0.2 M ammonium bicarbonate buffer, pH 8.2. Then 5 μl of the 3% mercaptoethanol solution and 0.3 mg of aminopeptidase M (Röhm and Haas, GmbH, Darmstadt, Germany) in 15 μl of water were added. After incubation for 3 hours at 37° another 0.3-mg portion of aminopeptidase M in 15 μl of water was added and the reaction was continued for 3 more hours at 37°. Then the digestion mixture, acidified with acetic acid, was lyophilized. Amino acid analyses of the enzymic digests of natural and synthetic RNase A were in good agreement and, with the exception of high values for glycine and leucine, corresponded to the values from acid hydrolysates. This result showed that the digestion of the synthetic product by papain and aminopeptidase M was essentially complete and, therefore, that no major racemization had occurred during the assembly of the polypeptide chain of

[5] H. Keutman and J. T. Potts, Jr., personal communication.

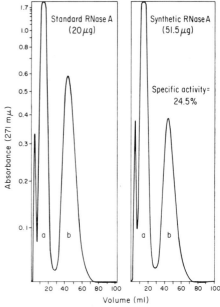

FIG. 9. RNase A assay toward 2′,3′-cyclic cytidine phosphate as substrate. The assay was performed according to the procedure of Fruchter and Crestfield (36). Three samples of natural RNase A (20 μg each) and three samples of synthetic RNase A (51.5 μg each from IRC-50 Peak I, Run A) were mixed with 1.4 mg of 2′,3′-cyclic cytidine phosphate in 2 ml of Tris-NaCl buffer, pH 7.48. After 10 min at 0° the reactions were stopped by adding 50 μl of 0.2 M mercuric chloride solution to each sample. Then the digestion mixtures were resolved on IRA-400 and the optical densities of the column effluents were recorded at 271 nm. *Peak a*, excess undigested 2′,3′-cyclic cytidine phosphate; *Peak b*, 3′-cytidine phosphate hydrolysis product. For synthetic RNase A an average activity of 24.5% was calculated when the area under *Peak b* was compared with that of the natural RNase A control.

RNase A or during the various work-up procedures. It also revealed that 79% of the methionine sulfoxide residues had been reconverted to methionine residues during the mercaptoethanol reduction of the synthetic RNase(S-SO$_3^-$)$_8$. The remainder was recovered as methionine sulfone. The Boc-methionine sulfoxide (35) used in the synthesis of the polypeptide chain of RNase A was chromatographically pure and did not contain any detectable amount of the sulfone derivative. Since air was not excluded and no special precautions against peroxide formation were taken it is assumed that the sulfone arose by a gradual oxidation during the synthesis.

Substrate Specificity of Synthetic Ribonuclease A

Since RNase A has a very high specificity for the cleavage of the 5′-ester bond of 3′,5′-phosphodiesters in which a pyrimidine ribonucleotide provides the 3′-ester, it was necessary to demonstrate that the synthetic enzyme possessed the same substrate specificity. It should split RNA, pyrimidine oligoribonucleotides, and pyrimidine nucleoside 2′,3′-cyclic phosphates, but not DNA, deoxyribonucleotides, purine ribonucleotides, or purine

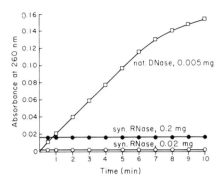

FIG. 10. DNase assay with synthetic RNase A. The DNase assay, using calf thymus DNA as substrate, followed the procedure of Kunitz (37). Three samples of DNA (0.2 mg each) in 5 ml of solution, pH 5, containing 0.005 M Mg^{++}, were mixed at 25° with 0.005 mg of DNase I, 0.2 mg or 0.02 mg of synthetic RNase A in 1 ml of water. The change of the optical densities was followed spectrophotometrically at 260 nm for 10 min. The absorbance of the samples containing synthetic RNase was recorded up to 2 hours and showed no increase, demonstrating that no reaction occurred between synthetic RNase A and DNA during that time.

cyclic phosphates. Most of the data on substrate specificity were obtained on material from IRC-50 Peak I, Run A (Fig. 4A).

RNA as Substrate—When yeast RNA was the substrate RNase A activities were determined either by the Kunitz method (19) from the initial velocities of the hypochromic shift at 300 nm or by spectrophotometric measurement at 260 nm of the acid-soluble oligonucleotides of the digestion mixtures (20, 21). The specific activities when measured by the two methods have been in good agreement. For example, the material from IRC-50 Peak I, Run A was 13% as active as natural RNase A by the Kunitz assay and also 13% by the Anfinsen assay (Fig. 5). Depending on the work-up conditions and the extent of purification, the various synthetic RNase A samples have shown enzymic activities against RNA that have been up to 78% as high as pure native RNase A (Fig. 7).

2′,3′-Cyclic Cytidine Phosphate (C>p) as Substrate—The synthetic enzyme was assayed against C>p by the procedure of Fruchter and Crestfield (36). The C>p (1.4 mg) was incubated at 0° with 20 μg of RNase in 2 ml of Tris-NaCl buffer, pH 7.48, and after 10 min the reaction was stopped with 50 μl of 0.2 M mercuric chloride solution. The hydrolysis product (3′-cytidine phosphate) and excess substrate were separated on an IRA-400 column by elution with Tris-NaCl buffer, pH 7.48. The extent of hydrolysis was quantitatively determined by integration of the peaks recorded at 271 nm on a Zeiss PMQ II spectrophotometer. The specific activities of the synthetic samples were calculated by comparison with natural RNase A standards after correcting for the blank hydrolysis values. The synthetic enzyme from IRC-50 Peak I, Run A showed a specific activity of 24.5% (Fig. 9). The $(NH_4)_2SO_4$-purified preparation had a specific activity of 65% against this substrate.

DNA as Substrate—DNase activity was measured by the method developed by Kunitz (37), based upon the increase of absorption at 260 nm during the depolymerization of deoxyribonucleic acid by DNase at 25°. To four samples of the substrate each containing 0.2 mg of DNA (calf thymus, Worthington) in

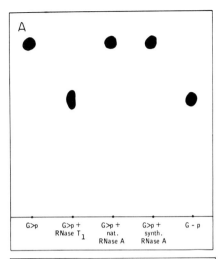

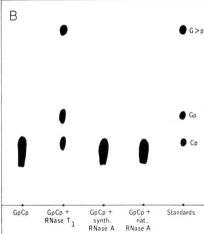

FIG. 11. Assay of synthetic RNase A for RNase T_1 activity. 2′,3′-Cyclic guanosine phosphate (*top*) and 5′-(3′-guanylyl)-cytidylic acid (*bottom*) were used as substrates for the RNase T_1 assay. Aliquots of the digests were chromatographed on Silica Gel GF-coated thin layer plates using isopropyl alcohol-water-ammonium hydroxide (70:25:5) as solvent. The spots were detected under an ultraviolet lamp. *A*, 10 μg each of RNase T_1 and natural and synthetic RNase A were added to solutions of 40 μg of 2′,3′-cyclic guanosine phosphate (G>p) in 0.2 ml of 0.1 M ammonium bicarbonate buffer. The samples were incubated 48 hours at 25°. G>p was hydrolyzed in the presence of RNase T_1 to give 3′-guanosine phosphate (Gp), but not by the RNase A samples. *B*, 10 μg each of RNase T_1 and natural and synthetic RNase A were mixed with solutions of 100 μg of 5′-(3′-guanylyl)-cytidylic acid (GpCp) in 0.1 ml of 0.1 M ammonium bicarbonate buffer and incubated 48 hours at 25°. The substrate was hydrolyzed by RNase T_1 to give G>p as an intermediate and Gp and 3′-cytidine phosphate (Cp). The RNase A samples had no effect on this substrate.

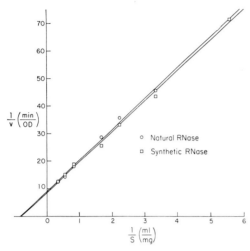

FIG. 12. Calculation of the K_m values for natural and synthetic RNase A from a Lineweaver-Burk plot. Initial velocities were measured spectrophotometrically at 300 nm toward yeast RNA as substrate by the Kunitz method. Samples (2 ml) of natural RNase A (2.5 µg per ml) and synthetic RNase A (from IRC-50 Peak I, Run A, 16.7 µg per ml) were mixed with 2 ml of RNA solutions (3.0, 1.8, 1.2, 0.6, 0.45, 0.3, and 0.18 mg per ml in 0.1 M sodium acetate buffer, pH 5). The final volume was 4 ml. The optical density curve of each sample was recorded at 300 nm for 10 min and the initial velocities were measured. The K_m values obtained from the Lineweaver-Burk plot (v = initial velocity, S = substrate concentration) were 1.20 mg per ml and 1.24 mg per ml for natural and synthetic RNase A, respectively.

5 ml of the Kunitz buffer were added 1 ml of water (blank), 0.005 mg of DNase I (2260 units per mg, Worthington) in 1 ml of water, and 0.2 mg and 0.02 mg of synthetic RNase A in 1 ml of water each. Absorbances at 260 nm were recorded every minute (Fig. 10). The results of the assay indicated complete resistance for at least 2 hours of DNA to a concentration of the synthetic RNase A 40 times higher than that of natural DNase which caused a rapid digestion within a few minutes.

2′,3′-Cyclic Guanosine Phosphate (G>p) as Substrate—RNase T_1 (38) (protein concentration 1.0 mg per ml, 600,000 units per mg, Schwarz BioResearch), natural RNase A, and synthetic RNase A (10 µg each) were added to solutions of 0.04 mg of G>p monobarium salt (Schwarz BioResearch) in 0.2 ml of 0.1 M ammonium bicarbonate buffer. After incubation at 25° for 48 hours, thin layer chromatograms of the samples were prepared on Silica Gel GF-coated plates (Analtech), using isopropyl alcohol-water-NH$_4$OH (70:25:5) as solvent. Fig. 11A shows the position of the spots detected with ultraviolet light. Under these conditions, RNase T_1 completely hydrolyzed the substrate to guanylic acid, whereas neither natural nor synthetic RNase A had a detectable effect on the G>p substrate.

5′-(3′-Guanylyl)-cytidylic Acid (GpCp) as Substrate—Natural RNase A, synthetic RNase A, and RNase T_1 (10 µg each) were added to solutions of 0.1 mg of GpCp (39) in 1 ml of 0.1 M ammonium bicarbonate buffer. After an incubation time of 48 hours at 25°, aliquots were analyzed by thin layer chromatography on silica gel plates in isopropyl alcohol-water-NH$_4$OH (70:25:5). The spots were detected under ultraviolet light (Fig. 11B). Under these conditions RNase T_1 completely cleaved this dinucleotide into cyclic guanylic acid, guanylic acid, and cytidylic acid, whereas the natural and synthetic RNase A had no effect whatever and GpCp was recovered unchanged.

5′-(3′-Adenylyl)-adenylic Acid (ApAp) as Substrate—Finally the specificity of the synthetic enzyme for the 3′,5′-phosphodiester bond adjacent to a pyrimidine ribonucleotide was confirmed by assaying it with ApAp (39) as substrate. RNase T_1, natural RNase A, or synthetic RNase A (10 µg each) were dissolved in solutions of 0.2 mg of ApAp in 0.2 ml of 0.1 M ammonium bicarbonate buffer and incubated for 48 hours at 25°. Upon thin layer chromatography in the above system the samples containing either RNase T_1, natural RNase A, or synthetic RNase A had the same R_F as the blank (ApAp), and no spot for Ap was visible, indicating the expected resistance of this dinucleotide to all three enzymes.

Determination of Michaelis Constant

Initial velocities of RNA hydrolysis were measured spectrophotometrically at 300 nm by the Kunitz method.

Each sample of the substrate was dissolved in 2 ml of 0.1 M sodium acetate buffer, pH 5, and then mixed with either natural or synthetic RNase A in 2 ml of water. The final volume was 4 ml. Seven different substrate concentrations were used: 3.0, 1.8, 1.2, 0.60, 0.45, 0.30, and 0.18 mg per ml. The enzyme concentrations were constant in all experiments: 2.5 µg per ml for natural and 16.7 µg per ml for synthetic RNase A (from IRC-50 Peak I, Run A). The absorbance at 300 nm was followed for 10 min. The K_m values, calculated from Lineweaver-Burk plots of initial velocities, were found to be 1.20 mg per ml for natural RNase A and 1.24 mg per ml for the synthetic enzyme (Fig. 12). The values that we originally reported (16) were both in error by a factor of 2 because of an incorrect dilution calculation. Edelhoch and Coleman (40) have reported a K_m value of 1.25 mg per ml for natural RNase A acting on yeast RNA.

Synthesis of Ribonuclease S-peptide (RNase(1-20))

S-peptide was synthesized by the solid phase method using the same procedures as described for the total synthesis of RNase A. In this synthesis Boc-Met was used instead of Boc-Met(O), but the other amino acid derivatives were the same as before. The eicosapeptide was cleaved from the resin with anhydrous HF in the presence of excess anisole (yield, 44%) and was purified by free-flow electrophoresis (Elphor FF, Brinkmann Instruments) in 0.5 M acetic acid (1500 volts, 200 ma) and by gel filtration on Sephadex G-25 in 0.1 M acetic acid. The R_{Arg} of both synthetic and natural S-peptide was 0.39 on paper electrophoresis in 0.1 M pyridine acetate, pH 5.0. The amino acid analysis of an acid hydrolysate of synthetic S-peptide is shown in Table I.

Preparation of Synthetic S-protein and Des-(21-25)S-protein and Their Combination with Natural or Synthetic S-peptide to Give Ribonuclease S and Des-(21-25)ribonuclease S

During the course of the RNase A synthesis, samples (213 and 385 mg, respectively) were removed at intermediate stages containing 99 and 104 amino acid residues. They were dried in a vacuum and stored for later work-up.

Cleavage from Resin and Deprotection of Side Chains—A 213-mg sample of dried RNase (26-124)-resin, containing 99 amino acid

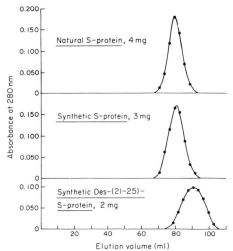

FIG. 13. Chromatography of synthetic S-protein, synthetic des-(21-25)S-protein, and natural S-protein on a Sephadex G-75 column (2.0 × 46.5 cm). The buffer was 0.05 M ammonium bicarbonate.

FIG. 14. Paper electrophoresis of synthetic S-protein and synthetic des-(21-25)S-protein in 2.4 M formic acid, 4 M in urea, pH 2.25, 780 volts, 23 ma, 90 min. The synthetic S-protein and des-(21-25)S-protein had been purified by fractionation on Sephadex G-75 (Fig. 13). Natural RNase A, natural S-protein, and histidine were run as controls.

TABLE IV

Combination of S-peptide with S-protein or des-(21-25)S-protein

The protein components were reduced with β-mercaptoethanol in 8 M urea. The isolated, reduced proteins were then reoxidized in the presence of S-peptide in dilute solution (0.028 mg per ml) at pH 8.25 and room temperature for 20 hours to give RNase S and des-(21-25)RNase S.

Components	Mole ratio	RNase A activity
		%
Natural S-protein + natural S-peptide	1:1.6	37
Synthetic S-protein + natural S-peptide	1:2	5
Synthetic des-(21-25)S-protein + natural S-peptide	1:2	9

residues, was mixed with 2 ml of anisole and then treated with 10 ml of anhydrous HF with stirring for 100 min while the temperature was allowed to rise slowly from 0–20°. At the end of the reaction, HF and anisole were evaporated in a vacuum. After traces of anisole had been extracted with ether, the peptide was dissolved in 15 ml of trifluoroacetic acid and filtered from the resin. Then TFA was removed on the rotary evaporator at room temperature, and the remaining peptide was treated for 2 hours with 10 ml of 5% sodium bicarbonate solution at pH 7.5. This mixture was dialyzed 5 hours and the peptide was isolated by lyophilization.

The 104-amino acid residue peptide was removed from the resin under the same conditions. A 185-mg sample of dried RNase(21–124)-resin was reacted with 10 ml of anhydrous HF in the presence of 2 ml of anisole for 100 min at 0–20°. The cleaved and deprotected peptide was dialyzed and lyophilized.

Fractionation of Synthetic S-protein and Des-(21-25)S-peptide on Sephadex G-75—The lyophilized polypeptides were each dissolved in 1 ml of 0.05 M ammonium bicarbonate buffer. After removal of some insoluble material by centrifugation, the solutions were applied on a Sephadex G-75 column (2.1 × 46.5 cm) and eluted with 0.05 M ammonium bicarbonate. The general patterns of the elution curves for the synthetic 99- and 104-residue peptides were very similar to those obtained with crude synthetic RNase A (Fig. 3). In each case a polymer fraction consisting of disulfide-linked aggregates was observed and a slower moving peak consisting of a mixture of smaller peptides was also obtained in addition to the main peak.

The fractions under the main peaks were combined, lyophilized, and rechromatographed on the Sephadex G-75 column. The yields obtained were 8.1 mg for the synthetic S-protein and 6.5 mg for the synthetic des-(21-25)S-protein. The elution volumes on Sephadex G-75 of samples of the synthetic S-protein and synthetic des-(21-25)S-protein are compared with that of natural S-protein in Fig. 13.

The partially purified products were then submitted to paper electrophoresis in 2.4 M formic acid, 4 M in urea, at pH 2.25 together with natural S-protein, natural RNase A, and histidine as a control (Fig. 14). The R_{His} values were: synthetic S-protein, 0.57; natural S-protein, 0.57; synthetic des-(21-25)S-protein, 0.58; and natural RNase A, 0.54. Amino acid analyses of natural and synthetic S-protein and synthetic des-(21-25)S-protein are given in Table I and show satisfactory agreement when the latter is corrected for the Asp (1), Tyr (1), and Ser (3) residues that were omitted in the synthesis.

Generation of Ribonuclease A Activity from Synthetic S-protein and Synthetic des-(21-25)S-protein upon Addition of Natural or Synthetic S-peptide—First, the synthetic proteins were unfolded by reduction. Synthetic S-protein or des-(21-25)S-protein (1.0 mg) was dissolved in 1 ml of 8 M urea solution, 0.01 ml of mercaptoethanol was added, and the sample was flushed with nitrogen for 10 min and then adjusted to pH 8.25 with 5% methylamine. Reduction was carried out for 20 hours at room temperature. After this time the mixture was acidified with 0.3 ml of glacial acetic acid, and the reduced protein was separated from urea and excess mercaptoethanol on a Sephadex G-25 column (2.1 × 36 cm) in 0.1 M acetic acid. The protein-containing fractions were pooled and diluted with distilled water to 50 ml in which 0.71 g of dibasic sodium phosphate was dissolved (0.1 M). The pH of the solution was adjusted to 8.25. Then 0.4 mg of natural or synthetic S-peptide was added, and the mixture was reoxidized with air for 20 hours at room temperature to allow the chains to refold and the four disulfide bonds to re-form. The protein concentration of the solution was 0.028 mg per ml and the mole ratio between the synthetic proteins and S-peptide was 1:2. As a control, natural S-protein was reduced and reoxidized in the presence of natural S-peptide.

The RNase A activities, measured spectrophotometrically by the Kunitz method (Table IV), were verified with assays by the Anfinsen method. Synthetic S-protein produced RNase S with 5% specific activity when mixed with either natural S-peptide or synthetic S-peptide. Synthetic des-(21-25)S-protein produced des-(21-25)RNaseS with 9% specific activity when mixed with natural S-peptide and 7% with synthetic S-peptide. The fully synthetic RNase A(1-124) when worked up under the same conditions had 9% specific activity.

DISCUSSION

Solid Phase Synthesis of Protected Linear 124-Amino Acid Residue Polypeptide Chain of Ribonuclease A

The successful synthesis of ribonuclease A by the solid phase method required that there be enough space within the resin beads to accommodate a protected polypeptide of 19,791 molecular weight (Fig. 1). For that reason a polystyrene resin with only 1% divinylbenzene was selected. This low cross-linking allowed good swelling in organic solvents and was the minimum that was still compatible with good physical stability of the beads. Earlier electron micrograph data (41) had indicated that under these conditions the average space available to each peptide chain probably would be large enough to accommodate the ribonuclease A molecule.

With growing chain length of the peptide the volume of the resin increased considerably, but its swelling and filtering behavior as well as its mechanical properties were not noticeably affected. During their synthesis of a cytochrome c analogue Sano and Kurihara (42) observed an extremely severe breakdown of the resin, which in the end resulted in the loss of about 80% of the polymer support from the reaction vessel. This did not occur at all in the solid phase synthesis of RNase A described here. The quality of the resin support is clearly of great importance in solid phase peptide synthesis. We have observed some variability in the properties of commercial styrene-divinylbenzene resins and of chloromethylated resins with regard to uniformity of size, mechanical stability, and chemical reactivity. It seems probable that some of the reported discrepancies between the results from various laboratories can be attributed to differences in resins.

Protecting groups for the amino acid side chains were selected which could be removed in one step together with the cleavage of the polypeptide chain from the polymer support. In general benzyl derivatives were used. However, since preliminary coupling experiments on small peptides with imidazole-unprotected histidine had been quite successful, Boc-histidine with an unblocked side chain (instead of Boc-N^{im}-benzylhistidine) was used to incorporate the 4 residues of histidine into the amino acid sequence of RNase A. Thus, the treatment of the final product with sodium in liquid ammonia was not required and side reactions, involving desulfurization (43) and peptide bond cleavage (44, 45) which have been reported for this deprotection method, could be avoided.

Three- to five-fold excesses of amino acid reagents, and coupling times of 5 hours for dicyclohexylcarbodiimide reactions and 10 hours for active esters were chosen for this synthesis on the basis of past experience with the solid phase method. It should be noted that none of the concentrations or reaction times were optimized. We believe, however, that they were in excess of those actually required and that they were adequate to ensure nearly quantitative reactions. Since the quantity of Boc-amino acid for each coupling step was calculated from the concentration of the 1st valine residue, the concentration of Boc-amino acids remained constant during the run, but the concentration of peptide amino groups continually decreased due to the loss of peptide chains from the resin as the synthesis proceeded. Thus, the actual molar excess of reagent increased at each step. The effect of this drift in the ratio of the reactants on the rate and extent of the coupling reactions is not known.

In this synthesis the individual coupling steps and deprotection steps were not monitored because of the excessive time required to carry out these procedures with our present methodology. However, the desirability of applying rapid monitoring methods is clearly recognized (46), since it is of utmost importance to minimize incomplete coupling reactions and deprotection steps that would lead to deletion peptides missing amino acid residues.

The course of the synthesis was followed by sampling at intervals of 6 to 10 residues throughout the synthesis and determining amino acid ratios on acid hydrolysates of the peptide-resins. The precision of such analyses is limited, but the data served to indicate the progress of the chain growth. During the elongation of the polypeptide chain from 99 to 104 amino acid residues a sample for hydrolysis and amino acid analysis was removed after each coupling step. These data indicated plainly that even when the peptide chain had reached a length of the order of 100 residues it still reacted rapidly and in high yield with added Boc-amino acids.

Quantitative amino acid analysis of the final fully protected 124-residue polypeptide-resin indicated a yield of 17% based on the amount of COOH-terminal valine originally esterified to the polymer support, which means that an average of about 1.4% of the peptide chains were lost from the resin at each of the 123 cycles of the synthesis. This slow cleavage of the ester bond linking the growing polypeptide to the resin is attributed to the deprotection step, in which 50% (v/v) TFA-CH_2Cl_2 was used, but it was not established that the loss was uniform at every step. It must be emphasized that these data are a measure of the retention of the peptide chain on the resin but that they are not to be taken as an indication of the coupling yields. The partial losses of the benzyloxycarbonyl group from the ϵ-amino group of lysine (47, 48) and of the β- and γ-benzyl esters of aspartic and glutamic acids during the acidic removal of the α-Boc protecting groups are also recognized as possible side reactions during this synthesis. There are no definitive data showing that these reactions actually occurred, but we assume that some of the observed heterogeneity of the crude product resulted from this cause. The introduction of more acid-labile amino-blocking groups like the biphenylisopropyloxycarbonyl (Bpoc) group (49), which can be cleaved with 0.5% (v/v) TFA-CH_2Cl_2 (50), should effectively eliminate these side reactions.

Cleavage of Ribonuclease A from Resin

After the synthesis of the fully protected RNase A sequence, the peptide was removed from the resin support and was deprotected under three different conditions which have been described as Runs A, B, and C. The first experiments (Run A) on the cleavage of the protected 124-residue peptide-resin were with anhydrous HF (26, 27) in TFA (51). However, the effect on the natural RNase A control of the HF-TFA mixture (90 min at 0–20°) was rather severe. There were differences in the amino acid analyses (Table I) and activities (Table V). When natural RNase A was treated with anhydrous HF in the presence of anisole, but with no TFA, the amino acid analysis was in good agreement with that of a sample not treated with HF. However, even under these conditions there probably was slight denaturation since the specific activity dropped to about 90%.

A third sample of protected RNase-resin was cleaved and deprotected first in HBr-TFA (29) and then in HF because there was reason to believe that Tyr(Bzl) might be difficult to deprotect completely in HF. Past experience had shown that the benzyl ether was cleaved completely from tyrosine in HBr-TFA. It was observed (Run C) that fewer components were actually present after such treatment, but the total number of units of RNase A activity was essentially the same as when only HF had been used.

Upon cleavage of the 124-residue peptide-resin with anhydrous HF-TFA, it was found that only 41% of the peptide chains were removed from the resin within 90 min at 0–20°. Comparison of the analysis of cleaved peptide with that of the peptide still bound to the resin showed no significant differences in the amino acid composition. It has been found that a second treatment with HF cleaved 22% more of the peptide, making a total of 63%. However, in the original work-up the longer contact with HF was avoided because of the possible side reactions that this strong acid is known to promote (28, 52). Several precautions were taken to minimize undesired reactions which might occur in HF. The HF was dried over CoF_3 (53) and redistilled onto the frozen sample. Anisole (26) was used as a trap for NO_2^+ and benzyl$^+$ ions to prevent them from attacking tyrosine and other susceptible amino acid residues. Methionine was introduced as the sulfoxide to avoid sulfonium ion formation and also to prevent the known specific chain cleavage at this amino acid (52). However, in the solid phase synthesis of S-peptide, with the use of Boc-Met and the cleavage of the eicosapeptide from the resin with anhydrous HF under the conditions described above, no rupture of methionine peptide bonds was observed and recovery of methionine was good. This suggested that blocking of the thioether group of methionine by partial oxidation is not necessary to prevent side reactions during the HF cleavage step. Since tryptophan is absent from RNase A, the known side reaction of this amino acid with HF was not a problem. Prolonged treatment of proteins with HF has given rise to the $N \rightarrow O$ acyl shift at serine or threonine residues (28). In case this had occurred to an appreciable extent during the 90-min treatment used here the product was held at pH 7.5 for 2 hours, a condition known to reverse the shift effectively. An important change which will be incorporated into future work will be the replacement of S-benzylcysteine by the more easily removed S-p-methoxybenzylcysteine (51). Preliminary experiments agree well with the findings of Sakakibara et al. (51) that Cys(MeOBzl) is readily deprotected at 0° in 30 min by HF.

Purification of Cleaved and Deprotected Ribonuclease A

When a long linear synthetic polypeptide chain containing cysteine residues must be purified, it is desirable to stabilize

TABLE V
Effect of various treatments on specific activity of ribonuclease A

Treatment	Specific activity
	%
Natural enzyme	
None	100
Reduced-reoxidized	100
HF	90
HF-TFA	40–50
HF-TFA, SSO_3^-, reduced, reoxidized	12
HF-TFA, SSO_3^-, reduced, reoxidized, IRC-50	48
Synthetic enzyme	
HF-TFA, SSO_3^-, reduced, reoxidized, IRC-50	13

the —SH groups to prevent formation of randomly oxidized monomers or of polymers. The S-sulfonates were used with such great success in all of the insulin syntheses (45, 54–56) that they were a logical choice here, too. Accordingly, RNase-$(SSO_3^-)_8$ was prepared from the synthetic protein and from the natural enzyme by oxidative sulfitolysis with sodium sulfite and sodium tetrathionate (30, 31). The crude synthetic product showed, in addition to the major component, two minor spots on paper electrophoresis (Fig. 2). One of these was removed by chromatography on Dowex 1-X2 and the other by gel filtration on Sephadex G-50. The partly purified main fraction, which moved as a single component on paper electrophoresis, was obtained in 50% yield from the crude cleaved product. The conditions used for the S-sulfonate synthesis were clearly not optimal. Thus, a sample of natural RNase A after HF-TFA treatment, S-sulfonation, reduction, and reoxidation exhibited only 12% activity compared with 40 to 50% when the S-sulfonation step was omitted (Table V). Material from the later runs (B and C) was purified by gel filtration on Sephadex G-75 (Fig. 3). This procedure separated polymers (Peak I), randomly oxidized monomers (Peak II), and low molecular weight peptides (Peak III). The position of Peak II agreed closely with that from natural RNase A. More monomer could be obtained from the polymer peak by passing it through another reduction-reoxidation cycle.

Reduction and Reoxidation of Synthetic Ribonuclease A Preparations

It was necessary to reduce both the RNase$(S-SO_3^-)_8$ preparation and the randomly oxidized Sephadex G-75 fractions to the sulfhydryl derivatives and then to oxidize in air, in dilute solution, to generate ribonuclease A, which presumably contained the proper disulfide bonds and was folded into the native three-dimensional conformation. This synthesis is, of course, totally dependent on the spontaneous reoxidation and refolding of the chain to give the native structure, as originally found for natural RNase A by White (11), Anfinsen and Haber (12), and Epstein et al. (13). As a control for the experiments with synthetic enzyme it was established that full biological activity could be recovered in this laboratory when natural RNase A was reduced and reoxidized under these conditions.

There is no way at this time to direct the pairing of the proper cysteine residues or the folding of the molecule by selective chemical methods. Efforts which are being made to develop selective protection procedures for cysteine residues (57, 58) have not yet reached the degree of refinement required for a synthesis of ribonuclease.

Fractionation of Synthetic Enzyme on IRC-50

This weakly acidic cation exchange resin has been used frequently for the fractionation and purification of natural RNase A (32). It was therefore chosen as a selective way in which to purify the synthetic enzyme. The number of peaks obtained when the synthetic material was fractionated on this column (Fig. 4) depended on the previous cleavage and purification procedures which had been applied to the sample.

The polypeptide (Run A) that had been cleaved in HF-TFA and carried through the S-sulfonation procedure gave rise to five peaks, whereas the material (Run B) cleaved in HF and purified by gel filtration produced three peaks, and the peptide (Run C) that had been cleaved in HBr-TFA and further deprotected in HF before purifying by gel filtration gave rise to only a single peak. In each case the main activity was associated with Peak I, and the elution position of Peak I was exactly the same in all three runs.

In Run A, material with a specific activity of 1 to 2% was applied to the IRC-50 column and a 40% yield of product with a specific activity of 13% was obtained. The total number of units of RNase A activity recovered in this run (including that from Peak II) increased approximately 4- to 7-fold. In Run B, starting material with 5% specific activity led to a 60% yield of product with 13% specific activity (1.6-fold increase in total units). In Run C, the specific activity rose from 9 to 16% and the purification yield was 72%. Therefore, the total number of units increased only about 1.3-fold. The results of the IRC-50 fractionation experiments showed that, by this method, the specific activity of the synthetic RNase A could be raised to between 13 and 16%, regardless of whether the crude material applied to the column had an activity of 9% or only 1 to 2%. The fact that the total number of enzyme units increased after the fractionation suggests that inhibitory materials which had arisen during the synthesis and work-up procedures were being removed by the ion exchange column.

During these fractionation experiments on IRC-50, it was found that the elution volume of natural reduced-reoxidized RNase A agreed well with that of the synthetic enzyme, but that the position of both differed from that of untreated, pure natural RNase A on this ion exchange column. Fig. 4 (A and B) shows that despite different work-up conditions the two samples of synthetic RNase A both eluted from a short IRC-50 column at a volume of 12 ml (maximum of peak). Reduced-reoxidized natural RNase A had the same chromatographic behavior whereas the peak of untreated natural enzyme did not appear until 18 ml. On a longer IRC-50 column both reduced-reoxidized natural RNase A and synthetic RNase A (Run C, HBr + HF cleaved, fractionated on Sephadex G-75, reduced, reoxidized) emerged between 31 and 45 ml with the peak at 37 ml and were indistinguishable from one another (Fig. 4, C and D). Again, pure untreated natural RNase A eluted later. It emerged between 47 and 62 ml with the peak at 55 ml. A mixture of reduced-reoxidized natural RNase A and untreated natural RNase A separated into two fully active peaks at 37 and 55 ml, respectively. It has become evident that very small differences in the reduction and reoxidation conditions for natural RNase A can give rise to variable proportions of the two peaks at 37 and 55 ml on IRC-50 chromatograms. Whereas the earlier experiments gave only the 37-ml peak, later runs have given reconstituted natural RNase A containing the two peaks in a ratio as high as 1:2. Although IRC-50 columns have been used many times for the chromatographic purification of native RNase A, there does not appear to be a report in the literature on the chromatography of reduced-reoxidized RNase A on IRC-50. It is not clear, therefore, whether the reduced-reoxidized natural RNase A prepared in other laboratories would also have been distinguishable from the native enzyme by chromatography in this system. It was also shown by Veber et al. (59) that reoxidation of reduced natural S-protein in the presence of S-peptide at pH 8 gave rise to more than one component on a Sephadex CG-50 column, whereas reconstitution at pH 6.5 gave active material that chromatographed largely as a single component.

In contrast, reduced-reoxidized natural and synthetic RNase A were indistinguishable from each other and from pure native RNase A upon chromatography on CM-cellulose (Fig. 6). This is the chromatographic system that was used in the past to demonstrate the similarity between native and refolded RNase A (11). By this criterion our synthetic RNase A was identical with the natural enzyme.

Further Purification of Synthetic Ribonuclease A by Treatment with Trypsin and by Fractional Precipitation with Ammonium Sulfate

RNase A is known to be very resistant to tryptic digestion (60), whereas S-protein, RNase S, and denatured RNase A (61, 62) are susceptible to tryptic hydrolysis. This meant for the synthetic enzyme that molecules with a conformation which resembled closely that of native RNase A should resist tryptic attack whereas impurities, even with the same charge and molecular weight, but with different tertiary structures than that of natural RNase A might be digested by trypsin. The degradation products then would be easily separable from the intact active enzyme by gel filtration. Therefore, if the synthetic RNase A that showed a low specific activity were a mixture of a small amount of fully active enzyme and a large amount of inert protein having an incorrect primary and tertiary structure the improperly folded chains would be expected to be more susceptible to cleavage by trypsin than those chains with the native conformation of biologically active RNase A, and the specific activity of the enzyme should increase. If, on the other hand, the synthetic enzyme were a mixture of proteins each of which had a low specific activity due to deviations from the native species in amino acid sequence and three-dimensional structure, it seemed likely that trypsin would destroy virtually all the enzymic activity in the sample. If low activity were caused by the binding of a peptide inhibitor to the active site of the synthetic RNase A, trypsin might digest the peptide chain of the inhibitor and thus reactivate the enzyme.

When a sample of synthetic enzyme with about 8% specific activity was incubated with trypsin, then acidified and fractionated on Sephadex G-50 two peaks were observed. The position of Peak T-I agreed closely with that of untreated natural RNase A, whereas Peak T-II appeared to be a mixture of smaller peptides. The protein in Peak T-I was found to have a specific activity of 61%.

The amount of synthetic material digested by trypsin and removed by separation on the Sephadex column was only 25% of the starting material. This, together with the very appreciable 7.6-fold increase in the specific activity, means that the total number of units of enzyme activity increased 5.3-fold and suggests strongly that an inhibitory peptide had been present in the crude synthetic product and was removed by tryptic hydrolysis.

The specific activity of the synthetic product was further raised by fractional precipitation. To an aqueous solution of the synthetic RNase A with 61% specific activity a saturated solution of ammonium sulfate was added in an amount just below that necessary to precipitate natural RNase A (63). After standing 2 days at 3°, some amorphous material had precipitated, and it was removed by centrifugation. The protein in the supernatant solution showed a specific activity against RNA of 78%. This is the most active synthetic RNase A we have prepared at this time.

Yields

It is helpful to examine the yields of the various steps of this synthesis in order to have a better feeling for what could be done. Since this first synthesis was experimental and exploratory the various steps were each examined in several ways and many samples were withdrawn both for analytical and preparative purposes. For that reason all of the starting material was not carried through to the final purification step. For the presentation of yields, therefore, we have selected Run B and the data are given in two ways (Table VI). The actual amounts obtained in Run B are recorded, and then the amounts that would have been expected if all of the original 2 g of Boc-valine-resin had been carried through these same steps were calculated and listed. Several conclusions can be drawn.

1. During the solid phase peptide synthesis of the protected RNase-resin, 83% of the chains were lost. This resulted from an average cleavage of 1.4% of the anchoring benzyl ester bond from the resin at each step.

2. A total of 1.59 g of protected RNase-resin was obtained. This number would have been 3.43 g if several samples had not been withdrawn during the synthesis. It means that the final resin-bound product was 29% protein, 13% protecting groups, and 58% polystyrene and contained the equivalent of 992 mg of unprotected protein on 2 g of polystyrene.

3. The cleavage yield in HF was 70%, and 41.5 mg of crude protein were obtained in the 203-mg run.

4. A total of 53% (19.5 mg) of the crude protein was recovered as RNase A monomer (Peak II) from Sephadex G-75.

5. A 1-mg sample was carried through the trypsin digestion and $(NH_4)_2SO_4$ fractionation to give 0.41 mg of purified synthetic RNase A with 78% specific activity.

6. Of the protein in the protected RNase-resin, 17% was recovered in the $(NH_4)_2SO_4$-soluble fraction as purified RNase A.

7. Of the crude, cleaved protein 24% was recovered in the $(NH_4)_2SO_4$-soluble fraction.

8. The crude, cleaved synthetic protein contained the equivalent of 22% of native RNase A by enzyme assay.

Evidence for Chemical and Physical Purity of Synthetic Enzyme and for Its Similarity to Native Ribonuclease A

The various data recorded and discussed here generally support the view that our synthetic protein is reasonably homogeneous and that it bears a close resemblance to the natural molecule. However, a rigorous proof of purity of a synthetic protein is difficult and the use of total synthesis to prove the structure of a protein in the classic organic chemical sense becomes very difficult indeed. One can establish certain identities, but always within the limits of the particular technique being applied. We do not suggest that our work, in its present state of development, constitutes a structure proof of ribonuclease A, only that the synthetic molecule bears a close chemical and physical resemblance to the natural protein and that it is a true enzyme with a specific activity not far from that of native ribonuclease A.

The chemical and physical comparisons were based on amino acid analyses, enzyme digestions, antibody neutralization, peptide maps, paper electrotrophoresis, gel filtration, and ion exchange chromatography.

The purified synthetic enzyme had the over-all amino acid composition (Table I) expected of RNase A with only minor deviations from that of the untreated natural control. Enzymic digestion, first by papain and then by aminopeptidase M,[5] completely degraded the synthetic molecule to free amino acids. This

TABLE VI
Summary of yields and activities obtained in Run B

Stage	Amount[a] of material carried through Run B	Yield[b] calculated from 2 g of resin	Specific activity[c]	Total RNase A activity[d]	
				In Run B	Calculated to 2 g of resin
	mg	mg	%	mg	
Boc-valine-resin	2000	2000	100		
Protected RNase-resin	1588 → 203	3430	17		
Crude cleaved protein	41.5 → 37	697	12		
RNase A monomer from Sephadex G-75 (Peak II)	19.5 → 1	373	6.4	9.0	1.73	33
Trypsin-resistant RNase A (Peak T-I)	0.69 → 0.62	256	4.4	61	0.42	156
Purified RNase A $((NH_4)_2SO_4$-soluble)	0.41	169	2.9	78	0.32	132

[a] This is the weight of protein or protein-resin that was actually carried through Run B. The number on the right is the amount used for each step and the number at the left below is the yield of that step. Thus, 203 mg of protected RNase-resin were used for the HF cleavage in Run B and it gave 41.5 mg of crude cleaved protein, 37 mg of which were used for the Sephadex G-75 fractionation.

[b] This is the amount of protein or of peptide-resin that would have been obtained if all of the starting 2 g of Boc-valine-resin had been carried through the steps described for Run B.

[c] The percentage of activity relative to pure native RNase A.

[d] The weight of pure RNase A required to give the activity observed in the synthetic preparations.

indicated that no major racemization had occurred during the synthesis or the various work-up procedures and is in agreement with the accumulated data from several laboratories (15) that the usual procedures of solid phase synthesis do not cause significant racemization. Although the synthetic RNase A was resistant to trypsin it was readily digested after performic acid oxidation (18, 33) of the 4 cystine residues to cysteic acid residues. Peptide maps of the digest showed, as expected, 14 spots that corresponded quite well with the positions of the peptides derived from natural RNase A, and a small additional spot near the position of lysine (Fig. 8). Since the separated peptides were not further identified there is no evidence that the corresponding spots were identical. However, major differences in protein composition would have been expected to give rise to quite different peptide patterns. Peptides arising from small amounts of chains with different sequences would probably have gone undetected.

The purified synthetic RNase A was indistinguishable from native RNase A on paper electrophoresis (Fig. 2), which was evidence for similarity of net charge and size. The synthetic and natural molecules also exhibited the same size by gel filtration on Sephadex G-75.

One of the original and principal criteria (11) for the identity of reduced-reoxidized natural RNase A with the untreated native protein was the equivalence of elution volume on a CM-cellulose column in sodium phosphate buffer. This, together with the recovery of full enzymic activity was the basis for the conclusion that the primary structure of RNase A determined its tertiary structure (11).

Our synthetic enzyme agreed perfectly with native RNase A on CM-cellulose and it possessed 78% of the enzymic activity of the pure natural enzyme.

The data on IRC-50 columns support the conclusion that the synthetic enzyme, after reduction and reoxidation, is similar to the natural enzyme that had been reduced and reoxidized under the same conditions. However, for reasons not understood, there was a discrepancy between the position on this column of native RNase A and natural or synthetic reduced-reoxidized RNase A. Whether this is an artifact or actually represents a real difference in conformation is not known at this time.

Enzymic Activity and Substrate Specificity of Synthetic Ribonuclease A

From the activity data on RNA and C>p there is no doubt that the synthetic RNase A catalyzed the chemical reactions that are characteristic for native RNase A. The most highly purified preparation of the synthetic enzyme was 78% as active against an RNA substrate as the pure, untreated natural enzyme, and when C>p was the substrate the specific activity was 65%. This means that it is possible to synthesize in the laboratory an extremely potent and specific organic catalyst starting with ordinary amino acids.

The synthetic RNase A preparation that had not been extensively purified did not have the same specific activity when it was tested against the two substrates, RNA and C>p. This might have been due to small deviations in its conformation from that of the natural enzyme, since it is known that in other instances slight changes in the conformation of RNase can alter the relative activity of the enzyme toward different substrates. Another possible cause might have been differences in the sensitivity of the substrates to the presence of inhibitors of the enzyme. Evidence for inhibitors had come from the data showing an increase in total number of units of activity upon purification of the enzyme and, in addition, the finding that the specific activity against RNA and C>p were more nearly equal for the purified synthetic enzyme than for the cruder preparation is also compatible with the idea that inhibitors had been present.

In a series of assays, DNase and RNase T_1 activities of the synthetic RNase A were excluded. The results of the DNase assay (Fig. 10) clearly showed that synthetic RNase A at two different concentrations (4-fold and 40-fold amount of DNase used) had no effect on DNA. With the dinucleotide GpCp as substrate, RNase T_1 catalyzed hydrolysis in accordance with the specificity of this enzyme for the 3′,5′-phosphodiester bond following a guanosine residue (Fig. 11). As expected, natural RNase A was not active against GpCp and also the synthetic RNase A was completely inactive in this assay. The results of the preceding assay could be confirmed with G>p as substrate. RNase T_1 catalyzed the hydrolysis of the 2′,3′-cyclic phosphate to give the 3′-phosphate, whereas natural and synthetic RNase A showed no reaction with this substrate. With ApAp as substrate all three enzymes, synthetic RNase A, natural RNase A, and RNase T_1, were without effect.

Although this is by no means a complete study of substrate specificity the data were all consistent with the specificity to be expected for RNase A. They demonstrated the ability of the synthetic enzyme to cleave both large (RNA) and small (C>p) substrates and to catalyze both the transphosphorylation and hydrolysis steps. They showed the expected requirement for D-ribose instead of D-deoxyribose and for a pyrimidine instead of a purine.

Calculation of Michaelis Constant—The close agreement of the K_m values of natural and synthetic RNase A (1.20 mg per ml and 1.24 mg per ml, respectively), as calculated from a Lineweaver-Burk plot (Fig. 12), provided good evidence that the enzymic activity exhibited by the synthetic product was a true RNase A activity.

Preparation of Synthetic S-protein and Synthetic Des-(21–25)S-protein and Their Combination with Natural or Synthetic S-peptide to Give Ribonuclease S and Des-(21–25)-ribonuclease S

Since the discovery of Richards (64) and Richards and Vithayathiol (65) that RNase A is cleaved under certain conditions by subtilisin, primarily at the peptide bond between alanine residue 20 and serine residue 21, several laboratories have been involved in chemical, physicochemical, and x-ray crystallographic studies on the complex, ribonuclease S, formed between S-peptide(1–20) and S-protein(21–124). Recently Wyckoff et al. (10) elucidated the **x-ray** structure of RNase S at a 2-Å resolution and constructed a three-dimensional model, which shows clearly the folding and interactions of the two polypeptide chains. But even such a detailed model cannot answer every question about the relationship between structure and function of the enzyme molecule. Many of these problems can best be attacked through the synthetic approach by replacement or omission of amino acid residues or by specific labeling of the molecule. Extensive studies with synthetic analogues and fragments of the S-peptide have already been performed by two groups (66–68). The first work on investigations of this type with synthetic S-protein is described in this paper. We were interested in the question whether an S-protein fragment lacking the first 5 residues (Residues 21 to 25) at its NH$_2$ terminus (Fig. 15) would still bind with S-peptide and generate ribonuclease activity.

For this purpose S-protein, des-(21-25)S-protein and S-peptide were prepared by the solid phase method as described under "Experimental Procedure and Results." The three peptides were partially purified by gel filtration on Sephadex, and S-peptide was further purified by free-flow electrophoresis. The three peptides then migrated as single components on paper electrophoresis and the electrophoretic mobilities of S-peptide and S-protein then agreed well with the corresponding natural peptides. The products were further characterized by amino acid analyses (Table I).

It has been demonstrated that the presence of S-peptide not only stabilizes the conformation of the S-protein but also increases the efficiency of the formation of the correct disulfide bonds upon reoxidation of reduced S-protein in dilute solution (69, 70). For this reason synthetic S-protein and des-(21-25)S-protein were reduced with mercaptoethanol and thereby unfolded. After isolation of the sulfhydryl forms of the two polypeptides by gel filtration on Sephadex G-25 in weakly acidic medium, natural or synthetic S-peptide was added. The mixtures were then reoxidized with air at low protein concentration to allow the chains to refold into the active conformation and the four disulfide bonds to close.

These studies have shown that approximately equivalent amounts of RNase activity were generated when the denatured, reduced form of either synthetic S-protein or synthetic des-(21-25)S-protein was allowed to refold and reoxidize in the presence of S-peptide. Although the specific activities ranged between 5 and 9% (Table IV) they are not considered to be significantly different from one another because of the several experimental operations involved. The activities are low for two reasons. First, this series of experiments was done on synthetic proteins that had not been highly purified. After cleavage they were simply passed through a Sephadex G-75 column. Second, when natural S-protein and S-peptide are reduced and reoxidized under these conditions they give only about 35% recovery of activity. This value which was reported by Haber and Anfinsen (69) in their original work was also found here. Thus, the crude synthetic S-protein and des-(21-25)S-protein gave between 15 and 25% as much activity as the natural S-protein. The control of synthetic RNase A, which had been partially purified in the same way, also gave 9% activity after reoxidation.

From their model of RNase S, Wyckoff et al. (71) predicted that the 3 NH$_2$-terminal serine residues, 21 to 23, of the S-protein could be omitted without serious consequences to conformation or activity of RNase S. There was no indication from the x-ray data of noncovalent interactions between these 3 serine residues and other parts of the complex. The situation for the next 2 amino acid residues, asparagine 24 and tyrosine 25, is different. In RNase A and RNase S, tyrosine 25 is one of the 3 "buried" tyrosyl residues and may be hydrogen-bonded with the buried side chain carboxylate group of aspartic acid 14 (72). Upon removal of the S-peptide this interaction is disrupted, and tyrosine 25 then titrates normally. In the RNase S model (10) the main chain of tyrosine 25 can be hydrogen-bonded with methionine 29, whereas for the phenolic hydroxyl group of the tyrosine side chain, interaction with the imidazole ring of histidine 48 has been suggested. For asparagine 24 three hydrogen bridges have been proposed linking it to tyrosine 97, asparagine 27, and glutamine 28.

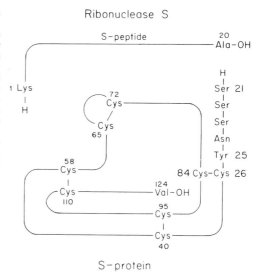

FIG. 15. Ribonuclease S. In des-(21-25)RNase S the first 5 amino acid residues at the NH$_2$ terminus of the S-protein (Residues 21 to 25) are missing.

In des-(21-25)S-protein those five hydrogen bonds are missing. The data now described (73) have demonstrated experimentally that on adding S-peptide to either S-protein or des-(21-25)S-protein essentially the same ribonuclease activity is generated. It can be concluded that the first 5 residues, 21 to 25, at the NH$_2$ terminus of S-protein are not required for the formation of the S-protein-S-peptide complex and the generation of enzymic activity. Thus, the proposed noncovalent interactions of asparagine 24 or tyrosine 25 are neither essential for binding the S-peptide nor for stabilizing the active conformation.

CONCLUSIONS

We wish to draw three general conclusions from these experiments.

1. A new and independent kind of evidence is provided for the view that the primary structure of a protein determines its tertiary structure. The original hypothesis (11, 13) was based on the observation that native ribonuclease could be reduced and unfolded in urea and that upon removal of the urea and reoxidation in air it would spontaneously refold into its original conformation, would re-form the proper disulfide bonds, and would regenerate its full enzymic activity. The conclusion that this response was determined solely by the primary structure of the protein depended on establishing that the unfolded chain had a completely random form and had lost all of its secondary and tertiary structure. Otherwise, if a small region of the molecule were to have retained some information about its original conformation it might serve as a nucleus to direct the remainder of the folding of the protein. Most of the data (1-3) have supported the random coil view, but some (74) have indicated a structural rigidity in the reduced protein. Whether the reduced protein retains any of its original structure in urea solution remains unclear.

In the case of a synthetic protein there is no possibility that any pre-existing information about the natural conformation could be present. The only information put into the synthetic protein is its amino acid sequence. Therefore, if an active enzyme is produced it must be solely a consequence of its primary structure. This was observed for the synthetic ribonuclease A.

2. The synthesis of des-(21-25)S-protein is a beginning toward structure-function studies on whole proteins by use of the synthetic approach. Through chemical synthesis it could be directly demonstrated that amino acid residues 21 to 25 were not required for S-protein and S-peptide to combine, noncovalently, and to fold into an active enzyme. This kind of conclusion can be made with some confidence because these five residues could not be present in the synthetic analogue in any amount and yet it was as active as the parent compound. Structure-activity data on synthetic analogues that prove to be inactive or of very low activity are much more difficult to interpret.

3. Finally we can conclude that it is possible to begin with free amino acids and to assemble them in the laboratory to give a real protein that possesses true enzymic activity.

Note Added in Proof—An antibody neutralization test has provided further evidence for the similarity between the synthetic and natural RNase A. Samples of synthetic and natural RNase A were incubated with an antibody prepared from rabbits immunized against crystalline bovine pancreatic RNase A and the extent of inactivation was determined by comparing the quantity of acid-soluble nucleotides (20, 21) produced with that in the controls without antibody. The incubation mixture contained: sodium acetate buffer (0.05 M, pH 5.0), 0.34 ml; RNA (1 mg per ml), 1.00 ml; antiserum, 0.12 ml; RNase (0.1 mg per ml), 0.01 ml. The presence of the antiserum caused a drop of absorbance at 260 nm (corrected for the blank) from 0.350 to 0.290 for the natural RNase A and from 0.277 to 0.217 for the synthetic RNase A. Since the assay is linear in this range the data indicate that under the conditions of this single experiment the antibody neutralized essentially equal quantities of the synthetic and natural enzymes.

We are most grateful to Dr. Cecil Yip, Banting and Best Department of Medical Research, University of Toronto, for suggesting this experiment and for supplying the antiserum and the procedure for the assay.

Acknowledgments—We wish to thank Miss Anita Bach for her excellent technical assistance and Miss Ursula Birkenmaier for the amino acid analyses.

REFERENCES

1. ANFINSEN, C. B., AND WHITE, F. H., JR., in P. D. BOYER, H. LARDY, AND K. MYRBÄCK (Editors), *The enzymes*, Vol. V, Part B, Academic Press, New York, 1961, p. 95.
2. SCHERAGA, H. A., AND RUPLEY, J. A., *Advan. Enzymol.*, **24**, 161 (1962).
3. BARNARD, E. A., *Annu. Rev. Biochem.*, **38**, 677 (1969).
4. HIRS, C. H. W., MOORE, S., AND STEIN, W. H., *J. Biol. Chem.*, **235**, 633 (1960).
5. SMYTH, D. G., STEIN, W. H., AND MOORE, S., *J. Biol. Chem.*, **237**, 1845 (1962).
6. POTTS, J. T., BERGER, A., COOKE, J., AND ANFINSEN, C. B., *J. Biol. Chem.*, **237**, 1851 (1962).
7. GROSS, E., AND WITKOP, B., *J. Biol. Chem.*, **237**, 1856 (1962).
8. SMYTH, D. G., STEIN, W. H., AND MOORE, S., *J. Biol. Chem.*, **238**, 227 (1963).
9. KARTHA, G., BELLO, J., AND HARKER, D., *Nature*, **213**, 862 (1967).
10. WYCKOFF, H. W., TSERNOGLOU, D., HANSON, A. W., KNOX, J. R., LEE, B., AND RICHARDS, F. M., *J. Biol. Chem.*, **245**, 305 (1970).
11. WHITE, F. H., JR., *J. Biol. Chem.*, **236**, 1353 (1961).
12. ANFINSEN, C. B., AND HABER, E., *J. Biol. Chem.*, **236**, 1361 (1961).
13. EPSTEIN, C. J., GOLDBERGER, R. F., YOUNG, D. M., AND ANFINSEN, C. B., *Arch. Biochem. Biophys. Suppl.*, **1**, 223 (1962).
14. MERRIFIELD, R. B., *J. Amer. Chem. Soc.*, **85**, 2149 (1963).
15. MERRIFIELD, R. B., *Advan. Enzymol.*, **32**, 221 (1969).
16. GUTTE, B., AND MERRIFIELD, R. B., *J. Amer. Chem. Soc.*, **91**, 501 (1969).
17. HIRSCHMANN, R., NUTT, R. F., VEBER, D. F., VITALI, R. A., VARGA, S. L., JACOB, T. A., HOLLY, F. W., AND DENKEWALTER, R. G., *J. Amer. Chem. Soc.*, **91**, 507 (1969).
18. HIRS, C. H. W., MOORE, S., AND STEIN, W. H., *J. Biol. Chem.*, **219**, 623 (1956).
19. KUNITZ, M., *J. Biol. Chem.*, **164**, 563 (1946).
20. ANFINSEN, C. B., REDFIELD, R. R., CHOATE, W. L., PAGE, J., AND CARROLL, W. R., *J. Biol. Chem.*, **207**, 201 (1954).
21. EGAMI, F., TAKAHASHI, K., AND UCHIDA, T., *Progr. Nucleic Acid Res. Mol. Biol.*, **3**, 59 (1964).
22. MERRIFIELD, R. B., STEWART, J. M., AND JERNBERG, N., *Anal. Chem.*, **38**, 1905 (1966).
23. STEWART, J. M., AND YOUNG, J. D., *Solid phase peptide synthesis*, Freeman and Company, San Francisco, 1969.
24. SHEEHAN, J. C., AND HESS, G. P., *J. Amer. Chem. Soc.*, **77**, 1067 (1955).
25. BODANSZKY, M., *Nature*, **175**, 685 (1955).
26. SAKAKIBARA, S., SHIMONISHI, Y., KISHIDA, Y., OKADA, M., AND SUGIHARA, H., *Bull. Chem. Soc. Japan*, **40**, 2164 (1967).
27. LENARD, J., AND ROBINSON, A. B., *J. Amer. Chem. Soc.*, **89**, 181 (1967).
28. LENARD, J., AND HESS, G. P., *J. Biol. Chem.*, **239**, 3275 (1964).
29. MERRIFIELD, R. B., *Biochemistry*, **3**, 1385 (1964); *Recent Progr. Hormone Res.*, **23**, 451 (1967).
30. BAILEY, J. L., *Biochem. J.*, **67**, 21P (1957).
31. BAILEY, J. L., AND COLE, R. D., *J. Biol. Chem.*, **234**, 1733 (1959).
32. HIRS, C. H. W., MOORE, S., AND STEIN, W. H., *J. Biol. Chem.*, **200**, 493 (1953).
33. HIRS, C. H. W., *J. Biol. Chem.*, **219**, 611 (1956).
34. BARROLLIER, J., HERLMANN, J., AND WATZKE, E., *Hoppe-Seyler's Z. Physiol. Chem.*, **309**, 219 (1957).
35. ISELIN, B., *Helv. Chim. Acta*, **44**, 61 (1961).
36. FRUCHTER, R. G., AND CRESTFIELD, A. M., *J. Biol. Chem.*, **240**, 3868 (1965).
37. KUNITZ, M., *J. Gen. Physiol.*, **33**, 349 (1950).
38. TAKAHASHI, K., STEIN, W. H., AND MOORE, S., *J. Biol. Chem.*, **242**, 4682 (1967).
39. MERRIFIELD, R. B., AND WOOLLEY, D. W., *J. Biol. Chem.*, **197**, 521 (1952).
40. EDELHOCH, H., AND COLEMAN, J., *J. Biol. Chem.*, **219**, 351 (1956).
41. MERRIFIELD, R. B., AND LITTAU, V., in E. BRICAS (Editor), *Peptides*, North-Holland, Amsterdam, 1968, p. 179.
42. SANO, S., AND KURIHARA, M., *Hoppe-Seyler's Z. Physiol. Chem.*, **350**, 1183 (1969).
43. KATSOYANNIS, P. G., *Amer. J. Med.*, **40**, 652 (1966).
44. BENESEK, W. F., AND COLE, R. D., *Biochem. Biophys. Res. Commun.*, **20**, 655 (1965).
45. MARGLIN, A., AND MERRIFIELD, R. B., *J. Amer. Chem. Soc.*, **88**, 5051 (1966).
46. MARGLIN, A., AND MERRIFIELD, R. B., *Annu. Rev. Biochem.*, **39**, 841 (1970).
47. YARON, A., AND SCHLOSSMAN, F., *Biochemistry*, **7**, 2673 (1968).
48. GRAHL-NIELSEN, O., AND TRITSCH, G. L., *Biochemistry*, **8**, 187 (1969).
49. SIEBER, P., AND ISELIN, B., *Helv. Chim. Acta*, **51**, 614, 622 (1968).
50. WANG, S. S., AND MERRIFIELD, R. B., *Int. J. Protein Res.*, **1**, 235 (1969).
51. SAKAKIBARA, S., SHIMONISHI, Y., OKADA, M., AND KISHIDA, Y., in H. C. BEYERMAN, A. VAN DE LINDE, AND W. MAAS-

SEN VAN DEN BRINK (Editors), *Peptides*, North-Holland, Amsterdam, 1967, p. 44.
52. LENARD, J., SCHALLY, A. V., AND HESS, G. P., *Biochem. Biophys. Res. Commun.*, **14**, 498 (1964).
53. JARRY, R. L., AND DAVIS, W., JR., *J. Phys. Chem.*, **57**, 600 (1953).
54. KATSOYANNIS, P. G., TOMETSKO, A., AND FUKUDA, K., *J. Amer. Chem. Soc.*, **85**, 2863 (1963).
55. KUNG, Y. T., DU, Y. C., HUANG, W. T., CHEN, C. C., KE, L. T., HU, S. C., JIANG, R. Q., CHU, S. Q., NIU, C. I., HSU, J. Z., CHANG, W. C., CHENG, L. L., LI, H. S., WANG, Y., LOH, T. P., CHI, A. H., LI, C. H., SHI, P. T., YIEH, Y. H., TANG, K. L., AND HSING, C. Y., *Sci. Sinica (Peking)*, **14**, 1710 (1965).
56. ZAHN, H., DANHO, W., AND GUTTE, B., *Z. Naturforsch.*, **21b**, 763 (1966).
57. ZERVAS, L., PHOTAKI, I., YOVANIDIS, C., TAYLOR, J., PHOCAS, I., AND BARDAKOS, V., in H. BEYERMAN, A. VAN DE LINDE, AND W. MAASSEN VAN DEN BRINK (Editors), *Peptides*, North-Holland, Amsterdam, 1967, p. 28.
58. HISKEY, R. G., THOMAS, A. M., SMITH, R. L., AND JONES, W. C., JR., *J. Amer. Chem. Soc.*, **91**, 7525 (1969).
59. VEBER, D. F., VARGA, S. L., MILKOWSKI, J. D., JOSHUA, H., CONN, J. B., HIRSCHMANN, R., AND DENKEWALTER, R. G., *J. Amer. Chem. Soc.*, **91**, 506 (1969).
60. DUBOS, R. J., AND THOMPSON, R. H. S., *J. Biol. Chem.*, **124**, 501 (1938).
61. ALLENDE, J. E., AND RICHARDS, F. M., *Biochemistry*, **1**, 295 (1962).
62. UZIEL, M., STEIN, W. H., AND MOORE, S., *Fed. Proc.*, **16**, 263 (1957).
63. KUNITZ, M., *J. Gen. Physiol.*, **24**, 15 (1940).
64. RICHARDS, F. M., *Compt. Rend. Trav. Lab. Carlsberg Ser. Chim.*, **29**, 329 (1955); *Proc. Nat. Acad. Sci. U. S. A.*, **44**, 162 (1958).
65. RICHARDS, F. M., AND VITHAYATHIL, P. J., *J. Biol. Chem.*, **234**, 1459 (1959).
66. FINN, F. M., AND HOFMANN, K., *J. Amer. Chem. Soc.*, **87**, 645 (1965).
67. HOFMANN, K., VISSER, J. P., AND FINN, F. M., *J. Amer. Chem. Soc.*, **92**, 2900 (1970).
68. ROCCHI, R., MARCHIORI, F., MORODER, L., BORIN, G., AND SCOFFONE, E., *J. Amer. Chem. Soc.*, **91**, 3927 (1969).
69. HABER, E., AND ANFINSEN, C. B., *J. Biol. Chem.*, **236**, 422 (1961).
70. KATO, I., AND ANFINSEN, C. B., *J. Biol. Chem.*, **244**, 1004 (1969).
71. WYCKOFF, H. W., HARDMAN, K. D., ALLEWELL, N. M., INAGAMI, T., JOHNSON, L. N., AND RICHARDS, F. M., *J. Biol. Chem.*, **242**, 3984 (1967).
72. LI, L., RIEHM, J. P., AND SCHERAGA, H. A., *Biochemistry*, **5**, 2043 (1966).
73. GUTTE, B., AND MERRIFIELD, R. B., *Fed. Proc.*, **29**, 727 (1970).
74. YOUNG, D. M., AND POTTS, J. T., JR., *J. Biol. Chem.*, **238**, 1995 (1963).

Reprinted from *Intra-Sci. Chem. Rept.,* **5**(3), 184–198 (1971), with permission from the Intra-Science Research Foundation

The Solid Phase Synthesis of Ribonuclease A

R. B. MERRIFIELD

ABSTRACT

In an effort to simplify and accelerate the synthesis of peptides a new approach to the problem was devised. It was called "solid phase peptide synthesis," and was based on the idea that peptides could be assembled in a stepwise manner while attached at one end to an insoluble solid particle. With the peptide securely bound in the solid phase it was possible to purify each of the intermediates simply and quickly by thorough washing, rather than by recrystallization or other tedious procedures. The method was applied to the synthesis of bradykinin, angiotensin, deamino-oxytocin, and some related compounds. The products were obtained in good yield and were shown to be chemically pure and biologically active. Solid phase peptide synthesis has now been automated so that all the reactions are carried out in a single reaction vessel, and each of the manipulations is performed in the proper sequence under the control of a simple programmer. Insulin and the dodecadepsipeptide valinomycin were synthesized by the automated system. The original aim of this work was the development of a method whereby large, complex polypeptides might become accessible to synthesis. We have recently succeeded in synthesizing a protein with the sequence and activity of bovine pancreatic ribonuclease.

I. INTRODUCTION

The idea of chemically synthesizing an enzyme must have occurred to many people over the years — it first occurred to me about twelve years ago. There was a time when such a thought would have been untenable even on philosophical grounds, but from the period when enzymes were shown to be proteins and proteins were shown to be discrete organic molecules it was a goal which chemists could begin to think about. If an enzyme could be made in the laboratory, then surely it would become possible to learn new things about how these very large and enormously complex compounds function. Specific changes could be made in their structures which could not readily be made by working with the native molecule and data should be forthcoming which would supplement that obtained from the natural enzymes themselves.

The obstacles to the idea, of course, were the complexity of the enzyme molecule and the inadequacies in the available synthetic techniques. Quite clearly new approaches to the synthesis of peptides and proteins were necessary before a realistic attack on the problem could begin. One of the new approaches that has evolved over the past decade is called "solid phase peptide synthesis." I want to describe and discuss this technique and to show how it was eventually applied to the total synthesis of the enzyme, ribonuclease A.

Anyone who has synthesized a peptide — even a small one — will know that two main kinds of problems are encountered: The first involves the chemistry of the reactions; the second involves the physical manipulations associated with the synthesis, the work-up, and the purifications. Very selective methods for protecting the functional groups reversibily are required, and efficient methods for forming the peptide bond in high yield and without racemization are also needed. Beginning with Curtius and Fischer and continuing to the present day these problems have been extensively studied, and very highly refined chemical reactions have evolved.[1] By the mid fifties it had become possible to define general procedures for the effective synthesis of small peptides, and du Vigneaud provided the dramatic demonstration with his synthesis of oxytocin.[2] From that time a rapid flow of new synthetic peptides followed.

It spite of this progress it was a difficult and time-consuming task to prepare a peptide. Many manipulations involving multiple transfers of materials, with their attendant losses, were needed, and each reaction mixture posed a new and unique purification problem. Crystallization conditions for each intermediate had to be devised, and where the product would not crystallize, another procedure such as chromatography, electrophoresis, or countercurrent distribution was usually needed. It seemed obvious that polypeptides containing over 100 amino acid residues could not be made in this way without an enormous expenditure of time and energy.

II. SOLID PHASE PEPTIDE SYNTHESIS

A. The Idea

With these problems in mind a new approach to the synthesis of peptides was devised which was expected to overcome some of the difficulties. It was called "solid phase peptide synthesis"[3] and was based on the idea that a peptide chain could be assembled, one unit at a time, while one end was anchored to an insoluble solid support. If the reactions could be driven very near to completion by the use of excess reagents, it should be possible to purify the intermediate products simply by a filtration and washing operation. In that way the manipulations in the work-up and purification steps could be greatly simplified and accelerated, and the synthesis of large molecules might become feasible. The work in our laboratory and in many others has shown that this approach to peptide synthesis was feasible and that it could lead to pure products in high yields.[4]

B. The General Plan

The general plan for solid phase peptide synthesis is outlined in FIGURE I. It indicates, first, the covalent attachment of a protected amino acid to a functional group on a solid particle, followed by the selective removal of the protecting group to liberate a reactive end of the amino acid. A second protected amino acid is then activated and coupled to the first to give a peptide bond. The deprotection and coupling steps are then repeated, alternately, until the peptide chain is assembled. Finally, the bond holding the peptide chain to the solid support is cleaved, and the completed peptide is liberated from the solid phase into the liquid phase.

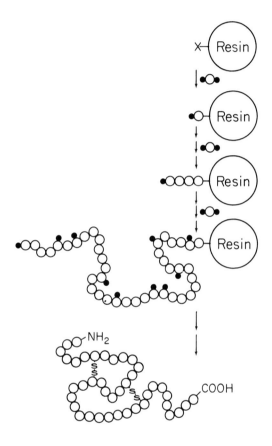

FIGURE I

Schematic View of Solid Phase Peptide Synthesis

The small open circles represent amino acid residues.
The solid circles represent protecting groups.

C. The Solid Support

A solid support was required which was chemically inert, completely insoluble in all solvents to be used, and readily filterable. For this purpose a polystyrene resin in which the chains were loosely cross-linked with 1 or 2% divinylbenzene was selected. It was in the form of small spheres (FIGURE IIa). The beads swell freely in organic

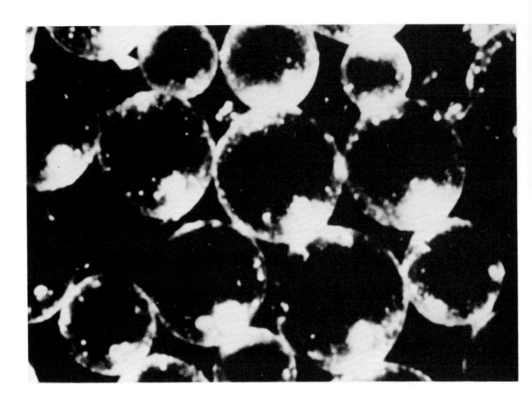

FIGURE IIa

The Resin Support

Photograph of Styrene–1%-Divinylbenzene Spheres

solvents and become permeable to reagents so that reactions can occur not only on the surface but also within the interstices of the gel-like matrix. The uniform distribution of the synthesized peptide chains within these resin beads was established by autoradiography[5] (FIGURE IIb).

D. A Chemical Scheme for the Stepwise Approach

The several chemical reactions involved in these operations were drawn from the reservoir of information developed for classical peptide chemistry. Many combinations are possible, provided they lead to high coupling yields and very selective protection and deprotection conditions. One of the early schemes is outlined in FIGURE III. It depended on the use of a relatively stable benzyl ester attachment to the resin support and relatively labile t-butyloxycarbonyl (Boc) protection for the α-amino groups. Side chain protection was also based on the benzyl group. The Boc group could, therefore, be removed repeatedly as the chain continued to grow without serious loss of the stable groups which were required to remain attached until the end of the synthesis. The latter were then removed with strong anhydrous acid, and the peptide was liberated into solution for final purification. The coupling reaction shown in this scheme makes use of dicyclohexylcarbodiimide. It provides a remarkably reactive derivative

Intra-Science Chemistry Reports

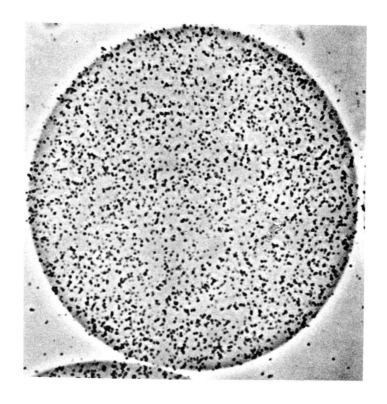

FIGURE IIb

The Resin Support

Autoradiograph of a Thin Section of a Sphere of Styrene–1%-Divinylbenzene Containing a Tritium-Labeled Peptide

which, when used in excess, leads to nearly quantitative reactions. Most of the known activating reagents for peptide synthesis have been tested for solid phase synthesis, but only the carbodiimide and active esters — particularly nitrophenyl esters — have been used extensively.

E. Applications

The new method was first tested in the synthesis of a model tetrapeptide[3] and then was applied to the synthesis of the nonapeptide, bradykinin.[6] The product was shown to be indistinguishable from the natural hormone both chemically and biologically. The general applicability of the technique to peptides of this size was established when it was extended to the synthesis of methionyl-lysyl-bradykinin, to angiotensin II by G. R. Marshall, to the A and B chains of insulin by Arnold Marglin, and eventually through the efforts of many other laboratories to a rather sizeable list of over 300 peptides.[7]

The experimental details have varied considerably in some of these syntheses, but in general they have been based on the original principle of the solid phase idea. Several changes in solid supports have evolved, and each of the chemical steps has been modified in numerous ways.[4]

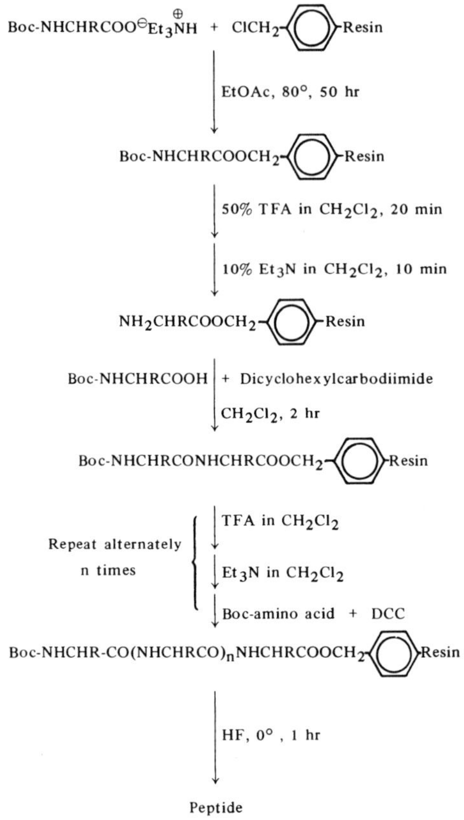

Boc-NHCHRCOO⁻Et₃NH⁺ + ClCH₂–⟨Ph⟩–Resin

↓ EtOAc, 80°, 50 hr

Boc-NHCHRCOOCH₂–⟨Ph⟩–Resin

↓ 50% TFA in CH₂Cl₂, 20 min

↓ 10% Et₃N in CH₂Cl₂, 10 min

NH₂CHRCOOCH₂–⟨Ph⟩–Resin

Boc-NHCHRCOOH + Dicyclohexylcarbodiimide

↓ CH₂Cl₂, 2 hr

Boc-NHCHRCONHCHRCOOCH₂–⟨Ph⟩–Resin

Repeat alternately n times
  ↓ TFA in CH₂Cl₂
  ↓ Et₃N in CH₂Cl₂
  ↓ Boc-amino acid + DCC

Boc-NHCHR-CO(NHCHRCO)ₙNHCHRCOOCH₂–⟨Ph⟩–Resin

↓ HF, 0°, 1 hr

Peptide

FIGURE III

A General Scheme for Stepwise Solid Phase Peptide Synthesis

F. A Solid Phase-Fragment Scheme

A new general scheme has recently been developed by Dr. Wang.[8] It was designed to give protected peptides of intermediate size for subsequent incorporation into larger polypeptides by fragment condensation methods. The scheme, shown in FIGURE IV, makes use of three classes of protecting groups with different degrees of sensitivity to acidic reagents. The α-amino protection is with a very acid-labile biphenylisopropyloxycarbonyl (Bpoc) group, the side chain protection is based on the relatively stable benzyl protection as before and the carboxyl attachment to the resin support is through a t-butyl derivative of intermediate stability. The latter required the synthesis of a new polystyrene resin derivative and its conversion into a t-alkyloxycarbonyl-hydrazide, which could then be acylated by the first Bpoc-amino acid. The relative stabilities to acid of the three protecting groups were approximately 1:3,000:1,000,000, which was sufficient to allow the synthesis of fully protected peptide fragments.

G. Automation

In order to extend the solid phase method to the synthesis of large polypeptides or proteins, it was important to take full advantage of the most important feature of this synthetic approach, namely, its suitability to mechanization. This potential feature was the real driving force between the original idea and early developmental work. It is interesting to recall the referee's comment to my first manuscript in which this goal was expressed. He wrote, "It is unfortunate that the author talks about the possibility of automation ... ". In spite of this discouraging reaction it did become possible to automate the process. Together with Dr. John Stewart a machine was built which will carry out all the steps involved in the synthesis of a peptide.[9] I believe the machine demonstrates very well the advantages of using a solid support for a multistage synthesis of this kind. The manipulations are greatly simplified, the times are shortened, and the yields are high.

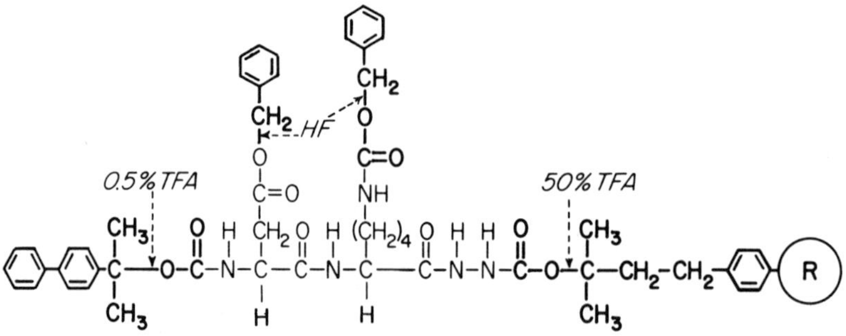

FIGURE IV

A Plan for a Solid Phase-Fragment Synthesis

III. THE SYNTHESIS OF RIBONUCLEASE A

A. Introduction to RNase A

With this brief historical introduction to the solid phase method, I would now like to turn to a description of our application of the technique to the synthesis of the enzyme ribonuclease A and to the use of the synthetic approach to answer some questions about the relation of the structure of the enzyme to its activity. This has been a collaborative effort with **Dr. Bernd Gutte** who most certainly should share this award with me.

Ribonuclease is one of the most thoroughly studied of all enzymes. Although its actual discovery is credited to Jones,[10] its presence in tissues was recognized or at least suspected many years earlier. The first isolation of ribonuclease was by Dubos and Thompson[11] in 1938, and two years later Kunitz[12] succeeded in crystallizing the protein. This was the important step that made this enzyme the object of the intensive research that has followed. Through the work of Hirs, Spackmann, Smyth, and others in the laboratories of Moore and Stein the sequence (FIGURE V) was established.[13] The three dimensional structure of crystalline ribonuclease A was determined

cise position in space of the polypeptide backbone and most of the 2000 individual atoms became known. The function of this enzyme in nature is to hydrolyze and depolymerize ribonucleic acids, which it accomplishes in two stages: first by a transphosphorylation and then by a hydrolysis step. Much of the detailed mechanism of these reactions is now understood, based on extensive chemical experiments from many laboratories and supplemented by data from x-ray, nmr, and other physical methods.

The purpose of a chemical synthesis of this molecule was, first, simply to demonstrate that a protein with the enormous catalytic activity and specificity of a naturally occurring enzyme could be synthesized in the laboratory. More importantly in the long range, the purpose was to provide a new approach to the study of enzymes. We believed that it should be possible to synthesize enzyme molecules with modified amino acid sequences and other chemical changes which would possess new structures with altered activities and possibly even altered substrate specificities. In that event it would be possible to obtain new information about the folding and stabilization of the protein structure, about the extent and specificity of binding of substrates, and about the mechanism of the catalytic action.

B. The Synthesis

The chemical synthesis of ribonuclease A[16,17] began with the attachment of the carboxyl-terminal amino acid, valine, to the resin support and continued with the automatic stepwise addition of protected amino acids until the 124-amino acid chain was assembled on the resin. The peptide was then cleaved from the resin, deprotected, purified, isolated, and characterized.

The resin support was a chloromethylated styrene–1%-divinylbenzene copolymer bead, which was selected because of its high swelling capacity in methylene chloride and other organic solvents. The first amino acid was attached as a benzyl ester by standard procedures and was present at a concentration of 0.21 mmole per gram of resin.

Lys-Glu-Thr-Ala-Ala-Ala-Lys-Phe-Glu-Arg-Gln-His-Met-Asp-Ser-
Ser-Thr-Ser-Ala-Ala-Ser-Ser-Ser-Asn-Tyr-Cys-Asn-Gln-Met-Met-
Lys-Ser-Arg-Asn-Leu-Thr-Lys-Asp-Arg-Cys-Lys-Pro-Val-Asn-
Thr-Phe-Val-His-Glu-Ser-Leu-Ala-Asp-Val-Gln-Ala-Val-Cys-Ser-
Gln-Lys-Asn-Val-Ala-Cys-Lys-Asn-Gly-Gln-Thr-Asn-Cys-Tyr-
Gln-Ser-Tyr-Ser-Thr-Met-Ser-Ile-Thr-Asp-Cys-Arg-Glu-Thr-Gly-
Ser-Ser-Lys-Tyr-Pro-Asn-Cys-Ala-Tyr-Lys-Thr-Thr-Gln-Ala-Asn-
Lys-His-Ile-Ile-Val-Ala-Cys-Glu-Gly-Asn-Pro-Tyr-Val-Pro-Val-
His-Phe-Asp-Ala-Ser-Val

FIGURE V

The Linear Sequence of Bovine Pancreatic Ribonuclease A

by Kartha, Bello, and Harker[14] and that for ribonuclease S was determined by Wyckoff and Richards and their colleagues[15] so that the pre-

All of the amino acids were protected in the α-amino position by the *t*-butyloxycarbonyl (Boc) group and were deprotected after each cycle of the synthesis by 50% trifluoroacetic acid in methylene chloride. The side chains of the trifunctional amino acids were protected with relatively acid-stable groups, *e.g.*, Asp(OBzl), Glu(OBzl), Cys-(Bzl), Ser(Bzl), Thr(Bzl), Tyr(Bzl), Lys(Z), Arg-(NO_2), Met(O). Only the imidazole side chain of histidine was unprotected.

The couplings were usually mediated with N,N'-dicyclohexylcarbodiimide (DCC) in methylene chloride. A three-fold excess of Boc-amino acid and DCC was used in all cases except for Boc-His were a five-fold excess was used in order to drive the reactions to completion. The coupling times were five hours. For the amides Boc-Asn and Boc-Gln it was necessary to activate with *p*-nitrophenyl esters to avoid nitrile formation and to allow a 10 hr coupling time.

At the end of the assembly of the 124-residue polypeptide on the resin the over-all yield was 17%, based on the amount of C-terminal valine originally esterified to the polymer support, which means an average loss of about 1.4% of the peptide chain at each cycle of the synthesis. In addition, several analytical and preparative samples were removed during the synthesis, leaving a total of 1.59 g of fully protected RNase-resin at the end of the run, of which 664 mg was protected peptide and 926 mg was polystyrene. The product contained 4 NO_2^-, 49 Bzl-, 10 Z-, and 4 sulfoxide groups and had a calculated molecular weight of 19,791.

C. Cleavage from the Resin

After the synthesis of the fully protected ribonuclease A sequence the peptide was removed from the resin support and was deprotected under three different conditions. The first experiments were with anhydrous HF in TFA. However, the effect of this reagent on the natural RNase A control was rather severe. It was then found that much less damage to the molecule occurred if TFA was omitted and the cleavage reaction was carried out for 90 min at 0–20° in liquid HF containing anisole as a trap for reactive cations. Both of these procedures removed all of the protecting groups at the same time. A third sample of protected RNase-resin was cleaved by passing a stream of dry HBr through a suspension of the resin in TFA containing anisole. This reagent removed all protecting groups except the S-benzyls from the eight cysteine residues. They were then removed in a separate step with HF. In all runs the methionine sulfoxide was reduced with mercaptoethanol. Several precautions against side reactions were taken during these cleavage experiments. The samples and the HF were thoroughly dried before the reaction, anisole was used as a trap, and the time and temperature were kept to a minimum. Methionine was incorporated as the sulfoxide to prevent sulfonium formation and to avoid the known specific chain cleavage at this amino acid. A treatment at pH 7.5 was also introduced to reverse any $N \rightarrow O$ acyl shift which might have occurred during the HF treatment. The yields of crude liberated protein ranged between 40 and 70% of the protein attached to the resin.

D. Purification Procedures

During the subsequent purification procedure the SH groups of the cysteine residues were stabilized by conversion to S-sulfonates through oxidative sulfitolysis. The resulting crude RNase-$(S-SO_3^-)_8$ product showed a major component on paper electrophoresis and two minor spots. One of these was removed by chromatography on Dowex 1-X2 and the other was separated by gel filtration on Sephadex G-50. The main fraction then migrated on paper electrophoresis as a single spot with the same mobility as the S-sulfonate prepared from natural RNase A. It was discovered, however, that the specific activity of natural RNase A, recovered from its S-sulfonate by reduction with mercaptoethanol and reoxidation with air, was substantially reduced, and this procedure was omitted from the later runs.

The material obtained from the runs in which the cleavage was by HF or HBr-TFA was placed directly on Sephadex G-75 columns and eluted with dilute ammonium bicarbonate. In each case three major components were obtained, which were the result of random oxidations during the work-up procedures. The first was a polymeric fraction produced by intermolecular disulfide bond formation, the second was a randomly oxidized monomer fraction, and the third, slower moving fraction contained a mixture of short-chain impurities. The monomer fraction was then reduced again to the sulfhydryl form and was carefully re-oxidized by air in very dilute solution to generate ribonuclease A that presumably contained the proper disulfide bonds and was folded into the native three-dimensional conformation. This synthesis is entirely dependent on the spontaneous refolding and re-pairing of the proper cysteine residues to give the native structure, as originally found for natural RNase A by White,[18] and Anfinsen and colleagues.[19] There is no way at this time to direct the pairing of the cysteines or the folding of the molecule by selective chemical methods.

The cation exchange resin, IRC-50, was used for the further fractionation and purification of the reoxidized synthetic RNase A preparations. The number of peaks obtained depended on the previous cleavage and purification procedures that had been applied to the sample. The protein that had been cleaved in HF-TFA and carried through the S-sulfonation procedure gave rise to five peaks, whereas the material cleaved in HF and purified by gel filtration produced three peaks, and the protein that had been cleaved in HBr-TFA and further deprotected in HF before purifying by gel filtration gave rise to only a single peak. In each case the main activity eluted at the same position as natural reduced-reoxidized RNase A.

Reduced-reoxidized natural and synthetic RNase A were indistinguishable from each other and from pure native RNase A upon chromatography on CM-cellulose (FIGURE VI). This is the system that was used previously to demonstrate the similarity between native and refolded RNase A. By this criterion our synthetic RNase A was identical with the natural enzyme. However, since the enzyme at this stage of purification was only 13 to 16% as active against RNA as the natural enzyme, it was clear that it was not pure.

E. Tryptic Digestion and Ammonium Sulfate Precipitation

RNase A is known to be very resistant to tryptic digestion, whereas S-protein, RNase S, and denatured RNase A are susceptible to tryptic hydrolysis.

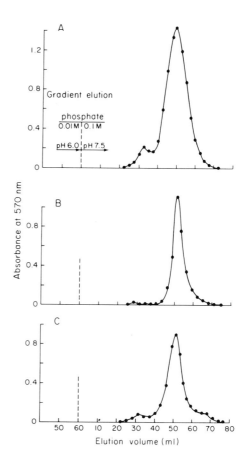

FIGURE VI

Ion Exchange Chromatography of

(A)   Natural RNase A

(B)   Reduced-Reoxidized Natural RNase A

(C)   Synthetic RNase A on CM-Cellulose

Elution was with 0.01 M phosphate buffer, pH 6.0, followed by a gradient with 0.1 M phosphate buffer, pH 7.5.

It was possible to take advantage of this property of the native enzyme in the further purification of the synthetic protein. It was reasoned that if the synthetic RNase A were a mixture of a small amount of fully active enzyme and a large amount of inert protein having an incorrect primary and tertiary structure, the improperly folded chains would be expected to be more susceptible to cleavage by trypsin and should be degraded to small fragments that could be readily separated from the active molecule, and the specific activity should increase. This argument could also apply to inhibitory peptides. If, however, the synthetic enzyme were a mixture of proteins, each of which had a low specific activity due to deviations from the native species in amino acid sequence and three-dimensional structure, it seemed likely that trypsin would destroy virtually all the enzymic activity.

When a sample of the synthetic enzyme with about 8% specific activity was incubated with trypsin and then fractionated on Sephadex G-50, two peaks were observed. One behaved as a mixture of small peptides, and the other eluted in the same position as untreated natural RNase A. It has a specific activity of 61%. Although the specific activity increased by 7.6-fold, the amount of protein digested and removed was only 25% of the starting material. Thus, the total number of measureable units of RNase A activity increased 5.3-fold, and the results suggest that an inhibitory peptide that had been present in the crude synthetic product had been removed by the tryptic hydrolysis.

The specific activity of the synthetic product was further raised by fractional precipitation. After standing two days at 3° in an amount of ammonium sulfate just below that required to precipitate natural RNase A an amorphous solid formed, which was removed by centrifugation. The protein in the supernatant solution showed a specific activity of 78%. This is the most active synthetic RNase A we have prepared at this time. TABLE I summarizes some of the yields and activities obtained in these experiments.

F. **Evidence for Chemical Purity and Identity**

The chemical and physical comparisons of synthetic and natural RNase A were based on amino acid analyses, enzyme digestions, peptide maps, paper electrophoresis, gel filtration and ion exchange chromatography. The purified synthetic enzyme had the over-all amino acid composition expected of RNase A with only small deviations from that of the untreated natural control (TABLE II). Such analyses indicate close similarities and exclude the existence

TABLE I

Yields and Activities

Stage of Synthesis	Over-all Yield (%)	Specific Activity
Boc-Val-Resin	100	
Protected RNase-Resin	17	
Crude cleaved protein	12	2
RNase A, after G-75	6.4	9
RNase A, after trypsin	4.4	61
RNase A, after $(NH_4)_2SO_4$	2.9	78

of gross differences in composition but, of course, do not eliminate the possibility of small differences or of the presence of a mixture of proteins with the average amino acid composition of natural RNase A. Enzymatic digestion with papain followed by aminopeptidase M completely degraded the synthetic molecule to free amino acids, indicating that no major degree of racemization had occurred during the synthesis or the various work-up procedures. On the other hand, the synthetic RNase A was very resistant to trypsin, as expected for this enzyme. After performic acid oxidation of the four cystine residues to cysteic acid residues, the product was readily digested by trypsin.[20] Peptide maps of the digest showed 14 spots that corresponded quite well with the positions of the peptides derived from natural RNase A. No further data were obtained to show whether the corresponding spots were identical. However, major differences in protein composition would have been expected to give rise to detectably different peptide patterns.

Both the S-sulfonate and the purified RNase A migrated on paper electrophoresis with the same mobilities as the corresponding natural controls. The S-sulfonate, however, gave an elongated spot, indicating some heterogeneity. Further evidence for similarity in size came from the results of gel filtration on Sephadex G-75 and for similarity in

TABLE II

Amino Acid Analysis of Synthetic RNase A[a]

Amino acid	Theory	Natural RNase A	Synthetic RNase A
Lys	10	10.0	9.9
His	4	3.8	3.7
Arg	4	4.1	4.4
Asp	15	15.7	14.9
Thr	10	9.8	9.6
Ser	15	14.0	15.0
Glu	12	12.1	11.8
Pro	4	3.8	3.5
Gly	3	3.1	3.6
Ala	12	12.0	12.0
Cys	8	7.8	7.3
Val	9	8.6	9.3
Met	4	4.1	4.1
Ile	3	2.7	2.8
Leu	2	2.2	2.3
Tyr	6	5.6	5.4
Phe	3	3.3	3.2

[a]Values are expressed as moles of amino acid per mole of RNase A, with alanine set at 12.0.

net charge came from ion-exchange chromatography on on IRC-50 and CM-cellulose.

G. Enzymatic Activity and Substrate Specificity

RNase A activity of the synthetic enzyme preparations was determined on both large and small substrates. When RNA was used the activity was measured either by the Kunitz[21] method from the initial velocities of the hypochromic shift at 300 nm or spectrophotometrically by estimation of the acid-soluble oligonucleotides in the digestion mixtures.[22] The specific activities from the two methods have been in good agreement. 2′,3′-cyclic cytidine phosphate (C>p) was used as the low molecular weight substrate. The cytidylic acid liberated by the hydrolysis was separated chromatographically and measured spectrophotometrically.[23]

The specific activity of the crude synthetic preparations was not the same by the two methods. We attribute the differences to different degrees of susceptibility to inhibitors which were present in the synthetic product. The polymeric substrate, RNA, apparently was more sensitive in inhibitors and also may have had a more rigorous requirement for a precise conformation of the enzyme than the smaller substrate.

The indication that inhibitory peptides were present in the crude synthetic protein comes from the observation that subsequent purification procedures not only increased the specific activity of the synthetic enzyme but also increased the total number of units of activity that were detectable by the assay methods. The simplest explanation is that both inert and inhibitory peptides were being removed. In the HF-TFA run the specific activity rose from 1 or 2% to 13% and the total units increased from 4 to 7-fold during the IRC-50 step. In the HF run there was a 1.6-fold increase in total units at that stage, and there was a further 5.3-fold increase in total measureable units after the trypsin digestion. From these data it could be calculated that 22% of the crude, cleaved synthetic protein was RNase A. At the end of the purification procedures the product was 78% pure RNase A by enzyme assay.

DNase and RNase T_1 activities of the synthetic RNase A were excluded. Even at high enzyme levels DNA was not at all affected by incubation with the RNase A under conditions where DNase caused complete hydrolysis in a few minutes. Cyclic guanylic acid and the dinucleotide GpCp were good substrates for RNase T_1 but were recovered unchanged when incubated with the synthetic RNase A.

The data were all consistent with the specificity to be expected for RNase A. They demonstrated the ability of the product to cleave both the transphosphorylation and hydrolysis steps.

They showed the expected requirement for D-ribose instead of D-deoxyribose and for a pyrimidine- instead of a purine-nucleoside 3'-phosphate.

Further evidence that the enzymatic activity exhibited by the synthetic product was a true RNase A activity was provided by the Michaelis constant. The K_m values with RNA as substrate were 1.20 mg/ml for natural RNase A and 1.24 mg/ml for the synthetic RNase A.

IV. PREPARATION OF SYNTHETIC S-PROTEIN AND SYNTHETIC DES(21-25)S-PROTEIN AND THEIR COMBINATION WITH NATURAL OR SYNTHETIC S-PEPTIDE TO GIVE RIBONUCLEASE S AND DES(21-25)RIBONUCLEASE S

As a beginning toward the application of the synthetic approach to the study of the relation of structure to function in enzymes we undertook to answer the question of whether an S-protein lacking the first five residues (21-25) at its NH_2-terminus would still bind S-peptide and generate ribonuclease activity.[24]

For this purpose S-protein, des(21-25)S-protein, and S-peptide were synthesized by the solid phase method according to the general procedures already described. They were partially purified on Sephadex, and the S-peptide was further purified by free flow electrophoresis. The three peptides then migrated as single components on paper electrophoresis and the mobilities of S-peptide and S-protein agreed well with those of the corresponding natural peptides. S-protein and des(21-25)S-protein were reduced and unfolded in mercaptoethanol, and each was mixed with either natural or synthetic S-peptide and then reoxidized in dilute solution. It was found that approximately equivalent amounts of RNase activity were generated from either synthetic S-protein or des-(21-25)S-protein by this procedure, and the activity was between 15 and 25% of that obtained in the same way from natural S-protein. It could be concluded that the first five residues (21-25) of S-protein are not required for the formation of the S-peptide–S-protein complex and the generation of enzymic activity. The x-ray structure of Wyckoff et al.[15,25] (FIGURE VII) had indicated that there were no noncovalent interactions between the three N-terminal serine residues (21-23) of S-protein and other parts of the complex and that they probably could be omitted without serious consequences to the conformation or activity of RNase S. There was, however, a strong indication that asparagine 24 forms three hydrogen bonds and tyrosine 25 forms two hydrogen bonds with other residues in the protein. We conclude that neither the three serines nor the proposed noncovalent interactions of the asparagine or tyrosine are essential for binding the S-peptide or stabilizing the active conformation of the enzyme.

V. CARBOXYL-TERMINAL TETRADECAPEPTIDE OF RIBONUCLEASE A

The remarkable discovery of the S-peptide–S-protein system by Richards[26] prompted us to look for a similar system at the carboxyl end of ribonuclease.[27] Since an enzymatic cleavage that would yield both a large protein and a smaller C-terminal peptide was not known, it seemed that the synthetic approach was again the way to examine this problem. For that purpose the C-terminal tetradecapeptide, H-Glu-Gly-Asn-Pro-Tyr-Val-Pro-Val-His-Phe-Asp-Ala-Ser-Val-OH, was synthesized. This was chosen because it was the longest chain that did not contain a cysteine residue. The peptide was synthesized by the same solid phase method that was used later for the RNase A synthesis. The product was purified by gel filtration on Sephadex and by free flow electrophoresis and was then found to be homogeneous by paper electrophoresis.

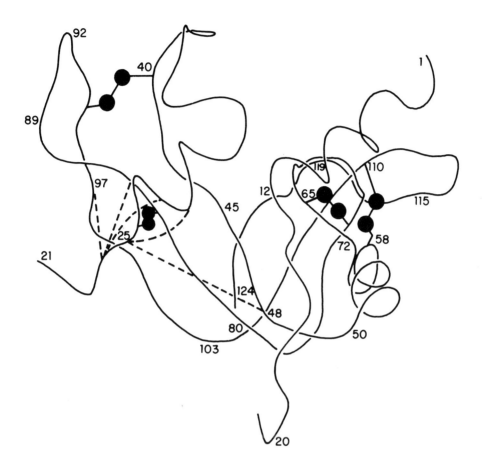

FIGURE VII

The Structure of Ribonuclease S, Based on the X-Ray Data of Wyckoff *et al.*, Showing the Probable Non-Covalent Interactions of Residues 21-25

The initial plan was to inactivate natural RNase A by carboxymethylation of the essential histidine 119 and to attempt to displace the carboxyl terminal portion of the chain by the synthetic peptide in the hope that an enzymatically active molecule would be regenerated. When the peptide was mixed with N^1-His(Cm)-119-RNase in a ratio up to 100:1 and was refolded after denaturation either by acid, alkali, urea, or heat, no activity was found. The carboxyl-terminal portion of the chain appears to have been too strongly bound by noncovalent forces to allow displacement by the free tetradecapeptide.

It was discovered, however, that when the last four residues of RNase A were removed by pepsin, producing an inactive enzyme, the tetradecapeptide could then bind to the protein with the regeneration of about 50% of the original activity. This is interpreted to mean that the intramolecular forces binding the shortened chain to the rest of the protein are not as strong as in the native enzyme. Furthermore, it was found that removal of one or two more residues from the protein allowed even more efficient binding and recovery of activity upon addition of the peptide. Thus, the removal of phenylalanine-120 and histidine-119 by carboxypeptidase A gave RNase(1-119) and RNase(1-118) which were reactivated to the extent of about 80% by two to three equivalents of the tetradecapeptide (111-124). This is remarkably effective binding considering that the peptide supplies not only the missing residues but also an overlap of eight to nine residues. In the case of RNase(1-120) or RNase(1-119) it is not clear whether the functional histidine is being supplied by the peptide or the protein, but in the complex with RNase(1-118) the catalytically active histidine must be from the peptide. From these experiments it was concluded that there does exist a system at the carboxyl end of ribonuclease that is quite analogous to the S-peptide–S-protein system at the amino end of the enzyme.

Finally, it has been found that peptides representing both ends of the RNase molecule, the N-terminal S-peptide (1-20) and the C-terminal tetradecapeptide (111-124), containing both of the catalytically essential histidine residues, could be bound noncovalently to an inactive core (21-118) to regenerate enzymatic activity, FIGURE VIII. This is a beginning toward the physical separation of the catalytic sites of an enzyme from its substrate binding and specificity sites which may someday lead to the synthesis of enzymes with simplified structures and new or modified activities.

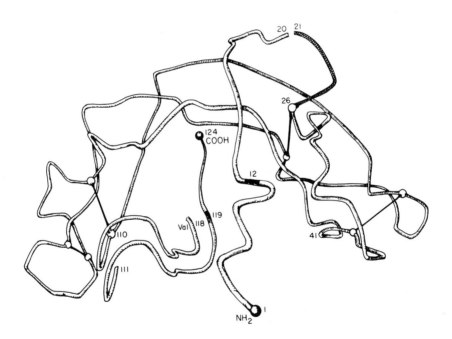

FIGURE VIII

Recombination of the N-terminal S-Peptide(1-20) and C-Terminal Tetradecapeptide(111-124) with the Protein Core of RNase(21-118)

The relative positions in space are not known. The drawing is intented only to illustrate the observation that the three fragments reassociate in solution in such a way that ribonuclease activity is regenerated. Based in part on the X-ray structure of RNase by Kartha, Bello, and Harker.[14]

VI. CONCLUSIONS

In summary I would like to draw the following conclusions from the material presented here:

1. Solid phase peptide synthesis has provided an alternate approach to the classical methods for the preparation of peptides. It has simplified and accelerated the synthesis of a large number of small peptides and is beginning to be applied to the synthesis of large molecules.

2. The synthesis of ribonuclease A demonstrates that it is possible to begin with simple amino acids and to assemble them in the laboratory into a real protein with true enzymatic activity.

3. A new kind of evidence was provided for the conclusion that the primary structure of a protein determines it tertiary structure. Since the only information put into the synthetic protein was its amino acid sequence the finding that an active enzyme was produced must have been a consequence of its primary structure.

4. Through the synthesis of S-protein and des(21-25)S-protein it was shown that amino acid residues 21 to 25 were not required for S-protein and S-peptide to combine, non-covalently, and to fold into an active enzyme.

5. The synthetic C-terminal tetradecapeptide of RNase A was shown to reactivate RNase A which had been shortened at the carboxyl end by 4, 5, or 6 amino acid residues. The activity of the non-covalently bonded complex between these components demonstrated that a system exists at the carboxyl end of the molecule which is quite analogous to the S-peptide–S-protein system at the amino end.

6. It was discovered that small peptides from each end of ribonuclease A, containing both of the catalytically essential histidine residues, could be combined non-covalently with a protein core to give an active enzyme.

7. The chemical synthesis of proteins, and especially of enzymes, is a new field which can reasonably be expected to continue to expand and to provide useful new information about this very important class of compounds.

REFERENCES

1. E. Schröder and K. Lübke, "The Peptides," Academic Press, New York, 1965.

2. V. du Vigneaud, C. Ressler, J. M. Swan, C. W. Roberts, P. G. Katsoyannis, and S. Gordon, *J. Amer. Chem. Soc.*, **75**, 4879 (1953). The Synthesis of an Octapeptide Amide with the Hormonal Activity of Oxytocin.

3. R. B. Merrifield, *J. Amer. Chem. Soc.*, **85**, 2149 (1963). Solid Phase Peptide Synthesis. I. The Synthesis of a Tetrapeptide.

4. R. B. Merrifield, *Advan. Enzymol.*, **32**, 221 (1969). Solid Phase Peptide Synthesis.

5. R. B. Merrifield and V. Littau in "Peptides," E. Bricas, Ed., North Holland, Amsterdam, 1968, p 179. The Distribution of Peptide Chains on the Solid Support.

6. R. B. Merrifield, *Biochemistry*, **3**, 1385 (1964). Solid Phase Peptide Synthesis. III. An Improved Synthesis of Bradykinin.

7. G. R. Marshall and R. B. Merrifield in

Handbook of Biochemistry," 2nd ed, H. A. Sober, Ed., The Chemical Rubber Co., Cleveland, 1970, p C-145. Peptides Prepared by Solid Phase Synthesis.

8. S. S. Wang and R. B. Merrifield, *J. Amer. Chem. Soc.*, **91**, 6488 (1969). Preparation of a Tertiaryalkyloxycarbonylhydrazide Resin and its Application to Solid Phase Peptide Synthesis.

9. R. B. Merrifield, J. M. Stewart, and N. Jernberg, *Anal. Chem.*, **38**, 1905 (1966). Instrument for Automated Synthesis of Peptides.

10. W. Jones, *Amer. Physiol.*, **52**, 203 (1920). The Action of Boiled Pancreas Extract on Yeast Nucleic Acid.

11. R. J. Dubos and R. H. S. Thompson, *J. Biol. Chem.*, **124**, 501 (1938). The Decomposition of Yeast Nucleic Acid by a Heat Resistant Enzyme.

12. M. Kunitz, *J. Gen. Physiol.*, **24**, 15 (1940). Crystalline Ribonuclease.

13. D. G. Smyth, W. H. Stein, and S. Moore, *J. Biol. Chem.*, **238**, 227 (1963). The Sequence of Amino Acid Residues in Bovine Pancreatic Ribonuclease.

14. G. Kartha, J. Bello, and D. Harker, *Nature*, **213**, 862 (1967). Tertiary Structure of Ribonuclease.

15. H. W. Wyckoff, D. Tsernoglou, A. W. Hanson, J. R. Knox, B. Lee, and F. M. Richards, *J. Biol. Chem.*, **245**, 305 (1970). The Three-Dimensional Structure of Ribonuclease-S.

16. B. Gutte and R. B. Merrifield, *J. Amer. Chem. Soc.*, **91**, 501 (1969). The Total Synthesis of an Enzyme with Ribonuclease A Activity.

17. B. Gutte and R. B. Merrifield, *J. Biol. Chem.*, **246**, 1922 (1971). The Synthesis of Ribonuclease A.

18. F. H. White, Jr., *J. Biol. Chem.*, **236**, 1353 (1961). Regeneration of Native Secondary and Tertiary Structures by Air Oxidation of Reduced Ribonuclease.

19. C. J. Epstein, R. F. Goldberger, D. M. Yound, Young, and C. B. Anfinsen, *Arch. Biochem. Biophys.*, Suppl., **1**, 223 (1962). A Study of the Factors Influencing the Rate and Extent of Enzymic Reactivation During Reactivation of Reduced Ribonuclease.

20. C. H. W. Hirs, *J. Biol. Chem.*, **219**, 611 (1956). The Oxidation of Ribonuclease with Performic Acid.

21. M. Kunitz, *J. Biol. Chem.*, **164**, 563 (1946). A Spectrophotometric Method for the Measurement of Ribonuclease Activity.

22. C. B. Anfinsen, R. R. Redfield, W. L. Choate, J. Page, and W. R. Carroll, *J. Biol. Chem.*, **207**, 201 (1954). Studies on the Gross Structure, Cross-linkages and Terminal Sequence in Ribonuclease.

23. R. G. Fruchter and A. M. Crestfield, *J. Biol. Chem.*, **240**, 3868 (1965). Preparation and Properties of Two Active Forms of Ribonuclease Dimer.

24. B. Gutte and R. B. Merrifield, *Fed. Proc.*, **29**, 727 (1970). Generation of Ribonuclease Activity from Synthetic Des(21-25)S-protein on Addition of S-Peptide.

25. H. W. Wyckoff, K. D. Hardman, N. M. Allewell, T. Inagami, L. N. Johnson, and F. M. Richards, *J. Biol. Chem.*, **242**, 3984 (1967). The Structure of Ribonuclease S at 3.5 Å Resolution.

26. F. M. Richards, *C. R. Trav. Lab. Carlsberg*, **29**, 322, 329 (1955). On an Active Intermediate Produced During the Digestion of Ribonuclease by Subtilisin.

27. M. C. Lin, B. Gutte, S. Moore, and R. B. Merrifield, *J. Biol. Chem.*, **245**, 5169 (1970). Regeneration of Activity by Mixture of Ribonuclease Enzymatically Degraded from the COOH Terminus and a Synthetic COOH-Terminal Tetradecapeptide.

6

Copyright © 1965 by the American Chemical Society

Reprinted from *Biochemistry,* **4**(11), 2394–2401 (1965)

Synthesis of Angiotensins by the Solid-Phase Method[*]

Garland R. Marshall and R. B. Merrifield

ABSTRACT: The new method of solid-phase synthesis was applied to the preparation of isoleucine[5]-angiotensin II. These syntheses started with *t*-butyloxycarbonylphenylalanylcopolystyrene–divinylbenzene, ended with chromatographically pure octapeptide, and gave an over-all yield of 56%. The synthetic angiotensin possessed full oxytocic activity. Since the rearrangement of the α-aspartyl bond to the β-aspartyl bond had been reported under conditions similar to those used in the synthesis, it was then ascertained by a variety of methods that this rearrangement did not occur in solid-phase synthesis. The analogs, asparagine[1]-isoleucine[5]-angiotensin II and β-aspartic[1]-isoleucine[5]-angiotensin II, were also synthesized.

The solid-phase method has recently been introduced to speed, simplify, and automate the synthesis of peptides and, ultimately, proteins (Merrifield, 1962, 1965). Both bradykinin and methionyllysylbradykinin have been successfully synthesized in this way (Merrifield, 1963, 1964a,b). This report on the syntheses of the biologically active octapeptide, isoleucine[5]-angiotensin II, and two of its analogs presents further evidence in support of the benefits and applicability of the solid-phase method.

Page and Helmer (1939) and Braun-Menendez *et al.* (1939) announced simultaneously the discovery of a pressor substance resulting from the action of the renal proteolytic enzyme, renin, on plasma. It was later shown that the product of the reaction of renin with plasma is an inactive decapeptide (angiotensin I) which is further degraded by a plasma enzyme to the biologically active octapeptide, angiotensin II (Skeggs *et al.*, 1954, 1956a). The structure of the active peptide from the horse (Skeggs *et al.*, 1956b) has been confirmed by synthesis (Schwarz *et al.*, 1957; Schwyzer *et al.*, 1957; Arakawa and Bumpus, 1961). It is designated isoleucine[5]-angiotensin II in order to distinguish it from the corresponding peptide of bovine origin which contains valine in position five.

The synthesis of isoleucine[5]-angiotensin II by the solid-phase method was undertaken for two reasons. First, it was of interest as a further test of the applicability of this new method of peptide synthesis. Since angiotensin contained four amino acids, aspartic acid, histidine, isoleucine, and tyrosine, which had not previously been introduced into peptides in this way, the synthesis provided additional evidence for the general applicability of solid-phase synthesis. Second, it was expected to provide a simplified synthetic route to this important compound and various derivatives. The synthesis followed the basic concept of solid-phase peptide synthesis as outlined previously (Merrifield, 1964a), but differed in that the coupling steps were carried out in methylene chloride where possible. The steps involving

---

[*] The Rockefeller University, New York, N. Y. 10021. This study was made in partial fulfillment of the requirement for the Ph.D. degree by G. R. Marshall. An abstract of this work was presented at the 49th Annual Meeting of the Federation of American Societies for Experimental Biology, April 1965 (Marshall and Merrifield, 1965). This is paper number V of this series.

t-BOC-nitro-L-arginine[1] and t-BOC-im-benzyl-L-histidine were performed in dimethylformamide due to greater solubility of these compounds in this solvent. Methylene chloride is preferred as a solvent when dicyclohexylcarbodiimide is used as a coupling reagent, since both rearrangement of the activated amino acid intermediate to the N-acylurea (Sheehan *et al.*, 1956) and racemization are minimized in this solvent (Smart *et al.*, 1960). Racemization has not been detectable with solid-phase synthesis and is presumed not to occur to any appreciable degree.

Synthesis of Angiotensin II. The synthesis consisted of attachment of the C-terminal amino acid of angiotensin to a solid polymer by a covalent ester linkage, the addition of the succeeding amino acids one at a time as shown in Figure 1 until the desired sequence was assembled, and finally the removal of the peptide from the solid support as shown in Figure 2. Through the attachment of the growing peptide chain to an insoluble particle, the intermediate products were easily purified by filtration and washing.

After the t-BOC-L-phenylalanine triethylammonium salt was esterified in ethanol with the chloromethylated copolystyrene-2% divinylbenzene, the polymer was introduced into the reaction vessel (Merrifield, 1963), where all of the steps up to and including the cleavage of the peptide were carried out. The cycle for each amino acid consisted of removal of the acyl group, of neutralization of the resulting hydrochloride with triethylamine in dimethylformamide, and then of coupling of the free base with a protected amino acid by the aid of dicyclohexylcarbodiimide (Sheehan and Hess, 1955). Excess reagents and by-products were removed by washing with methylene chloride, ethanol, and acetic acid. This cycle was repeated with the appropriate t-BOC-amino acid derivative seven times until the desired octapeptide, t-BOC-β-benzyl-L-aspartyl-nitro-L-arginyl-L-valyl-O-benzyl-L-tyrosyl-L-isoleucyl-im-benzyl-L-histidyl-L-prolyl-L-phenylalanine, was completed on the resin. The fact that the desired sequence was actually being produced was verified by analyses of samples taken after each new amino acid was coupled. The protected octapeptide was cleaved from the resin by bubbling hydrogen bromide through a suspension of the peptide resin in trifluoroacetic acid. This treatment also removed the benzyl groups from both the aspartyl and tyrosyl residues. The protecting im-benzyl group of histidine and nitro group of arginine were then removed by catalytic hydrogenation as shown in Figure 2. This alternative procedure to the usual sodium–liquid ammonia method for removal of im-benzyl groups has been used successfully in certain cases (Theodoropoulos, 1956; Li *et al.*, 1960), but the difficulties involved in other cases (Bricas and Nicot-Gutton, 1960) would seem to depend on the nature of the individual peptide (Kopple *et al.*, 1963).

The crude synthetic isoleucine⁵-angiotensin II contained one component and two minor contaminants

---

[1] Abbreviation used: t-BOC, t-butyloxycarbonyl.

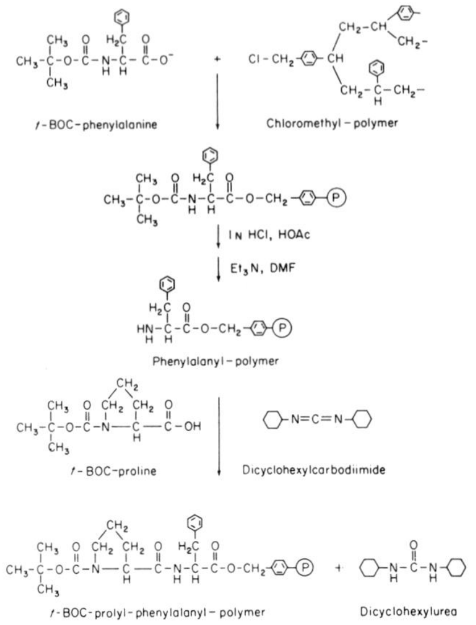

FIGURE 1: The solid-phase method. Synthesis of the initial dipeptide in angiotensin.

which were readily separated by countercurrent distribution (Schwarz *et al.*, 1957). The purified product was determined to be homogeneous by electrophoresis and paper chromatography. Amino acid composition and biological activity in the rat uterus assay (Schwarz *et al.*, 1955) were determined and found to be within the expected range.

Advantages. The major advantages of this new method of synthesis of isoleucine⁵-angiotensin II are its speed, simplicity, and yield. One person can easily lengthen the amino acid chain at the rate of two residues/day. Thus it becomes feasible to attempt the syntheses of a large number of analogs, *e.g.*, bradykinin (Stewart and Woolley, 1965), as well as large polypeptides which have not been accessible by conventional methods. The over-all yield of purified peptide was 56% when compared with the amount of the first phenylalanyl residue introduced into the reaction vessel. This was greater than reported previously (Schwarz *et al.*, 1957; Arakawa and Bumpus, 1961).

Rearrangement of the Aspartyl Residue. Since a rearrangement of the peptide bond in angiotensin involving the α-carboxyl group of aspartic acid to one containing the β-carboxyl has been shown to occur under certain acidic conditions and at elevated temperatures (Page, 1964; Riniker, 1964), several experiments were designed to determine if the conditions used in solid-phase synthesis would cause such a rearrangement. The problem was first studied by the synthesis of the model tri- and tetrapeptides, t-BOC-β-benzyl-L-aspartyl-L-alanyl-L-

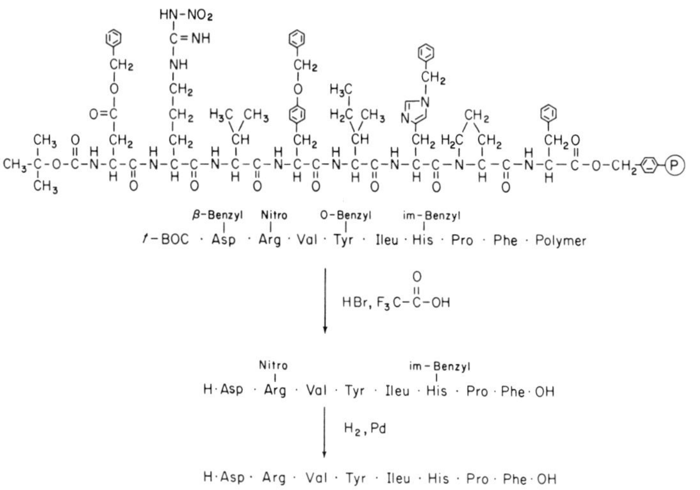

FIGURE 2: Cleavage of angiotensin from the polymer support and removal of protecting groups.

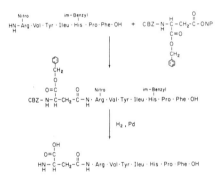

FIGURE 3: Preparation of β-aspartic[1]-isoleucine[5]-angiotensin II by a route which avoids conditions conducive to rearrangement. CBZ = carbobenzoxy, ONP = p-nitrophenyl ester.

valyl resin and t-BOC-L-leucyl-β-benzyl-L-aspartyl-L-alanyl-L-valyl resin, which were then deprotected by HCl–HOAc. The peptides were cleaved from the resin by the standard treatment with HBr in trifluoroacetic and also by saponification. Release of less than the theoretical amount of aspartic acid from the peptides upon digestion by leucine amino peptidase was assumed to be due to α–β rearrangement since the β-aspartyl bond has been shown resistant to leucine amino peptidase (Khairallah et al., 1963). Neither the HCl–HOAc deprotection step nor the HBr–trifluoroacetic acid cleavage caused detectable rearraagement in these representative peptides, regardless of whether or not the aspartyl residue was N terminal or centrally located. On the other hand, when the cleavage was by saponification, there was established an equilibrium ratio of approximately 4β to 1α in the tripeptide and 6β to 1α in the tetrapeptide. These results agreed with the well-known alkaline rearrangement of aspartic acid esters and amides (Sondheimer and Holley, 1954; Battersby and Robinson, 1955).

Although the above experiments indicated that no such rearrangement was to be expected in the standard procedure, it was nevertheless felt necessary to prove that none had actually occurred in the synthesis of angiotensin itself. For this purpose, standard reference angiotensins were prepared under conditions which were not conducive to the rearrangement. α- and β-aspartic[1]-isoleucine[5]-angiotensin II standards were synthesized by reaction of the appropriate aspartic acid derivative by conventional techniques with a heptapeptide. The latter was prepared by solid-phase synthesis and cleaved from the resin before the N-terminal aspartyl residue of angiotensin had been attached, as shown in Figure 3. Thus, nitro-L-arginyl-L-valyl-L-tyrosyl-L-isoleucyl-im-benzyl-L-histidyl-L-prolyl-L-phenylalanine was coupled with N-carbobenzoxy-β-benzyl-L-aspartic p-nitrophenyl ester and N-carbobenzoxy-α-benzyl-L-aspartic p-nitrophenyl ester which gave, upon hydrogenation, α- and β-aspartic[1]-isoleucine⁵-angiotensin II, respectively. These two syntheses illustrated that blocks of polypeptide could be made by solid-phase synthesis for subsequent incorporation into larger peptides by classical techniques. This method should be applicable to other special cases where the solid-phase synthetic method itself might not be easily adaptable.

Leucine amino peptidase digestion of the reference angiotensins confirmed the reports by others that the α-bond was degraded while the β-aspartyl bond was resistant. The product prepared totally by the solid-phase method in which the cleavage was performed by a short exposure of 5 min to HBr in trifluoroacetic acid was also degraded by leucine amino peptidase in accord with its α-aspartyl linkage.[2] The angiotensins synthesized on the resin and the standards were also compared by electrophoresis in 1 M acetic acid, pH 2.4. Table I shows that the α- and β-aspartic standards were

TABLE I: Electrophoretic Mobility of Angiotensin Derivatives.

Angiotensin II Derivative	Relative Mobility[a]
α-Aspartic[1]-valine[5b]	0.88
β-Aspartic[1]-valine[5b]	0.76
α-Aspartic[1]-isoleucine[5]	0.84
β-Aspartic[1]-isoleucine[5]	0.66
Synthesized on resin	0.84

[a] The electrophoretic mobility was measured relative to the mobility of asparagine[1]-valine[5]-angiotensin II on paper in 1 M HOAc, pH 2.4, at 17.4 v/cm for 1 hr.
[b] Calculated from the data of Riniker and Schwyzer (1964).

easily separated at this pH and that their mobilities did not differ markedly from those of the valine⁵ analogs as reported by Riniker and Schwyzer (1964). Table I also shows that the angiotensin synthesized on the resin had the mobility of the α-aspartic[1] derivative. There was no detectable material present with the mobility of the β-aspartic[1] derivative.

In a second experiment, the α-peptide linkage was shown by chemical modification of the free carboxyl group of the N-terminal aspartyl residue and detection of the resulting derivatives of aspartic acid after hydrolysis, as shown in Figure 4. The dimethyl ester of the resin-

[2] The product which resulted from exposure of the resin to HBr–trifluoroacetic acid for an hour or longer was resistant to leucine amino peptidase despite the fact that the α-aspartic[1]-derivative was shown to be present and no β-aspartic[1]-derivative was detectable by any of the other methods used. Addition of the β-aspartic standard to the α-aspartic standard did not inhibit its degradation by leucine amino peptidase.

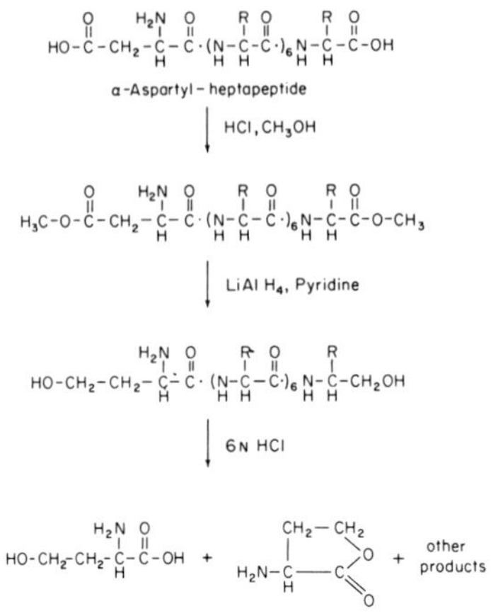

FIGURE 4: Method for determination of α or β linkage by modification of aspartyl residue. The derivatives formed with α linkage are shown.

prepared angiotensin was synthesized (Chibnall et al., 1958) and reduced with lithium aluminum hydride in pyridine (Lansbury, 1961). The resulting derivatives from the aspartyl residue were detected by hydrolysis and amino acid analysis. Reduction of aspartic acid, in which the α-carboxyl group is in peptide linkage and the β-carboxyl group is present as a methyl ester, would be expected to give homoserine and α-aminobutyrolactone upon hydrolysis. In the case where the α-carboxyl is present as a methyl ester and the β-carboxyl is in peptide linkage, one would expect to detect β-homoserine and β-aminobutyrolactone after esterification, reduction, and hydrolysis. The angiotensin prepared on the resin gave only homoserine and α-aminobutyrolactone. Thus, the results of chemical modification of the free carboxyl group of the aspartyl residue confirmed the previous results, in that the presence of the α-linkage was demonstrated and no β-aspartyl bonds were detected.

The α-aspartic[1] standard and the solid-phase preparation were both heated at 100° in neutral solution for 18 hr and then examined by electrophoresis. Both the α-aspartic standard and the angiotensin prepared on the resin were converted to mixtures of the α- and β-aspartic[1] derivatives by this treatment, as has been reported for the valine⁵ analog (Riniker, 1964). This experiment also lends support to the conclusion that no α–β rearrangement had occurred in the synthesis of angiotensin by the solid-phase method.

Synthesis of Asparagine¹-isoleucine⁵-angiotensin II. In order to investigate the use of asparagine in solid-phase synthesis, asparagine¹-isoleucine⁵-angiotensin II (Rittel et al., 1957) was prepared. The procedure was the same as described above except that in the last step *t*-BOC-L-asparagine *p*-nitrophenyl ester was coupled to the heptapeptide resin instead of *t*-BOC-β-benzyl-L-aspartic acid. The nitrophenyl ester method is preferred in the incorporation of asparagine into the growing peptide chain since formation of the anhydro derivative, the nitrile, has been observed with the diimide reagent (Gish *et al.*, 1956), but not with the nitrophenyl ester method (Bodanszky and du Vigneaud, 1959). Once the asparaginyl residue has been incorporated, the diimide method may be used without fear of formation of the nitrile derivative, since formation has been shown to require the "activated" carboxyl group of the intermediate formed with the diimide reagent (Paul and Kende, 1964; Kashelikar and Ressler, 1964). A similar approach with glutamine should also avoid formation of its nitrile derivative.

Experimental Section

Materials. The *t*-butyloxycarbonyl amino acids were either purchased or synthesized according to the procedure of Schwyzer *et al.* (1959). Purity was checked by thin layer chromatography. Dimethylformamide was purified by the barium oxide technique of Thomas and Rochow (1957). The asparagine[1]-valine[5]-angiotensin octapeptide used in this work was purchased from Ciba, Inc.

Amino Acid Analyses. Samples (50 mg) of the peptide resin were refluxed with a mixture of equal parts of 12 N HCl and dioxane for 18 hr. They were then filtered, evaporated, and rehydrolyzed in 6 N HCl for an additional 18 hr. Free peptides were hydrolyzed in 6 N HCl alone. The analyses were performed on the Spinco amino acid analyzer, Model 120B.

im-Benzylhistidine Analyses. Since *im*-benzylhistidine was too strongly held by the normal columns of the analyzer, a separate 3-cm column of Aminex MS blend Q15 was prepared. Elution by pH 5.28, 0.35 N sodium citrate buffer at a rate of 68 ml/hr produced a peak for *im*-benzylhistidine at 61 min. The color yield with ninhydrin was 0.88 that of leucine.

t-BOC-L-asparagine p-Nitrophenyl Ester. A solution of 4.6 g (22 mmoles) of dicyclohexylcarbodiimide in 5 ml of purified dimethylformamide was slowly added from a dropping funnel to a mixture of 4.6 g (20 mmoles) of *t*-BOC-L-asparagine (Stewart and Woolley, unpublished data) and 11.1 g (80 mmoles) of *p*-nitrophenol dissolved in 15 ml of dimethylformamide at 0°. The mixture was allowed to react for 48 hr at 4° and the dicyclohexylurea formed was filtered off and discarded. The dimethylformamide was removed under high vacuum on a rotary evaporator and the resulting crystalline material was washed with ether. The product was recrystallized from ethyl acetate to yield 3.78 g (54%), mp 163°. Additional product was separated from the ether wash, which also contained *p*-nitrophenol, by countercurrent distribution in ethyl acetate–hexane–ethanol–water (1:1.4:1.4:1). The separation was followed by measuring the optical density at 270 mμ and was complete after 192 transfers. The distribution constant for the product was 0.31 and for nitrophenol, k = 0.61. The solvent mixture was removed by evaporation and the additional product crystallized from ethyl acetate; yield 0.79 g. This, plus the previously obtained product, gave a total yield of 4.57 g (65%).

Anal. Calcd for $C_{15}H_{19}N_3O_7$: C, 50.99; H, 5.42; N, 11.89. Found: C, 50.63; H, 5.11; N, 11.99.

t-BOC-L-phenylalanyl Resin. A solution of 1.25 g (4.7 mmoles) of *t*-BOC-L-phenylalanine and 0.59 ml (4.7 mmoles) of triethylamine in 10 ml of ethanol was added to 5.3 g of the chloromethyated copolystyrene–2% divinylbenzene (Merrifield, 1963) which contained 1.3 mmoles of Cl/g and the mixture was stirred at 75° for 24 hr. The esterified resin was filtered off, washed with ethanol, water, and methanol, and dried under vacuum. Amino acid analysis showed the substituted polymer to contain 0.192 mmole of phenylalanine/g.

t-BOC-β-benzyl-L-aspartyl-nitro-L-arginyl-L-valyl-O-benzyl-L-tyrosyl-L-isoleucyl-im-benzyl-L-histidyl-L-prolyl-L-phenylalanyl Resin. Five grams of the *t*-BOC-L-phenylalanyl resin was introduced into the reaction vessel (Merrifield, 1963). The following cycle of reactions was used to introduce each new residue: (1) washed with three 30-ml portions of glacial acetic acid; (2) *t*-BOC group cleaved by 1 N HCl in glacial acetic acid (30 ml) for 30 min; (3) washed with three 30-ml portions of glacial acetic acid; (4) washed with three 30-ml portions of absolute ethanol; (5) washed with three 30-ml portions of dimethylformamide; (6) neutralized the hydrochloride with 3 ml of triethylamine in 30 ml of dimethylformamide for 10 min; (7) washed with three 30-ml portions of dimethylformamide; (8) washed with three 30-ml portions of methylene chloride; (9) introduced 3.83 mmoles of the appropriate *t*-BOC amino acid in 20 ml of methylene chloride and allowed to mix for 10 min; (10) introduced 3.83 mmoles of dicyclohexylcarbodiimide in 2 ml of methylene chloride and allowed to react for 2 hr; (11) washed with three 30-ml portions of methylene chloride; (12) washed with three 30-ml portions of absolute ethanol. For the *im*-benzyl-L-histidine and nitro-L-arginine cycles, step 8 was deleted, and dimethylformamide was substituted for methylene chloride in steps 9–11. Amino acid analysis showed the average value of the eight amino acid residues to be 0.13 mmole/g of peptide resin, or 0.16 mmole/g of unsubstituted copolymer.

L-*Aspartyl-nitro-L-arginyl-L-valyl-L-tyrosyl-L-isoleucyl-im-benzyl-L-histidyl-L-prolyl-L-phenylalanine.* The protected peptide polymer was suspended in 20 ml of anhydrous trifluoroacetic acid and a slow stream of HBr was bubbled through the fritted disk of the reaction vessel into the suspension for various lengths of time at 25°, with exclusion of water. The suspension was filtered and the resin was washed three times with 10-ml portions of trifluoroacetic acid. The filtrates were evaporated on a rotary evaporator at 25° under reduced pressure. The product was redissolved in trifluoroacetic acid and re-evaporated. The syrupy product was then dissolved in acetic acid and lyophilized; yield, 1.47 g; amino acid ratios: Asp, 0.94; Arg, 0.80; Val, 1.05; Tyr, 0.18; Ileu, 0.95; *im*-benzyl-His, 1.17; Pro, 0.83; and

Phe, 1.19. The value for tyrosine is known to be low in the presence of nitroarginine (Riniker and Schwyzer, 1961). Calculating from the average amino acid content, excluding tyrosine, a total of 0.80 mmole of protected peptide was recovered. The theoretical amount of protected peptide, based on the amount of phenylalanine initially on the resin, was 0.89 mmole which gave an 89% yield for all of the coupling steps and the cleavage step which were performed in the reaction vessel.

L-*Aspartyl*-L-*arginyl*-L-*valyl*-L-*tyrosyl*-L-*isoleucyl*-L-*histidyl*-L-*prolyl*-L-*phenylalanine* (*Isoleucine*[5]-*angiotensin II*). A portion (500 mg) of the protected peptide and 500 mg of 5% palladium oxide on barium sulfate (Engelhard Industries) were hydrogenated in 30 ml of a mixture of methanol, acetic acid, and water (10:1:1) at 50 psi for 24 hr. An aliquot was removed, hydrolyzed, and analyzed for amino acids. At the same time, 250 mg of additional catalyst was added, and the mixture was rehydrogenated for an extra 24 hr. After the first 24-hr period, the reaction mixture showed complete reduction of nitroarginine to arginine, but only a 65% conversion of *im*-benzylhistidine to histidine. After 48 hr of hydrogenation, there was less than 2% *im*-benzylhistidine remaining and the amount of histidine present was that expected. The tyrosine value had increased to a ratio of 0.91, compared with 0.18 in the presence of nitroarginine. The solution was filtered, evaporated, and lyophilized from acetic acid; yield, 445 mg. Paper electrophoresis in 0.1 M pyridine acetate, pH 5.0, showed a major spot at R_{arg} 0.29 when sprayed with ninhydrin or Sakaguchi reagents, and traces of material at R_{arg} 0.51 and at the origin.

Purification by Countercurrent Distribution. A 300-mg portion of the crude octapeptide was purified by 100 transfers in a 1-butanol–acetic acid–water (4:1:5) system. Over 80% of the Sakaguchi-positive material was located in one peak which matched closely a theoretical curve with distribution constant, $k = 0.30$. The material in the peak was collected and the organic phase was removed by evaporation. The residual aqueous phase was removed by lyophilization; yield, 193 mg. This was equivalent to an over-all yield of 56% from *t*-BOC-phenylalanyl resin; distribution constant in 1-butanol–1-propanol–0.1 N HCl (1:1:2), $k = 0.64$; reported value for this system, $k = 0.73$ (Schwarz *et al.*, 1957). The homogeneous product had R_f 0.28 on paper chromatography in 1-butanol–acetic acid–water (4:1:5); reported value, R_f 0.29 (Arakawa and Bumpus, 1961); $[\alpha]_D^{25}$ −0.66° (*c* 0.8, 1 N HCl); reported value, $[\alpha]_D^{21}$ −0.67° (Schwyzer and Turrian, 1960; Arakawa and Bumpus, 1961); amino acid ratios: Asp, 1.05; Arg, 1.00; Val, 1.12; Tyr, 0.99; Ileu, 0.98; His, 0.95; Pro, 1.00; and Phe, 1.09.

α- and β-Aspartic[1]-*isoleucine*[5]-*angiotensin II Standards.* The heptapeptide resin, *t*-BOC-nitro-L-arginyl-L-valyl-*O*-benzyl-L-tyrosyl-L-isoleucyl-*im*-benzyl-L-histidyl-L-prolyl-L-phenylalanyl resin, was synthesized as described previously, but was subjected to cleavage before the N-terminal aspartyl residue was coupled. Thus, the heptapeptide, nitro-L-arginyl-L-valyl-L-tyrosyl-L-isoleucyl-*im*-benzyl-L-histidyl-L-prolyl-L-phenylalanine, was prepared. Two 50-mg portions of this material were combined with an equimolar amount of triethylamine in dimethylformamide. One of the two portions was then treated with a fourfold molar excess of *N*-carbobenzoxy-β-benzyl-L-aspartic *p*-nitrophenyl ester for 4 hr at room temperature in a total volume of 5 ml of dimethylformamide. An equal volume of water was then added, and the mixture was evaporated to dryness under high vacuum on a rotary evaporator. After extraction with water, the residue was hydrogenated and purified by countercurrent distribution as described previously. There still was contamination by *p*-aminophenol, however, which was removed by gel filtration on Sephadex G-25 in 0.2 N acetic acid. The purified α-aspartic[1]-isoleucine[5]-angiotensin II standard had a trace contaminant with electrophoretic mobility of the free heptapeptide.

Similarly, the β-aspartic[1]-isoleucine[5]-angiotensin II standard was prepared by treating the second portion of the cleaved heptapeptide with *N*-carbobenzoxy-α-benzyl-L-aspartic *p*-nitrophenyl ester, followed by hydrogenation and purification as described above. This preparation also showed trace contamination by heptapeptide.

Esterification and Reduction of Angiotensin. A sample of 10 mg of the aspartic[1]-isoleucine[5]-angiotensin II which had been synthesized entirely on the resin was esterified in 3 ml of 0.85 N HCl in anhydrous methanol for various lengths of time (Chibnall *et al.*, 1958). The amount of esterification was determined by paper electrophoresis in 0.1 M pyridine acetate, pH 5.0 (detected by ninhydrin and Pauli reagents). After 6 hr, two additional components besides the starting angiotensin (R_{his} 0.28) could be detected and were assumed to be the monoester (R_{his} 0.46) and the diester (R_{his} 0.73). After 24 hr, the angiotensin had been entirely converted to diester with a trace of monoester. The esterified angiotensin was then precipitated by ether, and the precipitate was washed with ether and dried. Aliquots of this material were then used in the following experiments on reduction.

Several solvents were used in the attempt to reduce the esters with lithium aluminum hydride. Pyridine (Lansbury, 1961) was finally selected. A sample of 1 mg of esterified angiotensin was dissolved in 5 ml of pyridine which had been dried over barium oxide. A 10 molar excess of lithium aluminum hydride in pyridine was introduced, and the mixture was allowed to react at room temperature for 4 hr. A light green color developed during the reaction as it did in the reduction of phenylalanine benzyl ester. The reaction was stopped by the addition of 6 N HCl. An aliquot was removed, evaporated to dryness, and hydrolyzed for 18 hr in 6 N HCl. Amino acid analysis showed the loss of aspartic acid, and the appearance of two new peaks in positions corresponding to homoserine and α-aminobutyrolactone and no detectable material in the elution position of either β-homoserine or β-aminobutyrolactone (Blumenfeld and Gallop, 1962). These two new peaks were also detected upon reduction and hydrolysis of β-benzylaspartic acid. Reduction and hydrolysis of the α-benzyl ester of aspartic acid, corresponding to the ester in the product which would be formed with a β-aspartyl peptide, gave two peaks in the elution positions of β-homoserine and β-aminobutyrolactone.

Leucine Amino Peptidase Digestion. To determine the amount of rearrangement of the α-aspartyl bond to the β-aspartyl bond, the peptides were examined for susceptibility to leucine amino peptidase since the peptide with the β-linkage had been reported resistant (Khairallah *et al.*, 1963). A method similar to that of Hofmann and Yajima (1961) was used, and the peptide was incubated with Worthington leucine amino peptidase (1 mg/20 mg of peptide) for 2 days at 37°. The digest was compared on amino acid analysis with that found on acid hydrolysis. Incomplete release of aspartic acid was assumed to be due to the presence of the β-linkage. The β-aspartic[1] standard was found to be completely resistant while the α-aspartic[1] standard released 1 mole equiv of aspartic acid upon digestion. Release of amino acids in angiotensin stopped when histidine became N terminal as has been reported (Riniker, 1964).

Asparagine[1]-isoleucine[5]-angiotensin II. A 1.3-g portion of *t*-BOC-nitro-L-arginyl-L-valyl-*O*-benzyl-L-tyrosyl-L-isoleucyl-*im*-benzyl-L-histidyl-L-prolyl-L-phenylalanyl resin, synthesized as described previously, was added to the reaction vessel. Steps 1–7 of the reaction cycle were performed with one-third the volume of solvents (10 ml) reported previously. A 3 molar excess of *t*-BOC-L-asparagine *p*-nitrophenyl ester in 5 ml of dimethylformamide was added and allowed to react for 4 hr. The resin was then washed and cleaved for 1 hr in HBr–trifluoroacetic acid; yield, 273 mg.

A 100-mg portion was then hydrogenated and purified by countercurrent distribution as previously described; yield, 62 mg. The product appeared homogeneous and had an electrophoretic mobility equal to that of authentic asparagine[1]-valine[5]-angiotensin II in 1 M HOAc, pH 2.4. The amino acid ratios were Asp, 0.92; Arg, 1.00; Val, 1.00; Tyr, 1.01; Ileu, 0.82; His, 0.92; Pro, 1.08; Phe, 1.08; and NH$_3$, 0.93.

Bioassay. The oxytocic activity of the synthesized materials was compared with that of asparagine[1]-valine[5]-angiotensin II (Hypertensin, Ciba) in the rat uterus assay. Based on the time of response (Schwarz *et al.*, 1955), the isoleucine[5]-angiotensin II and asparagine[1]-isoleucine[5]-angiotensin II prepared by solid-phase synthesis were fully active, being equivalent within experimental error in their oxytocic activity with the asparagine[1]-valine[5]-angiotensin as reported by Page and Bumpus (1961). The three peptides were compared over a range of 2.6×10^{-10} to 3.3×10^{-9} g/ml.

Acknowledgment

The authors wish to express their sincere appreciation to Dr. D. W. Woolley for his interest and encouragement during this work.

References

Arakawa, K., and Bumpus, F. M. (1961), *J. Am. Chem. Soc. 83*, 728.
Battersby, A., and Robinson, J. C. (1955), *J. Chem. Soc.*, 259.
Blumenfeld, O. O., and Gallop, P. M. (1962), *Biochemistry 1*, 947.
Bodanszky, M., and du Vigneaud, V. (1959), *J. Am. Chem. Soc. 81*, 5688.
Braun-Menendez, E., Fasciolo, J. C., Leloir, L. F., and Munoz, J. M. (1939), *Rev. Soc. Arg. Biol. 15*, 420.
Bricas, E., and Nicot-Gutton, C. (1960), *Bull. Soc. Chim. France 26*, 466.
Chibnall, A. C., Mangan, J. L., and Rees, M. W. (1958), *Biochem. J. 68*, 114.
Gish, D. T., Katsoyannis, P. G., Hess, G. P., and Stedman, R. J. (1956), *J. Am. Chem. Soc. 78*, 5954.
Hofmann, K., and Yajima, H. (1961), *J. Am. Chem. Soc. 83*, 2289.
Kashelikar, D. V., and Ressler, C. (1964), *J. Am. Chem. Soc. 86*, 2467.
Khairallah, P. A., Bumpus, F. M., Page, I. H., and Smeby, R. R. (1963), *Science 140*, 672.
Kopple, K. D., Jarabak, R. R., and Bhatia, P. L. (1963), *Biochemistry 2*, 958.
Lansbury, P. T. (1961), *J. Am. Chem. Soc. 83*, 429.
Li, C. H., Schnabel, E., and Chung, D. (1960), *J. Am. Chem. Soc. 82*, 2062.
Marshall, G. R., and Merrifield, R. B. (1965), *Federation Proc. 24*, 224.
Merrifield, R. B. (1962), *Federation Proc. 21*, 412.
Merrifield, R. B. (1963), *J. Am. Chem. Soc. 85*, 2149.
Merrifield, R. B. (1964a), *Biochemistry 3*, 1385.
Merrifield, R. B. (1964b), *J. Org. Chem. 29*, 3100.
Merrifield, R. B. (1965), *Endeavour 24*, 3.
Page, I. H. (1964), *Federation Proc. 23*, 693.
Page, I. H., and Bumpus, F. M. (1961), *Physiol. Rev. 41*, 331.
Page, I. H., and Helmer, O. H. (1939), *Proc. Central Soc. Clin. Res. 12*, 17.
Paul, R., and Kende, A. S. (1964), *J. Am. Chem. Soc. 86*, 741.
Riniker, B. (1964), *Metabolism 13*, 1247.
Riniker, B., and Schwyzer, R. (1961), *Helv. Chim. Acta 44*, 674.
Riniker, B., and Schwyzer, R. (1964), *Helv. Chim. Acta 47*, 2357.
Rittel, W., Iselin, B., Kappeler, H., Riniker, B., and Schwyzer, R. (1957), *Helv. Chim. Acta 40*, 614.
Schwarz, H., Bumpus, F. M., and Page, I. H. (1957), *J. Am. Chem. Soc. 79*, 5697.
Schwarz, H., Masson, G. M. C., and Page, I. H. (1955), *J. Pharm. Exptl. Therap. 114*, 418.
Schwyzer, R., Iselin, B., Kappeler, H., Rittel, W., and Zuber, H. (1957), *Chimia (Aarau) 11*, 335.
Schwyzer, R., Sieber, P., and Kappeler, H. (1959), *Helv. Chim. Acta 42*, 2622.
Schwyzer, R., and Turrian, H. (1960), *Vitamins Hormones 18*, 237.
Sheehan, J. C., Goodman, M., and Hess, G. P. (1956), *J. Am. Chem. Soc. 78*, 1367.
Sheehan, J. C., and Hess, G. P. (1955), *J. Am. Chem. Soc. 77*, 1067.
Skeggs, L. T., Jr., Kahn, J. R., and Shumway, N. P.

(1956a), *J. Exptl. Med. 103*, 301.

Skeggs, L. T., Jr., Lentz, K. E., Kahn, J. R., and Shumway, N. P. (1956b), *J. Exptl. Med. 104*, 193.

Skeggs, L. T., Jr., Marsh, W. H., Kahn, J. R., and Shumway, N. P. (1954), *J. Exptl. Med. 100*, 363.

Smart, N. A., Young, G. T., and Williams, M. W. (1960), *J. Chem. Soc.*, 3902.

Sondheimer, E., and Holley, R. W. (1954), *J. Am. Chem. Soc. 76*, 2467.

Stewart, J. M., and Woolley, D. W. (1965), *Federation Proc. 24*, 657.

Theodoropoulos, D. (1956), *J. Org. Chem. 21*, 1550.

Thomas, A. B., and Rochow, E. G. (1957), *J. Am. Chem. Soc. 79*, 1843.

A Synthesis of Cyclic Peptides Utilizing High Molecular Weight Carriers

MATI FRIDKIN, ABRAHAM PATCHORNIK, and EPHRAIM KATCHALSKI

Sir:

Cyclic peptides are usually prepared from linear peptides by intramolecular cyclization. The carboxylic end of the peptide is as a rule activated by the formation of active esters, anhydrides, azides, or chlorides, and the free terminal amino end allowed to react with the active terminal carbonyl group at high dilution.[1-4] Because of intermolecular condensation occurring even under these conditions, linear oligopeptides are formed in addition to the desired cyclic peptide. The techniques available so far thus lead to reaction mixtures from which cyclic peptides are usually isolated only in relatively low yields. In the following we report on the development of a new method for the synthesis of cyclic peptides in which high molecular weight peptide active esters of type II (see Figure 1), in which the peptide is bound to a high molecular weight polyalcohol carrier, are used as intermediates. When insoluble esters of this type are employed condensation between the activated peptide moieties is suppressed, and internal aminolysis leads to the formation of the desired cyclic peptide (IV) which is released from the insoluble polyhydroxy carrier (III). Intermolecular condensation might be expected to be markedly reduced even when soluble high molecular weight peptide esters of type II are used.

Two high molecular weight poly(nitrophenol) derivatives have been used in the preparation of the peptide active esters: cross-linked poly-4-hydroxy-3-nitrostyrene (IIIa) and a branched copolymer of DL-lysine and 3-nitro-L-tyrosine (IIIb) in which free amino groups have been blocked by acetylation. The former has been prepared according to the literature[5] and is insoluble in the usual organic solvents. The latter has been prepared by total acetylation, with acetic anhydride, of a branched copolymer of DL-lysine and L-tyrosine (molar residue ratio 3:1),[6] removal of the O-acetyl groups in alkali, and nitration in concentrated nitric acid at 0°. IIIb is insoluble in dioxane, ether, and acetone, but is soluble in dimethylformamide (DMF),

(1) R. Schwyzer and P. Sieber, *Helv. Chim. Acta*, **40**, 624 (1957).
(2) T. Wieland and K. W. Ohly, *Ann. Chem.*, **605**, 179 (1957).
(3) K. Vogler, R. O. Studer, W. Lergier, and P. Lanz, *Helv. Chim. Acta*, **43**, 1751 (1960).
(4) H. Gerlach, J. A. Owtschinnikow, and V. Prelog, *ibid.*, **47**, 2294 (1964).
(5) D. I. Packham, *J. Chem. Soc.*, 2617 (1964).
(6) M. Sela, S. Fuchs, and R. Arnon, *Biochem. J.*, **85**, 223 (1962).

ethanol, and water at pH values above 10. N-Benzyloxycarbonyl derivatives of peptides were coupled with both poly(nitrophenol) compounds in DMF using N,N'-dicyclohexylcarbodiimide as the coupling reagent. Highest yields of the N-benzyloxycarbonyl active esters (I) were obtained when the reaction mixture contained at least a threefold excess of the N-benzyloxycarbonyl peptide in comparison with the nitrophenol content of the polymer. Excess of unreacted peptide was removed from Ia by washing and from Ib by dialysis. The protecting groups of I were removed with anhydrous hydrogen bromide in glacial acetic acid,[7] and the high molecular weight peptide ester hydrobromides were isolated by filtration (esters derived from IIIa) or by precipitation with ether (esters derived from IIIb). The free amino groups of the peptide esters (II) were liberated from the corresponding hydrobromides by neutralization with triethylamine in DMF, and the cyclization reaction was allowed to take place at room temperature in the same solvent. The DMF-insoluble free peptide esters studied (IIa), derived from IIIa, yielded under the experimental conditions used chromatographically pure cyclopeptides in 60–80% yield. The latter could readily be isolated from the DMF solution. The DMF-soluble peptide esters studied (IIb), derived from IIIb, yielded the corresponding cyclic peptides, also in 60–80% yield. The cyclic peptides in this case, however, which were only identified chromatographically, were contaminated with some peptide oligomers.

The synthesis of some representative cyclic peptides by our new method is described: Cyclo(Gly-Gly). Z-Gly-Gly was coupled with IIIa in DMF to yield the corresponding ester Ia containing 0.8 mmole of peptide/g. The hydrobromide formed after removal of the benzyloxycarbonyl group (1 hr. in 30% HBr in anhydrous acetic acid) was filtered, washed with ether, suspended in DMF, and neutralized with triethylamine. After 12 hr. at room temperature the polymer was filtered off and the filtrate brought to dryness *in vacuo*. Crystalline cyclo(Gly-Gly) was obtained on trituration with ether; yield 75% of the theoretical; m.p. 310° (fast heating). The identity of the product obtained as diketopiperazine was ascertained by the appearance of one spot on thin layer partition chromatography (t.l.p.c.)

(7) D. Ben-Ishai and A. Berger, *J. Org. Chem.*, **17**, 1564 (1952).

with R_f values identical with those of an authentic sample when developed with *t*-butyl hypochlorite–starch–KI[8] using 1-butanol–acetic acid–water (BAW, 4:1:1, v./v.) (R_f 0.40) or 1-butanol–pyridine–acetic acid–water (BPAW) (15:10:3:12, v./v.) (R_f 0.56) as solvents. As expected, cyclo(Gly-Gly) gives a negative ninhydrin test.

On coupling Z-Gly-Gly with IIIb in DMF a quantitative yield of the corresponding ester Ib was obtained. Excess Z-Gly-Gly was removed by dialysis against water. Removal of the protecting groups and neutralization with triethylamine was carried out as above. The amount of cyclo(Gly-Gly) formed in DMF after 12 hr. at room temperature corresponded to 75% of the theoretical as assayed by quantitative t.l.p.c.

By a procedure analogous to the above *cyclo*(L-*Ala-Gly*) [R_f 0.43 (BAW), 0.64 (BPAW); m.p. 240° dec. (lit. 240°[9]; 241°[10]); $[\alpha]^{23}$D $-2.5°$ (c 1.2, water)], *cyclo*(L-*Ala*-L-*Ala*) [R_f 0.53 (BAW)], and *cyclo*(L-*Phe-Gly*) [R_f 0.77 (BAW); m.p. 260–265° dec. (lit. 260–265°[11]); $[\alpha]^{23}$D $-12.4°$ (c 3.68, dichloroacetic acid)] were prepared from the corresponding high molecular weight esters IIa and IIb.

Cyclo(tetra-L-*Ala*) [R_f 0.60 (BAW), 0.68 (BPAW); m.p. 250°; $[\alpha]^{22}_{484}$ $-68°$, $[\alpha]^{22}_{366}$ $-157°$ (c 0.5, DMF)] was prepared from three different tetra-Ala active esters: the active esters derived from IIIa and IIIb, as well as from tetra-L-Ala-*p*-nitrophenyl ester. Cyclization of the high molecular weight tetra-L-Ala esters IIa and IIb was carried out in DMF solution at room temperature (12 hr.). Cyclization of tetra-L-Ala *p*-nitrophenyl ester was carried out in dilute pyridine solution at 90–100° (12 hr.). A chromatographically pure cyclic peptide in 50–65% yield was obtained from the high molecular weight active esters. Considerably lower yields (38–42%) were obtained from the low molecular weight tetra-L-Ala ester. Furthermore, the pyridine reaction mixture was found to contain linear tetra-Ala oligomers in addition to the cyclic peptide. Cyclo(tetra-L-Ala) gave on mild alkaline hydrolysis (N/15 LiOH, 1 hr., 100°)[12] tetra-, tri-, di-, and monoalanine, which could be separated by high voltage paper electrophoresis at pH 1.5.

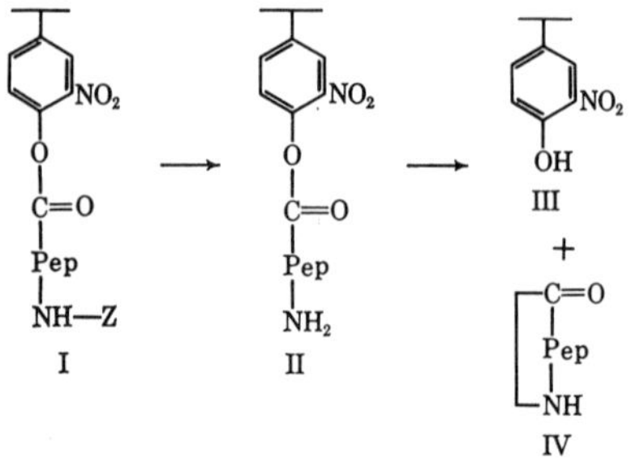

Cyclo(L-*Ala-Gly*-L-*Ala*-L-*Ala*) (R_f 0.54 (BAW), 0.64 (BPAW)] was prepared from the high molecular weight active ester of L-Ala-Gly-L-Ala-L-Ala with IIIa. In its chromatographic behavior it was found to be identical with a sample of the cyclic peptide obtained from Gly-L-Ala-L-Ala-L-Ala *p*-nitrophenyl ester [R_f 0.54 (BAW), 0.64 (BPAW)].

The above examples illustrate the use of high molecular weight polyalcohol carriers of the types IIa and IIb in the synthesis of cyclic peptides. It should be mentioned, however, that although the yields of the desired cyclic peptides on intramolecular aminolysis of II (II → III + IV) are good, the amount of the N-benzyloxycarbonyl peptide which can be bound to the insoluble carrier IIIa is rather low. Finally, it is pertinent to note that we were able to utilize the high molecular weight active esters of N-masked peptides and amino acids described here in the synthesis of linear peptides by their interaction with the suitable amino acid or peptide esters.

Figure 1. Scheme describing the method employed for the synthesis of cyclic peptides. IIIa stands for poly-4-hydroxy-3-nitrostyrene; Ia and IIa are the corresponding N-benzyloxycarbonyl (Z) peptide and free amino peptide ester derivatives of IIIa. IIIb stands for a branched copolymer of DL-lysine and 3-nitro-L-tyrosine in which free amino groups have been blocked by acetylation; Ib and IIb are the corresponding N-benzyloxycarbonyl (Z) peptide and free amino peptide ester derivatives of IIIb.

(8) R. H. Mazur, B. W. Ellis, and P. S. Cammarata, *J. Biol. Chem.*, **237**, 1619 (1962).
(9) J. L. Bailey, *J. Chem. Soc.*, 3461 (1950).
(10) T. I. Orlova and N. I. Gavilov, *Zh. Obshch. Khim.*, **27**, 3314 (1957).
(11) H. T. Huang and C. Niemann, *J. Am. Chem. Soc.*, **72**, 921 (1950).
(12) R. Schwyzer, B. Iselin, W. Rittel, and P. Sieber, *Helv. Chim. Acta*, **39**, 872 (1956).

Department of Biophysics, The Weizmann Institute of Science
Rehovoth, Israel
Received August 24, 1965

8

Reprinted from *J. Amer. Chem. Soc.*, **88**(13), 3164–3165 (1966)

Use of Polymers as Chemical Reagents: I. Preparation of Peptides

MATI FRIDKIN, ABRAHAM PATCHORNIK, and EPHRAIM KATCHALSKI

Sir:

Polymers of the type Ⓟ–A containing covalently bound groups A, which readily react with a low molecular weight reagent B, can be used to synthesize compound A–B according to eq 1. To increase yields

$$\text{Ⓟ-A} + \text{B} \longrightarrow \text{A-B} + \text{Ⓟ}$$

and to facilitate the synthetic procedure insoluble polymers of the above type may be added in large excess to a solution of B in a suitable solvent. At the end of the reaction the insoluble polymer can be removed by filtration or centrifugation. The filtrate which is devoid of A should thus contain only A–B and unreacted B. The most suitable polymers Ⓟ–A to be used as chemical reagents should contain a relatively large amount of A, should show high stability on storage, and should possess the suitable mechanical properties.

To test the possible use of chemically reactive polymers in acylation reactions we prepared insoluble, high molecular weight active polyesters of acetic acid (IIa) and benzoic acid (IIb) by allowing the corresponding chlorides to react with poly-4-hydroxy-3-nitrostyrene cross-linked with 4% divinylbenzene (I) in dimethylformamide (DMF), in the presence of pyridine. The high molecular weight active esters (II) contained approximately 5 mmoles of acyl residues/g of insoluble polymer and could be stored at room temperature in powder form without decomposition. Treating IIa or IIb (1.0 g) in suspension in DMF (15 ml) with 0.5 mmole of tri-L-alanyl-*p*-nitrobenzyl ester (L-Ala_3-PNB) was carried out at room temperature for 4–5 hr. The polymer was removed by centrifugation and the supernatants evaporated to dryness *in vacuo*. The solid residues were found to contain a practically quantitative yield of the corresponding acyl derivatives: acetyl-L-Ala_3-PNB, mp 226°, $[\alpha]^{25}\text{D}$ −28.9° (*c* 0.58, DMF); benzoyl-L-Ala_3-PNB, mp 216–218°, $[\alpha]^{25}\text{D}$ +2.9° (*c* 0.70, DMF).

The successful preparation of active esters of N-blocked amino acids of type II enabled their utiliza-

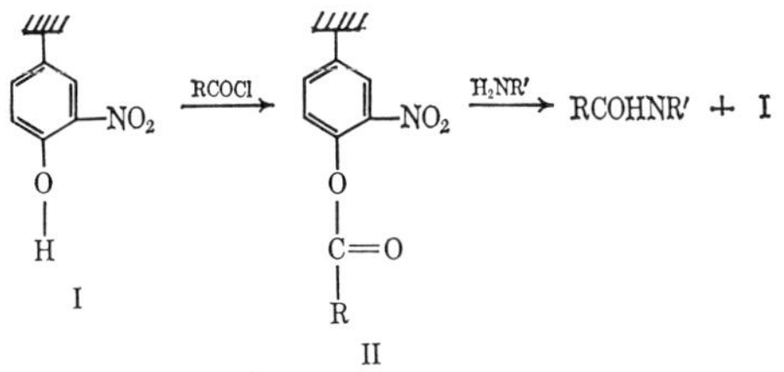

II: esters of: acetic acid (IIa); benzoic acid (IIb); Z-L-Phe (IIc); Z-L-Ileu (IId); Z-L-Pro (IIe); N,S-Di-Z-L-Cys (IIf); N-Z-α,-OBz-L-Glu (IIg), where Z = C_7H_7OCO and Bz = C_7H_7

tion in peptide synthesis. Thus peptides with N- and C-blocked terminal groups were obtained on coupling the insoluble active esters IIc to IIg with desired soluble amino acid or peptide esters containing a free α-amino group. Preferential removal of the N-blocking group from the newly formed peptide enabled the repetition of the coupling reaction with an insoluble active ester of another N-blocked amino acid. Further repetitions of this set of reactions lead obviously to the elongation of the peptide chain and formation of a peptide with a predetermined amino acid sequence.

The insoluble active esters IIc–g were derived from the following corresponding benzyloxycarbonyl amino acid derivatives: benzyloxycarbonyl-L-phenylalanine, benzyloxycarbonyl-L-isoleucine, benzyloxycarbonyl-L-proline, N,S-dibenzyloxycarbonyl-L-cysteine, and N-benzyloxycarbonyl-α-benzyl-L-glutamate, by their coupling with polymer I in DMF by the DCC method.[1,2] The insoluble polymers IIc to IIg contained per gram approximately 1.0–1.5 mmoles of amino acid and could be stored at room temperature without decomposition, similarly to IIa and IIb. In suspension in inert organic solvents polymers IIc to IIg showed chemical behavior similar to that of the cor-

(1) J. C. Sheehan and G. P. Hess, *J. Am. Chem. Soc.*, **77**, 1067 (1955); M. Rothe and F. W. Kunitz, *Ann.*, **609**, 88 (1957); D. F. Elliott and D. W. Russell, *Biochem. J.*, **66**, 499 (1957).

(2) M. Fridkin, A. Patchornik, and E. Katchalski, *J. Am. Chem. Soc.*, **87**, 4646 (1965).

responding low molecular weight amino acid active esters.

N,S-Di-Z-L-Cys-Gly-OBz was obtained by allowing IIf (1 g containing 1 mmole of cysteine) to react with glycine benzyl ester (0.5 mmole) in DMF (20 ml) with stirring for 5–8 hr at room temperature. The polymer was removed by centrifugation and washed with DMF, and the combined DMF solutions were evaporated to dryness *in vacuo*. The residue was dissolved in wet ethyl acetate and washed with 1 N HCl, 5% NaHCO$_3$, and water. On evaporation of the ethyl acetate a chromatographically pure solid product was obtained; yield 98%, based on the amount of glycine benzyl ester employed; mp 116–118°, $[\alpha]^{25}$D −45.3° (c 1.3, DMF) [lit. mp 118–119°,[3] $[\alpha]^{25}$D −45.5° (c 2.0, DMF)[3]]. Z-L-Pro-Gly-OBz was obtained analogously from IIe and glycine benzyl ester; yield 90%, mp 87–88° (lit. mp 88–89°[4]). Z-L-Ileu-(L-Ala)$_5$-PNB was obtained by coupling IId with H$_2$N-(L-Ala)$_5$-PNB in DMF at room temperature; yield 71%, mp 258–260°, $[\alpha]^{25}$D −91.1° (c 0.34, dichloroacetic acid). An identical peptide was obtained by the DCC method; mp 261–265°, $[\alpha]^{25}$D −90.5° (c 0.46, dichloroacetic acid). N,S-Di-Z-glutathione dibenzyl ester (N-Z-L-Glu-(α-OBz)-L-Cys-(-S-Z)-Gly-OBz), mp 156°, $[\alpha]^{25}$D −35.5° (c 1.0, methanol) [lit. mp 158–159°,[3] $[\alpha]^{25}$D −35.5° (c 1.0, methanol)[3]] was obtained in 85% yield from IIg and S-Z-L-Cys-Gly-OBz. The latter was derived from the N,S-di-Z-L-Cys-Gly-OBz preparation obtained from IIf and Gly-OBz on treatment with HBr in glacial acetic acid. Treating L-alanine *p*-nitrobenzyl ester with excess IIc, removal of the Z-group (by HBr-CH$_3$COOH) from the dipeptide formed (Z-L-Phe-L-Ala-PNB), neutralization with triethylamine in DMF, and treating the dipeptide ester with a new excess of IIc gave Z-L-Phe-L-Phe-L-Ala-PNB in an over-all 84% yield; mp 159°, $[\alpha]^{25}$D −26.8° (c 0.9, DMF). The same peptide prepared by the DCC method gave mp 160°, $[\alpha]^{25}$D −26.2° (c 1.39, DMF). Z-L-Phe-L-Phe-L-Phe-NH$_2$ was prepared analogously in 81% yield, mp 227°, $[\alpha]^{25}$D −31.5° (c 0.8, DMF). The same peptide prepared by the DCC method gave mp 226°, $[\alpha]^{25}$D −32.1°(c 1.2, DMF).

A comparison of the method for peptide synthesis described here with that of Merrifield[5] reveals that whereas in the "solid-phase peptide synthesis"[5] it is the peptide which is bound to the insoluble carrier and the N-blocked amino acid active ester is added while in solution, in our case a solution of free peptide ester is added to an insoluble N-blocked amino acid active ester. Furthermore, purification of the intermediate peptides formed during synthesis can be readily effected in our method, since these peptides are liberated into solution. In the Merrifield synthesis, on the other hand, peptide purification can be carried out only after detachment of the final product from the polymeric carrier. In the procedure described here the reaction between peptide ester and active amino acid ester can be carried out in the presence of a large excess of insoluble active ester which can be readily removed at the end of the reaction. It is thus possible to increase yields and to shorten the time of the coupling reaction.

The application of the new method for the synthesis of various low and high molecular weight peptides is under investigation.

Department of Biophysics, The Weizmann Institute of Science
Rehovoth, Israel
Received April 29, 1966

---

(3) M. Sokolovsky, M. Wilchek, and A. Patchornik, *J. Am. Chem. Soc.*, **86**, 1202 (1964).
(4) J. Kurtz, unpublished results.
(5) R. B. Merrifield, *J. Am. Chem. Soc.*, **85**, 2149 (1963); R. B. Merrifield, *Biochemistry*, 3, 1385 (1964); R. B. Merrifield, *J. Org. Chem.*, **29**, 3100 (1964); G. R. Marshall and R. B. Merrifield, *Biochemistry*, **4**, 2394 (1965).

Copyright © 1968 by the American Chemical Society

Reprinted from *J. Amer. Chem. Soc.*, **90**(11), 2953–2957 (1968)

Use of Polymers as Chemical Reagents. II. Synthesis of Bradykinin

Mati Fridkin, Abraham Patchornik, and Ephraim Katchalski

Contribution from the Department of Biophysics, The Weizmann Institute of Science, Rehovoth, Israel. Received December 13, 1967

Abstract: Coupling of cross-linked poly-4-hydroxy-3-nitrostyrene by the dicyclohexylcarbodiimide method with the following N-blocked amino acids or peptide: benzyloxycarbonyl-L-phenylalanine, benzyloxycarbonyl-L-proline, N-o-nitrophenylsulfenyl-O-benzyl-L-serine, N-o-nitrophenylsulfenyl-L-phenylalanine, N-o-nitrophenylsulfenylglycine, N-o-nitrophenylsulfenyl-L-prolyl-L-proline, and benzyloxycarbonyl-nitro-L-arginine, yielded the corresponding polymeric insoluble active esters. When these were treated in dimethylformamide with a soluble amino acid or peptide ester, containing a free α-amino group, blocked peptides were formed. Removal of the N-blocking groups from the newly formed peptides and repetition of the coupling reaction with the appropriate polymeric active ester enabled the stepwise elongation of the peptide chain. The use of the insoluble active amino acid esters in excess ascertained the synthesis of the desired peptides in high yield and enabled the removal of excess of polymeric reagent by filtration or centrifugation. By the method developed it was possible to synthesize the blocked nonapeptide: benzyloxycarbonylnitro-L-arginyl-L-prolyl-L-prolylglycyl-L-phenylalanyl-O-benzyl-L-seryl-L-prolyl-L-phenylalanylnitro-L-arginine *p*-nitrobenzyl ester in 65% yield from nitro-L-arginine *p*-nitrobenzyl ester and the polymeric active esters listed above. Removal of the blocking groups by treatment with liquid HF followed by catalytic reduction yielded fully biologically active bradykinin (L-arginyl-L-prolyl-L-prolylglycyl-L-phenylalanyl-L-seryl-L-prolyl-L-phenylalanyl-L-arginine) in 39% yield.

The possible use of polymeric active esters of N-blocked amino acids in the synthesis of peptides has been illustrated in a previous communication.[1] An N-blocked amino acid ester of cross-linked poly-4-hydroxy-3-nitrostyrene (I) was treated in dimethylformamide with a soluble amino acid or peptide ester (II) containing a free α-amino group to give the blocked peptide III. Upon removal of the N-blocking group from the newly formed peptide it could be allowed to react with an insoluble active ester of another N-blocked amino acid. Repetition of this set of reactions led, by stepwise elongation of the peptide chain, to the formation of a peptide with the desired amino acid sequence (Scheme I). The use of the insoluble active amino acid ester in excess ascertained the synthesis of the desired peptides in high yield (85–100%). The excess of the polymeric reagent could be quantitatively removed by filtration or centrifugation. The C- and N-blocked peptide (III) in the filtrate could be readily purified by extraction of the unreacted amino acid or peptide ester (II) with aqueous acid, after evaporation of solvent and

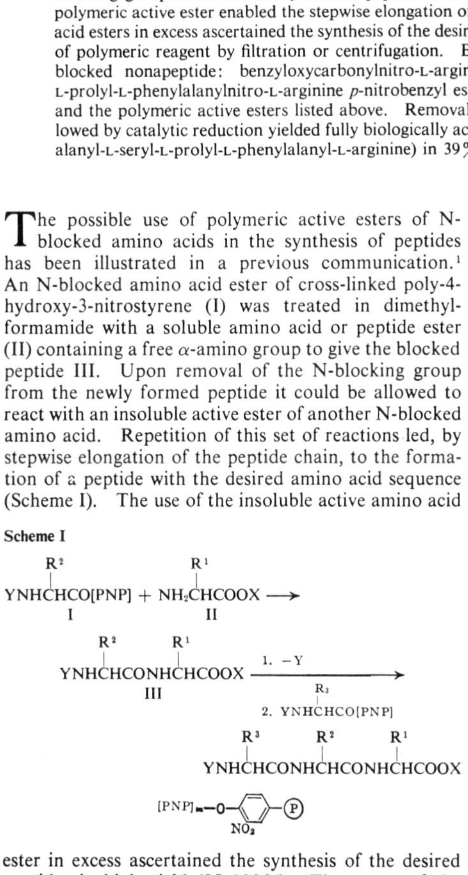

Scheme I

$$YNHCHCO[PNP] + NH_2CHCOOX \longrightarrow$$
$$\overset{R^2}{|}\overset{R^1}{|}$$
$$III$$

$$\underset{III}{YNHCHCONHCHCOOX} \xrightarrow[2.\ YNHCHCO[PNP]]{1.\ -Y}$$

$$YNHCHCONHCHCONHCHCOOX$$

$$[PNP] = -O-\!\!\bigcirc\!\!-\!\!(P)$$
$$NO_2$$

dissolution in ethyl acetate. In our previous communication[1] we described the synthesis by the new method of several dipeptides and tripeptides. In the present article we describe the use of polymeric active esters of N-blocked amino acids in the synthesis of the nonapeptide hormone, bradykinin.

Bradykinin has been synthesized by classical techniques[2–5] as well as by the recently developed solid-phase method.[6] Our synthesis of the nonapeptide with the aid of the corresponding active amino acid esters of the insoluble poly-4-hydroxy-3-nitrostyrene is summarized in Scheme II. Nitro-L-arginine *p*-nitrobenzyl ester was coupled in dimethylformamide with excess polymeric insoluble active ester of benzyloxycarbonyl-L-phenylalanine (Z-Phe-[PNP]) to yield the dimethylformamide-soluble benzyloxycarbonyl-L-phenylalaninenitro-L-arginine *p*-nitrobenzyl ester (IV). Removal of its benzyloxycarbonyl-protecting group with HBr in glacial acetic acid and neutralization with triethylamine yielded the dipeptide ester L-phenylalanylnitro-L-arginine *p*-nitrobenzyl ester (IVb). Compound IVb was coupled with Z-Pro-[PNP] to give the dimethylformamide-soluble protected tripeptide benzyloxycarbonyl-L-prolyl-L-phenylalanylnitro-L-arginine *p*-nitrobenzyl ester (V). Removal of the benzyloxycarbonyl group and neutralization with triethylamine as above yielded the tripeptide ester L-prolyl-L-phenylalanylnitro-L-arginine *p*-nitrobenzyl ester (Vb). Compound Vb was coupled in dimethylformamide with excess polymeric insoluble active ester of *o*-nitrophenylsulfenyl-O-benzyl-L-serine (NPS-Ser(Bzl)-[PNP]) to give *o*-nitrophenylsulfenyl-O-benzyl-L-seryl-L-prolyl-L-

(1) M. Fridkin, A. Patchornik, and E. Katchalski, *J. Am. Chem. Soc.*, **88**, 3164 (1966).
(2) R. A. Boissonnas, S. Guttmann, and P. A. Jaquenoud, *Helv. Chim. Acta*, **43**, 1349 (1960).
(3) E. D. Nicolaides and H. A. Dewald, *J. Org. Chem.*, **26**, 3872 (1961).
(4) S. Guttmann, J. Pless, and R. A. Boissannas, *Helv. Chim. Acta*, **45**, 170 (1962).
(5) S. Sakakibara and N. Inukai, *Bull. Chem. Soc. Japan*, **39**, 1567 (1966).
(6) R. B. Merrifield, *Biochemistry*, **3**, 1385 (1964).

Scheme II

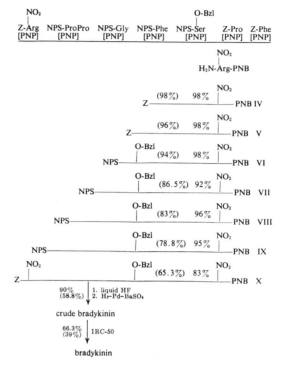

phenylalanylnitro-L-arginine *p*-nitrobenzyl ester (VI). Removal of the *o*-nitrophenylsulfenyl-protecting group with 1 N HCl in methanol and neutralization gave the tetrapeptide ester O-benzyl-L-seryl-L-prolyl-L-phenylalanylnitro-L-arginine *p*-nitrobenzyl ester (VIb). Coupling of VIb with NPS-Phe-[PNP], isolation of the dimethylformamide-soluble blocked pentapeptide VII, and removal of the *o*-nitrophenylsulfenyl-protecting group gave the pentapeptide ester L-phenylalanyl-O-benzyl-L-seryl-L-prolyl-L-phenylalanylnitro-L-arginine *p*-nitrobenzyl ester (VIIb). Coupling of VIIb with NPS-Gly-[PNP] and treatment of the product liberated as above yielded glycyl-L-phenylalanyl-O-benzyl-L-seryl-L-prolyl-L-phenylalanylnitro-L-arginine *p*-nitrobenzyl ester (VIIIb). The hexapeptide ester VIIIb was treated with NPS-Pro-Pro-[PNP] to give *o*-nitrophenylsulfenyl-L-prolyl-L-prolylglycyl-L-phenylalanyl-O-benzyl-L-seryl-L-prolyl-L-phenylalanylnitro-L-arginine *p*-nitrobenzyl ester (IX). The *o*-nitrophenylsulfenyl-protecting group was removed as usual and the free ester was treated with Z-Arg(NO$_2$)-[PNP] to give the blocked derivative of bradykinin, benzyloxycarbonylnitro-L-arginyl-L-prolyl-L-prolyl-glycyl-L-phenylalanyl-O-benzyl-L-seryl-L-prolyl-L-phenylalanylnitro-L-arginine *p*-nitrobenzyl ester (X), in an over-all yield of 65%. Removal of the benzyloxycarbonyl-, nitro- and benzyl-protecting groups was effected with liquid HF.[7] The remaining protecting ester group (*p*-nitrobenzyl) was subsequently removed by catalytic hydrogenation. The crude nonapeptide thus obtained was purified on an IRC-50 ion-exchange resin to yield a bradykinin preparation possessing full biological activity in 39% yield.[8,9] The material obtained showed chromatographic and electrophoretic behavior identical with that of an authentic sample of bradykinin.

In the stepwise synthesis of bradykinin described above, terminal *o*-amino groups were blocked in the first two steps by benzyloxycarbonyl groups and thereafter by *o*-nitrophenylsulfenyl groups. The guanidino group of arginine was masked by a nitro group, whereas the OH of serine was blocked by a benzyl group. The *p*-nitrobenzyl group was used as a protecting group of the terminal carboxyl. The use of *o*-nitrophenylsulfenyl as a reversible masking group in the different polymeric active amino acid esters, starting with O-benzyl-L-serine (see Scheme II), was found necessary since removal of the benzyloxycarbonyl group with HBr in glacial acetic acid leads to partial debenzylation of O-benzyl-L-serine accompanied by acetylation.[10] The hydroxyl of the side chain of serine was blocked with a benzyl group since preliminary experiments have shown that when synthesis with polymeric active esters is carried out with an unmasked serine, marked losses in serine content occur during the removal of the *o*-nitrophenylsulfenyl-protecting groups with methanolic hydrochloric acid. The nitro group was chosen as a protecting group of the guanidine moiety of arginine because of its quantitative

(7) S. Sakakibara, Y. Shimonishi, Y. Kishida, M. Okada, and H. Sugihara, *Bull. Chem. Soc. Japan*, **40**, 2164 (1967).
(8) G. P. Lewis, *Nature*, **192**, 596 (1961).
(9) H. Edery, *Brit. J. Pharmacol.*, **24**, 485 (1965).
(10) S. Guttmann and R. A. Boissonnas, *Helv. Chim. Acta*, **41**, 1852 (1958).

removal by treatment with liquid HF.[7] The *p*-nitrobenzyl masking group of the terminal carboxyl was retained on removal of the benzyloxycarbonyl- and the *o*-nitrophenylsulfenyl-protecting groups with HBr in glacial acetic acid and methanolic hydrochloric acid, respectively. Since treatment with liquid HF did not remove the *p*-nitrobenzyl group from the blocked nonapeptide it was necessary to unmask the carboxyl by catalytic hydrogenation to obtain the final product.

All of the polymeric active esters used in the synthesis of bradykinin, listed in Table I, were prepared by coupling the corresponding N-blocked amino acids or N-blocked peptide (NPS-Pro-Pro) with cross-linked poly-4-hydroxy-3-nitrostyrene, after activation by the dicyclohexylcarbodiimide (DCC) method.[11,12] The insoluble polymeric reagents obtained contained 1.0–2.0 mmoles of activated amino acid/g of polyester. Dimethylformamide and acetonitrile were found to be most useful solvents for the coupling reaction as they cause marked swelling of the polymer carrier. The amount of covalently bound amino acid or peptide in a polymeric active ester could be readily assayed by: (a) the increase in weight of the insoluble polymer as a result of the coupling reaction; (b) spectrophotometric determination of the amount of N-blocked amino acid liberated on short treatment with sodium methoxide, or on exhaustive acid hydrolysis; and (c) elemental analysis. Sulfur analysis was of particular use for the *o*-nitrophenylsulfenyl derivatives, as well as for the polymeric reagents containing cysteine and methionine. All three methods yielded similar results within the limits of experimental error.

The coupling reactions with polymeric active esters summarized in Scheme II were carried out in dimethylformamide for 8–12 hr, at room temperature. A fourfold molar excess of the polymeric reagent was used throughout. The purity of all of the intermediate peptides synthesized was checked on thin layer chromatography using at least two different solvent systems. The intermediate peptides were obtained in high yield (83–98%). Amino acid analysis of the peptides synthesized was carried out after exhaustive acid hydrolysis. A molar residue ratio close to the theoretical was found in all cases.

A comparison of the method for peptide synthesis described here with that of Merrifield[13] reveals that whereas in the "solid-phase peptide synthesis" it is the peptide which is bound to the insoluble carrier and the N-blocked amino acid active ester is added while in solution, in our case a solution of free peptide ester is added to an insoluble N-blocked amino acid active ester. Furthermore, purification of the intermediate peptides during synthesis can be readily effected in our method since these peptides are liberated into solution. In the Merrifield synthesis, on the other hand, peptide purification can be carried out only after detachment of the final product from the polymeric carrier.

Finally it is pertinent to note that in the new technique described, one can carry out the reaction between peptide ester and active amino acid ester in the presence of a large excess of insoluble active ester without markedly contaminating the reaction mixture, as the excess of the insoluble active ester can be readily removed by filtration or centrifugation. It is the excess of the polymeric amino acid active ester which ascertains the high yields of the peptides synthesized. The use of the amino acid polymer reagents in peptide synthesis obviously requires that the reacting amino acid or peptide esters be soluble in the solvents used. One might, therefore, expect that the new method developed will be of particular use in the synthesis of peptides, the blocked derivatives of which dissolve readily in the suitable organic solvents.

Experimental Section[14]

Synthesis of Cross-Linked Poly-4-hydroxy-3-nitrostyrene. 4-Acetoxystyrene and divinylbenzene, at a molar ratio of 100:4, were copolymerized according to the literature,[16] and the cross-linked polymer obtained (10 g) was nitrated by stirring with concentrated nitric acid (density 1.42, 200 ml) for 5 hr at 25°. The yellow suspension was poured into cold, distilled water (500 ml), and the nitrated polymer was filtered and washed with distilled water until the washings were neutral. The wet polymer was washed in addition with methanol and ether and dried *in vacuo* over sodium hydroxide pellets; yield, 12 g. The dried polymer gave on analysis N, 8.9; CH_3CO, 4.3. Removal of the remaining acetyl groups was effected by treatment with ammonia. The polymer (12 g) was suspended in dimethylformamide (100 ml) and concentrated aqueous ammonia (13.7 *M*, 5 ml) was added to the gel formed. The reaction mixture was kept at room temperature for 2 hr and poured into cold water with stirring, and the suspension formed was acidified with 3 *N* hydrochloric acid. The cross-linked nitro polymer was finally filtered, washed with water, methanol, and ether, and dried *in vacuo* over sodium hydroxide pellets; yield of brown polymer, 11.9 g.

Anal. Calcd for a 4% cross-linked (divinylbenzene) poly-4-hydroxy-3-nitrostyrene: N, 8.2; CH_3CO, 0.0. Found: N, 9.0; CH_3CO, 0.29.

N-Benzyloxycarbonyl-L-phenylalanyl Polymer (Z-Phe-[PNP]). Benzyloxycarbonyl-L-phenylalanine (12 g, 40 mmoles) in dimethylformamide (15 ml) or in acetonitrile (25 ml) was added to a suspension of cross-linked poly-4-hydroxy-3-nitrostyrene (5 g, containing 5.9 mmoles of OH groups/g) in dimethylformamide (50 ml) or in acetonitrile (30 ml). The reaction mixture was stirred for 10 min at 0°, whereupon DCC (8.24 g, 40 mmoles) in dimethylformamide (15 ml) or in acetonitrile (20 ml) was added. The reaction mixture was stirred for 1 hr at 0°, followed by 5 hr at room temperature. The Z-Phe-[PNP] formed was filtered and washed with three portions of hot methanol (250 ml each) and with ether; yield after drying *in vacuo*, 8.75 g. The Z-Phe-[PNP] synthesized contained 1.50 mmoles of N-benzyloxycarbonyl-L-phenylalanine/g, as determined by the various assays described below.

Polymer-Active Esters of Other N-Blocked Amino Acids. High molecular weight insoluble active esters of different N-blocked amino acids and peptides used in the synthesis of bradykinin were obtained by the coupling procedure given above for the preparation of Z-Phe-[PNP]. These are listed in Table I. It should be noted that under the experimental conditions of coupling used, the polymer was found to bind also *o*-nitrophenylsulfenyl-L-prolyl-L-proline.

Assay of the Content of Acyl Groups in the Insoluble Polymer-Active Esters Synthesized. The content of the various acyl residues in the high molecular weight active esters given in Table I could be determined by the following methods: (a) determination of the increase in weight of the cross-linked poly-4-hydroxy-3-nitrostyrene as a result of the coupling reaction; (b) determination of the amounts of free amino acids liberated on exhaustive acid hydrolysis

---

(11) J. C. Sheehan and G. P. Hess, *J. Am. Chem. Soc.*, **77**, 1067 (1955).
(12) M. Fridkin, A. Patchornik, and E. Katchalski, *J. Am. Chem. Soc.*, **87**, 4646 (1965).
(13) R. B. Merrifield, *ibid.*, **85**, 2149 (1963).

(14) All melting points were taken on a capillary melting point apparatus and are uncorrected. Dimethylformamide was dried over molecular sieves (Fischer Scientific Co., type 5A) and fractionally distilled *in vacuo*. Thin-layer chromatography of the peptides synthesized was carried out on Merck's silica gel G. The following solvent system were used: system I, chloroform–methanol (9:1); system II, *n*-butyl alcohol–acetic acid–water (4:1:1); system III, *n*-butyl alcohol–pyridine–acetic acid: water (15:10:3:12); system IV, *n*-propyl alcohol–aqueous NH_4OH (13.7 *M*) (67:33). R_f values are uncorrected. The chlorine method[15] was used for the detection of N-blocked peptides.
(15) H. Zahn and E. Rexroth, *Z. Anal. Chem.*, **148**, 181 (1955).
(16) D. I. Packham, *J. Chem. Soc.*, 2617 (1964).

Synthesis of Bradykinin

Table I. N-Blocked Amino Acid Esters of Cross-Linked Poly-4-hydroxy-3-nitrostyrene

Compound bound to polymer[a]	mmoles of amino acid or peptide bound per g of polyester[b]
Z-Phe	1.5
NPS-Phe	1.5
NPS-Gly	1.4
Z-Pro	1.4
NPS-Ser | Bzl	1.3
NPS-Pro-Pro	1.0
Z-Arg | NO₂	0.9

[a] Binding was effected in dimethylformamide by the DCC method.[11,12] [b] Assayed by the analytical methods described in the Experimental Section.

(The insoluble active ester (20–30 mg) was suspended in a mixture of acetic acid (1 ml) and 12 N hydrochloric acid (1 ml), and the suspension was refluxed for 24 hr. The insoluble polymer was filtered, and the amino acid content in the filtrate was determined by quantitative paper chromatography with the aid of the amino acid analyzer, Beckman–Spinco Model 120C); (c) spectrophotometric determination of the benzyloxycarbonyl group in an alkaline hydrolysate (The insoluble active ester (20–30 mg) was suspended in ethanolic sodium hydroxide (15 ml, consisting of 5 ml of aqueous 1 N NaOH and 10 ml of ethanol), and the mixture shaken for 2 min at room temperature. The insoluble polymer was filtered and the amount of benzyloxycarbonyl in the filtrate was derived from optical density measurements at 257 mμ. The molar extinction coefficient of this group is ϵ_{257} 200); and (d) sulfur content. The content of o-nitrophenylsulfenyl groups or of sulfur-containing amino acids could be derived from elemental sulfur analysis.[17]

General Procedure for the Preparation of Blocked Peptides. The required amino acid ester or peptide ester (0.5–1.0 mmole) in dimethylformamide (5 ml) was added to one of the acyl polymers listed in Table I (2 g, containing 2.0–4.0 mmoles of the acyl derivative) suspended in dimethylformamide (20 ml) in the form of highly swollen particles, and the reaction mixture was stirred for 8 hr at room temperature. The polymer was filtered and washed twice with 30-ml portions of dimethylformamide, and the filtrate and washings were combined and evaporated at low pressure (1 mm). The residue left was dissolved in wet ethyl acetate (150 ml), and the solution was washed with 1 N hydrochloric acid, 5% aqueous NaHCO₃, and water. Drying was effected with anhydrous sodium sulfate. The required, solid, pure product was obtained as a rule after evaporation of solvent *in vacuo*.

Benzyloxycarbonyl-L-phenylalanylnitro-L-arginine p-Nitrobenzyl Ester (IV). Nitro-L-arginine p-nitrobenzyl ester (354 mg, 1 mmole) in dimethylformamide (5 ml) was added to a suspension of Z-Phe-[PNP] (2.66 g containing 4 mmoles of phenylalanine in dimethylformamide (30 ml) and the reaction was stirred for 8 hr at room temperature. The mixture was then treated according to the general procedure above for the preparation of blocked peptides. Benzyloxycarbonyl-L-phenylalanylnitro-L-arginine p-nitrobenzyl ester (620 mg, 98%) was obtained as a colorless powder after trituration with petroleum ether (bp 30–60°): mp 169°, $[\alpha]^{25}$D −11° (c 0.4, methanol); lit.[17] mp 174°, $[\alpha]^{22}$D −9.9 ± 1° (c 1.0 methanol); R_{fI} 0.83, R_{fIII} 0.79.

Benzyloxycarbonyl-L-prolyl-L-phenylalanylnitro-L-arginine p-Nitrobenzyl Ester (V). L-Phenylalanylnitro-L-arginine p-nitrobenzyl ester hydrobromide was obtained on treatment of IV (620 mg) with 35% HBr in glacial acetic acid (2 ml) for 30 min at room temperature and precipitation with ether (750 ml). The hydrobromide was dissolved in dimethylformamide (5 ml) and neutralized with triethylamine, and the solution was added to a suspension of Z-Pro-[PNP] (2.85 g containing 4 mmoles of proline in dimethylformamide (30 ml). The mixture was stirred for 8 hr at room temperature and treated as above. Benzyloxycarbonyl-L-propyl-L-phenylalanylnitro-L-arginine p-nitrobenzyl ester (700 mg) was obtained in 98% yield: mp 112° dec; lit.[2] mp 115° dec; R_{fII} 0.90,

(17) W. Schöniger, *Z. Anal. Chem.*, **181**, 28 (1961).

R_{fIII} 0.89, R_{fIIII} 0.90. Compound V yielded L-prolyl-L-phenylalanylnitro-L-arginine p-nitrobenzyl ester hydrobromide (Va) on treatment with HBr in glacial acetic acid as above: R_{fII} 0.61, R_{fIV} 0.85. The tripeptide ester gave on acid hydrolysis (6 N HCl, 22 hr) a molar amino acid ratio of Arg, 0.90; Phe, 1.00; and Pro, 1.10.

N-o-Nitrophenylsulfenyl-O-benzyl-L-seryl-L-prolyl-L-phenylalanylnitro-L-arginine p-Nitrobenzyl Ester (VI). The coupling of L-prolyl-L-phenylalanylnitro-L-arginine p-nitrobenzyl ester, derived from Va on neutralization with triethylamine, with NPS-Ser-(Bzl)-[PNP] (3.1 g containing 4 mmoles of serine) and treatment of the product formed were carried out according to the general procedure described. Compound VI was obtained in semisolid form: yield, 870 mg (98%); R_{fI} 0.86, R_{fII} 0.91, R_{fIII} 0.93. O-Benzyl-L-seryl-L-prolyl-L-phenylalanylnitro-L-arginine p-nitrobenzyl ester hydrochloride (VIa) was obtained on removal of the o-nitrophenylsulfenyl-protecting group with 1 N HCl in methanol (3 ml, 5–10 min at room temperature) and precipitation with ether (750 ml); R_{fIII} 0.63, R_{fIV} 0.76. A molar amino acid ratio of Arg, 1.00; Phe, 1.04; Pro, 1.00; and Ser, 0.85 was found after acid hydrolysis (6 N HCl, 22 hr).

o-Nitrophenylsulfenyl-L-phenylalanyl-O-benzyl-L-seryl-L-prolyl-L-phenylalanylnitro-L-arginine p-Nitrobenzyl Ester (VII). The coupling of O-benzyl-L-seryl-L-prolyl-L-phenylalanylnitro-L-arginine p-nitrobenzyl ester, derived from VIa on neutralization with triethylamine, with NPS-Phe-[PNP] (2.66 g, containing 4 mmoles of phenylalanine) and treatment of the product formed was carried out as usual. Compound VII was obtained as a yellow powder: yield, 930 mg (92%); R_{fI} 0.84, R_{fIII} 0.89, R_{fIIII} 0.91. L-Phenylalanyl-O-benzyl-L-seryl-L-prolyl-L-phenylalanylnitro-L-arginine p-nitrobenzyl ester hydrochloride (VIIa) was obtained on removal of the o-nitrophenylsulfenyl-protecting group with 1 N HCl in methanol (3 ml) and precipitation with ether: R_{fIII} 0.59, R_{fIIII} 0.67. The molar amino acid ratio was Arg, 1.00; Phe, 2.36; Pro, 0.96; and Ser, 1.00.

o-Nitrophenylsulfenylglycyl-L-phenylalanyl-O-benzyl-L-seryl-L-prolyl-L-phenylalanylnitro-L-arginine p-nitrobenzyl Ester (VIII). The coupling of L-phenylalanyl-O-benzyl-L-seryl-L-prolyl-L-phenylalanylnitro-L-arginine p-nitrobenzyl ester, derived from VIIa, with NPS-Gly-[PNP] (2.85 g, containing 4 mmoles of glycine) and purification of the product formed was carried out as usual. Compound VIII was obtained as a yellow powder: yield, 938 mg (96%); R_{fI} 0.80, R_{fII} 0.82, R_{fIIII} 0.84. Glycyl-L-phenylalanyl-O-benzyl-L-seryl-L-prolyl-L-phenylalanylnitro-L-arginine p-nitrobenzyl ester hydrochloride (VIIIa) was obtained on removal of the o-nitrophenylsulfenyl-protecting group: R_{fIII} 0.61, R_{fIIII} 0.63.

o-Nitrophenylsulfenyl-L-prolyl-L-prolylglycyl-L-phenylalanyl-O-benzyl-L-seryl-L-prolyl-L-phenylalanylnitro-L-arginine p-Nitrobenzyl Ester (IX). The coupling of glycyl-L-phenylalanyl-O-benzyl-L-seryl-L-prolyl-L-phenylalanylnitro-L-arginine p-nitrobenzyl ester, derived from VIIIa, with NPS-Pro-Pro-[PNP] (4 g, containing 4 mmoles of prolylproline) and purification of the product formed was carried out as usual: yield of oily yellow product, 1040 mg (95%); R_{fI} 0.88, R_{fIII} 0.79, R_{fIIII} 0.85; after crystallization from acetonitrile–ether, mp 118–125°, $[\alpha]^{25}$D −72.1° (c 1.17, dimethylformamide). L-Prolyl-L-prolylglycyl-L-phenylalanyl-O-benzyl-L-seryl-L-prolyl-L-phenylalanylnitro-L-arginine p-nitrobenzyl ester hydrochloride (IXa) was obtained as a solid powder after removal of the o-nitrophenylsulfenyl-protecting group with 1 N HCl in methanol (3 ml), precipitation with ether, and trituration with isopropyl alcohol and ether: R_{fII} 0.61, R_{fIIII} 0.68. The molar amino acid ratio was Arg, 1.00; Phe, 2.00; Pro, 3.00; Ser, 0.92; and Gly, 1.10.

Benzyloxycarbonylnitro-L-arginyl-L-prolyl-L-prolylglycyl-L-phenylalanyl-O-benzyl-L-seryl-L-prolyl-L-phenylalanylnitro-L-arginine p-Nitrobenzyl Ester (X). The coupling of L-prolyl-L-prolylglycyl-L-phenylalanyl-O-benzyl-L-seryl-L-prolyl-L-phenylalanylnitro-L-arginine p-nitrobenzyl ester, derived from IXa, with Z-Arg-(NO₂)-[PNP] (4.4 g, containing 4 mmoles of arginine) and purification of the product formed were carried out as usual. The oily product obtained was crystallized from acetonitrile–ether: yield, 985 mg (83%); R_{fII} 0.75, R_{fIIII} 0.85; mp 110–125°, $[\alpha]^{25}$D −54.1° (c 1.08, dimethylformamide). Compound X gave on acid hydrolysis (6 N HCl, 22 hr) a molar amino acid ratio of Arg, 2.07; Phe, 2.00; Pro, 3.00; Ser, 0.94; and Gly, 1.00.

L-Arginyl-L-prolyl-L-prolylglycyl-L-phenylalanyl-L-seryl-L-prolyl-L-phenylalanyl-L-arginine (Bradykinin). The benzyloxycarbonyl-, nitro-, and benzyl-protecting groups of X were removed with anhydrous hydrofluoric acid according to Sakakibara.[7] Compound X (50 mg) was placed in a polyethylene test tube to which anisole

(0.05 ml) and anhydrous HF (1.5 ml) were added. The colorless solution turned brown-red within several minutes. It was kept at room temperature for 60 min whereupon the hydrogen fluoride was evaporated in a stream of dry nitrogen gas. The residue was triturated with dry ether and the solvent was removed by centrifugation. The powder which had sedimented was dried under vacuum (1 mm).

Final removal of the *p*-nitrobenzyl-protecting group was effected by catalytic hydrogenation. The dry powder was dissolved in acetic acid (90%, 5 ml), 100 mg of 5% Pd on BaSO$_4$ was added, and the hydrogenation was carried out for 2 days at atmospheric pressure. An additional amount of catalyst (100 mg) was added and the hydrogenation was continued for another 2 days. The yellow solution became gradually colorless during the reduction. The catalyst was then filtered and washed with 85% acetic acid. The filtrate and washings were combined and evaporated to dryness under vacuum (1 mm). The residue was dissolved in a few milliliters of water and lyophilized to give a colorless solid preparation of crude bradykinin (48 mg). The crude nonapeptide yielded on acid hydrolysis (6 *N* HCl, 22 hr) a molar amino acid ratio of Arg, 2.05; Phe, 1.98; Pro, 2.98; Ser, 0.96; and Gly, 1.00. The homogeneity of the product obtained was checked by paper electrophoresis (pH 1.5, 3.5, and 6.5) and by paper chromatography using *n*-butyl alcohol-acetic acid-water (4:1:1), 80% pyridine, or 1-propyl alcohol-water (2:1) as developers. Comparison of the patterns obtained, using ninhydrin and Sagakuchi reagents, respectively, with the corresponding patterns given by an authentic sample of bradykinin (kindly supplied by Dr. Sakakibara) revealed that the crude bradykinin synthesized contained traces of arginine and *p*-toluidine and probably traces of unreduced nonapeptide ester. Final purification was effected on IRC-50 ion-exchange resin which had been equilibrated with 1 *M* acetic acid according to Merrifield.[6]

The pure bradykinin obtained (31 mg) yielded on acid hydrolysis: Arg, 2.12; Phe, 1.85; Pro, 3.05; Ser, 0.90; and Gly, 1.10. Its biological activity as assayed by the guinea pig ileum contraction test,[8] or by the Edery sensitization test,[9] was found identical with that of an authentic sample of the biologically active nonapeptide.

Acknowledgment. The authors express their gratitude to Mrs. S. Ehrlich-Rogozinsky for the microanalyses, to Mr. I. Jacobson for technical help, and to Dr. H. Edery for the biological assays.

SYNTHESIS OF CYCLIC PEPTIDES ON DUAL FUNCTION SUPPORTS*

E. Flanigan** and G. R. Marshall
Departments of Physiology and Biophysics and of Biological Chemistry,
Washington University School of Medicine, St. Louis, Missouri 63110

(Received in USA 9 March 1970; received in UK for publication 15 May 1970)

There are two problems encountered in the synthesis of cyclic peptides [1]. The open chain must be built up from its constituent units. The linear chain is then activated, usually at the carboxyl end, and cyclization is achieved through intramolecular amide-bond formation. Yields in this step are frequently low due to the formation of linear oligomers even when conditions of high dilution are used.

Polymeric supports have been demonstrated to be extremely useful in peptide synthesis both for carboxyl protection as exemplified by the work of Merrifield [2] and for activation as shown by Fridkin [3] and Laufer [4]. Fridkin, using a 4-hydroxy-3-nitrostyrene polymer was able to cyclize peptides in high yields. The modification of a support during the synthesis to change its functional characteristics such as lability is now reported.

The 4-(methylthio)phenyl (MTP) group of Johnson and Jacobs [5] has been incorporated into two polymers, IA and IB, (see Fig. 1). Polymer IA was prepared by polymerization of 4-(methylthio)phenol with a 1.5 molar excess of formaldehyde. The mixture was refluxed 14 hours in the presence of a catalytic amount of 6N NaOH. The resulting brown amorphous solid showed characteristic bands in the infrared at 3450 cm^{-1} (b) and 750 cm^{-1} (s) due to the phenolic hydroxy moiety and aromatic ring, respectively. Polymer IB (R=H, R_1=C_6H_4-Polymer) was prepared [6] by the

* Supported by NIH Grants AM13025 and GM714.
** Predoctoral Trainee in Molecular Biology, 1967 - present.

reaction of 40 mmoles of p-mercaptophenol and an equivalent amount of KOH with 10 gms of chloromethylpolystyrene-2%-divinylbenzene (0.9 meq Cl/gm). The mixture was refluxed in dimethyformamide [7] (DMF) for 4 hours. Bands in the infrared were found at 3500 cm^{-1} (s), 1600 (m) and 750 cm^{-1} (s) due, again, to the phenolic hydroxy and aromatic moieties. Both polymers were insoluble in all solvents commonly used in peptide synthesis. The attractive feature of IA and IB is their facile conversion to the activated sulfonyl esters.

Polymeric MTP esters (see reaction 1) of Boc-α-amino acids were prepared by the dicyclohexylcarbodiimide (DDC), carbonyldiimidazole (CDI), mixed anhydride and N-ethoxycarboxyl-2-ethoxy-1,2-dihydroquinoline (EEDQ) methods using CH_2Cl_2 or tetrahydrofuran as solvents. When equimolar quantities of all reactants were used, the coupling efficiency was as follows: DCC>CDI>EEDQ> mixed anhydride.

Figure 1. Synthesis on MTP Supports

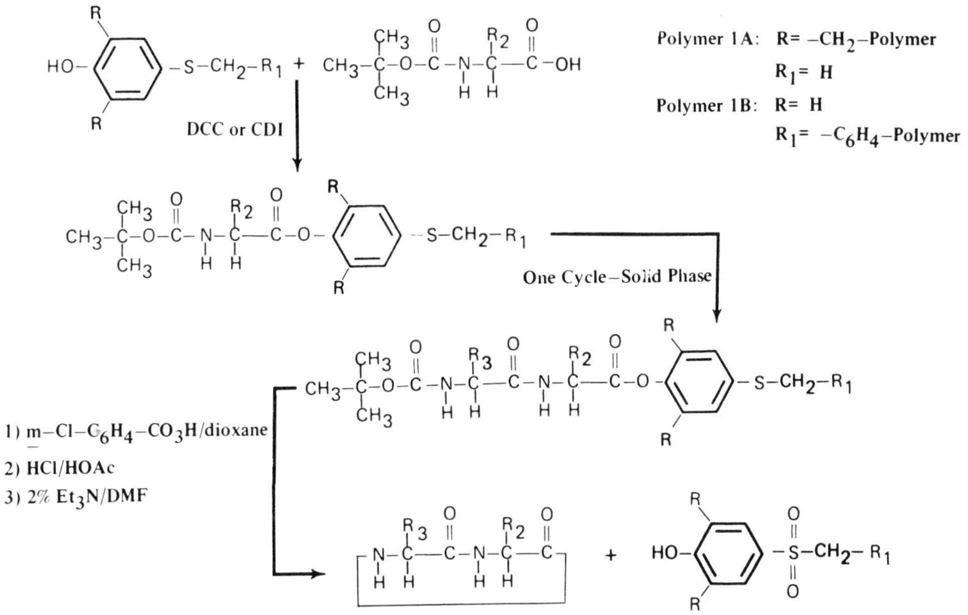

As shown in Fig. 1, chain elongation to desired lengths was carried out according to Merrifield [2] and Marshall and Merrifield [8]. The polymer was activated by oxidation to the sulfone using 3 equivalents of m-chloroperbenzoic acid [9] in dioxane. Complete conversion to the sulfone [I.R. bands at 1310 cm^{-1} (s) and 1138 cm^{-1} (s)] was accomplished in 24 hours. No loss of peptide was detected at this step. The terminal Boc-group was removed and the polymer was washed well with HOAc, EtOH and $MeCl_2$. The activated peptide ester was allowed to condense, intramolecularly, to the cyclic product by shaking the polymer 18 hours in 2% Et_3N/DMF. Evaporation of the solvent and re-crystallization of the solid from appropriate solvents yielded chromatographically pure cyclic peptides. The cyclic products gave ninhydrin negative tests in tlc, but were identified with hypochlorite-starch-KI [10]. Results of several preparations are shown in Table I. Low yields on IA may reflect the presence of the two ortho methylene groups or the degree of cross-linking in the polymer.

Table I. Cyclic Peptide Products*

Product	mp. °C.	R_f (I)	R_f (II)	% Yield IA	IB
c-(Ala-Gly)	238-239	.59	.46	23	52
c-(Gly-Gly)	308-309	.51	.35	40	63
c-(Gly-Phe)	261-265	.71	.66	21	42
c-(Ala-Ala-Ala-Gly)	279-280**	.57	.49	-	40
c-(Gly-Val-Ala-Phe-Ala-Gly)	209-210	.64	.58	-	50

* All melting points are uncorrected. Thin layer chromatography was carried out on Silica Gel G. Solvent Systems: (I) n-Butanol:HOAc:HOH (4:1:1); (II) sec-Butanol:EtOAc:HOH (65:60:25). Yields are based on amino acids initially esterified to the polymer. Amino acid analyses and molecular ions from mass spectrometry are consistent with the proposed structures.

** Decomposition

While the oxidative conditions of this method limits its application, a number of biologically-active cyclic peptides do not contain methionine, cysteine and cystine. The syntheses of valinomycin and antamanide are currently under investigation. The primary advantages of this system include higher overall yields, a minimization of racemization during activation and subsequent cyclization as well as the normal convenience of the solid phase method.

Literature

(1) M. Bodanszky and M. A. Ondetti, Peptide Synthesis, Interscience Publishers, New York, 1966.

(2) R. B. Merrifield, J. Am. Chem. Soc., 85, 2149 (1963). B. Gutte and R. B. Merrifield, ibid., 91, 501 (1969).

(3) M. Fridkin, A. Patchornik and E. Katchalski, J. Am. Chem. Soc., 87, 4646 (1965).

(4) D. A. Laufer, T. M. Chapman, D. I. Marlborough, V. M. Vaidya and E. R. Blout, J. Amer. Chem. Soc., 90, 2696 (1968).

(5) B. J. Johnson and P. M. Jacobs, Chem. Comm., 73 (1968).

(6) Preliminary work on one of these supports has been reported while this work was in progress. D. L. Marshall and I. E. Liener, Abstr. 158th. Natl. ACS Meeting, New York, 1969.

(7) J. R. Campbell, J. Org. Chem., 29, 1830 (1964).

(8) G. R. Marshall and R. B. Merrifield, Biochemistry, 4, 2394 (1965).

(9) B. J. Johnson, J. Org. Chem., 34, 1178 (1969). B. J. Johnson and T. A. Ruettinger, J. Org. Chem., 35, 255 (1970).

(10) D. E. Nitechi and J. W. Goodman, Biochemistry, 5, 665 (1966).

Synthesis of a Hydrophobic Potassium Binding Peptide[1]

B. F. Gisin and R. B. Merrifield

Contribution from The Rockefeller University, New York, New York 10021, and from the Department of Physiology, Duke University Medical Center, Durham, North Carolina 27710. Received March 22, 1972

Abstract: Based on the known structures of alkali ion complexing agents the design of a homodetic cyclopeptide that would be able to bind the potassium ion was undertaken. A molecular model of the peptide cyclo-[-L-Val-D-Pro-D-Val-L-Pro-]$_3$ exhibited that property as judged by the oxygen-lined cavity it could provide in one of its most probable conformations. The linear peptide was synthesized by the solid-phase method starting with L-proline at the C terminus. After cleavage from the resin and cyclization the neutral cyclododecapeptide was found to form a crystalline, hydrophobic 1:1 complex with potassium picrate.

Among the ligands for metal ions the so-called "ion carriers" constitute a group of compounds which are able to complex with alkali ions and make them soluble in nonpolar media. This quality has generated great interest mainly for three reasons. First, some of the ion carriers exhibit antibiotic activity[2-9] and they show dramatic effects on the ionic balance in mitochondria[10,11] and in red blood cells.[12,13] Second, they have been found to produce similar effects in lipid membranes[13,14] (lipid bilayers) and in other artificial systems related to membranes.[3,4,11,15-18] All of these properties appear to be a consequence of the ion complexing ability which is thought to be a necessary though not sufficient condition for activity in natural membranes.[4] Third, spectroscopic investigation of some of these compounds has contributed significantly to the understanding of conformational principles in molecules of biological origin.[3-5,15,19,20]

Alkali ion carriers of known structure from natural sources are depsipeptides (valinomycin,[21] the en-

---

(1) This investigation was supported by NIH Grant HE 12 157, U. S. Public Health Service Grant AM 1260, and by the Hoffman-La Roche Foundation. It was presented at the 163rd National Meeting of the American Chemical Society, Boston, Mass., April 1972.
(2) H. Brockmann and G. Schmidt-Kastner, *Chem. Ber.*, **88**, 57 (1955); C. E. Meyer and F. Reusser, *Experientia*, **23**, 85 (1967); K. Bevan, J. S. Davies, C. H. Hassall, R. B. Morton, and D. A. S. Phillips, *J. Chem. Soc. C*, 514 (1971).
(3) M. M. Shemyakin, Yu. A. Ovchinnikov, and V. T. Ivanov, *Angew. Chem.*, **81**, 53 (1969).
(4) M. M. Shemyakin, Yu. A. Ovchinnikov, V. T. Ivanov, V. K. Antonov, E. I. Vinogradova, A. M. Shkrob, G. G. Malenkov, A. V. Evstratov, I. A. Laine, E. I. Melnik, and I. D. Ryabova, *J. Membrane Biol.*, **1**, 402 (1969).
(5) Yu. A. Ovchinnikov, V. T. Ivanov, and I. I. Mikhaleva, *Tetrahedron Lett.*, 159 (1971).
(6) W. Mechlinski, C. P. Schaffner, P. Ganis, and G. Avitabile, *ibid.*, 3873 (1970).
(7) A. Agtarap, J. W. Chamberlin, M. Pinkerton, and L. K. Steinrauf, *J. Amer. Chem. Soc.*, **89**, 5737 (1967).
(8) L. K. Steinrauf, M. Pinkerton, and J. W. Chamberlin, *Biochem. Biophys. Res. Commun.*, **33**, 29 (1968).
(9) C. A. Maier and I. C. Paul, *Chem. Commun.*, 181 (1971).
(10) W. C. Murray and R. W. Begg, *Arch. Biochem. Biophys.*, **84**, 546 (1959); C. Moore and B. C. Pressman, *Biochem. Biophys. Res. Commun.*, **15**, 562 (1964); R. S. Cockrell, E. J. Harris, and B. C. Pressman, *Biochemistry*, **5**, 2326 (1966); H. Lardy, *Fed. Proc., Fed. Amer. Soc. Exp. Biol.*, **27**, 1278 (1968).
(11) B. C. Pressman, *Antimicrob. Ag. Chemother.*, 28 (1969).
(12) D. C. Tosteson, P. Cook, T. Andreoli, and M. Tieffenberg, *J. Gen. Physiol.*, **50**, 2513 (1967).
(13) D. C. Tosteson, *Fed. Proc., Fed. Amer. Soc. Exp. Biol.*, **27**, 1269 (1968).
(14) A. A. Lev and E. P. Buzhinsky, *Tsitologiya*, **9**, 102 (1967); P. Mueller and D. O. Rudin, *Biochem. Biophys. Res. Commun.*, **26**, 398 (1967); T. E. Andreoli, M. Tieffenberg, and D. C. Tosteson, *J. Gen. Physiol.*, **50**, 2527 (1967); P. Mueller and D. O. Rudin, *Nature (London)*, **217**, 713 (1968); A. Finkelstein and A. Cass, *J. Gen. Physiol.*, **51**, 145

(1968); G. Eisenman, S. M. Ciani, and G. Szabo, *Fed. Proc., Fed. Amer. Soc. Exp. Biol.*, **27**, 1289 (1968); P. Läuger and G. Stark, *Biochim. Biophys. Acta*, **211**, 458 (1970).
(15) M. M. Shemyakin, Yu. A. Ovchinnikov, V. T. Ivanov, V. K. Antonov, A. M. Shkrob, I. I. Mikhaleva, A. V. Estratov, and G. G. Malenkov, *Biochem. Biophys. Res. Commun.*, **29**, 834 (1967).
(16) M. Pinkerton, L. K. Steinrauf, and P. Dawkins, *ibid.*, **35**, 512 (1969).
(17) B. Dietrich, J. M. Lehn, and J. P. Sauvage, *Tetrahedron Lett.*, 2885, 2889 (1969); *Chem. Commun.*, 1055 (1970).
(18) D. C. Tosteson, T. E. Andreoli, M. Tieffenberg, and P. Cook, *J. Gen. Physiol.*, **51**, 373 (1968); Z. Stefanac and W. Simon, *Microchem. J.*, **12**, 125 (1967); L. A. R. Pioda, H. A. Wachter, R. E. Dohner, and W. Simon, *Helv. Chim. Acta*, **50**, 1373 (1967); H. K. Wipf, L. A. R. Pioda, Z. Stefanac, and W. Simon, *ibid.*, **51**, 377 (1968); H. K. Wipf, W. Pache, P. Jordan, H. Zähner, W. Keller-Schierlein, and W. Simon, *Biochem. Biophys. Res. Commun.*, **36**, 387, 1969; L. A. R. Pioda, H. K. Wipf, and W. Simon, *Chimia*, **22**, 189 (1968); R. P. Scholer and W. Simon, *ibid.*, **24**, 372 (1970); R. M. Izatt, J. H. Rytting, D. P. Nelson, B. L. Haymore, and J. J. Christensen, *Science*, **164**, 433 (1969); M. J. Hall, *Biochem. Biophys. Res. Commun.*, **38**, 590 (1970); M. S. Frant and J. W. Ross, Jr., *Science*, **167**, 987 (1970).
(19) V. T. Ivanov, I. A. Laine, N. D. Abdulaev, L. B. Senyavina, E. M. Popov, Yu. A. Ovchinnikov, and M. M. Shemyakin, *Biochem. Biophys. Res. Commun.*, **34**, 803 (1969); M. Ohnishi and D. W. Urry, *ibid.*, **36**, 194 (1969); D. F. Mayers and D. W. Urry, *J. Amer. Chem. Soc.*, **94**, 77 (1972).
(20) H. Diebler, M. Eigen, G. Ilgenfritz, G. Maas, and R. Winkler, *Pure Appl. Chem.*, **20**, 93 (1969); A. I. McMullen, *Biochem. J.*, **119**, 10 (1970); Th. Wieland, H. Faulstich, W. Burgermeister, W. Otting, W. Möhle, M. M. Shemyakin, Yu. A. Ovchinnikov, V. T. Ivanov, and G. G. Malenkov, *FEBS Lett.*, **9**, 89 (1970); V. T. Ivanov, A. I. Miroshnikov, N. D. Abdullaev, L. B. Senyavina, S. F. Arkhipova, N. N. Uvarova, K. Kh. Khalilulina, V. F. Bystrov, and Yu. A. Ovchinnikov, *Biochem. Biophys. Res. Commun.*, **42**, 654 (1971).
(21) M. M. Shemyakin, N. A. Aldanova, E. I. Vinogradova, and M. Yu. Feigina, *Tetrahedron Lett.*, **28**, 1921 (1963).

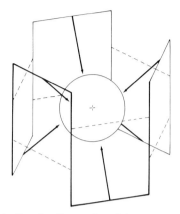

Figure 1. General architecture of a cyclododecapeptide folded in a way to provide a cavity lined with six amide carbonyl dipoles (arrows) to accommodate a potassium ion. The remaining amide groups form hydrogen bonds (broken lines) to stabilize the six U-shaped loops of the backbone (cf. Figure 2).

niatins,[22] beauvericin,[23] and monamycin[24]), peptides (alamethicin[25] and antamanide[26]), polyenes (amphotericin B[6]), and polyethers (the actins,[27] monensin,[7] nigericin,[8] and X-537A[9]). Some synthetic carriers have also become known, among them several depsipeptides,[3,4] a peptide,[28] and cyclic polyethers ("crown" compounds) or polythioethers.[17,29]

The inertness of most of the ion carriers which does not allow much room for chemical modification is a disadvantage. If one is to attempt to shed some light on the unanswered questions about the requirements of a molecule to exhibit alkali ion complexing one is therefore forced to resort to the total synthesis of compounds. The present study describes the design and synthesis of a neutral cyclododecapeptide which solubilizes alkali salts in an organic phase through complexation with the cation.

Despite their diversity in origin and chemical composition there remain a few structural features that are common to all of the alkali ion ligands and which are believed to be the roots of their activity. They all contain polar as well as nonpolar groups which are generally arranged in a cyclic structure. The folding of these molecules in their complexing conformation is such that the polar groups point toward the center to provide a polar cavity for the cation. The nonpolar groups, on the other hand, point outward, forming a lipophilic exterior. The result is a charged lipophilic complex which is large enough to allow its penetration into nonpolar media that normally exclude small charged particles. These general features of molecular architecture have been found to be realized in several instances by X-ray crystallography.[7-9,16,30]

Additional clues for the design of a carrier that would consist of α-amino acids exclusively are found in the well-investigated structure of a valinomycin–potassium complex.[16,19] This molecule is folded in a way that all of the six ester carbonyl oxygens point toward the center where they form the corners of an octahedron that encompasses the potassium ion (Figure 1). The backbone of the cyclododecadepsipeptide consisting of alternating hydroxy and amino acid residues encircles the cation in three complete sine waves, thus forming six loops. Each peptide carbonyl is engaged in a hydrogen bond with the closest peptide N-H in the sequence to form a bridge across the loop. These hydrogen bonds are thought to be essential in stabilizing this particular folding of the backbone. All of the side chains of the hydroxy and amino acid residues (methyl and isopropyl groups) point outward, thus shielding the potassium ion and the hydrogen bonds from the solvent.

From studies with space-filling models it becomes clear that the choice of changes in the nature or chirality of the hydroxy and amino acid residues that would not distort the symmetrical arrangement of the coordinating atoms is very limited.[4,31] There are two rules in particular that should not be violated. First, every other residue should provide an N-H which can, through hydrogen bonding, contribute to a stabilization of the "active conformation." Second, the orientation of the side chains, which is dictated by the optical configuration of the individual hydroxy or amino acid residues, should be so chosen that they could not prevent the formation of these hydrogen bonds.[31] However, there is no indication whatsoever that would require the carbonyl oxygens that coordinate to the potassium ion to be part of an ester group. In space-filling molecular models, substituting the ester bonds by amide bonds does not significantly alter the over-all geometry of the molecule. Nevertheless, in order not to introduce additional N-H groups that could conceivably stabilize other conformations or decrease the lipophilic character of the compound, these amide bonds should be part of an imino acid, such as proline.

Based on these considerations we chose to synthesize the cyclododecapeptide cyclo-[-L-Val-D-Pro-D-Val-L-Pro-]$_3$.[32] It may be considered an analog of valinomycin in that the L-lactic acid residues were replaced by L-proline and the D-α-hydroxyisovaleric acid residues by D-proline. In this molecule, should it assume a conformation analogous to valinomycin (Figure 2), the six carbonyl oxygens coordinating to the potassium ion still belong to the valine residues but they are now part of an amide rather than an ester bond.

(22) P. A. Plattner, K. Vogler, R. O. Studer, P. Quitt, and W. Keller-Schierlein, Helv. Chim. Acta, 46, 927 (1963); P. Quitt, R. O. Studer, and K. Vogler, ibid., 46, 1715 (1963).
(23) R. L. Hammill, C. E. Higgens, H. E. Boaz, and M. Gorman, Tetrahedron Lett., 4255 (1969).
(24) C. H. Hassall, R. B. Morton, Y. Ogihara, and D. A. S. Phillips, J. Chem. Soc. C, 526 (1971).
(25) J. W. Payne, R. Jakes, and B. S. Hartley, Biochem. J., 117, 757 (1970).
(26) Th. Wieland, G. Lüben, H. Ottenheim, J. Faesel, J. X. de Vries, W. Konz, A. Prox, and J. Schmid, Angew. Chem., 80, 209 (1968).
(27) J. Dominguez, J. D. Dunitz, H. Gerlach, and V. Prelog, Helv. Chim. Acta, 45, 129 (1962); J. Beck, H. Gerlach, V. Prelog, and W. Voser, ibid., 45, 620 (1962).
(28) R. Schwyzer, A. Tun-Kyi, M. Caviezel, and P. Moser, ibid., 53, 15 (1970).
(29) C. J. Pedersen, J. Amer. Chem. Soc., 89, 7017 (1967).

(30) B. T. Kilbourn, J. D. Dunitz, L. A. R. Pioda, and W. Simon, J. Mol. Biol., 30, 559 (1967); M. Dobler, J. D. Dunitz, and J. Krajewski, ibid., 42, 603 (1969); M. A. Bush and M. R. Truter, Chem. Commun., 1439 (1970); B. Metz, D. Moras, and R. Weiss, ibid., 444 (1971).
(31) B. F. Gisin, R. B. Merrifield, M. Tieffenberg, P. Cook, and D. C. Tosteson, unpublished work.
(32) Abbreviations according to the recommendations by the IUPAC-IUB Commission on Biochemical Nomenclature: J. Biol. Chem., 214, 2491 (1966); 242, 555 (1967).

Synthesis of a Hydrophobic Potassium Binding Peptide

Results and Discussion

The synthesis was performed essentially according to the established procedures[33] of solid-phase peptide synthesis[34] with the aid of a Beckman peptide synthesizer Model 990. Polystyrene-co-1% divinylbenzene resin was chloromethylated with chloromethyl methyl ether and stannic chloride[34,35] which was converted, first, to acetoxymethyl resin[36,37] and then aminolyzed with diethylamine to yield hydroxymethyl resin. Boc-L-proline was coupled to the resin with N,N'-carbonyldiimidazole[36,38] and the remaining hydroxymethyl groups were acetylated with acetic anhydride. Proline was chosen as the C-terminal residue in order to minimize the chance of racemization in the ultimate cyclization step, for imino acids are known not to undergo this side reaction upon activation of the carboxyl group.[39] During the synthesis of the dodecapeptide-resin[40] 50% trifluoroacetic acid in methylene chloride was used for deprotection and 5% diisopropylethylamine in methylene chloride for neutralization. Coupling was with 2 equiv of tert-butyloxycarbonyl amino acids and N,N'-dicyclohexylcarbodiimide (DCC)[41] in methylene chloride for 2 hr followed by another 2 hr with 1 additional equiv of each reagent.

The unusual tendency of the peptide ester, H-D-Val-L-Pro-resin, to cyclize to give D-valyl-L-proline diketopiperazine called for a modification of the coupling procedure with DCC.[42] In methylene chloride the intramolecular aminolysis was found to be catalyzed by carboxylic acids, e.g., by Boc-D-proline which was added to the dipeptide-resin prior to the addition of DCC. In this "regular" DCC coupling the loss of dipeptide amounted to approximately 70% while, using a "reversed" coupling (adding DCC prior to Boc-D-Pro-OH), the loss was only 10-20%. Furthermore, the sequence D-valyl-L-prolyl- has recently been shown to be cleaved from a tripeptide (H-D-Val-L-Pro-Sar-OH) to give D-valyl-L-proline diketopiperazine and sarcosine.[43] Since our peptide contains this sequence and its chemically equivalent optical antipode (-L-Val-D-Pro-) exclusively, taking measures to suppress diketopiperazine formation seemed advisable. For that reason the reversed DCC coupling procedure[42] was used throughout the synthesis.

Nevertheless, the loss of peptide chains in the first stages of the synthesis was not prevented completely as determined with the picric acid method.[44] The amine content of the resin was 72% of its original value at the tripeptide and 62% at the pentapeptide stage. There was no significant further decrease during the synthesis and the amount of amine found in the dode-

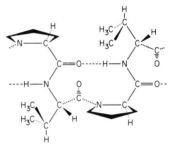

Figure 2. Probable conformation of one of the hydrogen bonded loops in the potassium complex of cyclo-[L-Val-D-Pro-D-Val-L-Pro-]$_3$. The sequence shown represents one-third of the molecule.

capeptide-resin was 60% of the value for the proline-resin at the beginning of the synthesis. Thus, the yield of peptide decreased more rapidly during the first five to six coupling cycles than in the later ones, an observation that has also been made during the synthesis of peptide sequences unrelated to this one.[45]

The peptide was cleaved from the resin with hydrobromic acid in acetic acid or TFA and purified by Sephadex LH-20 chromatography. Cyclization was with Woodward's reagent K.[46] The crystalline cyclododecapeptide gave the expected elemental analysis and, after hydrolysis, amino acid analysis by the method of Manning and Moore[47] indicated equimolar amounts of L-Val, D-Pro, D-Val, and L-Pro. The compound showed the calculated molecular weight of 1176 by mass spectrometry.[48]

The cyclic dodecapeptide was demonstrated to bind potassium ions in the following experiments. A known amount of the peptide was dissolved in methylene chloride and solid potassium picrate (which is insoluble in this solvent) was added. The yellow color of the picrate was immediately taken up by the solvent and spectrophotometric determination of the solubilized picrate indicated a 1:1 complex with the peptide. Upon evaporation of the solvent the compound was obtained in crystalline form. Judging from the two-phase dissociation constants (K_{D2})[49] in the system methylene chloride–water the peptide ($K_{D2} = 7 \times 10^{-6}$ M) showed a sevenfold higher affinity for potassium picrate than valinomycin ($K_{D2} = 5 \times 10^{-5}$ M).

The ir spectra of the cyclododecapeptide and its potassium complex are shown in Figure 3. The two N–H stretch bands of the uncomplexed peptide indicate the presence of both free (3397 cm^{-1}) and hydrogen-bonded (3317 cm^{-1}) amide hydrogens. Upon complexation these bands merge and undergo a bathochromic shift to form a single band at a frequency of 3280 cm^{-1}. The amide I band which is broad and structured in the free peptide (1630–1670 cm^{-1}) is considerably narrower in the complex. Except for a sharp ab-

---

(33) R. B. Merrifield, *Advan. Enzymol.*, **32**, 221 (1969).
(34) R. B. Merrifield, *J. Amer. Chem. Soc.*, **85**, 2149 (1963).
(35) K. W. Pepper, H. M. Paisley, and M. A. Young, *J. Chem. Soc.*, 4097 (1953).
(36) M. Bodanszky and J. T. Sheehan, *Chem. Ind. (London)*, 1597 (1966).
(37) J. M. Stewart and J. D. Young, "Solid-Phase Peptide Synthesis," W. H. Freeman, San Francisco, Calif., 1969, p 9.
(38) H. A. Staab, *Angew. Chem.*, **71**, 194 (1959).
(39) E. Schröder and K. Lübke, "The Peptides," Vol. I, Academic Press, New York, N. Y., 1965, p 147.
(40) "Peptide-resin" denotes a peptide esterified through the C-terminal carboxyl group to a polymeric benzyl alcohol.
(41) J. C. Sheehan and G. P. Hess, *J. Amer. Chem. Soc.*, **77**, 1067 (1955).
(42) B. F. Gisin and R. B. Merrifield, *ibid.*, **94**, 3102 (1972).
(43) J. Meienhofer, *ibid.*, **92**, 3771 (1970).
(44) B. F. Gisin, *Anal. Chim. Acta*, **58**, 248 (1972).

(45) B. Gutte, personal communication.
(46) R. B. Woodward, R. A. Olofson, and H. Mayer, *Tetrahedron, Suppl.*, **8**, 321 (1966); K. Blaha and J. Rudinger, *Collect. Czech. Chem. Commun.*, **30**, 3325 (1965).
(47) J. M. Manning and S. Moore, *J. Biol. Chem.*, **243**, 5591 (1968).
(48) We are very grateful to Professor F. Field of Rockefeller University for performing this analysis.
(49) B. C. Pressman, *Fed. Proc., Fed. Amer. Soc. Exp. Biol.*, **27**, 1283 (1968).

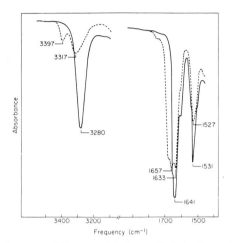

Figure 3. Partial ir spectra of the free peptide (broken line) and of its potassium picrate complex (solid line) in methylene chloride (concn, 2.5×10^{-3} M).

sorption band at a frequency of 2060 cm^{-1} due to the thiocyanate ion the potassium thiocyanate complex gave the same spectrum as the potassium picrate complex of Figure 3. The data for the free peptide are consistent either with a backbone structure indicated in Figure 1 in which most of the intramolecular hydrogen bonds are intact or with a mixture of conformers with a varying degree of intramolecular hydrogen bonding. Complexation appears to eliminate non-hydrogen-bonded amide hydrogens in favor of one single type of hydrogen-bonded N–H (Figure 2) as would be required for a symmetrical, compact complex of the type shown in Figure 1. The narrowing of the amide I band upon complexation is thought to be due to the superposition of the hydrogen-bonded proline carbonyl band by the band of the valine carbonyls.

The present work is an example of the possibility to program[50] the sequence of a polypeptide in a way that it contains all the information necessary to express a simple predetermined function, in this case to bind an alkali ion and form a hydrophobic complex.

Experimental Section

Amino acid analyses (Beckman Spinco amino acid analyzer Model 121) were performed by Miss L. Apacible and elemental analyses by Mr. T. Bella of Rockefeller University. Infrared spectra were taken on a Perkin-Elmer 237B or 621 ir spectrophotometer which were made available to us through the courtesy of Professor L. C. Craig of Rockefeller University. The melting points were determined in open capillaries and are not corrected. A Schmidt & Haensch polarimeter was used to measure the optical rotations. Solid-phase reactions were carried out on an automatic Beckman peptide synthesizer Model 990 or on a mechanical shaker[35] in screw cap vessels equipped with a fritted disk and a stopcock.[42] All chemicals and solvents were reagent grade. Trifluoroacetic acid, acetic acid, diisopropylethylamine, triethylamine, pyridine, acetic anhydride, and methylene chloride were redistilled prior to use.

Substitutions of resins are expressed in μequivalents per gram of benzyl polymer ($\cdot$CH$_2$–C$_6$H$_4$–resin) according to the formula,[51] $s = a/(1 - ae)$, where s = substitution in μequivalents per gram; a = analytical concentration of substituent on the resin in μequivalents per gram, and e = equivalent weight of the substituent bound to the benzyl polymer in gram per μequivalent. Peptide-resins were hydrolyzed with propionic acid–concentrated HCl (1:1, v/v),[52] and peptides with 6 N HCl, at 140° for 3–6 hr in sealed vessels.

Chloromethyl Resin. Polystyrene-*co*-1% divinylbenzene resin (100 g, Bio-Rad SX-1)[53a] was gently stirred at 90–100° in each of the following solvents for 30 min: benzene, methanol, dimethylformamide, dioxane–2 N aqueous NaOH (1:1, v/v), and dioxane–2 N aqueous HCl (1:1, v/v).[54] After each heating period the resin was filtered and washed thoroughly with the subsequent solvent. The last washes before air drying were with hot methanol, hot benzene, methanol, and methylene chloride. The resin was stirred in 600 ml of freshly distilled chloromethyl methyl ether at room temperature for 30 min and then cooled to -2 to 0° (ice-acetone bath). While, under exclusion of moisture, over a period of approximately 10 min a solution of 12 ml of stannic chloride in 100 ml of hexane was added from a dropping funnel to the stirred solution, the temperature remained within the range of -2 to $+2$°. After another 30 min at 0° 1000 ml of ice cold chloroform was added. The resin was filtered, washed thoroughly (dioxane–water (3:1, v/v), dioxane–2 N aqueous hydrochloric acid (3:1), dioxane–water (3:1), dioxane, water, methanol, and water), allowed to stand in methanol for 0.5 day, and floated in CH$_2$Cl$_2$ to remove small particles. The chloride content was 1950 μequiv/g (by combustion).

Hydroxymethyl Resin. Chloromethyl resin (20 g, substitution 1950 μequiv/g) was stirred and kept at 100–110° in 150 ml of methyl Cellosolve–potassium acetate (9:1, v/w) for 16 hr, filtered, and washed with methyl Cellosolve and CH$_2$Cl$_2$. This treatment was repeated for two more 16-hr periods. Washings included methyl Cellosolve, methyl Cellosolve–water (2:1, v/v), methyl Cellosolve, CH$_2$Cl$_2$, and methanol. Chloride was below the detectable limit (<30 μequiv/g by combustion). The resin was refluxed in 100 ml of diethylamine overnight, washed with CH$_2$Cl$_2$ and methanol, and dried. This procedure for converting acetoxymethyl resin to hydroxymethyl resin was adopted in order to convert, at the same time, any remaining traces of chloromethyl groups into diethylaminomethyl groups, thus eliminating the chance of the formation of quaternary ammonium sites later on. There was no ester band detectable at 1750 cm^{-1} indicating that essentially all of the acetoxy groups had been converted into hydroxy groups.

tert-**Butyloxycarbonyl-L-prolyl-Resin.** To a cold (-5°, acetone-ice bath) suspension of 5.85 g (36 mmol) of carbonyldiimidazole in 100 ml of CH$_2$Cl$_2$ Boc-L-Pro-OH[53b] (7.75 g, 36 mmol) was added. After 30 min of stirring at -5° all of the components had gone into solution. Hydroxymethyl resin (30 g, 54 mmol) was added and more solvent to make a volume of 250 ml. The suspension was stirred at room temperature for 3 days, filtered, and thoroughly washed with CH$_2$Cl$_2$, dimethylformamide, and methanol. The remaining hydroxymethyl groups were esterified by a treatment of the resin with acetic anhydride–pyridine (1:1, v/v) at room temperature for 30 min. Washings were with benzene, CH$_2$Cl$_2$, and methanol. Hydrolysis of the resin indicated a substitution of 630 μmol of proline per gram of resin. The substitution of a sample withdrawn after 18 hr reaction time was 265 μmol/g.

tert-**Butyloxycarbonyl-D-proline.** D-Proline[53c] was N-protected with the Boc group according to a procedure of Schnabel:[55] mp 133–134°; [α]^{26}D +63.6 (*c* 1, acetic acid) [lit.[55] (L isomer) mp 134–136°, [α]$^{18-25}$D -68.5° (*c* 1, acetic acid)].

Anal. Calcd for C$_{10}$H$_{17}$NO$_4$: C, 55.80; H, 7.96; N, 6.51. Found: C, 56.02; H, 7.95; N, 6.32.

Trifluoroacetate of L-valyl-D-prolyl-D-valyl-L-prolyl-L-valyl-D-prolyl-D-valyl-L-prolyl-L-valyl-D-prolyl-D-valyl-L-prolyl-Resin. First Run. The starting material was 4.52 g of *tert*-butyloxycarbonyl-L-prolyl-resin (substitution, 590 μequiv/g by picrate determination[44] and 595 μequiv/g by amino acid analysis). The peptide chain was built up using the following cycle: (a) deprotection with 50% (v/v) TFA in CH$_2$Cl$_2$, 2 × 15 min; (b) neutralization with 5% (v/v) diisopropylamine in CH$_2$Cl$_2$; (c) coupling with a twofold

---

(50) R. Schwyzer, *Experientia*, **26**, 577 (1970).
(51) V. A. Najjar and R. B. Merrifield, *Biochemistry*, **5**, 3765 (1966).

(52) J. Scotchler, R. Lozier, and A. B. Robinson, *J. Org. Chem.*, **35**, 3151 (1970).
(53) (a) Bio-Rad Laboratories, Richmond, Calif.; (b) Schwarz Bioresearch, Orangeburg, N. Y.; (c) Fox Chemical Co., Los Angeles, Calif.; (d) Aldrich Chemical Co., Milwaukee, Wis.; (e) Analtech, Wilmington, Del.
(54) R. B. Merrifield, unpublished work.
(55) E. Schnabel, *Justus Liebig Ann. Chem.*, **702**, (1967).

Synthesis of a Hydrophobic Potassium Binding Peptide

Table I. Analytical Data of Boc-Peptide-Resins (Second Run)

		Peptide					
		Tri	Penta	Hepta	Nona	Undeca	Dodeca
Pro:Val	Calcd	2:1	3:2	4:3	5:4	6:5	6:6
	Found	1.88:1	2.90:2	3.88:2	5.08:4	6.14:5	5.88:6
Free amine[a]		0.3	0.5	0.6	0.7	0.6	0.7

[a] As determined with picric acid,[44] expressed in per cent of the amount of peptide found by amino acid analysis. These values have not been corrected for the blank value of Boc-Pro-resin (0.4% or 2.6 μmol/g) and may therefore in fact be too high.

excess of DCC and acyl component (Boc-D-Val-OH,[56] Boc-D-Pro-OH, Boc-L-Val-OH,[53c] or Boc-L-Pro-OH,[53b] respectively) and a second coupling with 1 equiv each for 2 hr. The resin was washed thoroughly with CH_2Cl_2 after each step and with dimethylformamide and CH_2Cl_2 after each cycle, and was deprotected at the end of the synthesis. Because of the loss of peptide through intramolecular aminolysis that was observed when the dipeptide-resin was exposed to the acyl component a reversed DCC coupling[42] was used throughout: the DCC was added first followed by the Boc-amino acid in several batches. In order to inactivate unreacted amino groups and also hydroxy groups that may have formed upon loss of peptide from the resin, acetylation steps (acetic anhydride-pyridine, 1:1, 20 min) were added after the tri-, tetra-, penta-, and nonapeptide stages. The total amount of amine found in the deprotected peptide resins as determined with picrate was as follows: H-L-Pro-resin, 2.36 mequiv (100%); tripeptide-resin, 1.70 mequiv (72%); pentapeptide-resin, 1.47 mequiv (62%); heptapeptide-resin, 1.48 mequiv (62%); nonapeptide-resin, 1.39 mequiv (59%); undecapeptide-resin, 1.37 mequiv (58%); dodecapeptide-resin, 1.43 mequiv (60%). The dried dodecapeptide-resin trifluoroacetate had a substitution of 370 μequiv/g by picrate determination, thus indicating that 63% of the peptide chains were retained on the resin; amino acid analysis, Pro:Val = 6.6:6.0 (calcd 6:6).

Second Run. Starting with 5.0 g (2.8 mmol) of tert-butyloxycarbonyl-L-prolyl-resin this batch of peptide was synthesized on an automatic Beckman peptide synthesizer Model 990 which had been programmed to perform the same operations as described for the first run, with the following exceptions. The first couplings to the di- and tripeptide were done manually to allow addition of the Boc-amino acid in small portions and the acetylation steps were omitted. Samples of Boc-peptide-resins were taken and analyzed for free amino groups and amino acid content (Table I). These results indicate a better than 99.3% coupling yield at the stages examined. The yield of dodecapeptide-resin was 58% based on the substitution (340 μequiv/g by amino acid analysis).

L-**Valyl**-L-**prolyl**-D-**valyl**-L-**prolyl**-L-**valyl**-D-**prolyl**-D-**valyl**-L-**prolyl**-L-**valyl**-D-**prolyl**-D-**valyl**-L-**proline** Hydrochloride. Dodecapeptide-resin trifluoroacetate of the first run (2.4 g, 0.60 mmol) was allowed to swell in CH_2Cl_2 (10 ml) and was treated with 30% HBr in acetic acid for 90 min at room temperature. The resin was filtered and washed with glacial acetic acid and the filtrates were combined, evaporated, and lyophilized from glacial acetic acid (470 mg). This crude product was chromatographed on a 2.6 × 90 cm column of Sephadex LH-20 in 10^{-3} N HCl and 100 5-ml fractions were collected. The main component emerged as a symmetrical peak (by ninhydrin reaction with an aliquot of the fractions[57]) in the fractions 50 through 60. Evaporation and lyophilization from water yielded 335 mg (46%). The corresponding yield for material from the second run of the previous paragraph was 43%. Titration of 4.77 mg of this lyophilized product with 0.01 M NaOH in 0.1 M NaCl showed it to be a mixture of 2.10 μmol of peptide hydrochloride and 1.96 μmol of free peptide. These data indicate a molecular weight of 1154 for the peptide (calcd 1194). The apparent pK_a of 7.6 was within the expected range for the N-terminal amino group of a peptide. Hydrolysis followed by quantitative amino acid analysis gave a ratio of Pro:Val = 6.0:5.6 (calcd 6:6).

cyclo-[-L-**Valyl**-D-**prolyl**-D-**valyl**-L-**prolyl**-L-**valyl**-D-**prolyl**-D-**valyl**-L-**prolyl**-L-**valyl**-D-**prolyl**-D-**valyl**-L-**prolyl**]. HCl·H-[-L-Val-D-Pro-D-Val-L-Pro-]₃-OH (1.0 g, 0.8 mmol) was neutralized in aqueous solution with triethylamine to pH 5.5 and lyophilized. According to a procedure by Blaha and Rudinger[46] it was activated in 50 ml o-dimethylformamide with Woodward's reagent K[53d] (270 mg, 1.0 mmol) at −10° to +10° for 120 min. The solution was diluted with 800 ml of CH_2Cl_2 and cyclization was effected by addition of triethylamine (210 μl, 1.5 mmol). After 120 hr at room temperature the solvent was evaporated at reduced pressure and the oily residue was taken up in 75% methanol and passed through a 200-ml column of mixed bed ion exchange resin (AG 501-X8 (D)[53a]). The effluent (150 ml) was collected and evaporated at reduced pressure with gradual replacement of the solvent by glacial acetic acid. This procedure was used because the peptide tended to precipitate from aqueous or alcoholic solutions in an insoluble form. After lyophilization from acetic acid the product was homogeneous by tlc (silica gel G plates[53r]) in three systems (1-butanol-acetic acid-pyridine-water, 15:10:3:2, R_f = 0.73; 1-butanol-acetic acid-water, 4:1:1, R_f = 0.47; 1-butanol, R_f = 0.18), ninhydrin-negative, and gave a yellow-white spot with the iodine-tolidine reaction.[58] Crystallization by slow evaporation of a solution of the peptide in $MeCl_2$-benzene (1:1) gave 138 mg (16%) of needles: mp >330° dec (with browning at 300°); $[\alpha]^{29}D$ −2.2 ± 1.2° (c 0.2, TFA). The peptide retained 1 equiv of acetic acid.

Anal. Calcd for $C_{60}H_{96}N_{12}O_{12} \cdot C_2H_3O_2$: C, 60.20; H, 8.09; N, 13.60. Found: C, 60.39; H, 8.05; N, 13.77.

The molecular weight by mass spectrometry was 1175 ± 1 (calcd 1176).[48]

The amino acid composition of the peptide was determined after hydrolysis (6 N HCl, 130–140°, 48 hr in a sealed vessel) with a method of Manning and Moore[47] which involves conversion of the amino acids to L-Leu-dipeptides followed by chromatography on an automatic amino acid analyzer.[59] Under the conditions used (Beckman-Spinco amino acid analyzer 120B, 0.9 × 55 cm column with Beckman-Spinco AA-15 resin, T = 56°, buffer pH 4.25, flow rate, 70 ml/hr) the elution times were L-Leu-D-Pro 87 min, L-Leu-L-Pro and L-Leu-D-Val 115 min, and L-Leu-L-Val 129 min. The peak areas found indicated amino acid ratios of D-Pro:(L-Pro and D-Val):L-Val = 3.1:5.6:3.0 (calcd 3:6:3).

Potassium Picrate Complex of cyclo-[-L-Val-D-Pro-D-Val-L-Pro]₃. cyclo-[-L-Val-D-Pro-D-Val-L-Pro]₃ (20 mg) was dissolved in 1.5 ml of CH_2Cl_2. When crystalline potassium picrate was added (10 mg) the solvent quickly turned yellow indicating an uptake of picrate. (Substituting valinomycin[56] for the cyclododecapeptide the same observation was made. However, there was no picrate solubilized in CH_2Cl_2 in the absence of either compound.) After 16-hr agitation at room temperature the solution was filtered, diluted with an equal volume of benzene, and allowed to evaporate. A residue with 5-mm long, fern-like yellow crystals remained; amino acid ratio after hydrolysis, proline:valine = 1.03:1.00; mp 335° dec.

Anal. Calcd for $C_{66}H_{95}N_{15}O_{19}K + 2C_6H_6$: C, 58.52; H, 6.93; N, 13.19. Found: C, 58.50; H, 6.95; N, 13.52.

The complex retained 2 mol of benzene per mole which was removed by redrying at 140° (0.1 mm) for 70 hr.

Anal. Calcd for $C_{66}H_{95}N_{15}O_{19}$: C, 54.87; H, 6.84; N, 14.54. Found: C, 54.55; H, 6.76; N, 14.26.

An aliquot was dissolved in 95% ethanol and the picrate content of this solution was measured spectrophotometrically at 358 nm. It contained 104% of the amount expected for a 1:1 complex.

Similar to Pressman[49] the two-phase dissociation constant of the complex was determined by equilibrating a 10^{-4} M solution of the peptide in CH_2Cl_2 with various concentrations of aqueous potassium

---

(56) B. F. Gisin, R. B. Merrifield, and D. C. Tosteson, *J. Amer. Chem. Soc.*, **91**, 2691 (1969).

(57) S. Moore and W. H. Stein, *J. Biol. Chem.*, **211**, 907 (1954).

(58) Tlc according to G. Pataki, "Dünnschichtchromatographie in der Aminosäure und Peptidchemie," Walter de Gruyter & Co., Berlin, 1966.

(59) D. H. Spackman, W. H. Stein, and S. Moore, *Anal. Chem.*, **30**, 1190 (1958).

picrate and measuring the uptake of picrate into the organic phase spectrophotometrically. According to this constant (*i.e.*, the concentration of potassium picrate in the aqueous phase that caused half-saturation of the peptide with potassium picrate in the organic phase) the peptide ($K_{D_2} = 7 \times 10^{-6}$ M) has a sevenfold higher affinity for potassium picrate than valinomycin ($K_{D_2} = 5 \times 10^{-5}$ M).

Acknowledgments. We wish to thank Professor D. C. Tosteson of Duke University who contributed to this work through many inspiring discussions. The technical assistance of Mr. Arun Dhundale is gratefully acknowledged.

Preparation of a *t*-Alkyloxycarbonylhydrazide Resin and Its Application to Solid Phase Peptide Synthesis[1]

Su-sun Wang[2] and R. B. Merrifield

Contribution from The Rockefeller University, New York, New York 10021. Received May 24, 1969

Abstract: A new type of resin with *t*-alkyloxycarbonylhydrazide functional groups was prepared: $H_2NNH\text{-}COOC(CH_3)_2CH_2CH_2C_6H_5$ polymer. With this resin as solid support, a procedure was developed which should be useful for preparation of protected peptide hydrazides that can be purified, converted to the azides, and then condensed to other fragments to yield longer polypeptides. Combination of the conventional and solid phase approaches with retention of the best features of each is now possible. The procedure was tested by the synthesis of the crystalline tetrapeptide Z-Phe-Val-Ala-Leu-NHNH$_2$.

In solid phase peptide synthesis[3] the anchoring bond holding the peptide chain to the resin support has usually been a benzyl ester. The combination of this benzyl ester linkage with α-Boc[4] amino protection has allowed the convenient stepwise synthesis of a number of biologically active peptides by this method.[5] However, in order to release the desired peptide chain from the resin after completion of the synthesis treatment with strong acids such as HBr–TFA or anhydrous HF[6] is required and these reagents also remove most of the side chain protecting groups that are commonly used. It has, therefore, been difficult to obtain protected peptides that could be purified and then coupled to other peptide chains by the fragment condensation method to yield larger polypeptides. The opportunity for isolation and purification of the small peptides at intermediary stages during the synthesis is an important advantage of the fragment approach. Efforts to combine the conventional and solid phase approaches, with retention of the best features of each, have been made by Anfinsen, *et al.*,[7] and by Weygand.[8] The former employed hydrazinolysis of the α-benzyl ester linkage to prepare peptide hydrazides while the latter made use of sulfhydryl-sensitive phenacyl esters to anchor the peptide to the resin. Though both of these methods are attractive, they are not without some undesirable complications and limitations and a more generally applicable method is desirable.

Recent progress in the development of acid labile urethan amino protecting groups in peptide chemistry[9] has enabled us to consider the application of *t*-alkyloxycarbonyl hydrazides as the anchoring bonds in solid phase peptide synthesis. In the following, the preparation of a new type of substituted styrene–divinylbenzene copolymer resin, $H_2NNHCOOC(CH_3)_2CH_2CH_2$-resin, is described.

Peptides can be synthesized stepwise on this type of polymer support by using the 2-(*p*-biphenyl)isopropyloxycarbonyl (Bpoc) group[10,11] for α-amino protection. The Bpoc group can be removed at each cycle of the synthesis with very mild acid under conditions where the anchoring bond is stable. On the other hand the final peptide chain can be released as the hydrazide under more acidic conditions where side chain protecting groups remain unaffected, thus providing fragments suitable for further condensations. The feasibility of this approach was demonstrated by the synthesis of the protected tetrapeptide hydrazide, Z-Phe-Val-Ala-Leu-NHNH$_2$. The product was obtained in good yield in crystalline form with satisfactory amino acid and elementary analyses.

As indicated in Scheme I, a ketone functional group $CH_3COCH_2CH_2\text{-}$ was introduced into copolystyrene–2% divinylbenzene resin beads (I) by a Friedel–Crafts reaction with methyl vinyl ketone[12] using HF as catalyst. The ketone-containing resin (II) showed an intense carbonyl absorption at 1725 cm^{-1} in the ir spectrum (see Figure 1). This functional group was then converted to a tertiary alcohol by a Grignard reaction. The absorption band at 1725 cm^{-1} disappeared completely at this stage. To prepare the *t*-alkyloxycarbonyl

---

(1) This work was supported in part by Grant A 1260 from the U. S. Public Health Service.
(2) Postdoctoral Fellow, U. S. Public Health Service.
(3) R. B. Merrifield, *J. Amer. Chem. Soc.*, **85**, 2149 (1963); *Biochemistry*, **3**, 1385 (1964).
(4) Abbreviations used: Boc, *t*-butyloxycarbonyl; Bpoc, 2-(*p*-biphenyl)isopropyloxycarbonyl; Z, benzyloxycarbonyl; DCC, dicyclohexylcarbodiimide; TFA, trifluoroacetic acid; DIEA, diisopropylethylamine; DMF, dimethylformamide.
(5) R. B. Merrifield, *Advan. Enzymol.*, **32**, 221 (1969).
(6) J. Lenard and A. B. Robinson, *J. Amer. Chem. Soc.*, **89**, 181 (1967).
(7) C. B. Anfinsen, D. Ontjes, M. Ohno, M. Corley, and A. Eastlake, *Proc. Natl. Acad. Sci., U. S.*, **58**, 1806 (1967).
(8) F. Weygand, *Proc. European Peptide Symposium, 9th*, Orsay, 1968, p 183.

(9) P. Sieber and B. Iselin, *Helv. Chim. Acta*, **51**, 614 (1968).
(10) P. Sieber and B. Iselin, *ibid.*, **51**, 622 (1968).
(11) S. S. Wang and R. B. Merrifield, *Intern. J. Protein Res.*, in press.
(12) J. Colonge and L. Pichat, *Bull. Soc. Chim. France*, 177 (1949).

Scheme I

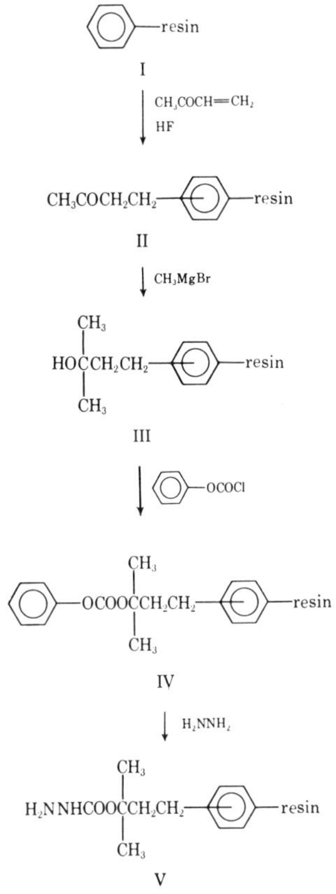

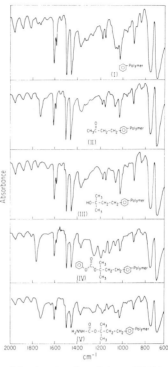

Figure 1. Infrared spectra of copolystyrene–2% divinylbenzene resin (I), 3-oxobutyl-resin (II), *t*-alkyl alcohol-resin (III), *t*-alkyl phenyl carbonate–resin (IV), and *t*-alkyloxycarbonylhydrazide–resin (V).

hydrazide resin, the alcohol resin (III) was allowed to react with phenyl chloroformate to form a phenyl carbonate resin (IV) which was then hydrazinolyzed to give *t*-alkyloxycarbonylhydrazide resin (V). Microanalysis shows that resin V contained 1.15% nitrogen while there was no detectable amount of nitrogen in resin IV. Both of these resins absorbed strongly near 1750 cm^{-1} as can be seen in Figure 1. From the nitrogen analysis, it could be calculated that the resin was substituted to the extent of 0.41 mmole of the functional group per gram of resin.

Bpoc-L-Leu[10,11] was condensed with the resin hydrazide V, using dicyclohexylcarbodiimide as coupling agent,[13] to form a protected amino acyl resin (VI) (see Scheme II). The Bpoc amino protecting group was found to be removed completely by 0.5% TFA in methylene chloride within a few minutes at room temperature while the anchoring bond was largely stable under these conditions for over 10 hr (see Figure 2). The stability of the anchoring bond under the conditions of Bpoc deprotection was such that less than a 6% loss of peptide chain would be expected even if the peptide synthetic cycle were repeated 40 times (see below).

A resin bound protected tetrapeptide hydrazide was synthesized from VI according to the general procedures

(13) J. C. Sheehan and G. P. Hess, *J. Amer. Chem. Soc.,* 77, 1067 (1955).

of solid phase peptide synthesis.[5,14] For that purpose Bpoc-L-Ala, Bpoc-L-Val, and Z-L-Phe were coupled sequentially to the growing peptide chain by the carbodiimide method. TFA (0.5%) in methylene chloride was utilized for the deprotecting agent, thus avoiding the strong acidic conditions that have previously been required. The protected peptide hydrazide (VIII) was released by treatment of the peptide resin VII with 50% TFA in methylene chloride at room temperature for 30 min. The product was obtained in pure, crystalline form in 76% over-all yield calculated from the leucine content of VI.

Experimental Section

Melting points were taken on a Thomas–Hoover capillary melting point apparatus and are uncorrected. Infrared spectra were taken on a Perkin-Elmer infrared spectrophotometer with KBr pellets. A Gilford Model 2000 was used to measure uv absorption. Amino acid analyses were carried out on a Spinco/Beckman Model 120B amino acid analyzer according to the accelerated[15] procedure of Spackman, Stein, and Moore.[16] Elementary analyses were performed by the Microanalysis Laboratory, Rockefeller University.

(14) J. M. Stewart and J. D. Young, "Solid Phase Peptide Synthesis," W. H. Freeman and Co., San Francisco, Calif., 1969.
(15) D. H. Spackman, "Methods in Enzymology," Vol. XI, C. H. W. Hirs, Ed., Academic Press, New York, N. Y., 1967, p 3.
(16) D. H. Spackman, W. H. Stein, and S. Moore, *Anal. Chem.,* 30, 1190 (1958).

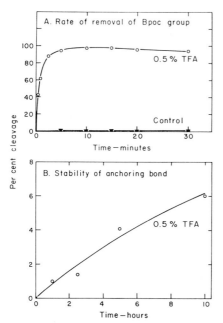

Figure 2. (A) Rate of Bpoc deprotection by 0.5% TFA in methylene chloride from Bpoc-Leu-HNNH–resin VI at room temperature. (B) The stability of the anchoring *t*-alkyloxycarbonyl bond in VI toward 0.5% TFA in methylene chloride at room temperature.

Copolymers of styrene–2% divinylbenzene (200–400 mesh) resin beads were obtained from Dow Chemical Co. Methyl vinyl ketone, TFA, and dicyclohexylcarbodiimide were purchased from Aldrich Chemical Co., Inc. Phenyl chloroformate and hydrazine hydrate were products of Distillation Products Industries. Magnesium turnings were obtained from Amend Drug and Chemical Co., Inc. Methyl bromide and anhydrous HF were supplied by Matheson Co., Inc. Bpoc-L-amino acid derivatives used were prepared in this laboratory.[11]

3-Oxobutyl–Resin (II). Styrene–2% divinylbenzene (200–400 mesh) beads were washed thoroughly with benzene, methylene chloride, 10% TFA in methylene chloride, 10% tributylamine in methylene chloride, DMF, and ethanol to remove styrene and other low molecular weight products. A 10-g sample of the washed resin (I) was then placed in a 200-ml Kel F vessel of the HF apparatus described by Sakakibara, *et al.*[17] After cooling the vessel in Dry Ice–acetone, 12.5 ml of methyl vinyl ketone was added and about 40 ml of anhydrous HF was distilled in. The mixture was stirred magnetically at room temperature for 30 min and the reaction mixture poured into 2 l. of absolute ethanol with caution. The resin particles were then collected on a glass filter and washed with ethanol, methylene chloride, benzene, DMF, methylene chloride, 3% TFA in methylene chloride, 10% tributylamine in methylene chloride, methylene chloride, dioxane, and ethanol to give 10.2 g of ketone-containing resin (II). This material absorbed strongly at 1725 cm^{-1} indicating the presence of a carbonyl function (see Figure 1). From a series of experiments, 30 min of reaction time was chosen since it gave the most desirable degree of substitution.

t-Alkyl Alcohol–Resin (III). Magnesium turnings (1 g) were suspended in 200 ml of dry ether and methyl bromide was bubbled through for about 20 min until all the metal had dissolved. A suspension of ketone-containing resin (II) in benzene (10 g in 50 ml) was then added in several portions to the freshly prepared Grignard reagent. Boiling occurred each time the suspension was added. The reaction mixture was left standing at room temperature for 1 hr and then the resin was collected on a glass filter, washed with alcohol and dioxane–water, and stirred with 300 ml of an equal mixture of dioxane–1 N H$_2$SO$_4$ for another hour. The resin was then collected and washed with dioxane–water, dioxane, and the solvents mentioned above. The material obtained (9.5 g) showed no carbonyl absorption in the ir spectrum (see Figure 1). Microanalysis indicated that this resin had no nitrogen but had 3.84% oxygen estimated from C, H, N analyses.

t-Alkyloxycarbonylhydrazide–Resin (V). The alcohol resin (III, 8 g) prepared above was suspended in 70 ml of methylene chloride and cooled to 0° while 5.5 ml of pyridine was added. To this suspension, 7.8 ml of phenyl chloroformate was added dropwise and then the mixture was stirred at 4° overnight. The suspension was poured onto a small amount of crushed ice and the resin particles were collected on a glass filter. After washing, a cream colored resin (IV) weighing 8.3 g was obtained. This material was hydrazinolyzed in 65 ml of DMF with 6.6 ml of hydrazine hydrate for 6 hr at room temperature. The resultant product was collected and washed to yield 8.2 g of the desired hydrazide resin (V). By microanalysis, it was shown that this resin contained 1.15% of nitrogen which corresponded to 0.41 mmole of functional group per gram of resin. The ir spectra of IV and V showed strong absorption at 1765 and 1725 cm^{-1}, respectively.

Bpoc-Aminoacylhydrazide–Resin (VI). Bpoc-L-leucine (600 mg, 1.64 mmoles) was dissolved in 8 ml of CH$_2$Cl$_2$ and was shaken with 1 g of hydrazide resin (V) for 10 min. DCC (350 mg in 8 ml of CH$_2$Cl$_2$) was then added and coupling was allowed to continue for 90 min (see Scheme II). After washing as usual, 1.18 g of Bpoc-Leu hydrazide resin (VI) was obtained which contained 0.289 mmole

---

(17) S. Sakakibara, Y. Shiminishi, M. Okada, Y. Kishida, *Proc. European Peptide Symposium, 8th,* Noordwijk, 1967, p 44.

of leucine per gram of resin according to amino acid analysis after hydrolysis in a mixture of dioxane–HCl.[18]

Stability of N-Bpoc Amino Protecting Group and the *t*-Alkyloxycarbonylhydrazide Anchoring Bond. A sample of Bpoc-Leu-resin (53 mg) was suspended in 1 ml of CH_2Cl_2 for a few minutes to allow the resin to swell. An equal volume of 1% TFA in the same solvent was then added and 0.1-ml samples were withdrawn at different time intervals. The resin particles were removed by filtration and washed several times with small volumes of CH_2Cl_2. The combined filtrate and washings were evaporated to dryness, dissolved in 24 ml of ethanol, and their absorbance at 254 mμ was measured to determine the amount of 2-(*p*-biphenyl)isopropyl alcohol present[10,11] with the results shown in Figure 2a. Trifluoroacetic acid was omitted from the control experiments. It can be seen that the removal of the Bpoc group was complete in about 10 min.

To see how stable the anchoring bond was under these conditions, several resin samples (VI) were prepared and suspended in 0.5% TFA in CH_2Cl_2 (30 mg/ml) in tightly capped small vials. At different times, one of the samples was taken and filtered and washed as described above. The liberated leucine hydrazide was then hydrolyzed with 6 N HCl at 105° for 24 hr. The leucine content of these samples is inversely related to the stability of the anchoring bond. Results of such experiments showed that there was only about 6% loss of the anchoring bond in 10 hr (see Figure 2b) which corresponded to 40 cycles of Bpoc deprotection according to the synthetic scheme described in the next section.

Z-Phe-Val-Ala-Leu-HNNH$_2$ (VIII). Bpoc-Leu-HNNH-resin (VI) (550 mg, 0.159 mmol) was placed in a peptide reaction vessel on a shaker[19] and treated as follows with 15-ml portions of solvents: (1) wash three times with CH_2Cl_2, (2) wash once with 0.5% TFA in CH_2Cl_2, (3) shake 10 min with 0.5% TFA in CH_2Cl_2, (4) wash three times with CH_2Cl_2, (5) wash once with 10% DIEA[4] in CH_2Cl_2, (6) wash three times with CH_2Cl_2, (7) wash three times with EtOH, (8) shake 10 min with CH_2Cl_2, (9) shake 10 min with 10% DIEA in CH_2Cl_2, (10) wash three times with CH_2Cl_2, (11) soak 10 min with 210 mg (0.64 mmole) of Bpoc-L-Ala in 6 ml of CH_2Cl_2, then add 132 mg (0.64 mmole) of DCC in 6 ml of CH_2Cl_2 and shake 90 min, (12) wash three times with CH_2Cl_2, (13) wash three times with DMF, (14) wash three times with EtOH. The cycle was repeated with 228 mg (0.64 mmol) of Bpoc-L-Val in step 11, and again with 185 mg (0.64 mmole) of Z-L-Phe. In each cycle a fourfold excess of amino acid derivative was used. The protected peptide hydrazide resin (VII) thus obtained weighed 610 mg after drying. According to the amino acid analysis, this material contained 0.232 mmole of peptide per gram of resin and had an amino acid composition of $Ala_{0.97}$ $Val_{1.00}$ $Leu_{0.97}$ $Phe_{1.05}$. The yield of the peptide at this stage was 89% calculated from the leucine content of VI.

The protected peptide hydrazide was liberated from the solid support by shaking 500 mg (0.116 mmole) of VII with 12 ml of 50% TFA in CH_2Cl_2 (v/v) at room temperature for 30 min. The resin was removed by filtration and washed with a few milliliters of CH_2Cl_2. After removal of the solvent by evaporation under reduced pressure, the peptide hydrazide was obtained as a white powder. It was then crystallized from methanol by addition of ether to give 60 mg (0.101 mmol) of the product melting at 255–257°. It had an amino acid composition of $Ala_{0.99}$ $Val_{1.00}$ $Leu_{1.04}$ $Phe_{1.00}$.

Anal. Calcd for $C_{37}H_{44}N_6O_6$ (596.71): C, 62.30; H, 7.43; N, 14.08. Found: C, 62.21; H, 7.40; N, 14.10.

Acknowledgments. We wish to thank Miss D. M. Cohen for her technical assistance in the preparation of Bpoc-amino acid derivatives, Miss U. Birkenmaier for carrying out the amino acid analyses, and Mr. S. T. Bella for performing the microanalysis.

---

(18) G. R. Marshall and R. B. Merrifield, *Biochemistry*, **4**, 2394 (1965).

(19) R. B. Merrifield and M. A. Corigliano, "Biochemical Preparations," Vol. 12, W. E. M. Lands, Ed., John Wiley & Sons, Inc., New York, N. Y., 1968, p 98.

p-Alkoxybenzyl Alcohol Resin and *p*-Alkoxybenzyloxycarbonylhydrazide Resin for Solid Phase Synthesis of Protected Peptide Fragments

Su-Sun Wang

Contribution from the Chemical Research Department, Hoffmann-La Roche Inc., Nutley, New Jersey 07110. Received August 9, 1972

Abstract: Two new resins—*p*-alkoxybenzyl alcohol resin ($HOCH_2C_6H_4OCH_2C_6H_4$ resin) and *p*-alkoxybenzyloxycarbonylhydrazide resin ($H_2NNHCOOCH_2C_6H_4OCH_2C_6H_4$ resin)—were prepared. The former resin is suitable for the synthesis of protected peptide fragments possessing a free carboxyl group while the latter is useful for the synthesis of protected peptide hydrazides. Applications of these resins in the syntheses of Z-Leu-Leu-Val-Phe, Z-Phe-Val-Ala-Leu-HNNH$_2$, Asp-Arg-Val-Tyr-Val-His-Pro-Phe, Z-Lys(Z)-Phe-Phe-Gly, and Z-Lys(Z)-Phe-Phe-Gly-Leu-Met-NH$_2$ are described.

Recent developments in solid phase peptide synthesis have been reviewed by Merrifield[1,2] and discussed by others.[3,4] The method has been widely and quite successfully utilized for the rapid and convenient synthesis of numerous polypeptides. However, the products obtained by this technique are, in general, rather difficult to purify. Although in certain instances, effective purification can be achieved by selective proteolysis[5] or affinity chromatography,[6]

(1) R. B. Merrifield, *Advan. Enzymol.*, **32**, 221 (1969).
(2) G. R. Marshall and R. B. Merrifield in "Biochemical Aspects of Reactions on Solid Supports," G. R. Stark, Ed., Academic Press, New York, N. Y., 1971, p 111.
(3) E. Wünsch, *Angew. Chem.*, **83**, 773 (1971).
(4) J. Meienhoffer, 163rd National Meeting of the American Chemical Society, Boston, Mass., April 1972, M15.

(5) B. Gutte and R. B. Merrifield, *J. Biol. Chem.*, **246**, 1922 (1971).
(6) H. Taniuchi and C. B. Anfinsen, *ibid.*, **244**, 3864 (1969).

those techniques are not universally applicable to all synthetic poypeptides. In view of obtaining the final peptides in greater purity, the use of modified resins suitable for solid phase synthesis of protected peptide fragments has been described by several authors.[7-14] In a previous communication[15] we have described the preparation of a *tert*-alkyloxycarbonylhydrazide resin that is useful for the synthesis of protected peptide hydrazides. In this paper, the preparation of the *p*-alkoxybenzyl alcohol resin (III) and the *p*-alkoxybenzyloxycarbonylhydrazide resin (V) and their application in the synthesis of a few protected peptide fragments are described.

The anchoring bonds employed are structurally similar to the *p*-methoxybenzyl ester[16] or *p*-methoxybenzyloxycarbonylhydrazide carboxyl protection of the classical peptide synthesis. These bonds can be selectively cleaved by trifluoroacetic acid under conditions where the side-chain protecting groups normally used are stable. When the very acid-labile 2-(*p*-biphenylyl)-2-propyloxycarbonyl[17,18] (Bpoc)[19] group is utilized for amino protection, resins III and V can be used satisfactorily for solid phase synthesis of protected peptide free acids and protected peptide hydrazides, respectively.

As outlined in Scheme I, Merrifield resin (I) (0.90 mmol of Cl/g) was allowed to react with methyl 4-hydroxybenzoate to form resin II. The product contained 0.87 mmol of OCH_3/g but no detectable amount of chlorine. An intense ester band at 1712 cm^{-1} and an aryl alkyl ether band at 1220 cm^{-1} appeared in the ir spectrum (Figure 1). Reduction of II with $LiAlH_4$ gave the desired *p*-alkoxybenzyl alcohol resin IIIa. In the ir spectrum, the ester band disappeared completely whereas the aryl alkyl ether band remained essentially unchanged. The same resin can also be conveniently prepared by reacting Merrifield resin with 4-hydroxybenzyl alcohol in the presence of $NaOCH_3$. The product obtained by this route (IIIb) is identical with IIIa as can be seen in Figure 1. Reaction of the alcohol resin III with phenyl chloroformate yielded IV which on hydrazinolysis gave *p*-alkoxybenzyloxycarbonylhydrazide resin V (see Figure 2). Nitrogen analysis indicated that there was 0.88 mmol/g of hydrazide on the resin. All of the reactions seem to have proceeded nearly quantitatively.

Bpoc-amino acids can be esterified to the alcohol

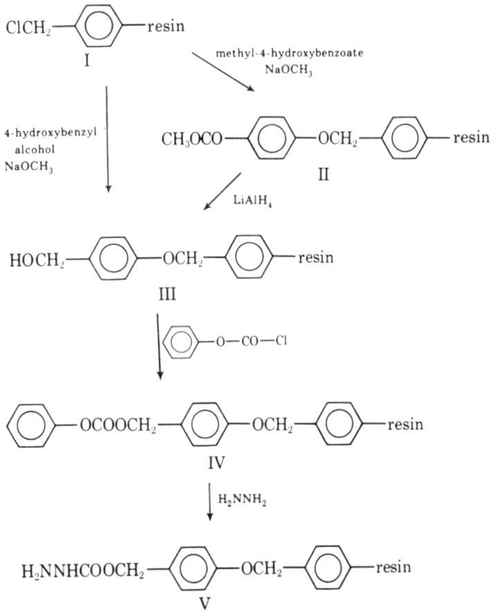

(7) T. Mizoguchi, K. Shigezane, and Takamura, *Chem. Pharm. Bull.*, **18**, 1465 (1970).
(8) G. L. Southard, G. S. Brooke, and J. M. Pettee, *Tetrahedron*, **27**, 1701 (1970).
(9) T. Wieland, J. Lewalter, and C. Birr, *Justus Liebigs Ann. Chem.*, **740**, 31 (1970).
(10) D. L. Marshall and I. E. Liener, *J. Org. Chem.*, **35**, 867 (1970).
(11) E. Flanigan and G. R. Marshall, *Tetrahedron Lett.*, 2403 (1970).
(12) S. S. Wang and R. B. Merrifield, *Proc. 10th Eur. Peptide Symp.*, 1969, 74 (1971).
(13) G. Losse and K. Neubert, *Tetrahedron Lett.*, 1267 (1970).
(14) G. W. Kenner, J. R. McDermott, and R. C. Sheppard, *Chem. Commun.*, 636 (1971).
(15) S. S. Wang and R. B. Merrifield, *J. Amer. Chem. Soc.*, **91**, 6488 (1969).
(16) F. Weygand and K. Hunger, *Chem. Ber.*, **95**, 1 (1962).
(17) P. Sieber and B. Iselin, *Helv. Chim. Acta*, **51**, 614, 622 (1968).
(18) S. S. Wang and R. B. Merrifield, *Int. J. Protein Res.*, **1**, 235 (1969).
(19) Abbreviations used are: Bpoc, 2-(*p*-biphenylyl)-2-propyloxycarbonyl; Boc, *tert*-butyloxycarbonyl; Z, benzyloxycarbonyl; Tos, *p*-toluenesulfonyl; Bzl, benzyl; But, *tert*-butyl; ONP, *p*-nitrophenyl ester; TFA, trifluoroacetic acid; NMM, *N*-methylmorpholine; DCC, dicyclohexylcarbodiimide; THF, tetrahydrofuran.

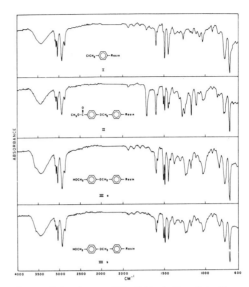

Figure 1. Infrared spectra of Merrifield resin (I); methyl *p*-alkoxybenzoate resin (II); *p*-alkoxybenzyl alcohol resin prepared from resin II by $LiAlH_4$ reduction (IIIa); *p*-alkoxybenzyl alcohol resin prepared directly from resin I and 4-hydroxybenzyl alcohol plus $NaOCH_3$ (IIIb).

Scheme I

resin III by DCC or active ester procedure. The extent of substitution was normally 0.2–0.6 mmol/g. To eliminate the unsubstituted free hydroxyl groups left on the resin, benzoylation or acetylation was found to be necessary. Attachment of Bpoc-amino acids to hydrazide resin V proceeded smoothly with the DCC method. To ascertain that no significant racemization

Synthesis of Protected Peptide Fragments

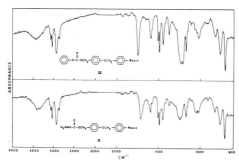

Figure 2. Infrared spectra of phenyl carbonate resin (IV); p-alkoxybenzyloxycarbonylhydrazide resin (V).

took place during esterification of Bpoc-amino acids to resin III, Bpoc-Phe-resin (VII) was deblocked, neutralized, and coupled with Boc-L-Glu(But) to give Boc-L-Glu(But)-Phe-resin. The dipeptide was released from the resin with 50% TFA (30 min) and then examined with an amino acid analyzer by a procedure similar to that described by Manning and Moore.[20] The product chromatographed identically with L-Glu-L-Phe. Less than 0.1% of contaminant L-Glu-D-Phe was found. The optical purity of Bpoc-Leu-resin prepared from Bpoc-L-Leu, DCC, pyridine, and resin III was also examined. The free amino acid was cleaved from the resin by 50% TFA and allowed to react with L-Leu-N-carboxyanhydride at pH 10.2. The resultant dipeptide was chromatographed with an amino acid analyzer. In this case, too, very little (less than 0.1%) L-Leu-D-Leu was detected. The product chromatographed identically with L-Leu-L-Leu.

In agreement with our previous observations,[15,18] the Bpoc group on VII can be removed by 0.5% TFA CH$_2$Cl$_2$ within a few minutes, while the anchoring bond is stable even after VII was treated with 0.5% TFA for 10 hr, as ascertained by ir spectroscopy. No free phenylalanine was liberated into the solution during this period of time. The anchoring linkage on the other hand can be very efficiently cleaved by 50% TFA in 30 min. As illustrated in Figure 3, the carbonyl band in the ir spectrum disappeared completely after this treatment and more than 85% of free phenylalanine was recovered from the filtrate. The experiments with Bpoc-Leu-HNNH-resin (XIV) gave similar results.

The synthesis of the protected tetrapeptide XI is outlined in Scheme II. Esterification of Bpoc-L-Phe to resin III was achieved by the DCC method using an equivalent amount of pyridine as catalyst. After benzoylation, Bpoc-L-Val, Bpoc-L-Leu, and Z-L-Leu were successively coupled to VII according to the general principle of solid phase peptide synthesis[1,2] with the necessary changes described before.[15] Treatment of X with 50% TFA gave XI as a pure crystalline compound in 68% overall yield calculated from the phenylalanine content of VII. Compound XI was further converted into its methyl ester XII and hydrogenated to give the tetrapeptide methyl ester XIII.

In Scheme III, the synthesis of the protected peptide

(20) J. Manning and S. Moore, *J. Biol. Chem.*, **243**, 5591 (1968).

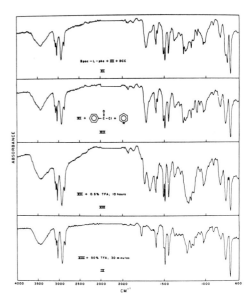

Figure 3. Infrared spectra of Bpoc-phenylalanyl resin (VI); resin VI after benzoylation (VII); resin VII that was treated with 0.5% TFA for 10 hr (VIII); resin VIII that was treated with 50% TFA for 30 min (IX).

Scheme II

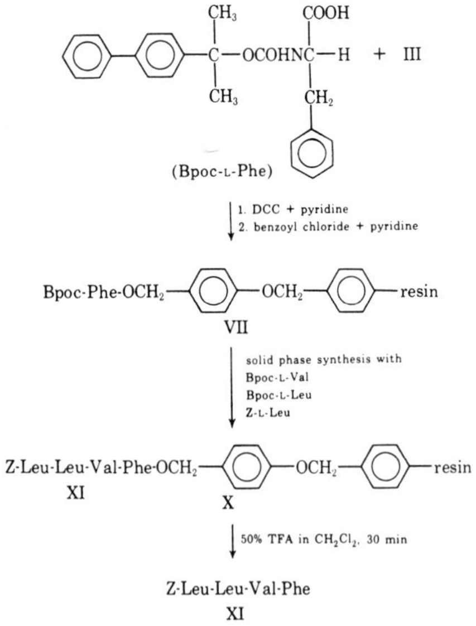

hydrazide XVI is outlined. A crystalline, pure product was obtained in 42% overall yield.

The alcohol resin III was utilized as the solid support for the synthesis of Val^5-angiotensin II (XVIII) (Scheme IV). In this synthesis, Bpoc-L-His(Tos) was used since partial racemization takes place when N^{Im}-benzyl-

Scheme III

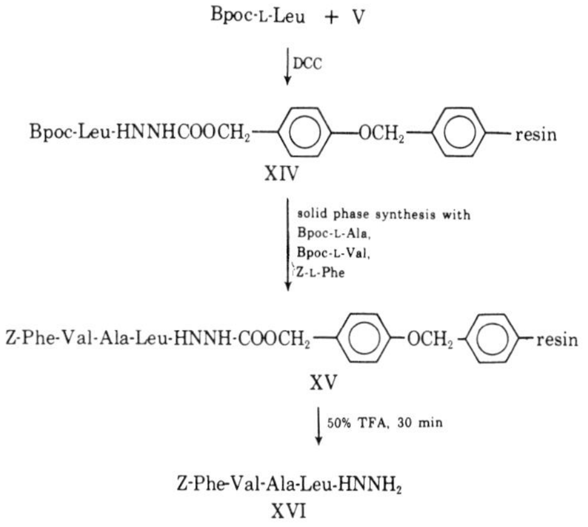

Bpoc·L·Leu + V
↓ DCC
Bpoc-Leu-HNNHCOOCH$_2$—⟨◯⟩—OCH$_2$—⟨◯⟩—resin
XIV
↓ solid phase synthesis with
Bpoc·L·Ala,
Bpoc·L·Val,
Z·L·Phe
Z-Phe-Val-Ala-Leu-HNNH·COOCH$_2$—⟨◯⟩—OCH$_2$—⟨◯⟩—resin
XV
↓ 50% TFA, 30 min
Z-Phe-Val-Ala-Leu-HNNH$_2$
XVI

Scheme IV

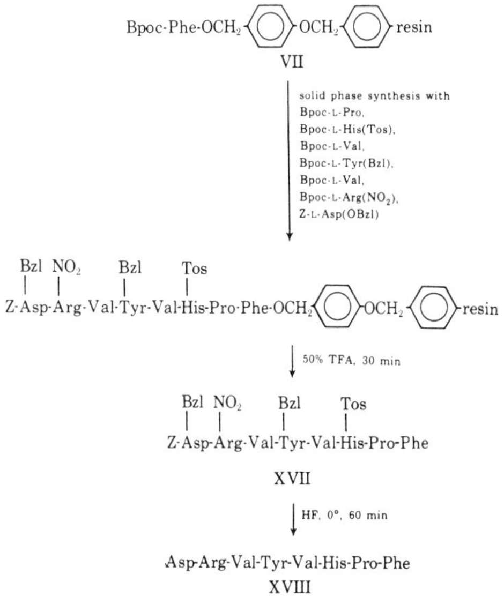

Bpoc-Phe-OCH$_2$—⟨◯⟩—OCH$_2$—⟨◯⟩—resin
VII
↓ solid phase synthesis with
Bpoc·L·Pro,
Bpoc·L·His(Tos),
Bpoc·L·Val,
Bpoc·L·Tyr(Bzl),
Bpoc·L·Val,
Bpoc·L·Arg(NO$_2$),
Z·L·Asp(OBzl)

     Bzl NO$_2$ Bzl   Tos
     | | | |
Z-Asp-Arg-Val-Tyr-Val-His-Pro-Phe-OCH$_2$—⟨◯⟩—OCH$_2$—⟨◯⟩—resin
↓ 50% TFA, 30 min

     Bzl NO$_2$ Bzl   Tos
     | | | |
Z-Asp-Arg-Val-Tyr-Val-His-Pro-Phe
XVII
↓ HF, 0°, 60 min
Asp-Arg-Val-Tyr-Val-His-Pro-Phe
XVIII

histidine derivatives are used in solid phase synthesis.[21] The protected octapeptide intermediate XVII was released from the resin by 50% TFA. All of the protecting groups were then removed by anhydrous hydrogen fluoride[22] and the crude product was purified by countercurrent distribution followed by gel filtration. Analytically pure material was obtained in 15% overall yield.

Resin III was also used in the synthesis of the eledoisin analog Lys-Phe-Phe-Gly-Leu-Met-NH$_2$ (XXIII)[23] by a combination of solid phase method and fragment condensation procedures (see Scheme V). The crys-

Scheme V

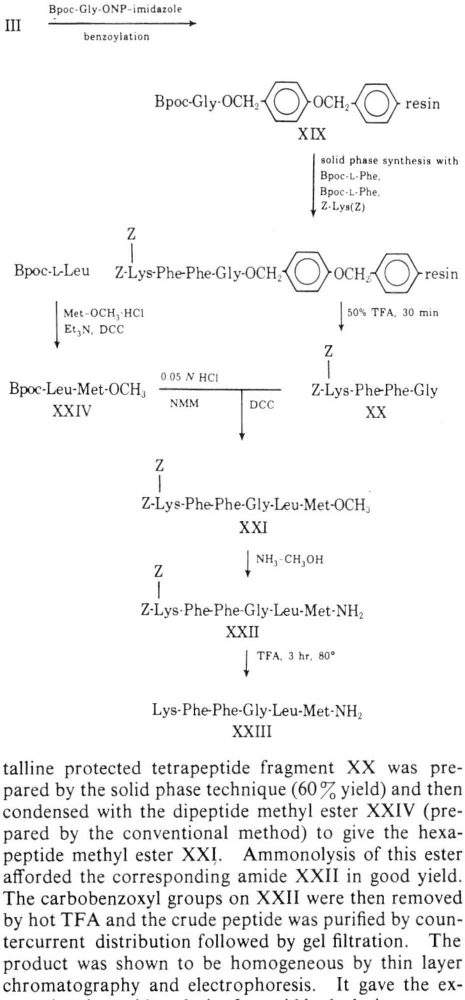

III —Bpoc-Gly-ONP-imidazole→ (benzoylation)
Bpoc-Gly-OCH$_2$—⟨◯⟩—OCH$_2$—⟨◯⟩—resin
XIX
↓ solid phase synthesis with
Bpoc·L·Phe,
Bpoc·L·Phe,
Z·L·Lys(Z)

    Z
    |
Bpoc·L·Leu   Z-Lys-Phe-Phe-Gly-OCH$_2$—⟨◯⟩—OCH$_2$—⟨◯⟩—resin
↓ Met-OCH$_3$·HCl   ↓ 50% TFA, 30 min
  Et$_3$N, DCC
                                     Z
                                     |
Bpoc-Leu-Met-OCH$_3$ —0.05 N HCl / NMM / DCC→ Z-Lys-Phe-Phe-Gly
XXIV   XX

    Z
    |
Z-Lys-Phe-Phe-Gly-Leu-Met-OCH$_3$
XXI
↓ NH$_3$–CH$_3$OH

    Z
    |
Z-Lys-Phe-Phe-Gly-Leu-Met-NH$_2$
XXII
↓ TFA, 3 hr, 80°
Lys-Phe-Phe-Gly-Leu-Met-NH$_2$
XXIII

talline protected tetrapeptide fragment XX was prepared by the solid phase technique (60% yield) and then condensed with the dipeptide methyl ester XXIV (prepared by the conventional method) to give the hexapeptide methyl ester XXI. Ammonolysis of this ester afforded the corresponding amide XXII in good yield. The carbobenzoxyl groups on XXII were then removed by hot TFA and the crude peptide was purified by countercurrent distribution followed by gel filtration. The product was shown to be homogeneous by thin layer chromatography and electrophoresis. It gave the expected amino acid analysis after acid hydrolysis.

Experimental Section

Melting points are uncorrected. Infrared spectra were taken on a Beckman IR-8 spectrophotometer with KBr pellets. Amino acid analyses were carried out on a Jeolco 5AH amino acid analyzer or Beckman Model 120B amino acid analyzer. A 500-tube Post countercurrent distribution apparatus was used for purification purposes. Paper electrophoresis was performed on a Camag high voltage electrophoresis apparatus and thin layer chromatography was carried out on the precoated silica gel plates (Merck, F-254).

The Merrifield resin (chloromethylated copolystyrene–1% divinylbenzene, 0.90 mequiv/g, 200–400 mesh) used in this work was purchased from Cyclo Chemical Co. Amino acid derivatives were obtained from Fox Chemical Co. or Schwarz/Mann. All the Bpoc-amino acids were prepared in this laboratory and were of L configuration unless otherwise stated.

Methyl p-Alkoxybenzoate Resin (II). Merrifield resin (I) (45 g, 40.5 mmol) in 250 ml of dimethylacetamide was allowed to react with 16 g of methyl 4-hydroxybenzoate (105 mmol) and 5.8 g of

(21) E. C. Jorgensen, G. C. Windridge, and T. C. Lee, *J. Med. Chem.*, **13**, 352 (1970).
(22) S. Sakakibara in "Chemistry and Biochemistry of Amino Acids, Peptides and Proteins," Vol. 1, B. Weinstein, Ed., Marcel Dekker, New York, N. Y., 1971, p 51.
(23) L. Bernardi, G. Bosisio, F. Chillemi, G. de Caro, R. de Castriglione, V. Erspamer, A. Glaeser, and O. Goffredo, *Experientia*, **20**, 306 (1964).

NaOCH₃ (107 mmol) at 80° for 24 hr. The resin was collected and washed with DMF, dioxane, CH₂Cl₂, and methanol to give 49.5 g of II. The product absorbed strongly at 1712 and 1220 cm⁻¹ (see Figure 1). It contained 2.69% (0.87 mmol/g) of OCH₃ but no detectable amount of Cl.

p-Alkoxybenzyl Alcohol Resin (III). Resin II (45 g) was treated with 4.6 g of LiAlH₄ in 600 ml of dry ether for 6 hr. The resin was then collected and washed carefully with ethyl acetate, methanol, CH₂Cl₂, and methanol to give a slightly grayish product. The color was removed by stirring in 2 l. of a 1:1 mixture of dioxane and 1 N H₂SO₄ for 45 hr. The snow white product (IIIa, 43 g) contained 0.3% OCH₃ (0.1 mmol/g). The ester band in the ir spectrum disappeared completely.

The same resin can also be prepared directly by treating Merrifield resin (5.1 g, 4.6 mmol) with 4-hydroxybenzyl alcohol (0.74 g, 5.9 mmol) and NaOCH₃ (0.32 g, 6.1 mmol) under similar conditions. The product obtained (IIIb, 5.2 g) gave an identical ir spectrum as that of IIIa. Microanalysis indicated that there was less than 0.07% (0.02 mmol/g) of chlorine.

p-Alkoxybenzyloxycarbonylhydrazide Resin (V). Resin III (8 g) in 70 ml of CH₂Cl₂ was treated with 7.8 ml of phenyl chloroformate and 5.5 ml of pyridine at 0° overnight. The reaction mixture was suction filtered and the resin washed consecutively with cold water, dioxane-water, DMF, CH₂Cl₂, and methanol to give 8.1 g of phenyl carbonate resin IV. It was immediately suspended in 70 ml of DMF and treated with anhydrous hydrazine (6.6 ml). After 6 hr of gentle stirring, the resin was collected and washed to give 6.8 g of the desired compound. Nitrogen analysis indicated that there was 0.88 mmol/g (2.45% N) of hydrazide. The ir spectra of IV and V are shown in Figure 2.

Bpoc-Phenylalanyl p-Alkoxybenzyl Alcohol Resin (VII). p-Alkoxybenzyl alcohol resin IIIa (5 g, 4.4 mmol) was washed several times with CH₂Cl₂ and then allowed to react with 2.5 g of Bpoc-L-Phe (6.3 mmol) and 1.3 g of DCC in the presence of 0.51 ml of pyridine for 150 min. After washings, 5.8 g of Bpoc-Phe-resin (VI) was obtained. It was then treated with 1.65 ml of pyridine and 1.95 ml of benzoyl chloride in 58 ml of CH₂Cl₂ at 0° for 15 min (see next section) to give 6.1 g of VII. The ir spectra of these resins are shown in Figure 3. There was 0.410 mmol of phenylalanine/g of resin according to amino acid analysis. Microanalysis indicated that there was 0.60% N (0.428 mmol/g) in this resin.

When IIIb was esterified with Bpoc-Phe in the same manner as above, the Bpoc-Phe-resin (VII) obtained was found to have 0.406 mmol/g of phenylalanine. The ir spectrum was identical with the corresponding aminoacyl resin described above.

Elimination of the Unreacted Hydroxyl Groups on the Resin. It was found necessary to eliminate the excess unreacted hydroxyl groups left on the resin after the first Bpoc-amino acid was attached. The rate of acylation of resin III by benzoyl chloride or acetic anhydride was therefore studied. A sample of resin III (0.3 g) was suspended in 3 ml of CH₂Cl₂ that contained 0.1 ml of pyridine. Benzoyl chloride (0.11 ml) was added at 0° while the mixture was kept stirred magnetically. Aliquots (0.1 ml) were withdrawn at different intervals and washed immediately with CH₂Cl₂ and methanol. The rate of increase in the ratio of the ester band at 1720 cm⁻¹ to the polystyrene band at 1600 cm⁻¹ in the ir spectra of these samples was taken as the rate of benzoylation. The reaction proceeded quite rapidly to completion with a half-life time of 40 sec. When similar experiments were carried out with tenfold excess of acetic anhydride with triethylamine as base, the reaction was found to proceed only to 70% completion after 60 min at 25°. Addition of a catalytic amount (0.05 equiv) of 4-dimethylaminopyridine[24] to the reaction mixture did bring the reaction to completion within 1 hr. Thus, acetylation with acetic anhydride in the presence of the catalyst appears to be satisfactory also.

Racemization Tests for Bpoc-Amino Acid Resins. Amino acid resin VII (0.5 g) was deprotected with 0.5% TFA (10 min), neutralized, and then coupled with 0.25 g of Boc-L-Glu(But) in the presence of 0.16 g of DCC to form Boc-L-Glu(But)-Phe-resin. The dipeptide was cleaved from the resin with 50% TFA (30 min) and the filtrate chromatographed on an amino acid analyzer in a system similar to that reported by Manning and Moore.[20] The product chromatographed identically with L-Glu-L-Phe. Only a trace amount of L-Glu-D-Phe (less than 0.1%) was detected.

In another experiment, 66 mg of Bpoc-Leu-OCH₂C₆H₄OCH₂-C₆H₄-resin was treated with 2 ml of 50% TFA (30 min) and the liberated free leucine was allowed to react with 4 mg of L-Leu-N-carboxyanhydrate at pH 10.2, 0°. The product (L-Leu-L-Leu) contained very little (less than 0.1%) L-Leu-D-Leu as revealed by the chromatographic analysis.

Bpoc-Glycyl p-Alkoxybenzyl Alcohol Resin (XIX). The alcohol resin III (6.1 g) was washed a few times with dioxane and stirred with 3.3 g of Bpoc-Gly-ONP (7.5 mmol) plus 5.1 g of imidazole (75 mmol) in 50 ml of dioxane for 18 hr. The esterified resin was benzoylated as above to give 7.0 g of XIX. Amino acid analysis indicated that there was 0.36 mmol/g of glycine.

Z-Leu-Leu-Val-Phe (XI). Phenylalanyl resin VII (3.2 g, 1.3 mmol) was placed in the peptide synthesis flask[25] and the solid phase synthesis carried out with 65-ml portions of solvents according to the procedures described previously[15] using a threefold excess (4.2 mmol) of amino acid derivative and DCC in each cycle. Thus, Bpoc-L-Val (1.5 g), Bpoc-L-Leu (1.55 g), and Z-L-Leu (1.1 g) were sequentially coupled to the resin to give 3.8 g of protected tetrapeptide resin X (see Scheme II). According to amino acid analysis, X contained 0.38 mmol/g of peptide and had an amino acid composition of Val₀.₉₈Leu₂.₀₀Phe₁.₂₁. To liberate the protected peptide from the resin, 3.47 g of X was stirred in 70 ml of 50% TFA for 30 min. After removal of the resin particles by filtration and the solvents by evaporation, an oily residue obtained was triturated with petroleum ether. The white powder was taken up in 200 ml of ethyl acetate-ether mixture and left standing overnight at 4°. Crystalline white solid formed slowly during this time: yield, 0.54 g (68.3%); mp 216-219°; nmr spectrum consistent with the structure. It had the amino acid composition of Val₀.₉₃Leu₁.₉₄-Phe₁.₀₈.

Anal. Calcd for C₃₄H₄₈N₄O₇ (624.8): C, 65.36; H, 7.74; N, 8.97. Found: C, 65.32; H, 7.59; N, 8.77.

Z-Leu-Leu-Val-Phe-OCH₃ (XII). Compound XI (0.42 g) was dissolved in 300 ml of 0.2 N methanolic HCl and kept overnight at 4°. Evaporation of the solvent left an oil which was crystallized from ethyl acetate with petroleum ether: yield, 0.35 g (82%); mp 204-206°.

Anal. Calcd for C₃₅H₅₀N₄O₇ (638.8): C, 65.80; H, 7.89; N, 8.77. Found: C, 65.22; H, 7.80; N, 8.78.

Leu-Leu-Val-Phe-OCH₃ (XIII). Hydrogenation of the above compound (XII, 0.57 g) in 20 ml of solvent mixture containing MeOH-THF-acetic acid (10:10:1) at 48 psi for 20 hr in the presence of 0.2 g of catalyst (5% Pd on BaSO₄) gave the tetrapeptide methyl ester free amine in a moderate yield (0.19 g, 43%), mp 157-159°.

Anal. Calcd for C₂₇H₄₄N₄O₅ (504.7): C, 64.20; H, 8.77; N, 11.10; OCH₃, 6.19. Found: C, 63.98; H, 8.72; N, 10.96; OCH₃, 6.09.

Z-Phe-Val-Ala-Leu-HNNH₂ (XVI). Bpoc-L-Leu (0.69 g, 1.8 mmol) was acylated to resin V (1.1 g, 0.96 mmol) by stirring in 15 ml of CH₂Cl₂ for 90 min in the presence of 0.37 g of DCC. The product (XIV, 1.2 g) was found to have 0.86 mmol/g of leucine according to amino acid analysis. A small sample was treated with 50% TFA and the filtrate examined on thin layer chromatography. Only Leu-HNNH₂ was found. There was no free leucine present. These results indicated that the conversion of resin III into resin V had proceeded quite completely. All of the hydroxyl groups on resin III appeared to have been transferred to hydrazide function. Solid phase synthesis[1,15] was carried out on 0.6 g (0.52 mmol) of XIV using 12-ml portions of solvents with a threefold excess of amino acid derivatives and DCC in each cycle. As outlined in Scheme III, Bpoc-L-Ala (0.52 g), Bpoc-L-Val (0.57 g), and Z-L-Phe (0.47 g) were sequentially coupled to the resin. The protected tetrapeptide hydrazide resin XV weighed 0.63 g after drying. Part of the material (0.5 g) was stirred in 10 ml of 50% TFA for 30 min and the peptide released was crystallized from methanol with ether: yield, 0.11 g (42%); mp 252-254° (lit.[15] mp 255-257°).

Anal. Calcd for C₃₁H₄₄N₆O₆ (596.7): C, 62.39; H, 7.43; N, 14.09. Found: C, 62.29; H, 7.33; N, 13.73.

Asp-Arg-Val-Tyr-Val-His-Pro-Phe (XVIII). Bpoc-Phe-resin VII (1.0 g, 0.41 mmol) was placed in the peptide synthesis flask and the solid phase synthesis[15] carried out with 18-ml portions of solvents. In each cycle, fourfold excesses of amino acid derivative and DCC were used. Thus, Bpoc-L-Pro (0.56 g), Bpoc-L-His(Tos) (1.08 g), Bpoc-L-Val (0.57 g), Bpoc-L-Tyr(Bzl) (0.82 g), Bpoc-L-Val (0.57 g), Bpoc-L-Arg(NO₂), and Z-L-Asp(OBzl) (0.57 g) were successively coupled to the resin. The resultant octapeptide resin (1.36 g) was stirred in 26 ml of 50% TFA for 30 min. Protected octapeptide (0.42 g) was isolated as amorphous white solid. It had an amino

---

(24) W. Steglich and G. Höfle, *Angew. Chem.*, **81**, 1001 (1969).

(25) R. B. Merrifield and M. A. Corigliano, *Biochem. Prep.*, **12**, 98 (1968).

acid composition of $Asp_{0.96}Pro_{0.99}Val_{2.34}Tyr_{0.25}Phe_{1.14}His_{0.80}Arg_{1.00}$. Part of the sample (0.35 g) was then dissolved in 2 ml of TFA. Anisole (0.35 ml) was added and the mixture treated with 8 ml of anhydrous HF for 60 min at 0°. Removal of the excess acids left an oily residue which was taken up in 60 ml of water, washed several times with ether, and lyophilized to give 0.17 g of crude peptide. It was then purified by countercurrent distribution in a solvent system made up from n-BuOH–HOAc–H_2O (4:1:5) for 200 transfers ($K = 0.27$) followed by gel filtration on a Sephadex G-10 column (2.5 × 85 cm) using 0.2 M acetic acid as eluent. The material in the major peak was collected and lyophilized to give 56 mg of pure product. It was shown to be homogeneous on thin layer chromatography and paper electrophoresis. On acid hydrolysis, the compound gave the correct amino acid analysis: $Asp_{1.06}$-$Pro_{1.03}Val_{1.93}Tyr_{1.02}Phe_{1.01}His_{0.93}Arg_{1.00}$.

Anal. Calcd for $C_{51}H_{83}N_{13}O_{19}$ (1182.3): C, 51.81; H, 7.08; N, 15.40. Found: C, 51.84; H, 7.06; N, 15.47.

Bpoc-Leu-Met-OCH₃ (XXIV). Bpoc-L-Leu (6.9 g, 18.6 mmol) was allowed to react with 3.73 g of Met-OCH₃·HCl (18.7 mmol) and 3.9 g of DCC in 80 ml of CH_2Cl_2 containing 2.6 ml of triethylamine at 0° for 2 hr. The insoluble by-product formed was filtered off and the filtrate washed a few times with water. The solution was dried over Na_2SO_4 and then concentrated to an oil. It was dissolved in a small volume of CH_2Cl_2 and treated with petroleum ether. Upon cooling, the product started to crystallized slowly: yield, 7.5 g (78%); mp 80–82°; $[\alpha]^{25}D -38.09°$ (c 1, MeOH).

Anal. Calcd for $C_{26}H_{38}N_2O_5S$ (514.7): C, 65.34; H, 7.44; N, 5.44. Found: C, 65.73; H, 7.83; N, 5.38.

Bpoc-Leu-Met-NH₂. Ammonolysis of the above compound (XXIV, 3 g) in 80 ml of methanol that had been saturated with dry ammonia resulted in the formation of the corresponding peptide amide (1.96 g, 67%): mp 99–102°; $[\alpha]^{26}D -37.46°$ (c 1, MeOH).

Anal. Calcd for $C_{17}H_{37}N_3O_4S$ (499.7): C, 64.90; H, 7.46; N, 8.41. Found: C, 64.90; H, 7.78; N, 8.27.

Z-Lys(Z)-Phe-Phe-Gly (XX). Bpoc-Gly-resin XIX (7.0 g, 2.52 mmol) was placed in the peptide synthesis flask and the solid phase synthesis[15] carried out with 150-ml portions of solvents using a 2.4-fold excess of amino acid derivative and DCC in each cycle. As outlined in Scheme V, Bpoc-L-Phe (2.42 g), Bpoc-L-Phe (2.42 g), and Z-L-Lys(Z) (2.43 g) were sequentially coupled to the resin to give 7.5 g of the protected tetrapeptide resin. The peptide was then released from the resin by stirring with 150 ml of 50% TFA for 30 min. After removing the resin particles and the solvents, the oily residue left was treated with 50 ml of ethyl acetate. The solid obtained was dissolved in THF and crystallized by addition of water: yield, 1.02 g (60%); mp 220–222°; $[\alpha]^{25}D -25.55°$ (c 1, DMF); nmr spectrum consistent with the structure.

Anal. Calcd for $C_{42}H_{47}N_5O_9$ (765.9): C, 65.87; H, 6.19; N, 9.14. Found: C, 65.81; H, 6.19; N, 9.14.

Z-Lys(Z)-Phe-Phe-Gly-Leu-Met-OCH₃ (XXI). Bpoc-Leu-Met-OCH₃ (XXIV) (0.52 g, 1.0 mmol) was dissolved in a mixture of 1 ml of 2.4 N HCl in ethyl acetate and 47.5 ml of CH_2Cl_2. After 10-min standing, the solvents were removed at 25° under reduced pressure and the oily residue of the dipeptide hydrochloride was taken up in 20 ml of DMF–CH_2Cl_2 mixture. The solution was cooled to 0° while 0.77 g of Z-Lys(Z)-Phe-Phe-Gly (1 mmol) was added followed immediately by 0.3 ml of N-methylmorpholine and 0.23 g of DCC. The mixture was stirred at 0° overnight. The insoluble material formed was filtered off and the filtrate washed a few times with water, dried over Na_2SO_4, and then evaporated to an oil. It was dissolved in DMF–CH_2Cl_2 mixture and precipitated with ether. The product was crystallized from THF by slow addition of water: yield, 0.85 g (83%), mp 180–184°.

Anal. Calcd for $C_{54}H_{69}N_7O_{11}S$ (1024.3): C, 63.32; H, 6.79; N, 9.57. Found: C, 63.87; H, 6.74; N, 9.54.

Z-Lys(Z)-Phe-Phe-Gly-Leu-Met-NH₂ (XXII). The above compound XXI (0.75 g, 0.73 mmol) was suspended in 100 ml of dry methanol and bubbled with dry ammonia gas for 2 hr at 0°. The compound became soluble in the solution but started to crystallize out slowly during overnight standing at room temperature. The product was collected and washed with ether to give 0.58 g of the desired compound: mp 238–242°; $[\alpha]^{26}D -39.28°$ (c 1, DMF).

Anal. Calcd for $C_{53}H_{68}N_8O_{10}S$ (1009.3): C, 63.08; H, 6.70; N, 11.10; S, 3.18. Found: C, 62.79; H, 6.70; N, 11.25; S, 2.90.

Lys-Phe-Phe-Gly-Leu-Met-NH₂ (XXIII). Compound XXII (0.15 g) was dissolved in 10 ml of TFA containing 0.5 ml of mercaptoethanol as well as 1 ml of anisole. The mixture was warmed at 80° for 3 hr during which time some white insoluble material came out of the solution. It was filtered off and the filtrate was treated with a large volume of ether to precipitate the product. The crude peptide was then purified by countercurrent distribution in a solvent system of n-BuOH–HOAc–pyridine–H_2O (8:2:2:9) for 300 transfers ($K = 2.8$) followed by gel filtration on a Sephadex G-10 column (2.5 × 85 cm) using 0.2 M acetic acid as eluent. The material in the main fraction was collected and lyophilized to give 33 mg of pure product. It gave correct amino acid analyses upon acid hydrolysis: $Gly_{1.00}$-$Met_{0.93}Leu_{1.03}Phe_{2.11}Lys_{0.92}$. The product was shown to be homogeneous on thin layer chromatography and paper electrophoresis.

Anal. Calcd for $C_{37}H_{56}N_8O_6S \cdot 2CH_3COOH$ (861.1): C, 57.19; H, 7.49; N, 13.01. Found: C, 57.38; H, 7.61; N, 13.00.

Acknowledgment. The author wishes to thank Dr. R. B. Merrifield for helpful discussions and Drs. G. Saucy and A. M. Felix for valuable suggestions.

Azide Solid Phase Peptide Synthesis[1]

Arthur M. Felix[2] and R. B. Merrifield

Contribution from the Rockefeller University, New York, New York 10021. Received June 19, 1969

Abstract: The stepwise synthesis of peptides by the initial attachment of an amino acid *t*-butyloxycarbonylhydrazide through its α-amino group to a polystyrene resin was investigated. The resin–amino acid azide was generated quantitatively and coupled with another amino acid *t*-butyloxycarbonylhydrazide. The peptide chain was elongated by further azide couplings. Finally, the C-terminal amino acid was added as a *t*-butyl ester. The peptide was deprotected and removed from the resin in one step with HBr. The feasibility of the approach was demonstrated by the synthesis of L-leucylglycine and L-leucyl-L-alanylglycyl-L-valine by this stepwise procedure and also by a fragment condensation. Studies on the stability and reactivity of the intermediate azides were carried out.

Modifications in the solid phase peptides synthesis procedure[3] have been directed toward the resin,[4] the coupling reaction,[4a,c,5] the N-protecting group,[1e,5e,6] and the cleavage step.[1d,5a,6a,7] In general the method has involved the attachment of the carboxyl group of the first amino acid to a resin, followed by peptide chain elongation at the amino end.[3] There has been one instance[8,9] in which an amino acid was attached to the resin through its amino group while the carboxyl was protected as an ethyl ester. However, the procedure described in those early experiments was somewhat limited because of the danger of racemization during the subsequent deprotection and coupling steps. In order to develop this general approach further it is necessary to find suitable carboxyl protection and efficient, racemization-free coupling methods. The acid azide route of classical peptide synthesis[10] is known to yield optically pure products[11] not only when amino acids are coupled, but even when the carboxyl groups of peptides are activated. The modified approach of Rudinger and Honzl[12] to the azide reaction utilizing a nonaqueous system, and the use of 1-aminoacyl-2-*t*-butyloxy-

---

(1) Supported in part by Grant No. A-1260 from the U. S. Public Health Service.
(2) Guest investigator (1968–1969) from Hoffmann-La Roche, Inc., Nutley, N. J.
(3) (a) R. B. Merrifield, *J. Amer. Chem. Soc.*, **85**, 2149 (1963); (b) R. B. Merrifield, *Advan. Enzym.*, **32**, 221 (1969).
(4) (a) M. M. Shemyakin, Y. A. Ovchinnikov, A. A. Kiryushkin, and I. V. Kozhevnikova, *Tetrahedron Lett.*, **27**, 2323 (1965); (b) G. I. Tesser and B. W. J. Ellenbroek, *Proc. Eur. Peptide Symp., 8th, Noordwijk* 124 (1967); (c) M. A. Tilak and C. S. Hollinden, *Tetrahedron Lett.*, 1297 (1968); (d) F. Weygand, *Proc. Eur. Peptide Symp., 9th, Orsay*, 183 (1968); (e) N. Inukai, K. Nokano, and M. Murakami, *Bull. Chem. Soc., Jap.*, **211**, 182 (1968); (f) F. Camble, R. Garrier, and G. T. Young, *Nature*, **217**, 248 (1968); (g) B. Green and L. R. Garson, *J. Chem. Soc., C*, 401 (1969).
(5) (a) M. Bodanszky and J. T. Sheehan, *Chem. Ind.*, 1597 (1966); (b) F. M. Bumpus, M. C. Khosla, and R. R. Smeby, Abstracts, 153rd Meeting of the American Chemical Society, Division of Medicinal Chemistry, Miami, Fla., 1967; (c) M. Rothe and H. Schneider, *Angew. Chem. Intern. Ed. Engl.*, **5**, 417 (1966); (d) H. Klostermeyer, H. Halstrøm, P. Kusch, J. Föhles, and W. Lunkenheimer, *Proc. Eur. Peptide Symp., 8th, Noordwijk*, 113 (1967); (e) F. Weygand and U. Ragnarsson, *Z. Naturforsch.*, **21b**, 1141 (1966).
(6) (a) Kessler and B. Iselin, *Helv. Chim. Acta*, **49**, 1330 (1966); (b) V. A. Najjar and R. B. Merrifield, *Biochemistry*, **5**, 3765 (1966); (c) A. Deér, *Angew. Chem. Intern. Ed. Engl.*, **5**, 1041 (1966); (d) G. Losse and K. Noubert, *Z. Physiol.*, **8**, 278 (1968); (e) P. Sieber and B. Iselin, *Helv. Chim. Acta*, **51**, 614 (1968).
(7) (a) J. Lenard and A. B. Robinson, *J. Amer. Chem. Soc.*, **89**, 181 (1967); (b) A. Loffet, *Experientia*, **23**, 406 (1967); (c) L. Stryer and R. P. Haugland, *Proc. Natl. Acad. Sci. U. S.*, **58**, 719 (1967); (d) M.

Ohno and C. B. Anfinsen, *J. Amer. Chem. Soc.*, **89**, 5994 (1967); (e) H. C. Beyerman, H. Hindriks, and E. W. B. DeLeer, *Chem. Commun.*, 1668 (1968); (f) B. Halpern, L. Chew, V. Close, and W. Patton, *Tetrahedron Lett.*, 5163 (1968).
(8) R. L. Letsinger and M. J. Kornet, *J. Amer. Chem. Soc.*, **85**, 3045 (1963).
(9) R. L. Letsinger, M. J. Kornet, V. Mahadevan, and D. M. Jerina, *ibid.*, **86**, 5163 (1964).
(10) T. Curtius, *Ber.*, **35**, 3226 (1902).
(11) N. H. Smart, G. T. Young, and M. W. Williams, *J. Chem. Soc.*, 3902 (1960).
(12) J. Honzl and J. Rudinger, *Collect. Czech. Chem. Commun.*, **26**, 2333 (1961).

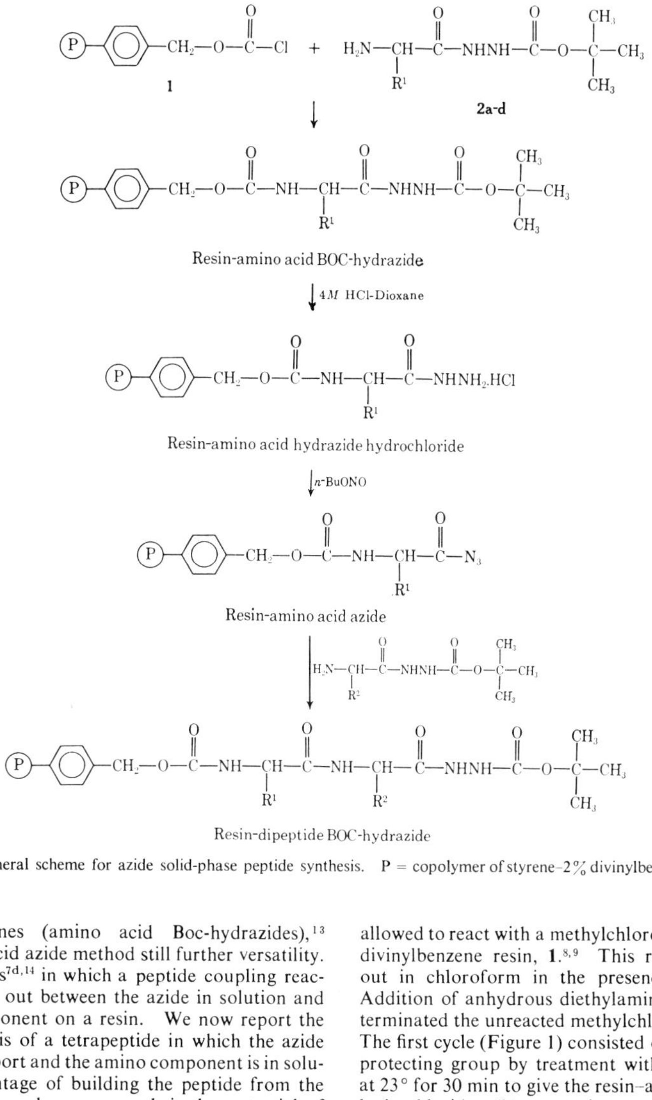

Figure 1. The general scheme for azide solid-phase peptide synthesis. P = copolymer of styrene-2% divinylbenzene.

carbonylhydrazines (amino acid Boc-hydrazides),[13] have given the acid azide method still further versatility. There are reports[7d,14] in which a peptide coupling reaction was carried out between the azide in solution and the amino component on a resin. We now report the stepwise synthesis of a tetrapeptide in which the azide is on a resin support and the amino component is in solution. An advantage of building the peptide from the carboxyl end by such an approach is the potential of coupling fragments without racemization. Moreover, this approach has application in special cases where a series of analogs with variations near the carboxyl end of the peptide is desired.

Synthesis of Peptides

The general procedure for the azide solid phase peptide synthesis is outlined in Figures 1 and 2. In the first step, 0.5 equiv of L-leucine Boc-hydrazide,[15] was

(13) (a) R. Schwyzer, Angew. Chem., 71, 742 (1959); (b) R. Schwyzer, E. Surbeck-Wegmann, and H. Dietrich, Chimia (Aarau), 14, 366 (1960); (c) R. A. Boissonnas, S. Guttmann, and P. A. Jacquenoud, Helv., 43, 1349 (1960).
(14) D. W. Woolley, unpublished results. See also ref 3b.
(15) E. Wünsch, F. Drees, and J. Jentsch, Ber., 98, 803 (1965).

allowed to react with a methylchloroformylated styrene–divinylbenzene resin, 1.[8,9] This reaction was carried out in chloroform in the presence of triethylamine. Addition of anhydrous diethylamine or dimethylamine terminated the unreacted methylchloroformylated resin. The first cycle (Figure 1) consisted of (a) removal of the protecting group by treatment with 4 M HCl–dioxane at 23° for 30 min to give the resin–amino acid hydrazide hydrochloride; (b) conversion to the resin–amino acid azide by treatment with n-butylnitrite in tetrahydrofuran at $-30°$ for 1 hr; and (c) coupling with an amino acid Boc-hydrazide in tetrahydrofuran using temperature steps of -30, 0, and 25°. Excess reagents and byproducts were removed by washing with tetrahydrofuran and methanol. This coupling cycle was repeated with other amino acid Boc-hydrazides. The last cycle of the synthesis (Figure 2) consisted of (a) removal of the protecting group, (b) conversion to the resin–peptide azide, and (c) coupling with the amino acid t-butyl ester corresponding to the C-terminal residue of the peptide. The t-butyl ester protecting group gave better yields than the benzyl ester in our system and was removed equally well by the HBr–TFA used for the cleavage of the peptide

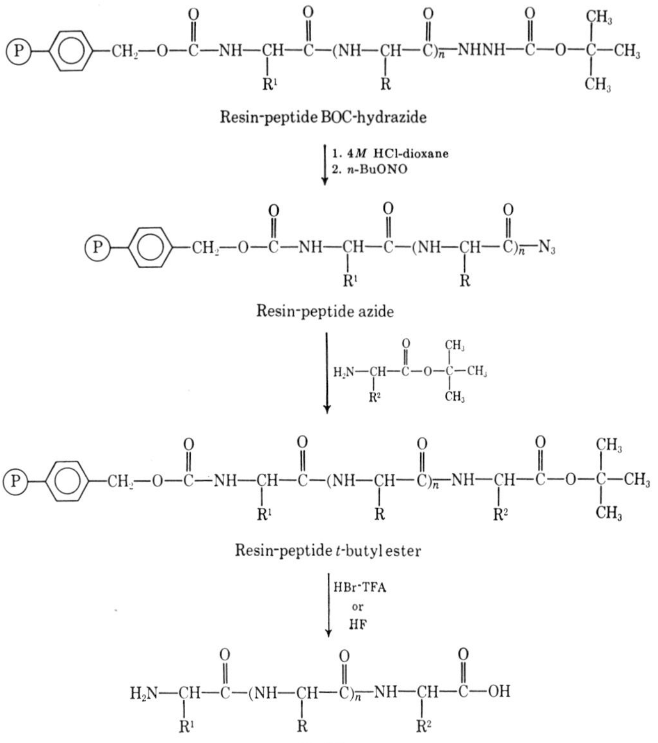

Azide Solid Phase Peptide Synthesis

Resin-peptide BOC-hydrazide

1. 4M HCl-dioxane
2. n-BuONO

Resin-peptide azide

Resin-peptide t-butyl ester

HBr·TFA
or
HF

Peptide

Figure 2. The termination and cleavage steps for azide solid-phase peptide synthesis.

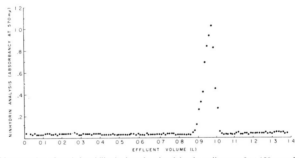

Figure 3. Chromatographic separation of crude lyophilized L-leucyl-L-alanylglycyl-L-valine on a 2 × 120 cm column of Dowex 50-X4 resin. Elution was with pyridine acetate buffer, pH 4.0, 0.5 M in acetate, with a gradient of 104 ml/hr. Aliquots (0.2 ml) from 5.5-ml fractions were analyzed by the ninhydrin test.

from the resin.[16] In order to demonstrate the feasibility of this approach the tetrapeptide, L-leucyl-L-alanylglycyl-L-valine, was synthesized. The crude product, after cleavage from the resin contained a minor contaminant which could be separated by column chromatography on Dowex 50-X4. An overall yield of 30% of purified tetrapeptide was obtained by this stepwise procedure.

The versatility of this method was illustrated by the synthesis of the same tetrapeptide, in better yield, by fragment condensation. In this instance, resin-L-leucyl-L-alanine azide prepared as described above, was coupled with glycyl-L-valine t-butyl ester. Cleavage of peptide from resin afforded the tetrapeptide. Chromatography of this product on a Dowex 50-X4 resin by elution with 0.5 M pyridine acetate buffer (pH 4) as shown in Figure 3 gave a single, homogeneous peak. The product gave one spot on thin-layer chromatography in each of three solvent systems. It was also

(16) R. B. Merrifield, *J. Amer. Chem. Soc.*, **86**, 304 (1964).

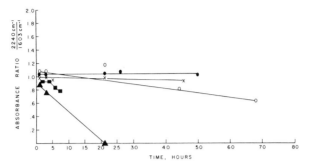

Figure 4. Stability in tetrahydrofuran: ●, resin-L-leucine azide at −30°; ×, resin-L-leucine azide at −10°; ○, resin-L-leucine azide at 4°; ▲, resin-L-leucine azide at 23°; ■, resin-L-cysteine azide at 23°.

judged to be pure by amino acid analysis and by comparison with an authentic sample.[3a] The overall yield of purified tetrapeptide was 60% (calculated from the amount of resin-L-leucine Boc-hydrazide). Fragment condensation of resin-L-leucine azide was also carried out successfully with L-alanylglycine Boc-hydrazide. This intermediate was essentially identical with that prepared by stepwise condensation.

L-Leucylglycine was prepared in a similar manner, i.e., resin-L-leucine azide was coupled with glycine t-butyl ester. Cleavage of peptide from resin and chromatography on Dowex 50-X4 afforded a product that was identical with a commercial sample of L-leucylglycine.[17] The overall yield of purified dipeptide was 57% (calculated from the resin-L-leucine Boc-hydrazide). It was judged pure by thin-layer chromatography, amino acid analysis, and elemental analysis.

Kinetic Studies

The major problems that have been encountered in the past with the acid azide method have been the formation of isocyanates by the Curtius rearrangement and the formation of amides. The former can lead to urea-containing peptides and the latter results in chain termination. The stability of peptide azides in dimethylformamide has been studied by Schwyzer[18] and Katsoyannis[19] who showed that the azides gave isocyanates at ambient temperature but were stable at 0°. Rudinger and Honzl[12] demonstrated that although the side reactions of carbobenzoxy-S-benzyl-L-cysteine azide occurred they could be suppressed in nonaqueous homogeneous systems. It was then necessary to determine if these side reactions occurred in the solid phase system and if they could also be suppressed.

Resin-L-leucine azide was prepared as outlined above at −30° in tetrahydrofuran. It was filtered, washed with cold tetrahydrofuran, isolated in the "dry state," and allowed to come to room temperature. An infrared spectrum of this pale yellow resin in a KBr pellet revealed a strong band at 2240 cm^{-1}. The position and intensity of this band (relative to a reference band at 1603 cm^{-1} from the same spectrum) was unchanged even when the resin was left exposed in the laboratory for many days. Grinding and reforming the pellet also had no effect on the spectrum. When this resin–azide was reacted with L-alanine Boc-hydrazide, using the temperature gradient described above, resin-L-leucyl-L-alanine Boc-hydrazide was obtained in high yield. Since there was no evidence of urea formation[20] the band at 2240 cm^{-1} was assigned to the azide function, rather than to isocyanate. It was concluded that the resin-L-leucine azide was stable for more than 10 days at room temperature in the dry state.

A new sample of the resin-L-leucine azide was prepared and allowed to stand at −30° in tetrahydrofuran. Aliquots were removed at various times and infrared spectra were measured as outlined above. It was determined that resin-L-leucine azide was stable in tetrahydrofuran at −30° for at least 50 hr. Since derivatives of cysteine are known to undergo side reactions readily,[21] resin S-p-methoxybenzyl-L-cysteine azide was also prepared and its stability examined in the same manner. These measurements indicated that the stability of resin-S-p-methoxybenzyl-L-cysteine azide was parallel with that of resin-L-leucine azide (i.e., no change was observed in the 2240 cm^{-1}:1603 cm^{-1} ratio for at least 24 hr). Studies were also carried out on the stability of resin-L-leucine azide and resin-S-p-methoxybenzyl-L-cysteine azide at higher temperature in tetrahydrofuran using the same parameters as outlined above (Figure 4). It was determined that the resin–azide was stable at −30 and −10° for longer than 40 hr but that there was some decomposition at 4° over a prolonged period of time. This decomposition was pronounced at 23° and the azide band completely disappeared within 24 hr. Since we failed to observe the formation of a new band in the 2240-cm^{-1} region, characteristic for the isocyanate, it was concluded that a Curtius rearrangement had not occurred even at the higher temperature but, rather, that the resin-azide had decomposed to give the amide.[12] It should be noted that preliminary experiments with resin-L-asparagine azide have revealed the presence of a second band at 2140 cm^{-1}, which is under further study.

The rate of reaction of resin-L-leucine azide and resin-S-p-methoxybenzyl-L-cysteine azide with L-alanine Boc-

---

(17) Mann Research Laboratories, New York, N. Y.
(18) R. Schwyzer and H. Kappeler, Helv. Chem. Acta, 44, 1991 (1961).
(19) P. G. Katsoyannis, A. M. Tometsko, C. Zalut, and K. Fukuda, J. Amer. Chem. Soc., 88, 5625 (1966).

(20) Amino acid analysis of the resin hydrolysate showed equimolar amounts of leucine and alanine. If urea formation had occurred via an isocyanate intermediate only alanine would have been found.
(21) (a) R. Roeske, F. M. C. Stewart, R. J. Stedman, and V. du Vigneaud, J. Amer. Chem. Soc., 78, 5883 (1956); (b) R. B. Merrifield and D. W. Woolley, ibid., 80, 6635 (1958).

Azide Solid Phase Peptide Synthesis

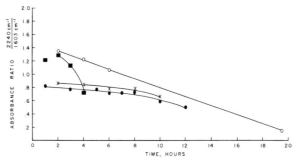

Figure 5. Reactivity in tetrahydrofuran with L-alanine Boc-hydrazide: ●, resin-L-leucine azide at $-30°$; ×, resin-L-leucine azide at $-10°$ ○, resin-L-cysteine azide at 4°; ■, resin-L-cysteine azide at 23°.

hydrazide (Figure 5) was also examined by following the disappearance of the azide band at 2240 cm^{-1}. These studies revealed that the coupling reaction proceeded very slowly at -30 and $-10°$ but that at 4° it was almost complete within 20 hr. Although coupling was very rapid at 23°, the rate of decomposition at this temperature (Figure 4) was appreciable. It was concluded from this kinetic study that the resin–azide should ideally be generated at $-30°$ and coupled at temperatures between -10 and 4°. Under these conditions we have no evidence for rearrangement of the resin-azide.

The generation of resin-L-leucine azide and the subsequent coupling reactions were also studied in other solvent systems (dimethylformamide, dioxane, methylene chloride, and chloroform). It was found that preparation and coupling of resin azides were best carried out in tetrahydrofuran. Variable results were obtained with dimethylformamide. Dioxane, methylene chloride, and chloroform were poor solvents in this system. The overall yields are presently marginal and we do not recommend this new approach as a general replacement for solid phase peptide synthesis. However, it may be a valuable supplement to peptide synthesis and demonstrates the versatility of the solid phase method.

Experimental Section

All analytical samples were dried for 24 hr at 56° under high vacuum. All melting points are uncorrected. Infrared spectra were recorded on a Perkin-Elmer 621 grating spectrophotometer. Nuclear magnetic resonance (nmr) spectra were recorded for all compounds on a Varian Associates A-60 spectrometer with 2% tetramethylsilane as internal standard. Optical rotations were taken in a jacketed 1-dm cell on a Perkin-Elmer Model 140 polarimeter. Thin-layer chromatography was carried out on plates prepared with silica gel G and developed with buffered ninhydrin reagent.[22] Mass spectra were determined with a (CEC) 21-110 spectrometer. Tetrahydrofuran (reagent grade) was freed of peroxides by passage through a column of alumina. All other reagents were obtained commercially and used without further purification. The reaction vessel for peptide synthesis, as described previously,[3a] was jacketed and cooled by a Lauda TK-30 circulator to maintain the required temperature. Amino acid analyses were performed on the spinco Amino Acid Analyzer, Model 120B. Resin hydrolysates were carried out on 10–20-mg samples of resin with concentrated HCl, glacial acetic acid, and phenol (2:1:1) at 110° for 18–24 hr.

Methylchloroformylated Resin (1).[8,9] Chloromethylated copolystyrene-2% divinylbenzene (1.04 mmol/g; 25 g, 26 mmol), suspended in 200 ml of methyl Cellosolve, was treated with potassium acetate (7.0 g, 71.3 mmol) and heated at 130° for 64 hr. The reaction mixture was filtered, and washed with water and methanol. The product was converted to the hydroxymethylated resin by saponification with 150 ml of 0.5 M NaOH (75 mmol) at 23° for 47 hr. It was filtered, washed with water and methanol, and dried in vacuo. Treatment of this resin with 200 ml of 1.27 M phosgene in benzene (254 mmol) at 23° for 4 hr followed by filtration and washing with benzene and ether, afforded methylchloroformylated resin, 1 (0.718 mmol of chloride/g of resin). The infrared spectra of the hydroxymethylated and chloroformylated resins were identical with those reported by Letsinger.[9]

1-L-Leucyl-2-t-butyloxycarbonylhydrazine (L-Leucine Boc-hydrazide) (2a). N-Carbobenzoxy-L-leucine (15.0 g, 0.056 mol) and t-butylcarbazate (9.6 g, 0.072 mol) in 250 ml of ethyl acetate at 0° were treated with dicyclohexylcarbodiimide (12.4 g, 0.06 mol). The reaction mixture was stirred for 2 hr at 0° and 4 hr at 23°. It was filtered, extracted in turn with 1 M citric acid, saturated NaHCO$_3$, saturated KCl, and evaporated to dryness. The residue was taken up in 160 ml of methanol, treated with 4 g of 5% Pd-BaSO$_4$, and hydrogenated for 19 hr at 40 psi. Evaporation of solvent and recrystallization from ethyl acetate–petroleum ether afforded 6.6 g (47%) of product: mp 112.0–114.0°; $[\alpha]^{20}$D 21.77° (c, 2.35; methanol); lit.[15] mp 114–116°; $[\alpha]^{20}$D 21.54° (c, 2.2; methanol).

Anal. Calcd for C$_{11}$H$_{23}$N$_3$O$_3$: C, 53.86; H, 9.45; N, 17.13. Found: C, 53.95; H, 9.46; N, 17.10.

L-Alanine Boc-hydrazide (2b). N-Carbobenzoxy-L-alanine (6.7 g, 0.03 mol) and t-butylcarbazate (4.8 g, 0.036 mol) in 100 ml of ethyl acetate at 0° were treated with dicyclohexylcarbodiimide (6.2 g, 0.03 mol) and worked up as described above for L-leucine. Hydrogenation for 3 hr in methanol over 2 g of 5% Pd-BaSO$_4$, followed by crystallization from ethyl acetate–petroleum ether, afforded 4.0 g (66%) of product: mp 86.0–90.5° $[\alpha]^{20}$D 6.29° (c, 2.07; methanol).

Anal. Calcd for C$_8$H$_{17}$N$_3$O$_3$: C, 47.28; H, 8.43; N, 20.67. Found: C, 47.50; H, 8.38; N, 20.48.

Glycine Boc-hydrazide. N-Carbobenzoxyglycine (6.3 g, 0.03 mol) and t-butylcarbazate (4.8 g, 0.036 mole) in 120 ml of ethyl acetate at 0° were treated with dicyclohexylcarbodiimide (6.2 g, 0.03 mol) and worked up in the usual manner. Hydrogenation for 12 hr in methanol over 2 g of 5% Pd-BaSO$_4$, followed by crystallization from ethyl acetate–petroleum gave 3.2 g (56%) of crystalline product, mp 130–132.5°.

Anal. Calcd for C$_7$H$_{15}$N$_3$O$_3$: C, 44.43; H, 7.99; N, 22.21. Found: C, 44.69; H, 8.03; N, 21.98.

L-Alanylglycine Boc-hydrazide. N-Carbobenzoxy-L-alanine (1.34 g, 6.0 mmol) and glycine Boc-hydrazide (2c) (1.13 g, 6.0 mmole) in 50 ml of ethyl acetate at 0° were allowed to react with dicyclohexylcarbodiimide (1.24 g, 6.0 mmol). The reaction mixture was stirred for 1.5 hr at 0° and 17 hr at 23°. It was filtered, evaporated to dryness, taken up in 50 ml of methanol, treated with 2.5 g of 5% Pd-BaSO$_4$, and hydrogenated for 5 hr at 40 psi. Evaporation of solvent and crystallization from methanol–ethyl acetate–petroleum ether gave 920 mg (59%) of crystalline product: mp 129.0–132.0° dec; $[\alpha]^{20}$D 7.52° (c 1.93; methanol).

Anal. Calcd for C$_{10}$H$_{20}$N$_4$O$_4 \cdot \frac{1}{2}$ H$_2$O: C, 44.65; H, 7.86; N, 20.88. Found: C, 44.88; H, 7.89; N, 20.46.

N-Trifluoroacetyl-S-p-methoxybenzyl-L-cysteine. This compound was prepared by the procedure of Schallenberg and Calvin.[23]

---

(22) J. M. Stewart and J. D. Young, "Solid Phase Peptide Synthesis," W. H. Freeman and Co., San Francisco, Calif., 1969.

A mixture of S-*p*-methoxybenzyl-L-cysteine[24] (4.82 g, 20 mmol), 1.0 N NaOH (20 ml, 20 mmol), saturated sodium borate (80 ml), and ethyl thioltrifluoroacetate (5.54 g, 31.2 mmol) was shaken at 23° for 7 hr. The reaction mixture was extracted three times with ether, acidified with 1 M HCl, and reextracted three times with ether. This latter ethereal extract was dried over $MgSO_4$, evaporated to dryness, and crystallized from benzene–petroleum ether. There was obtained 2.75 g (41%) of product: mp 94.0–95.0° dec; $[\alpha]^{20}D$ −76.01° (c 1.82; methanol).

Anal. Calcd for $C_{13}H_{14}NO_4F_3S$: C, 46.29; H, 4.18; N, 4.15. Found: C, 46.00; H, 4.08; N, 4.12.

N-Trifluoroacetyl-S-*p*-methoxybenzyl-L-cysteine Boc-hydrazide. N-Trifluoroacetyl-S-*p*-methoxybenzyl-L-cysteine (5.63 g, 16.7 mmol) and *t*-butylcarbazate (2.21 g, 16.7 mmol) in 75 ml of ethyl acetate at 0° were treated with dicyclohexylcarbodiimide (3.45 g, 16.7 mmol). The reaction mixture was stirred for 2 hr at 0° and 20 hr at 23°. Filtration, evaporation to dryness, and crystallization from ethyl acetate–petroleum ether afforded 5.8 g (77%) of white crystals: mp 110.0–111.5°; $[\alpha]^{20}D$ −41.65° (c 1.82; methanol).

Anal. Calcd for $C_{18}H_{24}N_3O_5F_3S$: C, 47.89; H, 5.36; N, 9.31. Found: C, 47.59; H, 5.39; N, 9.27.

S-*p*-Methoxybenzyl-L-cysteine Boc-hydrazide (2d). Deprotection of the precursor was carried out by a modification of the procedure of Weygand and Reiher.[25,26] N-Trifluoroacetyl-S-*p*-methoxybenzyl-L-cysteine Boc-hydrazide (5.75 g, 12.7 mmol) in 60 ml of absolute ethanol was treated with 135 ml of 6.3 M NH_4OH and stirred for 2 hr at 23°. The reaction mixture was evaporated to dryness, treated with 25 ml of water, and reevaporated. This residue was evaporated from another 25 ml of water, taken up in ethyl acetate, and extracted three times with cold saturated aqueous Na_2CO_3. The ethyl acetate layer was dried over $MgSO_4$, filtered, evaporated to dryness, and pumped out *in vacuo*. There was obtained 3.79 g (84%) of an amorphous solid $[\alpha]^{20}D$ −1.49° (c 2.01; methanol); tlc showed 1 ninhydrin positive spot with R_f 0.54 (methanol:chloroform; 1:9). The infrared spectrum of **2d** in KBr was consistent with expectation. Characteristic bands were observed at 1620, 1590, 1520, 1310, and 1265 cm^{-1}. The nmr spectrum (NaOD–D$_2$O) was also consistent with expectation giving peaks at 3.26 (m, 2 H), 3.88 (m, 1 H), 4.14 (s, 2 H), 4.18 (s, 3 H), and 7.61 ppm (aromatic A_2B_2 type, 4 H).

Glycyl-L-valine *t*-Butyl Ester. N-Carbobenzoxyglycine (2.09 g, 10 mmol) and L-valine *t*-butyl ester[27] (1.73 g, 10 mmol) in 50 ml of tetrahydrofuran at 0° were treated with dicyclohexylcarbodiimide (2.06 g, 10 mmol). The reaction mixture was stirred for 2 hr at 0° and 48 hr at 23°. It was filtered, evaporated to dryness, taken up in ethyl acetate, and extracted in turn with saturated $NaHCO_3$, saturated KCl, and 1 M citric acid. The ethyl acetate layer was dried over $MgSO_4$, filtered, evaporated to dryness and pumped out *in vacuo*. The resultant oil was taken up in 100 ml of methanol, treated with 2 g of 5% Pd–C, and hydrogenated at 45 psi for 12 hr. Filtration, and evaporation, afforded 1.83 g (84%) of a viscous pale yellow oil which resisted all attempts at crystallization: tlc (CHCl$_3$) gave one major spot at R_f 0.13; $[\alpha]^{20}D$ −21.55° (c 2.00; methanol). The infrared spectrum in CHCl$_3$ was consistent with expectation. Characteristic bands were observed at 3415, 3340, 1735, 1675, and 1525 cm^{-1}. The nmr spectrum (CDCl$_3$) was also consistent with expectation giving peaks at 0.94 (d, 6 H), 1.48 (s, 9 H), 2.25 (m, 1 H), 3.66 (m, 1 H), 4.17 (m, 2 H), 4.67 (m, 2 H, exchangeable), and 7.92 ppm (1 H, exchangeable).

Resin-L-leucine Boc-hydrazide. The methylchloroformylated resin (**1**) (5.0 g, 3.6 mmol of Cl) was suspended in 75 ml of chloroform and treated with L-leucine Boc-hydrazide (**2a**) (442 mg, 1.8 mmol), followed by triethylamine (0.25 ml, 1.8 mmol). The reaction mixture was stirred for 2 hr at 23° and was treated with dimethylamine (1.6 g, 36 mmol) in 15 ml of chloroform. After stirring for 3 more hours at 23°, the resin was filtered, washed with chloroform, and dried *in vacuo*. Amino acid analysis on a resin hydrolysate revealed a substitution of 0.297 mmol of L-leucine per gram of resin.

Addition of larger amounts of L-leucine Boc-hydrazide gave more highly substituted resin. Therefore, methylchloroformylated resin (**1**) (5.0 g, 3.6 mmol) was suspended in 50 ml of chloroform and treated with L-leucine Boc-hydrazide (**2a**) (884 mg, 3.6 mmol), followed by triethylamine (0.50 ml, 3.6 mmol). The reaction mixture was stirred for 2 hr at 23°, treated with diethylamine (2.6 g, 36 mmol) in 15 ml of chloroform, and stirred overnight at 23°. The resin was filtered, washed with chloroform, and dried *in vacuo*. Amino acid analysis of a resin hydrolysate revealed a substitution of 0.503 mmol of L-leucine per gram of resin.

Resin-S-*p*-methoxybenzyl-L-cysteine Boc-hydrazide. The methylchloroformylated resin (**1**) (3.0 g, 2.15 mmol) was suspended in 50 ml of chloroform and treated with S-*p*-methoxybenzyl-L-cysteine Boc-hydrazide (**2d**) (2.14 g, 6.02 mmol) followed by triethylamine (0.84 ml, 6.00 mmol). The reaction mixture was stirred for 42 hr at 23°. Dimethylamine (2.2 g, 49 mmol) in 10 ml of chloroform was added and stirring continued for 2 hr. The product was filtered, washed with chloroform, and dried *in vacuo*. Amino acid analysis on a resin hydrolysate revealed total cysteine substitution of 0.273 mmol/g of resin.

Resin-L-leucine Azide for Kinetic Studies. Resin-L-leucine Boc-hydrazide (100 mg, 0.03 mmol) was placed in the reaction vessel and shaken with 50 ml of 4 M HCl–dioxane for 30 min at 23°. The resin was filtered, and washed three times with 50-ml portions of tetrahydrofuran. Tetrahydrofuran (50 ml) was added and the reaction vessel was cooled to −30°. It was opened momentarily and *n*-butyl nitrite (0.02 ml, 0.16 mmol) followed by 4 M HCl–dioxane (0.04 ml, 0.16 mmol) was added. The reaction mixture was rocked at −30° for 1 hr, filtered and washed three times with 50-ml portions of precooled tetrahydrofuran at −30°. The resultant resin-L-leucine azide was used directly in the kinetic stability studies. For the kinetic reactivity studies, resin-L-leucine azide was washed three times with 50-ml portions of precooled 10% Et$_3$N–THF at −30°. The resin was washed with 50-ml portions of precooled tetrahydrofuran at −30°. A solution of L-alanine Boc-hydrazide (**2b**) (31 mg, 0.15 mmol) in 50 ml of precooled tetrahydrofuran at −30° was added. The required temperature for the kinetic reactivity study was attained, aliquots were removed, rapidly filtered, and washed with precooled tetrahydrofuran.

Resin-S-*p*-methoxybenzyl-L-cysteine Azide for Kinetic Studies. Resin-S-*p*-methoxybenzyl-L-cysteine Boc-hydrazide (400 mg, 0.11 mmol) was placed in the reaction vessel and shaken with 50 ml of 4 M HCl–dioxane for 30 min at 23°. The resin was filtered and washed three times with 50-ml portions of tetrahydrofuran. Tetrahydrofuran (50 ml) was added and the reaction vessel was cooled to −30°. It was opened momentarily and *n*-butyl nitrite (0.08 ml, 0.60 mmol) followed by 4 M HCl–dioxane (0.15 ml, 0.60 mmol) was added. The reaction mixture was rocked at −30° for 1 hr, filtered and washed three times with 50-ml portions of precooled tetrahydrofuran at −30°. The resultant resin-S-*p*-methoxybenzyl-L-cysteine azide was used directly in the kinetic stability studies. For the kinetic reactivity studies, resin-S-*p*-methoxybenzyl-L-cysteine azide (150 mg, 0.041 mmol) was washed three times with 50-ml portions of precooled 10% Et$_3$N–THF at −30°. The resin was then washed with 50-ml portions of precooled tetrahydrofuran at −30°. A solution of L-alanine Boc-hydrazide (**2b**) (41 mg, 0.20 mmol) in 50 ml of precooled tetrahydrofuran at −30° was added. The required temperature was attained, aliquots were removed, rapidly filtered, and washed with precooled tetrahydrofuran.

L-Leucylglycine. Resin-L-leucine Boc-hydrazide (1.5 g, 0.446 mmol) was introduced into the jacketed reaction vessel. The following cycle of reactions was employed: (1) *t*-Boc group cleaved by 4 M HCl–dioxane (50 ml) for 30 min; (2) washed with three 50-ml portions of tetrahydrofuran; (3) added 50 ml of tetrahydrofuran and cooled the vessel to −30°; (4) generated the azide by addition of 4 M HCl–dioxane (0.56 ml, 2.23 mmol; 5 meq), followed by *n*-butylnitrite (0.28 ml, 2.23 mmol, 5 meq) for 1 hr; (5) washed with three 50-ml portions of precooled 10% Et$_3$N–THF at −30°; (6) washed with three 50-ml portions of precooled tetrahydrofuran at −30°; (7) added glycine *t*-butyl ester[28] (292 mg, 2.23 mmol, 5 meq) in 50 ml of tetrahydrofuran precooled to −30°; (8) rocked at −30° for 2 hr, 0° for 16 hr and 23° for 3 hr; (9) washed with three 50-ml portions of tetrahydrofuran; (10) washed with three 50-ml portions of methanol. The resin was cleaved with HBr in trifluoroacetic acid as described previously.[16] The crude peptide was lyophilized twice from water and purified on a 2 × 120 cm Dowex 50-X4 column in a pH 4.0 pyridine acetate cycle. The sample was applied to the column in 1 ml of water. Elution

---

(23) E. E. Schallenberg and M. Calvin, *J. Amer. Chem. Soc.*, **77**, 2779 (1955).

(24) S. Akabori, S. Sakakibara, Y. Shimonishi, and Y. Nobuhara, *Bull. Chem. Soc. Jap.*, **37**, 433 (1964).

(25) F. Weygand and M. Reiher, *Chem. Ber.*, **88**, 26 (1955).

(26) J. P. Greenstein and M. Winitz, "Chemistry of the Amino Acids," John Wiley & Sons, Inc., New York, N. Y., 1961, p 1251.

(27) R. W. Roeske, *Chem. Ind.*, 1121 (1959).

(28) G. W. Anderson and F. M. Callahan, *J. Amer. Chem. Soc.*, **82**, 3359 (1960).

with pH 4.0 pyridine acetate (0.5 M in acetic acid) proceeded at a rate of 118 ml/hr for a total of 16 hr. Aliquots (0.2 ml) from 5.5-ml fractions were analyzed by the ninhydrin test. The fractions between 0.88 and 1.16 l. were combined and lyophilized; yield 52 mg (57% calculated from the amount of resin-L-leucine Boc-hydrazide); $[M]^{25}$D 170.7° (c 1.40; water); lit.[29] $[M]^{25}$D 161.3° (c 2.4; water); amino acid ratios: Leu, 1.00; Gly, 0.98.

Anal. Calcd for $C_8H_{16}N_2O_3 \cdot H_2O$: C, 46.59; H, 8.80; N, 13.58. Found: C, 46.45; H, 8.52; N, 13.77.

Infrared and nmr spectra of the product were identical with that of a commercial sample.[17]

L-Leucyl-L-alanylglycyl-L-valine (Stepwise Condensation). Resin-L-leucine Boc-hydrazide (2.0 g, 0.544 mmol) was placed in the reaction vessel. The following cycle of reactions was employed: (1) t-Boc group cleaved by 4 M HCl–dioxane (50 ml) for 30 min; (2) washed with three 50-ml portions of tetrahydrofuran; (3) added 50 ml of tetrahydrofuran and cooled the vessel to $-30°$; (4) generated the azide by addition of 4 M HCl–dioxane (0.68 ml, 2.72 mmol, 5 meq) followed by n-butylnitrite (0.34 ml, 2.72 mmol, 5 meq) for 1 hr; (5) washed with three 50-ml portions of precooled 10% Et$_3$N–THF at $-30°$; (6) washed with three 50-ml portions of precooled tetrahydrofuran at $-30°$; (7) added L-alanine Boc-hydrazide (**2b**) (552 mg, 2.72 mmol, 5 meq) in 50 ml of tetrahydrofuran precooled to $-30°$; (8) rocked at $-30°$ for 2 hr, 0° for 4 hr and 23° for 16 hr; (9) washed with three 50-ml portions of tetrahydrofuran; (10) washed with three 50-ml portions of methanol. The second and third cycles were carried out exactly as described above with the exception that in step 7 glycine Boc-hydrazide (**2c**) (481 mg, 2.72 mmol, 5 meq) or L-valine t-butyl ester (471 mg, 2.72 mmol, 5 meq) was used. A portion of the peptide-resin (1.13 g) was cleaved with HBr in trifluoroacetic acid.[16] The crude peptide was lyophilized twice from water and purified on a 2 × 120 cm Dowex 50-X4 column in a pH 4.0 pyridine acetate cycle. The sample was applied to the column in 1 ml of water. Elution with pH 4 pyridine acetate proceeded at a rate of 81 ml/hr for a total of 25 hr. Aliquots (0.2 ml) from 5.5-ml fractions were analyzed by the ninhydrin test. The fractions between 0.85 and 0.96 l. were combined and lyophilized: yield, 33 mg (30% calculated from amount of resin-L-leucine Boc-hydrazide); $[\alpha]^{25}$D 23.55° (c 0.15%, ethanol); lit.[30] $[\alpha]^{20}$D 24.55° (c 0.68%; ethanol); lit.[4f] $[\alpha]^{20}$D 22.5–23.7° (c 1–2%; ethanol); amino acid ratios: Leu, 0.97; Ala, 0.96; Gly, 1.07; Val, 1.00.

Anal. Calcd for $C_{15}H_{30}N_4O_5 \cdot 1.5H_2O$: C, 49.89; H, 8.63; N, 14.54. Found: C, 49.81; H, 8.37; N, 14.40.

---

(29) M. Bergmann, L. Zervas, and J. S. Fruton, *J. Biol. Chem.*, **111**, 225 (1935).

(30) This value was obtained on the sample prepared by Merrifield (ref 3a) when remeasured on the Perkin-Elmer Model 140 polarimeter.

Tlc of the tetrapeptide in three systems was identical with that of a sample previously synthesized by the standard solid phase procedure.[3a] It moved with R_f 0.64 (butanol–acetic acid–water; 8:2:2); R_f 0.71 (butanol–acetic acid–water–pyridine, 15:3:-12:10); R_4 0.71 (butanol–acetic acid–ethyl acetate–water; 1:1:1).

The peptide was homogeneous when run as an analytical sample on the amino acid analyzer. The pH 3.28 buffer was changed to pH 4.25 buffer at 120 min. The peptide peak emerged 17 min after the buffer artifact (187 min total time).

Leucyl-L-alanylglycyl-L-valine (Fragment Condensation). Resin-L-leucine Boc-hydrazide (1.5 g, 0.446 mmol) was placed in the reaction vessel. The following cycle of reactions was employed: (1) Boc group cleaved by 4 M HCl–dioxane (50 ml) for 30 min; (2) washed with three 50-ml portions of tetrahydrofuran; (3) added 50 ml of tetrahydrofuran and cooled the vessel to $-30°$; (4) generated the azide by addition of 4 M HCl–dioxane (0.56 ml, 2.23 mmol) followed by n-butylnitrite (0.28 ml, 2.23 mmol) for 1 hr; (5) washed with three 50-ml portions of precooled 10% Et$_3$N–THF at $-30°$; (6) washed with three 50-ml portions of precooled tetrahydrofuran at $-30°$; (7) added L-alanine Boc-hydrazide (**2b**) (453 mg, 2.23 mmol) in 50 ml of tetrahydrofuran precooled to $-30°$; (8) rocked at $-30°$ for 4 hr, 0° for 12 hr and 23° for 2 hr; (9) washed with three 50-ml portions of tetrahydrofuran; (10) washed with three 50-ml portions of methanol. The second cycle was carried out exactly as described above with the exception of step 7 in which glycyl-L-valine t-butyl ester (514 mg, 2.23 mmol) was added. The resin was cleaved with HBr in trifluoroacetic acid.[16] The crude peptide was lyophilized from water and purified on a 2 × 120 cm Dowex 50-X4 column in a pH 4.0 pyridine acetate cycle. The sample was applied to the column in 1 ml of water. Elution with pH 4 pyridine acetate proceeded at a rate of 103 ml/hr for a total of 16 hr. Aliquots (0.2 ml) from 5.5-ml fractions were analyzed by the ninhydrin test (Figure 3). The fractions between 0.87 and 1.06 l. were combined and lyophilized: yield, 102 mg (60% calculated from the amount of resin-L-leucine Boc-hydrazide); $[\alpha]^{20}$D 26.77° (c 0.96%, ethanol); lit.[30] $[\alpha]^{20}$D 24.55° (c, 0.68%; ethanol), lit.[4f] $[\alpha]^{20}$D 22.5–23.7° (c 1–2%; ethanol); amino acid ratios: Leu, 0.97; Ala, 1.00; Gly, 0.95; Val, 1.01.

Anal. Found: C, 50.51; H, 8.01; N, 14.40.

This tetrapeptide moved with the same R_f in three tlc solvent systems as the peptide made by the stepwise route. This peptide also emerged on the amino acid analyzer at 17 min after the buffer artifact. A small peak containing 1.0% of the ninhydrin color of the main peak was found at 9 min.

Acknowledgment. The authors would like to thank Miss Anita Bach for her technical assistance, Miss Ursula Birkenmaier for the amino acid analyses, and Mr. S. Traiman for the infrared spectra.

15

Reprinted from *Helv. Chim. Acta,* **56**(5), 1476–1482 (1973)

143. The Preparation of Merrifield-Resins Through Total Esterification With Cesium Salts

by **B. F. Gisin**[1])

Dept. of Physiology, Duke University Medical Center, Durham, N.C. 27710, USA

(26. II. 73)

Summary. The reaction of chloromethylated polystyrene-*co*-1%-divinylbenzene resin with the cesium salts of N-protected amino acids proceeds fast and without side reactions to give N-protected amino acyl resin esters free of quaternary ammonium sites or reactive chloride.

[1]) Present address: The Rockefeller University, New York, N.Y. 10021, USA.

In solid-phase peptide synthesis [1] the most widely used starting material is an N-protected amino acid bound *via* a benzyl ester linkage to an insoluble copolymer of styrene and divinylbenzene (*Merrifield*-resin)[2]. This derivative is accessible either in a one-step reaction or through intermediates, both approaches starting with chloromethyl resin. In the former scheme the chloromethyl resin is reacted with the triethylammonium salt of a Boc-amino acid[3]) in ethyl acetate [1], ethanol [3], *t*-butyl alcohol [4] or DMF [5]. Other organic bases have also proven useful in this reaction, namely, tetramethyl ammonium hydroxide [6], diisopropyl ethylamine [7], and dicyclohexyl amine [8]. Indirect schemes of synthesis involve the preparation of acetoxymethyl resin [9] followed by saponification or aminolysis [10] and Boc-amino acylation of the resulting hydroxymethyl resin with the aid of a coupling agent, such as N,N'-carbonyldiimidazole [11], triphenylphosphine/2,2'-dipyridyl disulfide [12], DCCI [11], or N,N-dimethylformamide dineopentyl acetal [13]. Boc-amino acid resin esters have also been obtained from chloromethyl resin in a two-step procedure *via* a dimethyl methylene sulfonium resin salt as an intermediate [14][4]).

Although all of these methods have been demonstrated to yield useful products, each one of them also suffers one or more of the following drawbacks:

1. The use of triethylammonium salts gives rise to a sizeable amount of quaternary ammonium sites on the resin [17]. The ion-exchange property of such groups can interfere [18] with those monitoring methods that depend on elution of anions from amine-containing resins, and they could conceivably retain TFA from the deprotection step and release it in a subsequent coupling step, resulting in the termination of peptide chains.

2. If there are unsubstituted chloromethyl groups left, they can react with triethyl amine to form additional quaternary ammonium groups in each neutralization step [19] giving rise to the problems mentioned above. In addition, they might alkylate other available nucleophiles, *e.g.*, the N-terminal amino groups of a peptide chain or alkylatable functional groups (His, Cys, Met) [20]. These are irreversible and yield-diminishing side-reactions.

3. Because of side-reactions or incomplete reaction it is difficult with presently available procedures to accurately determine the degree of substitution in advance.

For these reasons a search for a one-step procedure for obtaining Boc-amino acid resin esters devoid of quaternary ammonium sites or unreacted chloromethyl groups was undertaken, the result of which is presented here.

Although the lithium salt of a Boc-amino acid has been reported not to react with chloromethyl resin [21] the potassium salts of acetic acid [9] and of sebacid acid monomethyl ester [22] react at elevated temperatures. However, the conditions used in the latter procedures (125°, 24 h and 150°, 10 h respectively) appear too drastic for

---

[2]) For a recent review on solid-phase peptide synthesis see ref. [2].
[3]) The abbreviations recommended by the IUPAC-IUB Commission of Biomedical Nomenclature (J. biol. Chemistry *241*, 2491 (1966); *242*, 555 (1967)) have been used throughout. In addition, Boc = *t*-butyloxycarbonyl, TFA = trifluoroacetic acid, DMF = N,N-dimethylformamide, DCCI = N,N'-dicyclohexylcarbodiimide.
[4]) Not discussed here are special-purpose resins, such as 'activateable' resins [15] and a resin with a 'spacer' between polymer and anchoring point of the first amino acid [16].

Boc-amino acids. The sodium salt of a Boc-amino acid has been used to attach an amino acid to a soluble chloromethyl polymer [23]. In this instance, when a two-fold excess of salt and three days reaction time was used the product still contained over 10% of unreacted chloromethyl groups, thus indicating a rather sluggish reaction rate.

A cesium salt, containing a cation larger than lithium, sodium or potassium might be expected to be more lipophilic and therefore more compatible with the resin. In addition, in a polar solvent, such as DMF, the salt of a carboxylic acid should be dissociated to a greater extent if the cation is large than if it is small. Since the carboxylate rather than the ion pair is the nucleophile that displaces the chloride, a high degree of dissociation is very desirable: It will result in a high concentration of the reacting species and consequently increase the reaction rate. In agreement with this reasoning, such a dependence of the esterification rate on the size of the cation was found to exist when various salts of a Boc-amino acid were allowed to react with chloromethyl resin (Fig. 1). Under otherwise identical conditions (DMF,

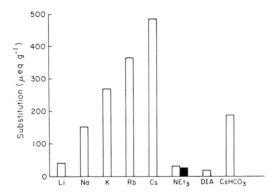

Fig. 1. *Incorporation of Boc-valine into chloromethyl resin* (substitution, 1950 µeq/g) *using various salts under otherwise identical conditions* (DMF, room temp., 5 h). The black bar represents quaternary ammonium groups obtained with the triethylammonium salt. When the cesium salt was prepared *in situ* from Boc-valine and cesium hydrogencarbonate, thus introducing one equivalent of water, the substitution was only half that found when dry Boc-Val-OCs was used (DIA = diisopropyl ethylamine).

room temperature, 5 h, 20% molar excess of salt over choloromethyl groups on the resin) the cesium salt yielded the highest incorporation of Boc-valine into the resin. It was more than one order of magnitude greater than for the lithium salt and nearly double that of the potassium salt. In order to demonstrate the absence of amino or ammonium groups the resins were treated with pyridinium picrate and washed thoroughly [24]. There was no (<1 µeq/g) picrate retained in the resins that had been prepared with alkali salts or with diisopropyl ethylamine, while in the case of the triethyl ammonium salt 25 µeq of picrate per gram of resin were found to be ionically bound to quaternary ammonium sites.

Because of their superior reaction rate cesium salts were used through the remainder of this study. When the temperature was raised to 50°, esterification with

Table 1. *Reaction of cesium salts with chloromethyl resins in DMF at 50°*

Salt Name	Equivalents	Chloromethyl Resin Equivalents	Chloromethyl Resin Substitution (µeq/g)	Total Volume per g of Resin (ml)	Reaction Time (h)	Substitution[a] (µeq/g)		Yield[b]
Boc-Phe-OCs	1.0	5.0	1950	6	16	Phe,	400	100%
Boc-Val-OCs	1.2	1.0	1950	20	1.5	Val,	1230	63%
Boc-Val-OCs	1.2	1.0	1950	20	6	Val,	1670	85%
Boc-Val-OCs	1.2	1.0	1950	20	20	Val,	1830	94%
Bpoc-Val-OCs[c]	1.0	4.5	530	8	15	Val,	120[d]	100%
Boc-D-Val-L-Lac-OCs [25][c]	1.0	3.5	1950	7	16	Val,	540	96%
Boc-D-Val-L-Lac-OCs [25][c]	1.0	1.3	820	6	24	Val,	590	94%
Boc-D-Val-L-Lac-OCs [25][c]	1.0	1.0	550	7	16	Val,	490	89%
Boc-Gly-OCs	1.0	1.0	820	7	16	Gly,	590	72%
Boc-Ala-OCs	1.0	1.0	820	7	16	Ala,	630	77%
Boc-Pro-OCs	1.0	1.0	820	7	16	Pro,	670	82%
Boc-Leu-OCs	1.0	1.0	820	7	16	Leu,	710	87%
Boc-Phe-OCs	1.0	1.0	820	7	16	Phe,	700	85%
Boc-Val-OCs	1.0	1.0	820	7	16	Val,	780	95%

[a] By amino acid analysis after hydrolysis.
[b] Based on limiting reactant and substitution, thus, discounting mechanical losses.
[c] Bpoc = 2-(*p*-biphenyl)isopropyloxycarbonyl; Lac = lactyl.
[d] Dr. *R. S. Feinberg*, The Rockefeller University, personal communication.

Boc-Val-OCs was complete in less than twelve hours (Fig. 2). The chloride content of the resin decreased concomitantly with the incorporation of valine. Furthermore, there was no deprotection of the amino group ($< 0.05\%$ in 22 h) under the reaction conditions. It is therefore concluded that essentially no side reactions occur.

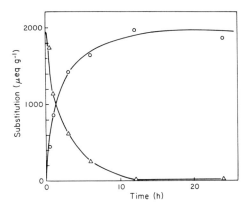

Fig. 2. *The esterification of chloromethyl resin with Boc-valine cesium salt as monitored by the decrease of the chloride content of the resin* ($\triangle$) *and by the increase in valine substitution* ($\circ$). Conditions: DMF, 50°, 20% molar excess of salt over chloromethyl resin.

Since the reaction is quantitative the degree of substitution is conveniently predetermined by choosing either the cesium salt or the chloromethyl resin as the limiting reactant. Thus, when one equivalent of Boc-L-phenylalanine cesium salt was reacted at 50° with five equivalents of chloromethyl resin a quantitative yield of Boc-Phe-resin was obtained (Table 1). Conversely, with an excess (1.2 to 3-fold) of cesium salt over chloromethyl groups almost all of the chloride was displaced (Table 2).

Table 2. *Reaction of chloromethyl resins with varying excesses of Boc-Val-OCs in DMF* (6–8 ml per g of resin). Reaction time, 16 h, temperature, 50°.

Substitution of Chloromethyl-Resin (µeq/g)	Molar Excess of Boc-Val-OCs (Resin = 1.0)	Residual Chloride[a] (µeq/g)
550	3.0	0
150	3.0	2
50	3.0	0
550	1.5	8
150	1.5	3
50	1.5	2
1950	1.2	6
50	1.2	8

[a] By potentiometric titration with $AgNO_3$ after quantitative displacement with hot pyridine. These values are corrected for the blank value (4 µeq/g) obtained with not chloromethylated polystyrene-*co*-1%-divinylbenzene resin.

The residual chloride represents a small fraction of chloromethyl groups that react only under far more drastic conditions (pyridine, 110°). Having such limited reactivity, however, these chloromethyl groups are of little concern, for they will most likely also be inert to substitution by other nucleophiles, such as tertiary amines or functional groups of the peptide chains.

For the preparation of resins to be used for peptide syntheses the scheme which achieves total esterification and leaves only little chloride unreacted is recommended. In a typical run 1.0 g of chloromethyl resin (substitution, 550 µeq/g), 232 mg of Boc-Val-OCs (660 µmol, 20% excess) and 9 ml of DMF were combined and stirred in a sealed vessel at 50° overnight. The resin was washed thoroughly with DMF, DMF/water 9:1 (v/v), DMF and ethanol, then dried and a sample was hydrolyzed. Amino acid analysis of the hydrolyzate indicated a substitution of 550 µequivalents of Boc-valine per gram of resin which corresponds to a yield of 98% based on the initial amount of chloromethyl groups on the resin.

The present method represents a smooth way to prepare *Merrifield*-resins in good yield and of a specific substitution. The high purity of these resins may help to reduce the number of unexplained side-products encountered in solid-phase peptide synthesis.

The author wishes to express his gratitude to Prof. *R. B. Merrifield* who suggested the need for a total esterification procedure for his helpful advice and to Mr. *Arunkumar Dhundale* and Mrs. *Maureen Sanz* for their technical assistance. This investigation was supported by NIH Grant HE 12 157, *U. S. Public Health Service* Grant AM 1260 and by the *Hoffmann-La Roche Foundation*.

Experimental Part

In order to correct for the weight change during reactions substitutions of resins are expressed in µ-equivalents per gram of benzyl polymer ($\cdot CH_2C_6H_4$-resin) according to the formula [26] $s = a/(1 - a \cdot e)$ where s = substitution in µ-equivalents per gram; a = analytical concentration of substituent on the resin in µ-equivalents per gram, and e = equivalent weight of the substituent bound to the benzyl polymer in grams per µ-equivalent.

All chemicals and solvents used were reagent grade. DMF ('Spectroquality', *Matheson, Coleman & Bell*, East Rutherford, N.J.) was stored over Molecular Sieve Type 4A (same supplier). Alkali hydroxides and hydrogencarbonates were purchased from *ROC/RIC* Corp., Sun Valley, Calif., Boc-amino acids from *Fox Chemical Co.*, Los Angeles, Calif., *Beckman* Instruments, Inc., Palo Alto, Calif. or Protein Research Foundation, Japan; Polystyrene-*co*-1%-divinylbenzene resin (*Bio-Beads* S-X1, 200–400 Mesh) from *Bio-Rad Laboratories*, Richmond, Calif. Amino acid analyses (*Beckman* Spinco amino acid analyzer Model 21) were by Miss *L. Apacible* of Rockefeller University.

Chloromethyl Resins. Polystyrene-*co*-1%-divinylbenzene resin was chloromethylated with chloromethylmethylether and stannic chloride [27] [1] as described previously [10]. The degree of chloromethylation was varied by using different amounts of catalyst. Thus, when 10 g samples of resin in 60 ml of chloromethylmethylether were combined with 0.3, 0.12 or 0.06 ml of $SnCl_4$ in 10 ml hexane and allowed to react for 1 h at 0° substitutions of 550, 150 and 50 µ-equivalents of chloride per g of resin respectively were obtained. The substitution of chloride was determined by quantitative displacement with pyridine at 100–110° followed by potentiometric titration of the released chloride with $AgNO_3$ [28].

Cesium Salts of Boc-amino Acids. 2 g of a Boc-amino acid was dissolved in 10–15 ml of ethanol and diluted with 3–5 ml water. The pH of this solution was adjusted to 7.0 (pH-meter) by adding aqueous cesium hydrogencarbonate. The neutral solution was then flash evaporated. After repeated evaporation to dryness withn bezene the Boc-amino acid cesium salts were obtained as white powders or solids. They were dried over P_2O_5 for 5 h and used without further purification.

Other Salts. The lithium, sodium, potassium and rubidium salts (Fig. 1) were prepared similarly to the cesium salt using the corresponding hydroxides or hydrogen carbonates. The triethylammonium and diisopropyl ethylammonium salts were prepared *in situ* by combining equivalent amounts of Boc-valine and base.

Boc-Amino Acyl Resin Esters. One millimol of dry Boc-amino acid cesium salt, one milliequivalent of chloromethyl resin and 6–8 ml of DMF per g of resin were placed in a screw-capped vial provided with a magnetic stirring bar. The suspension was stirred overnight while kept at 50° by means of a thermostated water bath. The resin was filtered, washed thoroughly with DMF, DMF/H_2O 9:1 (v/v), DMF and ethanol, and dried. The substitution of the resin was determined by amino acid analysis after hydrolysis with conc. HCl/propionic acid 1:1 (v/v) at 140° for 3 h [29]. The results of this and other experiments are summarized in Table 1 and 2.

REFERENCES

[1] R. B. Merrifield, J. Amer. chem. Soc. *85*, 2149 (1963).
[2] R. B. Merrifield, Advan. Enzymol. *32*, 221 (1969).
[3] R. B. Merrifield, J. Amer. chem. Soc. *86*, 304 (1964).
[4] Th. Wieland, B. Penke & Ch. Birr, Liebigs Ann. Chem. *759*, 71 (1972).
[5] A. Marglin, Tetrahedron Letters *33*, 3145 (1971).
[6] A. Loffet, Int. J. Protein Research *III*, 297 (1971).
[7] B. Mehlis & W. Fisher, Chem. Abstr. *72*, 450 (1970).
[8] H. Yajima, H. Kawatani & H. Watanabe, Chem. Pharm. Bull. *18*, 1333 (1970).
[9] J. M. Stewart & J. D. Young, 'Solid Phase Peptide Synthesis', W. H. Freeman, San Francisco, Calif., 1969, p. 9.
[10] B. F. Gisin & R. B. Merrifield, J. Amer. chem. Soc. *94*, 6165 (1972).
[11] M. Bodanszky & J. T. Sheehan, Chem. Ind. (London) *1966*, 1597.
[12] T. Mukaiyama, M. Ueki & R. Matsueda, Proc. 3rd Amer. Peptide Symposium, Ann Arbor Science Publishers Inc., Ann Arbor, 1972; J. Meienhofer, Editor, p. 209.
[13] J. Schreiber, Proc. 8th Europ. Peptide Symposium, North-Holland Publishing Co., Amsterdam, 1967; H. C. Beyerman *et al.*, Editors, p. 107.
[14] L. C. Dorman & L. D. Markley, J. med. Chemistry *14*, 5 (1971).
[15] E. Flanigan & G. R. Marshall, Tetrahedron Letters *1970*, 2403; D. L. Marshall & I. E. Liener, J. org. Chemistry *35*, 867 (1970); Th. Wieland, Ch. Birr & P. Fleckenstein, Liebigs Ann. Chem. *756*, 14 (1972).
[16] M. Buka & R. Zagats, Chem. Abstr. *72*, 448 (1970).
[17] R. B. Merrifield, 1966, personal communication.
[18] B. F. Gisin & R. B. Merrifield, J. Amer. chem. Soc. *94*, 3102 (1972).
[19] R. B. Merrifield, 1970, personal communication.
[20] Ref. [9], p. 8.
[21] J. Rudinger, Proc. 8th Europ. Peptide Symposium, North-Holland Publishing Co., Amsterdam, 1967; H. C. Beyerman *et al.*, Editors; p. 89.
[22] J. I. Crowley & H. Rapoport, J. Amer. chem. Soc. *92*, 6363 (1970).
[23] B. Green & R. Garson, J. chem. Soc. (C) *1969*, 401.
[24] B. F. Gisin, Anal. Chim. Acta *58*, 248 (1972).
[25] B. F. Gisin, R. B. Merrifield & D. C. Tosteson, J. Amer. chem. Soc. *91*, 2691 (1969).
[26] V. A. Najjar & R. B. Merrifield, Biochemistry *5*, 3765 (1966).
[27] K. W. Pepper, H. M. Paisley & M. A. Young, J. chem. Soc. 4097 (1953).
[28] L. C. Dorman, Tetrahedron Letters *1969*, 2319.
[29] J. Scotchler, R. Lozier & A. B. Robinson, J. org. Chemistry *35*, 3151 (1970).

Copyright © 1971 by Microforms International Marketing Corp.

Reprinted from *Tetrahedron Letters,* No. 28, 2633–2636 (1971), with permission of Microforms International Marketing Corporation as exclusive copyright licensee of Pergamon Press journal back files

A NEW SOLID SUPPORT FOR POLYPEPTIDE SYNTHESIS

W. Parr and K. Grohmann
Chemistry Department
University of Houston
Houston, Texas

(Received in USA 21 May 1971; received in UK for publication 15 June 1971)

The discovery of the solid-phase synthesis by Merrifield (1) opened a new way in the synthesis of polypeptides. However, since the chlorobenzyl-groups are uniformly distributed throughout the polystyrene-resin particles (2), all reaction rates are unfavorably controlled by diffusion. This distribution cannot be overcome by using resins with lower concentration of chlorobenzyl groups. A second difficulty arises from using different solvents during reaction and washing steps which cause different swelling of resin particles, thus affecting reaction rate, yields in coupling steps, and purity of synthesized peptides (3,4). Differential swelling of the polymer particles also causes difficulties in the automization of this process, and therefore batch procedures must be used instead of column procedures which would be more convenient and more easily controlled (5).

All of these difficulties can be overcome by using inorganic support material coated with a thin layer of molecules which are chemically bonded to an inorganic surface and have reactive groups located only on the surface. Esterification with 1,4-dihydroxydimethyl benzene was chosen by Bayer et al. (5,6) for formation of the chemically bonded monomolecular layer on which peptides can be built by stepwise synthesis. The advantage of this approach were proven by syntheses of the dodecapetide (leu-Ala)$_6$ and the tetrapeptide (leu-leu-glu-gly). All reaction times were reduced in comparison with the polystyrene-based resins. No failure sequences could be detected in the test peptide (leu-Ala)$_6$.

The main disadvantage with this support is that the Si-O-C bond is highly polarized and thus very sensitive to attack by all reagents containing free hydroxy groups, but especially towards water. This difficulty can be overcome by bonding organic molecules to siliceous surfaces through Si-O-Si-C bonds which are more stable against an attack by electrophilic or nucleophilic agents than are the Si-O-C bonds.

Si-O-Si-C bonds can be formed by reaction of surface silanol groups with compounds of formula R_xSiY_{4-x} (R = alkyl, aryl, Y = halogen, X = 1-3) (7). These reactions have been used for formation of chemically bonded layers of organic molecules on surface of siliceous materials in the field of gas and liquid chromatography (8,9). In order to obtain such material for the synthesis of peptides, we prepared p-bromomethyl-phenyl-silica. P-tolyl-magnesium bromide was reacted with silocon tetrachloride according to the method described by Chvalovsky et al. (10). The p-tolyl trichlorosilane obtained was brominated in CCl_4 with B-bromosuccinimide (NBS) catalyzed by benzoylperoxide according to:

$$CH_3-\text{C}_6H_4-SiCl_3 + NBS \rightarrow Br-CH_2-\text{C}_6H_4-SiCl_3$$

After removal of succinimide and the solvent, the oil was twice distilled *in vacuo* (B.P. 125°/2mm Hg). The purity and structure of the p-bromomethyl-phenyl-trichlorosilane was determined by NMR-spectroscopy.

The above compound was reacted with the silanol groups on the glass surface (Bio-Glass 2500, Bio. Rad Lab) by shaking the beads overnight with 5% solution of p-bromomethyl-phenyl-trichlorosilane in benzene. After filtration the coated glass beads were washed with benzene, a mixture of benzene and ethanol (1:1) and finally with benzene.

$$Br-CH_2-\text{C}_6H_4-SiCl_3 + 3\,HO-Si\equiv \rightarrow Br-CH_2-\text{C}_6H_4-Si(O-Si\equiv)_3$$

The remaining Si-Cl bonds were hydrolyzed by mixture Water: Ethanol: Benzene (5:45:50). The silicone layer was then polymerized by heating to 100°C for 24 hours in vacuo. The silanized glass beads were washed with ethanol and benzene in order to remove absorbed impurities. The capacity of silanized glass beads was determined by modified Volhard analysis for bromine (0.08 mmoles Br/g glass). At this point no chlorine could be detected. In order to demonstrate suitability of this support for solid phase synthesis of peptides, we have synthesized a model peptide H-Pro-Gly-Phe-Ala-OH.

The first amino acid, alanine, was connected with the carboxyl group to benzylic group of silicone layer by refluxing the silanzied glass beads in dioxane with a solution of triethylammonium salt of t-Boc amino acid for 24 hours. The amino acid thus bonded to a support were not affected by either trifluoroacetic acid and mixture of triethylamine/chloroform for a period of longer than 20 hours. No free amino acid could be detected in either reagent by thin layer chromatography. Further, the amino acid was not affected by the action of all common solvents used in solid phase synthesis. N,N'-dicyclohexylcarbodiimide (DCC) and CH_2Cl_2 was used for coupling of t-Boc amino acids to alanine amino attached to glass support. Reaction time for each coupling step was 2 hours at room temperature. For deprotection and washing we have used standard procedure described by Merrifield (1). The finished tetrapeptide was cleaved from the support by HBr/trifluoroacetic acid for 2 hours at room temperature. The identity of finished peptide was confirmed by amino acid analysis. Acid hydrolysis with 6NHCl for 24 hours at 110°C of the isolated tetrapeptide showed the following molar ratio: Pro 0.95, Gly 1.00, Ala 1.00, Phe 1.05 (Gly being taken as 1.00). The tetrapeptide showed a single spot by thin layer chromatography.

However, cleavage of the peptide from the support did not proceed to completion. Total hydrolysis of peptide remaining on the support, by amino acid analysis showed that cleavage step proceeded only to about 50% conversion. This is caused by negative effect of electron withdrawing SiO_3 group in para position toward a benzylic group. In order to overcome those difficulties we have synthesized several silanes of $X-CH_2-\langle\rangle-(CH_2)_n-SiCl_3$ type (n = 2,3,4), X = halogen. The applications of these compounds for peptide synthesis will be studied.

REFERENCES

1. R. B. Merrifield, J. Amer. Chem. Soc. 85, 2149 (1963); 86, 304 (1964). Advan. Enzymol. 32, 221 (1969).

2. R. B. Merrifield and V. Littau in E. Bricas, "Peptides 1968" North Holland Pub. Co., Holland, 1968, 179.

3. E. Bayer, H. Eckstein, K. Hägele, W. König, W. Brüning, H. Hagenmaier, and W. Parr, J. Amer. Chem. Soc. 92, 1735 (1970).

4. E. Bayer, H. Hagenmaier, W. Parr, H. Eckstein, G. Jung and P. Hunzicker in E. Scoffone, Peptides 1969, North Holland Pub. Col, Holland, 1970 in print.

5. E. Bayer, G. Jung, E. Breitmeier and W. Parr, 2nd American Peptide Symposium, Cleveland, Ohio (1970).

6. E. Bayer, G. Jung, I. Halasz, and I. Sebastian, Tetrahedron Letters 51, 4503 (1970).

7. W. Nool, "Chemistry and Technology of Silicones," Academic Press, New York and London (1968), 582.

8. J. J. Kirkland, J. J. Destefano in A. Zlatkis, Advances in Chromatography 1970, publ. by Chromatography Symposium, University of Houston, Department of Chemistry, Houston, Texas 77004, 1970, 397 and J. Chrom. Sci. 8, 309 (1970).

9. E. W. Abel, Z. H. Pollard, P. C. Uder, and G. Nickless, J. Chromatog. 22, 23 (1966).

10. V. Chvalovsky and V. Bazant, Coll. Czech. Chem. Comm. 16, 580 (1951).

A NEW SUPPORT FOR POLYPEPTIDE SYNTHESIS IN COLUMNS

E. Bayer and G. Jung

Chemisches Institut der Universität Tübingen, Germany

I. Halász and I. Sebestian

Institut für Physikalische Chemie der Universität Frankfurt /Main,

Germany

(Received in UK 17 September 1970; accepted for publication 8 October 1970)

An automated synthesis of polypeptides by the solid phase method (1) would be much easier if the support would always stay in a column. With the resins presently used, this is not possible and a batch process has to be applied because the different solvents necessary in a solid phase synthesis step cause variable swelling of the resin. Since the gel nature of the polystyrene-based resin is responsible for this property, we have tried to overcome these difficulties by using inorganic materials as supports. Alcohols can be esterified to the silanol groups on the surface of any type of silica (e. g. porous glass beads, silicagel, quartz) (2, 3). These modified supports have been successfully used in high speed chromatography (2, 3) and their properties show that the organic groups are located on the surface of the inorganic support, e.g., considered to stick out, somewhat like a brush having the desired functional group at the free end of the bristle. The mass transfer has to be faster than the rate of diffusion in the resins used up to now. In the case of polystyrene resins, it has been demonstrated that statistical failure sequences occur because the coupling of individual amino acids is incomplete (4-6). Every factor that increases the rate of reaction can improve the synthesis and extend the use of the solid phase method for synthesis of longer peptide chains with reasonable purity. The increase in speed of the mass transfer would be an important factor in this view. An increase of diffusion would enable shorter coupling times leading to faster synthesis. A real solid support material additionally prevents any ion exchange mechanism before or after the coupling step, which might be disadvantageous.

In order to obtain such materials suitable for column synthesis of polypeptides, we prepared the mono ester of 1.4-dihydroxy-methyl-benzene of silica (7). The ester bond is chemically stable in all anhydrous solvent systems used during a solid phase synthesis. The carboxylic group of the first amino acid is connected to the free benzylic alcohol group by 1,1'-carbonyldiimidazole (CDI). Both the N,N'-dicyclohexyl carbodiimide (DCCI) procedure and the activated ester method (ONP) have been successfully used for the further coupling steps.

Brush type supports (Biopak (7)) for the column solid phase synthesis have been prepared with different contents of benzyl alcohol groups varying from 0.006 - 0.06 mMol/g.

The synthesis of various peptides in a specially designed synthesizer for column procedure shows that the reaction time can be reduced by a factor of at least two and that the products can be obtained with the same purity as by the common procedures. The time can most likely be reduced much more than by a factor of two if an instrument is designed allowing higher flow rates. The column synthesizer used in these investigations did not allow faster flow rates. In Table 1, some conditions used in these experiments are compared with the conditions in solid phase synthesis on resins.

TABLE 1 Comparison of Conditions for Peptide Synthesis with Biopak (7) and Resins

	Polystyrene Resin (Bio-Beads S-X2)	Biopak (7)
Esterification	48 hrs, 80 °C, Triethylamine separate reaction vessel	2 hrs, 20 °C, CDI column procedure
Capacities	0.1 - 0.5 mMol/g	0.006 - 0.06 mMol/g
Time required for one cycle	DCCI : 4 hrs ONP : 10 hrs	2 hrs 6 hrs
Automation	Batch procedure	Column procedure
Cleavage	$HBr-CF_3COOH$, 90 min	$HBr-CH_3COOH$, 20 min or $HBr-CF_3COOH$

For the synthesis of the dodecapeptide (leu-ala)$_6$ on Biopak (7), the same derivatives and reagents have been used as described (4) for the synthesis on resins. 4 g each of each three supports with different capacities (I. 0.006 mMol/g; II. 0.02 mMol/g; III. 0.04 mMol/g) were utilized and the following yields obtained after hydrolysis from the support: I. 12 mg, II. 85 mg, III. 210 mg. These peptides were purified by precipitation from trifluoroacetic acid with ether. The products obtained have been found to be identical to samples prepared by common resin procedure (4). The partial hydrolysate was checked for ala-ala and leu-leu sequences indicative of any failure sequences by the described combined gas chromatography-mass spectrometry procedure (8). They were well beyond the figures found for resin synthesis (4, 8).

The tetrapeptide leu-leu-gln-gly, a partial sequence of secretin, was synthesized by the p-nitrophenyl ester method using a Biopak support with 0.05 mMol/g capacity.

These results suggest that the new support material for the solid phase method shortens the reaction time and enables an easier operation in a continuous or discontinuous column procedure. These real solid supports enable an easier scaling up of the synthesis in comparison to a batch procedure. Further advantages are: 1) no shaking is required, consequently no mechanical corrosion of the solid phase appears; and 2) a considerably smaller excess of the reagents is necessary to obtain the same equilibrium. Also, monitoring of the reaction might be possible by inserting thermistors into the column in order to measure the reaction heat.

We gratefully acknowledge the support of this work by the Deutsche Forschungsgemeinschaft and the Alexander von Humboldt-Stiftung.

REFERENCES

(1) R.B. Merrifield, Federation Proc. 21, 412 (1962)

(2) I. Halász, I. Sebestian, Lecture at the 5th Int. Symposium on Advances in Chromatography, Las Vegas, Nev., Jan. 1969

(3) I. Halász, I. Sebestian, Angewandte Chemie, Internatl. Ed., 8, 453 (1969)

(4) E. Bayer, H. Eckstein, K. Hägele, W. König, W. Brüning, H. Hagenmaier, and W. Parr, J. Amer. Chem. Soc. 92, 1735 (1970)

(5) E. Bayer, H. Hagenmaier, W. Parr, H. Eckstein, G. Jung, and P. Hunzicker, in E. Scoffone, Peptides 1969, North Holland Publ., in print

(6) E. Bayer, Chemie f. Labor u. Betrieb 3, (1969)

(7) The material is available under the name Biopak from Waters Associates, Framingham, Mass

(8) E. Bayer, and W. König, J. Chrom. Sci. 7, 95 (1969)

Removal of Protected Peptides from an *ortho*-Nitrobenzyl Resin by Photolysis

By Daniel H. Rich and S. K. Gurwara

(*School of Pharmacy, University of Wisconsin, Madison, Wisconsin* 53706)

Summary Protected peptides can be removed from an *ortho*-nitrobenzyl resin by photolysis.

Recently several modified resins suitable for the solid-phase synthesis of protected peptide fragments have been developed.[1] We report a method for the preparation of N-t-butoxycarbonylpeptide free acids by solid-phase peptide synthesis. The protected peptides are synthesized stepwise on an *ortho*-nitrobenzyl resin then removed from the resin by photolysis under conditions which do not cleave acid-labile protecting groups nor decompose aromatic amino-acids. The use of the photolabile *o*-nitrobenzyl group for protection of aldehyde, amino-, and carboxy-groups has been reported.[2]

An *o*-nitrochloromethyl resin was prepared by nitration of chloromethylated polystyrene beads (1% divinylbenzene) according to the procedure of Merrifield.[3] N-Butoxycarbonylamino-acids were attached to the nitro-resin by heating under reflux with triethylamine in ethyl acetate.[4]

To remove the N-protected amino-acids from the nitro-resin, the N-t-butoxycarbonylamino-acid nitro-resins were suspended in methanol, and irradiated under anaerobic conditions for 12—17 h with stirring in an RPR-100 apparatus (Rayonet, The Southern Co., Middletown, Conn.) equipped with RPR-3500 Å lamps. Wavelengths below 3200 Å were filtered out.[5] The resin was removed by filtration and the solvent evaporated. After purification by chromatography followed by crystallization, the N-t-butoxycarbonylamino-acids were isolated in good yield

Table

Photolysis of N-t-butoxycarbonylamino-acid o-nitrobenzyl resins in methanol

N-Protected amino-acid on resin	Yield of N-protected amino-acid (%)	M.p. (°) (reported)[7]	Photolysis time (h)
Gly	71·2	88—90 (89—90)	12
Leu	64·7	86—87 (86—87)	14
Phe (D,L)	66·4	145—147	17
Phe (L)	59·6	87—89 (88—88·5)	15
Tyr-OCH₂Ph	52·7	108—110	15
Trp	57·3	137—138 (138·5—139·5)	17

(see Table). No racemization of the amino-acids was detected, and no N-t-butoxycarbonylamino-acid remained on the resin.

N-t-Butoxycarbonyl-O-benzyl-L-seryl-O-benzyl-L-tyrosylglycyl-*o*-nitrobenzyl resin (1a) was synthesized using N-t-butoxycarbonylglycyl-*o*-nitro-resin (1·05 mmol/g) according to the general procedure of Merrifield.[3,6] Deblocking was achieved by treatment with 50% trifluoroacetic acid in methylene chloride. Dicyclohexylcarbodi-imide was used as the coupling reagent. The tripeptide (1b) was removed from the *o*-nitro-resin by irradiation at 3500 Å for 12 h as described and was isolated in 62% yield.† The tripeptide (1b) prepared in this way was identical to a sample prepared by solution procedures.†

Removal of protected peptides from the *o*-nitrobenzyl resin by irradiation provides a method for the synthesis of protected peptide fragments suitable for coupling in solution or on a solid support.

This work was supported by a PHS Research Grant from the National Institute of Arthritis, Metabolism, and Digestive Diseases.

(*Received*, 19*th June* 1973; *Com.* 859.)

† Satisfactory amino-acid analysis, microanalysis, t.l.c., n.m.r., and i.r. data were obtained.

[1] S. S. Wang, *J. Amer. Chem. Soc.*, 1973, **95**, 1328 and refs. 7—15 therein.
[2] U. Zehavi, B. Amit, and A. Patchornik, *J. Org. Chem.*, 1972, **37**, 2281; U. Zehavi and A. Patchornik, *ibid.*, p. 2285; A. Patchornik, B. Amit, and R. B. Woodward, 'Peptides,' ed. E. Scoffone, North-Holland Publishing Co., Amsterdam, 1969, p. 12; J. A. Barltrop, P. J. Plant, and P. Schofield, *Chem. Comm.*, 1966, 822.
[3] R. B. Merrifield, *Biochemistry*, 1964, **3**, 1385; *J. Amer. Chem. Soc.*, 1963, **85**, 2149.
[4] J. M. Stewart and J. D. Young, 'Solid Phase Peptide Synthesis,' W. H. Freeman and Co., San Francisco, 1969, p. 32.
[5] A. Singh, E. R. Thornton, and F. M. Westheimer, *J. Biol. Chem.*, 1962, **237**, 3006.
[6] R. B. Merrifield, *Intra-Sci. Chem. Reports*, 1971, **5**, 183.
[7] G. R. Pettit, 'Synthetic Peptides,' vol. I, Van Nostrand Reinhold Co., New York, 1970, pp. 11—78.

Retention of Configuration in the Solid Phase Synthesis of Peptides

E. Bayer, E. Gil-Av,[1] W. A. König, S. Nakaparksin, J. Oró, and W. Parr

Contribution from the Department of Chemistry and Biophysical Sciences, University of Houston, Houston, Texas. Received August 2, 1969

Abstract: Amino acid enantiomers can be resolved by the use of optically active stationary phases in a gas chromatographic system. This technique has allowed the study of racemization in the solid phase synthesis of polypeptides. The results indicate that almost complete retention of configuration is obtained.

Peptide synthesis by the solid phase method is assumed to proceed without racemization. Merrifield,[2] in his first publication on the subject has reported that the L-leucyl-L-alanyl-glycyl-L-valine prepared by his method was completely digested by leucine aminopeptidase. Furthermore, the high biological activity (450 units/mg, potency on the isolated rat uterus), coupled with high synthetic yield (79%) of the oxytocin prepared by Bayer and Hagenmaier[3] indicates that racemization did not occur to any appreciable extent. Nevertheless, this aspect of solid phase synthesis of peptides has not as yet been thoroughly investigated and the question has been raised whether racemization is completely excluded in all cases. With the advent of new, sensitive gas chromatographic methods for the resolution of enantiomeric amino acids[4-6]

(1) On sabbatical leave from the Weizmann Institute of Science, Rehovoth, Israel.
(2) R. B. Merrifield, *J. Amer. Chem. Soc.*, **85**, 2149 (1963).
(3) E. Bayer and H. Hagenmaier, *Tetrahedron Lett.*, 2037 (1968).
(4) E. Gil-Av, R. Charles-Sigler, G. Fischer, and D. Nurok, *J. Gas Chromatog.*, **4**, 51 (1966).
(5) J. W. Westley and B. Halpern in "Gas Chromatography 1968," S. L. A. Harbourn, Ed., Institute of Petroleum, London, 1969, p 119.
(6) E. Gil-Av, B. Feibush, and R. Charles-Sigler, *Tetrahedron Lett.*,

118

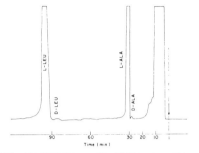

Figure 1. Gas chromatogram of the N-TFA-amino acid isopropyl esters of the (Leu-Ala)₆ hydrolysate. Column: 500 ft × 0.02 in. stainless steel capillary column, coated with N-TFA-L-valyl-L-valine cyclohexyl ester, temperature 110°; injector, 180°; detector FID, 280°; carrier gas He, pressure 20 psi.

procedures became available for an investigation of this problem.

Enantiomeric amino acids can be separated by gas chromatography using two different approaches. The mixtures of enantiomers to be analyzed can be transformed into diastereoisomers by reaction with an asymmetric reagent, such as (+)-2-butanol[4] or N-trifluoroacetyl-L-prolyl chloride,[5] or, alternatively, separation is achieved by chromatography on an optically active stationary phase.[6]

In the present study the second approach was used since it is more suitable for the detection of small amounts of antipodes. In fact, the method based on separation of diastereoisomers requires a derivatizing agent of high optical purity which cannot always be readily attained. The antipodal impurity in the reagent gives rise to a second peak which has to be corrected for in the calculation of the results, and this may introduce a large error when only small amounts of antipodes are present in the sample. Furthermore, one of the diastereoisomers can be formed preferentially so that the correction is not necessarily simply proportional to the amount of antipodal impurity in the reagent. In the second approach, on the other hand, symmetric reagents are used for derivatization and the only correction to be made is for the amount of racemized product formed during the hydrolysis of the peptide.

Results and Discussion

Retention of configuration in the dodecapeptide (Leu-Ala)₅-Leu-Ala and in desamidosecretin, a heptacosapeptide, both synthesized by the solid phase method, was examined. The compounds were hydrolyzed in 6 N HCl at 100° for 20 hr, and the amino acids obtained transformed into trifluoroacetyl-O-methyl esters and/or trifluoroacetyl-O-isopropyl esters which were then chromatographed on capillary columns coated with N-TFA-L-valyl-L-valine cyclohexyl ester.[7]

As can be seen in Figure 1, the dodecapeptide showed only traces of D isomers, determined as N-TFA-O-isopropyl esters, in amounts below 0.1%. The sensi-

1009 (1966); "Gas Chromatography 1966," A. B. Littlewood, Ed., Institute of Petroleum, London, 1967, p 227.
(7) P. Birrell, E. Gil-Av, S. Nakaparksin, and J. Oró, presented at the Southwest Regional Meeting of the American Chemical Society, Dec 4-6, 1968.

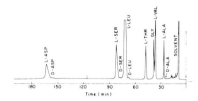

Figure 2. Gas chromatogram of the N-TFA-amino acid methyl esters of the desamidosecretin hydrolysate. Conditions as in Figure 1.

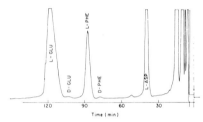

Figure 3. Gas chromatogram of the N-TFA-amino acid isopropyl esters of the desamidosecretin hydrolysate. Column: 100 ft × 0.02 in. stainless steel capillary, coated with N-TFA-L-valyl-L-valine cyclohexyl ester, temperature 110°; injector, 180°; detector FID, 280°; carrier gas He, pressure 5 psi.

tivity of the gas chromatographic method is between 0.05 and 0.1% according to the position of the peak in the chromatogram.

Desamidosecretin contains the following amino acids: alanine 1, valine 1, glycine 2, threonine 2, leucine 6, serine 4, aspartic acid 2, phenylalanine 1, glutamic acid 3, histidine 1, and arginine 4. The enantiomeric analysis of this hydrolysate is understandably somewhat less simple than in the first case. Histidine and arginine do not at present come within the scope of our gas chromatographic method of enantiomer separation. Alanine, valine, threonine, leucine, serine, and aspartic acid were analyzed for antipodal impurities in the form of their methyl esters on a 500-ft column (Figure 2). For leucine, 0.81% of D isomer was found, for alanine 0.76%, for serine 1.1%, and for aspartic acid 2.2%, whereas for valine and threonine the amount was below the detection limit of the gas chromatographic method of (0.1%).

Finally phenylalanine and glutamic acid were analyzed for the presence of D enantiomers by chromatographing their N-TFA-O-isopropyl esters on a 100-ft capillary column (Figure 3). For phenylalanine 0.39% D isomer was detected and for glutamic acid 0.73%.

The small amounts of D isomers found in the hydrolysate of (Leu-Ala)₆ are not caused by optical impurities in the Boc-amino acid derivatives used for the synthesis, as a control reaction showed less than 0.05% of the D isomers. Therefore they must be accounted for by racemization of the liberated amino acids in acid solution. In fact it has been shown on a series of neutral amino acids[8] that slow inversion occurs upon heating L-amino acids under the conditions of peptide hydrolysis. Thus, about 0.2% of D-valine, 0.5% of

(8) S. Nakaparksin, E. Gil-Av, and J. Oró, *Anal. Biochem.*, in press.

Solid Phase Synthesis of Peptides

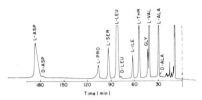

Figure 4. Gas chromatogram of the N-TFA-amino acid methyl esters of the bovine albumin hydrolysate. Conditions as in Figure 1.

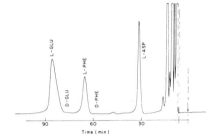

Figure 5. Gas chromatogram of the N-TFA-amino acid isopropyl esters of the albumin hydrolysate. Conditions as in Figure 3.

D-alloisoleucine, and 1.0% D-alanine were detected, when the corresponding L-amino acids were heated at 105° for 24 hr; that is, under somewhat more drastic conditions than those used in the present investigations.

The higher amounts of D-isomers found desaminosecretin, as compared with (Ala-Leu)$_6$, can in part also be explained by racemization of the liberated amino acids. In particular, it should be mentioned that aspartic acid has been reported[9] to racemize in 18 hr to the extent of 3.7% in 6 N HCl at 110°. In addition, however, inversion seems to occur, when opening certain peptide bonds, to a degree depending on the nature of the neighboring amino acid. Thus bovine albumin, which being a natural peptide, should not contain any D-isomer, gave by the identical procedure even larger amounts of some D-amino acids, than desaminosecretin, namely 1.26% D-leucine, 0.95% D-alanine, 0.9% D-glutamic acid, 1.6% D-aspartic acid, 0.8% serine, and 0.46% phenylalanine (Figures 4 and 5). Further evidence is given in the literature.[9] In conclusion it would appear that the D-amino acids found in the hydrolysates of the two synthetic peptides are essentially artifacts.

As far as can be judged by the present analyses, the solid phase synthesis proceeds with practically complete retention of configuration under the conditions specified.

Experimental Section

Desamidosecretin. The synthesis was carried out in a manner similar to that described by Bayer, et al.[3,10,11] t-Butyloxycarbonyl amino acids were prepared from t-butylcarbazate and the appropriate amino acid according to the pH-stat method of Schnabel.[12] The N-terminal amino acid L-histidine was used as the bisbenzoxycarbonyl-L-histidine-p-nitrophenyl ester,[13] because all the protected groups could then simultaneously be removed by catalytic hydrogenation.

t-Boc-L-valine (1.52 g, 7 mmol) was treated with 5 g of chloromethylated polymer (Bio-Beads S-X2 200–400 mesh, Bio-Rad Laboratories). The amino acid analysis of a dried sample showed that the Boc-L-valyl polymer contained 0.11 mmol of L-valine/g. The following cycles of reactions were used to introduce each new amino acid: cleavage of the Boc groups with 30 ml of N HCl in glacial acetic acid for 30 min; washing three times with 30-ml portions of glacial acetic acid, absolute ethanol, and N,N-dimethylformamide (DMF); neutralization of the resulting hydrochloride with 30 ml of triethylamine solution (10% triethylamine in DMF) for 10 min; washing three times with 30-ml portions of DMF and methylene chloride; and coupling of the new amino acid (2.2 mmol) with the free amino groups on the polymer with 2.2 mmol of dicyclohexylcarbodiimide (DCC) as condensation agent in methylene chloride. For the t-BOC-nitro-L-arginine cycles DMF was substituted for methylene chloride. His, Asp, Gln, and Gly were introduced as p-nitrophenyl esters of the Boc-amino acids in DMF as solvent. Boc-amino acids and Boc-amino acid p-nitrophenyl esters were used in 4-fold excess. The reaction time was 2 hr in the case of the DCC method and 8 hr in the case of the p-nitrophenyl esters. Excess reagents were removed by washing successively with methylene chloride (DMF in the case of t-Boc-nitro-L-arginine and t-Boc-amino acid p-nitrophenyl esters), ethanol, glacial acetic acid, ethanol, and methylene chloride.

The fully protected peptide, esterified to the resin, was suspended in 30 ml of trifluoroacetic acid, and a slow stream of anhydrous hydrogen bromide was bubbled through a fritted disk into the suspension for 90 min. The resin was filtered off and washed three times with 20 ml of trifluoroacetic acid. The combined filtrates were evaporated under reduced pressure. The resulting oil was dissolved in glacial acetic acid and lyophilized: yield 1.6 g.

A portion of the protected heptacosapeptide (750 mg) and 750 mg of 10% palladium on charcoal were hydrogenated in 50 ml of 90% acetic acid at 40 psi. After 24 hr the hydrogenation was stopped and 250 mg of additional catalyst was added and the mixture rehydrogenated for an additional 24 hr. Then, the catalyst was removed by filtration and the filtrate freeze dried: yield 500 mg. The amino acid analysis of this product was as follows: His 0.8, Ser 3.8, Asp 1.9, Gly 2.0, Thr 1.8, Phe 0.9, Ala 1.0, Glu 3.0, Leu 5.9, Arg 3.4, and Val 1.2. To detect racemization the peptide was hydrolyzed without further purification.

(Leu-Ala)$_6$ was synthesized as described in ref 13.

Bovine albumin was purchased from Mann Research Laboratories.

N-TFA-L-valyl-L-valine cyclohexyl ester was purchased from the Research Products Division, Miles Laboratories Ltd.

Hydrolysis of Peptides. The peptide (10 mg) was hydrolyzed in 5 ml of 6 N HCl at 100° for 20 hr in a sealed glass tube.

Esterification and Trifluoroacetylation of the Amino Acids. After evaporation of the hydrolysate, 3 ml of the desired alcohol, 1.25 N in HCl, was added per 5 mg of amino acid, and the mixture heated in a sealed tube at 100° for 3 hr. The product was evaporated in vacuo, and 3 ml of methylene chloride added per 5 mg of amino acid in a flask provided with a drying tube. The flask was then cooled to −20° in an acetone Dry Ice mixture, and 3 ml of trifluoroacetic anhydride added per 5 mg of amino acid. The mixture was allowed to heat up to room temperature and after 1 hr, was evaporated to dryness and dissolved in approximately 1 ml of chloroform.

Gas Chromatographic Analysis. The gas chromatographic conditions are described in Figures 1 and 3. Peak areas were measured with an Infotronics integrator. Since optical isomers have identical detector response, calibration factors are not required for the determination of the ratio of antipodes.

Acknowledgments. This investigation was supported by the Robert A. Welch Foundation (Grant No. E-227) and by NASA (Grant No. NGR-44-005-002, NGR-44-005-020, and NASA Contract NAS-9-8012).

---

(9) J. Manning and S. Moore, *J. Biol. Chem.*, **243**, 3591 (1968).
(10) E. Bayer, G. Jung, and H. Hagenmaier, *Tetrahedron*, **24**, 4853 (1968).
(11) E. Bayer, H. Eckstein, K. Hägele, W. A. König, W. Brüning, H. Hagenmaier, and W. Parr, submitted for publication.
(12) E. Schnabel, *Liebigs Ann. Chem.*, **702**, 188 (1967).
(13) K. Inoue, *J. Org. Chem.*, **30**, 1151 (1965).

Instrument for Automated Synthesis of Peptides

R. B. MERRIFIELD, JOHN MORROW STEWART, and NILS JERNBERG
The Rockefeller University, New York, N. Y. 10021

▶ An instrument which can perform automatically all of the operations involved in the stepwise synthesis of peptides by the solid phase method is described in detail. The synthesis of the peptide chain takes place on a solid polymer support and all of the reactions are conducted within a single vessel. The apparatus is composed of two main parts—the reaction vessel and the components required to store, select, and transfer reagents, and the programmer which controls and sequences the operation of the various components. The operation of the instrument and its application to the synthesis of several peptides are described.

EXTENSIVE ADVANCES in methods of isolation, purification, analysis, and structure determination of peptides and proteins have outdistanced our synthetic achievements in this area. To cope with many of the new problems which have arisen, a greatly accelerated and simplified approach to peptide synthesis was required. Solid phase peptide synthesis was devised (13), and developed (9, 10) with these objectives as guides. The principles of the method and the special features which make it adaptable to an automated process have been reviewed (11, 14), and an apparatus designed for automated peptide synthesis was constructed and briefly discussed (15). This article describes in detail the instrument which can perform automatically all of the operations involved in the stepwise synthesis of polypeptides.

GENERAL PRINCIPLES

The method is based on the fact that a peptide chain can be synthesized in a stepwise manner while one end of the chain is covalently attached to an insoluble solid support. During the intermediate synthetic stages the peptide remains in the solid phase and can therefore be manipulated conveniently without significant losses. All of the reactions, including the intermediate purification procedures, are conducted within a single reaction vessel. It is this feature which permits convenient automation of the process. The problem in essence is simply to introduce the proper reagents and solvents into the vessel in the proper sequence at the proper times.

The solid support is a chloromethylated styrene–divinylbenzene copolymer bead. The C-terminal amino acid is coupled as a benzyl ester to the resin and the peptide chain grows one residue at a time by condensation at the amino end with N-acylated amino acids. The tert-butyloxycarbonyl group has been

Figure 1. Apparatus for automated peptide synthesis

the protecting group of choice and activation has usually been by the carbodiimide or active ester routes. Since each of the reactions in the synthesis can be modified in a variety of ways it was important to design the apparatus with sufficient flexibility to cope with a wide range of reactions and conditions.

APPARATUS

The apparatus is composed of two main parts, the first being the reaction vessel with the components required to store and select reagents and to transfer them into and out of the vessel, and the second being the programmer which automatically controls and sequences the operation of the various components. A photograph of the complete instrument is shown in Figure 1, a schematic drawing is given in Figure 2, and the wiring diagram of the pro-

Instrument for Automated Synthesis of Peptides

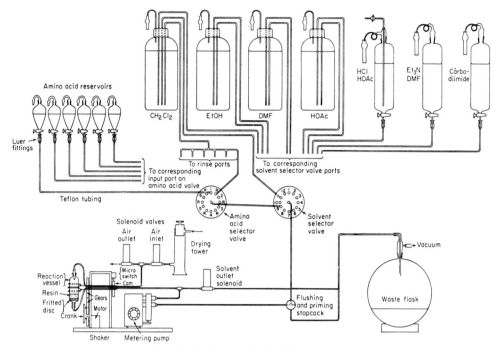

Figure 2. A schematic drawing of the apparatus

grammer is shown in Figure 3. Parts for the instrument are listed below.

Programmer. Stepping drum programmer model A-31-EZ-30, Tenor Co., Butler, Wis.

C1. Electrolytic capacitor, 100 mf., 50 volts.

C2. Paper capacitor, 1 mf., 600 volts.

D1. Silicon diode rectifier, 5 amperes, 100 piv.; RCA 1N1613.

P. Pilot lamp, neon, Drake HR117.

P1–P3. Pilot lamp, Dialco 812210, with GE 1829 bulbs (28 volts) and green jewels.

Pump. Beckman metering pump model 74603, 0 to 20 ml. per minute, modified as described.

R1–R3. 32-Pole shorting relay, Guardian IR-805-S, 24-volt d.c. coil.

R4. 3-Pole double throw relay, Potter-Brumfield KRP14AG, 115-volt a.c. coil.

R5. Time - delay relay, Amperite 115N010 (thermal, 115 volts, 10-second delay).

R6. Stepping relay, Guardian Rotomite IR-705-12P-24D, 24 volts d.c.

S1, S14. Switch, DPDT, 6 amperes, Cutler-Hammer 8373K7.

S2. Switch, SPST, 6 amperes, Cutler-Hammer 8381K8.

S3, S4, S5. Push-button switch, momentary contact, NO, Arrow 80541E.

S6–S10. Unit control switches, SPDT, center off, Cutler-Hammer 7503K13.

S11. Pump limit switch. Miniature microswitch, SPDT, Micro 1SM1.

S12. Shaker limit switch. Microswitch, SPDT, roller type, Micro BZ-2RM22-A2.

S13. End-of-run microswitch on amino acid selector valve, SPDT, Micro BZ-2RM22-A2, NC contacts used.

S15. Tap switch, rotary, 12-position, nonshorting, Mallory 32112J, on shaft of solvent valve.

S16. Tap switch, rotary, 2-gang, 12-position, nonshorting, Centralab 2005, on shaft of amino acid valve.

Drum Switches. Drum - controlled switches of Tenor programmer. NO contacts of all switches are used except Home switch, of which NC contacts are used.

Solvent Valve; Amino Acid Valve. 12-Position all-Teflon motor-driven rotary selector valves (see text).

Shaker. Motor-driven, to invert reaction vessel (see text).

SV1–SV3. Solenoid-operated valves, miniature diaphragm type, all-Teflon, normally closed, Mace EDV-122, Mace Corporation, San Gabriel, Calif.

T1. Transformer, primary 117 volts, secondary 25 volts, 2 amperes, Stancor P8357.

Timers. Cutler-Hammer 10336H46A. No. 1, 1 min; Nos. 2 and 3, 3 min.; No. 4, 30 min.; No. 5, 300 min.

Z1. Resistor, wire-wound, 200 ohms, 5 watts.

For the operation of the instrument the proper reagents and solvents are selected by the amino acid– and solvent selector valves and are transferred by the metering pump from the reservoirs to the reaction vessel which contains the peptide–resin. After the desired period of mixing by the shaker the solvents, excess reagents, and by-products are removed to the waste flask by vacuum filtration. These basic operations are repeated in a prearranged sequence under the control of the programmer until the synthesis of the desired peptide chain is complete. All parts of the apparatus which come into contact with the solvents and reagents are made of glass or chemically resistant polymers.

Reaction Vessel. A modification of the reaction vessel previously described (13) for use in the manual method has been designed and is shown in Figure 4. It consists of a glass cylinder (total volume 45 ml.) with a coarse grade fritted disk (Corning) at the lower end. The bottom is constructed to leave a minimum of space below the filter and is sealed to a male Luer connector. The top end is fitted with a ST 14/20 female joint which holds a specially ground stopper containing a 1-mm. i.d. tube extending 1.5 cm. into the vessel and ending with a 5-mm. coarse fritted disk. The outer end of the stopper is terminated in a male Luer connector. The vessel is held by a three-fingered, hollow-stemmed clamp (Fisher No. 5-742). Solvent and air lines are attached to the reaction vessel by means of Kel-F female Luer fittings (fittings with attached 0.076-inch i.d. tubing of Teflon are available from Hamilton Co., Whittier, Calif.). The tubing of Teflon passes through the hollow clampshaft to avoid entanglement during the shaking operation. Solvents and reagents are pumped into the bottom of the vessel while air is dis-

R. B. Merrifield, John Morrow Stewart, and Nils Jernberg

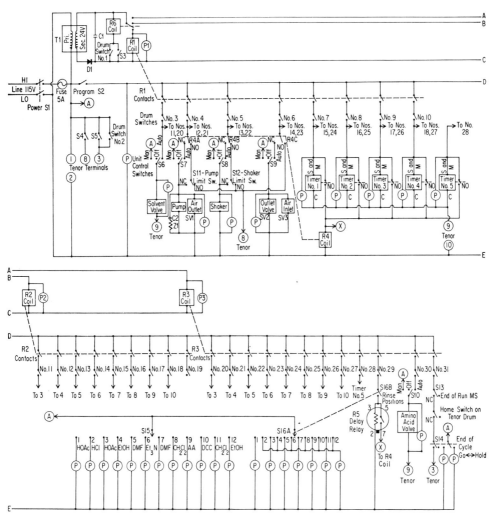

Figure 3. Wiring diagram

placed at the top (air outlet solenoid SV1 open). Solvents are removed at the bottom by vacuum (solvent outlet solenoid SV2 open) while air is drawn in at the top through a tower of Drierite (air inlet solenoid SV3 open). The volume of solvent is adjusted to fill the vessel more than half full but to keep the level below the tip of the air outlet. Thus, when the vessel is inverted all of the inner surface is washed with solvent, and any resin adhering to the walls is brought into contact with the reagents. With solenoid valves SV1 and SV3 closed, an air lock is maintained in the capillary tubing which prevents solvent from escaping at the top when the vessel is inverted. This vessel will accommodate 2 to 4 grams of resin. Two larger vessels of 80- and 120-ml. capacity but otherwise of similar design have been constructed for use with approximately 7- and 10-gram batches of resin.

Shaker. This is a device for producing a gentle mixing of the resin and solvents in the reaction vessel. An eccentric drive from a Hurst synchronous motor (Model PC-DA, 10 r.p.m., with clutch and brake) moves a gear through a 90° arc. This gear drives a second, smaller gear through 180°. The latter is attached to the clamp which holds the vessel, and thus repeatedly inverts the vessel to mix its contents during the shaking periods. A cam mounted on the rear end of the clampshaft is so positioned that it activates a microswitch (S12, shaker limit switch) each time the vessel comes to the upright position. This microswitch is energized by the programmer only at the end of each shaking cycle, and then serves to stop the vessel in the vertical position and to step the programmer to the next operation (see description below).

Pump. The Beckman metering pump (0 to 20 ml. per minute) is used for all of the pumping operations. It is modified in three ways. The standard Viton diaphragms are replaced with Teflon-coated diaphragms (No. 70992) to withstand the effects of dimethylformamide and other organic solvents. The holdup volume of the pump is minimized by insertion of threaded Kel-F plugs into the inlet and outlet ports. The holdup is also minimized by stopping the pump each time at the end of an exhaust stroke. This is done by means of a microswitch (S11, pump limit switch) mounted on the pump backplate so that it is actuated by the lever arm which drives the piston. The total volume pumped is controlled by adjusting both the volume per stroke and the pumping time.

When the pump was turned off high voltage transients were produced. These ruptured the silicon diodes of a bridge rectifier within SV1, which is in parallel with the pump. Addition of the surge filter (C2, Z1) corrected this difficulty.

Solvent Outlet System. Solvents are removed from the reaction vessel by vacuum filtration through the fritted disk at the bottom. The solvent passes out the Teflon tubing through a T connector (machined from Kel-F) to the solenoid valve SV2 and into the 12-liter round-bottomed waste flask. To accelerate the filtration rate and to avoid precipitation of materials and possible obstruction of the waste line, $1/_8$-inch i.d. polyethylene tubing is used for the line between the outlet solenoid (SV2) and the waste flask. A pressure of about 100 mm. is maintained in the waste flask with the laboratory vacuum line. Under these conditions solvent is removed at the rate of approximately 100 ml. per minute. To ensure complete removal of solvents, an excess of time is allowed for each filtration step in the automatic program. Since the Mace diaphragm solenoid valves are not designed to open against a vacuum, satisfactory operation was obtained only by connecting SV2 in the reverse direction—i.e., with the waste flask connected to the "in" port of the valve.

Reagent and Solvent Reservoirs. The solvents such as methylene chloride, dimethylformamide, and ethanol are stored in 4-liter brown bottles fitted with No. 38 polyethylene screw caps. The glacial acetic acid is stored in the original commercial 5-lb. bottles to avoid picking up moisture during transfer. Solvent is withdrawn through lines of Teflon tubing (AWG No. 16, standard natural, Pennsylvania Fluorocarbon Co., Clifton Heights, Pa.) which are inserted through tight-fitting holes in the caps and extend to the bottom of the bottles. Two or three such lines from each bottle are then attached to the appropriate inlet ports of the solvent selector valve (Figure 2). An additional line runs to a drying tube containing indicating Drierite to replace the solvents with dry air.

Solutions of *tert*-butyloxycarbonyl amino acids or esters are stored in 60-ml.

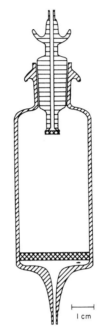

Figure 4. Reaction vessel

separatory funnels having Teflon stopcocks. The solutions are protected from moisture by small Drierite-packed drying tubes which are attached to the funnels with ST 14/20 connections. The stems of the funnels are fitted with Luer connectors and attached to Teflon lines (Hamilton) which run to the odd-numbered positions of the amino acid selector valve (Figure 2). The apparatus is fitted with six funnels which handle a 1-day supply of amino acids.

The other three reagents are contained in 1-liter cylindrical separatory funnels as shown in the upper right of Figure 2. To prevent excessive loss of the volatile solvents by diffusion during storage the drying tubes are connected to the funnels by coiled 3-ft. lengths of 0.076-inch i.d. Teflon tubing. These funnels are fitted with Teflon stopcocks and are connected by Luer joints to Teflon outlet lines which run to the solvent selector valve.

Rotary Selector Valves. To select the desired solvents and reagents rotary Teflon selector valves were constructed (Figure 5). This type of valve was chosen because it gives a very sharp cutoff of liquid, and prevents mixing of one reagent with another. With a manifold in which many reagents flow into a common chamber or line there is danger of cross-contamination of reagents. Two such valves are necessary, one to select the amino acid derivatives and a second one to select all of the other solvents and reagents.

The valves are constructed from two disks of Teflon with carefully machined faces which are mounted between stainless steel plates and held together under spring pressure. The front plate and disk are stationary and contain a center port and 12 evenly spaced ports around the circumference. The various inlet solvent tubes are connected to the outer ports and the single outlet tube is jointed to the center port. The connections are with a threaded nylon pressure screw and a tapered Teflon ferrule. The rear disk contains a center port and one circumferential port which are joined by a 1.5-mm. hole within the disk. As this disk is turned it connects, one at a time, the 12 inlet ports to the central outlet port. A leak-free seal between the two Teflon disks of the valve is obtained by fitting each orifice in the front disk with a specially machined X-ring of Kel-F (see Figure 5). The pressure spring applies sufficient force to the movable disk and pressure plate to seat the X-rings in their recesses and bring the two Teflon disks into light contact. The valve is advanced one position at a time by a Geneva-type intermittent gear drive mechanism (6) which is operated by a 23-r.p.m. Bodine capacitor motor No. KC1-23RM. This drive lacks the locking feature usually found on Geneva drives and thus allows the disk to be turned manually when the drive stud is not engaged. Spring pressure on the Teflon disks maintains the proper indexing in automatic operation. The rear end of the solvent selector valve shaft turns a 12-point tap switch (S15) which indicates the position of the valve at any time by pilot lamps on the programmer panel (Figure 1).

The valves are designed to move in only one direction and cannot select solvents at random. Therefore, the 12 inlet solution lines must be connected to the solvent selector valve in the sequence in which they are required during the synthesis (Figure 2). The outlet tube of the amino acid selector valve is connected to one of the outer ports (No. 9) of the solvent selector valve and the outlet tube of the solvent selector valve is connected to the pump through a three-way stopcock. The third arm of the stopcock is connected to the waste flask and is used to flush and prime the solvent and reagent lines.

The amino acid valve also contains 12 inlet ports, with the six amino acid reservoirs being connected to the odd-numbered positions, while alternate ports are connected to a rinse solvent to flush the line between the two valves and thereby prevent contamination by the previous amino acid solution. The six rinse lines are all supplied with methylene chloride from a six-arm glass

1908 • ANALYTICAL CHEMISTRY

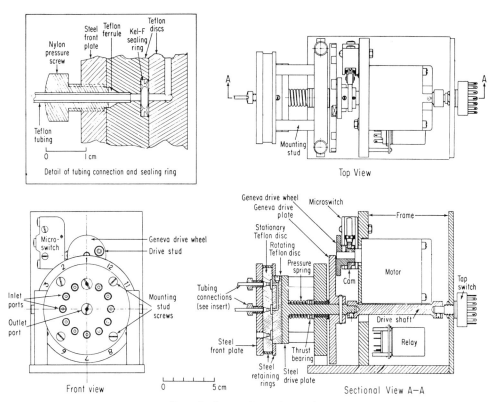

Figure 5. Rotary solvent selector valve

manifold. Instead of the single-gang tapswitch shown in Figure 5 for the solvent selector valve the amino acid valve shaft is fitted with a 12-point, 2-gang tap switch (S16A and B). The first gang (S16A) operates pilot lights to indicate the position of the valve. The odd-numbered positions are connected to the numbered amino acid pilot lights, while the even-numbered positions are all connected in parallel to a single "rinse" pilot (Figure 3). The second gang (S16B) has all of the even-numbered (rinse) positions wired together. These are used in conjunction with the time-delay relay (R5) to perform the rinse function after each amino acid is pumped. Operation of this function is described later. The odd-numbered contacts of S16B are not used. The amino acid valve also bears a microswitch (S13) so positioned that it is actuated by a pin in the Geneva drive plate when the valve is in position 12 (rinse after the sixth amino acid). This end-of-run switch serves to stop the entire instrument after the coupling of the sixth amino acid is completed.

The wiring diagram for the Geneva drive of the solvent and amino acid selector valves is given in Figure 6. The 23-r.p.m. motor shaft bears a cam which actuates a microswitch (Micro BZ 2RM22-A2, roller type, SPDT) once each revolution. The cam is positioned so that the switch is actuated and closes the NO contacts just as the drive stud is emerging from the slot in the Geneva drive plate after advancing the valve to the next position. The microswitch is actuated only momentarily, as the inertia of the motor carries the cam off the switch roller and restores the NC circuit. At the beginning of the Geneva drive cycle, the motor is energized through the NC contacts of the microswitch and the NC contacts of the relay (Potter-Brumfield KHP17A11). When the motor shaft has made one revolution to advance the valve to the next position, actuation of the microswitch by the cam stops the motor and energizes the relay. The NO contacts of the

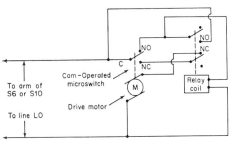

Figure 6. Wiring diagram for the selector valves

relay now serve to hold the relay energized and prevent the motor from advancing further until the source of power has been interrupted by the programmer. The return lead from the Geneva drives of the selector valves to the low side of the line passes through terminals 9 and 10 of the Tenor programmer. These terminals are connected to a switch which is opened each time the programmer steps. This allows the valve drive relay to open and makes it possible to operate the same selector valve on two successive steps if desired. This arrangement is necessary because the drum microswitches do not open during stepping when plugs are inserted in two successive positions.

Programmer. The proper sequential operation and timing of the previously described components is controlled by a Tenor stepping drum programmer. This programmer consists of a row of 32 roller-type microswitches positioned under a revolving drum. The drum bears rows of holes around its circumference, one row of 30 holes being positioned above each microswitch. The drum is driven by a Geneva-type drive which advances the drum by steps, 30 steps constituting one revolution of the drum. The desired program is established by the insertion of nylon plugs into the appropriate holes of the drum. These plugs actuate the corresponding microswitches which in turn control the various operating units (pump, shaker, or valves) and timers. The arrangement of the plugs in the diimide program drum is shown in Figure 7; that for the active ester program drum is shown in Figure 8. The two drums are readily interchangeable. Any future changes in the program necessitated by the use of different reagents or amino acid derivatives can be easily accommodated by changing the location of the plugs in the drums. At the end of the selected time for each step the timer furnishes a signal to cause the programmer drum to step to the next position where a new combination of switches will be actuated. The position of the programmer drum can be advanced manually one step at a time by depressing pushbutton switch S4, or continually by depressing S5. All of the operating units of the instrument can also be operated manually (independently of the automatic program) by means of the unit control switches (S6-S10).

One cycle in the automatic solid phase synthesis—i.e., the lengthening of the peptide chain by one amino acid residue—requires nearly 90 steps of the programmer drum (Figures 7 and 8) while the programmer model used provides only 30 steps per revolution. However, three times as many switches are available on the programmer as are needed for peptide synthesis and these switches

	Home	Bank relay	Step drum	Bank 1 Solvent valve	Pump	Shaker	Outlet	Timer No.1	Timer No.2	Timer No.3	Timer No.4	Bank 2 Solvent valve	Pump	Shaker	Outlet	Timer No.1	Timer No.2	Timer No.3	Timer No.4		Bank 3 Solvent valve	Pump	Shaker	Outlet	Timer No.1	Timer No.2	Timer No.3	Timer No.4	Timer No.5	Rinse timer	Amino acid valve	Return home
Drum step No.	H	1	2	3	4	5	6	7	8	9	10	11	12	13	14	15	16	17	18	19	20	21	22	23	24	25	26	27	28	29	30	31
1	X			X_1		X	X								X	X					X			X								
2					X		X				X				X								X	X								
3						X			X				X				X						X			X						
4							X	X				X_5		X	X						X				X							
5					X		X						X				X				X_9		X	X								
6						X		X				X						X			X		X									
7							X	X								X	X								X					X	X_R	
8					X			X				X						X			X_{10}				X		X					
9						X			X				X				X				X			X								
10				X_2		X	X								X	X					X_{11}	X							X			
11							X					X						X			X					X						
12						X			X				X				X							X	X							
13							X	X			X_6			X	X						X				X							
14							X	X				X	X								X					X						
15				X_3		X	X							X					X		X			X	X							
16					X			X			X_7		X		X						X				X							
17						X			X			X			X								X			X						
18						X	X					X						X					X	X								
19						X		X				X							X		X					X						
20					X		X						X					X					X		X							
21						X	X						X					X			X_{12}			X	X							
22					X		X					X						X					X			X						
23						X		X						X				X						X		X						
24				X_4		X	X								X		X							X	X							
25					X			X				X_8		X	X						X					X						
26						X	X								X		X							X			X				X_A	
27						X	X							X				X														X
28					X		X							X	X																	X
29						X		X						X				X														X
30	X	X																														

Figure 7. Drum program for use with dicyclohexylcarbodiimide.

The x marks indicate the positions of the nylon plugs on the programmer drum. Subscripts 1 to 12 indicate the port on the solvent selector valve which opens when the drum moves to the step shown. Subscripts A and R indicate that the amino acid valve is opened to an amino acid reservoir or to a rinse line, respectively

can therefore be divided into three banks. Each bank of switches is energized in turn with 115-volt power by means of a stepping relay and a multicontact program relay. In this way the drum makes three revolutions during the execution of the complete program and thus makes 90 separate steps available. This drum expansion and switch bank sequencing is controlled by a 24-volt d.c. system provided by T1, D1, and C1 (see Figure 3). The rotary stepping relay (R6) causes each of the three program relays (R1, R2, R3) to be energized in turn; the pilot lights (P1, P2, P3) indicate to the operator which relay is energized and thus show where in the program the instrument is operating. The Guardian stepping relay used (R6) makes 12 steps per revolution. Since this programmer needs only three steps, every third contact of R6 is wired together, thus converting it to a three-step device (contacts 1, 4, 7, and 10 to R1, contacts 2, 5, 8, and 11 to R2, and contacts 3, 6, 9, and 12 to R3). The stepping relay can be advanced manually by push-button switch S3, or automatically by drum switch No. 1, which is actuated by a plug in the drum at step 30. Drum switch 2 is also actuated at step 30 to advance the drum to step 1. Thus during automatic operation the program begins at step 1 of the drum with R1 and the first bank of drum switches (Nos. 3 to 10) energized. At step 30, R6 advances one step and the program drum makes another revolution with R2 and drum switches 11 to 19 energized. At step 30 on this second revolution R6 steps again and, during the third revolution of the program, drum R3 and drum switches 20 to 31 are energized. When the program is completed at step 26 of the third revolution (Figure 7), further operation is controlled by the position of the end-of-cycle switch, S14. If S14 is in the "hold" position (open) the programmer stops. If S14 is in the "go" position (closed), power is applied to terminal 3 of the Tenor programmer causing it to step continuously back to the "home" position (step 1) of the drum. The second pole of S14 energizes pilot lights which indicate the setting of S14. During this "return to home" phase, R6 is again stepped once, thus returning the instrument to the beginning of the program with R1 and drum switches 3 to 10 energized. The instrument then proceeds to carry out another cycle of the synthesis.

Full details of the wiring of the instrument are given in Figure 3. Connections to the Tenor programmer are indicated by circled numbers which correspond to numbered terminals on the Tenor unit (1 and 2, power for drive motor; 3, drum continuous step; 8, drum single step; 9 and 10, to NC microswitch which opens during each stepping operation). All points marked with circled A's are connected together to furnish power to operating units and pilot lights when the program is not in operation. A circled X indicates the connection of R5 switch to R4 coil.

Timers. All of the operations of the automatic instrument are time controlled. All of the operations except the rinse steps are controlled by one of the five Cutler-Hammer timers. Terminals S (solenoid) and M (motor) of the timers are connected together to the appropriate drum switch, while the terminals C (common) of all the timers are connected together and then to terminal 9 of the Tenor programmer. This use of the step-interrupted return circuit is necessary to reset the timer in those cases where the program requires use of the same timer on two successive steps. When a given timer is running and the preset time elapses, the normally open contacts of the timer-controlled switch close, and line voltage is applied to the coil of R4. Thus the timers do not cause the Tenor drum to step directly, but rather actuation of R4 causes the power being supplied to an operating unit to be switched to terminal 8 of the Tenor programmer, thus causing the drum to step. When the operating unit in use is the outlet valve, line voltage is switched by R4C directly to Tenor terminal 8, and the drum steps. When the operating unit is either the pump or shaker, however, the drum does not step until the operating unit limit switch (S11 or S12) is actuated—i.e., until the pump piston reaches the end of an exhaust stroke or the shaker brings the vessel to the upright position. In these cases the energization of R4 only brings the proper limit switch (S11 or S12) into the circuit; when the limit switch is actuated, the drum steps.

The rinse steps which flush the amino acid valve and tubing with methylene chloride to prevent cross-contamination of successive amino acid residues—i.e., step 7, bank 3 of the diimide program and step 3, bank 3 of the active ester program—are not timed by one of the Cutler-Hammer timers, but rather by the time-delay relay (R5). When the amino acid pumping step is completed—e.g., step 6, bank 3 of the diimide program—the drum advances to step 7, where drum switches for the pump (No. 21), amino acid selector valve (No. 30), and the rinse switch (No. 29) are actuated. Both the pump and the amino acid valve begin to operate. Both these devices are operated by 20-r.p.m. motors, and the cam on the amino acid valve Geneva drive is positioned so that the actual movement of the valve disk occurs during the second half of the motor rotation. Since the pump always stops at the end of an exhaust stroke, during the first intake stroke of the pump the valve disk will still be stationary and in the amino acid position. During the following exhaust stroke of the pump, the actual movement of the valve disk occurs, advancing it to the rinse position. This synchronization assures that the pump will not be on an intake stroke when the disk is in motion. When the amino acid valve reaches the rinse position the associated tap switch (gang 2, S16B) (see description above) closes and applies power from Tenor drum switch 29 to the delay relay, R5. After 10 seconds, R5 closes and applies power (through point X, Figure 3) to R4 coil, and the Tenor drum steps when the pump reaches the end of its next exhaust stroke. A similar scheme is used to provide a brief rinse after the diimide pumping step, although in this case (step 10, bank 3) the R5 timing operation begins immediately, since the amino acid valve is already in the rinse position, and S16B is closed. During the first pump exhaust stroke the solvent selector valve advances to position 11 to flush the line with methylene chloride.

The solvent and amino acid selector valves do not incorporate a device to

Drum step No.	Bank 1										Bank 2									Bank 3												
	Home	Bank relay	Step drum	Solvent valve	Pump	Shaker	Outlet	Timer No.1	Timer No.2	Timer No.3	Timer No.4	Solvent valve	Pump	Shaker	Outlet	Timer No.1	Timer No.2	Timer No.3	Timer No.4	Solvent valve	Pump	Shaker	Outlet	Timer No.1	Timer No.2	Timer No.3	Timer No.4	Timer No.5	Rinse timer	Amino acid valve	Return home
	H	1	2	3	4	5	6	7	8	9	10	11	12	13	14	15	16	17	18	19	20	21	22	23	24	25	26	27	28	29	30	31
1	X			X_1		X	X								X	X		X				X									X	X_R
2					X				X				X				X				X_{10}	X					X					
3					X			X					X					X				X					X					
4					X	X				X_5			X	X							X_{11}	X			X	X						
5				X_3	X				X				X				X				X				X							
6					X			X					X			X					X			X								
7						X	X						X	X							X	X										
8					X			X					X				X				X			X								
9					X				X				X			X					X			X								
10				X_2	X	X							X	X							X	X										
11					X			X					X			X					X			X								
12					X				X				X				X				X			X								
13					X			X	X_6			X	X							X_{12}	X	X										
14					X				X			X			X					X			X									
15				X_3	X	X						X				X					X			X								
16					X			X		X_7			X	X						X	X											
17					X			X					X			X					X			X								
18						X X							X			X					X			X								
19					X			X					X	X							X	X										
20					X				X				X			X					X			X								
21						X X						X			X					X			X								X_A	
22					X			X					X X			X					X			X								X
23					X				X				X			X					X			X								X
24				X_4	X X							X				X					X			X								X
25					X			X					X			X X					X			X								X
26					X				X				X			X					X			X								X
27						X X					X_8	X			X					X			X									X
28					X			X			X_9		X X							X			X									X
29					X				X				X			X					X			X								X
30	X X																															

Figure 8. Drum program for use with active esters

See Figure 7 for explanation

signal the programmer drum to step. Therefore these valves are always used in conjunction with a timer and another operating unit (pump, shaker, outlet) which will furnish the necessary stepping signal.

Housing and Interconnections. The programmer module is housed in a standard relay rack cabinet (Premier cabinet rack, DCR 210, $22^{3}/_{4}$ inches high). All of the switches and pilot lights are mounted on a standard $10^{1}/_{2}$-inch high panel (Figure 1) which is anchored to the Tenor programmer by sheet metal brackets. The remainder of the cabinet front is covered with a Lucite panel. A shelf-type chassis attached to the back of the front panel carries the low-voltage power supply (T1, D1, C1), the stepping relay R6, the program relays (R1, R2, R3), relay R4, and the delay relay R5. The Cutler-Hammer timers are mounted in a row on a panel above the back of the Tenor programmer (Figure 1). The operating units are connected to the programmer module by means of multiconductor cables which plug into the rear of the module. Sockets for these plugs are mounted on a small sheet metal chassis attached to the rear of the Tenor unit. The solvent and amino acid selector valves are connected by 18-conductor cables (Belden 8744), while the pump-shaker-solenoid valve unit is connected by means of a 9-conductor cable (Belden 8449).

OPERATION OF THE INSTRUMENT

Several preliminary operations are necessary before the synthesis of a peptide can be started. First, the supporting resin containing the C-terminal amino acid of the proposed peptide chain must be prepared and analyzed. This is done as previously described (9, 10) by esterification of a chloromethylated copolymer of sytrene and divinylbenzene with the *tert*-butyloxycarbonyl (*t*-BOC) amino acid (1, 4, 8, 17) (Cyclo Chemical Corp., Los Angeles). The product is freed of very fine particles of resin by flotation in methylene chloride to prevent subsequent clogging of the fritted disks of the reaction vessel. A sample of the vacuum-dried product is hydrolyzed in a 1:1 mixture of dioxane and 12N HCl (9) and the liberated amino acid is measured quantitatively on an amino acid analyzer. The amino acid content is used to calculate the amounts of subsequent amino acid derivatives and dicyclohexylcarbodiimide reagent which will be used in the synthesis. The best range of substitution has been 0.1 to 0.3 mmole per gram. *t*-BOC amino acid-resins are usually prepared in advance and are stored until needed.

The appropriate solvent reservoirs are filled with glacial acetic acid (Mallinckrodt analytical reagent), methylene chloride (dichloromethane, Matheson Coleman & Bell DX 835) and commercial (99.5%) absolute ethanol. *N,N*-Dimethylformamide (Matheson Coleman & Bell DX 1730) is freed of dimethylamine and formic acid by shaking with barium oxide and distillation under reduced pressure (20). The 1N HCl-acetic acid solution is prepared by adding 700 ml. of glacial acetic acid to the storage separatory funnel and passing in a slow stream of anhydrous hydrogen chloride. Samples are withdrawn at the bottom and titrated for chloride by the Volhard method. This solution, when protected by the long coil of capillary tubing and drying tube, is stable for several weeks without a significant decrease in concentration. The triethylamine reagent is prepared by mixing 50 ml. of triethylamine (Matheson Coleman & Bell TX 1200) with 450 ml. of purified dimethylformamide.

The solvent lines from the reservoirs are filled one at a time by turning the solvent selector valve to the corresponding position and applying suction through the flushing and priming stopcock.

The metering pump is calibrated by pumping methylene chloride into a graduated cylinder for a measured period. This operation is carried out under the control of the programmer with timer 2 set for 1 minute. The pump rate (approximately 20 ml. per minute) can be varied by changing the length of stroke of the piston.

The holdup volume of the system between the solvent selector valve and the bottom of the reaction vessel is determined by filling the line with the HCl-acetic acid reagent (port 2), turning the valve to the acetic acid line (port 3) and pumping until all of the HCl has been flushed out. The effluent is titrated for chloride and the holdup volume is calculated. In the system now in use it was 4.2 ml.

The reaction vessel is loaded with a weighed amount of the *t*-BOC amino acid–resin (2 to 4 grams for the small, 45-ml. capacity vessel). The stopper is lubricated with silicone high vacuum grease and secured in place with springs, and the inlet and outlet lines are attached. In the synthesis three equivalents of each *t*-BOC amino acid derivative are used per equivalent of the first amino acid on the resin. The calculated quantity of each of the first six amino acids is dissolved in 7 ml. of methylene chloride, filtered if necessary, and placed in the amino acid reservoirs in the proper sequence. Because of poor solubility in methylene chloride, *t*-BOC-nitro-L-arginine is first dissolved in 2 ml. of dimethylformamide and diluted with 5 ml. of methylene chloride, while *t*-BOC-*im*-benzyl-L-histidine is dissolved in 7 ml. of pure dimethylformamide. The *t*-BOC amino acid *p*-nitrophenyl esters are dissolved in 16 ml. of pure dimethylformamide. During the automated synthesis the amino acid solutions are pumped completely into the reaction vessel and a precise concentration therefore is not required.

The dicyclohexylcarbodiimide solution, on the other hand, is metered by the pump and the concentration of the reagent must be calculated for each run. Since the holdup volume and the total volume pumped are known the actual volume of diimide solution delivered into the vessel can be calculated. The required quantity of dicyclohexylcarbodiimide is dissolved in this volume of methylene chloride. The total volume of solution prepared at one time depends on the number of amino acids to be added.

The programmer is set for the run by inserting the proper program drum (diimide or active ester) and stepping it manually (S5) to step 2. Bank 1 of the drum switches is energized by using S3 to step R6 to the desired position. The timers are set as follows: No. 1, 30 seconds; No. 2, 60 seconds; No. 3, 90 seconds; No. 4, 10 minutes; No. 5, 120 minutes. For runs in the larger vessels, settings of timers 1 and 2 must be increased. The amino acid and solvent selector valves are set to position 1. The pump is set for 20 ml. per minute. The unit control switches (S6-S10) are placed in the auto position and the end-of-cycle switch (S14) is placed in the go position. Closing of the program switch (S2) starts the automatic synthesis.

Functioning of a Typical Diimide Cycle. The instrument first washes the resin three times with acetic acid by means of three sets of pumping, shaking, and outlet steps. As described above, the pump always stops at the end of an exhaust stroke to minimize solvent mixing, and the shaker always stops with the vessel in the upright position to make the following filtering (outlet) step possible. During the third of these outlet steps (step 10, bank 1), the solvent valve advances to position 2, and the HCl-acetic acid reagent is then pumped into the vessel. The 30-minute reaction period necessary for complete removal of the *tert*-butyloxycarbonyl protecting group is obtained by use of three successive 10-minute shaking steps.

After this deprotection step the resin is washed three times with acetic acid to remove hydrogen chloride, three times with ethanol to remove acetic acid, and three times with dimethylformamide. A 10-minute shaking period with triethylamine in dimethylformamide serves to neutralize the hydrochloride of the amino acid on the resin, thus liberating the free amine in preparation for coupling with the next protected amino acid. Triethylammonium chloride and excess triethylamine are removed by three washes with dimethylformamide, and three methylene chloride washes then prepare the resin for the coupling step. The t-BOC amino acid solution is then pumped into the vessel in a 30-second (timer 1) pumping step; the small amount of air pumped caused no harm. On the next step (rinse), the pump draws one more stroke of air, then three strokes of methylene chloride to flush the amino acid line.

The next step is a 10-minute shaking operation to allow the amino acid to soak into the resin beads. During this step, the solvent valve advances to the diimide (No. 10) position. At the next step, diimide solution is pumped for 30 seconds, and then the rinse step adds one more stroke of diimide solution and three strokes of methylene chloride. The coupling reaction then takes place during a 2-hour (timer 5) shaking cycle. After the coupling reaction, by-products and excess reagents are removed by three washes in methylene chloride and two washes in ethanol.

If the end-of-cycle switch (S14) was set in the hold position, the instrument stops after the third ethanol wash and the resin is left suspended in ethanol. If S14 is in the go position, the programmer returns to the beginning of the program and proceeds to carry out the next cycle of operation. The instrument will continue to operate for approximately 24 hours until the coupling cycle for the sixth amino acid has been completed. Then the end-of-run microswitch (S13) stops the instrument. To continue the run, the amino acid reservoirs are washed (solvents are added to the reservoirs and drawn through the amino acid valve and the solvent valve to the waste flask through the three-way stopcock). The amino acid reservoirs are then refilled with the proper new solutions, the reagent and solvent reservoirs are replenished if necessary, the amino acid valve is set to position 1, and the solvent valve is set to position 12. The programmer is then stepped manually (S5) back to step 1 to start the coupling of the next six amino acid residues.

The functioning of the active ester coupling cycle is similar in most respects to the diimide cycle, but with the following exceptions:

Since the active ester coupling reactions are done in dimethylformamide, the three methylene chloride washes preceding the coupling reaction are omitted, and instead the resin is washed six times with dimethylformamide. This is accomplished by programming the active ester drum to move the solvent valve past the unused methylene chloride position (No. 8). During the shaking period of the last dimethylformamide wash (step 27, bank 2), the solvent valve is advanced to position 8 (methylene chloride), and during the outlet step (step 28) the valve is advanced again to the amino acid position (No. 9). The instrument is thus ready to pump the amino acid active ester solution (16-ml., 1-minute pumping, timer 2) at step 29.

A 4-hour reaction period for the coupling reaction is provided by programming two successive 120-minute shaking steps.

Since the diimide reagent is not used, the program drum moves the solvent valve past this position without using it. During the coupling reaction the solvent valve is advanced to the diimide position, and during the next step (outlet, step 4) the valve is again stepped to bring it to the methylene chloride position (No. 11) for the washing operations.

The diimide reaction has been used routinely for the introduction of all amino acids except asparagine and glutamine. Since the diimide reagent causes an undesirable side reaction with these amino acids ($5, 7, 16$), they have been employed as their p-nitrophenyl esters ($2, 3$). When it is desired to change to the active ester program during the course of a synthetic sequence, the filling of the amino acid reservoirs is arranged so that the last amino acid before the asparagine or glutamine is placed in reservoir 6. This causes the instrument to stop at the point where the drum change must be made. A similar procedure may be followed for the reverse program change, or if only a single asparagine or glutamine is to be introduced, it may be placed in any reservoir and the end-of-cycle switch placed in the hold position.

Resin samples may be removed at any point during the synthesis for hydrolysis and amino acid analysis as described above. This allows the operator to ascertain that the synthesis is proceeding satisfactorily.

When the synthesis of the desired amino acid sequence has been completed, the peptide-resin is removed from the reaction vessel with the aid of ethanol, filtered, and dried. Weight gain of the resin during the synthesis provides an indication of the amount of peptide incorporated. The peptide is cleaved from the resin with HBr-trifluoroacetic acid as previously described (11), and subjected to a suitable purification procedure.

Reliability, Maintenance, and Possible Improvements. The automatic instrument described above has been used for over 400 coupling cycles. In general, a high degree of reliability of the instrument has been experienced. Certain precautions should be mentioned, however. Vacuum lines must be kept clear of accumulations of solid triethylammonium chloride which sometimes form from the vapors which pass through the system. The relays should be checked routinely to be sure that contacts are not fouled. The Tenor programmer unit incorporates three relays, one of which arcs badly and tends to transfer contact material; it failed after approximately 30,000 steps. The shaker limit microswitch (S12) has performed satisfactorily for over 1 million cycles, and the pump limit switch (S11) over 200,000 times. These values are in excess of the manufacturer's minimum life expectancy ratings.

While the Beckman pump has shown satisfactory resistance to the wide range of solvents used, some problems have been experienced resulting from failures of the valve actuating mechanism. One such failure of the outlet valve caused excessive cylinder pressure, leakage of HCl-acetic acid, and resultant dissolution of the nylon cylinder-retaining nut. This nylon nut has been replaced by one machined from Kel-F.

Experience in the use of the automatic instrument has suggested some possible improvements. A desirable change would be the incorporation of both the diimide and active ester programs on a single large drum, with a provision for automatic change of program as desired. In areas where power interruptions are common, a safety relay could be incorporated into the circuit to prevent the instrument from continuing upon restoration of power. This would be desirable because of the design of the amino acid and solvent selector valve circuits. These valves advance one position each time they receive power, requiring three seconds for this operation, and then hold until the power is removed. During automatic operation these valves are in the holding phase for considerable lengths of time. If even an instantaneous power failure should occur during such a holding period, upon restoration of power the valve would again advance and become completely out of phase with the remainder of the program. It would be preferable to stop the instrument entirely. An improved instrument embodying these modifications is currently under construction.

Other possible modifications would be the incorporation of other redundant or fail-safe controls. For example, a solvent level control for the reaction vessel to circumvent possible pump malfunction or exhaustion of solvents; interlocking controls to assure that the solvent and amino acid valves are in the proper position at any point in the program; a vacuum sensor in the waste receiver to prevent possible failure of the filtering operation; and a recorder to make a permanent record of the performance of each operation of the automatic cycle for help in trouble-shooting possible failures.

APPLICATIONS

The automated apparatus described here has been successfully utilized for the synthesis of several peptides, among which were bradykinin (15), several analogs of this nonapeptide plasma kinin (18), angiotensinylbradykinin (12), a decapeptide from tobacco mosaic virus protein (19) and insulin (8). There was, in each case, substantial saving of time and effort in the synthesis of these peptides and the overall yields were better than those usually achieved by conventional techniques. These advantages should become even more important as the synthesis of longer peptides is undertaken.

Since the various peptides which have been prepared have contained most of the common naturally occurring amino acids, it is believed that the method will have rather wide applicability to problems of peptide synthesis. However, several problems concerned with the chemistry of peptide synthesis by the solid phase method remain, and certain amino acids and combinations of amino acids still present difficulties. These questions are under active investigation.

It has already been suggested (12) that the principles of solid phase synthesis should be applicable to the synthesis of other polymers of defined structure. The flexibility which has been incorporated into the design of this instrument is expected to facilitate its application to the automated synthesis of such polymers in addition to the specific application described here for the synthesis of peptides.

LITERATURE CITED

(1) Anderson, G. W., McGregor, A. C., *J. Am. Chem. Soc.* **79**, 6180 (1957).
(2) Bodanszky, M., duVigneaud, V., *Ibid.*, **81**, 5688 (1959).
(3) Bodanszky, M., Sheehan, J. T., *Chem. Ind.* **1964**, p. 1423.
(4) Carpino, L. A., *J. Am. Chem. Soc.* **79**, 98 (1957).
(5) Gish, D. T., Katsoyannis, P. G., Hess, G. P., Stedman, R. J., *Ibid.*, **78**, 5954 (1956).
(6) Jones, F. D., "Ingenious Mechanisms for Designers and Inventors," Vol. **1**, p. 69, Industrial Press, New York, 1930.
(7) Kashelikar, D. V., Ressler, C., *J. Am. Chem. Soc.* **86**, 2467 (1964).
(8) McKay, F. C., Albertson, N. F., *Ibid.*, **79**, 4686 (1957).
(9) Marshall, G. R., Merrifield, R. B., *Biochemistry* **4**, 2394 (1965).
(10) Merrifield, R. B., *Biochemistry* **3**, 1385 (1964).
(11) Merrifield, R. B., *Endeavour* **24**, 3 (1965).
(12) Merrifield, R. B., in "Hypotensive Peptides," E. G. Erdös, N. Back, and Sicuteri, eds., Springer Verlag, New York, 1966.
(13) Merrifield, R. B., *J. Am. Chem. Soc.* **85**, 2149 (1963).
(14) Merrifield, R. B., *Science* **150**, 178 (1965).
(15) Merrifield, R. B., Stewart, J. M., *Nature* **207**, 522 (1965).
(16) Paul, R., Kende, D. S., *J. Am. Chem. Soc.* **86**, 741 (1964).
(17) Schwyzer, R., Sieber, P., Kappeler, H., *Helv. Chim. Acta* **42**, 2622 (1959).
(18) Stewart, J. M., Woolley, D. W., in "Hypotensive Peptides," E. G. Erdös, N. Back, and F. Sicuteri, eds., Springer Verlag, New York, 1966.
(19) Stewart, J. M., Young, J. D., Benjamini, E., Shimizu, M., Leung, C. Y., *Federation Proc.* **25**, 653 (1966).
(20) Thomas, A. B., Rochow, E. G., *J. Am. Chem. Soc.* **79**, 1843 (1957).

RECEIVED for review September 6, 1966. Accepted October 10, 1966. Work was supported in part by Grant A 1260 from the U. S. Public Health Service.

21

Copyright © 1972 by the American Society of Biological Chemists, Inc.

Reprinted from *J. Biol. Chem.*, **247**(19), 6224–6233 (1972)

Acyl Carrier Protein

XVIII. CHEMICAL SYNTHESIS AND CHARACTERIZATION OF A PROTEIN WITH ACYL CARRIER PROTEIN ACTIVITY*

(Received for publication, March 10, 1972)

WILLIAM S. HANCOCK,‡ DAVID J. PRESCOTT,§ GARLAND R. MARSHALL,¶ AND P. ROY VAGELOS

From the Washington University School of Medicine, Departments of Biological Chemistry and of Physiology and Biophysics, St. Louis, Missouri 63110

SUMMARY

A protected linear polypeptide of 74 amino acids with the sequence of *Escherichia coli* 1 to 74 apo-acyl carrier protein (ACP) was synthesized by the automated solid phase method. The polypeptide was removed from the solid support and partially deprotected by treatment of the peptide-resin with hydrogen bromide and trifluoroacetic acid, and the product was purified by gel filtration. The removal of protecting groups was completed by hydrogenation, and the prosthetic group, 4'-phosphopantetheine, was introduced enzymatically with holo-ACP synthetase. Ion exchange chromatography of the product yielded a preparation in which 55% of the protein in the purified fraction contained the prosthetic group. This synthetic 1 to 74 holo-ACP was as active as native holo-ACP in the malonyl pantetheine-CO_2 exchange reaction which is dependent upon malonyl-coenzyme A-ACP transacylase and β-ketoacyl-ACP synthetase.

The purified synthetic 1 to 74 holo-ACP preparation was found to be homogeneous and similar to native 1 to 74 holo-ACP as judged by co-chromatography on DEAE-cellulose and Sephadex G-50 and by sodium dodecyl sulfate disc gel electrophoresis. In addition the synthetic and native proteins were similar with respect to their ultraviolet spectra, amino acid compositions, and their immunological activity with antiserum prepared against native ACP.

Acyl carrier protein plays a central role in *de novo* biosynthesis of fatty acids in all biological systems that have been examined (2–12). It has the prosthetic group, 4'-phosphopantetheine, which is attached to the protein through a phosphodiester linkage to a serine residue (13, 14). The sulfhydryl group of the 4'-phosphopantetheine is the substrate binding site of ACP,[1] and the intermediates of fatty acid biosynthesis are bound as thioesters.

The enzymes that catalyze reactions involving ACP have a high degree of specificity for the protein component of this coenzyme, and little perturbation of its structure is tolerated (15–20). It has been shown, for example, that replacement of the ACP of a plant fatty acid-synthesizing system with *Escherichia coli* ACP causes a change in the spectrum of fatty acids produced (19). Structure-function studies have been initiated through investigations of ACP peptides (15–18) and investigations of the effects of chemical modifications of specific amino acid residues (16, 20). Both of these procedures, however, are limited in their ability to define the role of specific residues. It is only by chemical synthesis that specific amino acid substitutions can be systematically prepared, and for this reason the synthesis of ACP was undertaken as a necessary prelude to such a study. The synthesis of even a small protein, such as ACP containing 77 residues, presents a challenge since at the present time the syntheses of only a very limited number of proteins have been attempted, e.g. bovine insulin (21, 22), ferredoxin (23), ribonuclease A (24, 25), fragment P_2 of *Staphylococcus aureus* nuclease T (26), cytochrome *c* (27), a protein with growth hormone activity (28), and soybean trypsin inhibitor (29). It was decided to use the solid phase synthetic procedure of Merrifield because this method has been found to be much faster and to give a higher yield than the classical approach (30).

Several important prerequisites for the synthesis of ACP have been accomplished. The complete amino acid sequence of *E. coli* ACP was determined by Vanaman *et al.* (31), while Tagaki and Tanford (32) showed that denatured ACP can be readily renatured. Holo-ACP synthetase, which catalyzes the synthesis of holo-ACP from CoA and apo-ACP, as shown in Reaction 1 (see "Experimental Procedures") was purified from extracts of *E. coli* and characterized (18, 33). Thus if apo-ACP could be synthesized by the solid phase method, the enzyme could be utilized to add the prosthetic group to the deprotected product. *E. coli* 1 to 74 ACP, formed by digestion of native 1

---

* For the preceding report of this series see Reference 1. This investigation was supported in part by Grants 5-R01-HL 10406 and AM-13025 from the United States Public Health Service, Grant GB-5142X from the National Science Foundation, and the George Murray Scholarship (University of Adelaide, Australia).
‡ Present address, Department of Chemistry and Biochemistry, Massey University, Palmerston North, New Zealand.
§ Present address, Department of Biology, Bryn Mawr College, Bryn Mawr, Pennsylvania.
¶ Established Investigator, American Heart Association.

[1] The abbreviations used are: ACP, acyl carrier protein; holo-ACP, acyl carrier holoprotein; apo-ACP, acyl carrier protein lacking the 4'-phosphopantetheine prosthetic group. Nomenclature and abbreviations, where possible, follow the tentative rules of the IUPAC-IUM Commission on Biological Nomenclature ((1966) *J. Biol. Chem.* **241**, 2491; (1967) **242**, 555).

to 77 ACP with carboxypeptidase A, was shown to possess full biological activity with ACP synthetase (18) and in fatty acid biosynthesis (15). The synthesis of 1 to 74 apo-ACP would have an important advantage over that of 1 to 77 apo-ACP since ACP 1 to 74 does not contain histidine; histidine residues tend to complicate peptide syntheses (34).

The pentapeptide 33 to 37, around the active site of *E. coli* ACP, has been synthesized previously by Miura and Sato (35), while the manual synthesis of 1 to 74 ACP has been the subject of a previous communication (36). In this paper the preparation and purification of a protein with ACP activity is described, together with a comparison of the synthetic product with native ACP.

MATERIALS AND METHODS

Materials—*t*-Butoxycarbonyl-amino acids and *t*-butoxycarbonyl-glycine esterified to a styrene-1%-divinylbenzene resin were purchased from Schwarz BioResearch. The following side chain blocking groups were used: lysine, benzyloxycarbonyl; arginine, N^ϵ-nitro; aspartic acid, β-benzyl ester; threonine, serine and tyrosine, benzyl ethers; glutamic acid, γ-benzyl ester, while the *t*-butoxycarbonyl group was used for α-amino protection. Methionine was purchased as its dicyclohexylamine salt. Palladium (5%) on barium sulfate catalyst was purchased from Engelhard.

Sephadex G-25 and G-50 were purchased from Pharmacia; DEAE-cellulose was obtained from Whatman. CoA—SH (Chromatopure) was purchased from P-L Biochemicals. Sodium [^{14}C]bicarbonate, 20 mCi per mmole, was obtained from New England Nuclear. Dithiothreitol was obtained from Calbiochem. *E. coli* B, harvested in early exponential growth, was purchased from the Grain Processing Corporation, Muscatine, Iowa. All other chemicals of reagent grade or better were purchased from common sources.

Unlabeled *E. coli* ACP and *E. coli* ACP and CoA labeled with ^{3}H- or ^{14}C-labeled β-alanine in the prosthetic group were isolated as described previously (37, 38). Purified Fraction A and holo-ACP synthetase were prepared according to the method of Elovson and Vagelos (33) and desalted immediately before use.

Experimental Procedures—The two stage assay for apo-ACP utilized holo-ACP synthetase and measured the holo-ACP formed in a coupled system utilizing the malonyl pantetheine-CO$_2$ exchange reaction as described previously (33). The reactions involved are as follows:

CoA + apo-ACP $\xrightarrow{Mg^{2+}}$ holo-ACP

+ adenosine 3′,5′-bisphosphate (1)

Malonyl-pantetheine + holo-ACP

$\rightleftharpoons$ malonyl-holo-ACP + pantetheine (2)

Caproyl-pantetheine + holo-ACP

$\rightleftharpoons$ caproyl-holo-ACP + pantetheine (3)

Malonyl-holo-ACP + caproyl-holo-ACP

$\rightleftharpoons$ β-ketooctanoyl-holo-ACP + holo-ACP + CO$_2$ (4)

Reaction 1 is catalyzed by holo-ACP synthetase. The holo-ACP thus formed was measured by the holo-ACP-dependent incorporation of ^{14}CO$_2$ into malonyl pantetheine that involves the reversibility of Reactions 2 to 4. Reaction 2 is catalyzed by malonyl-CoA-ACP transacylase, and Reactions 3 and 4 are catalyzed by β-ketoacyl-ACP synthetase (39). Malonyl-CoA-ACP transacylase and β-ketoacyl-ACP synthetase are present in Fraction A, a crude enzyme preparation from *E. coli* (33). In the first stage of the assay the standard reaction mixtures contained 2 μmoles of Tris-HCl, pH 8.0, 0.2 μmole of dithiothreitol, 1 to 30 pmoles of apo-ACP, 24 nmoles of reduced CoA, 0.8 μmole of MgCl$_2$, and 0.001 unit of holo-ACP synthetase in a total volume of 0.04 ml. After 10 min at 33° the holo-ACP that was synthesized was assayed in the second stage of the assay by addition of 0.05 ml of a mixture containing 85 nmoles of malonyl pantetheine, 17 nmoles of caproyl pantetheine, 8.5 μmoles of imidazole-HCl, pH 6.2, 2 μmoles of EDTA, 4.5 μmoles of KH^{14}CO$_3$ (200 μCi per mmole), and 0.5 mg of Fraction A. The addition of EDTA immediately terminated the holo-ACP synthetase reaction, which requires Mg^{2+}. After 15 min at 33° the exchange reaction was stopped by the addition of 0.01 ml of 4 N HCl, the reaction mixtures were transferred to liquid scintillation counting vials and dried at 100° for 15 min. The samples were counted after the addition of 1 ml of water and 10 ml of Bray's solution (40). Native apo-ACP residues 1 to 77 and 1 to 74 gave identical values when assayed in this system.

Antibody against pure *E. coli* native ACP was prepared by Dr. D. A. K. Roncari from rabbits receiving subcutaneous injections of ACP mixed with Freund's complete adjuvant as described previously (41) and characterized by Ouchterlony double diffusion patterns (42) and quantitative precipitin tests (43).

Protein concentrations were determined by the method of Lowry *et al.* (44) and by amino acid analysis. Radioactive [β-alanine-^{14}C]ACP, 3.4 μCi per μmole, was used as a protein standard when the concentration of ACP was measured. All radioactivity measurements were carried out in a Packard 3380 scintillation counter.

Yields for coupling reactions in the synthesis were based on the limiting reactant, the peptide, and were calculated from the amount of the first amino acid, glycine, attached to the resin. Peptide-resin (2 mg) was hydrolyzed with 2 ml of 12 N HCl and propionic acid (1:1) for 2 hours at 130° according to the method of Scotchler *et al.* (45). Free peptides were hydrolyzed with 6 N HCl in sealed, evacuated tubes for 24 hours at 110°. Amino acid compositions of peptides were determined on a Beckman-Spinco amino acid analyzer.

Solid Phase Synthetic Procedure—A Schwarz automated synthesizer was used for the automated syntheses. The synthesis was based on the stepwise addition of protected amino acids to the carboxyl-terminal amino acid, glycine, which was esterified via a benzyl ester to a polystyrene resin. Most of the procedures common to the solid phase method (46) were used, although some significant modifications were introduced. The rationale for these changes will be described in a separate publication.[2] After the desired sequence of 74 amino acids had been assembled (Fig. 1), the peptide was cleaved from the resin, deprotected, and isolated as described below.

In a typical synthesis, 0.66 mmole of *t*-butoxycarbonyl-glycine esterified to 2 g of a 1% cross-linked polystyrene resin support was used. The sequence of washes utilized in the addition of a single amino acid residue is outlined in Table I. Since the number of washes was increased greatly in our modified procedure, a single manual synthesis of ACP required 3 months, while automation of the procedure decreased this time to 3 weeks. The results of the automated synthesis of ACP are described in this paper, while the manual synthesis was the subject of a previous communication (36). The coupling steps were carried out with

[2] W. S. Hancock, D. J. Prescott, P. R. Vagelos, and G. R. Marshall, manuscript in preparation.

```
                 1        6       10
NH₂-Ser-Thr-Ile-Glu-Glu-Arg-Val-Lys-Lys-Ile-Ile-Gly-Glu-
                         20
      Gln-Leu-Gly-Val-Lys-Gln-Glu-Glu-Val-Thr-Asp-Asn-Ala-Ser-
                    30                          (P)-Panetheine-SH
      Phe-Val-Glu-Asp-Leu-Gly-Ala-Asp-Ser-Leu-Asp-Thr-Val-Glu-
                                        36
           44                      50                 55
      Leu-Val-Met-Ala-Leu-Glu-Glu-Glu-Phe-Asp-Thr-Glu-Ile-Pro-
                              60
      Asp-Glu-Glu-Ala-Glu-Lys-Ile-Thr-Thr-Val-Gln-Ala-Ala-Ile-
           70                77
      Asp-Tyr-Ile-Asn-Gly-His-Gln-Ala-COOH
```

Fig. 1. The amino acid sequence of *E. coli* ACP (31).

Table I
Washes used during automated synthesis of acyl carrier protein

The cycle of washes constitutes one double addition of a residue and involves 84 washes of the resin. If 2 g of peptide-resin were used, then each wash was of 25 ml, except for trifluoroacetic acid washes (16 ml) and coupling reactions (14 ml). All amino acids were added by the dicyclohexylcarbodiimide method except glutamine and asparagine, which were added as the *p*-nitrophenyl esters. Coupling reactions with these two esters were carried out twice in dimethylformamide (10 ml) for 6 hours each time, and three washes of dimethylformamide were included before each coupling reaction.

Reagent	Number of applications	Time (each application)
		min
CH_2Cl_2	3	½
Trifluoroacetic acid-CH_2Cl_2 (1:1)	2	½, 10
CH_2Cl_2	3	½
3-Butanol	3	½
CH_2Cl_2	3	½
Trifluoroacetic acid-CH_2Cl_2 (1:1)	2	½, 10
CH_2Cl_2	3	½
3-Butanol	3	½
CH_2Cl_2	3	½
Et_3N-CH_2Cl_2	3	½
CH_2Cl_2	3	½
3-Butanol	3	½
CH_2Cl_2	3	½
t-Butoxycarbonyl-amino acid,[a] dicyclohexylcarbodiimide-CH_2Cl_2	1	120
CH_2Cl_2	3	½
3-Butanol	3	½
CH_2Cl_2	3	½
Et_3N-CH_2Cl_2	3	½
CH_2Cl_2	3	½
3-Butanol	3	½
CH_2Cl_2	3	½
t-Butoxycarbonyl-amino acid,[a] dicyclohexylcarbodiimide - CH_2Cl_2, dimethylformamide (1:1)	1	120
Dimethylformamide	3	½
CH_2Cl_2	3	½
3-Butanol	3	½
CH_2Cl_2	3	½
Et_3N-CH_2Cl_2	3	½
CH_2Cl_2	3	½
3-Butanol	3	½
CH_2Cl_2	3	½

[a] Before the addition of dicyclohexylcarbodiimide, the resin was shaken with the amino acid for 5 min.

a 4-fold excess of the appropriate amino acid, 0.15 M, with dicyclohexylcarbodiimide, 0.15 M, as the coupling reagent except for glutamine and asparagine. These two amino acids were added as the *p*-nitrophenyl esters. *t*-Butoxycarbonyl groups were removed by two treatments of the resin with 50% (v/v) trifluoroacetic acid in dichloromethane (CH_2Cl_2).

Cleavage of Peptide from Resin with Hydrogen Fluoride—The reaction was carried out in a polypropylene apparatus, which was based on the design of Pourchot and Johnson (47) and the procedure of Sakakibara *et al.* (48). The peptide-resin (0.2 g) was dried in a vacuum overnight, transferred into a polypropylene reaction vessel, and 0.2 ml of anisole and 3 mg of methionine were added to protect the product. The reaction and reservoir vessels were placed on the HF line and nitrogen gas was passed through the apparatus for 1 hour. HF (10 ml) was distilled into the reservoir vessel and then redistilled into the reaction vessel. The reaction mixture was stirred for 1 hour while the temperature was maintained at 10°. The HF was removed under a stream of nitrogen (3 hours), then the last traces were removed with the use of an oil pump. Residual anisole and its derivatives were removed by ether washes, and the product was dissolved in 5 ml of 0.01 M Tris-HCl, pH 7.3.

Cleavage of Peptide from Resin with Hydrogen Bromide—A 0.2-g sample of the peptide-resin was cleaved from the solid support by passing a stream of hydrogen bromide gas through a suspension of the resin in a mixture of 10 ml of trifluoroacetic acid, 2 mg of methionine, and 0.2 ml of anisole. The apparatus described by Stewart and Young (49) was used for the cleavage reaction. After ½ hour at 25°, the cleaved peptides were removed by filtration, the resin was washed with 20 ml of trifluoroacetic acid, and the combined filtrates were evaporated to dryness under reduced pressure. The product was dissolved in 5 ml of 0.01 M Tris-HCl, pH 7.3 (pH adjusted with base). The cleavage was then repeated on the peptide-resin under exactly the same conditions as before, except that the time of reaction was increased to 1 hour.

Deprotection of Cleaved Peptides by Hydrogenation—When HBr was used to cleave the peptide from the support, it was necessary to remove the protecting nitro group on Arg_6 by hydrogenation. A 20-mg sample of the cleaved product was dissolved in a mixture of 5 ml of acetic acid and 5 ml of water, and 22 mg of 5% Pd-barium sulfate catalyst was added. The sample was hydrogenated at atmospheric pressure for 2 hours. The catalyst was then removed by centrifugation at 2000 × *g* for 10 min, and the pellet was washed three times with 10-ml portions of an acetic acid-water mixture (1:1). The supernatants were combined, the solvent was removed under reduced pressure, and the residue was dissolved in 5 ml of 0.01 M Tris-HCl, pH 7.3.

RESULTS

Synthesis of Protected 74 Amino Acid Residue Polypeptide Chain of 1 to 74 Acyl Carrier Protein—The course of the synthesis was followed by the removal of samples at steps in the synthesis where the first of a given amino acid was added. The amino acid ratios were determined on the acid hydrolysate of the peptide-resin. Even with our modifications to the synthetic procedure, the yield of the growing peptide dropped considerably from Gly_{74} to Val_{65} (Table II). However, from that point, the yield of the synthesis remained constant at 40 to 50%, which indicated that the peptide chains that had become unavailable for coupling early in the synthesis did not reinitiate growth at a later stage.

The final weight of the fully protected 1 to 74 ACP-resin was

Acyl Carrier Protein

TABLE II

Incorporation of first residue of each amino acid in sequence of acyl carrier protein

A 2-mg sample of peptide-resin was taken at various stages of the synthesis and hydrolyzed with HCl and propionic acid. The incorporation of the amino acid was determined by amino acid analysis.

Amino acid	Residue number	Amount incorporated
Gly	74	1.0
Asp	73	0.85
Ile	72	0.80
Tyr	71	0.80
Ala	68	1.00
Glu	66	0.60
Val	65	0.40
Thr	64	0.50
Lys	61	0.50
Pro	55	0.50
Phe	50	0.50
Leu	46	0.40
Met	44	0.45
Ser	36	0.40
Phe[a]	28	0.45
Lys[a]	18	0.40
Arg	6	0.35

[a] Second residue of this amino acid.

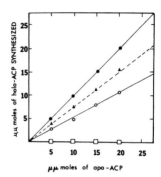

FIG. 2. Stability of native apo-ACP to treatment with HF, HF-trifluoroacetic acid, and HBr-trifluoroacetic acid. Apo-ACP was treated with either HF, HF-trifluoroacetic acid, or HBr-trifluoroacetic acid as described in the text and "Experimental Procedures." Native apo-ACP (●——●), HBr-trifluoroacetic acid-treated apo-ACP (▲---▲), HF-trifluoroacetic acid-treated apo-ACP (○——○) and HF-treated apo-ACP (□——□) were assayed in the two-stage assay for apo-ACP described in "Experimental Procedures."

3.6 g, while the product contained 1 NO_2^-, 31 benzyl-, and 4 benzyloxycarbonyl groups and had a calculated molecular weight of 11,685. The product contained 0.065 mmole of protein per g of protected ACP-resin as estimated by amino acid analysis and by the amount of peptide obtained from the HBr-trifluoroacetic acid cleavage reaction. This corresponds to 20% of the initial value for the amount of glycine esterified to the resin.

Stability of Apo-ACP in Conditions Necessary for Cleavage of Peptide-Resin—Before the removal of the protected peptide from the resin could be attempted, it was necessary to determine the stability of native apo-ACP in the conditions necessary for the cleavage of the benzyl ester. Treatment of apo-ACP with HF as described in "Experimental Procedures" led to complete inactivation of the protein (Fig. 2). When equal volumes of HF and trifluoroacetic acid were utilized under similar conditions, the recovered apo-ACP was 46% active. The best retention of biological activity (72%) was obtained when apo-ACP was treated with HBr and trifluoroacetic acid. Therefore, this procedure appeared to be suitable for use with the synthetic product.

As HF could not be used to cleave the synthetic apo-ACP from the resin, it was necessary to investigate the use of a second deprotection step, hydrogenation, which would remove the nitro group protecting Arg_6. Apo-ACP was hydrogenated using the conditions described in "Experimental Procedures." Samples (10%) were removed at different time intervals during the hydrogenation, and the activity of the peptide was measured in the two stage assay for apo-ACP (Fig. 3). In a parallel experiment, the efficiency of the palladium on barium sulfate catalyst was tested in the reduction of N-nitro-t-butoxycarbonyl-arginine. Although N-nitro-arginine residues in some proteins have been found to be resistant to hydrogenation (49), presumably due to steric hindrance, this study would at least give the minimum time necessary for complete removal of the nitro group. A sample of N-nitro-t-butoxycarbonyl-arginine was hydrogenated

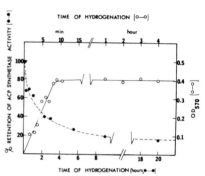

FIG. 3. Stability of native apo-ACP to hydrogenation and the rate of hydrogenolysis of N-nitro-t-butoxycarbonyl-arginine. A 2-mg sample of apo-ACP was dissolved in 2 ml of 50% acetic acid and 3 mg of Pd-BaSO$_4$ catalyst was added. The sample was hydrogenated at atmospheric pressure and samples (10%) were taken at various time intervals. The samples were processed as described in "Experimental Procedures," and the activity of the apo-ACP was measured in the two-stage assay. The apo-ACP activity was expressed as a percentage of the original apo-ACP activity (●---●). In a second experiment, N-nitro-t-butoxycarbonyl-arginine (2.8 mmoles), dissolved in 10 ml of 50% acetic acid, was hydrogenated under exactly the same conditions with 5 mg of Pd-BaSO$_4$ catalyst, and samples (0.1 ml) were withdrawn at different time intervals. The ammonia released in the reduction was trapped in the reaction mixture as ammonium acetate, which was measured by its reaction with ninhydrin. The catalyst was removed by centrifugation, the supernatant was diluted to 1 ml with water, and 0.5 ml of ninhydrin solution (50) and 2 ml of methanol were added. After the samples were heated at 100° for 10 min, 3 ml of water were added and the absorbance was read at 570 nm (○---○). *OD*, optical density.

under conditions identical with those used for apo-ACP deprotection; the reduction was rapid (Fig. 3) as no further release of ammonia was observed after 10 min. The preceding experiments indicated that native apo-ACP could not be quantitatively recovered with any of the cleavage or deprotection procedures.

TABLE III

Yield of product cleaved from peptide-resin by various methods

The reaction conditions that were used are described in "Experimental Procedures" and text. The amount of protein cleaved was measured by Lowry protein determination on an aliquot of the isolated protein or by amino acid analysis on the acid hydrolysate of a weighed sample of the residual resin. The yield of cleavage was based on the amount of peptide, as measured by amino acid analysis, present in the peptide-resin before the cleavage reaction.

Method of cleavage	Conditions	Weight of peptide cleaved from 50 mg of peptide resin	Yield[a]
		mg	%
HF	1 hr at 0°	12.5[b]	92
HF-trifluoroacetic acid (1:1)	30 min at 10–15°	9.2[c]	68
HBr-trifluoroacetic acid	30 min and 1 hr at 25°	10[c,d]	74
Methanol-Et₃N	40 hrs at 25°	3.0[c]	22
Benzyl alcohol-Et₃N	40 hrs at 60°	1.0[c]	7

[a] Yield of cleavage.
[b] Measured by amino acid analysis on acid hydrolysate of weighed sample of residual resin.
[c] Measured by Lowry protein determination on an aliquot of isolated protein.
[d] Material from both cleavages was combined.

It appeared that HBr-trifluoroacetic acid treatment for cleavage of the peptide from the resin, followed by hydrogenation to remove the N-nitro protecting group, might yield the best results.

Isolation and Purification of Synthetic Product—Samples of the peptide-resin were subjected to the HF, HF-trifluoroacetic acid and HBr-trifluoroacetic acid cleavage procedures in an attempt to verify the optimal procedure for cleavage of the peptide from the resin. The yield and conditions of cleavage are described in Table III. The general procedure used for the transesterification studies was based on the work of Beyerman et al. (51), but even with the most vigorous conditions only a small portion of the peptide was cleaved from the resin with methanol-Et₃N or benzyl alcohol-Et₃N (Table III).

From studies of the stability of apo-ACP (Figs. 2 and 3) and the extent of cleavage of the peptide-resin (Table III), it was decided to use HBr-trifluoroacetic acid as the method of cleavage and deprotection. The amino acid composition of the cleaved product was found to be very similar to that of the peptide-resin (Table IV). The ratio of the values for tyrosine (residue 71) to phenylalanine (residues 28 and 50), which should be 0.5, was 1.3 in the crude cleaved peptide, indicating the presence of small incomplete peptides in the product. For comparison, the amino acid analysis of the peptide-resin obtained in the manual synthesis (36) is also shown in Table IV.

The crude, partially deprotected material was purified by gel filtration on Sephadex G-25. The elution profile of the column is shown in Fig. 4. Peak I, which had the same elution volume as native ACP, contained 57% of the protein (Table V). Only material from Peak I had any biological activity in the assay with ACP synthetase (see below). The rest of the material presumably consisted of ACP fragments formed by incomplete coupling reactions. The amino acid analysis for Peak I is shown in Table IV, and the improvement in the composition, relative

TABLE IV

Amino acid analyses

Values are expressed as moles per mole of ACP (calculated from the average of the number of moles of all amino acids).

Amino acid	Native 1 to 74 ACP Expected[a]	Native 1 to 74 ACP Found[b]	Peptide-resin Manual[c] synthesis	Peptide-resin Automated synthesis	Peptides left on resin	Peptides cleaved from resin	Sephadex G-25 Peak I	After DEAE-cellulose
Lys	4	4.5	5.2	4.1	4.2	4.4	4.4	4.5
His	0	0.1						
Arg	1	0.8	1.0	0.35	0.8	0.4	0.6	0.6
Asp	9	10.2	11.5	13	14.2	13.4	10.1	9.0
Thr	6	6.4	6.2	4.7	5.7	4.4	5.6	6.4
Ser	3	3.1	3.2	1.5	3.3	1.6	2.0	3.0
Glu	17	16.3	19.5	14.1	22.2	13.8	14.3	14.6
Pro	1	1.2	0.8	1.4	1.6	1.1	1.3	1.2
Gly	4	4.5	4.4	6.3	7.6	6.5	6.4	4.9
Ala	6	6.3	7.2	9.0	11.1	8.8	9.5	8.8
Val	7	8.7	7.8	5.0	5.2	5.1	9.9	6.6[d]
Met[e]	1	0.9	0.6	0.5	0.5	0.5	0.5	0.6
Ile	7	6.9	7.6	8.3	8.2	8.3	10.0	8.2[d]
Leu	5	5.2	4.7	3.1	5.0	3.3	3.4	3.6[d]
Tyr	1	0.6	1.0	1.9	1.0	1.9	1.0	1.3
Phe	2	1.7	1.9	1.6	1.9	1.5	1.4	1.8
β-Ala	1	1.2						0.7

[a] Based on the published amino acid sequence of 1 to 77 ACP (31).
[b] 1 to 74 ACP was prepared by the carboxypeptidase A digestion of native 1 to 77 ACP (18).
[c] Manual synthesis was reported elsewhere (36).
[d] Values are from 72-hour hydrolyses.
[e] All values corrected for the formation of methionine sulfoxide and sulfone.

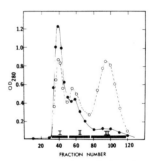

FIG. 4. Chromatography of the synthetic product on Sephadex G-25. A 20-mg sample of the partially deprotected synthetic apo-ACP, dissolved in 5 ml of 0.01 M Tris-HCl, pH 7.3, was chromatographed on a column (1.5 × 150 cm) of Sephadex G-25, with a flow rate of 2 ml per min and 0.01 M Tris-HCl, pH 7.3, as the eluting buffer. Fractions (3.8 ml) were collected and the protein content was measured by absorbance at 280 nm (●——●). The product, 4 mg, from the manual synthesis (36) was chromatographed with a similar procedure, (○- - -○). *OD*, optical density.

to the crude peptide mixture, can be attributed to the removal of short peptides, *e.g.* the ratio of tyrosine to phenylalanine decreased to 0.7. For comparison the elution profile obtained with peptide from the manual synthesis (Fig. 4) and the peptide yields in the different peaks of the chromatogram (Table V) are shown.

The deprotection of the synthetic product was then completed

Acyl Carrier Protein

TABLE V
Protein content of different peaks obtained from Sephadex G-25 chromatography of crude synthetic protein

	Manual synthesis			Automated synthesis		
	Protein	V_E	Total	Protein	V_E	Total
	mg^a	ml	%	mg^b	ml	%
Peak I (Fractions 30 to 50)	1.2	74^c	25	11.4	71.5^c	57
Peak II (Fractions 51 to 80)	0.7	112	23	7.4	110	37
Peak III (Fractions 81 to 120)	1.9	185	48	1.3	180	10

[a] A total of 4 mg was chromatographed on the column, but the optical densities were increased proportionally in Fig. 4 to match the other sample.

[b] A total of 20 mg was chromatographed on the column.

[c] The elution volume for native ACP was found to be 72 ml.

by removal of the N-nitro group protecting arginine as described under "Experimental Procedures." The recovery of protein from this reaction was 88% and the amino acid analysis of the product was not significantly different from that obtained with material of Peak I (under the acidic conditions required to hydrolyze the sample for amino acid analysis most of the nitro-arginine was converted to arginine).

An attempt was then made to convert the synthetic 1 to 74 apoprotein to 1 to 74 holo-ACP by reacting it with holo-ACP synthetase and CoA. To facilitate identification of the holo-ACP, [pantetheine-^{3}H]CoA was used so that the product would be radioactive, and the reaction mixture was chromatographed on DEAE-cellulose-52 (Fig. 5). Peak I was readily identified as the synthetic ACP, as the holo-ACP activity in the malonyl pantetheine-CO_2 exchange reaction was catalyzed by fractions in this peak. The fractions of Peak I were pooled, concentrated, and desalted on a Sephadex G-25 column. The desalted material contained 6×10^4 dpm of ^{3}H radioactivity, which corresponded to 1.5 μmoles of synthetic 1 to 74 holo-ACP. Lowry protein determination and amino acid analysis indicated that the sample contained 22 mg of protein (2.75 μmoles). Thus 55% of the protein in this sample contained the radioactive 4'-phosphopantetheine. This material was used without further purification in the following studies which attempted to establish the identity of the synthetic product.

The amino acid analysis of the material from Peak I gave a composition which agreed closely with the expected values for 1 to 74 ACP (Table IV). In addition, β-alanine, a constituent of the prosthetic group, could be identified in the analysis, and the value of 0.7 residue per molecule was in close agreement with the above finding that 55% of the protein contained 4'-phosphopantetheine.

Because of the small difference in properties of 1 to 74 ACP and CoA on DEAE-cellulose, a sample of Peak I (^{3}H radioactivity), unlabeled CoA, and native 1 to 77 ACP (^{14}C radioactivity) were chromatographed on DEAE-cellulose-52, and the peak of ^{3}H radioactivity was shown to be well resolved from the peak of CoA (Fig. 6). It is of interest that 1 to 77 ACP, which contains three additional residues including a histidine residue, was eluted at a much higher salt concentration than 1 to 74 ACP.

Comparison of Properties of Synthetic Product and Native Acyl Carrier Protein—Synthetic 1 to 74 [pantetheine-^{3}H]holo-ACP and native 1 to 74 [pantetheine-^{14}C]holo-ACP were shown to co-

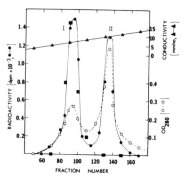

Fig. 5. Chromatography of products formed by holo-ACP synthetase from synthetic 1 to 74 apo-ACP and CoA. The incubation mixture contained Tris-HCl, pH 8.5, 0.2 mM; dithiothreitol, 0.1 mM; [pantetheine-^{3}H]CoA, 5 μmoles (2.0×10^5 dpm); synthetic apoprotein, 3.3 μmoles; and ACP synthetase, 30 μg (1.0 enzyme unit) in a total volume of 20 ml. The reaction mixture was incubated at 37° for 20 hours. The mixture was then diluted with water to a conductivity of 0.5 mmho and applied to a DEAE-cellulose-52 column (2 × 10 cm), which had been equilibrated with 0.01 M Tris-HCl, pH 7.3, 0.001 M dithiothreitol. The column was eluted with a linear gradient composed of 250 ml of 0.01 M Tris-HCl, pH 7.3, 0.001 M dithiothreitol, and 250 ml of the same buffer containing 0.3 M LiCl (18.2 mmhos). Fractions (3 ml) were collected and were assayed for ^{3}H radioactivity (●——●), absorbance at 280 nm (○- - -○), conductivity (▲——▲), and holo-ACP activity, ^{14}C radioactivity (■). Holo-ACP activity was measured in the malonyl pantetheine-CO_2 exchange reaction, described in "Experimental Procedures" as the second stage of the two-stage assay for apo-ACP. *OD*, optical density.

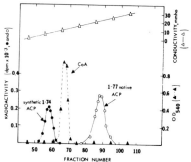

Fig. 6. Chromatography of synthetic 1 to 74 ACP, native 1 to 77 ACP, and CoA on DEAE-cellulose. A solution containing synthetic 1 to 74 ACP (^{3}H radioactivity, 2×10^4 dpm), unlabeled CoA (2 mg), and native 1 to 77 ACP (^{14}C radioactivity, 3×10^4 dpm) in 1 ml of 0.01 M Tris-HCl, pH 7.3, 0.001 M dithiothreitol was applied to a column (2 × 10-cm) of DEAE-cellulose-52 which had been equilibrated with 0.01 M Tris-HCl, pH 7.3, 0.001 M dithiothreitol. The column was eluted with a linear gradient composed of 100 ml of 0.01 M Tris-HCl, pH 7.3, 0.001 M dithiothreitol, and 100 ml of the same buffer containing 0.7 M LiCl (conductivity 29 mmhos). Fractions (1.8 ml) were collected and assayed for ^{3}H radioactivity (●——●), ^{14}C radioactivity (○——○), CoA (▲- - -▲), and conductivity (△——△). The CoA was measured by the phosphotransacetylase assay (52). *OD*, optical density.

chromatograph on a DEAE-cellulose column that was developed with a shallow lithium chloride gradient (36), indicating that the synthetic and native products had similar charge properties.

Synthetic 1 to 74 [pantetheine-^{3}H]ACP and native 1 to 74 [pan-

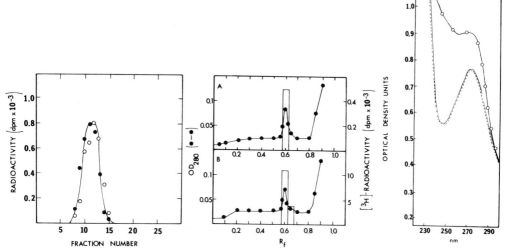

FIG. 7 (*left*). Cochromatography of synthetic and native 1 to 74 ACP on Sephadex G-50. A mixture of synthetic 1 to 74 [pantetheine-^{3}H]ACP (1.25 nmoles, 1×10^5 dpm) and native 1 to 74 [pantetheine-^{14}C]ACP (20 nmoles, 1×10^5 dpm) in 0.1 ml of 0.01 M Tris-HCl, pH 7.3, 0.001 M dithiothreitol was applied to a column (1×24 cm) of Sephadex G-50. The column was eluted with the same buffer and fractions (0.4 ml) were assayed for ^{3}H radioactivity (●——●) and ^{14}C radioactivity (○——○).

FIG. 8 (*center*). Sodium dodecyl sulfate polyacrylamide gel electrophoresis of synthetic 1 to 74 [pantetheine-^{3}H]holo-ACP and native 1 to 77 [pantetheine-^{3}H]holo-ACP. The samples were subjected to electrophoresis on 15% gels containing 0.1% sodium dodecyl sulfate at pH 8.9 according to the method of Laemmli (53). *A*, 80 µg of synthetic 1 to 74 ACP (0.6×10^3 dpm of ^{3}H radio-activity). *B*, 50 µg of native 1 to 77 ACP (19×10^3 dpm of ^{3}H radioactivity). The gels were scanned for absorbance at 280 nm (●——●) and were then sliced. The slices were digested in 0.5 ml of 33% hydrogen peroxide by heating at 100° for 10 min. Bray's solution (10 ml) was added and the ^{3}H radioactivity (*bar graphs*) was measured. *OD*, optical density.

FIG. 9 (*right*). Ultraviolet absorption spectra of ACP preparations. Absorption spectrum of native 1 to 74 ACP (———), synthetic 1 to 74 ACP (- - -) and product from HF-trifluoroacetic acid cleavage (○——○) in 0.01 M Tris-HCl, pH 7.3. The concentration of protein was 0.45 mg per ml. A Spectronic 600 recording spectrophotometer was used for the measurements of the spectra.

tetheine-^{14}C]ACP were chromatographed on a Sephadex G-50 column, and fractions were assayed for ^{3}H and ^{14}C radioactivity. The ^{3}H and ^{14}C radioactivity were found to be coincident (Fig. 7), indicating that the synthetic and native 1 to 74 ACP were of similar size.

Sodium dodecyl sulfate disc gel electrophoresis (Fig. 8) of synthetic 1 to 74 [*pantetheine*-^{3}H]holo-ACP and native 1 to 77 [*pantetheine*-^{3}H]holo-ACP indicated that both samples had identical electrophoretic mobility at pH 8.9. A single protein peak, obtained with the synthetic ACP preparation, contained all of the ^{3}H radioactivity, suggesting that the synthetic ACP was homogeneous as measured by this particular technique.

As 1 to 74 ACP contains only 1 tyrosine, 2 phenylalanine, and no histidine or tryptophan residues, there is little absorption in the ultraviolet spectrum of native ACP, and one could expect that the ultraviolet spectrum of synthetic ACP would be significantly different in the presence of protecting groups such as benzyloxycarbonyl and benzyl derivatives. It was found that the ultraviolet spectra of synthetic and native 1 to 74 ACP (Fig. 9) were completely superimposable at all wave lengths, which indicated that all of the protecting groups had been removed by the procedures used for deprotection. This was in distinct contrast to the crude material from the HF and trifluoroacetic acid cleavage which had a much greater ultraviolet absorption than the native protein.

Synthetic 1 to 74 holo-ACP was examined for activity in the malonyl pantetheine-CO$_2$ exchange reaction, a coupled assay based upon the activities of malonyl-CoA-ACP transacylase and β-ketoacyl-ACP synthetase (39). As seen in Fig. 10, synthetic 1 to 74 holo-ACP was as active as native 1 to 77 holo-ACP in this assay. Furthermore the activity of synthetic 1 to 74 holo-ACP in the exchange reaction (Table VI) was dependent upon

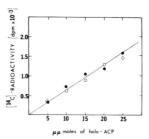

FIG. 10. Activity of synthetic 1 to 74 ACP and native 1 to 77 ACP in the malonyl pantetheine-CO$_2$ exchange reaction. The assay procedure for the determination of holo-ACP activity was identical with that described in Table VI and "Experimental Procedures." Similar concentrations of native 1 to 77 holo-ACP and synthetic 1 to 74 holo-ACP were tested. Only 55% of the protein in the synthetic sample was holo-ACP; the remaining protein contained no 4'-phosphopantetheine. ^{14}C radioactivity (●——●) incorporated with synthetic 1 to 74 holo-ACP and (○——○) incorporated with native 1 to 77 holo-ACP.

Acyl Carrier Protein

TABLE VI

Activity of synthetic 1 to 74 holo-ACP in malonyl pantetheine-CO_2 exchange reaction

Synthetic 1 to 74 holo-ACP, 4 µµmoles, was incubated with 2 µmoles of Tris-HCl, pH 8, and 0.2 µmole of dithiothreitol in a total volume of 0.04 ml for 10 min at 33° to ensure complete reduction of holo-ACP. The holo-ACP was then assayed in the malonyl pantetheine-CO_2 exchange reaction as described under "Experimental Procedures."

System	Amount of $^{14}CO_2$ fixed
	pmoles
Complete	2.6
Complete, minus synthetic ACP	0.75
Complete, minus caproyl pantetheine	0.44
Complete, minus malonyl pantetheine	0.64
Complete, minus caproyl pantetheine and malonyl pantetheine	0.41
Complete, minus Fraction A	0.01

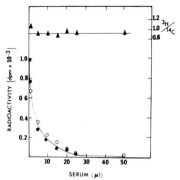

FIG. 11. Activity of synthetic 1 to 4 ACP with antibodies specific for native ACP. A mixture of synthetic 1 to 74 (0.13 mg, ^{3}H radioactivity, 2×10^4 dpm) and native 1 to 74 ACP (0.13 mg, ^{14}C radioactivity, 2×10^4 dpm) was incubated with increasing amounts of antiserum. The incubation mixtures contained the ACP mixture and 1 to 50 µl of serum in 0.2 ml of 0.02 M potassium phosphate buffer, pH 7.4, 0.15 M sodium chloride, and 0.001 M dithiothreitol. After incubation at room temperature for 2 hours and at 0° for 16 hours, the ACP-antibody complex was precipitated by the addition of 0.2 ml saturated ammonium sulfate solution (pH 7.0), and the precipitate was removed by centrifugation at $10,000 \times g$ for 10 min. A sample (5%) of the supernatant was assayed for ^{3}H radioactivity (●——●) and ^{14}C radioactivity (○——○). Ratio of ^{3}H:^{14}C radioactivity (▲——▲) in the supernatant.

the presence of caproyl pantetheine, malonyl pantetheine, and Fraction A, which contained both malonyl-CoA-ACP transacylase and β-ketoacyl-ACP synthetase, and these requirements are characteristic of this exchange reaction with native ACP (39).

Antiserum specific for ACP was prepared from rabbits immunized against native ACP (41). This antiserum was found to react equally well with native apo-ACP and holo-ACP. A mixture of synthetic 1 to 74 ACP (^{3}H radioactivity) and native 1 to 74 ACP (^{14}C radioactivity) was incubated with increasing amounts of antiserum. As is shown in Fig. 11, synthetic 1 to 74 and native 1 to 74 ACP were precipitated equally well by the antibodies to ACP independent of antibody concentration. This indicated that the synthetic protein and the native protein contained antigenic sites that had similar affinity for the antibodies.

DISCUSSION

The synthesis of ACP was not monitored, except by amino acid analysis, because none of the available analytical methods was found to be satisfactory for use with a rapid automated synthesis. Fortunately, the first residue of each amino acid is scattered throughout the ACP sequence (Table II), and the yield of the addition of these residues gave an estimate of the amount of peptide that was growing at any stage of the synthesis. The analyses indicated that the amino acid additions occurred in good yield even toward the end of the synthesis, so that an increase in size of the peptide chain did not cause additional steric hindrance for the coupling reactions. To ensure an adequate yield, however, it was found necessary to repeat all coupling and deprotection reactions and to include a much larger number of washes into the synthetic scheme; for example, extensive use was made of a 3-butanol wash to shrink the resin. Thus the resin was subjected to a swell-shrink-swell cycle before a deprotection or coupling reaction was repeated (Table I). Such a wash scheme has been shown to expose buried functional groups so that repetition of a given step will lead to further reaction with newly exposed terminal residues.[2] The extra steps, however, greatly increased the time that was required for a single synthesis, which now involved 6216 washes of the resin. If the synthesis of analogues was to be attempted, it was obvious that the process had to be automated so that the protein could be prepared with a reasonable expenditure of effort. The automation of the synthesis has the further advantage of much greater reproducibility as human error becomes a significant factor in long syntheses. A commercially available peptide synthesizer was adapted so that the synthesis of ACP could be carried out continuously for 24 hours a day. The automated synthesis gave a product, in a similar yield, indistinguishable from that prepared by the manual method (see "Appendix"). As both coupling and deprotection reactions were repeated, the time of each reaction was decreased to 2 hours and 10 min, respectively, except for active esters which were still coupled for 6 hours. There is considerable evidence that extended dicyclohexylcarbodiimide-mediated coupling reactions are not useful because of a rapid side reaction, the formation of N-acyl ureas, which consumes the reactants. One could expect, therefore, that two 2-hour couplings would be much more effective than one 4-hour coupling. As is common with other syntheses, the molar excess of the reagents increased during the synthesis, since the same quantity of reagents was used even though the number of available peptide chains decreased.

The side chain protecting groups were chosen so that cleavage of the benzyl ester linkage of the peptide chain to the resin would remove all of the protecting groups. This goal could not be achieved because of the instability of ACP to anhydrous HF, and this necessitated the use of HBr and trifluoroacetic acid as the cleavage procedure, with the consequent retention of N-nitro protecting groups. Although benzyloxycarbonyl groups were used to block the ε-amino groups of lysine, careful sizing of the cleaved product on Sephadex G-50 indicated that little growth had occurred on side chains because of premature removal of this derivative during the synthesis. Methionine was used without protection of the thioether, and amino acid analysis of the cleaved peptides (Table IV) indicated that little destruction of this amino acid had occurred during the synthesis or the cleavage step.

The use of HBr-trifluoroacetic acid was by far the best method for cleavage of the peptide from the resin, as apo-ACP had reasonable stability under the reaction conditions, a good yield of peptide was released (74%), and the product, after hydrogenation, was biologically active. It was found necessary to repeat the HBr-trifluoroacetic acid treatment as the first cleavage only released 50 to 70% of the available material. Amino acid analysis of the material left on the resin after two HBr-trifluoroacetic acid cleavages was in close agreement with the expected values for ACP, but repetition of the cleavage reaction did not release any more material.

The amino acid analysis of the crude mixture of cleaved peptides indicated that the product contained significant quantities of short peptides, which were readily removed by gel filtration on Sephadex G-25. The short peptides contained a preponderance of tyrosine, and as the only tyrosine in ACP occurs at residue 71, this amino acid was present in incomplete peptides in much higher molar quantities than amino acids such as phenylalanine, leucine, or serine, which were added much later in the synthesis when the percentage of chains growing was decreased. The presence of these short peptides, together with the data obtained from amino acid analysis of samples taken at different stages of the synthesis (Table II), indicated that residues which became unavailable at a given stage of synthesis did not reinitiate growth to a significant extent. The unavailability of truncated sequences for further reaction was consistent with the suggestion[2] that the solvation of the peptide-resin changes as the synthesis proceeds, because of the increasing protein character of the solid phase matrix. A decrease in the solvation could cause a loss of certain reaction sites that are situated in the less accessible regions of the resin beads. This problem could be expected to become intensified as the synthesis proceeded, so that, with minor exceptions, one would not expect buried sites to reinitiate growth at a later stage of the synthesis.

When native apo-ACP was subjected to the conditions of deprotection, i.e. HBr-trifluoroacetic acid treatment followed by hydrogenation, the protein retained only 40% of its biological activity. It is interesting to note that 55% of the purified synthetic product was active with holo-ACP synthetase. In addition, the synthetic product which accepted the prosthetic group in the reaction with holo-ACP synthetase was as active as native holo-ACP in the malonyl panteheine-CO_2 exchange reaction. Although a direct comparison between the activities may not be valid, as the stability of a denatured, partially deprotected peptide may be different from the native protein, it is clear that a very significant proportion of the product has biological activity very similar to that of native ACP. It has been shown that both holo-ACP synthetase (18) and β-ketoacyl-ACP synthetase (15) possess an unusually high degree of specificity for the intact protein chain. One can, therefore, assume that the synthetic product which is biologically active has a close resemblance to the native protein.

The unique nature of the prosthetic group, which contains the only sulfhydryl in the molecule, could serve as the basis for a further purification step to remove the remaining inactive protein. This approach, as well as the preparation of analogues of ACP, is being actively pursued as part of our studies of the structure-function relationships of ACP.

Acknowledgments—We wish to thank Dr. D. A. K. Roncari for the generous gift of *E. coli* ACP antiserum, Dr. R. Ray Fall for carrying out the polyacrylamide disc gel electrophoreses, and Dr. R. A. Bradshaw for numerous amino acid analyses.

TABLE VII
Comparison of yields obtained with manual and automated syntheses

	Manual synthesis		Automated synthesis	
	Yield of this step	Over-all yield	Yield of this step	Over-all yield
	%	%	%	%
Peptide resin[a]	15	15	20	20
Cleaved peptide	12	1.8	74	15
Peptide in Peak I of Sephadex G-25	25	0.45	57	8.5
Peptide after hydrogenation	94	0.42	88	7.5
Peptide active with holo-ACP synthetase	25	0.11	46	3.4

[a] Based on the amount of glycine originally esterified to the resin.

TABLE VIII
Extraction of cleaved peptides by various solvents

	Total extracted protein (%)	
Solvent used in extraction	Manual synthesis	Automated synthesis
1st trifluoroacetic acid wash	33	70
2nd trifluoroacetic acid wash	10	25
Dioxane		
Acetic acid-water (1:1)	57	5
Total protein (%)[a] extracted by these washes	12	74

[a] Measured by amino acid analysis of a weighed sample of peptide-resin.

We wish to thank Mrs. L. Warren and Mrs. K. Woodin for expert technical assistance.

APPENDIX

The manual solid phase synthesis of ACP, which was reported previously (36), utilized several modifications which were not used in the subsequent automated syntheses. As is shown in Table VII, the yield of the automated synthesis was significantly better than that of the manual synthesis, although it has been shown that automation of the procedure is not responsible for this increase in the yield.[3] It was of interest, therefore, to examine these modifications in the manual synthesis procedure and to determine the reasons for this lower yield. In the manual synthesis all coupling reactions were repeated, but the second coupling reaction was done with 1.5 M urea included in the solvent. The use of this reagent has been reported to increase the yield of difficult coupling reactions (54). It can be noted in Table IV that the amino acid analysis of the peptide-resin at the end of the manual synthesis was closer to the expected values for ACP than in the automated synthesis, which would suggest that the urea treatment increased the yield of the synthesis. Unfortunately, the properties of the resin after the urea treatment were quite different from the normal polystyrene resin; it had a "shriveled" appearance when dry, as if the urea treatment caused cross-linking of the resin. This was reflected in the difficulty experienced in the cleavage and extraction of the product (Table VIII); in fact, only 12% of the product could be

[3] W. S. Hancock, unpublished results.

extracted from the resin with a wide variety of solvents. A disadvantage of the use of urea is the presence of isocyanate impurities, which can cause undesirable side reactions such as chain termination by reaction with free amino groups to form thioureas (55).

Acetylation was also used in the first manual synthesis to block partially complete sequences after the addition of residues 2, 10, 20, 47, 62, and 70 (36). Subsequent studies, however, have shown that acetylation is often not effective in blocking unreactive amino groups[2] (56, 57) and, in fact, the product from the manual synthesis contained less higher molecular weight material (Tables V and VII and Fig. 4). The use of acetic anhydride has been found to cause side reactions (58), and if such reactions occurred in the manual synthesis, they would explain the greater amount of short peptides in the product.

It was clear that these modifications were harmful in the synthesis of ACP, and, therefore, the use of urea and acetic anhydride were abandoned.

REFERENCES

1. BIRGE, C. H., & VAGELOS, P. R. (1972) *J. Biol. Chem.* **247**, 4930-4938
2. GOLDMAN, P., ALBERTS, A. W., & VAGELOS, P. R. (1963) *J. Biol. Chem.* **238**, 3579-3583
3. MAJERUS, P. W., ALBERTS, A. W., & VAGELOS, P. R. (1964) *Proc. Nat. Acad. Sci. U. S. A.* **51**, 1231
4. WAKIL, S. J., PUGH, E. L., & SAUER, F. (1964) *Proc. Nat. Acad. Sci. U. S. A.* **52**, 106
5. OVERATH, P., & STUMPF, P. K. (1964) *J. Biol. Chem.* **239**, 4103-4110
6. VAGELOS, P. R., MAJERUS, P. W., ALBERTS, A. W., LARRABEE, A. R., & AILHAUD, G. P. (1966) *Fed. Proc.* **25**, 1485
7. MAJERUS, P. W., & VAGELOS, P. R. (1967) *Advan. Lipid Res.* **5**, 1
8. LYNEN, F. (1967) *Biochem. J.* **102**, 381-400
9. BUTTERWORTH, P. H. W., JACOB, E. J., DORSEY, J., & PORTER, J. W. (1967) *Fed. Proc.* **26**, 671
10. WILLECKE, K., RITTER, E., & LYNEN, F. (1969) *Eur. J. Biochem.* **8**, 503-509
11. VAGELOS, P. R. (1971) *Curr. Top. Cell Reg.* **4**, 119
12. PRESCOTT, D. J., & VAGELOS, P. R. (1972) *Advan. Enzymol.* **36**, 269-311
13. MAJERUS, P. W., ALBERTS, A. W., & VAGELOS, P. R. (1965) *J. Biol. Chem.* **240**, 4723-4726
14. PUGH, E. L., & WAKIL, S. J. (1965) *J. Biol. Chem.* **240**, 4727-4733
15. MAJERUS, P. W. (1967) *J. Biol. Chem.* **242**, 2325-2332
16. MAJERUS, P. W. (1968) *Science* **159**, 428
17. VAGELOS, P. R., & LARRABEE, A. R. (1967) *J. Biol. Chem.* **242**, 1776-1781
18. PRESCOTT, D. J., ELOVSON, J., & VAGELOS, P. R. (1969) *J. Biol. Chem.* **244**, 4517-4521
19. SIMONI, R. D., CRIDDLE, R. S., & STUMPF, P. R. (1967) *J. Biol. Chem.* **242**, 573-581
20. ABITA, J. P., LAZDUNSKI, M., & AILHAUD, G. P. (1971) *Eur. J. Biochem.* **23**, 412-420
21. MEIENHOFER, J., SCHNABEL, E., BREMER, H., BRINKHOFF, O., ZABEL, R., SROKA, W., KLOSTERMEYER, H., BRANDENBURG, D., OKUDA, T., & ZAHN, H. (1963) *Z. Naturforsch.* **18b**, 1130
22. MARGLIN, A., & MERRIFIELD, R. B. (1966) *J. Amer. Chem. Soc.* **88**, 5051
23. BAYER, E., JUNG, P., & HAGENMAIER, H. (1968) *Tetrahedron Lett.* **24**, 4853
24. GUTTE, B., & MERRIFIELD, R. B. (1969) *J. Amer. Chem. Soc.* **91**, 502
25. DENKEWALTER, R. G., VEBER, D. F., HOLLY, F. W., & HIRSCHMAN, R. (1969) *J. Amer. Chem. Soc.* **91**, 503
26. ONTJES, D. A., & ANFINSEN, C. B. (1969) *Proc. Nat. Acad. Sci. U. S. A.* **64**, 428
27. SANO, S., & KURIHARA, M. (1969) *Hoppe-Seyler's Z. Physiol. Chem.* **350**, 1183
28. LI, C. H., & YAMASHIRO, D. (1970) *J. Amer. Chem. Soc.* **92**, 7608
29. NODA, K., TERADA, S., MITSUYASU, N., WAKI, M., KATO, T., & IZUMIYA, N. (1971) *Naturwissenschaften* **58**, 147
30. MARSHALL, G. R., & MERRIFIELD, R. B. (1971) in *Biochemical Aspects of Reactions on Solid Supports* (STARK, G. R., ed) p. 111, Academic Press, New York
31. VANAMAN, T. C., WAKIL, S. J., & HILL, R. L. (1968) *J. Biol. Chem.* **243**, 6420-6431
32. TAKAGI, T., & TANFORD, C. (1968) *J. Biol. Chem.* **243**, 6432-6435
33. ELOVSON, J., & VAGELOS, P. R. (1968) *J. Biol. Chem.* **243**, 3603-3611
34. WINDRIDGE, G. C., & JORGENSEN, E. C. (1971) *Intra-Sci. Chem. Rep.* **5**, 375
35. MIURA, Y., & SATO, S. (1969) *Bull. Chem. Soc. Japan* **42**, 3592
36. HANCOCK, W. S., PRESCOTT, D. J., NULTY, W. L., WEINTRAUB, J., VAGELOS, P. R., & MARSHALL, G. R. (1971) *J. Amer. Chem. Soc.* **93**, 1799
37. ALBERTS, A. W., & VAGELOS, P. R. (1966) *J. Biol. Chem.* **241**, 5201-5204
38. MAJERUS, P. W., ALBERTS, A. W., & VAGELOS, P. R. (1969) *Methods Enzymol.* **14**, 43
39. ALBERTS, A. W., BELL, R. M., & VAGELOS, P. R. (1972) *J. Biol. Chem.* **247**, 3190-3198
40. BRAY, G. A. (1960) *Anal. Biochem.* **1**, 279-280
41. RONCARI, D. A. K. (1972) Doctoral dissertation, Washington University
42. OUCHTERLONY, O. (1966) in *Immunological Methods* (ACKROYD, J. F., ed) p. 55, F. A. Davis Company, Philadelphia
43. KABAT, B. A., & MAYER, M. M. (1961) *Experimental Immunochemistry*, Charles C Thomas, Springfield, Ill.
44. LOWRY, O. H., ROSEBROUGH, N. J., FARR, A. L., & RANDALL, R. J. (1951) *J. Biol. Chem.* **193**, 265-275
45. SCOTCHLER, J., LOZIER, R., & ROBINSON, A. B. (1970) *J. Org. Chem.* **35**, 3151
46. MERRIFIELD, R. B. (1963) *J. Amer. Chem. Soc.* **85**, 2149
47. POURCHOT, L. M., & JOHNSON, J. J. (1969) *Org. Prep. Proc.* **1**, 121
48. SAKAKIBARA, S., SHIMONISHI, Y., OKADA, M., & SUGIHARA, H. (1968) *Bull. Chem Soc. Japan* **41**, 438
49. STEWART, J. M., & YOUNG, J. D. (1969) *Solid Phase Peptide Synthesis*, p. 19, W. H. Freeman & Co., San Francisco
50. MOORE, S., SPACKMAN, D. H., & STEIN, W. H. (1958) *Anal. Chem.* **30**, 1185
51. BEYERMAN, H. C., HINDRIKS, H., & deLEER, E. W. B. (1968) *Chem. Commun.* **49**, 5163
52. STADTMAN, E. R. (1955) *Methods Enzymol.* **1**, 596
53. LAEMMLI, U. K. (1970) *Nature* **227**, 680-685
54. WESTALL, F. C., & ROBINSON, A. B. (1970) *J. Org. Chem.* **35**, 2842
55. STARK, G. R., STEIN, W. H., & MOORE, S. (1960) *J. Biol. Chem.* **235**, 3177
56. HAGENMAIER, H. (1970) *Tetrahedron Lett.* **4**, 283
57. BAYER, E., HAGENMAIER, H., JUNG, J., PARR, W., ECKSTEIN, H., HUNZIKER, P., & SIEVERS, R. E. (1971) in *Peptides 1969* (SCOFFONE, E., ed) p. 65, North-Holland Publishing Co., New York
58. HALSTRØM, J., BRUNFELDT, K., & KOVÁCS, K. (1970) *Experientia* **27**, 17-18

III

Problems with the Solid-Phase Method

Editors' Comments on Papers 22 Through 25

22 Bayer, Eckstein, Hägele, König, Brüning, Hagenmaier, and Parr: *Failure Sequences in the Solid Phase Synthesis of Polypeptides*

23 Hancock, Prescott, Vagelos, and Marshall: *Solvation of the Polymer Matrix: Source of Truncated and Deletion Sequences in Solid Phase Synthesis*

24 Gisin and Merrifield: *Carboxyl-Catalyzed Intramolecular Aminolysis: A Side Reaction in Solid-Phase Peptide Synthesis*

25 Sharp, Robinson, and Kamen: *Synthesis of a Polypeptide with Lysozyme Activity*

This section of papers delineates some of the recognized problems in the area of polypeptide synthesis. The topic is of general interest to anyone interested in the synthesis of macromolecules by means of resin supports. Many of the problems that have been defined in the area of polypeptide synthesis are also generally applicable to the synthesis of oligosaccharides and oligonucleotides.

The major problems identified to date fall into five categories:

1. *Resin:* the type of resin used must be compatible with the reagents and solvent. In particular, the resin must swell with the solvents used for all the reactions.

2. *Blocking groups:* the blocking groups, usually on the amino group, must be easily attached to the amino acid, remain intact during the coupling to the resin, and yet be readily removed from the attached amino acid. Obviously, the hydrolytic step must be sufficiently mild so that the peptide bonds are not hydrolyzed. A variety of blocking groups have been used in SPS with the *t*-BOC group receiving by far the most use.

3. *First coupling reaction:* a variety of problems may develop in the attachment of the first amino acid moiety. A particular amino acid such as proline (see Paper 11) may lead to difficulty. Another problem is the degree of completion of the first reaction. If the unreacted groups on the polymer are not masked, these groups may become reactive sites later in the synthesis. The result would be the formation of short sequences. One recent improvement has been the use of cesium salt of the amino acid (see Paper 15). There also may be some initial sites for reaction of the polymer that are not accessible. This effect is influenced by the extent of bead swelling, which in turn is affected by the solvent used.

4. *Truncation and sequence deletion:* two related problems can also occur in later steps of coupling. In the first, a growing chain is terminated and does not participate in further couplings (truncation). The second problem involves a truncated chain that does become involved in later coupling steps, giving rise to a sequence deletion.

5. *Cleavage reactions:* the papers found in the polypeptide section reflect the variety of reagents used to cleave the completed polypeptide from the resin. The problem simply stated is that a reagent must accomplish the cleavage reaction of the peptide bonds. The more recent publications show milder or shorter reaction times that can meet the basic requirements.

Even with the problems discussed in the following four papers (Papers 22–25), the SPS technique is probably best summarized by a quotation from the Hancock et al. (Paper 23) ". . . . this method has been found to be much faster and to give a higher yield than the classical approach."

22

Copyright © 1970 by the American Chemical Society
Reprinted from *J. Amer. Chem. Soc.*, **92**(6), 1735–1738 (1970)

Failure Sequences in the Solid Phase Synthesis of Polypeptides

E. Bayer, H. Eckstein, K. Hägele, W. A. König, W. Brüning, H. Hagenmaier, and W. Parr

Contribution from the Department of Chemistry, University of Houston, Houston, Texas and the Department of Chemistry, University of Tübingen, Tübingen, Germany. Received August 2, 1969

Abstract: Failure sequences occur during solid phase synthesis of polypeptides. The number of these failure sequences can be considerably decreased by acetylation of the amino groups which do not react or by the use of specially prepared, resin-coated glass beads.

One of the limitations of Merrifield's[1-3] solid phase approach to the synthesis of peptides and proteins is the possibility of creating failure sequences, which cannot be separated or even analytically distinguished from the desired sequence. Theoretically the number of failure peptides increases exponentially with increasing chain length. Only 100% yield in every coupling step could prevent the formation of failure sequences.

We can distinguish between the truncated sequences, in which amino acids are missing from the amino end, and the failure sequences, in which amino acids are missing from within the chain. If a truncated sequence cannot couple in later steps, no failure sequences are

(1) R. B. Merrifield, *J. Amer. Chem. Soc.*, **85**, 2149 (1963); **86**, 304 (1964).
(2) R. B. Merrifield, *Biochemistry*, **3**, 1385 (1964).
(3) R. B. Merrifield, *Science*, **150**, 178 (1965).

Synthesis of Polypeptides

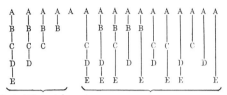

Figure 1. Possible wrong sequences during a solid phase synthesis of a pentapeptide A-B-C-D-E.

present, and the isolation of the desired peptide is much more likely to be achieved than in the case when all possible failure sequences are formed. The possible truncated and failure sequences for a pentapeptide are shown in Figure 1.

It had not been previously investigated whether only truncated sequences were occurring or whether all theoretically possible failure sequences were occurring. The main reason for this was that analytical methods are not sensitive enough to detect failure sequences in a solid phase synthesis conducted with high yields in the coupling steps.[4-6]

We therefore synthesized polypeptides of repeating dipeptide sequences $(A-B)_x$ to improve the detection limit. Failure sequences would be clearly indicated by the occurrence of peptides: A-A or B-B after partial hydrolysis. Truncated sequences and the correct sequence would result in peptides A-B or B-A. The detection limit increases with increasing chain length. The dipeptides obtained after partial hydrolysis were separated as N-trifluoroacetylmethyl esters by gas chromatography and the higher peptides were separated by liquid chromatography and both were then identified by mass spectrometry.[5,7,8]

Results and Discussion

The results for the solid phase synthesis of the dodecapeptides (Leu-Ala)$_6$[9] and (Ala-Phe)$_6$,[9] summarized in Tables I and II, clearly indicate that failure sequences occur despite the fact that there were no difficulties in coupling, as high yields have been obtained in every step. Since random hydrolysis does not occur, the values in Tables I and II can be considered only as semiquantitative. These values however are in agreement with the observed reaction rates of more than 99% in every coupling step. With peptide sequences such as Arg-His, Arg-Arg, or Arg-Lys, much lower yields and more failure sequences are detectable.[10] The results are definite experimental proof of failure sequences. In order to define the separation and identification problems caused by failure sequences, it was necessary to know the chain length and amount of contaminating peptides. One can calculate the amount of contaminating peptides for polypeptides of different chain lengths and estimate the optimum chain length to which the Merrifield procedure may be applied. Figure 2 and Table III show the expected yield of myoglobin with 153 amino acid residues, ribonuclease with 124, apoferredoxin with 55, insulin B chain, and a dodecapeptide. Figure 2 also shows the yields of the contaminating failure peptides, assuming 90 or 99% average yield in every coupling step. Unfortunately most of the contaminating peptides are nearly identical in chain length with the desired product, especially in the case of 99% yield. In addition, there is little possibility of separating these by existing separation methods when more than 20 amino acid residues are present.

Table I. Observed Sequences in Hydrolysates of (Leu-Ala)$_6$

Peptide sequence	Normal synthesis, %	Acetylating amino groups, %	Synthesis with pellicular resin, %
Ala-Ala	1.35	0.1	0
Leu-Ala	67.50	71.00	70.55
Ala-Leu	28.53	29.25	30.23
Leu-Leu	2.62	0.1	0

Table II. Observed Sequences in Hydrolysates of (Ala-Phe)$_6$

Sequence	Amount, %
Ala-Ala	Not detectable
Phe-Ala	39.7
Ala-Phe	51.0
Phe-Phe	9.3

Table III. Expected Yields of Several Polypeptides Assuming Average Coupling Yields of 90 and 99%

Peptide	Number of amino acid residues	Yield assuming 90% coupling rates, %	Yield assuming 99% coupling rates, %
Dodecapeptide	12	31.4	89.5
Insulin B chain	30	4.7	74.1
Ferredoxin	55	0.3	57.5
Ribonuclease	124	0.0002	29.0
Myoglobin	153	<0.00001	21.7

With 90% yield in every coupling step, ribonuclease would be obtained only in 0.0002% yield as is shown in Table III. Polypeptides with one amino acid less, however, would be synthesized in a 15-fold higher yield. Even with 99% yield in every coupling step, the yield of the main product would be 29% as opposed to 36% $n-1$ peptides which could not be separated. In the case of myoglobin, the possibility of a synthesis is even more remote. Assuming 99% yield in every coupling step (and that only the $n-1$ peptides cannot be separated), the limit of the chain length of a peptide, which can be synthesized to 80% purity by the Merrifield method is about 30 amino acid residues. Higher yields or better separation methods would increase this limit. The presence of amino acids that are difficult to couple decreases this value.

There are two possibilities of extending the application of the solid phase method: increasing the yields

---

(4) E. Bayer, H. Hagenmaier, G. Jung, and W. A. König in "Peptides 1968," E. Bricas, Ed., North-Holland Publishing Co., Amsterdam, 1968, p 162.
(5) E. Bayer and W. A. König, *J. Chrom. Science*, **7**, 95 (1969).
(6) E. Bayer in "Proceedings of the 1st American Peptide Symposium," Marcel Dekker, Inc., New York, N. Y., in press.
(7) F. Weygand, A. Prox, H. H. Fessel, and K. Z. Kun Sun, *Z. Naturforsch*, **20b**, 1169 (1965).
(8) E. Bayer, H. Hagenmaier, W. A. König, and H. Pauschmann, *Z. Anal. Chem.*, **243**, 670 (1968).
(9) (Leu-Ala)$_6$ and (Phe-Ala)$_6$ are used as abbreviations for H-L-Leu-(L-Ala-L-Leu)$_5$-L-Ala-OH and H-L-Ala-(L-Phe-L-Ala)$_5$-L-Phe-OH.
(10) E. Bayer and K. Hägele, unpublished results.

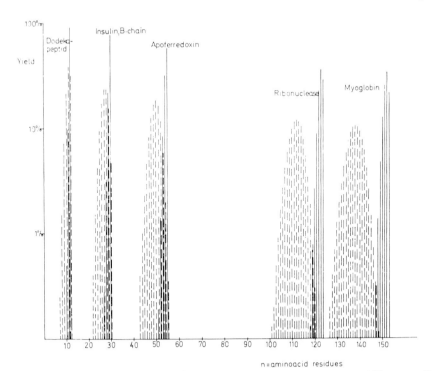

Figure 2. Yield of desired polypeptide and of failure sequences assuming 90% (dotted line) and 99% (solid line) yield in every coupling step.

or preventing truncated sequences from growing into failure sequences. The truncated sequences are easier to separate from the main product as has been shown in the example of the synthesis of the ACTH sequence 1–20 by the solid phase method.[10]

To increase the yield, one must consider that the solid phase method is a heterogeneous reaction which is controlled by the diffusion in the resin. Theoretically, the mass transfer in the resin bead is dependent upon the particle diameter.[11] It is difficult to reduce the particle size because of experimental difficulties in handling such small diameter resin beads. In liquid chromatography, pellicular resins have been introduced for the purpose of decreasing diffusion pathway and thus decreasing the time of chromatography. Such pellicular resins have been prepared by polymerizing a thin layer of resin onto inert glass beads.[12] Using these resins as support, the yield is considerably increased, as the investigation of failure sequences of (Leu-Ala)$_6$ in Table II shows in comparison to the conventional beads. An additional advantage of the pellicular resin is the low swelling ratio and the possibility of conducting the synthesis in column operations.

To reduce the failure sequences to truncated sequences, the amino groups not coupling in one step must be prevented from further coupling in later steps. Merrifield[1] introduced a procedure for this in his early publications; however, it has not been followed up generally. We acetylated after every coupling step. Table II demonstrates that in this case the failure sequences are considerably reduced.

By using these methods, and by synthesizing peptides with chain lengths up to twenty or thirty amino acids (which are then coupled by fragment techniques), the Merrifield method might be used to synthesize homogeneous proteins.

Experimental Section

Materials. Boc-amino acids were prepared according to the pH-stat method[13] and the purity established by thin-layer chromatography.

Boc-Ala Polymer. Chloromethylated Bio Beads S-X2 (2 g, capacity 1.1 mmolar equiv/g) were stirred in 50 ml of absolute ethanol with 0.34 g (1.8 mmol) of Boc-Ala-OH and 0.13 g (1.8 mmol) of triethylamine for 24 hr at 80°. The resin was then filtered off, and washed with ethanol, DMF, and CH$_2$Cl$_2$.

(Leu-Ala)$_6$ Polymer. The further synthesis of the peptide resin was conducted as described by Bayer, et al.[14] The loading on the resin represented 0.39 mmol. The number of free amino groups decreased by approximately 1% per synthesis step.

The following amounts of Boc-amino acids were used for the coupling step: 6 × 0.46 g (2 mmol) of Boc-Leu-OH dried *in vacuo* over P$_2$O$_5$, and 5 × 0.38 g (2 mmol) of Boc-Ala-OH.

(Leu-Ala)$_6$ Hydrobromide. The peptide resin was dried *in vacuo* and suspended in 25 ml of trifluoroacetic acid. The suspension was stirred and purified HBr was passed through for 90 min. The resin was then filtered off and washed twice with 20 ml of trifluoroacetic acid. The combined filtrates were concentrated;

(11) J. F. K. Guber and J. A. R. Halsman, *Anal. Chim. Acta*, **38**, 305 (1967).
(12) C. G. Horvath, B. A. Preiss, and S. R. Lipsky, *Anal. Chem.*, **39**, 1422 (1967).
(13) E. Schnabel, *Ann. Chem.*, **702**, 188 (1967).
(14) E. Bayer, G. Jung, and H. Hagenmaier, *Tetrahedron*, **24**, 4853 (1968).

the (Leu-Ala)₆-hydrobromide was precipitated with 200 ml of ether. The product was filtered off and washed with ether. After drying *in vacuo*, 415 mg of a white powder was obtained (0.345 mmol = 88% with respect to the amount of alanine esterified onto the resin). A sample of this product was hydrolyzed for 20 hr in 6 N HCl at 110°. The amino acid analysis of the hydrolysate indicated a ratio of Ala:Leu = 1:0.97.

(Ala-Phe)₆. The synthesis was carried out according to the description given for (Leu-Ala)₆. According to amino acid analysis the ratio of Ala-Phe was 0.94:1.0. (Ala-Phe)₆-HBr is practically insoluble in all conventional solvents. It is soluble only in trifluoroacetic acid.

Synthesis of (Leu-Ala)₆ with Pellicular Resin. Boc-Ala Beads. Glass beads (150 g, 88–105μ) coated with chloromethylated polystyrene[15] (capacity 0.15 mmolar equiv/g) were put into a column (1 × 100 cm) and a solution of 1.9 g of Boc-Ala-OH (10 mmol) and 1 g (10 mmol) of triethylamine in 60 ml of ethanol soaked in the resin bed and allowed to react for 48 hr at 60°. The coated glass beads were then washed with ethanol, DMF, and CH₂Cl₂. The capacity loading on the coated beads represented 0.07 mmolar equiv of Ala/g.

Boc-(Leu-Ala)₆ Beads. The further synthesis of the peptide beads was carried out in a column (3 × 10 cm) with 75 g of the Boc-Ala-OCH₂ beads using the basic steps and washing procedures as described for (Ala-Phe)₆. In every coupling step, 9 mmol of Boc-L-Ala-OH and Boc-L-Leu-OH and 9 mmol of dicyclohexylcarbodiimide, respectively, were used. The product obtained from the beads after hydrolysis showed a ratio of Leu-Ala of 1.0:1.0, and was contaminated with resin.

Synthesis of (Leu-Ala)₆ with Acetylation. The synthesis was carried out as described above with 2.5 g of chloromethylated resin (0.7 mmolar equiv/g Bio Beads S-X2, 200–400 mesh). The total quantity of the first amino acid attached to the resin was 0.4 mmol; the amount of the free amino groups after each cleaving step of the *t*-Boc group was constant during the entire synthesis. After each coupling step, the resin was washed three times with dimethylformamide, and the acetylation was carried out with 2.5 ml of acetic anhydride in a mixture of 25 ml of dimethylformamide and 0.75 ml of triethylamine for 20 min. Afterward, the resin was washed three times with dimethylformamide. The yield of the peptide hydrobromide was 321 mg (55%). According to the amino acid analysis, the ratio of Leu-Ala was 1.0:1.0.

Gas Chromatography–Mass Spectrometry and Sephadex Chromatography. The same methods, previously described in ref 5 were used. For gas chromatographic and mass spectrometric investigations, an LKB 9000 gas chromatograph–mass spectrometer was used. All mass spectra were taken at 70 eV. For solid samples, a heated and water-cooled direct inlet was used. The separation of the trifluoroacetyldipeptide methyl esters of the partial hydrolysate was achieved on SF 96 coated steel capillary columns 500 ft × 0.03 in., isothermal at 190°: carrier gas, 15 ml He/min; detector, FID, 300°; injector 280°. For the gas chromatography–mass spectrometry combination, the total ion current detector was used.

Partial Hydrolysis of Peptides. A peptide (10–100 mg) was sealed in a glass tube with 5 ml of concentrated HCl and left for 72 hr at 37°. After evaporation of the HCl *in vacuo* (18 Torr) the peptide mixture was esterified and trifluoroacetylated for gas chromatography or directly submitted to partition chromatography. Under these conditions approximately 50% (Leu-Ala)₆ was hydrolyzed to dipeptides, 30% to higher peptides, and 20% to amino acids. Due to this and to the fact that random hydrolysis was not occurring, the results in Table I and II, calculated on the basis of sequences found in di- to pentapeptides, are only semiquantitative.

Esterification of Peptides. The almost neutral peptide hydrolysate was dissolved in methanol. A concentrated solution of diazomethane in diethyl ether was added until the yellow color remained. If any precipitate formed, a few drops of water and more diazomethane were added until the precipitate dissolved. The yellow solution was evaporated to dryness *in vacuo* (18 Torr).

Trifluoroacetylation. The peptide esters were dissolved in 5 ml of dry methanol and brought to a pH of 7.5–8 by adding triethylamine. Methyl trifluoroacetate (2 ml) was added and the solution left at room temperature for 8 hr. After evaporation to dryness *in vacuo* (18 Torr), the residue was shaken with a mixture of equal amounts of ethyl acetate and water. The organic phase was evaporated *in vacuo* (18 Torr), resulting in the TFA–peptide-methyl esters.

Acknowledgment. This investigation was supported by the Robert A. Welch Foundation, Grant No. E-227, Houston, Texas, and the Dentsche Forschungsgemeinschaft.

---

(15) The sample of chloromethylated glass beads was provided by Dr. C. G. Horvath and S. R. Lipsky, New Haven, Conn.

Copyright © 1973 by the American Chemical Society

Reprinted from *J. Org. Chem.*, **38**(4), 774–781 (1973)

Solvation of the Polymer Matrix. Source of Truncated and Deletion Sequences in Solid Phase Synthesis[1]

WILLIAM S. HANCOCK,[2] DAVID J. PRESCOTT, P. ROY VAGELOS, AND GARLAND R. MARSHALL[3,4]

*Departments of Physiology and Biophysics and of Biological Chemistry,
Washington University Medical School, St. Louis, Missouri 63110*

Received November 9, 1972

An automated procedure utilizing hydrogen chloride-36 for monitoring the free amine in automated solid phase synthesis was developed. Discrepancies were found between the values determined by this procedure and those from amino acid analysis in the synthesis of a peptide, residues 63–74, of acyl carrier protein. These results led to a hypothesis of dynamic solvation changes of the polymer matrix as synthesis proceeds. The effects of chain termination by acetylation were also in agreement with the hypothesis. Dynamic solvation changes of the polymer matrix leads to the sequence-dependent problems of solid phase synthesis, both truncated and deletion sequences. It may also be responsible for difficulties encountered with monitoring procedures and with attempts to terminate unreacted peptide chains. Based on these observations, a modified procedure of solid phase peptide synthesis was developed which significantly improved the synthesis of residues 63–74 of acyl carrier protein.

Since the introduction of the solid phase method for peptide synthesis by Merrifield in 1962,[5] an enormous number of peptides have been prepared.[6] A more striking achievement, however, has been the synthesis by the solid phase procedure of several large proteins with high biological activity, *e.g.*, ribonuclease A,[7] fragment P$_2$ of *Staphylococcus aureus* nuclease T,[8] soybean trypsin inhibitor,[9] and acyl carrier protein.[10]

Despite these impressive achievements, many peptides have not been prepared in an adequate yield and the cause of such failures is still not clear despite extensive studies of the sequence-dependent[11] problems of solid phase synthesis.[6,12–15] Most failures in the method have been attributed to incomplete coupling and deprotection steps.[6,16–21] Fortunately, a wide variety of methods has been established for the formation of a peptide bond,[18] and recently several new resin supports have been developed.[6a] One could expect, therefore, that many of these synthetic problems could be overcome by a judicious choice of the polymeric support, side chain and α-amino protecting groups, the method of removal of the α-amino protecting group, and the reagent used for mediating the coupling reaction. Some recent studies[16] have indicated that difficult coupling reactions may be facilitated by the use of a mixed solvent or the addition of urea to the reaction mixture.

The formation of truncated peptides[22] by incomplete coupling and deprotection steps is a problem in that the overall yield of the synthesis is decreased. The regrowth of partially complete sequences, with the formation of deletion sequences,[23] poses a much greater threat to the success of the synthesis of a large molecule, as the separation of the desired peptide from a large number of very similar products may be beyond the scope of present methods of protein purification. Bayer[12] has established that such deletion sequences can occur in specialized cases, and by analogy it has been suggested that the larger synthetic proteins must contain such products. Also the formation of deletion sequences places a severe limit on the size of the peptide that can be synthesized.[13,24]

Two approaches which have been taken to reduce the occurrence of deletion sequences are either to increase the yield of the coupling reactions, by multiple couplings, for example, or to terminate truncated sequences by treatment of the resin, after completion of the coupling reaction, with a very reactive acylating reagent, *e.g.*, acetic anhydride[14,25] or other reagents.[26–28] It has yet to be demonstrated, however, that such reagents, because of their reactivity, will not cause side

---

(1) This work was supported by Grants No. HE-10406 and AM-13025 from the USPHS and Grant GB-5142X from the National Science Foundation.
(2) Recipient, George Murray Scholarship (University of Adelaide, Australia).
(3) To whom correspondence should be addressed.
(4) Established Investigator, American Heart Association.
(5) R. B. Merrifield, *Fed. Proc.*, **21**, 412 (1962).
(6) (a) G. R. Marshall and R. B. Merrifield, "Biochemical Aspects of Reactions on Solid Supports," Academic Press, New York, N. Y., 1971; (b) A. Marglin and R. B. Merrifield, *Ann. Rev. Biochem.*, **39**, 841 (1970).
(7) B. Gutte and R. B. Merrifield, *J. Biol. Chem.*, **246**, 1922 (1971).
(8) D. A. Ontjes and C. B. Anfinsen, *Proc. Nat. Acad. Sci. U. S.*, **64**, 428 (1969).
(9) K. Noda, J. S. Terada, N. Mitsuyasu, M. Waki, T. Kato, and N. Izumiya, *Naturwissenschaften*, **58**, 147 (1971).
(10) (a) W. S. Hancock, D. J. Prescott, W. L. Nulty, J. Weintraub, P. R. Vagelos, and G. R. Marshall, *J. Amer. Chem. Soc.*, **93**, 1799 (1971); (b) W. S. Hancock, D. J. Prescott, G. R. Marshall, and P. R. Vagelos, *J. Biol. Chem.*, **247**, 6224 (1972).
(11) It has been established that these failures are not necessarily associated with a particular amino acid residue, and therefore are described as sequence-dependent problems.
(12) E. Bayer, H. Eckstein, K. Hagele, W. A. Konig, W. Bruning, H. Hagenmaier, and W. Parr, *J. Amer. Chem. Soc.*, **92**, 1735 (1970).
(13) E. Bayer, H. Hagenmaier, G. Jung, W. Parr, H. Eckstein, P. Hunziker, and R. E. Sievers, "Peptides 1971," North-Holland Publishing Co., New York, N. Y., 1971, pp 65–73.
(14) H. Hagenmaier, *Tetrahedron Lett.*, 283 (1970).
(15) R. C. Sheppard, "Peptides 1971," North-Holland Publishing Co., New York, N. Y., 1971.
(16) F. C. Westall and A. B. Robinson, *J. Org. Chem.*, **35**, 2842 (1970).
(17) F. Chuen-Heh Chou, R. K. Chawla, R. F. Kibler, and R. Shapira, *J. Amer. Chem. Soc.*, **93** 267 (1971).
(18) R. B. Merrifield, *Advan. Enzymol.*, **32**, 221 (1969).
(19) S. Karlsson, G. Lindeberg, J. Porath, and U. Ragnarsson, *Acta Chem. Scand.*, **24**, 1010 (1970).
(20) J. Scotchler, R. Lozier, and A. B. Robinson, *J. Org. Chem.*, **35**, 3151 (1970).
(21) K. Lubke, "Peptides 1969," North-Holland Publishing Co., New York, N. Y., 1971, pp 154–156.

(22) A truncated peptide is defined as a peptide which becomes unavailable for reaction at some stage in the synthesis and does not add any further amino acids.
(23) A deletion sequence is defined as a truncated peptide which resumes growth at some later stage in the synthesis.
(24) (a) For example, the synthesis of human growth hormone (188 amino acids) would require an average coupling yield of 99.5% if the desired protein was to be the major product of the synthesis. Such calculations, however, are based on the assumption that all truncated sequences become deletion sequences and grow at the same rate as the correct sequence. (b) J. M. A. Baas, H. C. Beyerman, B. van de Graff, and E. W. B. de Leer, "Peptides 1969," North-Holland Publishing Co., New York, N. Y., 1971, pp 173–176.
(25) R. B. Merrifield, *J. Amer. Chem. Soc.*, **85**, 2149 (1963).
(26) L. B. Markley and L. C. Dorman, *Tetrahedron Lett.*, 1787 (1970).
(27) T. Wieland, C. Birr, and H. Wissenbach, *Angew. Chem., Int. Ed. Engl.*, **8**, 764 (1969).
(28) H. Wissmann and R. Geiger, *ibid.*, **9**, 909 (1970).

reactions that are deleterious to the synthesis, e.g., acetic anhydride.[29] Also it is unclear whether the amino groups which are resistant to coupling will react with the terminating reagent.

Before these variations in procedure can be rationally applied, it is necessary to have a reliable means of quantitation for both the coupling and deprotection steps, and in fact numerous assay procedures have been reported,[6a] most of which depend on monitoring the free amino group. The automation of the solid phase synthetic method has added the further requirements that the analytical procedure must be rapid enough in order to monitor and perhaps control the progress of the automated synthesizer. At this stage, the potentiometric method of Brunfeldt[30] is the procedure which seems most applicable for use with an automated synthesizer.

Little progress has been made in understanding the cause of the sequence-dependent problems of solid phase synthesis, although several possible explanations have been advanced. The effect of steric hindrance of the amino terminus of the growing peptide chain has often been discussed with relevance to solid phase synthesis.[15,17,21] The environment of the peptide can be dramatically effected by the solvent used, as it has been found that only solvents which swell the resin and provide a reasonably polar environment will allow an efficient coupling reaction, e.g., dimethylformamide and chloroform, but not benzene.[25] Since both the polymerization of the polystyrene support and the chloromethylation reaction are random processes, one could expect a heterogenous environment due to the random distribution of cross-links and reactive sites so that some reactive sites would be more sensitive than others to steric hindrance.

It has been noted[12] that solid phase synthesis is a heterogeneous reaction and as such depends on the rate of diffusion of reagents in the resin. One could expect, however, that diffusion should normally be a rapid process, as the swollen beads contain 80-90% solvent.[25]

In a recent review,[6a] it was stressed that as a synthesis proceeds the physical properties of the resin reflect the change from that of hydrophobic polystyrene to that of a mixed polystyrene-protein matrix. One might then expect a change, perhaps dramatic, in local solvation of the heterogeneous polymer matrix as the synthesis proceeds. If the solvation was decreased, then the accessibility of the heterogenous population of sites could also decrease with a consequent drop in the yield. It was difficult to evaluate such a concept because of the lack of published examples of this phenomenon; in fact only two well-documented examples have been described.[6a,20]

At the beginning of our investigations on acyl carrier protein, it became clear that the synthesis of the initial sequence, residues 74-63 (see Figure 1), was a good example of the sequence-dependent problems of solid phase synthesis—the yield of growing peptide at the end of the sequence was only 20% of the initial glycine value. It was decided to examine monitoring procedures and methods for minimizing truncated and deletion sequences before proceeding further. The synthesis was then repeated under a variety of conditions and followed by analysis of the yield of both the coupling and deprotection steps.

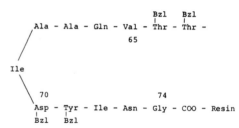

Figure 1.—Fully protected partial sequence 74-63 of acyl carrier protein.

For these analytical studies, the method developed by Dorman[31] was adapted so that the progress of an automated synthesis could be followed. From these studies, a modified synthetic procedure was derived that overcame to a considerable extent the problems originally observed in the synthesis of this peptide, and which was subsequently applied to the synthesis of acyl carrier protein. Of more general significance was the evidence which indicated that the difficulties observed in this synthesis can be attributed to changes in solvation of the peptide resin at different stages in the synthesis, which give rise to both truncated and deletion sequences. With change in the accessibility of the free amino group, monitoring methods must be used with extreme caution.

Results and Discussion

The sequence 74-63 of ACP (see Figure 1) was synthesized by the stepwise addition of suitably protected amino acids to 0.66 mmol of *tert*-butyloxycarbonyl (BOC)-glycine (0.33 mmol of amino acid per gram of peptide resin) esterified to a 1% cross-linked polystyrene resin support. The procedures used are similar to those described by Merrifield for solid phase synthesis,[32] and the specific details are described in the Experimental Section.

Although the synthesis of the initial C-terminal sequence of ACP, 74-63, would appear straightforward, early attempts to synthesize this peptide by standard procedures were discouraging (Figure 2). The addition of asparagine$_{73}$, isoleucine$_{72}$, and tyrosine$_{71}$ gave a progressive decrease in the yield of the growing peptide chain, which implies truncation. Steric hindrance of the coupling of amino acids with bulky protecting groups should be greatest at the beginning of the synthesis,[33] and this could account for these initial difficulties, especially in the case of isoleucine, with its bulky isobutyl side chain.

The most significant feature of the growth profile of the sequence 74-63 is the dramatic increase in yield for

---

(29) J. Halstrom, K. Brunfeldt, and K. Kovacs, *Experientia*, **27**, 17 (1971).

(30) K. Brunfeldt, P. Roepstorff, and J. Thomsen, "Peptides 1969." North-Holland Publishing Co., New York, N. Y., 1971, pp 148-153.

(31) L. C. Dorman, *Tetrahedron Lett.*, 2319 (1969).

(32) (a) R. B. Merrifield, *J. Amer. Chem. Soc.*, **85**, 2149 (1963); (b) R. B. Merrifield, *Advan. Enzymol.*, **32**, 221 (1969).

(33) The initial amino acid, glycine, was attached to the resin under conditions that are much more vigorous (100°, 24 hr) than one uses for peptide bond formation. It could be expected therefore, that some of the reaction sites would be located in the more hindered regions of the resin matrix. This conclusion is supported by the frequent observation that cleavage of the peptide from the resin releases significant quantities of the initial amino acid.

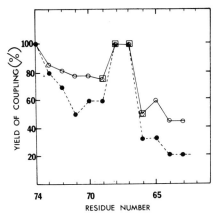

Figure 2.—The growth profile for the synthesis of the sequence 74–63 of ACP by two different synthetic procedures. The yield of each synthesis was determined by amino acid analysis of the peptide resin after the addition of each amino acid. The data present are the average of two syntheses in which the standard solid phase synthetic procedures were used (●) and four syntheses in which the modified procedures were used (○). All duplicate syntheses gave essentially the same results. If the resin peptide was acetylated after the addition of Ile$_{69}$, and then the synthesis was continued as for the modified procedure, the incorporation of the subsequent residues is shown by the points □.

the addition of alanine$_{68}$ and alanine$_{67}$, which implies regrowth and formation of deletion sequences. However, the next amino acid, glutamine, was incorporated with a much lower yield, which again implied truncation.

It was clear at this stage that the synthesis had to be followed by some analytical technique which allowed quantitation of the yield of both the coupling and deprotection steps. The Dorman[31] procedure was chosen, which measures the amount of chloride bound to the resin by conversion of free amino groups to their corresponding hydrochlorides on treatment of the resin with pyridine hydrochloride. The chloride is displaced from the resin with triethylamine and measured by titration. However, it was decided that the use of chloride-36 would greatly increase both the speed and sensitivity of the assay. Instead of a laborious titration procedure, the chloride could be simply measured by radioactivity, while a very small amount of bound chloride (as would be expected for an efficient coupling reaction) could be accurately determined by a suitable increase in the specific radioactivity of the chloride-36. Table I indicates that both methods of chloride determination gave the same values for BOC and deprotected glycine resin, and, therefore, the chloride-36 procedure was used in all subsequent assays.

It was found necessary to modify some of the washes that were used in the original Dorman procedure so that the chloride-36 could be accurately measured. In Table II, the two procedures are compared and it is clear that the modifications do not effect the amount of bound chloride. Dimethylformamide was not a suitable solvent to remove the excess pyridine hydrochloride, presumably due to traces of dimethylamine in the solvent which displaced some of the bound chloride, and as a consequence the level of chloride-36 in the

TABLE I
COMPARISON OF METHODS USED FOR THE MEASUREMENT OF CHLORIDE^a

	Chloridimeter	^{36}Cl radioactivity
BOC-Gly resin	30	26
Total recovery, %	98	103
Gly resin	762	725
Total recovery, %	98	97

^{a} Gly resin (0.66 mmol) was treated with pyridine hydrochloride (7.5 mmol). The excess chloride was removed by a series of washes and the bound chloride was displaced with triethylamine (see Experimental Section). The chloride was then determined by titration of the chloride with a chloridimeter or by measurement of the ^{36}Cl radioactivity as described in the Experimental Section. Each value in the table is an average of three determinations (error for both assays was ±5%).

TABLE II
DORMAN ANALYSIS OF DEPROTECTED GLYCINE RESIN^a

Original procedure			Modified procedure		
No. of wash	Reagent	^{36}Cl dpm in wash	No. of wash	Reagent	^{36}Cl dpm in wash
1	Pyr HCl^b	29,420	1a	Pyr HCl	28,580
2	CH$_2$Cl$_2$	12,643	2a	CH$_2$Cl$_2$	13,195
3	CH$_2$Cl$_2$	4,898	3a	CH$_2$Cl$_2$	4,956
4	CH$_2$Cl$_2$	933	4a	CH$_2$Cl$_2$	741
5	DMF	612	5a	t-BuOH	906
6	DMF	194	6a	t-BuOH	486
7	DMF	115	7a	t-BuOH	131
8	DMF	120	8a	t-BuOH	23
9	DMF	90	9a	CH$_2$Cl$_2$	20
10	DMF	70	10a	CH$_2$Cl$_2$	0
11	DMF	80	11a	CH$_2$Cl$_2$	0
12	Et$_3$N^c	4,906	12a	Et$_3$N^d	3,370
13	DMF	587	13a	Et$_3$N	1,835
14	DMF	306	14a	Et$_3$N	306
15	DMF	63	15a	CH$_2$Cl$_2$	257
			16a	CH$_2$Cl$_2$	61
			17a	CH$_2$Cl$_2$	23
Total dpm 1–15		55,037			54,890
Total dpm 12–17		5,862			5,852
Recovery of total ^{36}Cl, %		100			100

^{a} Gly resin (0.66 mmol) was treated with pyridine hydrochloride (7.5 mmol, 1.4 × 10^6 dpm). In two determinations the resin was treated with the washes 1–15 and 1a–17a, each wash was collected in a 25-ml volumetric flask, and the sample was made up to the mark with dichloromethane. A 1-ml sample was counted for ^{36}Cl radioactivity. ^{b} The abbreviations used in the diagram are Pyr HCl = pyridine hydrochloride, CH$_2$Cl$_2$ = dichloromethane, DMF = dimethylformamide, Et$_3$N = triethylamine. ^{c} 10% (v/v) triethylamine dissolved in dimethylformamide. ^{d} 3% (v/v) triethylamine dissolved in dichloromethane.

washes would not fall to zero (see washes 5–11). This problem was solved by the substitution of an alcohol wash in steps 5a–8a. The resin was then washed with dichloromethane (steps 9a–11a) to prevent loss of peptide from the resin by transesterification caused by traces of alcohol present contaminating the triethylamine wash.[34] Triethylamine, at the concentrations used by Dorman, was found to severely quench the counting of chlorine-36. However, it was found that three washes of a lower concentration of triethylamine were sufficient to displace all of the bound chloride, and allow satisfactory counting of the sample (steps 12a–14a).

(34) The possibility of this side reaction was reduced further by the substitution of the more sterically hindered *tert*-butyl alcohol for ethanol in steps 5a–8a.

The assay was found to be quite reproducible and, in all cases, the recovery of chloride-36 was excellent (98%). Moreover, the procedure was easily adapted to monitor an automated synthesis. An automated synthesizer[35] was modified to accommodate a fraction collector, which was used to collect the effluent from the reaction flask. In order to quantitate effluent collection, it was necessary to drain by nitrogen pressure rather than by vacuum filtration. The machine was readily programmed to execute a Dorman analysis after each coupling and deprotection step, especially as pyridine hydrochloride was the only reagent that was not already used in the synthetic procedure. Washes 1a–11a and 12a–17a (Table II) were collected in separate containers and a sample of each was then counted. The volume and pH of the fractions collected were used as a check on the operation of the synthesizer. The results of the analyses performed during the synthesis of the peptide 74–67 of ACP are shown in Table III.

TABLE III
CHLORIDE BINDING DATA[a] FOR THE SYNTHESIS OF THE PEPTIDE 74–67

Residue	1st Coupling in CH_2Cl_2	2nd Coupling in CH_2Cl_2	3rd Coupling in CH_2Cl_2–DMF (1:1)	Acetylation[b]	1st Deprotection[b]	2nd Deprotection[b]
74 Gly	38[c]				782[f]	
73 Asn[d]	300	240			675	
72 Ile	362	370	304		765	
71 Tyr[d,e]	431	411	430	380	390	415
70 Asp	84	88		178	360	
69 Ile	186	187			374	
68 Ala		47			267	320
67 Ala	106	92			275	

[a] Expressed as μmoles of free amino groups as measured by the Dorman analysis. [b] As described in the Experimental Section. [c] Result of analysis of the initial t-BOC–glycine resin. [d] Both active ester couplings were carried out in DMF. [e] Dorman analysis after a fourth coupling in benzene gave 425 μM of chloride bound. The peptide 74–67 of ACP (Figure 1) was synthesized using the standard procedure described in the Experimental Section. The yield of each coupling reaction was estimated by the procedure of Dorman. [f] The amino acid analysis of the BOC–glycine resin indicated that 660 μmol of amino acid was esterified to the resin.

The yield of each coupling step was also followed by amino acid analysis and the results of the two analytical techniques are compared in Figure 3.

The two measurements agreed well for the addition of the first three amino acids, but after the addition of aspartic$_{70}$ there was a sharp drop in the amount of bound chloride, although the incorporation of aspartic$_{70}$ had not increased. Furthermore, the analysis of each amino acid addition and deprotection was complicated by a background of bound chloride,[36] which was superimposed on the amount of *chloride bound to* the free amino groups, although the value would remain constant after repeated couplings of that particular residue (see Table III). A disturbing feature was that, as the synthesis proceeded, the amount of chloride bound after deprotection showed a steady decrease (see Figure 2). The monitoring of the synthesis of the sequence 74–63 was repeated four times on different preparations with

(35) Based on the design of R. B. Merrifield, J. M. Stewart, and N. Jernberg, *Anal. Chem.*, **38**, 1905 (1966).
(36) This background is presumably due to binding of chloride by the resin, perhaps, by continued formation of quaternary amine by exposure of the residual chloromethyl groups to triethylamine.

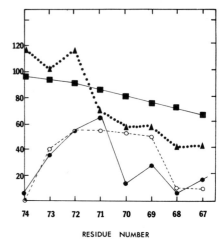

Figure 3.—The yield of the synthesis of peptide 74–63 as measured by amino acid analysis. The values recorded in the graph were an average of three separate syntheses which gave essentially the same results and are expressed as per cent of total free amine converted to the hydrochloride. The analyses were carried out as described in the Experimental Section. The amount of free amine after deprotection was determined by Dorman analysis (▲). The free amine by Dorman analysis (●) after double coupling is compared with the free amine which should be present (○) based on amino acid analysis. The per cent polystyrene (■) present in the polymer–peptide matrix is also shown.

essentially the same results. The constant background observed after several additions of a particular acid suggests that this effect is caused by a change in the properties of the polymer rather than a random accumulation of by-products from the coupling reaction.[37] A similar background effect due to the resin was also found by Beyerman[38] when N-(2-^{14}C-BOC) amino acids were used to follow the progress of a coupling reaction, or ^{35}S-sulfuric acid was used to determine the peptide content.

In Figure 3, it can be seen that the amount of chloride bound by the deprotected peptide decreased as the synthesis proceeded. Bayer noted a similar effect in the synthesis of ferredoxin.[39] This decrease could be caused by the cleavage of the peptide from the resin during the deprotection steps, or the pyridine hydrochloride wash in the case of the Dorman analysis, or by a decrease in the background of bound chloride as the properties of the resin change.

One of the problems of the Dorman analysis is that repeated treatment of the resin peptide with pyridine hydrochloride may cause cleavage of some of the peptide from the resin or other undesirable side reactions due to the acidic nature of this reagent. In Figure 4, the yield of the synthesis of the peptide 74–69 is compared with an identical synthesis that has been analyzed at each coupling and deprotection step with the Dor-

(37) For example, dicyclohexylurea and other by-products have been observed to give a positive test with ninhydrin reagent.
(38) H. C. Beyerman, P. R. M. van der Kamp, E. W. B. de Leer, W. Maassen van den Brink, J. H. Parmentier, and J. Westerling, "Peptides 1971," North-Holland Publishing Co., New York, N. Y., 1972.
(39) E. Bayer, G. Jung, and H. Hagenmeier, *Tetrahedron Lett.*, 4853 (1968).

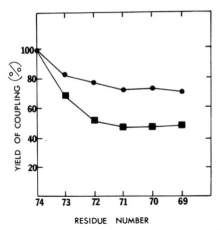

Figure 4.—The effect of the Dorman analysis on the yield of the synthesis of the sequence 74–69 of ACP. The yield of each synthesis was determined by amino acid analysis of the peptide resin after the addition of each amino acid. The data presented are the average of three syntheses that have been followed by the Dorman analysis (■) and four unmonitored syntheses (●). All duplicate syntheses gave essentially the same results. Both procedures involved the same number of coupling and deprotection steps.

man method. It is clear that the analytical procedure has caused a significant drop in the yield of the synthesis.

The variable background of bound chloride makes it difficult to use the Dorman analysis to follow the yield of a synthesis, unless the coupling of a particular amino acid is repeated until the amount of bound chloride remains constant. Although this background has been observed elsewhere,[38,40,41] several studies have not noted this problem.[14,31] This difference, however, may be due to the wide variation in properties that has been observed for different resin preparations.[7,42] Dorman[31] used a hydroxymethyl resin and did not comment on any background due to the resin binding chloride. Beyerman[38] noted that a chloromethyl resin, which had been treated with triethylamine, took up a considerable amount of ^{35}S-sulfuric acid, while a hydroxymethylated resin, or a chloromethylated resin in which the esterification of the amino acid was carried out with sodium carbonate instead of triethylamine, did not react with ^{35}S-sulfuric acid. Therefore, if the correct resin is chosen, the problem of a variable background may be overcome. The procedure should still be used with caution, however, as it can lower the yield of the synthesis (see above).

The variable background of bound chloride observed during the synthesis can be understood only if one considers that the solvation of the resin changes with the nature of the peptide–polystyrene matrix. For example, the sharp decrease in the amount of chloride bound by the resin after the addition of aspartic$_{70}$ (see Figure 3) can be explained by a decrease in solvation of

(40) R. B. Merrifield, unpublished observations.
(41) B. Mehlis, W. Fischer, and H. Niedrich, "Peptides 1969," North-Holland Publishing Co., New York, N. Y., 1971, pp 146–147.
(42) We have noted different backgrounds for glycine resins which were obtained commercially and those that were prepared in our laboratory.

the resin with a corresponding loss of chloride binding sites. These observations are consistent with studies on the rate of release of various amino acids, esterified to a chloromethyl resin, with acidic hydrolysis in a variety of solvents.[16] It was found that some sites on the resin were less accessible than others, and the effect of improper solvation during hydrolysis was to close off completely the less accessible sites, rather than generally decrease the rate of reaction at all sites by similar amounts.

The hydrophobic nature of the polystyrene resin requires that a relatively nonpolar solvent is used to allow swelling of the resin and penetration of the reagents into the matrix. As the length of the peptide chain increases, however, one would expect the requirement of a polar solvent for correct solvation of the peptide chains. In fact, in two published examples,[6a,20] a sudden drop in yield of the synthesis was overcome by the use of a polar solvent, and these studies led to the proposal that the use of a mixture of dichloromethane and dimethylformamide might be used for difficult amino acid additions. In Table III, the chloride binding data indicated that the use of this solvent mixture gave an increased incorporation of isoleucine$_{72}$.

This problem of a variable background of chloride (see Table III) could be explained if the resin contained a constant number of binding sites for chloride whose exposure was variable. If the degree of solvation of the resin changed significantly after the addition of a particular amino acid, then one would expect a change in the measured number of binding sites. If this assumption is correct, it should be possible to expose some of the buried sites by treatment of the resin with a series of washes which alternately shrink and swell the polymer matrix. The availability of previously buried sites could be readily tested by treatment of the resin with triethylamine, which should displace some further chloride from the resin. The development of the modified Dorman procedure was ideally suited for this purpose, as the use of chloride-36 allowed the measurement of very small quantities of chloride.

Three such studies at different stages of the synthesis are shown in Table IV and, in each case, it was found that further chloride-36 could be displaced from the resin after a series of shrink and swell washes. The amount of chloride trapped by the resin was not sufficient to affect the overall recovery of chloride in the assay (maximum amount of chloride released by the extra washes was only 1–5% of the total chloride added to the resin), but it was sufficient to prevent the accurate determination of small amounts of free amino groups. The same result was obtained if the original method of Dorman was followed (washes 1–15 in Table II) so that this effect was not a result of our modifications to the washing procedure.

It is clear from the growth profile of the peptide 74–63 (see Figure 2), synthesized by standard procedures for solid phase synthesis, that the synthesis has several problems which lead to a very low yield of the completed peptide. The analytical studies (see Table III) indicated, however, that variation in the conditions of coupling and deprotection could significantly improve the yield of the synthesis. More significantly, the studies described in Table IV indicated that a swell–shrink–swell wash cycle could expose buried functional

TABLE IV
REEXPOSURE OF CHLORIDE BINDING SITES AT DIFFERENT
STAGES OF THE SYNTHESIS OF THE PEPTIDE 74–67[a]

Number of wash	Reagent	BOC-Gly$_{74}$ ^{36}Cl dpm in washes	BOC-Ala$_{67}$ ^{36}Cl dpm in washes	Deprotected Ala$_{67}$ ^{36}Cl in washes
12	Et$_3$N	206	427	1,809
13	Et$_3$N	65	167	241
14	Et$_3$N	10	43	
15	CH$_2$Cl$_2$	5	25	
16	CH$_2$Cl$_2$		16	
17	CH$_2$Cl$_2$			
18	t-BuOH			
19	t-BuOH			
20	t-BuOH			
21	CH$_2$Cl$_2$			
22	CH$_2$Cl$_2$			
23	CH$_2$Cl$_2$			
24	Et$_3$N	73	213	630
25	Et$_3$N	16	75	112
26	CH$_2$Cl$_2$		20	19
Total dpm in wash 11–15		286	778	2,050
Total dpm in wash 23–25		89	308	761
% of dpm in wash 23–25 relative to wash 11–15		31	40	37

[a] The resin peptide was treated with pyridine hydrochloride (7.5 mmol, 1.4 × 10^6 dpm) and the excess pyridine hydrochloride was removed by the same washes (1a–11a) as described in Table II. All available chloride has been displaced by the washes 12–17 as shown in this table. The resin was then subjected to washes 18–23 in an attempt to expose buried regions of the resin that had bound chloride, and in fact the second triethylamine treatment washes 24–26 did liberate more chloride. Samples of the washes were counted for chloride-36 in the manner described for Table II.

groups. These considerations led to the development of a modified synthetic procedure.

The chloride binding data described in Table III suggested that the coupling of isoleucine$_{72}$ was improved if the coupling reaction was carried out in a mixed solvent of dichloromethane and dimethylformamide, while the yield of addition of asparagine$_{73}$ was also increased if the coupling reaction was repeated. A second deprotection after the addition of tyrosine$_{71}$ and alanine$_{68}$ increased the amount of free amine as detected by the chloride binding measurements. If these data were an accurate estimate of the amino groups that were available for a coupling reaction, then it was clear that persistent efforts to ensure reaction at the less accessible sites on the resin were necessary to achieve a good yield in the synthesis.

In an attempt to meet these conditions, all coupling reactions were repeated with a 1:1 mixture of dichloromethane and dimethylformamide as the solvent, and the deprotection step was repeated in an attempt to ensure complete deblocking of the peptide chain. Also the swell–shrink–swell wash was used between all coupling and deprotection steps (the exact sequence of washes used is described in the Experimental Section).

The sequence 74–63 of ACP was then resynthesized with this new procedure (see Experimental Section) and the progress of the synthesis was followed by amino acid analysis. As is shown in Figure 2, the modifications were successful, as the yield of completed peptide at the end of the synthesis was doubled.

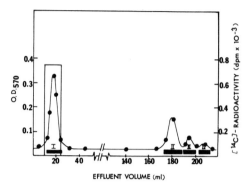

Figure 5.—Separation of the products from the synthesis of the sequence 74–69 of ACP. This figure shows the separation on a Dowex 50-X8 column (0.9 × 17 cm) of the products from the synthesis of the peptide 74–69. The column was developed at 30 ml/hr with a pyridine acetate gradient at 55°, and 6-ml fractions were collected. The progress of the column was monitored by ninhydrin analysis (570 nm) of the alkaline hydrolysate of a sample of each fraction. A portion (6%) of each fraction was checked for ^{14}C radioactivity.

However, the decrease in yield of the synthesis for the addition of asparagine$_{73}$, isoleucine$_{72}$, and tyrosine$_{71}$, as well as the regrowth of partially complete sequences after the addition of alanine$_{68}$ and alanine$_{67}$ was still evident (see Figure 2). It was decided that these problems would serve as an excellent test for the effectiveness of acetylation as a reagent for the termination of incomplete sequences. If the peptide resin was acetylated after the addition of tyrosine$_{71}$ and the peptide cleaved from the resin, one could expect to obtain N-acetylglycine, N-acetylaspartylglycine, and N-acetylisoleucylaspartylglycine. Alternatively, if the resin was acetylated after the addition of Ile$_{69}$, then the yield of addition for Ala$_{68}$ and Ala$_{67}$ should not be greater than that of isoleucine$_{69}$.

The peptide 74–63 (Gly-Asn-Ile-Tyr-Asp-Ile-Ala-Ala-Gln-Val-Thr-Thr) was synthesized with an identical procedure used for the synthesis depicted in Figure 2 except that the peptide resin (0.3 mmol) was acetylated with a large excess of ^{14}C-acetic anhydride (3 mmol, 5.4 × 10^6 dpm/mmol) for 20 min after the coupling of isoleucine$_{69}$. Despite this treatment, the synthesis gave the same regrowth of incomplete peptides with the coupling of alanine$_{68}$ and alanine$_{67}$.

Peptide 74–63 (Gly-Asn-Ile-Tyr-Asp-Ile) was synthesized as before, except that acetylation with ^{14}C-acetic anhydride was performed after the addition of tyrosine$_{71}$. A sample of the peptide resin (50 mg, 16 μmol of glycine esterified to the resin) was cleaved by a HBr and trifluoroacetic acid treatment and the product was isolated (see Experimental Section). The peptides were chromatographed on Dowex 50-X8 and the column was monitored by ninhydrin analysis after alkaline hydrolysis of a sample of each fraction (see Figure 5). The absorbance at 570 nm indicated that four fractions were present and these were numbered consecutively by Roman numerals. A portion (6%) was checked for ^{14}C radioactivity while the rest was subjected to an acid hydrolysis and the amino acid content was determined (see Table V). The analysis in-

TABLE V
COMPOSITION OF THE MIXTURE OF PEPTIDES FROM THE
SYNTHESIS OF THE SEQUENCE 74–69 OF ACP[a]

Peak no.	Gly	Asp	Ile	Tyr	Peptide[b]	[14]C dpm
I	1.0	2.0	0.81	0.79	125	1,500[c]
II	1.0	0.1			38	
III	1.0	0.95	0.75		17.5	
IV	1.0	0.98	0.15		3.1	
Resin peptide[d]	1.0	1.41	1.36	0.6	306	

[a] Peptide 74–69 (0.3 mmol), which had been acetylated with ^{14}C-acetic anhydride (3 mM, 5.2 × 10^3 dpm/μmol) after the addition of Ile$_{72}$, was cleaved from the resin and fractionated on Dowex 50-8 (see results and Figure 4). The amino acid composition and ^{14}C radioactivity of the pooled peaks were then determined. [b] Based on μmoles of glycine. [c] This corresponds to the acetylation of 2.3 μmol of peptide. [d] The amino acid analysis of the resin peptide at different stages of the synthesis indicated that amino acids were added with the following yields: Asn$_{73}$, 78%; Ile$_{72}$, 71%; Tyr$_{71}$, 59%; Asp$_{70}$, 63%; Ile$_{69}$, 65%.

dicated that peak I consisted of the complete peptide 74–69 and acetylated peptides while peaks II–IV were the products of incomplete coupling of asparagine$_{73}$, isoleucine$_{72}$, and tyrosine$_{71}$, i.e., Gly, Gly-Asx, and Gly-Asx-Ile, respectively. The quantities of these peptides corresponded closely to the values of truncated sequences expected from the amino acid analysis of the peptide resin, and the recovery of material was good (60% of the glycine esterified to the resin was recovered from the Dowex 50 column). The finding that only a small fraction (4%) of the incomplete peptides present in this sample had been terminated with the acetic anhydride treatment is consistent with the hypothesis that incomplete coupling and/or deprotection reflects sites which are not solvated by the reagents and may be inaccessible to chain-terminating reagents or to titration as well as to coupling.

As the yield of addition of isoleucine$_{69}$ is only 60% of the value for the initial amino acid, one would expect that up to 40% of peptide chains would be available for acetylation, but this was obviously not the case, as all of the incomplete peptide chains coupled with both alanine$_{68}$ and alanine$_{67}$. Therefore, on deprotection of the α-amino group of the isoleucyl peptides, the solvation of peptide–polymer matrix must undergo a dramatic change so that the amino groups that were unavailable for acetylation can readily form a peptide bond. On the addition of glutamine$_{66}$, the yield then dropped to the level of the synthesis before the addition of alanine$_{68}$ and alanine$_{67}$.

Acetylation of the peptide resin after the addition of tyrosine$_{71}$ occurred with only a small fraction (4%) of the truncated peptides formed by the incomplete coupling of asparagine$_{73}$, isoleucine$_{72}$, and tyrosine$_{71}$ despite the use of a large excess (tenfold) of acetic anhydride (see Figure 5 and Table V). Bayer made a similar observation from studies with model peptides,[12] where it was found that only part of the uncoupled free amino groups could be acetylated.

Therefore, one cannot expect that repeated couplings of an amino acid will increase the yield of a difficult step if the unreactive amino groups are buried. This conclusion is supported by the data presented in Table III, where it is clear that four couplings of tyrosine, carried out in a variety of solvents, did not decrease the amount of free amino groups (as measured by the amount of bound chloride), although amino acid analysis after the four couplings indicated that tyrosine had reacted with only 60% of the peptide chains. Similar observations about the ineffectiveness of repeated couplings have been made in the synthesis of (Leu-Ala)$_6$[14] and of a tetrapeptide.[25] In the synthesis of lysozyme, it was found that repeated couplings resulted in the growth of peptides on partially deprotected side chains.[43]

The use of an insoluble support allows the solid phase method to have considerable advantages over protein synthesis carried out in solution, such as the facile removal of excess reagents, tremendous savings in time, and avoidance of the problems of insolubility of large fragments. At the same time, the use of a polymeric support introduces a new set of problems which still require extensive investigation before the solid phase method can be used routinely for any particular peptide sequence. If the use of a polystyrene support is to be completely successful, then an analytical method will have to be developed which allows the rapid determination of the yield of both coupling and deprotection reactions. This goal may be difficult to achieve, as the properties of the polymeric support may distort the analytical results, either by masking some of the functional groups or by trapping by-products of the coupling reaction which will react with the assay reagents. By a similar argument, the use of terminating reagents, such as acetic anhydride, has only limited application in stopping the regrowth of partially complete sequences. The problems of the solid phase synthetic procedure, however, cannot be so general or so serious as the various analytical studies[13,14,17] would suggest, because of the enormous number of peptides that have been synthesized successfully by the method. In fact, the success of the method would suggest that in most syntheses, truncated peptides, when formed, do not regrow to any significant extent during the rest of the synthesis.

An alternative to the problems of a polystyrene resin is the use of a special polymer in which a thin layer of styrene is localized on the surface of an inert bead. Several successful syntheses have been carried out with such a support either on Teflon[44] or on glass[45] and these achievements point to a possible solution to the sequence-dependent problems of solid phase synthesis. Other physical forms of polystyrene, such as the macroreticular resins,[46] should also be further investigated. A different approach to this problem proposed by Sheppard[15] would be the use of a polymer support whose solvation properties would be similar to that of the protected polypeptide which was being synthesized in order to minimize solvation charges as synthesis progresses. Once the problem of dynamic solvation changes is overcome, then monitoring procedures can be rationally applied to overcome the sequence-dependent problems which hinder the application of solid phase synthesis.

(43) J. J. Sharp, A. B. Robinson, and M. D. Kamen, *Fed. Proc.*, **30**, 1286 (1971).
(44) G. W. Tregear, H. D. Niall, J. T. Potts, S. E. Leeman, and M. M. Chang, *Nature (London), New Biol.*, **232**, 87 (1971).
(45) (a) E. Bayer, G. Jung, I. Halasz, and I. Sebastian, *Tetrahedron Lett.*, 4503 (1970); (b) W. Parr and K. Grohmann, *ibid.*, 2633 (1971).
(46) S. Sano, R. Tokunaga, and K. A. Kun, *Biochim. Biophys. Acta*, **244**, 201 (1971).

Experimental Section

Reagents.—*tert*-Butyloxycarbonyl (BOC)-amino acids and BOC-glycine which was esterified to a polystyrene–1% divinylbenzene resin were purchased from Schwarz BioResearch. The following side-chain blocking groups were used—aspartic acid, β-benzyl ester; threonine and tyrosine, benzyl ethers; glutamic acid, γ-benzyl ester—while the BOC group was used for α-amino protection. Hydrogen chloride-36 and ^{14}C-acetic anhydride were purchased from New England Nuclear. All other chemicals, of reagent grade or better, were purchased from common sources.

A Packard liquid scintillation spectrometer, Model 544, was used for measurement of radioactivity. Amino acid analyses were measured with a Beckman Spinco amino acid analyzer, Model MS. Automated syntheses were performed on a synthesizer based on the design of Merrifield.[47] A Cotlove chloridometer was used in addition to Volhard titrations for chloride determinations.

Preparation of the Pyridine Hydrochloride Reagent.—Pyridine hydrochloride (0.3 mol) was dissolved in dichloromethane (1 l.) and the chloride content was checked by chloridimetry. Pyridine hydrochloride (^{36}Cl) was prepared in exactly the same manner except that $H^{36}Cl$ (25 μCi) was added to the solution to give a specific radioactivity of 183 dpm/μmol of chloride.

Estimation of Bound Chloride.—The peptide resin (2 g) was washed with dichloromethane (3 × 1 ml) and then treated for 15 min with 25 ml of the standard pyridine hydrochloride solution (0.3 M, measured with a volumetric pipette). The excess reagent was then removed by the following washes: (1) dichloromethane (3 × 20 ml); (2) ethanol or *tert*-butyl alcohol[48] (3 × 20 ml, each for 2 min); (3) dichloromethane (4 × 20 ml, each for 2 min). All washes were removed by drying the resin under nitrogen pressure (1 min). Washes 1–3, as well as the remainder of the pyridine hydrochloride solution, were collected in a 250-ml volumetric flask made up to volume with dichloromethane and a sample was taken for chloride measurements. The amine hydrochloride, that had been formed by the pyridine hydrochloride treatment, was neutralized by two triethylamine–dichloromethane washes (1.5% triethylamine, 2 × 20 ml, each for 10 min). The peptide resin was then washed with dichloromethane (3 × 20 ml, each for 2 min). These washes were combined with triethylamine washes in a 100-ml volumetric flask and a sample was taken for chloride measurements.

Measurement of Chloride with the Chloridimeter.—Although the chloridimeter was developed by Cotlove, *et al.*,[49] for the determination of chloride in serum and urine, it was found that the chloride was quantitatively extracted from dichloromethane into the aqueous phase under the conditions of the assay.

The sample (1 ml), dissolved in dichloromethane, was added to a mixture of 0.1 M nitric acid and 1.7 M acetic acid (3 ml) and gelatin reagent (0.2 ml). The assay mix was then added to the reaction vessel of the chloridimeter and the chloride concentration was determined. A sodium chloride solution (1.6 mg/l.) was used to calibrate the instrument, while the blank value was measured with dichloromethane (1 ml) and the assay mix (3.2 ml).

Calculation of Chloride by ^{36}Cl Radioactivity.—The sample (1 ml) was added to Bray's solution (10 ml) and the sample was counted in a Packard liquid scintillation spectrometer, Model 544, which had been programmed to present the data as disintegrations per minute of chloride-36. As several of the reagents used in the washes, particularly triethylamine and pyridine hydrochloride, caused strong quenching of the radioactivity, it was necessary to calibrate the program with quenching standards.

Procedure for Acetylation of Incomplete Peptides.—Peptide resin (2 g, 0.66 mmol) was treated with a solution of 3 mmol of ^{14}C-acetic anhydride (specific radioactivity 5.4 × 10³ dpm/μmol) and 3 mmol of triethylamine dissolved in dichloromethane (CH_2Cl_2, 20 ml) for 20 min. It was found that the following washing procedures were necessary for adequate removal of the excess reagent (all washes 20 ml and for 2 min unless otherwise specified): 3 × CH_2Cl_2, 3 × *tert*-butyl alcohol, 3 × CH_2Cl_2, 3 × triethylamine (1.5% v/v in CH_2Cl_2), 3 × CH_2Cl_2, 3 × *tert*-butyl alcohol, 3 × CH_2Cl_2. All the washes were combined and an aliquot was counted for ^{14}C radioactivity to check that all of the excess acetic anhydride had been removed.

Standard Procedure for the Coupling of BOC-Amino Acids.—As both triethylamine and ethanol were used in each cycle of the synthesis, a small loss of peptide chain by transesterification may occur repeatedly during a long synthesis. To minimize this possibility, the sterically hindered alcohol, *tert*-butyl alcohol, was substituted for ethanol and the concentration of triethylamine was reduced from 10 to 3% (v/v). The *tert*-butyl alcohol was found to be efficient in shrinking the resin even when 5% CH_2Cl_2 was added to prevent freezing of the alcohol.

The following sequence of reactions was used to prepare the peptide resin for a coupling reaction (all washes 20 ml and 2 min unless otherwise specified): 3 × CH_2Cl_2, trifluoroacetic acid–CH_2Cl_2, (1:1, v/v, 20 ml, 2 min), 6 × CH_2Cl_2, 3 × triethylamine (3% v/v in CH_2Cl_2), 3 × CH_2Cl_2. The coupling step was carried out with a threefold excess of the appropriate amino acid (0.2 M) and dicyclohexylcarbodiimide (DDC, 0.2 M) as the coupling reagent, except for glutamine and asparagine, which were added in the same concentration as the *p*-nitrophenyl ester. All couplings were left for 6 hr, except for active esters which were coupled for 12 hr. After coupling the excess reagents were removed by 3 × CH_2Cl_2 washes.

Modified Procedure for the Coupling of BOC-Amino Acids.—The BOC group was removed and the peptide resin was prepared for coupling by the following sequence of washes (20 ml and for 2 min unless specified): 3 × CH_2Cl_2, trifluoroacetic acid–CH_2Cl_2 (1:1, 2 × 10 ml, 2 and 20 min), 3 × CH_2Cl_2, 3 × *tert*-butyl alcohol, 3 × CH_2Cl_2, 2 × triethylamine (3%, v/v in CH_2Cl_2), 3 × CH_2Cl_2, 3 × *tert*-butyl alcohol, 3 × CH_2Cl_2, 3 × dimethylformamide (only for couplings in that solvent). The coupling procedure was the same as described above, except that the time of reaction was reduced to 2 hr for DCC-mediated couplings. First, couplings were routinely carried out with dichloromethane as the solvent. The by-products from the reaction were removed by the following washes: 3 × CH_2Cl_2, 3 × *tert*-butyl alcohol, 3 × CH_2Cl_2, 3 × triethylamine (3% v/v), 3 × CH_2Cl_2, 3 × *tert*-butyl alcohol, 3 × CH_2Cl_2, 3 × dimethylformamide. The second coupling was carried out in a mixed solvent of dichloromethane and dimethylformamide (1:1) and left for 2 hr. The following washes then completed the procedure: 3 × dimethylformamide, 3 × CH_2Cl_2, 3 × *tert*-butyl alcohol, 3 × CH_2Cl_2, 3 × triethylamine (3% v/v), 3 × CH_2Cl_2, 3 × *tert*-butyl alcohol, 3 × CH_2Cl_2.

Preparation of Samples for Amino Acid Analysis.—Peptide resin (2 mg) was hydrolyzed with a mixture of HCl (12 N) and propionic acid (1:1, 2 ml) at 130° according to the method of Scotchler, *et al.*[20] Free peptides were hydrolyzed with 6 N HCl in sealed, evacuated tubes for 24 hr at 110°.

HBr–Trifluoroacetic Acid Cleavage of the Peptide from the Resin.—Peptide resin (1 g) was added to a mixture of trifluoroacetic acid (30 ml) and anisole (0.3 ml) in a cleavage apparatus as described by Stewart, *et al.*[50] HBr bubbled through the solution for 30 min at 25°, and the trifluoroacetic acid was removed from the resin by filtration. The resin was then washed with trifluoroacetic acid (3 × 10 ml) and the filtrate was combined with these washes. The trifluoroacetic acid was immediately removed by evaporation under reduced pressure and the residue was dissolved in 0.01 M Tris-HCl, pH 7.3 (10 ml). The cleavage was then repeated on the peptide resin under exactly the same conditions as before, except that the time of reaction was increased to 1 hr. The two products were then combined.

Registry No.—Pyridine hydrochloride-^{36}Cl, 22069-61-0; hydrochloric-^{36}Cl acid, 36640-18-3; sequence 74–63, 37746-85-3; sequence 74–69, 37746-86-4.

---

(47) R. B. Merrifield and J. M. Stewart, *Nature* (London), **207**, 522 (1965).

(48) The *tert*-butyl alcohol used in this paper was 5% (v/v) dichloromethane added to prevent freezing.

(49) E. Cotlove, H. V. Trantham, and R. L. Bowman, *J. Lab. Clin. Med.*, **50**, 358 (1958).

(50) J. M. Stewart and J. D. Young, "Solid Phase Peptide Synthesis," W. H. Freeman, San Francisco, Calif., 1969, pp 40–41.

Carboxyl-Catalyzed Intramolecular Aminolysis. A Side Reaction in Solid-Phase Peptide Synthesis

B. F. Gisin and R. B. Merrifield

Contribution from The Rockefeller University, New York, New York 10021, and Department of Physiology, Duke University Medical Center, Durham, North Carolina 27706. Received October 2, 1971

Abstract: The polymer-supported peptide ester, D-valyl-L-prolyl-resin, was found to undergo intramolecular aminolysis which was catalyzed by carboxylic acids. The resulting loss of the dipeptide from the resin, which amounted to 70% during a regular coupling with N,N'-dicyclohexylcarbodiimide, was repressed by adding the carbodiimide reagent prior to the carboxyl component. The diketopiperazine of D-valyl-L-proline, the only detectable product of this side reaction, was isolated and characterized. The rate of the intramolecular aminolysis was dependent on the composition and configuration of the dipeptide. None of the other reagents tested were as efficient catalysts as the carboxylic acids.

In the course of the synthesis of the peptide sequence D-Pro-D-Val-L-Pro[1] by the solid-phase method,[2] we observed a considerable loss of peptide from the resin.[3] Although the yield of the protected dipeptide was nearly quantitative, only about 30% of the expected amount of tripeptide was found. A step-by-step monitoring of the synthesis indicated that the loss did not occur during deprotection or neutralization of the dipeptide-resin[4] but during the coupling with Boc-D-proline and DCC. This unexpected finding called for a closer investigation, some aspects of which are presented here.

The methods and procedures employed were essentially the established techniques of solid-phase peptide synthesis.[5] Polystyrene-co-1% divinylbenzene resin was chloromethylated with chloromethyl methyl ether and stannic chloride[2,6] which was converted, first, to acetoxymethyl resin[7,8] and then aminolyzed with diethylamine[3] to yield hydroxymethyl resin. Boc-L-proline was esterified to the resin by the N,N'-carbonyldiimidazole method[7,9] and the remaining hydroxy groups were blocked by esterification with acetic anhydride. This procedure was chosen in order to avoid the introduction of any quaternary ammonium groups into the polymer[10] which can interfere with the quantitative determination of amino groups as described below. The dipeptide-resins were prepared using two DCC couplings[11] with a twofold excess of Boc-amino acid and DCC reagent each time.

In order to monitor the loss of dipeptide from the resin, a procedure for the determination of amino groups on an insoluble polymer with picric acid[12] was adopted. The amine-containing resins were treated with a solution of picric acid to form the polymer supported amine picrate. After thorough washings to remove nonionically bound picric acid, the resins were treated with an excess of diisopropylethylamine which quantitatively released the picrate from the polymer into solution. The concentration of picrate in this solution, which was determined spectrophotometrically, reflected the amine content and therefore the amount of dipeptide on the resin. These values were used to compute apparent first-order rate constants for the decrease in amine content of dipeptide resins (Tables I and II[13]) as described in the Experimental Section.

Since diketopiperazines can be quantitatively determined by gas-liquid chromatography,[14] this method was used to measure the release of D-Val-L-Pro diketopiperazine[15] from the solid support. These experi-

---

(1) The abbreviations recommended by the IUPAC–IUB Commission on Biochemical Nomenclature (*J. Biol. Chem.*, **241**, 2491 (1966); **242**, 555 (1967) have been used throughout. In addition, TFA = trifluoroacetic acid, DMF = dimethylformamide, DCC = N,N'-dicyclohexylcarbodiimide.
(2) R. B. Merrifield, *J. Amer. Chem. Soc.*, **85**, 2149 (1963).
(3) B. F. Gisin and R. B. Merrifield, in preparation.
(4) The expression "peptide-resin" denotes a peptide, the C-terminal carboxyl group of which is esterified to a polymeric benzyl alcohol.
(5) R. B. Merrifield, *Advan. Enzymol.*, **32**, 221 (1969).
(6) K. W. Pepper, H. M. Paisley, and M. A. Young, *J. Chem. Soc.*, 4097 (1953).
(7) M. Bodanszky and J. T. Sheehan, *Chem. Ind. (London)*, 1597 (1966).
(8) J. M. Stewart and J. D. Young, "Solid Phase Peptide Synthesis," W. H. Freeman, San Francisco, Calif., 1969, p 9.
(9) H. A. Staab, *Angew. Chem.*, **71**, 194 (1959).
(10) R. B. Merrifield, 1966, unpublished.
(11) J. C. Sheehan and G. P. Hess, *J. Amer. Chem. Soc.*, **77**, 1067 (1955).
(12) B. F. Gisin, *Anal. Chim. Acta*, **58**, 248 (1972).
(13) C. E. Moore and R. Peck, *J. Org. Chem.*, **20**, 673 (1955).
(14) A. B. Mauger, *J. Chromatogr.* **37**, 315 (1968).

Table I. Apparent First-Order Rate Constants of the Disappearance of Amino Groups from H-D-Val-L-Pro-Resin with Various Reagents

Reagent	Concn,^{a} M	k_{app},^{b} min^{-1}	k_{rel}
None		2.5×10^{-4}	1.0^c
Diisopropylethylamine	0.3^d	6.0×10^{-4}	2.4
N-tert-Butyloxycarbonyl-D-proline^e	0.05	1.3×10^{-1}	520
Acetic acid	0.05	8.5×10^{-2}	340
Trimethylacetic acid	0.06	5.7×10^{-2}	230
Benzoic acid	0.06	7.8×10^{-2}	310
Trifluoroacetic acid	6.8^f	1.5×10^{-3}	6.1
	2.7^g	6.3×10^{-4}	2.5
	0.06	1.5×10^{-4}	0.6
Picric acid	0.1	6.9×10^{-4}	2.8
3,5-Dimethylpicric acid^h	0.06	2.7×10^{-3}	11
2,4-Dinitrophenol	0.06	2.3×10^{-3}	9.2
p-Nitrophenol	0.06	1.2×10^{-2}	48
Imidazole	0.06	4.5×10^{-3}	18
2-Hydroxypyridine	0.06	4.7×10^{-3}	19
N-Hydroxysuccinimide	0.03^i	8.9×10^{-4}	3.6
Pyridine picrate	0.03^i	4.6×10^{-4}	1.8
Triethylamine hydrochloride	0.06	$<1.3 \times 10^{-4}$	<0.5

^{a} In methylene chloride. ^{b} See Experimental Section. ^{c} Reference value. ^{d} 5% by volume. ^{e} Reference 3. ^{f} 50% by volume. ^{g} 20% by volume. ^{h} Reference 13. ^{i} Saturated solution.

Table II. Apparent First-Order Rate Constants of the Acetic Acid Catalyzed Disappearance of Amino Groups from Dipeptide-Resins^a

Compound	k_{app}, min^{-1}	k_{rel}	Half-time, min
H-D-Val-L-Pro-resin	8.5×10^{-2}	100^b	8.1
H-L-Val-L-Pro-resin	7.3×10^{-3}	8.6	95
H-D-Pro-L-Pro-resin	6.5×10^{-3}	7.6	107
H-L-Pro-L-Pro-resin	9.2×10^{-2}	108	7.5
H-L-Val-Gly-resin	4.7×10^{-3}	5.5	150
H-Gly-L-Val-resin	1.0×10^{-3}	1.2	690

^{a} 0.1 M HOAc in methylene chloride. 25°. ^{b} Reference value.

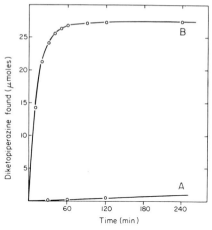

Figure 1. The cleavage of D-Val-L-Pro diketopiperazine from H-D-Val-L-Pro-resin with methylene chloride (A) and with 0.05 M acetic acid in methylene chloride (B). Experimental values (O) by glc.

ments (Figure 1) were performed in a thermostated vessel. After the cyclization reaction the released diketopiperazine was injected into the gas chromatograph without derivatization to yield the experimental data in Figure 1. Each point represents the total diketopiperazine found after the corresponding period of time.

Results

It was found (Table I) that H-D-Val-L-Pro-resin was stable as the trifluoroacetate and was nearly so in its free amine form ($k_{app} = 2.5 \times 10^{-4}$ min^{-1}, $k_{rel} = 1.0$) when suspended in methylene chloride. There was only a slight increase in the rate of loss of amino groups from the resin with 5% diisopropylethylamine in methylene chloride ($k_{rel} = 2.4$). With the carboxylic acid Boc-D-Pro-OH (0.05 M, in methylene chloride), however, amine was lost from the resin at a rate 520 times greater than with methylene chloride alone ($k_{app} = 1.3 \times 10^{-1}$ min^{-1}). We take this rate, corresponding to a half-time of 5.3 min, to account for the low yield in the preparation of D-Pro-D-Val-L-Pro-resin mentioned earlier. For, in that synthesis, the standard procedure for DCC coupling[2] was employed, which involves the equilibration of the amino-resin with the carboxyl component for 10 min prior to the addition of the coupling agent. When the order in which the reagents were added to the amine component was reversed, namely, DCC followed by Boc-Pro-OH in several small portions, the loss was reduced and the yield of tripeptide was over 90%.

The loss of dipeptide was not only accelerated by Boc-D-Pro-OH but also by other carboxylic acids such as acetic acid ($k_{rel} = 340$, Figure 1), benzoic acid ($k_{rel} = 310$), and trimethylacetic acid ($k_{rel} = 230$). In all three instances tlc of the supernatant showed, in addition to the spot corresponding to the reagent, a single new ninhydrin negative spot which stained blue in the iodine–tolidine reaction. This suggested to us a cleavage of the anchoring bond to the resin to form the diketopiperazine of D-valyl-L-proline. In order to test this hypothesis a sample of H-D-Val-L-Pro-resin was treated with 0.1 M acetic acid in methylene chloride for 1 hr at room temperature. The resin was filtered off and the filtrate was evaporated to give a crystalline residue which, with one crystallization, had the same melting point as that reported for L-Val-D-Pro diketopiperazine.[16] The product which was obtained in 98% yield gave a satisfactory C, H, and N analysis and was also found to give an ir spectrum identical with the one of the diketopiperazine synthesized by cyclization of D-valyl-L-proline p-nitrophenyl ester.

Table II shows the effect of 0.1 M acetic acid in methylene chloride on the disappearance of amino groups from six different dipeptide-resins. H-L-Pro-L-Pro-resin had a sensitivity comparable to H-D-Val-L-Pro-resin whereas H-D-Pro-L-Pro-resin and H-L-Val-L-Pro-resin both were about ten times more resistant to this reagent. H-L-Val-Gly-resin was 18 times and H-Gly-L-Val-resin was 80 times more stable than H-D-

(15) Synonyms: *trans*-1,6-trimethylene-3-isopropyl-2,5-piperazinedione; *cyclo*-[D-valyl-L-prolyl].

(16) A. Stoll, A. Hofmann, and B. Becker, *Helv. Chim. Acta*, **26**, 1602 (1943).

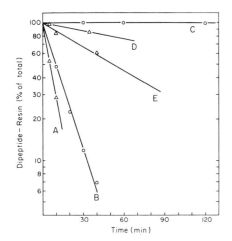

Figure 2. Semilogarithmic plot of the disappearance of amino groups from H-D-Val-L-Pro-resin upon treatment with different reagents in methylene chloride: (A) Boc-D-Pro-OH (0.05 M); (B) acetic acid (0.05 M); (C) methylene chloride; (D) imidazole (0.06 M); (E) p-nitrophenol (0.06 M). Experimental values by glc (O) and by picrate determination (△).

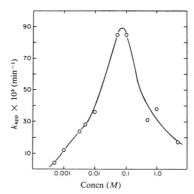

Figure 3. Semilogarithmic plot of the apparent first-order rate constants of the cleavage of D-Val-L-Pro diketopiperazine from H-D-Val-L-Pro-resins under the influence of acetic acid at various concentrations in methylene chloride determined by monitoring the disappearance of amino groups with picrate.

Val-L-Pro-resin. Based on these data we can expect losses in the order of 1-5% for a normal coupling procedure to a non-imino acid dipeptide-resin, in which the carboxyl and amine components are premixed for 10 min before the addition of DCC.

A semilogarithmic plot of the amount of diketopiperazine that was found by glc in the supernatant of an acid treated dipeptide-resin vs. time is consistent with pseudo-first-order kinetics (Figure 2, curve B).

The rate of cleavage is dependent on the concentration of the catalyst. Figure 3 shows how differing concentrations of acetic acid in methylene chloride affect the rate of cleavage of dipeptide from the resin at room temperature. The plot indicates a maximal efficiency of acetic acid in a concentration of approximately 0.08 M. By coincidence, similar concentrations are normally used in the presoaking step of the amine-resin with the Boc-amino acid prior to the addition of DCC for the coupling reaction.

Discussion

The side reaction which caused a low yield in the synthesis of Boc-D-Pro-D-Val-L-Pro-resin was a carboxylic acid catalyzed intramolecular aminolysis of the ester bond to the resin at the dipeptide stage. This conclusion was arrived at by interpretation of the following experimental results.

(a) It was shown that the cleavage occurred during the coupling step. The cleavage rates of H-D-Val-L-Pro-resin at other steps of the synthesis, $i.e.$, with 50% TFA in methylene chloride, 5% diisopropylamine in methylene chloride, or with the solvent alone, could not account for the magnitude of the loss of peptide that was observed.

(b) The yield of tripeptide was drastically improved by adding DCC to the dipeptide-resin prior to the carboxyl component.

(c) The dipeptide is also lost from the resin under the influence of acetic acid or other weak carboxylic acids at rates comparable to that with Boc-D-Pro-OH.

(d) The only detectable product of the cleavage, D-Val-L-Pro diketopiperazine, was isolated in high yield.

Experiment b was chosen with reference to the mechanism for DCC couplings which is generally thought to involve the very rapid formation of an activated derivative of the carboxyl component, which then aminolyzes more slowly to form the peptide bond. If, therefore, the DCC were added first, the subsequently added carboxyl component would be consumed almost immediately to form the active intermediate and would not catalyze the formation of diketopiperazine. With this "reversed DCC coupling" the exposure of the dipeptide-resin to carboxyl groups was expected to be minimal as compared to the standard procedure or to simultaneous addition of the two components to the resin. The increase in yield of Boc-D-Pro-D-Val-L-Pro-resin from 32% with the standard procedure to over 90% with reversed DCC coupling was compatible with that reasoning.

The cyclization can either be catalyzed or inhibited by acetic acid (Figure 3). At higher concentrations the amine becomes increasingly protonated and cannot participate as a nucleophile. This hypothesis is supported by the values for the cleavage rates with strong acids at low concentrations (picric acid, TFA) which are not significantly different from the reference rate with solvent alone (Table I). Higher concentrations of TFA (20% or 50%) lead to the linear dipeptide (verified by tlc) by the known acidolytic cleavage of the benzyl ester linkage that anchors the peptide on the resin[17] rather than to the diketopiperazine by intramolecular aminolysis.

A definite acceleration of the reaction was observed with other weak acids (3,5-dimethylpicric acid, 2,4-dinitrophenol, p-nitrophenol) and with "bifunctional catalysts"[18] (2-hydroxypyridine, imidazole). However,

(17) B. Gutte and R. B. Merrifield, $J. Biol. Chem.$, **246**, 1922 (1971).
(18) C. G. Swain and J. F. Brown, Jr., $J. Amer. Chem. Soc.$, **74**, 2538 (1952); M. L. Bender, $Chem. Rev.$, **60**, 53 (1960).

the rates with these acids were much smaller than with the carboxylic acids at similar concentrations. Whether this is due to different mechanisms or to the result of a different concentration dependence or whether it reflects an inherent property of these reagents awaits a more detailed investigation.

Owing to the ease with which acylimino acids can form cis peptide bonds,[19] the tendency of diketopiperazine formation is high in the case of peptides that contain proline[20–23] or sarcosine.[22] One is also reminded of the acceleration of the aminolysis of esters[24] and "active esters"[25,26] by "catalytic amounts" of a carboxylic acid. Similarly, the cyclization of glutamic esters is catalyzed by carboxylic acids.[27] However, this type of reaction is not restricted to the aminolysis of esters.[21b,22,23,28–31] Recently, the formation of pyroglutamyl peptides from Boc-glutaminyl peptides during the deprotection step has been linked to the presence of carboxylic acids[28a] and the spontaneous decomposition of the pure tripeptide, H-D-valyl-L-prolylsarcosine, to yield D-valyl-L-proline diketopiperazine and sarcosine has been reported.[23] An explanation offered for the latter reaction was based on the limited conformational freedom of the peptide due to the bulkiness of its substituents.[29] It might now be supplemented with the conceivable participation of the carboxyl group of sarcosine that was present under the prevailing conditions. Another carboxyl-catalyzed intramolecular aminolysis reaction is the rearrangement of N-acyl-N'-α-aminoacylhydrazines into acyl-α-aminoacylhydrazides.[30,31] There,[31] as well as in the cases of the cyclization of glutamic esters,[27] and of the aminolysis of active esters,[26] a catalysis-inhibition dependence on acid concentration was demonstrated, which was similar to that found here for the disappearance of dipeptide from the resin (Figure 3). The postulated mechanism[26,27,31] might therefore also obtain in our case (Scheme I). It involves a concerted reaction in which the un-ionized carboxyl group of the catalyst acts through a hydrogen-bonded cyclic intermediate.

Although many peptides have been prepared by the solid-phase method[32] (several of them with C-terminal proline),[21a,33] this carboxyl-catalyzed side reaction has not been reported before. We presume that it reflects an unusually susceptible structure of certain dipeptide-resins under the specified conditions rather than a general phenomenon. In those instances where the cyclization is quantitatively important this side reaction can now be effectively suppressed.

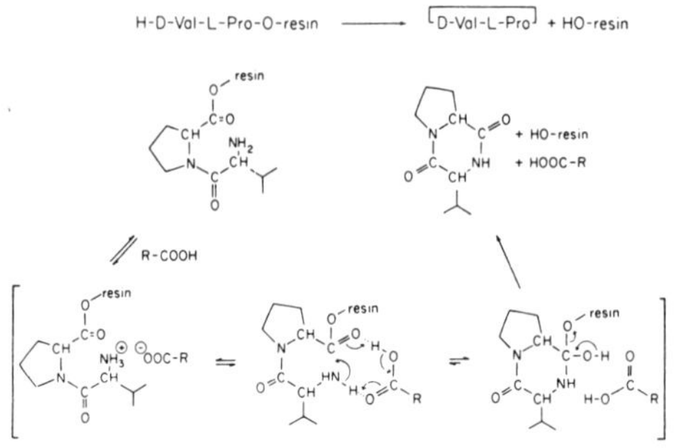

Scheme I

Experimental Section

Amino acid analyses (Beckman Spinco amino acid analyzers 120B and 121) were performed by Miss L. Apacible and elemental analyses by Mr. T. Bella of Rockefeller University. Infrared spectra were taken on a Perkin-Elmer 237B ir spectrophotometer through the courtesy of Dr. L. C. Craig of Rockefeller University. Melting points (not corrected) were determined in capillaries and optical rotations on a Schmidt & Haensch polarimeter. Solid-phase reactions were carried out in vessels that were made from screw-capped Pyrex culture tubes (screw cap fitted with Teflon interface, Scientific Glass Apparatus, Inc., Bloomfield, N. J., Catalogue No. T-2040-a). In order to obtain reactors of capacities ranging from 5 to 30 ml that could be used both for analytical and preparative purposes, the tubes were cut and fitted with a glass fritted disk (medium porosity) and a stopcock with a 1.5-mm bore Teflon plug. The volumes were adjusted so that the walls of the vessel were completely wetted during the mixing period of the standard mechanical shaker.[2] Hydrolyses of resins were with either 12 N HCl-dioxane (1:1, v/v) in sealed vessels at 110° for 18–20 hr followed by filtration and rehydrolysis in 6 N HCl (110°, 90 hr) or with propionic acid–12 N HCl (1:1, v/v) at 130–140° for 3–6 hr.[34]

tert-**Butyloxycarbonyl Dipeptide-Resins.** Starting with 1.0 mmol of Boc-L-Pro-resin[3] (substitution, 370 μmol/g), Boc-dipeptide-resins were prepared in the following way: (a) deprotection with TFA–CH$_2$Cl$_2$ (1:1, v/v) 2 × 15 min, CH$_2$Cl$_2$ 3 × 2 min; (b) neutralization with diisopropylethylamine–CH$_2$Cl$_2$ (1:19, v/v) 2 × 3 min, CH$_2$Cl$_2$ 3 × 2 min; (c) coupling with 2.0 mmol of DCC in CH$_2$Cl$_2$ for 2 min and 2.0 mmol of Boc-amino acid (Boc-D-Pro-OH,[3] Boc-L-Pro-OH,[35] Boc-D-Val-OH, Boc-L-Pro-OH,[36a] or Boc-L-Val-OH)[36b] for 2 hr, washing with alternating CH$_2$Cl$_2$ and DMF 3 × 2 min each; (d) coupling step c repeated for 5 hr. Amino acid analysis of the dried resins gave the following substitutions (in μeqiv/g): Boc-D-Pro-L-Pro-resin, Pro 650; Boc-D-Val-L-Pro-resin, Pro 350, Val 340; Boc-L-Pro-L-Pro-resin, Pro 710; Boc-L-Val-L-Pro-resin, Pro 332, Val 343.

D-Prolyl-D-valyl-L-prolyl-Resin. (a) By Regular DCC Coupling. Boc-D-Val-L-Pro-resin, 100 mg, was deprotected (2 × 15 min with TFA–CH$_2$Cl$_2$, 1:1) and neutralized (2 × 3 min with diisopropylethylamine–CH$_2$Cl$_2$, 5%). The amine content at this stage was

---

(19) G. N. Ramachandran and V. Sasisekharan, *Advan. Protein Chem.*, **23**, 283 (1968).
(20) E. Schröder and K. Lübke in "The Peptides," Vol. I, Academic Press, New York, N. Y., 1965, p 148.
(21) (a) M. Rothe, R. Theysohn, and K. D. Steffen, *Tetrahedron Lett.*, 4063 (1970); (b) R. H. Mazur and J. M. Schlatter, *J. Org. Chem.*, **28**, 1025 (1963).
(22) J. Dale and K. Titlestad, *J. Chem. Soc. D*, 656 (1969); 1403 (1970).
(23) J. Meienhofer, *J. Amer. Chem. Soc.*, **92**, 3771 (1970).
(24) P. K. Glasoe, J. Kleinberg, and L. F. Audrieth, *J. Amer. Chem. Soc.*, **61**, 2387 (1939).
(25) R. Schwyzer, M. Feurer, and B. Iselin, *Helv. Chim. Acta*, **38**, 83 (1955); E. Taschner, G. Blotny, B. Bator, and C. Wasielewski, *Bull. Acad. Pol. Sci.*, **12**, 755 (1964).
(26) N. Nakamizo, *Bull. Chem. Soc. Jap.*, **42**, 1071, 1078 (1969).
(27) A. J. Hubert, R. Buyle, and B. Hargitay, *Helv. Chim. Acta*, **46**, 1429 (1963).
(28) (a) H. C. Beyerman, personal communication 1971; (b) H. T. Huang and C. Niemann, *J. Amer. Chem. Soc.*, **72**, 921 (1950); D. G. Smyth, A. Nagamatsu, and J. S. Fruton, *ibid.*, **82**, 4600 (1960); J. C. Sheehan and D. N. McGregor, *ibid.*, **84**, 3000 (1962).
(29) J. Meienhofer, Y. Sano, and R. P. Patel in "Peptides: Chemistry and Biochemistry," B. Weinstein, Ed., Marcel Dekker, New York, N. Y., 1970, p 419.
(30) M. Brenner and W. Hofer, *Helv. Chim. Acta*, **44**, 1794, 1798 (1961); B. Gisin and M. Brenner, *ibid.*, **53**, 1030 (1970).
(31) W. Hofer and M. Brenner, *ibid.*, **47**, 1625 (1964).
(32) G. R. Marshall and R. B. Merrifield in "Handbook of Biochemistry," 2nd ed, H. A. Sober, Ed., Chemical Rubber Publishing Co., Cleveland, Ohio, 1970, p C-145.
(33) J. J. Hutton, *et al.*, *Arch. Biochem. Biophys.*, **125**, 779 (1968); L. Stryer and R. P. Haugland, *Proc. Nat. Acad. Sci. U. S.*, **58**, 719 (1967); G. S. Omenn and C. B. Anfinsen, *J. Amer. Chem. Soc.*, **90**, 6571 (1968); G. Gabor, *Biopolymers*, **6**, 809 (1968).
(34) J. Scotchler, R. Lozier, and A. B. Robinson, *J. Org. Chem.*, **35**, 3151 (1970).
(35) B. F. Gisin, R. B. Merrifield, and D. C. Tosteson, *J. Amer. Chem. Soc.*, **91**, 2691 (1969).
(36) (a) Schwarz Bioresearch, Orangeburg, N. Y.; (b) Fox Chemical Co., Los Angeles, Calif.; (c) Applied Science Labs, State College, Pa.; (d) Bio-Rad Laboratories, Richmond, Calif.

340 μmol/g (by picrate determination) and amino acid analysis indicated 350 μmol of proline and 340 μmol/g of valine per gram. The resin was soaked with a 0.1 M solution of Boc-D-Pro-OH in CH$_2$Cl$_2$ (2 ml, 200 μmol) for 10 min prior to the addition of DCC (42.5 mg, 200 μmol). After 3-hr agitation at room temperature and thorough washing (DMF, CH$_2$Cl$_2$), the resin had a picrate value of 5 μmol/g, indicating over 98% coupling. After deprotection (2 × 15 min, TFA) only 110 μmol of free amine, 229 μmol of proline, and 127 μmol of valine per gram were found. (All of the analytical values have been corrected for the weight change of the peptide-resin and are thus expressed as μmoles per gram of Boc-D-Val-L-Pro-resin.) Therefore, the yield was 32% based on the picrate values of H-D-Val-L-Pro-resin and H-D-Pro-D-Val-L-Pro-resin, and 35% based on the averaged amino acid substitutions.

(b) By "Reversed" DCC Coupling. This experiment was identical with (a) except that the sequence of the addition of the coupling reagents was reversed. The resin was soaked in a 0.1 M solution of DCC in CH$_2$Cl$_2$ (2 ml, 200 μmol) for 2 min prior to the addition of Boc-D-Pro-OH (43 mg, 200 μmol). Coupling by this scheme gave the following analytical results: before coupling, free amine 340 μmol, Pro 350 μmol, and Val 340 μmol per gram; picrate value after coupling, 2 μmol per gram; tripeptide-resin after deprotection, free amine 310 μmol, Pro 685 μmol, and Val 309 μmol per gram (cor). This amounts to a yield of 91% by picrate determination and 93% by amino acid analysis.

tert-Butyloxycarbonyl-D-valyl-L-proline Benzyl Ester. In 30 ml of CH$_2$Cl$_2$, L-proline benzyl ester hydrochloride[36b] (2.41 g, 10 mmol), triethylamine (1.4 ml, 10 mmol), Boc-D-Val-OH[35] (2.17 g, 10 mmol), and DCC (2.06 g, 10 mmol) were combined and agitated overnight. After addition of 1 ml of acetic acid the mixture was diluted with 100 ml of ether and filtered. The filtrate was washed (potassium bicarbonate, citric acid, water), dried (Na$_2$SO$_4$), evaporated, and crystallized from hexane: yield, 2.8 g (69%); mp 105–105.5°; [α]^{26}D + 16.4° (c 1, benzene).

Anal. Calcd for C$_{22}$H$_{32}$N$_2$O$_5$: C, 65.32; H, 7.97; N, 6.93. Found: C, 65.37; H, 7.92; N, 6.79.

tert-Butyloxycarbonyl-D-valyl-L-proline. In 20 ml of methanol, *tert*-butyloxycarbonyl-D-valyl-L-proline benzyl ester (2.0 g, 4.95 mmol) was hydrogenated with 10% Pd on BaSO$_4$ for 18 hr at 50 psi of H$_2$. After filtration and evaporation of the solvent, the crude acid was dissolved in aqueous bicarbonate, extracted with ether, acidified with citric acid, and extracted into ether. The ether layer was separated, dried (Na$_2$SO$_4$), and evaporated to give 0.55 g (35%) of a colorless solid: mp 69–72°; [α]^{27}D − 11.6° (c 1, ethanol); cyclohexylammonium salt, mp 182–184° dec.

Anal. Calcd for C$_{21}$H$_{39}$N$_3$O$_5$: C, 60.99; H, 9.51; N, 10.16. Found: C, 60.77; H, 9.88; N, 10.07.

D-Valyl-L-proline Diketopiperazine. (a) Solution Method. Boc-D-Valyl-L-proline (500 mg, 1.6 mmol), DCC (410 mg, 2.0 mmol), and *p*-nitrophenol (450 mg, 3.2 mmol) were added to 2.5 ml of CH$_2$Cl$_2$ at 5°. The mixture was agitated for 18 hr at 5° and, after addition of 1 ml of acetic acid, for 15 min at room temperature. The dicyclohexylurea was filtered off and the filtrate was diluted with ether, washed with aqueous citric acid, H$_2$O, aqueous bicarbonate, and H$_2$O, then dried (Na$_2$SO$_4$), and evaporated to dryness. By repeatedly dissolving the residue in acetone or ether, filtering, and evaporating the solvent, more dicyclohexylurea was removed. The yellow oil (Boc-D-Val-L-Pro *p*-nitrophenyl ester) was dissolved in 10 ml of TFA and after 10 min was evaporated to dryness. The residue (TFA–H-D-Val-L-Pro *p*-nitrophenyl ester) was dissolved in 200 ml of benzene and 100 ml of pyridine was added. After 18 hr at room temperature, the mixture was evaporated to dryness, dissolved in ethanol–water (1:1, v/v), and put on a 30-ml mixed-bed ion-exchange resin column (AG 501-X8 (D)).[36d] The neutral diketopiperazine was eluted from the column with 120 ml of H$_2$O. Evaporation of the solvent gave 190 mg (60%). After crystallization from ethyl acetate–hexane the diketopiperazine had mp 148–148.5°, [α]^{30}D −86° (c 0.9, H$_2$O) [lit.[16] for *cyclo*-L-Val-D-Pro: mp 147–149°, [α]^{20}D +88° (c 1, H$_2$O)].

Anal. Calcd for C$_{10}$H$_{16}$N$_2$O$_2$: C, 61.20; H, 8.22; N, 14.28. Found: C, 60.83; H, 8.26; N, 14.21.

(b) Solid-Phase Method. Boc-D-Val-L-Pro-resin (325 mg, 110 μequiv) was deprotected and neutralized as described above. It was treated with 0.1 M acetic acid in CH$_2$Cl$_2$ (5 ml) for 60 min, filtered, and washed with CH$_2$Cl$_2$. Filtrate and washes were combined and evaporated to give 21 mg (98%) of crystalline diketopiperazine. In aqueous solution it was passed through a short (3 ml) column of mixed-bed ion-exchange resin as above and, after evaporation of the solvent, crystallized from ethyl acetate–hexane: mp 147–148°, [α]^{30}D −95° (c 0.7, H$_2$O). The ir spectrum (KBr pellet) of this preparation was found to be identical with the one of the product of (a). Amino acid content after hydrolysis (6 N HCl, 135°, 3 hr in a sealed vessel) yielded Pro, 5.35 μmol/mg, Val, 4.85 μmol/mg (calcd 5.10 for each).

Anal. Calcd for C$_{10}$H$_{16}$N$_2$O$_2$: C, 61.20; H, 8.22; N, 14.28. Found: C, 61.20; H, 8.09; N, 14.26.

Determination of Amine Content of Resins. The following procedure was used.[12] The resin was allowed to swell in CH$_2$Cl$_2$ (1 × 5 min), neutralized with 5% (v/v) diisopropylethylamine (DIA) in CH$_2$Cl$_2$ (2 × 3 min), washed with CH$_2$Cl$_2$ (3 × 2 min), treated with 0.1 M picric acid in CH$_2$Cl$_2$ (2 × 3 min), and washed with CH$_2$Cl$_2$ (5 × 2 min). The picrate was eluted with 5% DIA in CH$_2$Cl$_2$ or 0.1 M pyridine hydrochloride in CH$_2$Cl$_2$ (2 × 3 min) and CH$_2$Cl$_2$ (3 × 2 min) and, after dilution with 95% ethanol, measured spectrophotometrically. The molar extinction coefficient of DIA picrate (E_{358} 14,500) was constant in the concentration range of (1–20) × 10^{-5} M if the ethanolic measuring solution contained less than 20% CH$_2$Cl$_2$. A convenient ratio of resin to solvent was 1:20 (w/v).

Determination of Cleavage Rates with Different Reagents. Method a (Tables I and II, Figure 2 and 3). Samples of Boc-dipeptide-resins (50–100 mg, 10–60 μequiv) were deprotected and neutralized as described. The picrate value representing the amine content of the resin was determined before and after treatment for periods of 5–60 min at room temperature with each of the reagents listed. (The chemicals were of reagent grade and were obtained from commercial sources unless specified.) Treatment with each reagent and picrate determination were repeated to give at least three points on the cleavage curve. After minor corrections for loss of peptide during the determination itself, these values were plotted semilogarithmically, and the apparent first-order rate constants[37] were deduced from the graphically averaged slope of the curve.

Method b (Figures 1 and 2). In a jacketed thermostated vessel (24 ± 0.2°) a sample of Boc-D-Val-L-Pro-resin was deprotected and neutralized. It was treated first with CH$_2$Cl$_2$ for 120 min and second with 0.05 M acetic acid in CH$_2$Cl$_2$ for 240 min. During both periods of time the resin was filtered periodically, washed with CH$_2$Cl$_2$, and resuspended in a fresh batch of reagent. In each case, filtrate and washes were combined, evaporated to dryness, and subjected to quantitative determination of the D-Val-L-Pro diketopiperazine by glc according to Mauger.[14] The analyses were performed on an F & M Model 402 gas chromatograph equipped with a flame detector and a U-shaped column (6 ft × 3.5 mm) containing 3% EGSP-Z on Gas Chrom Q, 100–200 mesh.[36] The flow rates of the gases were kept at 40 (H$_2$), 100 (He), and 250 cm^3/min (air), the temperature at 210°. The samples were dissolved in CH$_2$Cl$_2$ (c (2–5) × 10^{-3} M) which was 2.0 × 10^{-3} M in *p*-phenylphenol as internal standard and injected (without derivatization) in volumes of 2–5 μl. Under these conditions the diketopiperazine had an average retention time of 7.6 min with only slight tailing, allowing determination of the peak area by the height × half-width method. The amino acid content of aliquots of three samples after hydrolysis (6 N HCl, 130–140°, 4 hr in a sealed vessel) was determined and was found to correlate satisfactorily with the quantity of diketopiperazine found by glc (Table III).

Table III. Determination of the Concentration of Diketopiperazine by Gas–Liquid Chromatography (glc) and by Amino Acid Analysis after Hydrolysis (Val, Pro) in Three Different Samples (I, II, III)[a]

	I	II	III
Val	582	262	140
Pro	556	270	149
Glc	570	280	141

[a] nmole/ml.

Acknowledgments. The authors are grateful to Mr. Arun Dhundale for his technical assistance. This work was supported, in part, by U. S. Public Health Service Grant AM 1260, the Hoffman-La Roche Foundation, and NIH Grant HE 12 157.

(37) W. P. Jencks, "Catalysis in Chemistry and Enzymology," McGraw-Hill, New York, N. Y., 1969.

Copyright © 1973 by the American Chemical Society

Reprinted from *J. Amer. Chem. Soc.*, **95**(18), 6097–6108 (1973)

Synthesis of a Polypeptide with Lysozyme Activity

J. J. Sharp, [1] **A. B. Robinson, and M. D. Kamen**

Contribution from the Department of Chemistry, University of California, San Diego, La Jolla, California 92037. Received January 27, 1973

Abstract: The results of two Merrifield solid-phase syntheses of hen egg-white lysozyme are reported. The first synthesis essentially duplicated those procedures used by Gütte and Merrifield for the synthesis of ribonuclease A. 2-Mercaptoethanol was incorporated into the Boc removal solution (TFA–CH_2Cl_2–2-mercaptoethanol–anisole) for tryptophan protection but this reagent proved unsatisfactory. Further studies led to the choice of 1,2-ethanedithiol for tryptophan protection. Much of the synthetic product exhibited a molecular weight less than that of native lysozyme (as evidenced by Sephadex G-100 chromatography). The S-benzyl protecting group of cysteine proved difficult to remove with anhydrous HF, but once conditions were found for its removal, the resultant crude synthetic product showed ~0.05% specific enzymatic activity of the native molecule. Experiments were carried out which indicated poor coupling of amino acids resulted in the low molecular weight peptide. In addition, 93% of the initially esterified material had been removed during the course of the synthesis. Consideration of which chains were being lost led to the decision to use a milder Boc removal solution which removed less peptide from the resin (4 N HCl–dioxane–anisole–ethanedithiol). Examination of ε-NH_2 deprotection of lysine also led to the choice of this HCl–dioxane system for Boc removal. The synthetic procedures for a second synthesis were altered in accordance with these and other results. The changes resulted in a product in the correct molecular weight range (elution position on Sephadex G-75) as well as a nonreducible (by dithiothreitol or 2-mercaptoethanol) high molecular weight fraction. Only the former fraction had enzymatic activity. The specific activity for the crude product this was 0.5–1% that of native lysozyme. Following chromatography on chitin, the specific activity was 2–3% that of native lysozyme (or 9–25% that of native lysozyme subjected to the same HF conditions and purification as the synthetic product). Three HF cleavage conditions were compared and the effect of one of these conditions on native, reduced lysozyme was examined. The results of these two lysozyme syntheses are discussed.

The methodology of Merrifield solid phase peptide synthesis has gained wide acceptance and given rise to expectations that facile syntheses of large biologically active polypeptides might be in prospect.[2,3] Of the syntheses attempted, the most successful has been that of ribonuclease A by Gütte and Merrifield.[4]

However, at present there is no universal methodology which will guarantee such success.[5-7] Certain difficulties associated with a synthesis (poor coupling, removal of side chain blocking groups, etc.) assume increasing magnitude as the peptide length increases. The resulting problems of product purity, isolation, and yield also increase with increasing length of the desired product. Thus it may be expected that considerable developmental research will be required before the potential of solid phase peptide synthesis is realized.

We have attempted the total synthesis of hen egg-white lysozyme and have found it necessary to concentrate on definition of some of the problems involved in the synthesis of a large polypeptide. We report herewith details of the synthetic procedures and difficulties encountered in the hope that this information may prove useful for future syntheses of large molecules.

In this paper, we present the results of two attempts to synthesize lysozyme. The methodology employed in the first synthesis essentially duplicated that used by Gütte and Merrifield in their synthesis of ribonuclease A. Our experiences resulted in investigations of some of the solid-phase chemistry, results of which also are presented. These results led to the conclusion that the incorporation of several synthetic alterations would improve the possibility for a successful lysozyme synthesis. The improved strategy of this second attempt to synthesize hen egg-white lysozyme centered on improvements in the coupling procedures. Attempts were made to drive each coupling reaction to completion, and if these proved inadequate, to acetylate the remaining chains, thereby eliminating their participation in further reactions. There was no added consideration given to either the Boc removal or the neutralization steps.

A preliminary report of this work has appeared elsewhere.[8]

Experimental Section

Synthetic Conditions. First Synthesis. Crosslinked polystyrene beads (1%) (200–400 mesh) obtained from Bio-Rad (SX-1 Bio-Beads) were washed, chloromethylated, and esterified by the usual procedures.[9] The chloromethylated resin contained 0.7 mequiv of chloride/gram of resin. Resin-leucine (1.5 g) was used in the synfthesis. The esterification of Boc-leucine was 0.41 mmol/gram o-resin.

The *tert*-butyloxycarbonyl (Boc) amino acid derivatives used were the following: Ala, Arg (NO_2), Asp (β-OBzl), Asn-ONp, Cys (S-Bzl), Glu (γ-OBzl), Gln-ONp, Gly, His, Ile, Leu, Lys (ε-carbobenzoxy (Z)), Met, Phe, Pro, Ser (Bzl), Thr (Bzl), Trp (Bzl), and Val. These derivatives were either synthesized in our laboratory or purchased from Schwarz BioResearch Co. The purity of each derivative was checked by thin-layer chromatography and melting point determination.

Amino acid derivative (1.5 mmol) was used for each coupling

---

(1) Address correspondence to this author at the Department of Biochemistry, Dartmouth Medical School, Hanover, N. H. 03755. National Institutes of Health Research Fellow (1 F01 GM-38214-01A1) 1969–1971.
(2) R. B. Merrifield, *Fed. Proc., Fed. Amer. Soc. Exp. Biol.*, **21**, 412 (1962).
(3) R. B. Merrifield, *J. Amer. Chem. Soc.*, **85**, 2149 (1963).
(4) B. Gütte and R. B. Merrifield, *J. Biol. Chem.*, **246**, 1922 (1971).
(5) S. Sano and M. Kurihara, *Z. Physiol.*, **30**, 1183 (1969).
(6) C. H. Li and D. Yamashiro, *J. Amer. Chem. Soc.*, **92**, 7608 (1970).
(7) E. Bayer, G. Jung, and H. Hagenmaier, *Tetrahedron*, **24**, 4853 (1968).

(8) J. J. Sharp, A. B. Robinson, and M. D. Kamen, *Fed. Proc., Fed. Amer. Soc. Exp. Biol.*, **30**, Abstract 1287 (1971).
(9) J. M. Stewart and J. D. Young, "Solid Phase Peptide Synthesis," W. H. Freeman, San Francisco, Calif., 1968.

whether mediated by dicyclohexylcarbodiimide (DCC) or a *p*-nitrophenyl ester (ONp). Single 5-hr couplings were used in all cases except for later couplings of the ONp derivatives.

Boc removal was accomplished with a mixture of trifluoroacetic acid (TFA)–CH_2Cl_2–2-mercaptoethanol–anisole, 45:50:5:2 (this and future solution compositions are on a volume basis). This solution will be referred to as "TFA." Anisole was added to prevent possible alkylations of labile groups (*i.e.*, unprotected Met) by *tert*-butyl cations present during Boc removal. The TFA was injected into the reaction vessel twice: first it was rocked for 1 min and then for 30 min.

Neutralization was accomplished with a mixture of triethylamine (Et_3N)–$CHCl_3$, 12.5:87.5. This solution was injected into the reaction vessel twice: the first was rocked for 1 min and the second for 10 min.

Synthetic Conditions. Second Synthesis. Crosslinked polystyrene beads (1 %) (200–400 mesh) were obtained from Bio-Rad (SX-1 Bio-Beads). Following chloromethylation there was 0.9 mequiv of chloride/gram of resin. The esterification of Boc-leucine was 0.4 mmol/gram of resin; 4.0 grams was used for the synthesis, an amount anticipated to be sufficient to meet the requirement that resin-peptide would be removed for monitoring of coupling.

The Boc amino acid derivatives were the same as the first synthesis with the following exceptions: Cys (*S*-(*p*-methoxy)Bzl) and His (*im*-Tos). The derivatives were either synthesized in our laboratory or purchased from Fox Chemical Co. The purity of the derivatives was checked as before. Amino acid derivative (4.0 mmol) was used for each coupling. The DCC was injected in two 2-mmol increments. Each DCC injection was followed by a 3-hr reaction time. There were three coupling solvents used throughout the synthesis: CH_2Cl_2, dimethylformamide (DMF), and DMF saturated with recrystallized urea (DMF–urea). In all cases, the amino acid was dissolved in one of the three above solvents, the DCC solvent always being CH_2Cl_2. Recoupling times (when necessary) were variable and were generally between 1 and 6 hr. The *p*-nitrophenyl ester couplings (ONp) of glutamine and asparagine were between 16 and 24 hr in length and 5 mmol of the derivative was used for each coupling. Boc[^{14}C]glycine was purchased from Schwarz Bio-Research.

Acetylations were performed using 6 mmol of glacial acetic acid with two equal injections of 3 mmol of DCC. The reaction times were from 30 min to 2 hr. The solvent system for acetylation varied, being either CH_2Cl_2, DMF, or DMF–urea. When acetylimidazole was used for acetylation, 5 mmol of this compound in CH_2Cl_2 was injected directly into the reaction vessel. The reaction time varied between 1 and 3 hr.

Boc removal was accomplished with 4 *N* HCl–dioxane–anisole–ethanedithiol, 98:1:1 (v/v). This solution will be referred to as "HCl–dioxane." The HCl–dioxane injection procedure and reaction times were the same as for TFA in the first synthesis. Anisole was redistilled before use. Dioxane used for the washes prior to, and succeeding, the Boc removal reaction, as well as for the preparation of the HCl–dioxane, was purified by the method of Fieser and Fieser[10] and was stored under 1.2 atm of argon.[11]

Neutralization was accomplished as in the first synthesis. The triethylamine (Et_3N) used for neutralization was purified by the method of Fieser and Fieser.[12]

The DMF used in this synthesis was "Sequenal Grade" DMF purchased from Pierce Chemical Co. It was used without further purification. Saturated urea (recrystallized) in DMF was prepared weekly.

Sequence. The amino acid sequence used was that of Canfield[13] and Jollès.[14]

Automation. The design of the automated machine, as well as that of the reaction vessel (A. B. Robinson and P. Yeager), is described elsewhere.[11,15]

Hydrolysis and Amino Acid Analysis. Resin-peptide samples were hydrolyzed at 130° for 2 hr in 50:50 (v/v) concentrated hydrochloric acid–propionic acid in anaerobic, vacuum-sealed tubes.[16] This procedure removes all blocking groups except the benzyl group of cysteine. Before hydrolysis the resin-peptide sample was thoroughly washed with TFA–CH_2Cl_2, 50:50 (v/v), CH_2Cl_2, and ethanol to remove any ion-exchange adsorbed amino acids from the resin. All other peptide and protein samples were hydrolyzed at 110° for 20–24 hr in constant-boiling HCl in anaerobic, vacuum-sealed tubes.

Amino acid analyses were performed on a modified Beckman Spinco amino acid analyzer. Benzylcysteine was determined on the short column (elution at 20 min in the system used). Tryptophan was determined by titration with *N*-bromosuccinimide.[17]

Karl–Fisher titrations for water were determined by the method of Fieser and Fieser.[18]

Elemental analysis was performed by Huffman Laboratories in Wheaton, Colo.

HF Conditions. The specific HF conditions are given in the following text. The construction of the HF line is described elsewhere.[9,15]

Following treatment with anhydrous HF the peptide was taken up in TFA–ethanedithiol, 98:2, and filtered. This solution was cooled to 0° and 4° ether was added to precipitate the product. The precipitate was washed twice with 4° ether and dried under vacuum. Tryptophan destruction of native lysozyme in the TFA–ethanedithiol had previously been followed spectrally (decrease in A_{280} and increase in A_{335}). There was no evidence of tryptophan destruction after 5 hr at room temperature in this solvent. TFA–ethanedithiol, 98:2, was routinely used to recover the cleaved, deblocked peptide from the dried HF reaction mixture.

Reduction and Reoxidation of Disulfide Bonds. Reduction and reoxidation of disulfide bonds was carried out by one of three methods. In the first, the method of Epstein and Goldberger[19] was followed. In the second procedure, the protein was reduced in 6 *M* guanidine hydrochloride (Matheson), 0.03 *M* dithiothreitol (Pierce),[20] pH 8.0, for 1–3 hr at 37°. After reduction, the solution was passed through a Sephadex G-10 column equilibrated with 0.1 *M* acetic acid. Reoxidation was effected by diluting the protein–acetic acid solution 30- to 50-fold with 0.025 *M* Tris-HCl, 10^{-6} *M* $CuCl_2$, pH 8.0 buffer, and stirring at room temperature for 24 hr. Isolation of the oxidized protein was accomplished by adsorption onto a 1 × 4 cm carboxymethylcellulose (CMC) column equilibrated with 0.025 *M* Tris-HCl, pH 8.0, and elution from the column with 0.5 *M* ammonium bicarbonate. Thirdly, the mixed reoxidation procedure of Saxena and Wetlaufer[21] was used on occasion. All three procedures gave similar results with native lysozyme.

Enzymatic Assay for Lysozyme Activity. The assay for lysozyme activity was performed according to the method of Epstein and Goldberger,[19] which involved adding 1–500 μl of protein solution to 3.0 ml of a *Micrococcus lysodeikticus* cell suspension and monitoring the absorption decrease at 450 nm. The assay was carried out at 30° in a temperature controlled cuvette holder. A Gilford spectrophotometer (Model 2000) was used to follow the 450-nm absorbance decrease.

The specific activity of a protein solution was defined either as the 450-nm absorbance decrease per second per mg of protein (A_{450}/sec per mg) or the 450-nm absorbance decrease per second per A_{280} per milliliter of the solution assayed (A_{450}/sec per A_{280}/ml). The per cent lysozyme activity of a particular synthetic sample was defined as the per cent specific activity the sample exhibited when compared with the specific activity of native lysozyme (Miles Laboratories, 6X crystallized lysozyme, Lot no. 902) measured under the same conditions, *i.e.*, when the same amount of native lysozyme in the same volume was simultaneously assayed.

Determination of Protein Concentration. Protein concentration was determined by one of two methods. In the first, the concentration was determined by assuming an $E_{1\,cm}^{1\%}$ of 26.9 at 280 nm[22] for both native and reduced material. In the second, a portion of the sample was removed, dried, and hydrolyzed, and an amino acid analysis was performed. The concentration was determined from the amino acid recovery.

Mixed Disulfide of Lysozyme. The procedure of Bradshaw, *et al.*[23] (with minor modification), was used to form the cysteine-

---

(10) L. Fieser and M. Fieser, "Reagents for Organic Synthesis," Wiley, New York, N. Y., 1967, p 333.
(11) J. J. Sharp, Ph.D. Thesis, University of California, 1971.
(12) Reference 10, p 1198.
(13) R. E. Canfield and A. K. Liu, *J. Biol. Chem.*, **240**, 1997 (1965).
(14) J. Jauregui-Adell, J. Jollès, and P. Jollès, *Biochim. Biophys. Acta*, **107**, 97 (1965).
(15) A. B. Robinson, Ph.D. Thesis, University of California, 1967.
(16) J. Scotchler, R. Lozier, and A. B. Robinson, *J. Org. Chem.*, **35**, 3151 (1970).

(17) A. Patchoenik, W. B. Lawson, and B. Witkop, *J. Amer. Chem. Soc.*, **80**, 4747 (1958).
(18) Reference 10, p 528.
(19) C. J. Epstein and R. F. Goldberger, *J. Biol. Chem.*, **238**, 1380 (1963).
(20) W. W. Cleland, *Biochemistry*, **3**, 480 (1964).
(21) V. P. Saxena and D. B. Wetlaufer, *Biochemistry*, **9**, 5015 (1970).
(22) K. Imai, T. Takagi, and T. Isemura, *J. Biochem. (Toyko)*, **53**, 1 (1963).
(23) R. A. Bradshaw, L. Kanarek, and R. L. Hill, *J. Biol. Chem.*, **242**, 3789 (1967).

Synthesis of a Polypeptide with Lysozyme Activity

Table I. Amino Acid Ratios[a] of Selected Resin-Peptide Samples

	No. 1	No. 2	No. 3
Asp	1.0 (1)	6.7 (8)	17.8 (21)
Thr	0.8 (1)	1.3 (2)	4.6 (7)
Ser	0 (0)	3.0 (5)	6.5 (10)
Glu	1.0 (1)	1.5 (1)	5.6 (5)
Pro	0 (0)	Trace (1)	1.1 (2)
Gly	2.3 (2)	5.1 (4)	16.1 (12)
Ala	1.0 (1)	5.9 (6)	14.7 (12)
Cys		Trace	
Val	0.8 (1)	4.0 (4)	8.0 (6)
Met	0 (0)	0.4 (1)	1.2 (2)
Ile	0.8 (1)	3.1 (4)	5.9 (6)
Leu	1.3 (1)	2.9 (3)	9.1 (8)
Tyr	0 (0)	0 (0)	2.6 (3)
Phe	0 (0)	0 (0)	2.8 (3)
His	0 (0)	0 (0)	0.9 (1)
Lys	0 (0)	3.3 (3)	9.8 (6)
Arg	b (2)	4.6 (4)	12.3 (11)
Bzl-Cys	b (1)	b (4)	9.2 (8)
Trp	b (1)	b (3)	b (6)
Total residues	13	53	129

[a] The calculations are based on the theoretical length of the peptide. The theoretical number of each residue appears in parentheses beside the calculated value. [b] No determination made.

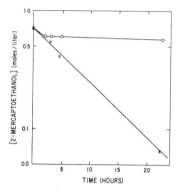

Figure 1. Kinetics of 2-mercaptoethanol decrease (water production) in TFA–CH_2Cl_2–2-mercaptoethanol–anisole, 45:50:5:2, at room temperature (×—×) and 0° (O—O). (The experimental points were determined by subtracting the concentration of water from the initial concentration of 2-mercaptoethanol; see text.)

mixed disulfide of lysozyme (lyso-S-Cys). Performic acid oxidation[24] and amino acid analysis of lyso-S-Cys and native lysozyme showed 14 and 7 residues, respectively, of cysteic acid. The difference spectrum (vs. native lysozyme) of the lyso-S-Cys agreed well with the published spectrum.[23]

Cellulose Acetate Electrophoresis. Barbitone buffer (pH 8.6) was used for the electrophoresis.[25] Cellulose acetate strips (1 × 12 in.) were employed. Electrophoresis time was 2 hr at 300 V. The strips were first stained in Ponceau S and then in Nigrosin.

HBr–TFA Cleavage. HBr–TFA cleavages were carried out in a 40-ml jacketed solid phase peptide synthesis reaction vessel.[8] The top of the vessel included a drying tube filled with Drierite. Gaseous HBr was dried by passage through P_2O_5 before introduction into the bottom of the reaction vessel. Constant temperature was maintained by pumping 0° or room temperature water through the jacket on the vessel.

Carboxymethylation was carried out by standard procedures.[26]

Chitin Column Chromatography. Chitin columns were run according to the procedures of Cherkasov and Kravchenko.[27,28] Chitin was purchased from Matheson Chemical Co. The material was screened and the 60–150 mesh particles were used.

Enzymatic digestion was performed by the procedure of Pisano, et al.[29] (with minor modification).

Synthetic Results. First Synthesis

The total synthesis time was 10 weeks. Amino acid analysis of the 24-residue resin-peptide showed that the first two asparagines were incompletely coupled (~60% complete). Hence, reaction times for the remaining ONp couplings were increased from 5 to 24 hr. A machine error was responsible for incomplete coupling of residue 51 (proline).[30]

After 2 weeks a precipitate formed in the Boc removal solution (TFA–CH_2Cl_2–anisole–2-mercaptoethanol). Elemental analysis of this precipitate was consistent with the composition ($-CH_2CH_2S-)_n$ which suggested that a reaction of the following type was occurring: $(HOCH_2CH_2SH)_n \xrightarrow{H^+} (-CH_2CH_2S-)_n + (H_2O)_{n-1}$. To test this possibility, water production in the Boc removal solution was followed by Karl-Fisher titrations. The results are shown in Figure 1. The apparent first-order rate constant for this reaction at room temperature is 0.063 hr^{-1}.[31]

It is obvious that water production in this solution is quite rapid and that water appears long before a precipitate becomes noticeable. To eliminate the potentially harmful conditions of water under acidic conditions, the reagents (TFA, CH_2Cl_2, anisole, 2-mercaptoethanol) were mixed at room temperature and then placed at 0°. Water production was again monitored and these results are shown in Figure 1. After an initial production of water, which took place while the solution was cooling to 0°, the rate was considerably retarded (first-order rate constant ≃0.003

---

(24) C. H. W. Hirs, *J. Biol. Chem.*, **219**, 611 (1956).
(25) C. B. Laurell, S. Laurell, and N. Skoog, *Clin. Chem.*, **2**, 59 (1956).
(26) F. Gurd in "Methods in Enzymology," Vol. 25, C. H. Hirs and S. N. Timasheff, Ed., Academic Press, New York, N. Y., 1972, p 424.
(27) I. A. Cherkasov and N. A. Kravchenko, *Biokhimiya*, **33**, 761 (1968).
(28) I. A. Cherkasov and N. A. Kravchenko, *Biokhimiya*, **34**, 1089 (1969).
(29) J. J. Pisano, J. S. Finlayson, and M. P. Peyton, *Biochemistry*, **8**, 871 (1969).
(30) For synthesis purposes, amino acids are numbered beginning with the C-terminal amino acid instead of the normal N-terminal amino acid.

(31) The first-order rate equation and considerations of carbonium ion stability are consistent with the following mechanism involving neighboring group participation:

$$HSCH_2CH_2OH \xrightleftharpoons{H^+} HSCH_2CH_2\overset{+}{O}H_2$$

$$\underset{HS}{\overset{CH_2-CH_2}{\diagdown\diagup}}\underset{\overset{+}{O}H_2}{} \xrightleftharpoons{slow} \underset{S}{\overset{CH_2-CH_2}{\diagdown\diagup}}\underset{H^+}{} + H_2O$$

$$\underset{\underset{H^+}{S}}{\overset{CH_2-CH_2}{\diagdown\diagup}} + HSCH_2CH_2OH \xrightleftharpoons{fast} CH_2CH_2\overset{+}{S}CH_2CH_2OH \\ | | \\ SH$$

$$\underset{SH}{CH_2CH_2SCH_2CH_2OH} \xrightleftharpoons{H^+} \underset{SH}{CH_2CH_2SCH_2CH_2\overset{+}{O}H_2}$$

$$\underset{SH}{CH_2CH_2SCH_2CH_2\overset{+}{O}H_2} \xrightleftharpoons{} \underset{S^+}{\overset{CH_2-CH_2}{\diagdown\diagup}} + H_2O \\ | \\ HSCH_2CH_2$$

etc.

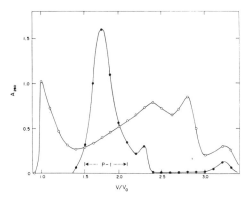

Figure 2. The A_{280} profile of native lysozyme (●—●) and the synthetic product (○—○) chromatographed on similar 2 × 60 cm Sephadex G-100 columns. The column was equilibrated with 8 M urea–0.1 M acetic acid. The lysozyme was reduced with 2-mercaptoethanol in 8 M urea before application. P-1 is the fraction of the product exhibiting the correct elution position. The high molecular weight material disappeared if the product peptide was reduced before application but the other characteristics of the profile remained constant. (V is the elution volume and V_0 the void volume of the column.)

hr^{-1}). On the basis of this result these reagents were stored separately at $-10°$ for the remainder of the synthesis. They were mixed at this temperature and then stored at 0°. A solution was kept no longer than 4 or 5 days. Immediately prior to Boc removal the correct volume was brought to room temperature and injected into the reaction vessel.

Amino acid analyses of the resin-peptide were performed throughout the synthesis and results of three of these appear in Table I. It was difficult to draw any conclusions from these data. However, a few points could be made. The amino acid ratios became poorer as the synthesis progressed. Because of the heterogeneity of the product and the method used to calculate these data, molar ratios greater than theoretical undoubtedly reflected, in part, the ease with which an amino acid coupled rather than multiple additions during the coupling reaction. Conversely, ratios less than theoretical could be ascribed to amino acids which coupled poorly. Of those which survive the hydrolysis conditions, the low aspartic acid value probably indicated the poor coupling of ONp esters (13 of the 21 "aspartic acids" are asparagine) as noted elsewhere.[32,33]

Upon completion of the synthesis the dried resin-product weighed 2.75 g, and amino acid analysis showed that the esterification of the peptide was 0.03 mmol/gram of resin, assuming a molecular weight of 19,000 daltons for the blocked peptide and an average molecular weight of 147 daltons for the blocked amino acids. On this basis, 93% of the initially esterified material had been removed during the course of the synthesis. The blocked peptide constituted 50% of the total weight of the resin-peptide, or 1.38 g.

The peptide was deblocked and removed from the resin by reacting 200 mg of the resin-product in 5 ml of anhydrous HF and 0.5 ml of anisole for 90 min at 0°. A large portion of the product (~80%) exhibited a molecular weight less than that of native lysozyme, as evidenced by Sephadex G-100 chromatography in 8 M urea–0.1 M acetic acid (Figure 2). The recovery of peptide with the correct molecular weight was always about 20% of the material recovered from the column and there was never a peak corresponding to the reduced native molecule. Following isolation of product with the correct molecular weight (Figure 2) and reoxidation, the peptide showed no enzymatic activity, based on the whole cell assay method (~0.05% native specific activity could have been observed). Amino acid analysis showed that 2.7 residues of benzylcysteine were present in the product and only a trace of cystine.

The same HF reaction conditions were repeated at room temperature in an attempt to completely deblock the benzylcysteine. The peptide product of the correct molecular weight again exhibited no enzymatic activity. Amino acid analysis indicated the level of benzylcysteine had been reduced to 0.5 residue but there was still only a trace of cystine present.

In an attempt to clarify the cystine recovery problem 0.1 mmol of Boc-Cys (S-Bzl) was treated with HF using the room temperature reaction conditions. After recovery from HF the material was subjected to normal HCl hydrolysis and amino acid analysis. The major product (~85% of the material recovered from the column assuming a normal proline analyzer constant for this material) was present as a peak and shoulder eluting at the normal proline position. The A_{440} was higher than the A_{560} value, as with proline, indicating it was possibly an imine. On reexamination of previous analyses this peak and shoulder were present and could, therefore, be used as an indication of cystine destruction during the HF reaction. About 5% of the total material recovered from the analysis was cystine and 10% was benzylcysteine.

In an attempt to improve the cystine recovery, 0.1 g of resin-product was reacted in 10 ml of anhydrous HF and 2 ml of anisole for 90 min at room temperature. This represented a fivefold dilution of peptide in HF and a twofold increase in anisole concentration. After collection of the correct molecular weight fraction from Sephadex G-100 (8 M urea–0.1 M acetic acid) chromatography, reduction, and reoxidation, the peptide-product had ~0.05–0.1% native lysozyme activity. Amino acid analysis showed 4.6 residues of cystine and 0.5 residue of benzylcysteine present in the active fraction. The cysteine degradation product was also considerably reduced. Peptide was removed from the resin several times by this method. The yield of material with 0.05–0.1% native activity was typically about 0.1%.[34] The low molecular weight fractions never exhibited any activity.

No attempt was made to further purify the product. The known errors that occurred during the chain assembly did not account for the poor amino acid ratios of the product or for the large percentage of low molecular weight peptide. Several lines of investigation were undertaken in an attempt to better understand the results of this synthesis.

---

(32) M. Bodansky and R. J. Bath, *Chem. Commun.*, 1259 (1969).
(33) S. Karlsson, G. Lindeberg, and U. Ragnarsson, *Acta Chem. Scand.*, 24, 337 (1970).

(34) The basis of the per cent yield calculation is the following: 100% yield would be a 129 amino acid peptide esterified to the resin (0.41 mmol/gram of resin). The yield prior to HF treatment was therefore 7%.

Results of Investigations into the Synthetic Procedure. The Coupling Reaction

Incomplete coupling of amino acids is the most obvious explanation for the large percentage of low molecular weight peptide and poor amino acid ratios found during the lysozyme synthesis. To investigate this possibility the coupling efficiency was examined by employing the ninhydrin test of Kaiser, et al.[35] This test indicated less than 90% coupling in several instances when coupling procedures of the lysozyme synthesis were used. Subsequent couplings could drive the reaction to completion.

The coupling reaction was also investigated under conditions in which a molar excess of Boc amino acid to DCC was used and under conditions which will be referred to as "incremental DCC additions." If an amino acid were to be coupled incrementally, the DCC was added to the reaction vessel in 2 or 3 equal increments, the final concentration being equimolar with the Boc amino acid. The reason for provision of a molar excess of Boc amino acid and for incremental DCC addition was to promote *in situ* symmetrical anhydride formation.[36]

The rates of O-acyl to N-acyl isomerization in DCC mediated couplings are not well known, and coupling rates most probably decrease as the resin-peptide increases in length. The coupling efficiency could therefore come under greater influence of O-acyl to N-acyl isomerization rates as the chain length increases. By the *in situ* generation of the symmetrical anhydride of the Boc amino acid the influence of isomerization is reduced. Since the soluble product of the solid phase aminolysis of a symmetrical Boc amino acid anhydride is the Boc amino acid, DCC could be added later to utilize this product. This was the basis for incremental DCC addition.

To determine the usefulness of a molar amino acid to DCC excess and of incremental DCC addition, several resin-peptide and Boc amino acid combinations that had previously been difficult to couple were examined. The results of one such experiment are given in Table II. There was a definite advantage when using a molar excess of Boc-Leu to DCC in these experiments. The result was quite reproducible for this and other resin-peptide-Boc amino acid combinations. In several instances (*i.e.*, Boc-Gly and Boc-Phe), a twofold excess of Boc amino acid–DCC, with and without incremental DCC injection, gave the same result as an equimolar coupling. However, an amino acid–DCC excess was never found to couple less completely than an equimolar coupling.

The effect of solvent on the O- to N-acyl migration is well known and the incremental addition of DCC might prove useful in increasing coupling efficiency when DCC couplings are carried out in DMF.[9]

Removal of Peptide from the Resin. Several investigations, including ours, have shown that peptide is removed from the resin throughout the course of a synthesis.[4,7,37] Resin-Leu-Ala-Val-[^{14}C]Gly was carried through several synthetic cycles and the washes were collected and examined for radioactivity. As expected, peptide was removed from the resin during the acidic Boc removal reaction. The amino acid ratios were unchanged even though there was an accompanying 40% reduction in esterification, indicating that intact peptide had been removed.

It is known that the resin sites are not homogeneous with respect to coupling rates.[38–40] This heterogeneity is presumably owing to the local environment created by the resin matrix.[37] Since there are heterogeneous coupling rates, one might also expect heterogeneity in rate of removal of peptide from the resin. It may be that the peptides at resin sites which participate faster in the synthetic reactions are also the peptides most easily removed from the resin sites. The resin would, therefore, gradually become enriched in poor coupling sites. This difficulty would be of greater importance as the chain length increased and would tend to favor the production of a heterogeneous low molecular weight product if accompanied by incomplete coupling. The possibility of heterogeneous rates of removal from the resin made us consider it advisable to reduce this source of loss by minimizing rates of removal.

The peptide removal rates of the acidic Boc removal solvents currently in use were compared. (Because of previous Boc removal difficulties, 1 N HCl–acetic acid was not included.[9,38]) The solvent systems compared were: TFA–CH_2Cl_2 (50:50); TFA–CH_2Cl_2–anisole–mercaptoethanol, 45:50:2:5; TFA–CH_2Cl_2–anisole–ethanedithiol, 45:50:2:5; and 4 N HCl–dioxane–anisole–ethanedithiol, 98:1:1. Resin-Leu-Ala-Val[^{14}C]Gly was reacted successively with the four solvents for 30-min periods. Following each reaction all radioactivity was washed from the resin with either CH_2Cl_2 or dioxane. The per cent of peptide removed from the resin was calculated from the cpm recovered and specific activity of the peptide. HCl–dioxane removed less peptide per 30-min reaction time (1.1% of the total peptide was removed in 30 min) (TFA–CH_2Cl_2 = 2.9%; TFA–CH_2Cl_2–anisole–mercaptoethanol = 1.6%; TFA–CH_2Cl_2–anisole–ethanedithiol = 2.1%) than any of the other solvents. These

Table II. The Coupling of Boc-Leu to Resin-Asp-Phe-Ala under Various Conditions

	Coupling conditions		Amino acid analysis[b]			
Expt	Mole ratio of Boc-Leu/DCC	Coupling time, hr	φ-Asp	-Phe	-Ala	-Leu
1	1:1	3	1.0	1.0	0.9	0.4
2	1.3:1	3	1.0	1.0	0.9	0.5
3	1.6:1	3	1.0	1.0	0.9	0.6
4	2:1	3	1.0	1.0	0.9	0.7
5	1:1	5	1.0	1.0	0.9	0.5
6	Incremental[a]	5	1.0	1.0	0.9	0.8

[a] The initial ratio was 1.6:1. After 3 hr DCC was added to the vessel to bring the ratio to 1:1, and coupling continued for the remaining 2 hr. [b] The amino acid analysis was performed after hydrolysis of the resin-peptide as described in the Experimental Section.

(35) E. Kaiser, R. L. Colescott, C. D. Bossinger, and D. I. Cook, *Ann. Biochem. Exp. Med.*, **34**, 595 (1970).
(36) H. Smith, J. G. Moffatt, and H. G. Khorana, *J. Amer. Chem. Soc.*, **80**, 6204 (1958).
(37) R. B. Merrifield, *Advan. Enzymol. Relat. Areas Mol. Biol.*, **32**, 221 (1969).
(38) R. B. Merrifield, *Recent Progr. Horm. Res.*, **23**, 451 (1967).
(39) J. Rudinger and U. Gut, *Peptides, Proc. Eur. Peptide Symp., 8th, 1966*, 89 (1967).
(40) M. Monahan, A. B. Robinson, and M. D. Kamen, unpublished results.

data could be used to calculate the yield of peptide following a 129 amino acid synthesis. For TFA–CH_2Cl_2–anisole–mercaptoethanol the theoretical yield would be 9.3% (found 7%). For 4 N HCl–dioxane–anisole–mercaptoethanol the theoretical yield would be 18.1% (found 30%).

The peptide removal rate might be reduced further by shortening the reaction time in the Boc removal solvent. However, because the kinetics of Boc removal in solid-phase synthesis are not well known (especially for large peptides), the 30-min reaction time was considered minimal for any of the solvents used.

Tryptophan Protection during Boc Removal. 2-Mercaptoethanol had previously been used for tryptophan protection during Boc removal,[41] but not when TFA was the deblocking agent. The polymer and water formation in TFA necessitated either the use of another tryptophan protecting agent and/or the use of an acid other than TFA. Even though no visible polymer formation had been observed when 2-mercaptoethanol was used in other acidic solvents, it was considered desirable to find another reagent for tryptophan protection. 1-Mercaptobutane and 1,2-ethanedithiol were tested for their ability to protect tryptophan under acidic conditions.

The acid catalyzed resin-tryptophan degradation product was used as an indicator of tryptophan destruction (the resin becomes dark purple). Resin-Trp (0.4 mmol/gram of resin) (0.2 g) and resin-Val-Asp-Met-Asn-Asn-Ile-Ala-Trp-Glu (0.4 mmol/gram of resin) (0.2 g) were stored covered at room temperature in 2 ml of the following solutions: TFA–CH_2Cl_2–anisole–ethanedithiol, 43:50:2:5; TFA–CH_2Cl_2–anisole–mercaptobutane, 43:50:2:5; TFA–CH_2Cl_2–anisole–mercaptoethanol, 43:50:2:5 (this solution contained the polymer); TFA–CH_2Cl_2, 50:50. Both resin samples turned dark purple after 8 hr in TFA–CH_2Cl_2. After 5 days only the samples containing ethanedithiol were uncolored. The samples containing mercaptoethanol were a moderate purple and those with mercaptobutane were dark purple.

1,2-Ethanedithiol was also tested for tryptophan protection in 4 N HCl–dioxane–anisole, 99:1, using the same resin-peptides; 0, 1, 2, 3, 4, and 5% ethanedithiol was examined. After 2 weeks at room temperature none of the samples containing 1,2-ethanedithiol were colored while those without this reagent were dark purple.

Lysine ε-Amino Protection. There have been reports that Z protection of the ε-amino group of lysine is inadequate[37,38] and that trifluoroacetyl (TFA) ε-amino protection of lysine is better.[42] However, removal of the TFA group requires strong basic conditions (1 M piperidine). We found that the cysteine mixed disulfide of native lysozyme, when treated with 1 M piperidine (5 mg in 0.5 ml), had lost considerable enzymatic activity after regeneration of the native molecule. An untreated sample had 63% native lysozyme activity, a sample treated at 0° 28%, and a room temperature treated sample 5% (2-hr reaction time). Because the conditions for removal of the TFA group

(41) G. R. Marshall, "Milan Symposium on Peptides and Proteins," N. Back, R. Padetti, and L. Martini, Ed., Plenum Press, New York, N. Y., 1968.
(42) M. Ohno and C. B. Anfinsen, J. Amer. Chem. Soc., **89**, 5994 (1967).

are variable[9] and might require room temperature treatment, TFA protection of lysine was considered unacceptable for the solid phase synthesis of lysozyme. Therefore, investigations of ε-Z stability were performed.

Initial experiments accomplished by incubation of α-Boc-ε-Z-lysine in TFA–CH_2Cl_2–anisole–ethanedithiol, 43:50:2:5, followed by silica gel thin-layer chromatography of an aliquot of the reaction mixture (developed with chloroform–methanol–acetic acid, 85:10:5), showed that free lysine appeared in the reaction mixture after 1.5 hr. After 75 hr, more than 75% of the lysine was present in the free form.

To obtain further data on the removal of the ε-Z group, α-Boc-ε-Z-lysine was incorporated into two peptides: φ-Asp-Leu-Val-Phe-Lys and φ-Asp-Leu-Val-Phe-Lys-Ala. These resin-peptides were reacted for 24 hr with TFA–CH_2Cl_2–anisole–ethanedithiol, 43:50:2:5, and 4 N HCl–dioxane–anisole–ethanedithiol, 98:1:1. Glycine was then coupled to completion (negative ninhydrin test). If a molar excess of glycine was found in the 24-hr treated samples, this reflected ε-amino deprotection (i.e., glycine would be coupled at both the α- and ε-amino positions). The results (Table III) show that the Z group was removed during

Table III. Deprotection of ε-Z-Lys in Acidic Solution

Peptide[a]	Treatment	Amino acid analysis		
		Lys	Ala	Gly
1	None[c]	1.0		1.0
1	HCl–dioxane;[b] 24 hr	1.0		1.0
1	TFA–CH_2Cl_2;[b] 24 hr	1.0		1.4
2	None[c]	1.0	1.0	1.0
2	HCl–dioxane;[b] 24 hr	1.0	1.1	0.9
2	TFA–CH_2Cl_2;[b] 24 hr	1.0	1.1	1.2

[a] Peptide 1 = φ-Asp-Leu-Val-Phe-Lys-Gly. Peptide 2 = φ-Asp-Leu-Val-Phe-Lys-Ala-Gly. [b] The composition of these solutions is given in the text. [c] These samples were deblocked in TFA–CH_2Cl_2 (50:50) for 30 min.

the 24-hr TFA treatment (about 40%). Within the sensitivity of this procedure there was no detectable removal of the ε-amino Z group with HCl–dioxane. In TFA there was perhaps greater removal of the Z group when lysine was N-terminal than when alanine was N-terminal.

Cystine Sulfhydryl Protection. Because the benzyl sulfhydryl protecting groups of cysteine proved difficult to remove during this synthesis, the S-(p-methoxy)-benzyl group was considered more applicable for cystine sulfhydryl protection.[43]

Boc-S-(p-methoxy)Bzl-L-cysteine was used to test the p-methoxy-Bzl group stability in TFA–CH_2Cl_2–anisole–ethanedithiol, 44:50:1:5, and 4 N HCl–dioxane–anisole–ethanedithiol, 98:1:1. The same thin layer chromatographic procedure that was used for α-Boc-ε-L-lysine was employed. There was no evidence of S-protection removal in either solvent after 75 hr.

(43) S. Sakakibara, Bull. Chem. Soc. Jap., **38**, 1412 (1965).

Synthesis of a Polypeptide with Lysozyme Activity

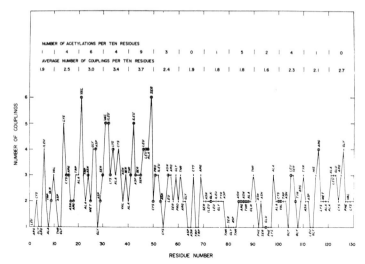

Figure 3. A graphical representation of the number of times each amino acid had to be coupled, the number of acetylations/every 10 amino acids, and the average number of couplings/every 10 amino acids. If an amino acid on the graph is represented by a · then CH_2Cl_2 and/or DMF was used as the coupling solvent, by an X then DMF–urea was used at least once, and if the · or X is circled then that particular amino acid was acetylated.

Synthetic Results. Second Synthesis

The strategy for a second synthesis was based on our investigations which indicated that the most probable cause of low molecular weight product was incomplete coupling of amino acids. It was suggested that this incompleteness of coupling was worsened because the fast coupling chains were also fastest removed from the resin. For this reason it was considered desirable to reduce the peptide removal rate through the choice of a milder Boc removal solution (HCl–dioxane). A consideration in reducing the loss of peptide from the resin is the effect of a resultant greater number of growing sites. It is possible that greater crowding and steric requirements would result from this situation and that coupling efficiency thus might be reduced, especially when synthesizing a large peptide. The fact that HCl–dioxane did not remove the Z ϵ-amino blocking group of lysine while the TFA solvent did favored the use of HCl–dioxane.

The synthetic alterations therefore included the use of 4 N HCl–dioxane for Boc removal, of 1,2-ethanedithiol for tryptophan protection in the 4 N HCl–dioxane, of incremental DCC injection during the coupling of amino acids, of the p-methoxybenzyl group for S-protection of cysteine, and of the ninhydrin test to monitor degree of completion of the coupling reaction. Also, in an attempt to drive each coupling reaction to completion, the coupling solvent (CH_2Cl_2) was varied when necessary. Residues which proved difficult to couple in CH_2Cl_2 (as evidenced by the ninhydrin test) were coupled in dimethylformamide (DMF) and if necessary in DMF saturated with urea (DMF–urea).[44] These solvent systems had been shown to be of value when coupling in CH_2Cl_2 was difficult.[11,44]

Residues which still proved difficult to react to a

(44) F. C. Westall and A. B. Robinson, *J. Org. Chem.*, **35**, 2842 (1970).

sufficient degree were acetylated (either by DCC-mediated coupling of acetic acid or reaction with acetylimidazole). Thus, the unreacted chains would be terminated and easily separated by gel filtration during purification. Obviously this modification of the synthetic procedure would be of decreasing value as the synthesis progressed to completion.

The total time spent on this synthesis was about 6 months. The numerous recouplings and often lengthy coupling times reduced the assembly rate to less than one residue per day.

A summary of the number of couplings per residue, coupling solvents, and acetylations is presented in Figure 3.

The most easily coupled portion, as expected, was from residues 1 to 13. These reactions were driven to completion in CH_2Cl_2 and acetylation was not necessary. An attempt was made to complete the coupling in CH_2Cl_2 before using DMF or DMF–urea. In general, coupling in CH_2Cl_2 became more difficult as the chain length increased. After 40 amino acid additions it was apparent that repeated couplings in CH_2Cl_2 failed to yield better than about 80% completion (as evidenced by the ninhydrin test). After residue 40 the initial coupling of an amino acid was the only one carried out in CH_2Cl_2. Subsequent couplings were done in either DMF or DMF–urea. (Exceptions to this were residues 64–68, 79–83, and 92–97 for which the initial coupling solvent was DMF and for which CH_2Cl_2 was never used.) The choice between DMF and DMF–urea as a second coupling solvent depended on the amino acid involved and on the results obtained in coupling the amino acids immediately prior to the one in question.

The summary in Table IV also indicates expected trends for the coupling of the individual amino acids. Although coupling conditions (solvent, time, etc.)

Table IV. Average Couplings/Residue and the Number of Acetylations Employed for Each Amino Acid during the Synthesis

Amino acid	No. of each amino acid	Average couplings/ residue	Total acetylations for each amino acid
Ala	12	2.4	2
Arg	11	2.4	4
Asp	8	2.1	3
Asn	13	2.2	6
Cys	8	2.4	3
Glu	2	2.0	1
Gln	3	2.0	1
Gly	12	1.6	0
His	1	3.0	0
Ile	6	3.5	2
Leu	8	2.7	3
Lys	6	2.8	2
Met	2	2.0	1
Phe	3	1.3	0
Pro	2	2.5	1
Ser	10	2.7	3
Thr	7	2.0	2
Trp	6	2.0	1
Tyr	3	2.3	1
Val	6	3.3	3
Total	129	2.4	39

ing for the fact that the hydrolysis conditions led to poor recoveries of serine, cysteine, methionine, and tyrosine, these analyses appeared very satisfactory through the 99 amino acid peptide. We did not consider these favorable amino acid ratios to indicate homogeneity in the product, however. Comparison of these data with similar data from the first synthesis (Table I) seemed to indicate a significant improvement in synthesis.

The amino acid ratios in subsequent analyses became gradually poorer. This was of minor concern because the same coupling and testing procedures were being used. It seemed more likely that the analyses through the 99 amino acid peptide gave a poor indication of product homogeneity than that the coupling and testing procedures suddenly deteriorated at residue 100. We suspect that a combination of product heterogeneity, some incomplete coupling and removal of peptide from the resin are responsible for the unsatisfactory amino acid ratios after residue 99.

The 126th amino acid to be added was incorporated as [^{14}C]Boc-glycine. This was used to examine the rate of removal of peptide from the resin. Both 4 N HCl–dioxane–anisole–ethanedithiol, 98:1:1, and TFA–

Table V. Amino Acid Ratios[a] of Selected Intermediate Resin-Peptide Samples

Amino acid	No. 1	No. 2	No. 3	No. 4	Lysozyme
Asp	0.9 (1)	6.9 (8)	18.6 (18)	18.1 (21)	20.1 (21)
Thr	0.7 (1)	1.9 (2)	6.4 (7)	5.1 (7)	6.5 (7)
Ser		3.9 (5)	5.5 (9)	5.0 (10)	6.5 (10)
Glu	0.8 (1)	0.9 (1)	4.4 (4)	6.2 (5)	5.3 (5)
Pro		1.2 (1)	2.4 (2)	1.9 (2)	2.1 (3)
Gly	2.0 (2)	3.7 (4)	8.4 (8)	14.2 (12)	12.0 (12)
Ala	1.1 (1)	6.5 (6)	11.7 (9)	15.6 (12)	12.5 (12)
Cys	1.5 (2)	3.2 (4)	3.4 (6)	4.1 (8)	4.0 (8)
Val	0.8 (1)	3.2 (4)	3.4 (6)	6.0 (6)	5.9 (6)
Met		0.2 (1)	0.5 (1)	1.5 (2)	1.5 (2)
Ile	1.1 (1)	3.9 (4)	6.4 (6)	5.1 (6)	5.1 (6)
Leu	1.3 (1)	4.0 (3)	6.4 (5)	10.6 (8)	7.7 (8)
Tyr			1.3 (1)	2.9 (3)	2.3 (3)
Phe			2.7 (2)	4.1 (3)	2.7 (3)
His				1.0 (1)	1.0 (1)
Lys	1.0 (1)	2.7 (3)	4.7 (4)	7.6 (6)	5.6 (6)
Arg	2.2 (2)	4.1 (4)	7.7 (8)	8.4 (11)	10.4 (11)
Trp	b (1)	b (3)	b (5)	b (6)	b (6)
Total residues	15	53	99	129	129

[a] The calculations were based on the theoretical length of the peptide. The theoretical number of each residue appears in parentheses beside the calculated value. [b] No determination made.

were not identical for each amino acid, glycine obviously coupled very easily (presumably owing to minimal steric requirements). Phenylalanine also coupled easily. Conversely, the sterically hindered amino acids (valine, isoleucine) were more resistive to complete reaction.

A total of 38 residues were acetylated during the synthesis. In this respect the ninhydrin test never gave a completely negative test after residue 15. Acetylation with either acetic acid–DCC or acetylimidazole in any of the three coupling solvents failed to completely eliminate the residual slightly positive ninhydrin value, as did a thorough washing of the resin before testing.

Amino acid analyses were performed on the resin-peptide every 10 residues throughout the synthesis. Representative samples appear in Table V. Allow-

CH_2Cl_2–anisole–ethanedithiol, 47:50:1:2, were reacted with the resin-peptide for 30 min and the washes collected and counted (three washes with each solution were carried out). The TFA under these conditions removed about five times as much peptide from the resin as did the HCl–dioxane. Using a specific activity of 0.01 mci/mmol and average molecular weight of 19,000 daltons for the blocked resin-peptide, about 0.4% of the peptide was removed by the HCl–dioxane and 1.9% by the TFA. The resin-peptide was also considerably more swollen in the TFA than the HCl–dioxane. This fact plus a comparison of the removal rate data obtained on the 4 and 126 residue peptide indicate TFA was better able to penetrate a large resin-peptide than was HCl–dioxane. It was unknown if the relative rates of Boc removal would be similarly affected and/or if the 30-min reaction time was sufficient for 4 N HCl–dioxane to deblock a 126 residue resin-peptide.

At the completion of the synthesis the resin-peptide was vacuum dried and weighed. Its weight was 14.8 g and amino acid analysis indicated that 7.4 g of this weight was peptide. This was a substantial yield but was complicated by the fact that the synthesis was begun with 4.0 g of resin-leucine. In addition it was estimated that an equivalent of about 1.5 g of the starting resin was removed throughout the synthesis for amino acid analysis, ninhydrin tests and removal at critical residues which later might be examined for activity and binding properties. This meant that approximately 35% of the final weight (~5 g) was due to material other than peptide or resin. In an attempt to identify this unknown material, HCl–propionic acid hydrolyses for extended periods of time were carried out but failed to release any more peptide from the resin. Extensive washing by various solvents (dimethyl sulfoxide, TFA, TFA–CH_2Cl_2 (50:50), chloroform, glacial acetic acid, methanol, ethanol) also failed to remove anything from the resin. Elemental analysis of the resin-peptide was inconclusive although it did indicate that no Cl, F, or S was contained in this unknown material. It did seem that the material was removed during the HF cleavage. Although no positive identification was made, it is possible the unknown material is tightly bound solvent (although this problem was not observed with the first resin-product) and/or urea (DMF–urea was used extensively as a coupling solvent). Perhaps related to the presence of this unknown material is the fact that the resin-product was resistant to all attempts at cleavage by HBr–TFA.

The final esterification of the peptide was 0.15 mmol/gram of resin (60% of the originally esterified material was removed during the synthesis). As shown below, a large portion of the product consisted of a nonreducible (by mercaptoethanol or dithiothreitol) high molecular weight peptide. Assuming one-half of the product had a molecular weight twice that of the native peptide (and the other half had the correct molecular weight) the corrected esterification was 0.12 mmol/gram of resin (70% of the original material was removed). In either case the amount of peptide removed was less than that predicted from the results on a 4 amino acid resin-peptide (82% removal would have been expected). This, in addition to the removal rate obtained on the 126 residue resin-peptide, suggests that the rate of removal of peptide from the resin decreased as the chain length increased when 4 N HCl–dioxane is used for Boc removal. Again it appeared that 4 N HCl–dioxane was less able to penetrate a large resin-peptide than was TFA (for TFA the theoretical removal rate based on a 4 amino acid resin-peptide agreed well with what was found during the first synthesis, 91% vs. 93%, respectively).

To obtain information on the optimal HF conditions, three different cleavage conditions were examined. In the first (A), 200 mg of resin-product was reacted in 20 ml of anhydrous HF and 4 ml of anisole for 1 hr at 0°. The reaction vessel was then allowed to warm to room temperature for 30 min, at which time the HF and other volatile products were removed under vacuum. The second conditions (B) were

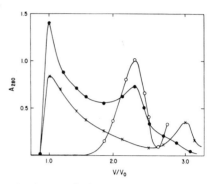

Figure 4. The A_{280} profile of native lysozyme (O—O), the synthetic product removed from the resin by the A cleavage conditions (●—●), and A-h following reduction with dithiothreitol (×—×) on Sephadex G-75 (see text). (V is the elution volume and V_0 the void volume of the column.)

identical with the first except that 200 mg of methionine was added to the reaction mixture. In the third (C), 200 mg of methionine, 200 mg of glutamine, and 200 mg of tryptophan were added to the reaction mixture. Evidence has been presented that deamidation of asparagine and glutamine occurs during the HF cleavage and that this deamidation can be eliminated by the addition of glutamine to the reaction mixture.[45] Hence we also examined the addition of methionine and tryptophan to the reaction mixture in order to protect these possibly labile amino acids.

The dried samples were taken up in 4 ml of 0.1 M acetic acid and any precipitate was removed by centrifugation. The peptide product was chromatographed on a 2 × 60 cm Sephadex G-75 column equilibrated with 0.1 M acetic acid. The chromatography profile of A appears in Figure 4 along with the profile of reduced native lysozyme chromatographed on a similar column.

The products from the three HF reaction conditions exhibited similar profiles on G-75. In contrast to the first synthesis, about 30–50% of the product has eluted at the same position as native reduced lysozyme. There was very little low molecular weight product. There was, however, a large amount (50–70%) of high molecular weight peptide.

Each of the samples was divided into a high molecular weight and a correct molecular weight (elution position) fraction and freeze dried. The weights of the A fractions were: high molecular weight (A-h) = 17 mg, correct molecular weight (A-c) = 14 mg (yield = 31%). The weights of the B fractions were: B-h = 19 mg, B-c = 13 mg (yield = 32%). The weights of the C fractions were C-h = 7 mg, C-c = 10 mg (yield = 17%). From amino acid analyses the amount of peptide expected to be recovered from each of these cleavages was 100 mg.

We attempted to resolve further the high molecular weight fractions by reduction with mercaptoethanol or dithiothreitol in 6 M guanidine–HCl. Following reduction the samples were rechromatographed on Sephadex G-75. The A_{280} profiles of the initial A re-

(45) J. McKerrow, Ph.D. Thesis, University of California, 1972.

action mixture and the A-h fraction after reduction are shown in Figure 4. Reduction failed to change the molecular weights (elution position) of any of the high molecular weight fractions.

Samples of all fractions from this experiment (A-h, A-c, B-h, B-c, C-h, and C-c) were taken for amino acid analysis following both HCl and enzymatic hydrolysis. These data failed to show any improvement in amide yield when glutamine was present in the HF reaction mixture. There was also no improvement in tryptophan recovery when free tryptophan was added to the HF. The same was true for methionine. The amino acid analyses of the enzymatic digests were examined for the presence of Glu (γ-OBzl), Asp (β-OBzl), Ser (Bzl), Thr (Bzl), and Tyr (Bzl). None of these derivatives were found indicating the peptide had been completely deblocked in all three HF reaction mixtures.

All fractions were reduced with dithiothreitol, reoxidized, concentrated on CMC, and assayed for enzymatic activity. None of the high molecular weight fractions were enzymatically active. The specific activity of A-c was 0.5% that of native lysozyme and 2.0% that of similarly reduced and reoxidized native lysozyme. Corresponding data for B-c were 0.06% of native and 0.2% of reduced and reoxidized native specific activity. For C-c the figures were 0.1% of native and 0.4% of reduced and reoxidized native specific activity. Free amino acid additions to the HF reaction mixture seemed to decrease the enzymatic activity of the product. Subsequent cleavage procedures were therefore performed using method A.

The results of the amino acid analyses of both the enzymatic and HCl hydrolysis for the A fractions and native lysozyme appear in Table VI. Few conclusions can be drawn from these data. As expected the amino acid ratios did not match native lysozyme, and were similar to those of the first synthesis (Table I). Most noticeable were the low recoveries of Asp, Thr, Ser, and Ile.

Following the information obtained on HF cleavage conditions we decided to examine milder conditions to see if a more active product could be recovered. In this case 200 mg of resin-product was reacted in 20 ml of anhydrous HF and 4 ml of anisole for 90 min at 0°. The HF and other volatile products were again removed under vacuum. The peptide was recovered and chromatographed as previously described. Under these milder conditions only the high molecular weight product was cleaved from the resin. These data offered support to our proposal that heterogeneity in resin esterification sites was directly related to heterogeneity in the coupling sites (i.e., chains more easily removed from the resin were also the fastest coupling sites). It seemed probable that the high molecular weight peptide was the result of overzealous concern for completing each coupling reaction. Deblocked amino acid side chains (incurred during the synthesis or from impure amino acid derivatives) branched as a result of the extreme coupling conditions. It followed that most of this branching occurred in the more accessible chains. The fact that these were the only chains removed by mild HF conditions supported our contention of a direct relationship between coupling and cleavage efficiency.

One further aspect of the HF cleavage conditions remains to be reported. Before beginning the first synthesis the stability of native lysozyme in HF was examined. The reaction conditions employed were essentially those of method A. After recovery from the HF this material had 70% of the native specific activity. On the basis of these data it was concluded that the molecule was sufficiently stable to withstand the cleavage conditions. Following this second synthesis the stability of *reduced* (i.e., unfolded) native lysozyme was tested when the A cleavage conditions were used. After recovery from HF, Sephadex G-75 chromatography, and reoxidation the product had 10% the specific activity of the native molecule. This result prompted us to examine the HF stability of the reduced native molecule in the presence of the correct molar ratios of the blocked free amino acids. We attempted to more closely simulate the actual conditions a blocked synthetic product would experience during the HF reaction. Reduced native lysozyme (3.5 μmol) was treated by the A cleavage conditions in the presence of 38 μmol of Arg (NO$_2$), 28 μmol of Cys (S-(p-methoxy)Bzl), 7 μmol of Glu (γ-OBzl), 28 μmol of Asp (β-OBzl), 21 μmol of Lys (ϵ-Z), 3.5 μmol of His (im-TOS), 35 μmol of Ser (Bzl), 25 μmol of Thr (Bzl), and 10 μmol of Tyr (Bzl). On the basis of results obtained with added amino acids in the HF reaction mixture (B and C cleavage conditions) it was requisite to be concerned with a similar effect of these amino acids. However, it should be noted that in the B and C cleavage conditions the amino acids were present in a two- and sixfold weight excess, respectively, over the product peptide while in this experiment the native lysozyme was present in a twofold weight excess over the amino acids.

Following recovery from the HF, Sephadex G-75 chromatography, and reoxidation, the product had 2–3% the specific activity of the native molecule. It

Table VI. Amino Acid Ratios[a] of the Synthetic Product

Amino acid	A-h HCl	A-h Enzymatic	A-c HCl	A-c Enzymatic	Native lysozyme HCl[b]	Native lysozyme Enzymatic
Asp	18.3	3.5	15.8	3.6	21.3 (21)	5.7 (8)
Thr	5.7	4.2	4.9	3.7	6.9 (7)	6.2 (6)
Ser	9.3		7.5		9.4 (10)	
Glu	5.5	3.1	6.5	4.6	5.3 (5)	2.3 (2)
Pro	2.0		1.9		2.1 (2)	
Gly	14.2	10.9	15.6	12.7	13.5 (12)	9.9 (12)
Ala	15.3	15.1	16.5	15.7	13.4 (12)	13.1 (12)
Cys	5.0	6.3	3.7	5.5	6.9 (8)	8.8 (8)
Val	5.9	6.4	6.3	6.5	5.6 (6)	6.6 (6)
Met	0.6	2.2	1.1		1.2 (2)	1.8 (2)
Ile	5.3	4.2	3.9	2.5	5.2 (6)	4.8 (5)
Leu	10.3	11.8	10.3	9.1	7.8 (8)	8.5 (8)
Tyr	1.9	3.4	2.7	3.7	2.4 (3)	4.0 (3)
Phe	3.6	4.6	4.4	5.3	2.9 (3)	3.3 (3)
His	1.3	2.1	1.8	2.8	1.0 (1)	1.2 (1)
Lys	6.4	7.4	8.6	7.1	5.8 (6)	6.1 (6)
Arg	12.5	10.8	14.3	10.7	12.3 (11)	10.6 (11)
Trp	c	c	4.8[d]	c	6.4[d] (6)	c

[a] The basis of the calculations was the same as those in Table II. [b] The numbers in parentheses are the theoretical number of each residue. [c] No determination made. [d] Determined by p-toluenesulfonic acid hydrolysis: T. Y. Liu and Y. H. Chang, *J. Biol. Chem.*, **246**, 2842 (1971).

is interesting that the specific activity of the crude product was 9–25% that of native lysozyme treated with HF in the presence of the blocked amino acids. Based on amino acid analysis and product characterization it seems unlikely that the crude product could have been so homogeneous. However, these results indicated that problems existed with the HF cleavage conditions and that better synthetic results might result from more thorough examination of the HF reaction.

The synthetic product (A–c) was examined for purity by CMC chromatography and by cellulose acetate electrophoresis. In both cases most of the product was more acidic than native lysozyme. It is possible that this increased acidity arose from an unfavorable balance between the ratios of glutamic acid (Glu) + aspartic acid (Asp) and glutamine (Gln) + asparagine (Asn). From the amino acid analyses in Table VI it was calculated that for native lysozyme the recovery of Glu + Asp was 42% of the total of Glu + Asp + Gln + Ans (theoretical 39%) while the same figure for the synthetic product was 46%. In addition, if some chain branching or acetylation did occur at the ε-amino group of lysine the apparent acidity of the product would be further increased.

We have previously reported[46] the results of fingerprinting of the crude product and native lysozyme (in cooperation with L. E. Barstow, V. J. Hruby, T. Shimoda, and J. A. Rupley at the University of Arizona). The synthetic product fingerprint showed a similarity to native lysozyme except for one peptide and low yields of several large peptides. Although encouraging, this type of analysis was inconclusive without an extensive characterization of the peptides.

Chitin column chromatography and ammonium sulfate fractionation were employed in an attempt to further purify the synthetic product. Following chitin chromatography the synthetic product had 4–6% the specific activity of native lysozyme that had been similarly reduced, reoxidized, and chromatographed (2–3% the specific activity of native lysozyme). Ammonium sulfate precipitation failed to improve the specific activity. Both native lysozyme and the synthetic product precipitated in saturated ammonium sulfate and in neither case did peptide precipitate before the saturating concentration was reached.

The specificity of the enzymatic bond breakage was examined and has previously been reported.[46,47] The synthetic product exhibited the proper specificity in the hydrolysis of a hexamer of N-acetylglucosamine $(GlcNAc)_6$ and in the glucosyl transfer from $(GlcNAc)_6$ to [^{14}C]GlcNAc. The specific activity of the crude product was 1% that of native lysozyme.

Further characterization of the product was not attempted. Since the completion of this synthesis, two further syntheses have been carried out by the Arizona group.[48] The crude activities of these products were approximately those of this second synthesis.

(46) L. E. Barstow, V. J. Hruby, A. B. Robinson, J. A. Rupley, J. J. Sharp, and T. Shimoda, *Fed. Proc., Fed. Amer. Soc. Exp. Biol.*, **30**, Abstr. 1292 (1971).
(47) This assay was carried out by L. E. Barstow, V. J. Hruby, T. Shimoda, and J. A. Rupley at the University of Arizona.
(48) L. Barstow, D. Cornelius, V. Hruby, T. Shimoda, J. Rupley, J. J. Sharp, A. B. Robinson, and M. D. Kamen in "Chemistry and Biology of Peptides," J. Meienhofer, Ed., Ann Arbor Science Publishers, Ann Arbor, Mich., 1972, p 231.

A product of high specific activity (~70%) has been obtained with the use of affinity chromatography, although with low yield. Neither of the products discussed in this paper were subjected to the affinity chromatographic procedures.

Discussion

A polypeptide(s) with lysozyme activity was synthesized using Merrifield solid-phase peptide procedures. The yield[34] after HF treatment and Sephadex G-75 chromatography was ~10%. The yield of the correct molecular weight fraction after reduction, reoxidation, and concentration on CMC was ~1%. The total possible recovery of peptide with the correct molecular weight was ~210 mg (if the total resin-product was reacted in HF by method A). The specific activity of this crude product was 0.5–1% that of native lysozyme (or 9–25% that of native lysozyme which had been similarly treated with HF in the presence of blocked amino acids). Following chitin chromatography the specific activity of the correct molecular weight fraction was 2–3% that of native lysozyme. This material exhibited the correct enzymatic specificity with regard to the hydrolysis of $(GlcNAc)_6$ and in the glucosyl transfer from $(GlcNAc)_6$ to [^{14}C]GlcNAc.

The problems encountered during these syntheses will be discussed below.

Coupling Reaction. From the results of the first synthesis and following experiments it was concluded that poor coupling and selective removal of esterified peptide were the reasons for the large excess of low molecular weight peptide. By using the ninhydrin test[4] for estimating the coupling efficiency, the amount of low molecular weight material was considerably reduced in the second synthesis. The possibility that the number of couplings performed were unnecessary and/or harmful has been discussed.

Boc Removal. The possibility that poor Boc removal during the first synthesis resulted in the low molecular weight product was never ruled out. Rather, the difficulty was attributed to poor and variable coupling. The results of the second synthesis indicate that Boc removal was not a problem. Even with a milder Boc removal reagent (HCl–dioxane) there was very little low molecular weight product in the second synthesis. (There is a report[49] that HCl–dioxane is a much poorer reagent for Boc removal than TFA–CH_2Cl_2, or even than HCl–acetic acid.)

It is possible that there was ε-amino deprotection of lysine during the second synthesis and that this served as a chain branching point resulting in the high molecular weight polymer. The lysine deprotection experiments were performed by rocking the appropriate resin-peptides in the acid solution for 24 hr and then examining the amount of ε-amino deprotection. To reproduce the synthetic conditions the acid should have been withdrawn and fresh acid injected in 30-min time intervals. These experiments did show, however, that TFA–CH_2Cl_2 was unacceptable for Boc removal if Z was used for the ε-amino protection of lysine. Certainly any Z removal was compounded by the fact that each coupling reaction was driven to completion.

(49) S. Karlsson, G. Lindeberg, J. Porath, and U. Ragnarsson, *Acta Chem. Scand.*, **24**, 1010 (1970).

HF Reaction. The HF cleavage conditions are not ideal as demonstrated by the activity measurements of native lysozyme after this reaction. Certainly ribonuclease was not affected so severely by the HF conditions. The severity of this reaction undoubtedly varies from protein to protein.

Aspartic Acid. The α-β peptide bond migration of aspartic acid is well known. This reaction has been reported when glycine is C-terminal to aspartic acid.[50] There are three Asp-Gly sequences in lysozyme but unfortunately no simple method exists to determine the extent of cyclization and/or β migration in a peptide of this size.

Asparagine and Glutamine. The deamidation of asparagine and glutamine during HF cleavage has been reported.[51] It is not known how much, if any, deamidation occurs during the chain assembly process. Although asparagine and glutamine are usually used unblocked in the solid-phase procedures, blocking groups for these amino acids are available (for example, ref 52). Nor is it known what effect these groups will have in reducing any HF caused deamidation.

From the results of these two lysozyme syntheses it follows that there is need for milder conditions and a better solid support when attempting a long-term solid-phase synthesis. It has been demonstrated that cleavage conditions milder than HF would be desirable. Milder cleavage conditions, however, necessitate more easily removable side chain blocking groups, which in turn require less severe conditions for α-amino deprotection removal.

If the goal of protein synthesis had been to obtain a lysozyme-like native protein with high specific activity, the purification of the product obtained from this synthesis could have been further pursued. However, the synthesis of the native molecule was only the initial stage of researches leading to the production of synthetic analogs for use in structure–function investigations. We feel that the heterogeneity exhibited by the products obtained from these two lysozyme syntheses (and from other syntheses) underlines the necessity for further development of solid-phase procedures as applied to the synthesis of large molecules before attempts based on analog syntheses can be mounted. In spite of the remarkable success in the syntheses of ribonuclease A[1] and acyl-carrier protein,[53] it seems that the method is, as yet, not reliable enough for general applicability to the synthesis of any large molecule and its significant structural analogs. A problem obviously exists in the purification of analogs (affinity chromatography purification of structural analogs may or may not be of value, depending on their substrate binding properties). At the present time thorough characterizations of the analog are required to establish its nature. For a large molecule with one or two amino acid replacements (the remainder of the molecule being identical with the "native" product) this is a significant undertaking. For this reason it is felt that the development of solid-phase procedures specifically aimed at improving the syntheses of large molecules is crucial.

Acknowledgments. Financial support provided by grants from the National Institutes of Health (HD-01262 to M. D. K. and AM-14879 to A. B. R.) and the National Science Foundation (GB-7033X to M. D. K.) is gratefully acknowledged.

(50) E. E. Haley, B. J. Corcoran, F. E. Dorer, and D. L. Buchanan, *Biochemistry*, **5**, 3229 (1966).
(51) A. B. Robinson, J. H. McKerrow, and P. Cary, *Proc. Nat. Acad. Sci. U. S.*, **66**, 753 (1970).
(52) P. G. Pietta and G. R. Marshall, *Chem. Commun.*, 650 (1970).
(53) W. S. Hancock, D. J. Prescott, G. R. Marshall, and P. R. Vagelos, *J. Biol. Chem.*, **247**, 6224 (1972).

IV

Solid-Phase Synthesis of Nucleotides

Editors' Comments on Papers 26 Through 33

26 **Letsinger and Mahadevan:** *Oligonucleotide Synthesis on a Polymer Support*

27 **Letsinger and Mahadevan:** *Stepwise Synthesis of Oligodeoxyribonucleotides on an Insoluble Polymer Support*

28 **Letsinger, Caruthers, and Jerina:** *Reactions of Nucleosides on Polymer Supports: Synthesis of Thymidylylthymidylylthymidine*

29 **Hayatsu and Khorana:** *Studies on Polynucleotides: LXXII. Deoxyribooligonucleotide Synthesis on a Polymer Support*

30 **Melby and Strobach:** *Oligonucleotide Syntheses on Insoluble Polymer Supports: I. Stepwise Synthesis of Trithymidine Diphosphate*

31 **Rubinstein and Patchornik:** *Polymers as Chemical Reagents: The Use of Poly (3, 5 Diethylstyrene) —Sulfonyl Chloride for the Synthesis of Internucleotide Bonds*

32 **Seliger and Aumann:** *Oligonucleotide Synthesis on a Polymer Support Soluble in Water and Pyridine*

33 **Chapman and Kleid:** *Oligonucleotide Synthesis on Polar Polymer Supports: The Use of a Polypeptide Support*

   In this section we have selected a number of papers that reflect the current status of polymer-based solid-phase synthesis. The fact that there are fewer papers in this section than in that on peptides reflects the greater difficulty encountered in nucleotide chemistry. The major difficulties lie in the necessity for protection of more groups (saccharide, base, and phosphate moieties) and solubility problems due to very large and polar molecules. This has been particularly true in the cross-linked polymers, where very slow reactions occur.
   It is an easy prediction that we will see greater future activity in this area—in particular the use of different methods and blocking groups coupled with the use of resins that would be more compatible with oligonucleotides (see Paper 33 for an example using a polypeptide as the resin).
   The first nucleotide synthesis on a "popcorn" polymer by Letsinger and Mahadevan (Paper 26) began with the attachment of the base to the polymer via an amide bond. The authors took full advantage of the steric differences found in attached units on a polystyrene to couple a nucleoside in which no protective groups were used in the deoxyribose moiety. Other oligonucleotide syntheses are reported by the Letsinger group in Papers 27 and 28. Hayatsu and Khorana (Paper 29) employed the common 5' blocking group (trityl) of nucleotide chemistry on a pyridine-soluble polymer. The usual steps of elongation of the oligonucleotide chain were used. Exceptionally high yields were obtained in each coupling step, owing in part, no doubt, to the solubility of the polymer. Melby and Strobach in Paper 30 used techniques similar to those of Hayatsu and Khorana, with the distinction that an *insoluble* support was investigated. The insoluble polymer gave yields comparable to those in the soluble system. Another approach to nucleotide synthesis by means of a soluble polymer used as the ribose blocking group is reported by Rubinstein and Patchornik (Paper 31). A highly sterically

hindered sulfinyl chloride polymer was investigated as a means of increasing the selectivity at the 5' position of the sugar moiety. This method can be used only in the condensation step of a nucleoside and a nucleotide. The paper is significant in demonstrating the many advantages of the SPS method over classical reagents. Paper 32 by Seliger and Aumann, and Paper 33 by Chapman and Kleid, describe the use of polymers different from the usual polystyrene. The former authors prepared a copolymer of N-vinylpyrolidone and vinyl acetate, performing the condensation steps in solution. Dialysis was used to remove excess reactants and by-products. The latter paper is unique in the use of a polypeptide as the backbone polymer.

Oligonucleotide Synthesis on a Polymer Support[1,2]

ROBERT L. LETSINGER and V. MAHADEVAN

Sir:

With the objective of simplifying procedures for the stepwise synthesis of complex substances such as polynucleotides and polypeptides, we undertook a study of chemical reactions on insoluble supports. In a previous paper it was shown that functional groups on styrene–divinylbenzene popcorn polymer are available for many conventional chemical reactions and that a dipeptide may be prepared on and removed from this polymer.[3] The present communication reports experiments illustrating the use of a polymer support in the synthesis of oligonucleotides.

The general scheme is typified by the preparation of deoxycytidylyl-(3'→5')-thymidine (V) outlined in Chart I. All reactions on the polymer support (a–f) were

(1) Part II in a series on Nucleotide Chemistry. Paper I: R. L. Letsinger, J. Fontaine, V. Mahadevan, D. A. Schexnayder, and R. E. Leone, *J. Org. Chem.*, **29**, 2615 (1964).
(2) This research was supported by the Division of General Medical Sciences, National Institutes of Health, Grant 10265.
(3) (a) R. L. Letsinger and M. J. Kornet, *J. Am. Chem. Soc.*, **85**, 3045 (1963); (b) R. L. Letsinger, M. J. Kornet, V. Mahadevan, and D. M. Jerina, *ibid.*, **86**, 5163 (1964).

carried out at room temperature. In the initial step (a) 5 g. of polymer acid chloride, ⓟ$_c$–COCl,[3b,4] was stirred with 2.8 g. (6.0 mmoles) of 5'-O-trityldeoxycytidine[5] in 50 ml. of dry pyridine for 2 days. Methanol was then added to convert residual acid chloride to ester, and the solid polymer (6.3 g.) was separated by filtration (1.2 g. of trityldeoxycytidine was recovered from the solvent). Phosphorylation (b) was effected by stirring 12 g. of I (from two experiments) with pyridinium β-cyanoethyl phosphate[6] (obtained from 5.1 g., 15 mmoles, of the barium salt trihydrate) and 6.2 g. of dicyclohexylcarbodiimide for 7 days. After filtration, treatment with 50% aqueous pyridine (c), and successive washing with methanol, ethanol, 1:1 ethanol–cyclohexane, and ether, a portion of the polymer (9.5 g.) was stirred with 2.5 g. (11 mmoles) of mesitylenesulfonyl chloride[7] (d) in 100 ml. of pyridine for 1 day. The polymer was then removed and mixed with 2.1 g. (9 mmoles) of dry thymidine (e)

(4) The polymer had 1.0 mmole of acid chloride groups/g.
(5) A. M. Michelson and A. R. Todd, *J. Chem. Soc.*, 34 (1954).
(6) G. M. Tener, *J. Am. Chem. Soc.*, **83**, 159 (1961).
(7) H. G. Khorana, J. P. Vizsolyi, and R. K. Ralph, *ibid.*, **84**, 414 (1962); T. M. Jacob and H. G. Khorana, *ibid.*, **86**, 1630 (1964).

Chart I

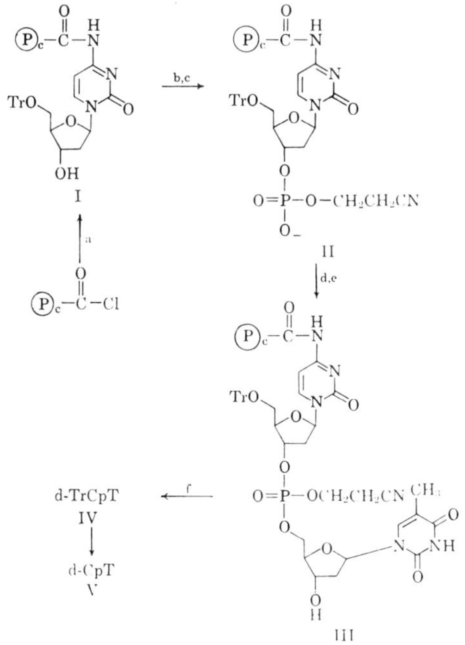

in 40 ml. of pyridine. After 2 days the solid was separated and washed carefully. Nucleotidic material was cleaved from the support by four successive 3-hr. treatments with 40-ml. portions of 0.2 M sodium hydroxide (f) in 1:1 dioxane–ethanol. The resulting alkaline solution was neutralized with Dowex-50 resin (pyridinium form), concentrated, and chromatographed on DEAE-cellulose with a linear gradient of triethylammonium bicarbonate. From 1.215 g. of III was obtained 1340 O.D.$_{267}$[8] units of 5′-O-trityldeoxycytidylylthymidine (IV), which was isolated as the triethylammonium salt (101 mg.) by concentration and lyophilization; R_f 0.78[9]; electrophoretic mobility relative to deoxycytidine 5′-phosphate at pH 10.8, 0.32; λ_{max} 267 mμ, λ_{min} 244 mμ. Some 5′-O-trityl-

[8] T. M. Jacob and H. G. Khorana, *J. Am. Chem. Soc.*, **87**, 372 (1965).
[9] The paper chromatograms were all run on Whatman 3MM paper with isopropyl alcohol–ammonium hydroxide–water (7:1:2).

deoxycytidine (32 mg.), deoxycytidine (10 mg.), thymidine (16 mg.), and a trace of a compound corresponding to trityldeoxycytidine 3′-phosphate were also obtained.

Detritylation of IV with 80% aqueous acetic acid afforded deoxycytidylylthymidine (V), which was isolated as the ammonium salt; R_f 0.33; electrophoretic mobility relative to deoxycytidine 5′-phosphate at pH 10.8, 0.58; ultraviolet at pH 6.9, λ_{max} 267 mμ, λ_{min} 240 mμ.; at pH 2.15, λ_{max} 275 mμ, λ_{min} 239 mμ. In the presence of phosphodiesterase from Russel's viper venom[10] IV hydrolyzed completely to thymidine 5′-phosphate (R_f 0.16) and 5′-O-trityldeoxycytidine (R_f 0.85). With spleen phosphodiesterase[11] V was hydrolyzed extensively (~95%) to deoxycytidine 3′-phosphate (R_f 0.13) and thymidine (R_f 0.70).

In addition to the polymer support aspect, this synthetic route has two other novel features: (1) mesitylenesulfonyl chloride was used to activate a phosphodiester rather than a phosphomonoester and (2) a nucleoside with both the 3′- and 5′-hydroxyl groups free was employed. As a check on the selectivity of the condensation step (e), deoxycytidylyl-(3′→5′)-thymidine was prepared independently by using 3′-O-2,4-dinitrobenzenesulfenylthymidine[1] in place of thymidine in the synthetic sequence. Following the condensation step and prior to cleavage (f), the dinitrobenzenesulfenyl group was removed by treating the insoluble polymer with excess thiophenol in pyridine at room temperature. The dinucleotide obtained had the same physical and chromatographic properties as V prepared directly from thymidine, and it behaved the same on enzymatic degradation, indicating that step (e) with thymidine involves attack at the 5′-hydroxyl. The selectivity probably depends upon the fact that the hydroxyl group must approach a relatively hindered phosphorus in the condensation step.

Following the general procedure outlined in the flow sheet, the 5′-O-trityl derivatives of deoxycytidylyldeoxycytidine, deoxycytidylylthymidylyldeoxyadenosine, and deoxycytidylylthymidylylthymidine have been prepared.

[10] Calbiochem, Los Angeles, Calif.
[11] Nutritional Biochemical Corp., Cleveland, Ohio.

Division of Biochemistry of the Department of Chemistry
Northwestern University, Evanston, Illinois
Received May 13, 1965

Stepwise Synthesis of Oligodeoxyribonucleotides on an Insoluble Polymer Support[1,2]

Robert L. Letsinger and V. Mahadevan

Contribution from the Department of Chemistry, Northwestern University, Evanston, Illinois 60201. Received August 5, 1966

Abstract: Procedures are described for synthesizing oligonucleotide derivatives (*e.g.*, TrdCpTpTpT) on an insoluble support polymer.

The comprehensive studies by Khorana and his co-workers have provided elegant and effective methods for the chemical synthesis of a wide variety of oligonucleotides.[3] Nevertheless, the stepwise synthesis of relatively long chain polynucleotides remains a formidable task. Some time ago it occurred to us that the labor involved in repetitive step syntheses of this type might be materially reduced if the syntheses were carried out on an insoluble polymer support. In the initial step a nucleoside would be joined covalently to the support. Nucleotide units would subsequently be added stepwise to this nucleoside, and in the final reaction the covalent bond joining the oligonucleotide chain to the support would be broken and the oligonucleotide eluted. This technique would enable one to separate the products in the building stages from the solvents, excess reagents, and soluble by-products simply by filtration, thus avoiding numerous time-consuming isolation steps. In testing this idea we first developed an insoluble, functionalized polymer and

---

(1) Part III, Nucleotide Chemistry. A preliminary account of some of this work was published in *J. Am. Chem. Soc.*, **87**, 3526 (1965).
(2) This research was supported by the Division of General Medical Sciences, National Institutes of Health, Grant 10265.
(3) For leading references see: (a) H. G. Khorana, T. M. Jacob, M. W. Moon, S. A. Narang, and E. Ohtsuka, *J. Am. Chem. Soc.*, **87**, 2954 (1965); (b) T. M. Jacob and H. G. Khorana, *ibid.*, **86**, 1630 (1964); (c) H. Schaller, G. Weimann, B. Lerch, and H. G. Khorana, *ibid.*, **85**, 3821 (1963); (d) P. T. Gilham and H. G. Khorana, *ibid.*, **80**, 6212 (1958). Since completion of the present paper a procedure for synthesizing oligonucleotides on a soluble polymer support has been described by H. Hayatsu and H. G. Khorana, *ibid.*, **88**, 3182 (1966).

demonstrated that chemical transformations could be carried out satisfactorily on it when suspended in an organic solvent.[4] The present paper reports the utilization of a support polymer in the synthesis of oligonucleotide derivatives.

The support chosen for the synthetic work was a popcorn copolymer prepared from styrene (88%), p-vinylbenzoic acid (12%), and p-divinylbenzene (0.2%). It was insoluble in water, alkaline solutions, and all organic solvents examined. Promising results have also been obtained recently with a polymer that contained 0.02% p-divinylbenzene as the cross-linking agent. It swelled considerably more in benzene and in pyridine than did the higher cross-linked polymer yet it remained insoluble and could be separated satisfactorily from the solvents by centrifugation or by filtration. Popcorn-type polymers were obtained from styrene and p-vinylbenzoic acid mixtures that contained as little as 0.01% p-divinylbenzene; however, these products were partially soluble in pyridine. Other materials considered as supports included a bead copolymer of acrylonitrile and acrylic acid and a popcorn copolymer of 2,3-dimethylbutadiene and p-vinylbenzoic acid.[5] While usable, these polymers were somewhat less stable than the styrene-based supports and offered no apparent advantage over them.

5'-O-Trityldeoxycytidine was utilized as the anchor group in these syntheses. It was joined to the support by reaction with the acid chloride form of the polymer (Ⓟ-COCl).[6] That the juncture was through the amino group of the cytosine moiety was shown by enzymatic degradation of the oligonucleotide derivatives synthesized on the support.[7] Approximately 40% of the acid chloride functional groups reacted when Ⓟ-COCl was stirred with excess 5'-O-trityldeoxycytidine in pyridine for 48 hr. Since the remaining acid chloride groups would have been undesirable in later synthetic steps, they were esterified by treatment with methanol in pyridine. The resulting polymer is indicated by I.

A number of experiments were carried out to establish conditions for removing the trityldeoxycytidine from the support. The reagent selected was 0.4 M sodium hydroxide in dioxane-ethanol-water. At room temperature it removed all the trityldeoxycytidine

within 36 hr. Solutions of ammonium hydroxide in pyridine were less satisfactory. When the polymer was heated in the ammonia mixture at atmospheric pressure, the cleavage was very slow; when the reaction was carried out in an autoclave at 100°, extensive degradation of oligonucleotides joined to the support took place.

Nucleotide units were built onto the terminal hydroxyl groups in polymer I in two steps: (1) phosphorylation and (2) condensation of the phosphoryl derivative with a nucleoside. Phosphorylation could be effected by a variety of reagents, including phosphorus oxychloride, p-nitrophenyl phosphorodichloridate,[8] β-cyanoethyl phosphorodichloridate, p-nitrophenyl phosphate plus dicyclohexylcarbodiimide, and β-cyanoethyl phosphate plus dicyclohexylcarbodiimide.[9] For construction of oligonucleotide chains the β-cyanoethyl phosphate-dicyclohexylcarbodiimide combination proved to be the most satisfactory. Step 2 was accomplished by activating the phosphodiester resulting from the β-cyanoethyl phosphate reaction (II) with mesitylenesulfonyl chloride and then adding the appropriate nucleoside derivative. The β-cyanoethyl group in the resulting phosphotriester was subsequently removed by the alkali used to cleave the oligonucleotides from the support polymer. Initially, thymidine blocked at the 3' position by the dinitrobenzenesulfenyl group[10] was used in order to assure attack at the hydroxyl at the 5' position; however, later experiments demonstrated that polymer II, activated by mesitylenesulfonyl chloride, attacked thymidine itself at the 5' position. Consequently thymidine was used directly in the preparative scale reactions.

The over-all sequence of reactions may be illustrated by the synthesis of 5'-O-trityldeoxycytidylyl-(3'-5')-thymidine (TrdCpT). Starting with Ⓟ-COCl, the insoluble polymer was subjected successively to the following reagents: 5'-O-trityldeoxycytidine, methanol (yielding I), β-cyanoethyl phosphate and dicyclohexylcarbodiimide, water (yielding II), mesitylenesulfonyl chloride, thymidine (yielding III), and sodium hydroxide (yielding TrdCpT). Pyridine was used as the liquid medium throughout except in the alkaline cleavage reaction. At the end of each step the polymer

---

(4) R. L. Letsinger and M. J. Kornet, *J. Am. Chem. Soc.*, **85**, 3045 (1963); R. L. Letsinger, M. J. Kornet, V. Mahadevan, and D. M. Jerina, *ibid.*, **86**, 5163 (1964).
(5) Work of Jerina and Becker.
(6) In this notation the relevant functional group is indicated explicitly and the remainder of the support is denoted by Ⓟ-. Other abbreviations used in this paper are: Tr, triphenylmethyl; dC, deoxycytidine; T, thymidine; dA, deoxyadenosine; p, the phosphate in a nucleotide or oligonucleotide derivative. The symbolism is typified by TrdCpTpdA, which refers to 5'-O-trityldeoxycytidylyl-(3'-5')-thymidylyl-(3'-5')-deoxyadenosine.
(7) The oligonucleotide derivatives obtained were degradable with snake venom phosphodiesterase. If the amino group had not been blocked, nondegradable products should have been obtained.

(8) A. F. Turner and H. G. Khorana, *J. Am. Chem. Soc.*, **81**, 4651 (1959).
(9) P. T. Gilham and G. M. Tener, *Chem. Ind.* (London), 542 (1959); G. M. Tener, *J. Am. Chem. Soc.*, **83**, 159 (1961).
(10) R. L. Letsinger, J. Fontaine, V. Mahadevan, D. A. Schexnayder, and R. E. Leone, *J. Org. Chem.*, **29**, 2615 (1964).

was separated from the liquid medium, washed, and resuspended in a fresh solution containing the next reagent. The final alkaline solution, which contained the nucleotidic material, was neutralized, concentrated, and chromatographed on DEAE-cellulose. From 2.2 g of polymer III was obtained 0.15 g of TrdCpT, a 53% yield based on the amount of trityldeoxycytidine bound to the support.

Since polymer III, like polymer II, possessed free 3'-hydroxyl groups, additional nucleotide units could be added to the chains by repetition of the phosphorylation, activation, and condensation steps utilized in building III. Thus, 5'-O-trityldeoxycytidylyl-(3'-5')-thymidylyl-(3'-5')-thymidine (TrdCpTpT) and 5'-O-trityldeoxycytidylyl-(3'-5')-thymidylyl-(3'-5')-thymidylyl-(3'-5')-thymidine (TrdCpTpTpT) were obtained in yields of 16 and 14%, respectively, based on the weights of the products isolated and the amount of trityldeoxycytidine bound to the support. 5'-O-Trityldeoxycytidylyl-(3'-5')-thymidylyl-(3'-5')-deoxyadenosine (TrdCpTpdA) was similarly prepared by substituting N-benzoyldeoxyadenosine for thymidine in the final condensation step.

These oligonucleotide derivatives behaved as pure compounds on electrophoresis and chromatography on paper, and the ultraviolet spectra in neutral, acidic, and basic solutions were in accord with the proposed structures. Pertinent data are summarized in Table I.

Table I. Properties of Synthetic Trityloligonucleotides[a]

	TrdCpT[b]	TrdCpTpT	TrdCpTpTpT
R_f solvent A	0.72	0.50	0.29
R_f solvent B	0.74	0.57	0.48
Electrophoretic mobility rel. to pdC	0.37	0.59	0.80
Ultraviolet in 0.01 M HCl[c]	275 (17900)	271 (24800)	269 (32000)
	243 (5500)	241 (9200)	237 (12800)
Ultraviolet in water[c]	270 (15600)	270 (23000)	269 (30600)
	247 (9600)	246 (13000)	242 (16400)
Ultraviolet in 0.01 M NaOH[c]	269 (14300)	269 (20200)	268 (26300)
	249 (10200)	249 (14400)	249 (19800)

[a] The values for ϵ depend upon the molecular weights assumed for the nucleotide derivatives. Since these products were hygroscopic, varying amounts of water could have been taken up; consequently the absolute values are not very precise. The molecular weights used in the calculations were 828 for TrdCpT, 1204 for TrdCpTpT, and 1436 for TrdCpTpTpT, corresponding to the formulas calculated from the combustion analyses (see Experimental Section). [b] The solvent for the ultraviolet determinations on TrdCpT contained a small amount of ethanol (5% v/v). [c] λ_{max} (ϵ) and λ_{min} (ϵ) are given.

That the nucleotides were joined in 3'–5' phosphodiester linkages was shown by degradation by snake venom phosphodiesterase, an enzyme which acts on oligonucleotides bearing a free 3'-OH group. In each case the trityloligonucleotide was hydrolyzed to 5'-O-trityldeoxycytidine and thymidine 5'-phosphate in the appropriate mole ratio.

These synthetic experiments demonstrate the utility of a polymer support in building oligonucleotide chains. The principal limitations stem from the yields, which are lower than desirable, and the time required in the reaction with β-cyanoethyl phosphate. Recent experiments have shown that both aspects may be somewhat improved. On the assumption that poor diffusion through the cross-linked support polymer was responsible for the inefficient conversions, we have repeated the synthesis of TrdCpT on a polymer support prepared from a monomer mixture that contained only 0.02% divinylbenzene. Manipulations with this support were quite satisfactory, and the yield of TrdCpT was 61% of the theroretical based on the millimoles of trityldeoxycytidine bound to the support. It has also been found that the condensation of β-cyanoethyl phosphate with a 3'-hydroxyl group on a nucleoside joined to the support may be effected with mesitylenesulfonyl chloride in place of dicyclohexylcarbodiimide. In this case the time required to join a full nucleotide unit to the chain on the support may be reduced to 2 days or less. By use of mesitylenesulfonyl chloride throughout with the low cross-linked support, TrdCpT and TrdCpTpT were prepared from polymer–TrdC in yields of 44 and 21%, respectively.

Experimental Section

Ultraviolet spectra were obtained with a Cary 14 or a Beckman DU spectrophotometer. Infrared spectra were obtained with a Baird Model AB2 spectrometer with the sample in a potassium bromide disk. Melting points were determined with a Fisher-Johns apparatus. Descending paper chromatography was carried out on Whatman 3MM paper with solvent A (isopropyl alcohol, concentrated ammonium hydroxide, and water in the proportions 7:1:2 by volume) and in solvent B (1-butanol, acetic acid, and water in the proportions 5:2:3 by volume). Nucleosides and nucleotides were observed by their fluorescence in ultraviolet light (~2537 A), and trityl-containing compounds were observed by the yellow color which developed when the paper was sprayed with 10% aqueous perchloric acid and subsequently warmed in a drying oven at 60° for 30 min.

Electrophoretic separations were made at pH 8.1 on Whatman 3MM paper strips with a Savant flat plate electrophoresis apparatus operated at 2000 v. The buffer solution contained 9.03 g of disodium hydrogen phosphate and 0.453 g of potassium hydrogen phosphate per liter.

Diethylaminoethyl- (DEAE) cellulose and Sephadex G-25 and G-15 were used in separating the reaction products. For the former, 80 g of DEAE-cellulose (Calbiochem, 0.93 mequiv/g) was washed[11] and added with stirring to a glass column (4 × 80 cm) as a slurry in 1.5 l. of 0.2 M ammonium bicarbonate buffer. The height, after a washing with 6 l. of 0.02 M ammonium bicarbonate, was approximately 35 cm. The Sephadex column was 1.5 × 120 cm and was packed with 0.02 M ammonium bicarbonate as the liquid phase. Effluent from the columns was monitored with a Gilson UV-2651F ultraviolet absorption meter and a Texas instrument rectiriter recorder; fractions (~13 ml) were collected automatically with a Gilson Model VL fractionator.

Chemicals. Nucleosides were purchased from Nutritional Biochemicals Corp. 5'-O-trityldeoxycytidine,[12] mp 249°, and N-benzoyldeoxyadenosine,[3c] mp 113–114°, were prepared by procedures described in the literature. Barium β-cyanoethyl phosphate was converted to the pyridinium salt[9] in aqueous solution by exchange with the pyridinium form of Dowex 50 resin. After evaporation of the resulting solution to dryness the salt was rendered anhydrous by repeated additions of dry pyridine and concentration of the solution; then a 0.8 M solution of the salt was prepared by addition of the appropriate amount of pyridine. Mesitylenesulfonyl chloride, mp 57°, was prepared by the method of Wang and Cohen.[13] Pyridine was dried over calcium hydride and fractionated through a 35-cm helices-packed column immediately before use.

General Procedure. The reactions were carried out at room temperature, unless otherwise specified, in glass-stoppered flasks equipped with a magnetic stirrer. Pyridine was employed as a solvent for the reagents except in the cleavage step. After the appropriate period of stirring, the polymer was separated from the

---

(11) G. M. Tener, H. G. Khorana, R. Markham, and E. H. Pol, *J. Am. Chem. Soc.*, **80**, 6223 (1958).
(12) A. M. Michelson and A. R. Todd, *J. Chem. Soc.*, **34** (1954).
(13) C. Wang and S. G. Cohen, *J. Am. Chem. Soc.*, **79**, 1924 (1957).

solution of reagents and by-products by filtration, centrifugation, or settling under gravity. Filtration was employed when exposure to air and moisture was not detrimental; otherwise, the centrifugation or settling technique was used. Since the support with adjoining nucleotides was insoluble in pyridine, the reaction products could in principle be recovered quantitatively, though in practice small mechanical losses were experienced.

Polymer I. A support made from styrene (88%), p-vinylbenzoic acid (12%), and p-divinylbenzene (0.2%) was converted to the acid chloride form by treatment with excess thionyl chloride in benzene[4] and stored in a vacuum desiccator over phosphorus pentoxide. The infrared spectrum had two strong bands in the carbonyl region at 5.65 and 5.75 μ.

A portion (12 g) of the polymer acid chloride was stirred with 7.0 g (14.9 mmoles) of 5′-O-trityldeoxycytidine in 100 ml of dry pyridine for 2 days. To esterify any acid chloride groups that remained, 50 ml of methanol was added and stirring continued for 12 hr. The polymer (I) was then collected by filtration, extracted with methanol-pyridine (9:1 v/v) in a Soxhlet extractor for 4 hr to remove residual trityldeoxycytidine, and dried under vacuum. At this stage the polymer weighed 13.3 g and exhibited a band at 5.9 μ in the infrared.

To recover excess trityldeoxycytidine, the filtrate from the polymer reaction was concentrated and mixed with 1 l. of water and ice containing 5 ml of ammonium hydroxide. Filtration afforded 5.6 g of 5′-O-trityldeoxycytidine, identified by its melting point and infrared spectrum.

The amount of trityldeoxycytidine which had been covalently bound to the support was determined from the optical density of a solution prepared from the alkaline cleavage products. For this purpose 100 mg of polymer (I) was stirred with approximately 30 ml of 0.4 M sodium hydroxide in dioxane-ethanol-water (10:10:1, v/v/v) for 36 hr and then separated from the solution by filtration. That all trityldeoxycytidine had been cleaved from the polymer under these conditions was indicated by the absence of absorption characteristic of this nucleoside in the infrared spectrum of the support. The alkaline solution was neutralized with Dowex 50 (pyridinium form) and evaporated to dryness in a rotary evaporator at about 35°. Ethanol (50 ml) and ammonia (2 ml) were added and the solution again taken to dryness *in vacuo* in order to remove residual pyridine. The process was repeated three more times; then the residue was dissolved in 100 ml of ethanol which was 0.01 M in hydrochloric acid. The optical density at 280 mμ of 2 ml of this solution, after dilution to 10 ml, was 0.44. On the basis that the extinction coefficient at 280 mμ is 1.33×10^4, this corresponds to 78 mg (0.166 mmole) of trityldeoxycytidine bound per gram of polymer I.

Polymer II. A mixture containing 12 g of polymer-trityldeoxycytidine (I) (2.0 mmoles of bound trityldeoxycytidine), 12.5 mmoles of pyridinium β-cyanoethyl phosphate, 5.16 g (25 mmoles) of dicyclohexylcarbodiimide, and 60 ml of dry pyridine was stirred for 8 days. Water (100 ml) was added and after an additional 24-hr period of stirring the solids (polymer plus dicyclohexylurea) were separated by filtration, washed with 20% aqueous pyridine, and stirred for 3 hr with 100 ml of 1:1 cyclohexane-ethanol. Filtration and two additional washes with cyclohexane-ethanol removed the dicyclohexylurea. The polymer was then washed with methanol and ether and dried in a desiccator over phosphorus pentoxide.

Polymer III. Activation of the phosphodiester was achieved by stirring 5.5 g of the polymer (II) with 3.27 g (15 mmoles) of mesitylenesulfonyl chloride in 50 ml of pyridine for 24 hr. An additional 50-ml portion of pyridine was then added, the polymer was allowed to settle out, and the supernatant liquid was withdrawn. This process was repeated two times to remove the major portion of the excess mesitylenesulfonyl chloride. Condensation with thymidine was effected by stirring the residual slurry with 0.600 g (2.48 mmoles) of thymidine and 25 ml of pyridine for 48 hr. Methanol (50 ml) was then added and after 3 hr of stirring the mixture was filtered and the polymer washed successively with 30% aqueous pyridine, methanol, and ether.

5′-O-Trityldeoxycytidylyl-(3′-5′)-thymidine. Nucleotidic material was cleaved from a 2.2-g portion of III by stirring the polymer successively with six 50-ml portions of 0.4 M sodium hydroxide in dioxane-ethanol-water (10:10:2.2, v/v/v). After approximately 6 hr of stirring each portion was filtered and the filtrate was neutralized with Dowex 50 resin in the pyridinium form (approximately 12 g of resin required for each portion). The resin was removed by filtration and washed with 50% aqueous pyridine and ethanol. The combined filtrates and washings were concentrated to 10–20 ml in a rotary evaporator. After addition of 50 ml of water and reconcentration, the pH was brought to 10 with ammonium hydroxide, and the mixture was centrifuged to remove the insoluble trityldeoxycytidine (56 mg, 0.12 mmole, identified by mp 232–235°, infrared spectrum, and R_f 0.86 in solvent A). The solution (~25 ml) was applied to the top of a DEAE-cellulose column and elution effected by a linear gradient of ammonium bicarbonate buffer (2.2 l. of 0.02 M ammonium bicarbonate buffer in the mixing vessel and 2.2 l. of 0.25 M ammonium bicarbonate in the reservoir). Fractions were collected approximately every 7 min. After an initial peak, due primarily to pyridine, two peaks were obtained. The tubes within each were pooled and the absorbance was determined (see Table II).

Table II. Products from Preparation of TrdCpT

Peak	Tube no.	Total vol., ml	$OD_{270}{}^a$	$OD_{270} \times$ vol.[b]
i	62–125	400	0.42	168
ii	135–310	2400	1.20	2900

[a] OD_{270} refers to the optical density measured at 270 mμ. [b] This product gives the number of optical density units present.

Paper chromatography of a sample of i in solvent A showed one spot, R_f 0.49, positive for the trityl group. The substance is probably TrdCp. Peak ii similarly gave a single spot on chromatography in solvent A, R_f 0.78, positive for trityl. Concentration and lyophilization of ii afforded 150 mg (0.181 mmole) of trityldeoxycytidylylthymidine as a fluffy white powder. It was homogeneous in solvents A and B and on electrophoresis at pH 8.1. *Anal.*[14] Calcd for $C_{38}H_{40}N_5O_{11}P \cdot 3H_2O$: C, 55.13; H, 5.60; N, 8.46; P, 3.75. Found: C, 54.85; H, 5.61; N, 8.11; P, 4.18.

The yield may be estimated from (a) the weight of TrdCpT recovered per gram of polymer hydrolyzed or (b) the OD_{270} units in fraction peak ii relative to the number of OD_{270} units eluted from the polymer by the alkaline treatment. On the weight basis, 150 mg of TrdCpT corresponds to ~53% of the amount calculated from the number of TrdC units bound to the polymer. In terms of OD_{270} units, the units obtained in peak ii amount to 60% of the total OD_{270} units eluted from the support, including the trityldeoxycytidine collected as a solid prior to fractionation. If the reactions had been quantitative, of course, all of the OD_{270} units would have been in fraction ii.

5′-O-Trityldeoxycytidylyl-(3′-5′)-thymidylyl-(3′-5′)-thymidine. A sample of polymer III (3 g) was stirred with 4 mmoles of pyridinium β-cyanoethyl phosphate and 1.65 g (8 mmoles) of dicyclohexylcarbodiimide in pyridine for 7 days. By the procedure followed in preparing III, the resulting polymer was activated with 1.32 g (6 mmoles) of mesitylenesulfonyl chloride and condensed with 0.300 g (1.24 mmoles) of dry thymidine. Alkaline cleavage of 2.85 g of the polymer and work-up as in the previous case afforded 70 mg of trityldeoxycytidine and the products indicated in Table III.

Table III. Products from Preparation of TrdCpTpT

Peak	Tube no.	Total vol., ml	OD_{270}	$OD_{270} \times$ vol.
i	50–101	420	0.59	248
ii	102–140	600	1.30	858
iii	145–160	290	0.96	278
iv	163–222	850	1.30	1110
v	225–325	1360	1.15	1565

A forerun from the fractionation, containing as ultraviolet absorbing material mainly pyridine with trace amounts of thymidine and a substance which appeared to be deoxycytidine, was also obtained.

---

(14) The compounds were prepared for analysis by drying under vacuum over phosphorus pentoxide at room temperature. After drying they were hygroscopic, taking up moisture slowly when exposed to air, and they gave acidic solutions when redissolved in water. The theoretical values for the analyses were calculated for hydrates of the nucleotides in the acid form. The C, H, and N analyses were made by the Micro-Tech Laboratories, Skokie, Ill. Phosphorus was determined by the method of P. S. Chen, T. Y. Toribara, and H. Warner, *Anal. Chem.*, **28**, 1756 (1956).

Synthesis of Oligodeoxyribonucleotides

Anal.[14] Calcd for $C_{48}H_{53}N_7O_{18}P_2 \cdot 7H_2O$: C, 47.88; H, 5.61; N, 8.14; P, 5.23. Found: C, 47.74; H, 5.58; N, 7.51; P, 5.47.

Paper chromatography showed that i contained thymidine as the single fluorescent material (R_f 0.70 in solvent A), ii had one component (R_f 0.48, trityl positive, probably TrdCp), iii had two components (R_f 0.1 and 0.86, both trityl free), iv had one component (R_f 0.77, trityl positive, probably TrdCpT), and v had one component (R_f 0.61, trityl positive). Concentration and lyophilization of v afforded 80 mg of TrdCpTpT as a white powder ($\sim$16% of the weight calculated on the basis of the weight of polymer hydrolyzed). The OD_{270} units in v correspond to 29% of the OD_{270} units removed from the support by the alkaline treatment.

5'-O-Trityldeoxycytidylyl-(3'-5')-thymidylyl-(3'-5')-thymidylyl-(3'-5')-thymidine. This substance was prepared from a fresh batch of polymer-trityldeoxycytidine. Polymer I (8.25 g.) was phosphorylated with 10 mmoles of β-cyanoethyl phosphate and 4.54 g (22 mmoles) of dicyclohexylcarbodiimide over a period of 7 days as in the previous synthesis. Subsequent activation with 4.38 g (20 mmoles) of mesitylenesulfonyl chloride and treatment with 0.85 g (3.5 mmoles) of thymidine afforded the dinucleotide bound to the polymer (polymer type III). This sequence was repeated twice to build up the polymer-tetranucleotide derivative (7.7 g), which was then cleaved with 0.4 M sodium hydroxide in dioxane–ethanol–water and worked up as before. The trityldeoxycytidine recovered amounted to 82 mg. From the fractionation on DEAE-cellulose was obtained an initial fraction, which contained pyridine and the products indicated in Table IV. Paper chromatography

Table IV. Products from Polymer Support Synthesis of TrdCpTpTpT

Peak	Tube no.	Total vol., ml	OD_{270}	$OD_{270} \times$ vol.
i	50–85	315	6.25	1970
ii	86–100	200	2.3	460
iii	101–140	560	8.5	4760
iv	141–245	1420	4.2	5960
v–viii	250–420	2340		5506
ix	425–550	1900	1.65	3140

with solvent A revealed a single ultraviolet-absorbing product in i (R_f 0.70, thymidine), ii (R_f 0.59, probably deoxycytidine), iv (R_f 0.27, trityl positive, possibly TrdCpTp), and ix (R_f 0.35, trityl positive, identified as TrdCpTpT). Fraction iii contained a substance with R_f 0.38 and a trace amount of a material with R_f 0.78. Portions v–viii contained a total of three components with R_f values of 0.61, 0.15, and 0.04, the former two being positive for trityl and the last negative for trityl.

The yield of TrdCpTpTpT calculated from the weight of isolated material (220 mg) corresponds to 14% of the amount calculated on the basis of the amount of trityldeoxycytidine bound per gram of polymer. For comparison, the OD_{270} units in fraction ix correspond to 15% of the total OD_{270} units removed from the polymer support.

Anal.[14] Calcd for $C_{58}H_{66}N_9O_{23}P_3 \cdot 3H_2O$: C, 48.50; H, 5.05; N, 8.78; P, 6.46. Found: C, 48.53; H, 5.29; N, 8.50; P, 6.10.

5'-O-Trityldeoxycytidylyl-(3'-5')-thymidylyl-(3'-5')-deoxyadenosine. Essentially the same procedure was used as in the preparation of TrdCpTpT except for the cleavage step. In this case 2 g of polymer at the second phosphorylation stage was activated by 0.750 g (3.63 mmoles) of mesitylenesulfonyl chloride (24 hr) and condensed with 0.711 g (2 mmoles) of N-benzoyldeoxyadenosine (2 days). Cleavage was effected by three successive 5-hr treatments of the polymer with 100 ml of 1:1 pyridine–concentrated ammonium hydroxide on a steam bath. During the heating periods, ammonia was replenished periodically (0.5-hr intervals) by addition of 10–20 ml of ammonium hydroxide. This work was done before the cleavage method utilizing sodium hydroxide was developed. The method is inefficient in that not all the nucleotidic material is removed from the polymer support. Fractionation of the products afforded 525 OD_{270} units of TrdCpTpA, $\sim$26% of the total of $\sim$2000 OD_{270} units eluted from the polymer support. This material was homogeneous by all criteria applied: R_f (solvent A) 0.60; R_f (solvent B) 0.64; electrophoretic mobility relative to pdC at pH 10.8, 0.67. In the ultraviolet in 0.01 M HCl, water, and 0.01 M NaOH λ_{max} ($\epsilon \times 10^{-4}$) was 272 (2.5), 269 (2.2), and 267 (2.0),

respectively, and λ_{min} ($\epsilon \times 10^{-4}$) was 239 (1.1), 246 (1.5), and 267 (1.6), respectively.

Preparation of TrdCpT on a Low-Cross-Linked Support. A 5.5-g sample of support polymer acid chloride (the support was a popcorn polymer made from styrene (90%), p-vinylbenzoic acid (10%), and p-divinylbenzene (0.02%)) was stirred with 3 g (6.5 mmoles) of 5'-O-trityldeoxycytidine in 50 ml of pyridine for 2 days. Methanol (50 ml) was added and the polymer was filtered and washed with 9:1 methanol-pyridine in a Soxhlet extractor. From the filtrate and washings was obtained 1.95 g of trityldeoxycytidine. The polymer, after drying in a desiccator over phosphorus pentoxide, weighed 6.48 g and contained 0.245 mmole of TrdC per gram of polymer.

One portion of this polymer (2.18 g) was used to build TrdCpT by the procedure previously described. In this case phosphorylation with 4 mmoles of pyridinium β-cyanoethyl phosphate and 12 mmoles of dicyclohexylcarbodiimide (7 days) afforded 2.3 g of a polymer–phosphorylated derivative. Activation (1 day) of 2.1 g of this product with 1.05 g (48 mmoles) of mesitylenesulfonyl chloride, condensation (2 days) with 0.480 g (2 mmoles) of thymidine, and cleavage of 2.12 g of the resulting derivative with 0.4 M sodium hydroxide in dioxane–ethanol–water (40 m) yielded 560 OD_{270} units ($\sim$26 mg) of trityldeoxycytidine and the fractionation products shown in Table V. Evaporation to dryness and analysis

Table V. Products from Preparation of TrdCpt on Low-Cross-Linked Support

Peak	Tube no.	Total vol., ml	OD_{270}	$OD_{270} \times$ vol.
i	8–41	450	2.5	1120
ii	60–80	320	0.6	192
iii	81–105	350	0.82	287
iv	125–410	4090	1.02	4170

of the residue from i indicated that i contained pyridine (370 OD units) and three substances with R_f 0.69 (thymidine), 0.56 (deoxycytidine), and 0.84 (probably mesitylenesulfonic acid). On paper chromatography in solvent A, ii showed three components (R_f 0.50, 0.60, 0.84), iii showed two components (R_f 0.40 and 0.84), and iv showed one component (R_f 0.73, TrdCpT). The yield of TrdCpT (240 mg of solid isolated) corresponded to 61% conversion of the polymer–TrdC to TrdCpT. Furthermore, the mixtures obtained by base treatment of the polymer were relatively simple, 77% of the OD_{270} units being in the TrdCpT fraction.

Another portion (1.0 g) of the low-crossed-linked support polymer with adjoined trityldeoxycytidine was phosphorylated by stirring with 3 mmoles of pyridinium β-cyanoethyl phosphate and 1.96 g (9 mmoles) of mesitylenesulfonyl chloride in 15 ml of pyridine for 24 hr. After centrifugation and removal of the supernatant liquid, the jelly-like polymer was hydrolyzed by stirring with 100 ml of 30% pyridine for 4 hr and then with 100 ml of 10% aqueous pyridine for 30 min on a steam bath. The polymer was then separated by filtration, washed with methanol and ether, and dried in a vacuum desiccator over phosphorus pentoxide; polymer weight, 1.01 g. Part of the resulting polymer (0.86 g) was then activated by 0.306 g (1.4 mmoles) of mesitylenesulfonyl chloride and condensed with thymidine as in the previous synthesis of TrdCpT; polymer weight recovered, 0.95 g. Cleavage of 0.90 g of this polymer and work-up as usual afforded 400 OD_{270} units of trityldeoxycytidine, 1390 OD_{270} units of TrdCpT (53% of total), and 817 OD_{270} units of other material absorbing in the ultraviolet region.

Preparation of TrdCpTpT on Low-Cross-Linked Support. The technique and procedures were the same as employed in the synthesis of TrdCpT. In the first stage 1.5 g of support polymer bearing trityldeoxycytidine (0.02% divinylbenzene in polymer; 0.245 mmole of trityldeoxycytidine per gram of polymer) was phosphorylated with β-cyanoethyl phosphate and mesitylenesulfonyl chloride in pyridine and then condensed with thymidine by means of mesitylenesulfonyl chloride. The resulting polymer slurry was mixed with 150 ml of methanol and stirred for 1 hr. The solid was then filtered off, stirred with 20% aqueous pyridine for 3 hr at room temperature and 1 hr at 90°, separated by filtration, and dried under vacuum. The second nucleotide was added to the chain by repetition of the steps used in the first stage, and the nucleotide products were removed from the final polymer (weight, 1.68 g) by conventional treatment with 0.4 M sodium hydroxide. Work-up of the products afforded 21 mg of trityldeoxycytidine and the fractions listed in Table VI.

Table VI. Products from Preparation of TrdCpTpT on Low Cross-Linked Support

Peak	Tube no.	Total vol., ml	OD_{270}	OD_{270} × vol.
i	6–38	430	1.62^a	700
ii	42–105	850	1.375	1170
iii	121–200	1040	1.175	1220
iv	212–240	375	0.67	251
v	242–345	1500	1.15	1720

^{a} OD after evaporation to remove pyridine and then redilution.

Paper chromatography in solvent A indicated that i was mainly thymidine with a trace of trityldeoxycytidine; ii contained two components, R_f 0.49 (probably TrdCp) and 0.71; iii was largely TrdCpT (R_f 0.73) with a trace of material with R_f 0.15; iv had one compound with R_f 0.29; and v showed one component with R_f 0.62, which was the desired TrdCpTpT. Fraction v contained 34% of the ultraviolet-absorbing material removed from the the support polymer. The OD units of TrdCpTpT correspond to 21% of the amount calculated assuming quantitative conversion of the TrdC on the polymer to TrdCpTpT.

Spectral Determinations and Enzyme Assays. For spectral determinations and enzyme assay, the oligonucleotide derivatives were dissolved in water and rechromatographed on DEAE-cellulose. Better than 90% of the solids were recovered in all cases (Table I). This purification procedure had little effect on the ϵ value for TrdCpT, but increased the ϵ values for TrdCpTpT and TrdCpTpTpT by about 10%.

Table VII. Products of Enzymatic Degradation

Compound	TrdC OD_{278} units	pT OD_{267} units	TrdC/pT Found, %	TrdC/pT Calcd, %
TrdCpt	2.45	1.9	1.1	1
TrdCpTpT	0.75	1.15	2.1	2
TrdCpTpTpT	0.60	1.35	3.1	2

Hydrolyses were carried out by incubating 0.4–1.0-mg samples of the nucleotide derivatives with 0.1 ml of an aqueous solution of snake venom phosphodiesterase for 7 hr at 37°. The enzyme solution was made up by dissolving 500 units of phosphodiesterase (Russel's viper venom, lyophilized)[15] in 2.5 ml of 0.33 M Tris buffer at pH 9.1. Turbidity which developed due to formation of trityldeoxycytidine was cleared by addition of two drops of pyridine and the clear solution was chromatographed on 3MM paper in solvent A. Product spots were cut out and eluted by soaking for 24 hr in 30% ethanol-water 0.01 M in HCl (for the phosphates) or 90% ethanol-water 0.01 M in HCl (for trityldeoxycytidine). Appropriate blanks were cut from other areas of the chromatograms and soaked in the same solvents. Concentrations of the nucleotides and trityldeoxycytidine were determined from differences between the absorbances of solutions resulting from elution of the spots and the blanks. Values used for extinction coefficients in these calculations were 13,200 for trityldeoxycytidine and 9600 for thymidine 5'-phosphate. A similar hydrolysis of TrdCpTpA proceeded well and afforded approximately equal amounts of trityldeoxycytidine, thymidine 5'-phosphate, and adenosine 5'-phosphate, as judged visually from the spots produced on paper chromatography.

(15) Calbiochem.

Reactions of Nucleosides on Polymer Supports. Synthesis of Thymidylylthymidylylthymidine[*]

Robert L. Letsinger, Marvin H. Caruthers, and Donald M. Jerina[†]

ABSTRACT: A procedure is described for synthesizing thymidylyl-(3'-5')-thymidylyl-(3'-5')-thymidine on an insoluble polymer support. The support is a carboxylated styrene popcorn polymer. In the acid chloride form it reacts with 5'-O-monomethoxytritylthymidine, forming an ester link at the 3'-O position of the nucleoside. Subsequent cleavage of the monomethoxytrityl group by acid and condensation of the liberated 5'-hydroxyl group with 5'-O-monomethoxytritylthymidine 3'-phosphate afford a dinucleoside phosphate derivative on the support. On repetition of these steps and treatment with alkali to break the ester link holding the nucleotide product to the support, thymidylyl-(3'-5')-thymidylyl-(3'-5')-thymidine is obtained in 51% yield based on thymidine originally bound to the support.

Since introduction of the technique of synthesizing oligonucleotides on polymer supports (Letsinger and Mahadevan, 1965, 1966) several laboratories have reported work in the area. A major advantage of the support technique is that the products in a multistep synthesis may be separated easily from excess reagents and soluble by-products at each step. In the original procedure 5'-O-trityldeoxycytidine was joined to an insoluble polystyrene-type popcorn polymer by reaction of the 4-amino group with an acid chloride function on the support. Nucleotide units were added by successive phosphorylation of the 3'-hydroxyl group with β-cyanoethyl phosphate and dicyclohexylcarbodiimide, activation of the phosphate with mesitylenesulfonyl chloride, and condensation of the active phosphate with a nucleoside at the 5'-oxygen position.

Recently Hayatsu and Khorana (1966) and Cramer et al. (1966) described a synthetic procedure in which the initial nucleoside was joined through the 5'-hydroxyl group to a triarylmethyl derivative of a polymer that was soluble in pyridine. The chain was lengthened from the 3' position by condensation with

---

[*] From the Department of Chemistry, Northwestern University, Evanston, Illinois. *Received January 20, 1967.* This research was supported by the Division of General Medical Sciences, National Institutes of Health (GM-10265), and by Public Health Service predoctoral fellowship awards to M. H. C. (5-F1-GM-23-558) and D. M. J. (5-F1-GM-23-577). Part VI in series on Nucleotide Chemistry. Part V: Letsinger and Ogilvie (1967).

[†] Present address: National Institute of Arthritis and Metabolic Diseases, National Institutes of Health, Bethesda, Md.

1379

SCHEME I: Reaction of Thymidine on Support Polymer.

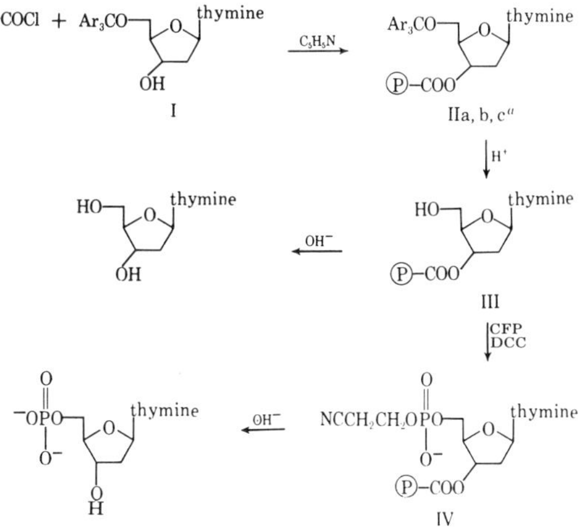

^{a} Ar$_3$C = (a) trityl, (b) monomethoxytrityl, and (c) dimethoxytrityl.

3'-O-acetylthymidine 5'-phosphate. On hydrolytic cleavage of the acetate, a new 3'-hydroxyl group was liberated for the next cycle. Separation of the polymer from the solvent was effected in this case by precipitation with water followed by filtration. Essentially the same chemical steps have also been used with an insoluble polymer (Melby and Strobach, 1967).

Concurrent with the synthetic experiments with deoxycytidine, a study was initiated to test the feasibility of building oligonucleotide chains on polymer supports from the 5'-hydroxyl terminus of the bound nucleoside. The approach that was envisaged involved reaction of the acid chloride form of the support polymer (Ⓟ–COCl)[1] with the 3'-hydroxyl of a 5'-O-protected nucleoside derivative, cleavage of the 5'-O-blocking group, and condensation of the 5'-hydroxyl with a suitable phosphate derivative. In developing this approach we first carried out a number of experiments to determine conditions for joining nucleoside derivatives through the 3'-hydroxyl position to the support and for removing nucleotidic material from the support. Triarylmethyl groups were selected to protect the 5'-hydroxyl positions since the derivatives are readily prepared and the triarylmethyl groups may be removed satisfactorily from the pyrimidine nucleotide derivatives.

[1] Abbreviations used: T, thymidine; pT, thymidine 5'-phosphate; Tp, thymidine 3'-phosphate; TpT, thymidylyl-(3'–5')-thymidine; TpTpT, thymidylyl-(3'–5')-thymidylyl-(3'–5')-thymidine; DCC, dicyclohexylcarbodiimide; CEP, β-cyanoethyl phosphate; Ⓟ–, inert portion the polymer support; MTr, monomethoxytrityl.

Although infrared spectroscopy (Letsinger et al., 1964) provides a means for determining when nucleosidic material has been added to or removed from the support polymer, the technique is not generally suitable for following the synthesis of an oligonucleotide quantitatively. The products for a given chemical step were, therefore, determined by cleaving the covalent bonds holding the nucleosidic and nucleotidic material to the support, washing the support, separating the soluble components by paper chromatography or electrophoresis, and assaying the materials spectrophotometrically. Cleavage was effected initially with mixtures of ammonium hydroxide and pyridine, the organic solvent being used to swell the polymer; however, this combination did not prove very effective. Much nucleotide material remained on the support even after several separate treatments. Attention was then turned to the use of sodium hydroxide solutions. It was found that 0.5 M sodium hydroxide in 50% aqueous dioxane at room temperature was quite satisfactory. With it 93% of the thymidine was removed from the polymer–thymidine sample (III) within 16 hr and all of the thymidine was liberated within 2 days. There is danger that deamination may occur when nucleotides bearing amino groups are subjected to alkaline treatment; however, no evidence for deamination was found when deoxycytidine derivatives were treated with 50% aqueous dioxane solutions of sodium hydroxide under these conditions (Letsinger and Mahadevan, 1966). Ethanol was a poor solvent for the cleavage reaction. Thus, only 43% of the thymidine was released when the polymer–thymidine was treated with 0.1 M sodium hydroxide in ethanol at reflux for 5 hr.

Reactions of Nucleosides on Polymer Supports

SCHEME II: Sequence for Addition of One Nucleotide Unit.

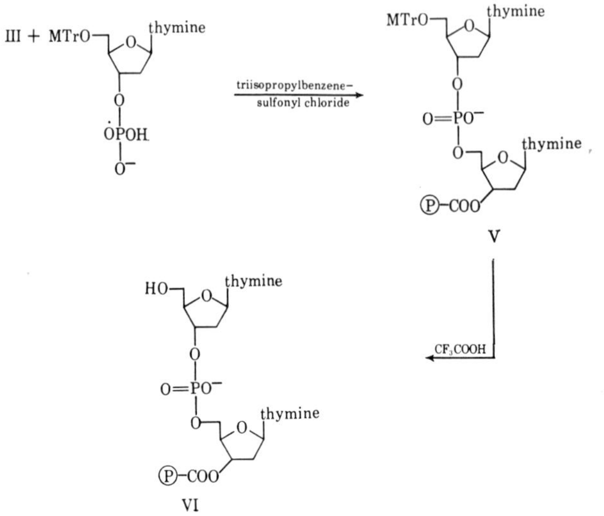

The hydroxyl groups in polymer III (Scheme I) reacted satisfactorily with the common phosphorylating reagents. Phosphorylations of III derived from the monomethoxytritylthymidine derivative IIb yielded thymidine 5'-phosphate as the sole mononucleotide; however, chromatographic data indicated that both thymidine 3'-phosphate and thymidine 5'-phosphate were obtained from samples of III that had been prepared from the dimethoxytritylthymidine derivative IIc. The latter result is consistent with the assumption that some of the dimethoxytrityl groups had been lost during the reaction with the acid chloride and that the resulting thymidine had coupled to the support through the 5'-O position. In support of this explanation is the observation that alkaline hydrolysis of IIc afforded thymidine as well as dimethoxytritylthymidine. Since the dimethoxytrityl group was not satisfactory in this system, all further work was carried out with samples of III prepared from the monomethoxytritylthymidine derivative IIb.

Ninety per cent of the thymidine in III was phosphorylated when the polymer was treated with β-cyanoethyl phosphate and dicyclohexylcarbodiimide for a period of 6 days. With mesitylenesulfonyl chloride as the activating agent in place of dicyclohexylcarbodiimide a 76% yield of thymidine 5'-phosphate was obtained from a reaction run for 4 hr. About the same yield (73%) of thymidine 5'-phosphate was realized by phosphorylating with phosphorus oxychloride in the presence of imidazole and triethylamine.

Attempts to condense the cyanoethyl derivative of the polymer-thymidine (IV) with 5'-O-monomethoxytritylthymidine were not promising. With mesitylenesulfonyl chloride as the condensing agent a mixture of products was obtained which contained methoxytritylthymidylylthymidine and thymidine 5'-phosphate in approximately a 5:3 mole ratio. The reaction of the activated 5'-phosphodiester with the 3'-hydroxyl group is, therefore, less satisfactory than the corresponding reaction of an activated 3'-phosphodiester with a 5'-hydroxyl group (Letsinger and Mahadevan, 1966).

Direct condensation of a nucleoside 3'-phosphate with the 5'-hydroxyl of the polymer derivative proved superior for adding nucleotide units to the chain. This method is illustrated by the synthesis of thymidylyl-(3'-5')-thymidylyl-(3'-5')-thymidine. The phosphorylation reaction is similar to that utilized in the preliminary studies, with 5'-O-monomethoxytritylthymidine 3'-phosphate being used in place of β-cyanoethyl phosphate. The nucleotide derivative was obtained by reaction of thymidine 3'-phosphate with monomethoxytriphenylmethyl chloride in pyridine. In pilot runs carried to the dinucleoside phosphate stage it was found that somewhat better yields of the phosphodiester were obtained by using 2,4,6-triisopropylbenzenesulfonyl chloride (72 ± 2% yield) in place of mesitylenesulfonyl chloride (66 ± 3% yield), and that the monomethoxytrityl group could be removed very rapidly from the nucleotide derivatives on the polymer with trifluoroacetic acid in benzene (see Hayatsu and Khorana, 1966). Consequently these modifications were introduced.

The sequence for adding a nucleotide unit to the chain is outlined in Scheme II. The two steps involved were condensation of 5'-O-monomethoxytritylthymidine 3'-phosphate with the 5'-hydroxyl of the terminal nucleoside on the support, and cleavage of the monomethoxytrityl ether to liberate a new 5'-hydroxyl group. The time required for addition of one nucleotide unit, including periods for washing the polymer, was 1 day. For preparation of the trinucleoside diphosphate the cycle was repeated and the products were stripped from the support by alkaline hydrolysis. The three major products obtained were thymidine, thymidylylthymidine, and thymidylylthymidylylthymidine. On the basis of the thymidine units originally bound to the support the over-all yields of the compounds for the two-cycle synthesis were 33, 16, and 51%, respectively. The absorbance of these materials (measured at 267 mμ) constituted 76% of the total absorbance of the substances eluted from the support. The other ultraviolet-absorbing material (in part, thymidine phosphate) was probably held to the insoluble polymer by pyrophosphate-type links.

Thymidylylthymidylylthymidine prepared in this manner was homogeneous on paper chromatography and electrophoresis. Complete degradation by snake venom and spleen phosphodiesterase indicated that the nucleotide units were joined in the natural 3'-5' linkage. The ultraviolet and infrared spectra and the elemental analysis were also consistent with the proposed structure.

This synthetic procedure is more rapid and gives better yields than the procedure which builds from trityldeoxycytidine bound to be an insoluble support (Letsinger and Mahadevan, 1965, 1966). The yields are better than those reported for support synthesis by Cramer et al. (1966) and by Melby and Strobach (1967) but are somewhat lower than the yields reported by Hayatsu and Khorana (1966) for syntheses conducted in solution.

At the present stage of development the procedure utilizing 5'-O-monomethoxytritylthymidine 3'-phosphate and an insoluble polymer support provides a convenient method for synthesizing short-chain oligonucleotides derived from thymidine. The method can probably be used for deoxycytidine derivatives as well. For extension to the synthesis of oligonucleotides containing deoxyguanosine and deoxyadenosine, however, blocking groups which do not require acidic conditions for removal will have to be used.

Experimental Section

General Methods. Paper chromatography was conducted by the descending technique on Whatman 3MM paper. The solvent systems were (A) isopropyl alcohol–concentrated ammonium hydroxide–water (7:1:2), (B) 1-butanol–acetic acid–water (5:2:3), and (C) 1-propanol–2 M hydrochloric acid (3:1). The solvents were prepared on a volume basis.

Paper electrophoresis was performed using Whatman 3MM paper with an applied potential of about 40 v/cm. A Savant flat-plate electrophoretic chamber and 2000-v power supply were used. The buffers were ammonium bicarbonate (0.05 M, pH 7.55), ammonium formate–formic acid (0.05 M in ammonium ion, pH 3.6), and potassium dihydrogen phosphate–disodium hydrogen phosphate (0.0033 and 0.063 M, respectively, pH 8.0). Nucleosides and nucleotides were observed by fluorescence on paper in ultraviolet light (254 mμ). Compounds containing monomethoxytrityl or dimethoxytrityl groups were detected by spraying the papers with 10% aqueous perchloric acid and allowing them to dry at room temperature. Compounds containing monomethoxytrityl appeared yellow-orange; those containing the dimethoxytrityl group appeared red-orange.

Preparative separation of nucleotide mixtures was accomplished using a DEAE-cellulose column (4 × 30 cm) containing about 60 g of DEAE-cellulose (Calbiochem, 0.93 mequiv/g). Column effluents were monitored with a Gilson UV-2651F absorption meter coupled to a Texas Instruments Rectiriter recorder. A Gilson Model VL fractionator was used to collect 12–13-ml fractions. The DEAE-cellulose column was run at a rate of 10 fractions/hr.

Infrared spectra were obtained with a Baird Associates Model AB2 infrared recording spectrophotometer. The samples were prepared in potassium bromide disks. Ultraviolet spectra were obtained with a Cary 14 recording spectrophotometer or a Beckman DU manual spectrophotometer. Elemental analyses were performed by the Micro-Tech Laboratories, Skokie, Ill.

Solvents and Reagents. Reagent grade pyridine was distilled from *p*-toluenesulfonyl chloride, redistilled from calcium hydride, and stored over Linde Molecular Sieves. Triethylamine was distilled from calcium hydride and stored over barium oxide. Benzene was dried by distillation and dioxane was purified according to Fieser (1957). Monomethoxytrityl chloride (mp 119–121°) (Marvel et al., 1944), dimethoxytrityl chloride (mp 112–114°) (Hogenkamp and Oikawa, 1964), 5'-O-tritylthymidine (mp 159–160°) (Weimann and Khorana, 1962), 5'-O-monomethoxytritylthymidine (mp 102–105°) (Schaller et al., 1963), 5'-O-dimethoxytritylthymidine (mp 124–126°) (Schaller et al., 1963), mesitylenesulfonyl chloride (mp 56°) (Wang and Cohen, 1957), and 2,4,6-triisopropylbenzenesulfonyl chloride (mp 95–96°) (Lohrman and Khorana, 1966) were prepared by procedures described in the literature. Barium β-cyanoethyl phosphate was converted to the pyridinium salt by ion exchange with Dowex 50 (pyridinium form) by the method of Tener (1961).

Alkaline Cleavage and Analysis. The standard procedure used in removing products from the support was to stir the polymer for 16 hr with 0.5 M sodium hydroxide in 50% aqueous dioxane. The mixture was then neutralized with Dowex 50 resin in the pyridinium form, the resin and support polymer were separated from the solution by filtration through glass wool, and the resin was washed several times with 50% aqueous pyridine. The combined filtrate and washings were evaporated to dryness below 35° and then redissolved in water. A portion of the solution was

TABLE I: Products from Cleavage of Partially Phosphorylated Thymidine-3′-Polymer.

Products	R_F in Solvent A	Fraction of Total Ultraviolet-Absorbing Products	% Removed in Alkaline Hydrolysis		
			1st Cycle[a]	2nd Cycle	3rd Cycle
Thymidine	0.67	0.56	93	4	3
Thymidine 5′-phosphate	0.13	0.36	96	2	2
A	0.41	0.048	67	16	17
B	0.16	0.032	81	12	9

[a] In each cycle the polymer was stirred for 16 hr in 0.5 M NaOH in 50% aqueous dioxane.

applied to a 42 × 57 cm sheet of Whatman 3MM paper on a 10-cm line and developed with solvent A or C. Bands visible in ultraviolet light were eluted with water and the optical densities of the resulting solutions were measured at 267 mμ. For calculation of the amounts of nucleosides or nucleotides present, the optical densities of solutions eluted from blanks cut from regions of the paper near the product spots were determined and substracted from the observed values. Corrections for the blanks were relatively small; e.g., for 5 μmoles of thymidine the absorbance of the blank was 2% that of the thymidine. As a test of the elution procedure five thymidine solutions containing from 0.3 to 9.9 μmoles of thymidine were spotted on Whatman 3MM paper and developed with solvent A. Elution and spectrophotometric analysis indicated recovery of 98.8 ± 1.4% of the thymidine for these samples (9700 was used as the extinction coefficient for thymidine).

The efficiency of the hydrolytic method may be illustrated with the cleavage of a sample of thymidine–3′-polymer (III) which had been partially phosphorylated by reaction with phosphorus oxychloride in the presence of imidazole. As shown in Table I all nucleosidic and nucleotidic material was eluted by three alkaline cleavage cycles, and 96% of the thymidine 5′-phosphate and 93% of the thymidine were removed in the first alkaline cycle. Two other materials (A and B) which were not identified but have R_F values corresponding to dithymidine pyrophosphate and dithymidine phosphate, respectively, were removed somewhat less readily. Reproducibility in the alkaline cleavage reactions was also satisfactory; duplicate cleavage experiments on two supports bearing thymidine and thymidine 3′-phosphate (60% of the thymidine had been phosphorylated in one case; 86% in the other) yielded consistent results (±0.7% for each component). It was also shown in a control reaction that neither thymidine 5′-phosphate nor thymidylyl-(3′–5′)-thymidine was degraded under the alkaline conditions used for the cleavage reactions.

Several experiments were carried out to test the practicality of cleaving nucleotidic material from the support with ammonical solutions. In a typical case a 5-g sample of thymidine-3′-polymer which had been phosphorylated with phosphorus oxychloride was suspended in 60 ml of pyridine–concentrated ammonium hydroxide (1:1, v/v) and heated on a steam bath with stirring for 4 hr. During this time 150 ml of concentrated ammonium hydroxide was slowly added dropwise. The polymer was separated by filtration, washed thoroughly with pyridine, methanol, and ether, and the combined solution was evaporated under vacuum. The nucleotidic material amounted to 0.116 g and by spectral analysis of paper chromatograms was 80% thymidine phosphate and 19% thymidine. In three subsequent treatments the amounts of material removed were 0.094, 0.044, and 0.060 g. Analysis of the third portion indicated 30% thymidine and 68% thymidine phosphate. Assay of another sample of the polymer by the sodium hydroxide method showed 0.778 g of nucleotidic material/5 g of polymer sample. The product analyzed as 56% thymidine and 38% thymidine phosphate. Clearly the ammonical cleavage reaction is very slow and results in preferential release of thymidine phosphate relative to thymidine.

Thymidine–3′-Polymer (III). The support polymer used in these studies was prepared by polymerizing styrene (52.0 g), p-vinylbenzoic acid (11.4 g), and p-divinylbenzene (0.12 g) in the absence of a catalyst at 50° (Letsinger and Mahadevan, 1966). Titration with alkali showed 1.38 mequiv of acid/g. The carboxy polymer was converted to the acid chloride form by treatment with excess thionyl chloride in benzene at reflux for 5 hr, following which the polymer was washed with benzene and stored over phosphorous pentoxide *in vacuo*. Infrared absorption attributable to the carboxyl group in Ⓟ–COOH was completely replaced by bands at 5.65 (strong) and 5.75 μ (weak) in Ⓟ–COCl.

5′-O-Triarylmethylthymidine was joined to the support by reaction in pyridine. In a typical case 7.15 g (13.9 mmoles) of 5′-O-monomethoxytritylthymidine was stirred with 10.0 g (13.8 mmoles) of acid chloride of the acid chloride polymer in 70 ml of dry pyridine for 2 days. The polymer was removed from the solution and stirred with 25 ml of pyridine and 5 ml of methanol for 9 hr in order to esterify the remaining acid chloride groups. At this stage the polymer weighed

1383

13.1 g (the weight increase corresponds to esterification of 65% of the acyl groups in the polymer) and showed a strong band at 5.8–6.1 μ in the infrared spectrum. Analysis of a small portion by the alkaline cleavage technique gave on chromatography in solvent A a single spot at R_F 0.87 corresponding to monomethoxytritylthymidine (positive test with acid spray).

A portion (1.2 g) of the 5'-O-monomethoxytrityl 3'-polymer was suspended in 30 ml of 80% aqueous acetic acid and heated at reflux for 3 hr. The insoluble polymer (0.96 g) was separated from the yellow solution, washed with acetic acid and benzene, and dried under vacuum. Chromatography in solvent A of the base cleavage product obtained from a small portion (~20 mg) of the resulting polymer showed a single spot (R_F 0.67), corresponding to thymidine. Quantitative assay of this material indicated 0.71 mequiv of thymidine/g of polymer. Considering the theoretical weight change of the polymer for conversion of the acid chloride to the thymidine derivative, this corresponds to 0.61 thymidine unit/initial carboxyl group on the support of polymer.

From a similar set of reactions carried out with 5'-O-dimethoxytritylthymidine, alkaline cleavage of the dimethoxytritylthymidine-3'-polymer (prior to treatment with acid) yielded thymidine as well as the expected dimethoxytritylthymidine, the ratio being ~1:3.

Phosphorylation Experiments. The results of three experiments on the phosphorylation of thymidine-3'-polymer are reported in Table II. Reaction conditions for each are described below.

(1) A mixture of 21 mmoles of pyridinium β-cyanoethyl phosphate and 8.65 g (42 mmoles) of dicyclohexylcarbodiimide in 40 ml of pyridine was stirred for 18 hr; then 3 g of thymidine-3'-polymer was added and the mixture was stirred for an additional 7 days at room temperature. Following filtration, the solid was washed with benzene and with absolute ethanol, heated on a steam bath for 8 hr with 80% aqueous pyridine, and separated by filtration. It was then washed successively with dioxane, methanol (1:1), benzene, and ether, and dried over phosphorus pentoxide. The final polymer weighed 2.92 g. Data for the product analysis for a sample that was cleaved by alkali are presented in Table II. The only phosphate obtained was thymidine 5'-phosphate (R_F 0.80 in solvent C). In a similar phosphorylation experiment in which the thymidine-3'-polymer sample had been prepared from 5'-O-dimethoxytritylthymidine rather than 5'-O-monomethoxytritylthymidine, a mixture of thymidine 5'-phosphate (R_F 0.80) and thymidine 3'-phosphate (R_F 0.86, solvent C) was obtained.

(2) A mixture of β-cyanoethyl phosphate (0.145 mmole), 0.095 g (0.432 mmole) of mesitylenesulfonyl chloride, and 0.043 g of thymidine-3'-polymer in 1 ml of pyridine was stirred for 4 hr at room temperature. The solid was separated by centrifugation and worked up as in part 1.

(3) Phosphorus oxychloride (0.87 ml, 9.4 mmoles in 3 ml of pyridine was added over a period of 15 min to 2.25 g of imidazole and 4 ml of triethylamine in 18 ml of pyridine at −16°. After 1.5 hr 0.5 g of thymidine-3'-polymer was added and the mixture was stirred at ~−16° for 2 days. The solid was separated by filtration, washed with pyridine, stirred with 50% aqueous pyridine for 13 hr, and worked up as in 1.

Condensation of 5'-O-Monomethoxytritylthymidine with Phosphorylated Polymer-Thymidine. A sample of polymer-thymidine III was phosphorylated with β-cyanoethyl phosphate and dicyclohexylcarbodiimide as in 1 of the previous experiment. Analysis by alkaline hydrolysis of a small portion showed that 82% of the thymidine was phosphorylated. The cyanoethyl phosphoryl derivative of the polymer thymidine IV (0.170 g, 0.088 mmole of thymidine units) was stirred with 0.082 g (0.36 mmole of mesitylenesulfonyl chloride in 2 ml of pyridine for 23 hr; then 0.183 g (0.35 mmole) of 5'-O-monomethoxytritylthymidine in 1.5 ml of pyridine was added and the mixture was stirred for an additional 76 hr. The red-brown polymer was separated, washed well, dried, and analyzed by the alkaline cleavage procedure. The results are shown in Table III. Minor products A–C were not characterized.

5'-O-Dimethoxytritylthymidine 3'-Phosphate. To an anhydrous mixture of 20 mmoles of pyridinium β-cyanoethyl phosphate and 1 g of pyridinium Dowex 50 resin in 20 ml of dry pyridine was added a solution of 12.4 g (60 mmoles) of dicyclohexylcarbodiimide and 2.18 g (4.00 mmoles) of 5'-O-dimethoxytritylthymidine in 20 ml of pyridine. The mixture was sealed and stirred in the dark for 2 days; then it was cooled to 0°, mixed with 40 ml of water, and stirred for another 12 hr. Precipitated dicyclohexylurea was filtered from the solution and washed with 50% aqueous

TABLE II: Phosphorylation Experiments on Thymidine-3'-Polymer.

Expt	Reagent	Time	Products	%[a]
1	CEP + DCC	6 days	pT	90
			T	9
			A[b]	~0.5
2	CEP + mesitylenesulfonyl chloride	4 hr	pT	76
			T	17
			A[b]	7
3	POCl₃ + imidazole	49 hr	pT	73
			T	15
			A[b]	1
			B[c]	11

[a] Mole %. [b] A, probably a dinucleoside monophosphate; R_F in solvent A, 0.41; R_E pH 8.1, relative to pT, 0.24. [c] B, probably a dinucleoside pyrophosphate. R_F in solvent A, 0.16; R_E pH 8.1, relative to pT, 0.75. B hydrolyzes in 1 M hydrochloric acid at 100° (30 min) to pT.

TABLE III: Products from Condensation of Thymidine with IV.

Product	R_F (solvent A)	% of Ultraviolet-Absorbing Material Eluted
MTr-TpT	0.81	50
Thymidine	0.70	20
A	0.56	6
B	0.44	3
C	0.30	6
Thymidine 5'-phosphate	0.18	15

pyridine. The combined filtrate and wash (about 125 ml) were extracted several times with hexane (50 ml) and then extracted four times with 100-ml portions of chloroform. On combination of the chloroform extracts and evaporation with repeated additions of dry pyridine, a gum was obtained which was dissolved in 25 ml of pyridine and reprecipitated by dropwise addition into 500 ml of rapidly stirred ether. The cyanoethyl 5'-O-dimethoxytritylthymidine 3'-phosphate thus obtained was collected as a white solid by centrifugation and washed twice with ether. For removal of the cyanoethyl group this material was dissolved in 30 ml of pyridine and added to 70 ml of 1 N aqueous sodium hydroxide at 0°. The solution was allowed to warm up and was stirred at room temperature for 30 min. Sodium ions were removed by treatment with pyridinium Dowex 50 resin and the solution was concentrated to a small volume under reduced pressure with several additions of pyridine. Dilution with water and lyophilization produced 1.71 g (61%) of crude 5'-O-dimethoxytritylthymidine 3'-phosphate. Chromatography in solvent A showed approximately 95% of the material at R_F 0.49 (positive for dimethoxytrityl) and trace amounts at R_F 0.10 (probably thymidine 3'-phosphate) and near the solvent front (dimethoxytritanol). On electrophoresis at pH 7.55 the major product (R_F 0.49) moved half the distance of thymidine 5'-phosphate. This preparation was repeated on a tenfold scale with the same results (61% yield).

5'-O-Monomethoxytritylthymidine 3'-Phosphate. Since a good quantity of 5'-O-dimethoxytritylthymidine 3'-phosphate was available from the previously described reaction, it was used as a source of thymidine 3'-phosphate for synthesis of 5'-O-monomethoxytritylthymidine 3'-phosphate. Thus, a solution of 2.00 g of 5'-O-dimethoxytritylthymidine 3'-phosphate in 15 ml of 80% aqueous acetic acid was allowed to stand for several hours at room temperature; then the solution was evaporated to dryness under vacuum. The gum was taken up in a mixture of 50 ml of water and 50 ml of ether and the layers were separated. The water layer was extracted several times with fresh portions of ether and then centrifuged to remove a small amount of suspended solid. On lyophilization, the water layer yielded a white solid, which was mixed with pyridine and evaporated to dryness two times. The product was homogeneous on electrophoresis and on chromatography in solvent C, which served to distinguish between thymidine 3'- and 5'-phosphates. For conversion to the monomethoxytrityl derivative, the gum was mixed with 3.0 g of monomethoxytrityl chloride in 20 ml of pyridine. The solution was stirred in the dark for 4 days. Then the precipitate was removed by centrifugation and the solution was added dropwise to 2 l. of ether with stirring. Pyridinium 5'-O-monomethoxytritylthymidine 3'-phosphate precipitated. It was collected by centrifugation, washed with ether, and dried *in vacuo*; weight 1.90 g. Electrophoresis at pH 8.0 showed a single spot (positive for the monomethoxytrityl group) with R_E 0.58 compared to thymidylic acid. Chromatography in solvent A showed the product to be essentially pure with R_F 0.51. A trace of impurity (negative for a trityl-type compound) moving slightly faster than thymidine and a trace of mono-*p*-methoxytriphenylcarbinol at the solvent front were also present.

Synthesis of Thymidylyl-(3'–5')-thymidylyl-(3'–5')-thymidine. A solution of pyridinium 5'-O-monomethoxytritylthymidine 3'-phosphate (600 mg) was prepared in dry pyridine (15 ml) and triethylamine (5 ml), and the solvents were removed at room temperature *in vacuo* to leave the triethylammonium salt of the blocked nucleotide as a pale yellow gum. The evaporation process was repeated twice to ensure dryness; then the gum was dissolved in a solution containing 840 mg of 2,4,6-triisopropylbenzenesulfonyl chloride in 15 ml of pyridine at 0°. After the solution had warmed to room temperature, 314 mg of thymidine–3'-polymer (0.71 mmole of thymidine/g, prepared from monomethoxytritylthymidine) was added and the mixture was stirred for 7 hr. The solid was then separated by centrifugation, washed twice with fresh pyridine, and stirred with pyridine–water (95:5) for ~16 hr. It was then washed with methanol and benzene and treated with a 5% solution of trifluoroacetic acid in benzene (15 ml) for 15 min to remove the monomethoxytrityl group. Formation of the triarylmethyl carbonium ion was indicated by appearance of a deep yellow color. After the polymer had been washed repeatedly with benzene and ether, a second acid treatment produced very little color. The polymer was washed with benzene, methanol, pyridine, and ether, and then was dried *in vacuo*.

For addition of the third nucleotide unit the polymer bearing the dinucleotide VI was added to a fresh pyridine solution of triethylammonium-5'-O-monomethoxytritylthymidine 3'-phosphate, prepared from 600 mg of the pyridinium salt as described above; and the entire sequence of steps used for the first coupling reaction was repeated. The final polymer weighed 390 mg.

Hydrolytic cleavage from the support was effected by stirring a portion of the polymer (351 mg) with 15

TABLE IV: Fractionation of Products from Preparation of TpTpT.

Peak	Fractions Pooled	ODU (267 mμ)	R_F Solvent A	R_F Solvent B	R_E (rel to TpTpT, pH 8)
I	29–45	370	0.67^a		
II	150–170	325	0.38	0.45	0.56
III	178–195	60	0.47	0.46	1.11
			0.56	0.62	0.56
IV	245–260	210	0.10	0.46	1.47
V	278–305	1630	0.16	0.32	1.0
VI	355–373	265	0.10^b	0.32^c	1.54 (1.09)^d
VII	400–416	135	0.047^b	0.21^c	1.20 (1.04)^d
VIII	432–445	68	0.013^b	0.21^c	1.58 (1.07)^e

$^{a-d}$ Very minor components were also found: (a) at R_F 0.85, 0.78, and 0.84; (b) at R_F 0.84; (c) near the solvent front; (d) at R_E indicated in parentheses. ^{e} About 40% of product at R_E 1.07 and 60% at R_E 1.58.

ml of 0.5 M sodium hydroxide in 50% aqueous dioxane for 12 hr. This process was repeated two times in order to assure complete removal of the nucleotides. At this stage the recovered polymer showed no carbonyl bands in the 5.6–6.0-μ region. The alkaline solution was initially red-violet; however, the color disappeared on standing. Sodium ions were removed by stirring the alkaline solution with pyridinium Dowex 50 resin. The resin was removed from the solution and washed with 50% aqueous pyridine, and the washings and neutralized solution were combined and stripped to a small volume below 25°. In this operation a small portion of the solution was inadvertently lost. Following treatment with 3 ml of concentrated ammonium hydroxide the remaining solution was evaporated. The residual gum was dissolved in 50 ml of 0.01 M ammonium bicarbonate in aqueous ethanol (10% ethanol) and applied to the top of the DEAE-cellulose column (4 × 30 cm, bicarbonate form). The sample was washed into the column with 650 ml of the 0.01 M bicarbonate buffer; then a linear salt gradient was begun using 2 l. of 0.01 M ammonium bicarbonate in 10% ethanol solution in the mixing chamber and 2 l. of 0.25 M ammonium bicarbonate in 20% ethanol in the reservoir. When this gradient expired, another was run using 2 l. of 0.25 M ammonium bicarbonate (20% ethanol) in the mixing chamber and 2 l. of 0.50 M ammonium bicarbonate (20% ethanol) in the reservoir. Common tubes from the fractionation were pooled and concentrated to dryness *in vacuo* below 35°. The resulting gums were dissolved in distilled water and the evaporation repeated to assist in the removal of ammonium bicarbonate; then the solids were dissolved in water and lyophilized to yield dry powders. Data for the fractionation and the products are collected in Table IV.

The products from peaks I, II, and IV correspond to thymidine, thymidylyl-(3'–5')-thymidine, and thymidylic acid, respectively. The material in peak V (61 mg) was homogeneous on electrophoresis and on paper chromatography, moving identically with a sample of thymidylyl-(3'–5')-thymidylyl-(3'–5')-thymidine prepared independently by Dr. Clifford L. Leznoff by detritylation of the 5'-O-trityl derivative which had been made by the method of Jacob and Khorana (1965). The material in peaks IV and VI–VIII may well have been held to the polymer through pyrophosphate-type linkages at the phosphate bridges in the trinucleoside diphosphate chain. In support of this idea is the fact that the same substances were slowly liberated into solution when the polymer was subjected to the action of aqueous pyridine prior to treatment with sodium hydroxide, and the material balance for the components in I, II, and V agreed well with that calculated on the basis of thymidine originally bound to the support (see below).

The white powder from peak V gave an acidic solution (pH ~2) when dissolved in water. The possibility that ammonia was lost during lyophilization or storage under vacuum was tested by titrating a solution of 12.30 mg of the solid in 1.0 ml of water with 1.99 × 10^{-2} M sodium hydroxide. The solution gave a sharp end point at 7.45 × 10^{-3} mequiv of base (pH of solution 6.5), corresponding to the loss of about 30% of the ammonia from the salt of TpTpT.

For analysis, a portion of the material from peak V was further purified by passage through a G-25 Sephadex column (1.5 × 120 cm, packed as a liquid suspension in 0.02 M ammonium bicarbonate). The column was eluted with 0.02 M ammonium bicarbonate, 12–13-ml fractions being collected. Most of the material (~99%) was eluted in fractions 16 and 17. A trace impurity was eluted in fraction 19. Fractions 16 and 17 were pooled, lyophilized, redissolved in water, and lyophilized again to ensure removal of residual ammonium bicarbonate, and the solid product was stored *in vacuo* over phosphorus pentoxide. The theoretical values for the analysis are calculated for the trihydrate of thymidylyl-(3'–5')-thymidylyl-(3'–5')-thymidine with 1.5 equiv of ammonia bound as the salt. The presence of less than 2 equiv of ammonia is consistent with the titrimetric data reported above.

Anal. Calcd for $C_{30}H_{40}N_6O_{19}P_2 \cdot 3H_2O \cdot 1.5NH_3$: C, 38.73; H, 5.47; N, 11.29. Found: C, 38.47; H, 5.17; N, 10.94.

Ultraviolet spectra were obtained in acidic, neutral, and alkaline solution for the sample purified by G-25 Sephadex gel filtration. The extinction coefficients (calculated on the basis of a molecular weight of 930, which corresponds to the formula consistent with the analytical data) are summarized in Table V. Values

TABLE V: Ultraviolet Spectra of Thymidylyl-(3′–5′)-thymidylyl-(3′–5′)-thymidine.

Solvent (M)	λ_{max}	ϵ_{max}	λ_{min}	ϵ_{min}
HCl (0.01)	266	25,200	235	6,600
Water	266	25,400	235	8,200
NaOH (0.01)	266	22,000	247	15,000

of 25,400 (Gilham and Khorana, 1958) and 25,800 (Jacob and Khorana, 1965) for λ_{max} 266 (neutral solution) have been reported previously.

The characterization of the compound in peak V was completed by a study of its hydrolysis. Material purified by gel filtration was incubated with snake venom phosphodiesterase and was found to hydrolyze completely to 5′-thymidylic acid and thymidine in a ratio of 1.86:1 (calcd ratio 2:1). Material obtained directly from the DEAE-cellulose separation was hydrolyzed completely by spleen phosphodiesterase to 3′-thymidylic acid and thymidine in a ratio of 1.7:1 (calcd ratio 2:1) The enzyme assays were conducted in a manner similar to that described by Razzell and Khorana (1959, 1961). Thus a solution of 0.2–0.5 mg of the oligonucleotide derivative was incubated with 0.2 mg of spleen phosphodiesterase (Nutritional Biochemicals Corp.) in a mixture of 0.05 ml of 1 M tetrasodium pyrophosphate (adjusted to pH 6.0 with phosphoric acid) and 0.05 ml of 1 M ammonium acetate (adjusted to pH 6.0 with acetic acid) at 37° for 4 hr. The resulting solution was spotted as a 5–10-cm band on a sheet of Whatman 3MM paper and developed in solvent A. Appropriate spots and blanks were cut from the paper and eluted with distilled water. The solutions were made up to specific volumes and the absorbances were measured at 267 mμ. For degradations by snake venom phosphodiesterase 500 units of Russel's viper venom phosphodiesterase (Calbiochem Co.) was dissolved in 2.5 ml of 0.33 M Tris buffer adjusted to pH 9.1 with hydrochloric acid. Samples of 0.2–0.5 mg of nucleotidic material were dissolved in 0.1 ml of the enzyme solution and incubated at 37° for 7 hr. Analyses were conducted by the method described for the spleen enzyme.

For determining the yield of TpTpT in the synthesis another portion of the trinucleotide–polymer derivative (27.6 mg, 7.1% of the total polymer from the synthesis) was hydrolyzed with alkali and the products were separated by chromatography on DEAE-cellulose as before. Since the amount of thymidine in the initial 314-mg sample of thymidine-3′-polymer amounted to 223 μmoles and the per cent of the final polymer analyzed was 7.1%, a total of 0.071 × 223 μmoles = 15.8 μmoles of products would be expected. In fact, 50 optical density units of thymidine, 44 optical density units of TpT, and 205 optical density units of TpTpT were obtained. In terms of extinction coefficients of 9,700, 18,500, and 25,400 for thymidine, TpT, and TpTpT, respectively, these values correspond to 5.16 μmoles of thymidine, 2.38 μmoles of TpT, and 8.06 μmoles of TpTpT. The total of 15.6 μmoles agrees well with the theoretical amount (15.8 μmoles). The other ultraviolet-absorbing products obtained from the reaction (see, *e.g.*, Table III) probably stemmed from the arenesulfonyl chloride and the excess triarylthymidine 3′-phosphate which had reacted with the polymer derivative at the nucleophilic oxygen of the phosphodiester groups (forming pyrophosphates or mixed anhydrides which were later hydrolyzed). On the basis of the reasonable assumption that these minor components did not contain thymidine from the original thymidine-3′-polymer, 51% of the thymidine in III was converted to TpTpT in this synthesis.

References

Cramer, F., Helbig, R., Hettler, H., Scheit, K. H., and Seliger, H. (1966), *Angew. Chem. 78*, 640.
Fieser, L. F. (1957), Experiments in Organic Chemistry, Boston, Mass., D. C. Heath, p 284.
Gilham, P. T., and Khorana, H. G. (1958), *J. Am. Chem. Soc. 80*, 6212.
Hayatsu, H., and Khorana, H. G. (1966), *J. Am. Chem. Soc. 88*, 3182.
Hogenkamp, P. C., and Oikawa, T. G. (1964), *J. Biol. Chem. 239*, 1911.
Jacob, T. M., and Khorana, H. G. (1965), *J. Am. Chem. Soc. 87*, 368.
Letsinger, R. L., Kornet, M. J., Mahadevan, V., and Jerina, D. M. (1964), *J. Am. Chem. Soc. 86*, 5163.
Letsinger, R. L., and Mahadevan, V. (1965), *J. Am. Chem. Soc. 87*, 3526.
Letsinger, R. L., and Mahadevan, V. (1966), *J. Am. Chem. Soc. 88*, 5319.
Letsinger, R. L., and Ogilvie, K. K. (1967), *J. Org. Chem. 32*, 296.
Lohrmann, R., and Khorana, H. G. (1966), *J. Am. Chem. Soc. 88*, 829.
Marvel, D. S., Whitson, J., and Johnston, H. W. (1944), *J. Am. Chem. Soc. 66*, 415.
Melby, L. R., and Strobach, D. R. (1967), *J. Am. Chem. Soc. 89*, 450.
Razzell, W. E., and Khorana, H. G. (1959), *J. Biol. Chem. 234*, 2105.
Razzell, W. E., and Khorana, H. G. (1961), *J. Biol. Chem. 236*, 1144.
Schaller, H., Weimann, G., Lerch, B., and Khorana,

H. G. (1963), *J. Am. Chem. Soc. 85*, 3821.
Tener, G. M. (1961), *J. Am. Chem. Soc. 83*, 159.
Wang, C. H., and Cohen, S. G. (1957), *J. Am. Chem. Soc. 79*, 1924.
Weimann, G., and Khorana, H. G. (1962), *J. Am. Chem. Soc. 84*, 419.

Studies on Polynucleotides. LXXII.[1] Deoxyribooligonucleotide Synthesis on a Polymer Support[2]

H. Hayatsu and H. G. Khorana

Contribution from the Institute for Enzyme Research, University of Wisconsin, Madison, Wisconsin 53706. Received February 10, 1967

Abstract: The concept of carrying out stepwise deoxyribooligonucleotide synthesis on a polystyrene support, which permits condensation reactions in completely homogeneous organic medium such as pyridine, has been successfully developed. A portion of the total phenyl groups in polystyrene was derivatized to yield monomethoxytrityl chloride groups. Condensation reactions with deoxyribonucleosides afforded polystyrene-supported 5'-monomethoxytrityl deoxyribonucleosides. Previously developed methods involving condensations with protected deoxyribonucleoside 5'-phosphates were now used for internucleotide bond synthesis. After condensations, polymer-supported deoxyribooligonucleotides were separated from the excess of reagents and mononucleotide by precipitation from an aqueous medium. Methods for the removal of protecting groups were adapted for work with polymer-supported compounds. The compounds prepared by the new method include: thymidylyl-(3'→5')-thymidine, thymidylyl-(3'→5')-deoxycytidine, thymidylyl-(3'→5')-deoxyadenosine, thymidylyl-(3'→5')-deoxyguanosine, and a trinucleotide, thymidylyl-(3'→5')-thymidylyl-(3'→5')-thymidine. The yield at each internucleotide bond synthesis ranged between 88 and 96%.

Chemical synthesis of polynucleotides containing defined nucleotide sequences has formed the subject of extended investigations in recent years and the methods developed have been used successfully in the synthesis of a large variety of ribo- and deoxyribopolynucleotides.[3] In all of the synthetic work hitherto reported, the products obtained after each condensation step to form internucleotide bonds are separated by time-consuming procedures involving mainly column chromatography. Marked rapidity in synthetic work would result if after the condensation reactions the product is present in a form readily separable from the remainder of the reaction components. This concept, while having been expressed in literature from time to time,[4] has recently been developed with striking success by Merrifield for the synthesis of polypeptides.[5] In the Merrifield procedure, a polypeptide chain is built up in a stepwise manner from one end while it is linked by a covalent bond at the other end to an insoluble polymeric support. Shemyakin and co-workers have more recently reported an alternative approach to polypeptide synthesis in which the polymer carrying the growing peptide chain is soluble in the medium of reaction, and therefore the repetitive condensations are performed in completely homogeneous phase.[6]

Clearly, it would be desirable to develop the use of similar concepts for work in the polynucleotide field and in this paper we report on an approach to deoxyribooligonucleotide synthesis by the use of a soluble polymer as a support. In concept, therefore, the approach is similar to that used previously by Shemyakin and co-workers in the peptide field. Preliminary reports of this work have already appeared.[7] A number of other laboratories have also reported activity in this area. Thus, Letsinger and Mahadevan[8] reported an

---

(1) Paper LXXI in this series is by R. D. Wells and J. Blair, *J. Mol. Biol.*, in press.
(2) This work has been supported by grants from the Life Insurance Medical Research Fund (Grant No. G-62-54), the National Science Foundation (Grant No. GB-3342), and the National Cancer Institute of the National Institutes of Health, U. S. Public Health Service (Grant No. CA-05178).
(3) Selected references which give literature citations are: H. G. Khorana, T. M. Jacob, M. W. Moon, S. A. Narang, and E. Ohtsuka, *J. Am. Chem. Soc.*, **87**, 2954 (1965); R. Lohrmann, D. Söll, H. Hayatsu, E. Ohtsuka, and H. G. Khorana, *ibid.*, **88**, 819 (1966); and H. G. Khorana, H. Büchi, T. M. Jacob, H. Kössel, S. A. Narang, and E. Ohtsuka, *ibid.*, **89**, 2154 (1967).
(4) I. H. Silman and E. Katchalski, *Ann. Rev. Biochem.*, **35**, 873 (1966).

(5) For a review see R. B. Merrifield, *Science*, **150**, 178 (1965).
(6) M. M. Shemyakin, Y. A. Ovchinnikov, A. A. Kinyushkin, and I. V. Kozhevnikova, *Tetrahedron Letters*, 2323 (1965).
(7) H. Hayatsu and H. G. Khorana, *J. Am. Chem. Soc.*, **88**, 3182 (1966); Abstracts, 152nd National Meeting of the American Chemical Society, New York, N. Y., Sept 1966, p 59C.
(8) R. L. Letsinger and V. Mahadevan, *J. Am. Chem. Soc.*, **87**, 3526 (1965); **88**, 5319 (1966).

approach which uses a popcorn type of insoluble polymer as the support. Cramer and co-workers[9] have briefly outlined an approach essentially identical with that reported here while a similar approach but using an insoluble polymer as the support has been developed by Melby and Strobach.[10]

Preparation of Polymer Support. Polystyrene having p-methoxytrityl groups as a part of its structure was chosen as the support for deoxyribooligonucleotide synthesis. Considerations for this choice were as follows. While a number of ways may be conceived for covalently linking the first nucleoside or nucleotide to a polymer support, a major consideration was the chemical principles and protecting groups which are in current use for deoxypolynucleotide synthesis. The use of a methoxytrityl group as a point of linkage of the first deoxyribonucleoside to a polymer was therefore an attractive possibility. Secondly, methoxytrityl derivatives of polystyrene and the protected oligo- and polynucleotides built onto this polymer were all expected to be freely soluble in anhydrous pyridine, which is used most often as the medium of reaction in polynucleotide synthesis. Thirdly, the introduction of the methoxytrityl groups was expected to be straightforward using the reaction sequence shown in Chart I. Finally,

Chart I. Preparation of Polystyrene-Supported p-Methoxytrityl Chloride

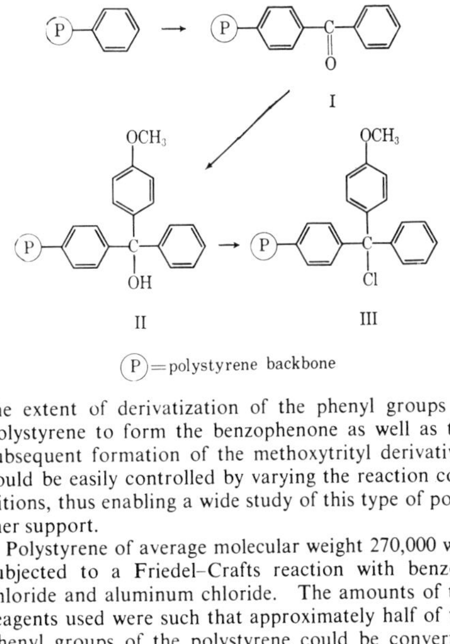

the extent of derivatization of the phenyl groups in polystyrene to form the benzophenone as well as the subsequent formation of the methoxytrityl derivatives could be easily controlled by varying the reaction conditions, thus enabling a wide study of this type of polymer support.

Polystyrene of average molecular weight 270,000 was subjected to a Friedel–Crafts reaction with benzoyl chloride and aluminum chloride. The amounts of the reagents used were such that approximately half of the phenyl groups of the polystyrene could be converted into the benzophenone derivative. Under the reaction conditions used probably optimal conversion to the ketone took place. As judged by the increase in weight of the product isolated, at least 30% of the total phenyl groups had in fact been derivatized. The ultraviolet and infrared absorption characteristics of the product were as expected for the benzophenone derivative I. This product was brought into reaction with the Grignard reagent, p-methoxyphenylmagnesium bromide, to give the trityl alcohol derivative II. Because of the insolubility of I in ether, benzene was used as the solvent for it, the Grignard reagent being introduced as its ethereal solution. Although the reaction mixture was heterogeneous, the reaction appeared to be extremely rapid. The product isolated after work-up gave a coloration on acidification which is characteristic of the p-methoxytrityl cation.[11] Treatment of II with acetyl chloride gave the trityl chloride III which was isolated as a fluffy white powder. Determination of the active trityl chloride groups as measured by the capacity of the preparation to react with methanol in pyridine showed a halide content of 0.4 mmole/g of the derivatized polymer. It would therefore appear that in the above preparation, out of every 100 styrene units, 25–35 have benzophenone groups and 5–7 have trityl chloride groups. The trityl content in this polymer was lower than that expected from the amount of the Grignard reagent used (2 mmoles/g of the benzophenone). In several repetitions of the Grignard reaction, the amounts of the reagent used were varied.[12] One preparation carried out using 3 mmoles of the Grignard reagent per gram of the polymer gave a trityl chloride with higher capacity to react with thymidine (see below). While in all these experiments the Grignard step was not easy to control because of the heterogeneity of the reaction mixture, the aim was to convert only a portion of the benzophenone groups to trityl groups.

All the work on deoxyribooligonucleotide synthesis described herein has been carried out with trityl chlorides which started with a polymer containing a high benzophenone content as described above. In more recent experiments, a polymer derivative with a low (3%) ketone content has been used. At the Grignard reaction step a homogeneous solution is obtained and a large excess of the reagent is used so as to convert essentially all of the ketone groups to the trityl groups. The resulting preparation of polystyrene methoxytrityl chloride appears to offer distinct advantages in synthetic work aimed at the construction of longer polynucleotide sequences. The preparation of the polymer-supported trityl chloride with the low ketone content is included in the Experimental Section. Synthetic work using these modified polymers will be reported upon in a forthcoming paper.

Deoxyribooligonucleotide Synthesis. The typical steps used are shown in Chart II. The first step was the attachment through a covalent linkage of deoxyribonucleosides to the derivatized polymer. ⓟ-Tr(OCH$_3$)-Cl[13] was allowed to react with an excess of dry thymidine in pyridine. This treatment was followed

---

(9) F. Cramer, R. Helbig, H. Hettler, K. H. Scheit, and H. Seliger, *Angew. Chem.*, 78, 640 (1966).
(10) L. R. Melby and D. R. Strobach, *J. Am. Chem. Soc.*, 89, 450 (1967).
(11) H. Schaller, G. Weimann, B. Lerch, and H. G. Khorana, *ibid.*, 85, 3821 (1963).
(12) In one experiment in which a larger amount (5 mmoles/g of I) of the Grignard reagent was used the product was found to have an extremely high viscosity and a low solubility in organic solvents such as chloroform or benzene.
(13) Abbreviations for protected mononucleotides are as have been used in previous papers [see, e.g., H. Schaller and H. G. Khorana, *J. Am. Chem. Soc.*, 85, 3841 (1963)]. The polymer-supported methoxytrityl chloride is abbreviated ⓟ-Tr(OCH$_3$)-Cl, and the oligonucleotides supported on this polymer are abbreviated in the usual way; thus ⓟ-Tr(OCH$_3$)-T for polymer-supported methoxytritylthymidine, ⓟ-Tr(OCH$_3$)-TpT for the corresponding thymidylylthymidine, and so on.

Deoxyribooligonucleotide Synthesis

Chart II. Deoxyribooligonucleotide Synthesis Using Polystyrene-Supported Methoxytrityl Chloride (III)

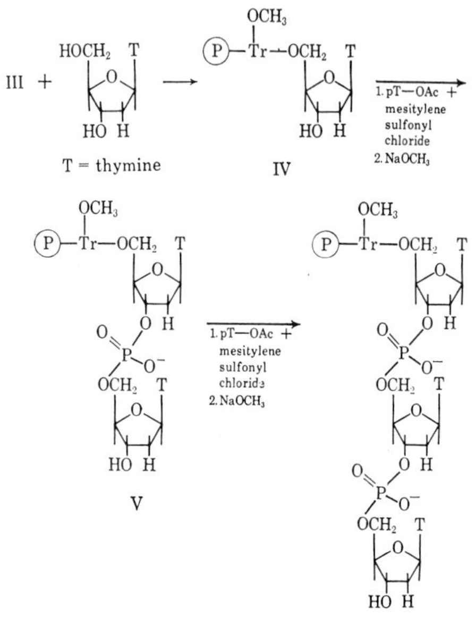

[Chart II: Reaction scheme showing III + thymidine → IV (Ⓟ-Tr(OCH₃)-T), then condensation with pT-OAc + mesitylene sulfonyl chloride, followed by NaOCH₃, giving V (Ⓟ-Tr(OCH₃)-TpT). T = thymine.]

by the addition of methanol so as to convert unreacted trityl chloride groups to the corresponding methyl ether. Since a large excess of the nucleoside was used, it was not necessary to protect the 3'-hydroxyl group for ensuring selective reaction with the 5'-hydroxyl group. The thymidine content of the polymer-supported methoxytritylthymidine was determined by release of thymidine with acid (see below) and extraction into an aqueous medium. Using a preparation of the polymer-supported methoxytrityl chloride, which showed a chloride content of 0.4 mmole/g of the polymer derivative, the thymidine content was found to be 60 μmoles/g of the Ⓟ-Tr(OCH₃)-T (IV). Thus, in this case, only about 15% of the total trityl chloride groups had accepted thymidine. This result suggests that there is considerable steric hindrance around the trityl groups in the polymer.

Using several other preparations of Ⓟ-Tr(OCH₃)-Cl, in which the Grignard reaction was carried out by using a larger proportion of the reagent, the thymidine content after reaction with an excess of thymidine was higher. Thus, in three preparations of Ⓟ-Tr(OCH₃)-T, thymidine content varied between 200 and 340 μmoles/g of the polymer derivative.

It was important to establish (1) that the attachment of thymidine to the polymer was through a covalent linkage, (2) that the linkage was acid labile as expected for a trityl ether, and (3) to develop suitable nonaqueous conditions for the acid-catalyzed release of the nucleoside. (The conventional method, aqueous acetic acid, for this purpose was not applicable because of the insolubility of the polymer in both water and glacial acetic acid.) In accordance with expectation, thymidine linked to the polymer was stable to repeated precipitation of Ⓟ-Tr(OCH₃)-T from aqueous medium and to alkaline treatment. Only by an acidic treatment was thymidine released in the free form. A variety of acidic conditions was investigated for this purpose. With very small amounts of trifluoroacetic acid in chloroform (1:6000, v/v) the release was complete in 10 min at room temperature; with 0.1% (v/v) dichloroacetic acid in chloroform it was complete in 1 hr; and with 20% acetic acid in chloroform (v/v) it was complete in 2 days.

For the analogous attachment of other deoxyribonucleosides to the polymer derivative, the N-protected compounds, N-benzoyldeoxyadenosine,[11] N-anisoyldeoxycytidine,[11] and N-benzoyldeoxyguanosine[14] were used. In one preparation of Ⓟ-Tr(OCH₃)-dA^{Bz}, deoxyadenosine content was found to be 35 μmoles/g of the polymer derivative. The experience with other nucleosides is very limited so far. One preparation of Ⓟ-Tr(OCH₃)-dC^{An} had a deoxycytidine content of 8.4 μmoles/g and the attachment of deoxyguanosine has so far been effected only in a very small amount (1 μmole/g of the polymer).

For internucleotide bond synthesis, a study was first performed on the rate of condensation of 3'-O-acetylthymidine 5'-phosphate (pT-OAc) with Ⓟ-Tr(OCH₃)-T. A large excess (about 40-fold with respect to thymidine on the polymer) of the protected mononucleotide and a proportionate excess of the condensing agent, mesitylenesulfonyl chloride, was used in order to maintain a concentration of activated phosphomonoester group comparable to those used in extensive previous syntheses. The condensation product Ⓟ-Tr(OCH₃)-TpT (V) was isolated simply by pouring the reaction mixture in an aqueous medium followed either by filtration or by centrifugation. The extent of reaction was determined by paper chromatographic analysis after releasing the nucleotidic products from the polymer with trifluoroacetic acid. As seen in Table II, the yield (93%) of TpT reached a plateau within 2 hr.

Applying similar reaction conditions, the reaction of Ⓟ-Tr(OCH₃)-T with the other three protected deoxyribonucleoside 5'-phosphates gave the protected products, Ⓟ-Tr(OCH₃)-TpA^{Bz}-OAc, Ⓟ-Tr(OCH₃)-TpC^{An}-OAc, and Ⓟ-Tr(OCH₃)-TpG^{Ac}-OAc. The yields in all cases were nearly quantitative.

From previous work, the removal of N-protecting groups, especially that of the benzoyl group on adenine, is required before the acidic treatment to cleave the trityl ether linkage. Treatment of the polymer-supported protected dinucleotides with either methanolic ammonia or aqueous ammonia could obviously not be used. Benzylamine, however, was found to dissolve the polymer and this amine was effective in removing the protecting groups. This method was actually used in the case of Ⓟ-Tr(OCH₃)-dA^{Bz} and Ⓟ-Tr(OCH₃)-TpA^{Bz}-OAc and the results were satisfactory.[15]

(14) Unpublished work of Dr. H. Weber in our laboratory. The N-benzoyldeoxyguanosine was prepared by benzoylation of deoxyguanosine with benzoyl chloride followed by selective de-O-benzoylation[11] of the fully acylated product. We thank Dr. Weber for a generous gift of this compound.

(15) In unpublished work, Dr. H. Weber, in our laboratory, has shown that when an alkylamine such as n-butylamine is used to remove the N-anisoyl group from N-anisoyldeoxycytidine, a side product is formed. A similar side product has also been observed when benzylamine was used to remove the N-anisoyl group. The benzylamine procedure therefore cannot be used when N-anisoyldeoxycytidine is present in the oligonucleotide chain to be constructed on the polymer support.

The next requirement for oligonucleotide chain elongation was the selective removal of the acetyl group at the 3'-hydroxyl end. A new procedure using organic solvents was developed for the purpose. In model experiments, treatment of the protected mononucleotides, pT-OAc, d-pA^{Bz}-OAc, d-pC^{An}-OAc, and d-pG^{Ac}-OAc, with a mixture of dimethyl sulfoxide, pyridine, and 1 M sodium methoxide in methanol was found to give at room temperature complete de-O-acetylation in 2 min. Under these conditions, the N-protecting groups were completely stable for 10 min, the maximum time period tested. Using this method, ⓟ-Tr(OCH$_3$)-TpT was prepared from the corresponding 3'-O-acetyl derivative and was subjected to a repeat condensation with pT-OAc. The product ⓟ-Tr(OCH$_3$)-TpTpT-OAc was treated with alkali and after precipitation was treated with acid to release the oligonucleotidic material. The latter was analyzed by a combination of the techniques of paper chromatography and paper electrophoresis. The yield at this second step of internucleotide bond synthesis was 88%.

All of the products, TpT, d-TpC, d-TpA, d-TpG, and TpTpT, were checked for their purity in paper chromatography (two solvents) and paper electrophoresis and were completely susceptible to the action of spleen phosphodiesterase, thus showing the exclusive presence of $C_{3'}$–$C_{5'}$ internucleotidic linkages in them.

Discussion

The approach described incorporates all of the principles previously developed for construction of deoxyribopolynucleotide chains containing predetermined sequences. Thus, in brief, the 5'-hydroxyl group of the terminal nucleoside is protected by a methoxytrityl group and chain elongation occurs by successive condensations with the 3'-hydroxyl group. Further, the use of a soluble polymer support such as developed here appears to be theoretically preferable to the use of an insoluble polymer support. Using the soluble polymer support, it would be hoped that the repetitive condensation steps would always be carried out in completely homogeneous medium and therefore no sharp or sudden deviation at any stage in the pattern of kinetics or yields would be expected. In contrast, in work with insoluble polymers the yields obtained as a function of polynucleotide chain length could be unpredictable. One conceivable concern about the use of soluble polymer support would be that as the size of the growing polynucleotide chain increases, the polymer may show a change in solubility properties. This would obviously be determined by the ratio of the total nucleotidic material carried by the polymer to the amount of the polystyrene backbone. It is, in fact, from a consideration of this reason that although it was possible to attach much higher amounts of thymidine to ⓟ-Tr(OCH$_3$)-Cl, the oligonucleotide syntheses described were carried out with a polymer derivative having a thymidine content of about 50 μmoles/g. It can be seen that if the chain built onto this polymer were the size of a decanucleotide, the increase in its weight would be about 20%.

A prerequisite for success of the concept of polymer support synthesis is that the yield at every repetition of internucleotide bond synthesis be as close to 100% as possible. The yields obtained in the present work were high (88–96%) but further improvement along this line would be desirable. Further work with this approach in the synthesis of longer oligonucleotide chains will certainly provide additional critical evaluation of the efficiency in repetitive condensations.

Finally, specific improvements and modifications at different steps, especially removal of N- and O-protecting groups, continue to be made. Further, the experience with the release of oligonucleotide chains containing purine nucleosides by acidic treatment is limited so far. It is possible that the methoxytrityl group may have to be replaced by the dimethoxytrityl group, which would be more acid sensitive. This modification could in principle be introduced simply by carrying out a Friedel–Crafts reaction with anisoyl chloride in place of benzoyl chloride. Work along the various lines indicated will be reported in subsequent papers.

Experimental Section[16]

Materials and Methods. Polystyrene used throughout this work was a preparation with an average molecular weight of 270,000 with atactic configuration of the skeleton. This preparation (lot no. 683) and several others with different molecular weights were kindly supplied by Dr. E. T. Dumitru of Dow Chemical Co., Midland, Mich. Reagent grade carbon disulfide was dried over calcium chloride prior to use. Dry pyridine was prepared by distillation of reagent grade pyridine over potassium hydroxide and keeping the distilled portion over Linde Molecular Sieve Type 4A. Benzoyl chloride and acetyl chloride were redistilled before use. Reagent grade benzene was dried by keeping it over calcium hydride for more than a week. Cyclohexane was dried over sodium sulfate. Chloroform, which was used as solvent for the acidic treatment of polymer-supported nucleotidic materials, was prepared by treatment of reagent grade chloroform with aqueous sodium carbonate, washing with water, and drying over sodium sulfate. Mesitylenesulfonyl chloride was recrystallized from pentane and stored in a desiccator over phosphorus pentoxide.

Paper chromatography was carried out by the descending technique using the following solvent systems: solvent 1, 2-propanol–concentrated ammonia–water (7:1:2, v/v); solvent 2, 1-butanol–acetic acid–water (5:2:3, v/v); solvent 3, ethanol–1 M ammonium acetate, pH 7.5 (7:3, v/v).

Paper electrophoresis was performed using 0.03 M potassium phosphate buffer, pH 7.1, at 80–100 v/cm for 20–30 min. For checking of purity of nucleotidic compounds by paper chromatography or paper electrophoresis, at least 3 OD units (260 mμ) of the compound was used per spot. R_f's and electrophoretic mobilities of different compounds are listed in Table I.

Table I. R_f Valus of Compounds in Paper Chromatography and in Paper Electrophoresis

Compound	Paper chromatography Solvent system 1	2	Paper electrophoresis at pH 7.1[a]
Thymidine	0.67	0.67	0
Thymidine 5'-phosphate	0.13	0.35	1.0
Thymidylyl-(3'→5')-thymidine	0.41	0.35	0.45
Thymidylyl-(3'→5')-deoxycytidine	0.34	0.33	0.47
Thymidylyl-(3'→5')-deoxyadenosine	0.40	0.37	0.47
Thymidylyl-(3'→5')-deoxyguanosine	0.22	0.26	0.46
Thymidylyl-(3'→5')-thymidylyl-(3'→5')-thymidine	0.17	0.27	0.75

[a] Relative mobilities compared to those of thymidine 5'-phosphate and thymidine.

(16) OD refers to the absorbance value obtained for a 1-ml solution of a compound measured in a quartz cuvette with a 1-cm light path. The number in subscript indicates the wavelength at which the measurement was taken.

All evaporations were carried out under reduced pressure at room temperature. For internucleotide bond synthesis great care was taken to exclude moisture from the reaction mixtures. For this purpose any release of vacuum or opening of the reaction vessel was carried out in the atmosphere of a drybox in which relative humidity was less than 10%. Adequate stirring was always maintained during precipitation of the polymer or polymer-supported nucleotidic material which was carried out by pouring a dilute solution of the polymer into a large volume of a second liquid medium in which the polymer was insoluble.

Spleen phosphodiesterase digestion of the oligodeoxynucleotides was carried out according to the previously reported procedure.[17]

The calculated OD_{280}/OD_{260} ratios are based on the assumption that the absorption spectra of the oligonucleotides are the sum of the spectra of constituent nucleosides.[18]

Removal of Protecting Groups. Removal of 3'-O-acetyl groups in polymer-supported oligonucleotides was carried out in a mixture of dimethyl sulfoxide–pyridine and 1 M sodium methoxide in methanol. The conditions used are described in individual experiments. In model experiments with protected mononucleotides (d-pC^{An}-OAc, d-pA^{Bz}-OAc, d-pG^{Bz}-OAc, and d-pG^{Ac}-OAc) when a 5-mg sample was treated with a mixture of dimethyl sulfoxide (0.25 ml), pyridine (0.20 ml), and 1 M sodium methoxide in methanol (0.05 ml) at room temperature, de-O-acetylation was found to be complete in 2 min, and in a period up to 10 min, no de-N-acylation was observed for any of the protected derivatives.

Removal of the N-protecting groups from polymer-supported nucleosides and oligonucleotides was carried out by keeping their solution in benzylamine at room temperature for periods shown in individual experiments.

Preparation of Monomethoxytrityl Chloride Derivative of Polystyrene (Ⓟ-Tr(OCH$_3$)-Cl). Preparation I. a. Benzophenone Derivative Ⓟ-COC$_6$H$_5$. To a mechanically stirred suspension of anhydrous aluminum chloride (6.8 g) in carbon disulfide (100 ml) was added a solution of polystyrene (10 g) and benzoyl chloride (6 ml, 51 mmoles)[19] in carbon disulfide (50 ml). The reaction was commenced by gentle warming in a glycerol bath when a vigorous evolution of hydrogen chloride was observed. Soon a very viscous solution resulted and then red polymeric material separated out from the reaction mixture. The mixture was heated under reflux for 3 hr. At the end of the period the evolution of hydrogen chloride was almost undetectable. After cooling, the mixture was poured into ice water (1.5 l.) along with chloroform (200 ml), and this mixture was stirred for 2.5 hr. The red semisolid complex which was initially present had now completely dissolved and only the two liquid phases were present. The mixture was allowed to stand until the phases separated, and the lower organic layer was collected. The aqueous phase was extracted with chloroform (150 ml). The combined organic phase was then washed with aqueous sodium carbonate and water. The organic phase was dried over sodium sulfate and evaporated to dryness. The residue was dissolved in benzene (180 ml) and the solution poured into pentane (2 l.) under stirring. The precipitate was collected by filtration and washed with pentane. The product was dried under vacuum over phosphorus pentoxide to give 13.2 g of a white powder. Theoretical yield (assuming quantitative utilization of benzoyl chloride and quantitative recovery of the resulting product) is 15.4 g. Therefore, as judged by the increase in weight, at least 30%[20] of the benzene residues were converted to benzophenone derivatives. The ultraviolet absorption spectrum showed λ_{max} 256–262 mμ in dioxane and the infrared spectrum taken in CHCl$_3$ showed an absorption band at 6.1 μ.

b. p-Monomethoxytrityl Alcohol Ⓟ-Tr(OCH$_3$)-OH. A Grignard reagent was prepared from p-bromoanisole (0.51 ml) and magnesium (100 mg) using dry ether (5 ml) as a solvent and by heating the mixture under reflux for 3 hr under exclusion of moisture. This Grignard reagent was added to a vigorously stirred solution of Ⓟ-COC$_6$H$_5$ (2 g) in dry benzene (50 ml). A yellow precipitate appeared immediately and stirring was continued for 30 min at room temperature. Concentrated hydrochloric acid (20 ml) was added to decompose the solid complex, and the mixture was stirred for 10 min. The resulting brown organic phase was separated and washed successively with water, aqueous sodium carbonate, and again water. If an emulsion resulted, which was often the case during these washings, it was broken by adding a small amount of ethanol to the mixture. The organic phase was finally dried over sodium sulfate and evaporated to dryness. The dry residue was then dissolved in benzene (40 ml) and the product was precipitated by pouring the solution into pentane (500 ml). The precipitate was collected by filtration and washed with pentane. The powder obtained after drying weighed 2 g. When a solution of this material in chloroform or benzene was treated with 60% aqueous perchloric acid or trifluoroacetic acid, an orange coloration characteristic of monomethoxytrityl derivatives[11] was observed in the organic phase (λ_{max} 256–262 mμ in dioxane).

c. p-Monomethoxytrityl Chloride Derivative Ⓟ-Tr(OCH$_3$)-Cl. A solution of Ⓟ-Tr(OCH$_3$)-OH (1.2 g) in a mixture of cyclohexane (4 ml) and acetyl chloride (8 ml) was heated under reflux for 30 min. After this time, evolution of hydrogen chloride, which was vigorous in the first 10 min of heating, had mostly subsided. After cooling, the reaction mixture was concentrated under reduced pressure to remove most of the acetyl chloride. The concentrated solution (one-third to one-fourth of the original volume) was diluted with dry benzene (30 ml) and the solution poured into pentane (0.5 l.) under stirring. The precipitate thus formed was collected by centrifugation and washed with pentane (three 50-ml portions). The white powder after drying weighed 1.1 g and gave a positive copper flame test.

Estimation of active chloride was performed in the following manner. A solution of the trityl chloride (Ⓟ-Tr(OCH$_3$)-Cl, 70 mg) in a mixture of pyridine (4 ml) and methanol (0.4 ml) was heated at 70–75° (bath temperature) for 1 hr. After cooling, the solution was poured into aqueous 2% sodium nitrate (75 ml) under stirring. The polymer thus precipitated was removed by filtration and the filtrate was concentrated to a volume of approximately 5 ml under reduced pressure to remove most of the pyridine and methanol. To the concentrated solution, water (45 ml) and 0.1 M potassium chromate (1 ml) were added. This solution was then titrated with freshly prepared 0.04 M silver nitrate. The consumption of the silver nitrate solution by this test solution was 0.70 ml whereas a blank (75 ml of 2% aqueous sodium nitrate) reached an end point on addition of no more than 0.05 ml of the silver nitrate solution. From this titration the active chloride was calculated to be 0.4 mmole/g of the derivatized polymer.

In another run of active chloride assay, which was carried out for another lot of Ⓟ-Tr(OCH$_3$)-Cl prepared from polystyrene in a similar fashion as described above, 300 mg of the Ⓟ-Tr(OCH$_3$)-Cl consumed 2.50 ml of 0.04 M silver nitrate, this value corresponding to 0.33 mmole of active chloride/g of the polymer.

Monomethoxytrityl Chloride Derivative of Polystyrene (Ⓟ-Tr(OCH$_3$)-Cl). Preparation II. A second preparation in which the proportion of benzophenone groups was much reduced was prepared as follows.

a. Benzophenone Derivative. To a stirred suspension of aluminum chloride (450 mg, 3 mmoles) in carbon disulfide (180 ml) was added a solution of polystyrene (9.4 g, 90 mmoles in styrene units) and benzoyl chloride (0.36 ml, 3 mmoles) in carbon disulfide (70 ml). The mixture was heated under reflux with stirring under exclusion of moisture for 18 hr. The reaction mixture was then cooled and the solution washed with water (two 200-ml portions), aqueous sodium carbonate (two 100-ml portions), and water (two 100-ml portions) successively. The organic phase was dried with sodium sulfate and evaporated to dryness. The residual oil was taken up in benzene (150 ml) and the benzene solution poured into pentane (3 l.). The precipitate was collected by filtration, washed with pentane, and dried. The white solid was precipitated again by pouring its dry benzene solution into an excess of pentane. This product was again collected and dried, yield being 8.9 g.

b. Monomethoxytrityl Alcohol Derivative. A Grignard reagent was prepared by heating under reflux a mixture of p-bromoanisole (0.63 ml, 5 mmoles), magnesium (125 mg, 5 g-atoms), and a small piece of iodine in dry ether (5 ml) for 7 hr under exclusion of moisture. The Grignard reagent in which a small amount (26 ml) of magnesium was still present was then added dropwise into a vigorously stirred solution of the ketone (3 g of the preceding preparation corresponding approximately to 1 mmole of the ketone) in dry benzene (100 ml). A yellow coloration immediately resulted and an increase in viscosity of the reaction mixture was observed. In the transfer of the Grignard reagent, the pieces of magnesium which remained in the original reaction vessel were washed with dry ether

---

(17) M. Smith, D. H. Rammler, I. H. Goldberg, and H. G. Khorana, *J. Am. Chem. Soc.*, **84**, 430 (1962).

(18) The spectral data of nucleosides used here were those reported in Schwarz BioResearch, Inc. Catalog, 1966, p 58.

(19) This amount corresponds to the benzoylation of 50% of the total of phenyl groups in 10 g of polystyrene.

(20) This figure does not take into account any loss during recovery of the product. Probably the per cent of benzene groups converted to benzophenone is higher.

(two 4-ml portions), and the ether washings were added to the reaction mixture. The reaction mixture was then heated under reflux for 20 min under exclusion of moisture. After cooling, concentrated hydrochloric acid (10 ml) was added giving a red mixture. After stirring for several minutes, when the color changed to yellow, the organic phase was separated and washed successively with water (two 100-ml portions), aqueous sodium carbonate (two 100-ml portions), and water (two 100-ml portions). The organic phase was then dried and evaporated to dryness. The residual oil was dissolved in benzene (100 ml) and the resulting solution poured into pentane (2 l.). The precipitate thus formed was collected by filtration, washed with pentane, and dried. A yellowish fluffy powder (2.92 g) was obtained as the product. A chloroform solution of this material, on addition of 70% perchloric acid, exhibited orange coloration indicating the presence of mono-p-methoxytrityl groups in the product.

c. **Monomethoxytrityl Chloride.** The above trityl alcohol (2 g) was treated with a mixture of cyclohexane (10 ml) and acetyl chloride (20 ml). After heating the solution under reflux for 4.5 hr, the mixture was diluted with dry benzene (70 ml) and the diluted solution was poured into anhydrous pentane (2 l.) under exclusion of moisture. The resulting fine precipitate was collected by centrifugation, washed with pentane, and dried under vacuum over phosphorus pentoxide. The yield was 1.8 g.

Polymer-Supported 5′-O-Monomethoxytritylthymidine (℗-Tr(OCH$_3$)-T). Preparation I of the polymer-supported monomethoxytrityl chloride was used in this and the subsequent preparations. A solution of ℗-Tr(OCH$_3$)-Cl (0.4 mmole of Cl/g) (500 mg) and thymidine (70 mg)[21] in pyridine (6 ml) was allowed to stand at room temperature for 12 hr and then heated at 65–75° (bath temperature) for 0.5 hr. Methanol (1 ml) was added and the mixture heated at 75° for 0.5 hr. Pyridine (5 ml) was then added and the solution poured into water[22] (200 ml) under stirring. The product (436 mg) was collected by filtration, washed with water, and dried.

The amount of thymidine in this product was determined by treatment with 10% (v/v) trifluoroacetic acid in chloroform at 0° for 2 min followed by extraction of thymidine into 0.5 N aqueous ammonium hydroxide and measurement of OD at 267 mμ after neutralization. Thymidine content was thus found to be 6 μmoles/100 mg of the product. A control experiment in which a chloroform solution of ℗-Tr(OCH$_3$)-T without addition of trifluoroacetic acid was extracted with 0.5 N ammonia in a similar manner gave no ultraviolet-absorbing material in the ammoniacal extract.

Determination of the thymidine content of the above preparation of polymer-supported 5′-O-methoxytritylthymidine was carried out after a repeat of the precipitation procedure. The latter involved pouring a pyridine solution of ℗-Tr(OCH$_3$)-T into an excess of 2% aqueous sodium chloride solution. Thymidine content was found to be unchanged. A sample (42 mg) of ℗-Tr(OCH$_3$)-T was treated with a mixture (1 ml) of dimethyl sulfoxide–pyridine–1 M sodium methoxide in methanol (5:4:1, v/v) at room temperature for 1 hr. On pouring the solution into water and examining the supernatant aqueous phase, no thymidine release in the medium was detected. Thymidine was thus concluded to be bound to the polymer with a covalent bond which is labile in acid but not in neutral or alkaline media.

Polymer-Supported 5′-O-Monomethoxytrityl-N-benzoyldeoxyadenosine. The preparation of ℗-Tr(OCH$_3$)-Cl used in this experiment was the one that resembled preparation I described above in ketone content but contained a much higher trityl chloride content as judged by reactivity toward thymidine. Thus, this sample of methoxytrityl chloride when treated with an excess of dry thymidine showed a methoxytritylthymidine content of 20–34 μmoles/100 mg of the polymer-supported methoxytritylthymidine. A solution of this preparation of ℗-Tr(OCH$_3$)-Cl (258 mg) and N-benzoyldeoxyadenosine (300 mg) in pyridine (5 ml) was allowed to stand at room temperature for 3 days. Pyridine (3 ml) and methanol (0.4 ml) were then added, and the solution was allowed to stand at room temperature further for 17 hr. The solution was then poured into 2% aqueous sodium chloride solution (250 ml). The precipitate was collected by filtration and washed with water, methanol, and pentane successively. The dried product weighed 247 mg.

---

(21) In other experiments, to increase thymidine content, 50–100 mg of dry thymidine per milliliter of pyridine was used.
(22) In later experiments water containing 2% sodium chloride was used instead of pure water to avoid the possible formation of colloidal solutions.

The amount of deoxyadenosine in this product was determined after removal of the N-benzoyl group with benzylamine (3 days). The product was precipitated, dried, and treated with 20% acetic acid in chloroform for 48 hr at room temperature. Paper chromatography (solvent 1) showed that this acid treatment caused 9% cleavage of the glycosidic linkage in deoxyadenosine. The amount of deoxyadenosine on the polymer was found to be 3.5 μmoles/100 mg.

Rate of Release of Thymidine from ℗-Tr(OCH$_3$)-T by Treatment with Acid. i. **With Trifluoroacetic Acid.** ℗-Tr(OCH$_3$)-T (20 mg which contained 0.5 μmole of thymidine) was dissolved in dry chloroform (5 ml). To this solution, trifluoroacetic acid solution [1 ml of 0.1% (v/v)] in chloroform was added and the reaction mixture kept at room temperature (25°) in a stoppered flask. Aliquots (1 ml) were removed from time to time and treated with 0.05–0.1 ml of triethylamine. Thymidine was extracted into water (two 1-ml portions), the water phase being separated by centrifugation. The pH of the aqueous phase was adjusted to 4–6 with 1 N hydrochloric acid and then optical density determinations were performed at 267 mμ. The solutions obtained from all of the aliquots showed ultraviolet absorption spectra characteristic of thymidine. Release of thymidine as a function of time was as follows: 2 min, 57%; 5 min, 84%; 10 min, 100%; 30 min, 105%.

ii. **With Dichloroacetic Acid.** Using a solution of ℗-Tr(OCH$_3$)-T (50 mg containing 1.25 μmoles of thymidine) in 10 ml of 0.1% dichloroacetic acid in chloroform the rate determination was carried out in a similar manner as for trifluoroacetic acid described above. Time of reaction and release of thymidine were as follows: 0 min (before the addition of acid), 0%; 5 min, 40%; 10 min, 62%; 20 min, 82%; 40 min, 94%; 1 hr, 100%; 1.5 hr, 96%; 3.5 hr, 101%. A control solution of P-Tr(OCH$_3$)-T in chloroform to which no acid was added showed no liberation of thymidine after 1.5 hr.

iii. **With Acetic Acid.** A solution of ℗-Tr(OCH$_3$)-T (50 mg containing 2.5 μmoles of thymidine) in 20% acetic acid in chloroform (v/v; 5.5 ml) was allowed to stand at room temperature in a tightly stoppered flask. Aliquots (1 ml) were withdrawn from time to time and the thymidine liberated was extracted into water (two 2-ml portions). The aqueous phase was concentrated under reduced pressure and submitted to paper chromatography using solvent system 1. The single ultraviolet-absorbing spot observed corresponded to thymidine. The spot was eluted with water and the amount of thymidine estimated. A blank solution was prepared by eluting a blank area, parallel in R_f, of approximately the same size as the thymidine spot. Time of reaction and release of thymidine were: 5 hr, 54%; 24 hr, 85%; 48 hr, 99%; 72 hr, 103%.

Kinetic Study of Internucleotide Bond Synthesis. A mixture of ℗-Tr(OCH$_3$)-T (300 mg containing 7.5 μmoles of thymidine) and pyridinium pT-OAc (132 mg, 300 μmoles) was dissolved in pyridine (3 ml) and the solution evaporated to dryness under reduced pressure. The evaporation with pyridine was repeated two additional times, and the final residue was dissolved in pyridine (2 ml). To this solution mesitylenesulfonyl chloride (128 mg, 600 μmoles) was added followed by addition of more pyridine (1 ml). The reaction mixture was allowed to stand at room temperature. Aliquots (0.6 ml) were taken up from time to time under exclusion of moisture and mixed with pyridine (1 ml) and water (0.1 ml). The aqueous pyridine solutions of aliquots were kept overnight and were each treated with more pyridine (1.5 ml) followed by dimethyl sulfoxide (4 ml) and 1 M sodium methoxide in methanol (0.8 ml). The mixtures were allowed to stand at room temperature for 10 min, then poured into 2% aqueous sodium chloride solution (50 ml) under stirring. The resulting precipitates were collected by centrifugation and washed with water and methanol successively. The products were then dried *in vacuo* over phosphorus pentoxide and submitted to analysis of the condensation reaction. For analysis a part of the product in every case was treated with 0.2% (v/v) trifluoroacetic acid in chloroform (2 ml) at room temperature for 10 min. Pyridine (0.1 ml) was added to stop the reaction and the solution evaporated to dryness. The residue was dissolved in pyridine (2 ml) and the solution poured into water (40 ml). The aqueous mixture including the colloidal precipitate was concentrated to a small volume and the precipitate was removed by centrifugation. The supernatant solution was evaporated and the residue submitted to paper chromatographic analysis. The results are given in Table II.

Thymidylyl-(3′→5′)-thymidine (TpT). A mixture of ℗-Tr(OCH$_3$)-T (190 mg containing 9 μmoles of thymidine) and pyridinium pT-OAc (80 mg, 180 μmoles) was dissolved in pyridine (3 ml). The pyridine solution was evaporated to dryness under reduced pressure.

Deoxyribooligonucleotide Synthesis

Table II. Kinetic Study of the Internucleotide Bond Synthesis Using ⓟ-Tr(OCH₃)-T and pT-OAc[a]

Reaction time, hr	Weight of the product submitted to analysis, mg	Amt of products obtained by paper chromatographic analysis ——— T ———		——— TpT ———		Recovery of nucleosidic and nucleotidic material, μmoles/300 mg[b]	Extent of internucleotide bond synthesis, %
		$OD_{267\,m\mu}$	μmole	$OD_{267\,m\mu}$	μmole		
0.5	35	3.54	0.37	8.62	0.45	7.0	55
1	39	1.62	0.17	15.15	0.79	7.4	82.5
2	34	0.52	0.05	13.99	0.73	7.0	93
4	35	0.39	0.04	12.59	0.65	6.2	94
10	38	0.21	0.02	14.20	0.74	6.0	97

[a] For details of procedure see text. [b] The starting material contained 7.5 μmoles of thymidine/300 mg.

Evaporation with pyridine was repeated four more times to remove any trace of moisture. The final residue was dissolved in pyridine (1 ml) and to the solution were added mesitylenesulfonyl chloride (77 mg, 0.360 mmole) and more pyridine (1.5 ml). The reaction mixture, which was homogeneous, was stirred using a magnetic stirrer at room temperature. After 1.5 hr, at which stage the reaction mixture was highly viscous, more pyridine (2 ml) was added and stirring was continued. After a total period of 3.5 hr, pyridine (2.5 ml) and water (0.4 ml) were added under cooling in ice. The solution was stirred for 15 hr at room temperature and was then poured into aqueous 2% sodium chloride solution (500 ml) under stirring. The resulting precipitate was collected by filtration, washed with water, and dried over phosphorus pentoxide under reduced pressure. This material was then dissolved in a mixture of pyridine (8 ml) and dimethyl sulfoxide (10 ml). To the solution was added 1 M sodium methoxide in methanol (2 ml). The slightly turbid solution was allowed to stand at room temperature for 5 min, and then more pyridine (8 ml) was added to obtain a completely clear solution. After a total period of 15 min, the solution was quickly (within 2 min) poured into aqueous 2% sodium chloride solution (500 ml). The precipitate was collected by filtration and washed with water. The product was dried *in vacuo* over phosphorus pentoxide overnight to give 178 mg of a slightly colored powder.

Subsequent treatment of this product (21 mg) with trifluoroacetic acid in chloroform followed by chromatographic analysis (solvent 1) gave only two spots, corresponding to thymidylyl-(3′→5′)-thymidine (22 OD_{267} units, 96%) and thymidine (0.46 OD_{267} unit, 4%) as the products. The recovery of TpT from the total ⓟ-Tr(OCH₃)-TpT therefore corresponded to 10 μmoles (110% of that present in the starting material). The TpT was chromatographically (solvents 1 and 2) and electrophoretically pure. R_f values are listed in Table I. The ultraviolet spectrum showed λ_{max} 267 mμ and λ_{min} 234 mμ in water; $OD_{280}/OD_{260} = 0.71$ (calculated ratio, 0.72). Upon treatment with spleen phosphodiesterase the preparation (5.5 OD_{267} units) was completely hydrolyzed to thymidylic acid and thymidine, the ratio of thymidylic acid to thymidine being 0.99.

Thymidylyl-(3′→5′)-deoxycytidine (d-TpC). ⓟ-Tr(OCH₃)-T (110 mg containing 21 μmoles of thymidine) and pyridinium d-pC^An-OAc (125 mg, 183 μmoles) were treated with pyridine (2 ml) in the presence of trihexylamine (0.06 ml) using mesitylenesulfonyl chloride (70 mg, 327 μmoles) as the condensing agent. The reaction mixture was stirred with a magnetic stirrer at room temperature for 1.5 hr. At this stage more pyridine (1 ml) was added because of the formation of a thick viscous solution. Stirring was continued further for 1 hr and then pyridine (4 ml) and water (0.4 ml) were added under cooling. The mixture was stirred further for 16 hr and was then poured into 2% aqueous sodium chloride solution (150 ml). The precipitate was collected by filtration and washed with water. The wet product, which was obtained in the form of a wet cake, weighed 170 mg. For analysis, a sample (22 mg) was immediately weighed and treated with 5 ml of 0.2% trifluoroacetic acid in chloroform at room temperature for 20 min. Pyridine (1 ml) was added to the solution and the solution evaporated to dryness under reduced pressure. The residue was dissolved in pyridine (3 ml) and the solution poured into water (20 ml). The precipitate thus formed was removed by centrifugation and the supernatant evaporated to dryness. The residue was treated with concentrated ammonium hydroxide (4 ml) for 24 hr and submitted to paper chromatographic analysis (solvent 1). Compounds detected and estimated on the chromatogram were: d-TpC, 35 OD_{260} units; T, 1.73 OD_{260}; and d-pC, 0.45 OD_{260}. Thus, the extent of reaction (T to d-TpC) was 91%, and the actual recovery as based on the amount of thymidine present in the starting material was 88%.

The preparation of d-TpC was chromatographically (solvents 1 and 2) and electrophoretically pure. R_f values are shown in Table I. The ultraviolet spectrum showed λ_{max} 268 (pH 8) and 273 mμ (pH 1) and λ_{min} 233 (pH 8) and 235 mμ (pH 1); $OD_{280}/OD_{260} = 0.81$ (pH 8) and 1.25 (pH 1) [calculated ratio, 0.83 (pH 8) and 1.31 (pH 1)]. Six OD_{260} units of d-TpC were digested with spleen phosphodiesterase. Complete hydrolysis was observed on chromatographic analysis and the products were Tp and dC, the molar ratio being 1.06.

Thymidylyl-(3′→5′)-deoxyadenosine (d-TpA). A solution of ⓟ-Tr(OCH₃)-T (300 mg, containing 18 μmoles of thymidine) and pyridinium d-pA^Bz-OAc (200 mg, 360 μmoles) in pyridine (4 ml) was rendered anhydrous by repeated evaporation of its pyridine solution. To the dry residue, pyridine (4 ml) and mesitylenesulfonyl chloride (170 mg, 800 μmoles) were added, and the solution was allowed to stand at room temperature for 3.5 hr. Then, more pyridine (4 ml) and water (0.4 ml) were added under ice cooling. The mixture was allowed to stand at room temperature for 18 hr and then poured into 2% aqueous sodium chloride solution (250 ml) under stirring. The resulting precipitate was collected by filtration, washed with water, methanol, and pentane successively, and dried *in vacuo* over phosphorus pentoxide overnight. The yield of the dry powder was 304 mg.

For analysis a portion (103 mg) of the product was treated with benzylamine (4 ml) at room temperature for 36 hr. The benzylamine solution was diluted with dioxane (6 ml) and poured into 2% aqueous sodium chloride solution (125 ml). The precipitate thus obtained was collected by centrifugation and washed with water, methanol, and pentane successively. The dried material weighed 99 mg. A portion (40 mg) of this product was then treated with a mixture of acetic acid (0.6 ml) and chloroform (2.4 ml) at room temperature for 41 hr. The solution was evaporated to dryness under reduced pressure, and the residue was dissolved in benzene and evaporated again to dryness to remove almost all of the acetic acid. The remaining oily material was dissolved in pyridine (2 ml) and the solution poured into water (40 ml). The precipitate which resulted was removed by centrifugation, and the supernatant, after concentration, was submitted to paper chromatography. The products detected on the chromatogram were as follows: d-TpA, 31.7 OD_{260} units (86%);[23] thymidine, 1.35 OD_{260} units (10%); adenine, 0.79 OD_{260} unit (4%);[23] deoxyadenylic acid, 0.34 OD_{260} unit; thymidylic acid, 0.47 OD_{260} unit (4%).[23] Thus the total recovery of the nucleotidic compounds was 76% as based on the amount of thymidine in the starting material.

The d-TpA was found pure in chromatography (solvents 1 and 2) and in paper electrophoresis. R_f values are given in Table I. The ultraviolet spectrum showed λ_{max} 260 mμ and λ_{min} 230 mμ in water; $OD_{280}/OD_{260} = 0.36$ (calculated ratio, 0.36). On digestion with spleen phosphodiesterase the d-TpA (6.5 OD_{260} units) was completely hydrolyzed to give Tp and dA, the ratio being 1.04.

Thymidylyl-(3′→5′)-deoxyguanosine (d-TpG). The internucleotide bond synthesis was carried out between ⓟ-Tr(OCH₃)-T (1.03 g, containing 25 μmoles of thymidine) and pyridinium d-pG^Ac-OAc (520 mg, 1 mmole) in pyridine (10 ml) using mesitylenesulfonyl chloride (420 mg, 2 mmoles) as the condensing agent. The reaction time was 3.5 hr. During this time, a viscous gel had resulted

---

(23) Adenine and thymidylic acid would have been produced from d-TpA during the work-up procedures. Thus the extent of internucleotide bond synthesis was believed to be approximately 90%.

which, after addition of pyridine (15 ml) and water (1.2 ml) and shaking, quickly went into solution. This aqueous pyridine solution was kept at room temperature for 17 hr. For analysis, an aliquot (1.0 ml) was poured into 2% aqueous sodium chloride solution (20 ml). The precipitated product from the aliquot was collected by filtration, washed with water, methanol, and pentane successively, and then dried to a slightly colored powder (33 mg). A portion (21 mg) of this product was treated with 0.1% trifluoroacetic acid in chloroform (3 ml) at room temperature for 6 min followed by treatment with concentrated ammonia for 20 hr. Paper chromatographic analysis (solvent 1) showed the following products: d-TpG, 9.1 OD_{260} units (94%); thymidine, 0.26 OD_{260} unit (6%); deoxyguanylic acid, 0.3 OD_{260} unit; and a compound traveling with R_f 0.3, 0.36 OD_{260} unit. Thus the recovery of the nucleotidic material was 94% as based on the amount of thymidine in the starting material. The d-TpG thus isolated from the paper chromatogram was checked for its purity by paper chromatography (solvents 1 and 2) and by paper electrophoresis. R_f values are given in Table I. The ultraviolet spectrum showed λ_{max} 254 mμ and 227 mμ in water, $OD_{280}/OD_{260} = 0.67$ (calculated ratio, 0.69). On digestion with spleen phosphodiesterase the d-TpG (6 OD_{260} units) was completely hydrolyzed to form thymidylic acid and deoxyguanosine, the ratio of these two products being 1.03.

To the main portion of the reaction mixture, which had been kept cold for 2 days while the aliquot was being worked up, pyridine (7 ml) and dimethyl sulfoxide (40 ml) were added. This solution was next made alkaline by addition of 1 M sodium methoxide in methanol (8 ml). After being kept at room temperature for 10 min, the solution was poured into 2% aqueous sodium chloride solution (2 l.). This transfer was completed in 5 min. The fine precipitate which formed was collected by filtration and washed with water, methanol, and pentane successively. The lightly colored powder weighed 1.00 g after drying overnight *in vacuo* over phosphorus pentoxide.

Thymidylyl-(3'→5')-thymidylyl-(3'→5')-thymidine (TpTpT). Polymer-supported 5'-monomethoxytritylthymidylyl-(3'→5')-thymidine[24] (125 mg which contained 7.4 μmoles of nucleotidic material composed of thymidylyl-(3'→5')-thymidine, 96%, and thymidine, 4%) was dissolved together with pyridinium 3'-O-acetylthy-

---

(24) This was present, presumably, as the sodium salt since the alkaline treatment for removal of the 3'-O-acetyl group involved sodium methoxide and subsequent precipitation was from aqueous sodium chloride.

midine 5'-phosphate (200 mg, 450 μmoles) in pyridine (2.5 ml). The solution was rendered anhydrous by repeated evaporation with pyridine. After final evaporation with pyridine, the residue was dissolved in pyridine (2 ml), and to the solution was added mesitylenesulfonyl chloride (200 mg, 900 μmoles) followed by more pyridine (0.5 ml). The solution was allowed to stand at room temperature for 2.75 hr. Pyridine (4.5 ml) and water (0.4 ml) were then added under ice cooling and the aqueous pyridine solution was left at room temperature for 17 hr. It was then poured into 2% aqueous sodium chloride solution (200 ml). To collect the resulting precipitate, most of the liquid was carefully sucked off after centrifugation and the resulting loosely packed product was now collected by filtration and washed with water. The semidried material was dissolved in pyridine (4 ml). To this solution were added dimethyl sulfoxide (5 ml) and 1 M sodium methoxide in methanol (1 ml). The solution was allowed to stand at room temperature for 12 min and then poured into 2% aqueous sodium chloride solution. The product thus precipitated was collected by centrifugation followed by filtration, as described above, and washed with water. Drying under vacuum over phosphorus pentoxide for a day gave a slightly colored powder (105 mg) as the product.

A portion of this material (21 mg) was treated with 1% (v/v) trifluoroacetic acid in chloroform (2 ml) at 0° for 15 min. The solution was then evaporated to dryness under reduced pressure, and the residue was rendered acid free by repeated evaporation with benzene. The oily residue was then taken up in pyridine (2 ml) and the pyridine solution poured into water (20 ml). The precipitate was removed by filtration and the filtrate concentrated and subjected to paper chromatography (solvent 1). The compounds detected on the chromatogram were: TpTpT, 33.8 OD_{267} units (84%); TpT, 3.28 OD_{267} units (12%); and thymidine, 0.55 OD_{267} unit (4%). The TpTpT fraction, upon paper electrophoresis, was shown to be contaminated with thymidylic acid (1.7%) and a material (1.9%) traveling slightly ahead of TpTpT. Thus, the extent of reaction from TpT to TpTpT was 88%, and the actual recovery of the total nucleotidic material was 95% as based on the amounts present in the starting material.

The TpTpT thus purified was homogeneous upon paper chromatography (solvents 1 and 2). The R_f values are given in Table I. The ultraviolet spectrum showed λ_{max} 267 mμ and λ_{min} 234 mμ in water; $OD_{280}/OD_{260} = 0.70$ (calculated ratio, 0.72). The preparation (10 OD_{260} units) was digested with spleen phosphodiesterase. A complete hydrolysis was observed as examined by paper chromatography (solvent 1) and the products, thymidylic acid and thymidine, were found to be in the ratio of 2.16:1.

Oligonucleotide Syntheses on Insoluble Polymer Supports. I. Stepwise Synthesis of Trithymidine Diphosphate

L. Russell Melby and Donald R. Strobach

Contribution No. 1249 from the Central Research Department, Experimental Station, E. I. du Pont de Nemours and Company, Wilmington, Delaware 19898. Received September 12, 1966

Abstract: An *insoluble* cross-linked polystyrene with pendant monomethoxytrityl chloride groups was prepared and condensed with thymidine to obtain a polymer containing approximately 500 μmoles of bound thymidine per gram of polymer. Condensation of the thymidine derivative with 3′-O-acetylthymidine 5′-phosphate (pTOAc) in the presence of dicyclohexylcarbodiimide (DCC) resulted in 54% conversion of bound thymidine to the dinucleoside phosphate. Deacetylation of the latter polymer and subsequent condensation with pTOAc and DCC gave the trinucleoside diphosphate derivative from which thymidylyl-(3′→5′)-thymidylyl-(3′→5′)-thymidine was isolated in 38% conversion based on polymer-bound dinucleoside phosphate.

The success achieved by Merrifield in stepwise synthesis of polypeptides on insoluble polymer supports[1] and the procedural advantages which the method affords prompted us to investigate the application of similar procedures to oligonucleotide synthesis. Reports of related studies in other laboratories have recently appeared.[2-4]

The support polymer under study in this laboratory consists of an *insoluble* styrene–divinylbenzene bead copolymer containing trityl chloride or 4-methoxytrityl chloride functional groups to which nucleosides are subsequently attached by trityl ether formation. Suitably protected nucleotides are then condensed with the polymer-bound nucleoside. The insolubility and form of this type of support confer physical and chemical characteristics which differentiate it from the soluble supports reported by Hayatsu and Khorana[3] and by Cramer, *et al.*[4] On the other hand, Letsinger and Mahadevan[2] utilized an insoluble popcorn polymer to which they attached amino-containing nucleosides through amide formation with polymer-borne carbonyl chloride groups.

Methoxytrityl Chloride Polymer Synthesis (Chart I). The synthesis of the supporting polymer was similar to procedures reported by Braun and Seelig[5] and is summarized in Chart I. Thus, a mixture of styrene and *p*-iodostyrene (mole ratio 4:1) containing 1% by weight of divinylbenzene (DVB) was polymerized in aqueous polyvinyl alcohol with benzoyl peroxide (Bz_2O_2) initiator to obtain cross-linked iodo copolymer (**1**) in the form of beads approximately 75 to 150 μ in diameter. The iodine content of the polymer corre-

(1) R. B. Merrifield, *Science*, **150**, 178 (1965).
(2) R. L. Letsinger and V. Mahadevan, *J. Am. Chem. Soc.*, **87**, 3526 (1965).
(3) H. Hayatsu and H. G. Khorana, *ibid.*, **88**, 3182 (1966).
(4) F. Cramer, R. Helbig, H. Hettler, K. H. Scheit, and H. Seliger, *Angew. Chem. Intern. Ed. Engl.*, **5**, 601 (1966).

(5) (a) D. Braun and E. Seelig, *Chem. Ber.*, **97**, 3098 (1964); (b) D. Braun, *Angew. Chem.*, **73**, 197 (1961).

Chart I. Cross-Linked Polystyrene with Pendant 4-Methoxytrityl Groups[a]

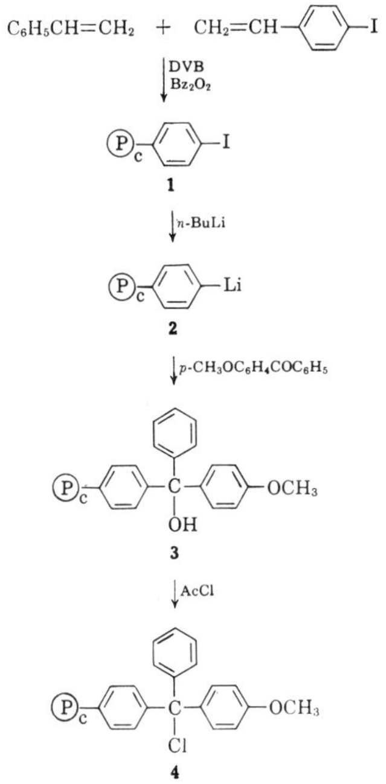

[Chart I shows reaction scheme: C₆H₅CH=CH₂ + CH₂=CH-C₆H₄-I, with DVB/Bz₂O₂, producing polymer 1 (Pc-C₆H₄-I); then n-BuLi gives 2 (Pc-C₆H₄-Li); then p-CH₃OC₆H₄COC₆H₅ gives 3 (methoxytrityl alcohol polymer with OH); then AcCl gives 4 (methoxytrityl chloride polymer with Cl).]

[a] Ⓟc- = cross-linked polystyrene backbone.

sponded to about 1600 μmoles/g and the degree of cross-linking was such that the polymer absorbed more than five times its weight of solvents such as benzene, pyridine, or dimethylformamide (DMF) before acquiring fluid mobility.

The iodo polymer, as a suspension in benzene, reacted with excess n-butyllithium to form a lithio polymer (2) which was not isolated, but was treated in situ with 4-methoxybenzophenone and then with a mixture of acetic and hydrochloric acids to liberate the methoxytrityl alcohol polymer 3. The infrared spectrum of this polymer (Nujol mull) exhibits characteristic bands at 2.95, 6.62, 7.70, 8.0, and 12.2 μ. The methoxytrityl alcohol polymer was virtually free of iodine, suggesting almost quantitative reaction with n-butyllithium, but considerable reduction had presumably taken place since conversion to the trityl alcohol was incomplete.[6] Treatment of the alcohol polymer with excess acetyl chloride in boiling benzene formed the methoxytrityl polymer 4 whose chlorine content corresponded to approximately 730–760 μmoles/g.[7] The infrared absorption spectrum is similar to that of the alcohol precursor but lacks the 2.95-μ hydroxylic absorption. Both trityl derivatives developed a deep red-brown color when warmed with 72% perchloric acid[8] and exhibited yellowish-white fluorescence under 3660-A ultraviolet irradiation, both qualities being characteristic of methoxytrityl chloride itself. The methoxytrityl chloride polymer is moderately stable to atmospheric moisture, but to ensure its integrity it was stored and handled in a drybox.

Polymer-Supported Thymidine (5, Chart II).[9] Reaction of the monomethoxytrityl chloride polymer with excess thymidine (T) in anhydrous pyridine or DMF–pyridine mixtures at room temperature for periods of 24–48 hr formed methoxytritylthymidine polymer (5) containing up to 490–550 μmoles of bound thymidine per gram of polymer, representing 75–85% conversions based on polymer-bound chloride.[10] Furthermore, by using a pyridine–benzene solvent mixture (1:2, v/v) conversions of over 95% were achieved to obtain polymer containing 610–620 μmoles of T/g. The thymidine content of the product was determined by exhaustive hydrolysis of the polymer in boiling acetic acid–hydrochloric acid. This procedure quantitatively liberated thymine which was assayed by ultraviolet spectral analysis of the hydrolysate. The infrared spectrum of the thymidine polymer exhibits a strong band at 5.92 μ characteristic of thymidine carbonyl group absorption.

Chart II. Insoluble Polymer-Supported Oligonucleotide Synthesis[a]

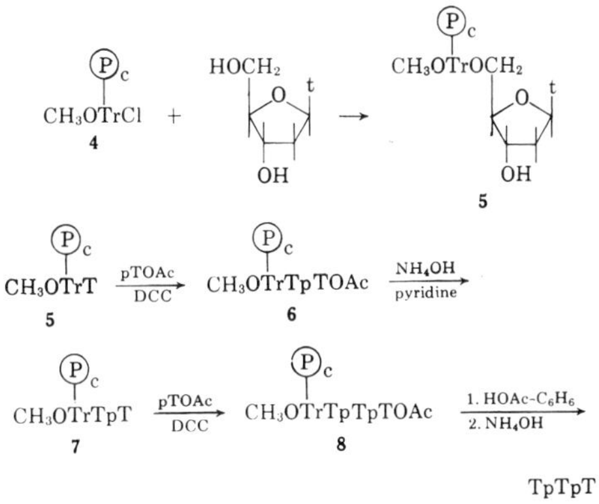

[Chart II shows: CH₃OTrCl (4) + HOCH₂-furanose-t-OH → CH₃OTrOCH₂-furanose-t-OH (5); 5 + pTOAc/DCC → CH₃OTrTpTOAc (6); 6 + NH₄OH/pyridine → CH₃OTrT (Ⓟc)(7); 7 + pTOAc/DCC → CH₃OTrTpTpTOAc (8); 8 with 1. HOAc-C₆H₆, 2. NH₄OH → TpTpT]

[a] T = thymidine; t = thymine.

Polymer-Supported Thymidylyl-(3′→5′)-thymidine 3′-O-Acetate (6, Chart II). Reaction of the methoxytritylthymidine polymer 5 with 3′-O-acetylthymidine 5′-phosphate[11] (pTOAc) and dicyclohexylcarbodiimide (DCC) in anhydrous pyridine–dimethylacetamide (DMAc) solvent mixture gave the supported thymidylyl-(3′→5′)-thymidine 3′-O-acetate (TpTOAc)[12] derivative 6. Treatment of the polymer with an acetic acid–water–benzene solution (32:8:10, v/v) at room temperature for 5 hr liberated TpTOAc as well as some unreacted thymidine. The TpTOAc was deacetylated to form free thymidylyl-(3′→5′)-thymidine (TpT) which was

---

(6) For example, methoxyl determination showed functionality much below that expected for quantitative conversion.

(7) This estimation was based on combustion analysis for chlorine. However, the chloride was quantitatively released from the polymer by hydrolysis in aqueous pyridine at room temperature during 18 hr. The chlorine content varied to a modest degree among several preparations.

(8) Also noted by Hayatsu and Khorana.[3]

(9) The symbol -Ⓟc will be used to denote the cross-linked polystyrene backbone according to the convention of R. L. Letsinger, et al. (see ref 2), and CH₃OTr- to denote the pendant methoxytrityl group.

(10) Percentage conversions take into account the weight gain of the polymer resulting from substitution of Cl by thymidine.

(11) H. G. Khorana and J. P. Vizsolyi, J. Am. Chem. Soc., **83**, 675 (1961).

(12) Abbreviations are as previously defined by E. Ohtsuka, M. W. Moon, and H. G. Khorana, ibid., **87**, 2956 (1965); see also H. Schaller and H. G. Khorana, ibid., **85**, 3841 (1963).

chromatographically identical with an authentic sample[13] and was completely hydrolyzed to Tp and T by spleen phosphodiesterase.[14] This hydrolysis verified the exclusiveness of $3' \rightarrow 5'$ phosphodiester bonds in the product and thus proved the 5'-O attachment of thymidine to the polymer. The TpT was also completely degraded to pT and T by snake venom phosphodiesterase.[15]

The yield of isolated TpT corresponded to 220 μmoles/g of polymer, representing 54% conversion of polymer-bound thymidine.[16]

The release of TpTOAc by the acetic acid–water–benzene reagent revealed a marked lability of the methoxytrityl ether bond associated with bound TpTOAc relative to the equivalent bond in unelaborated thymidine polymer. Thus, treatment of the polymer with the acetic acid reagent for 5 hr liberated substantially all of the TpTOAc and some thymidine equivalent to 251 μmoles of initially bound thymidine per gram of polymer. On the other hand, the thymidine precursor polymer liberated about 60 μmoles of T/g under identical conditions. As yet we have no adequate explanation for this trityl ether labilization subsequent to nucleotide condensation, but we are pursuing its study.

Polymer-Supported TpTpTOAc (8, Chart II). The dinucleoside phosphate 3'-O-acetate polymer **6** was deacetylated by treatment with concentrated ammonium hydroxide in pyridine for 3 days. Ammonium ion and occluded ammonium hydroxide were removed from the deacylated product by ion exchange with pyridinium acetate since cations derived from strong bases have been shown to inhibit DCC-promoted phosphorylations.[17] The derived TpT polymer **7** was condensed with pTOAc and DCC in pyridine to obtain the trinucleoside diphosphate polymer **8** from which TpTpT was isolated in amounts corresponding to 29 μmoles/g, representing 38% conversion of polymer-bound TpT. The trinucleoside diphosphate was accompanied by TpT and thymidine. The constitution of the TpTpT was confirmed by spleen and venom phosphodiesterase hydrolysis.

Addendum. We are continuing our study of this insoluble polymer support system for oligonucleotide synthesis with derivatives containing deoxycytidine, -adenosine, and -guanosine residues. This developing work will be reported in subsequent papers.

Experimental Section

General Methods. Paper chromatography was carried out by the descending technique using Whatman No. 40 paper for quantitative work and Whatman No. 1 paper for qualitative work. Solvent systems used were: A, 2-propanol–concentrated ammonium hydroxide–water (7:1:2, v/v); B, ethanol–1 M ammonium acetate (pH 7.5) (7:3, v/v); and C, 1-butanol–acetic acid–water (5:2:3, v/v).

Ultraviolet spectra were determined on a Cary Model 11 recording spectrophotometer. The expression OD_{267} unit (or OD_{264} unit) is defined as that amount of substance in 1 ml of solution which gives an optical density of 1.00 through a 1-cm path length at the indicated wavelength.

Pyridinium 3'-O-acetylthymidine 5'-phosphate was prepared according to the method of Khorana and Vizsolyi,[11] purified by ether precipitation,[18] and stored as a 0.2–0.3 M solution in purified anhydrous pyridine.[17] Its purity was checked by chromatography in solvent B.

Iodo Polymer 1. Into a 1-l., three-necked flask equipped with a mechanical stirrer and reflux condenser was placed 2.5 g of polyvinyl alcohol[19] and 250 ml of deaerated water, and the mixture was stirred under a nitrogen atomosphere until the alcohol had dissolved. A solution of 16.0 g (0.154 mole) of freshly distilled styrene, 9.0 g (0.04 mole) of p-iodostyrene,[20] 0.25 ml of divinylbenzene, and 0.25 g of benzoyl peroxide was added to the flask; the suspension was stirred under nitrogen and heated with a water bath (80–90°) for 5 hr. The snow-white bead polymer was collected and washed thoroughly with water and ethanol (or methanol). The dried polymer was suspended in 400 ml of benzene and agitated overnight to remove undesirable low molecular weight contaminants. The polymer was collected, washed with a large volume of ethanol, dried at 100° *in vacuo*, and sieved. The portion passing 100 mesh but retained on 200 mesh sieve[21] (19.5 g, 77% yield) was used in subsequent reactions. *Anal.* Calcd for $C_{40}H_{39}I$: C, 74.3; H, 6.1; I, 19.6. Found: C, 72.7; H, 6.0; I, 20.4.

The iodine content corresponded to 1610 μmoles/g. The infrared spectrum (Nujol mull) of the polymer exhibits a characteristic band at 12.2 μ assignable to *para*-disubstituted benzene. This band was absent in polystyrene homopolymer.

Methoxytrityl Alcohol Polymer 3. To a suspension of 65 g of the iodo polymer (equivalent to *ca.* 0.1 mole of iodine) in 1 l. of anhydrous benzene was added 140 ml of 1.6 M n-butyllithium in n-hexane[22] (0.22 mole), and the mixture was stirred under nitrogen at room temperature for 24 hr. During this time the polymer changed from swollen, translucent beads to an opaque white granular material. Most of the liquid was removed by suction through a filter stick and the solid was washed in the flask with two 500-ml portions of benzene, removing each wash in the same manner. To this material was added 1 l. of anhydrous benzene and 25 g (0.12 mole) of 4-methoxybenzophenone, and the mixture was stirred under nitrogen at room temperature for 24 hr. A mixture of 200 ml of glacial acetic acid and 50 ml of 6 N hydrochloric acid was added cautiously, and stirring was continued for 24 hr. The dark brown product was collected and washed on the filter successively with 1 l. each of benzene, ethanol, 3 N hydrochloric acid, and ethanol. The granular orange product was suspended in a mixture of 1500 ml of a benzene–ethanol mixture (2:1, v/v) and warmed gently on a steam bath until the polymer assumed a light yellow color. It was collected, washed with a large volume of ethanol, and vacuum dried at 100°. The yield was 67 g. *Anal.* Found: I, 0.26 and 0.23.

The infrared spectrum (Nujol mull) exhibits bands at 2.95, 8.0, and 12.2 μ assignable, respectively, to –OH, =COC–, and *para*-disubstituted benzene, as well as other characteristic bands at 6.62 and 7.70 μ.

Methoxytrityl Chloride Polymer 4. A suspension of *ca.* 60 g of the alcohol polymer **3** in 1 l. of benzene and 100 ml of acetyl chloride was boiled under reflux for 12 hr with exclusion of moisture. The product was collected in a dry atmosphere, washed with 1 l. of dry benzene and 1 l. of dry petroleum ether, and dried under high vacuum over phosphorus pentoxide at room temperature. It was stored and handled in a drybox. The product, a slightly yellow free-flowing solid, was obtained in essentially quantitative yield. *Anal.* Found: Cl, 2.6 and 2.6.

The chloride content of the polymer corresponded to *ca.* 730 μmoles/g.

A 1-g portion of the polymer was agitated for 20 hr in a mixture of 20 ml of pyridine containing 100 μl of water. The product was washed with pyridine, benzene, and hexane, and dried. It was virtually free of chlorine. *Anal.* Found: Cl, 0.12, 0.14; methoxyl, 1.8, 1.8.

Methoxytritylthymidine Polymer 5. A. To a solution of 100 mg (415 μmoles) of thymidine[23] in 10 ml of anhydrous pyridine[17]

---

(13) Prepared by the method of P. T. Gilham and H. G. Khorana, *J. Am. Chem. Soc.*, **80**, 6212 (1958).
(14) W. E. Razzel and H. G. Khorana, *J. Biol. Chem.*, **236**, 1144 (1961).
(15) W. E. Razzel and H. G. Khorana, *ibid.*, **234**, 2105 (1959).
(16) The conversion figure takes into account the weight gain of the polymer in going from bound thymidine to pyridinium TpTOAc.
(17) T. M. Jacob and H. G. Khorana, *J. Am. Chem. Soc.*, **86**, 1630 (1964), and footnote 17 therein.

(18) T. M. Jacob and H. G. Khorana, *ibid.*, **87**, 2971 (1965).
(19) Elvanol[R] 52-22, a medium viscosity grade of polyvinyl alcohol, was obtained from the Electrochemicals Department, E. I. du Pont de Nemours and Co., Inc., Wilmington, Del. 19898.
(20) Ionac Chemical Co., Birmingham, N. J.
(21) A.S.T.M. specifications.
(22) Foote Mineral Co., Exton, Pa.

was added 500 mg of monomethoxytrityl chloride polymer **4** (equivalent to 365 μmoles of chloride), and the mixture was agitated for 48 hr at room temperature. The insoluble polymer **5** was collected and washed exhaustively with dry pyridine until the washings were free of thymidine. A small portion (*ca.* 50 mg) of the polymer was removed under anhydrous conditions, washed with ethanol, and vacuum dried at 60°. A sample (10.0 mg) was refluxed overnight in glacial acetic acid–6 N hydrochloric acid (1:1, v/v, 10 ml) to liberate thymine. The hydrolysate was evaporated to dryness on a rotary evaporator, and the residue was shaken with 25.0 ml of 0.01 N hydrochloric acid and filtered. Ultraviolet spectral analysis of the filtrate indicated a thymine content of 4.94 μmoles (39.0 OD_{264} units using ϵ 7900) corresponding to 494 μmoles of T/g of polymer or 77.5% conversion of chloride.

To verify the absence of occluded thymidine a 20-mg portion of the polymer was agitated with 3 ml of pyridine for 20 hr. The extract was completely devoid of thymidine as judged by ultraviolet spectral analysis.

The infrared spectrum of the thymidine polymer exhibits a strong, sharp, slightly asymmetric band at 5.95 μ characteristic of thymidine carbonyl absorption.

B. In 10 ml of anhydrous pyridine was dissolved 400 mg (1650 μmoles) of thymidine, and 20 ml of dry benzene was added to the solution (the solution remained homogeneous). Two grams of the chloride polymer (equivalent to 1455 μg-atoms of Cl) was added, and the mixture was agitated for 40 hr. A white crystalline solid deposited on the wall of the reaction vessel, presumably pyridine hydrochloride. The polymer was collected, washed, and dried as described above, to obtain 2.13 g of product. Thymine analysis indicated 610 μmoles of T/g. The product was shown to be free of occluded thymidine.

TpTOAc Polymer 6. To the remaining pyridine-swollen thymidine polymer **5** from procedure A (*ca.* 450 mg, equivalent to 220 μmoles of bound thymidine) was added 900 μmoles of 3′-O-acetylthymidine 5′-phosphate[11] in 4 ml of anhydrous pyridine, 1.8 g of dicyclohexylcarbodiimide, and 4 ml of anhydrous dimethylacetamide, and the suspension was agitated for 5 days at room temperature. The polymer beads were collected and washed exhaustively with pyridine, then with ethanol, and the product (**6**) was dried under vacuum at room temperature. Exhaustive hydrolysis with acetic acid–hydrochloric acid indicated a thymine content of 706 μmoles/g (58.5 OD_{264} units from 10.47 mg of polymer). To liberate thymidylyl-(3′→5′)-thymidine a portion of the dried polymer (10.21 mg) was agitated in 2 ml of a solution of 80% aqueous acetic acid saturated with benzene (HOAc–H₂O–C₆H₆, 32:8:10, v/v) for 5 hr at room temperature. The hydrolysate was filtered, concentrated, treated with concentrated ammonium hydroxide for 1 hr, and subjected to paper chromatography in solvent A. Two bands were resolved, one with chromatographic mobility identical with that of authentic TpT[13,24] and the second corresponding to thymidine (representative R_f 0.67). The yield of isolated TpT (41.6 OD_{267} units or 2.25 μmoles using ϵ 18,500[13]) corresponded to 220 μmoles/g of polymer or 54% conversion of polymer-bound thymidine. The isolated thymidine amounted to 31 μmoles/g.

When the products released by acetic acid were chromatographed directly without ammonia treatment TpTOAc was isolated (R_f 0.55 in solvent C *vs.* R_f 0.42 for TpT).

TpTpTOAc Polymer 8. A TpTOAc polymer similar to **6** (*ca.* 300 mg, TpTOAc content 78 μmoles/g) was deacetylated by agitating it for 3 days at room temperature with 10 ml of pyridine–concentrated NH₄OH (3:1, v/v). The polymer was collected and washed with 10% acetic acid in pyridine for 30 min, and then with dry pyridine and DMAc. The swollen polymer was treated with 240 μmoles of pTOAc and 0.24 g of DCC in a solvent mixture of 1.2 ml of dry pyridine and 2 ml of DMAc for 4 days and worked up as described above. Aqueous acetic acid–benzene hydrolysis of 16.6 mg of polymer followed by ammonia treatment and paper chromatography in solvent A afforded three components corresponding to TpTpT (R_f 0.21, 13.5 OD_{267} units), TpT (21.6 OD_{267} units), and thymidine (6.6 OD_{267} units). The TpTpT isolated from polymer **8** thus corresponded to 29 μmoles/g,[25] representing 38% conversion of TpT.

The R_f value for TpTpT in solvent A varied between 0.21 and 0.25 among various runs and the mobility referred to that of pT (R_{pT}) varied between 1.65 and 1.85. In solvent C the R_f was 0.24 and the R_{pT} 0.69.[26]

Enzymic Hydrolyses. Spleen phosphodiesterase hydrolyses were conducted by treating lyophilized oligonucleotide preparations (*ca.* 0.5 μmole) with 30 μl of enzyme solution[27] and incubating at 37° for 5–6 hr. The hydrolysate was chromatographed in solvent A to reveal complete hydrolysis of TpTpT to Tp and T in a mole ratio of 1.75:1 in one experiment and 2.0:1 in another. In a similar manner TpT was completely hydrolyzed to Tp and T. Snake venom phosphodiesterase[28] (25 μl of solution/0.2 μmole of oligonucleotide, 5–6 hr) also completely degraded TpTpT and TpT to pT and T.

Acknowledgment. We wish to thank Miss Eleanor Applegate for technical assistance, and Misses B. K. Londergan and L. E. Williams and Messrs. B. R. Stevens and A. Vatvars for elemental analyses. Helpful discussions with Dr. R. E. Benson are also gratefully acknowledged.

---

(23) P-L Biochemicals, Inc., Milwaukee, Wis.

(24) The R_f for TpT in solvent A varied between 0.32 and 0.45 among several runs. In solvent C the R_f was 0.40–0.42.

(25) Using ϵ 28,100 for TpTpT, *i.e.*, the sum of those for pT (9600) and TpT (18,500).

(26) For TpTpT ref 14 reports R_f 0.20, R_{Tp} 1.35 in solvent A, and R_f 0.20, R_{Tp} 0.67 in solvent C. G. Weimann and H. G. Khorana, *J. Am. Chem. Soc.*, **84**, 419 (1962), report R_f 0.23, R_{Tp} 1.75 in solvent A.

(27) Worthington Biochemical Corp., Freehold, N. J. The lyophilized enzyme (10–15 units) was dissolved in 1 ml of 0.03 M succinate hydrochloride buffer, pH 6.5. The solution was stored at −20°.

(28) Calbiochem, New York, N. Y. The lyophilized enzyme from Russels viper was dissolved in 0.15 M Tris-HCl buffer, pH 8.9, to a concentration corresponding to 800 units of enzyme/ml.

31

Reprinted from *Tetrahedron Letters,* No. 28, 2881–2884 (1972)

POLYMERS AS CHEMICAL REAGENTS. THE USE OF POLY(3,5 DIETHYLSTYRENE) - SULFONYL CHLORIDE FOR THE SYNTHESIS OF INTERNUCLEOTIDE BONDS (1)

M. Rubinstein, A. Patchornik

Department of Organic Chemistry, The Weizmann Institute of Science

Rehovot, Israel.

(Received in UK 31 May 1972; accepted for publication 7 June 1972)

Insoluble polymeric reagents were shown to be effective in transfer reactions of the type Ⓟ- A + B → Ⓟ + A-B. Where Ⓟ - represents a polymeric leaving group and B a nucleophile. Ⓟ- A can be taken in large excess over B, thus increasing the yield of the product A-B. Then, A-B is isolated by filtration of the polymer. This priciple was demonstrated in peptide synthesis (2).

The chemical synthesis of oligonucleotides, namely condensing monoesters of phosphoric acid with alcohols to give diesters of phosphoric acid has been effected by two methods according to the following scheme:

a. $ROPO_3H^- + R'OH + ArSO_2Cl \xrightarrow[2.H_2O]{1.C_5H_5N} ROPO_2^- OR' + ArSO_3H + HCl$ (3), (4).

b. $ROPO_2^-OX + R'OH + ArSO_2Cl \xrightarrow[2.H_2O]{1.C_5H_5N} ROPO(OX)OR' + ArSO_3H + HCl \xrightarrow{OH^-} ROPO_2^-OR'$ (5).

(Dicyclohexylcarbodiimide can replace $ArSO_2Cl$ in method a. but with poorer results, X in method b. represents a base labile protecting group), the activation of the phosphates in both methods is effected by sterically hindered aromatic sulfonyl chloride like mesitylenesulfonyl chloride and 2,4,6-triisopropylbenzenesulfonyl chloride (TPS). These reagents have two disadvantages: 1. In spite of the steric hindrance, some sulfonation of R'OH does occur, leading to undesired sulfonate esters. 2. In many cases it is quite difficult to remove completely the sulfonic acid from the desired product.

It seemed to us that by using a polymeric arylsulfonyl chloride one might overcome these problems. The sulfonate esters would remain attached to the polymer and thus be easilly removable together with the polymeric sulfonic acid.

Thus the synthesis of poly (3,5 diethylstyrene) sulfonyl chloride was undertaken. When sym. triethylbenzene (I) was treated with N-bromosuccinimide (NBS), α bromoethyl 3,5 diethylbenzene (II) was formed. (B.P. 106°/1.5 mm. NMR δ 1.2, 2.6 for ethyl protons, 2.0 for methyl protons α to the bromine, 5.0 for the proton vicinal to the bromine and 6.85, 6.95 for the aromatic protons. Anal. calcd. for $C_{12}H_{17}Br$: Br. 33.3%. Found: Br. 33.0%). The bromide (II) was refluxed in pyridine to give 3,5 diethylstyrene (III). (B.P.85°/4mm. NMR δ 5.4, 6.6 for vinyl protons, the absorbtions of the ethyl and aromatic protons were the same as in II). The styrene (III) was copolymerised with 5% divinylbenzene as a cross linking agent. A 0.25% aqueous solution of polyvinylalcohol was used as a medium for the bead polymerisation. The cross linked polymer was obtained in bead form and was chlorosulfonated by chlorosulfonic acid in chloroform to give the totally chlorosulfonated polymer (V). (Anal. calcd. for $(C_{12}H_{15}SO_2Cl)_n$: S.12.3%, Cl 13.7%. Found: S.12.1% Cl. 13.2%). No attempt was made to determine the position of the SO_2Cl group on the aromatic ring. However in both possibilities the SO_2Cl group is sterically hindered.

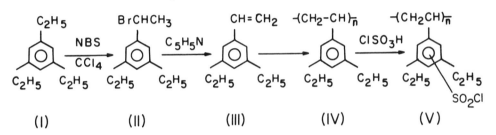

Scheme 1. The preparation of poly (3,5 diethylstyrene) sulfonyl chloride.

The synthesis of a dinucleoside phosphate by this polymer was then undertaken and compared with the conventional methods. 5' Trityl thymidine (1 eq.) and 3' acetyl thymidine 5' phosphate (2 eq.) were treated with 3 eq. of polymer V in dry pyridine overnight. Water then was added and after 24 hrs. the polymer was filtered. The protecting groups were removed according to the literature, the reaction mixture was chromatographed and the yield of TpT in the condensation step was 90% based on VI as determined spectrophotometrically. The synthesis is described in scheme 2.

TpT was obtained also by method b. in a two stage synthesis: VI (1 eq.) and pyridinium β cyanoethyl phosphate were treated with polymer V (3 eq.) in the usual manner. X was isolated by filtration, extracted with $CHCl_3$ and reacted in the second stage, with Thymidine (2 eq.) and polymer V (3 eq.) in dry pyridine overnight. The yield of TpT based on X was 70%, determined as before. The synthesis is described in scheme 3. TpT was degradable by snake venom phosphodiesterase to thymidine and 5' TMP; T/pT=1.03 .

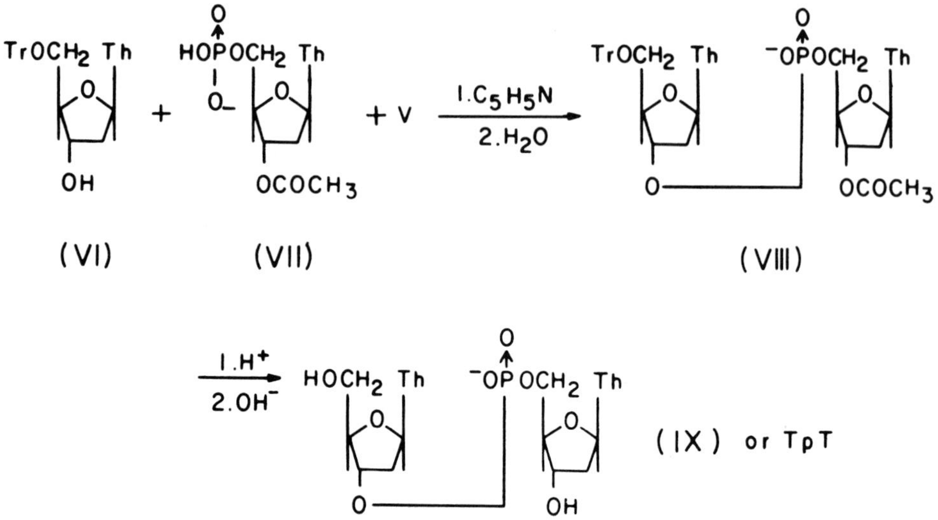

Scheme 2. The synthesis of TpT by polymeric reagent according to method a.
(Tr = trityl, Th = Thymin)

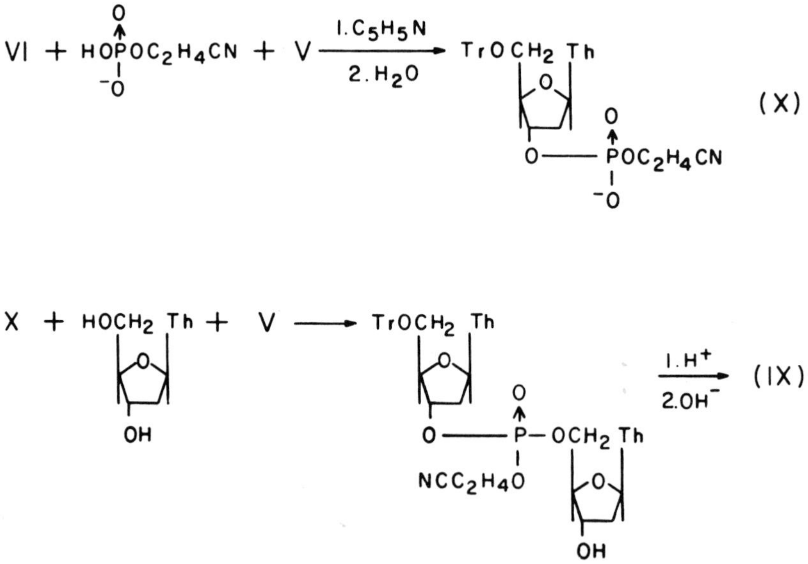

Scheme 3. The synthesis of TpT by polymeric reagent according to method b.

The results were compared with those of the conventional methods. The yields and rate of the reactions were found to be similar, but when TPS was used, triisoprophylbenzenesulfonic acid contaminated the products, as could be seen by TLC and spraying with an acid base indicator.

The sulfonic acid. owing to its detergency, caused stable emulsions with chloroform when the later was used for extracting the products. These problems were completely eliminated when polymer V was used instead of TPS. Moreover, contrary to the TPS procedure, the reaction mixtures described in scheme 3, remained colourless and no sulfonate esters or other impurities were detected on TLC.

Finally. the rate of sulfonation of primary alcohols by polymer V was measured and compared with other arylsulfonyl chlorides. The results of the sulfonation are summarised in Fig. 1.

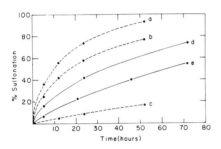

a. Tosyl chloride. b. Mesitylenesulfonyl chloride. c. TPS. d. Copolystyrene, 2% Divinylbenzenesulfonyl chloride. e. V.

Fig. 1. The rate of sulfonation of thymidine and 2'3'dibenzoyluridine (broken lines) by different arylsulfonyl chlorides. The conditions and results with 2'3'dibenzoyluridine are taken from (3).

The rate of sulfonation by polymer V was close to that of TPS. Therefore the polymer is suitable for oligo nucleotide synthesis. The synthesis of longer oligo nucleotides by this approach is under investigation.

References

1. This work has been supported by Grant No. AM 05098 from the National Institute of Health.
2. M. Fridkin, A. Patchornik and E. Katchalsky, J. Amer. Chem. Soc., 90, 2953 (1968).
3. R. Lohrmann and H.G. Khorana, J. Amer. Chem. Soc., 88, 829 (1966).
4. H.G. Khorana et, al., Nature, 227, 27 (1970) and references cited in.
5. R.L. Letsinger and K.K. Ogilvie, J. Amer. Chem. Soc., 91, 3350 (1969).

Copyright © 1973 by Pergamon Press

Reprinted from *Tetrahedron Letters*, No. 31, 2911–2914 (1973)

OLIGONUCLEOTIDE SYNTHESIS ON A POLYMER SUPPORT SOLUBLE IN WATER
AND PYRIDINE [1]

Hartmut Seliger and Gerd Aumann

Institut für Makromolekulare Chemie der Universität Freiburg,

D 78 Freiburg i. Br., Stefan-Meier-Str. 31

(Received in UK 26 April 1973; accepted for publication 20 June 1973)

Introduction: Early studies on oligonucleotide support synthesis have centered on the use of polystyrene based carriers soluble in organic solvents [2,3]. These permitted the elongation of the oligonucleotide graft chain in solvents for which a maximum rate of this reaction had been shown, such as pyridine, whereas all reactants and by-products not bound to the support could easily be removed after the condensation by precipitation of the polymer in water. Thus, by repetition of the condensation and precipitation steps oligonucleotides of short chain length were built up in yields which, in some cases, equalled those obtainable in carrier-free condensations. The disadvantage of these systems, however, was that some crosslinking and a change of solubility would take place after going through several chain elongation cycles [4]. Recent advances in the development of techniques for the separation of low molecular weight compounds from water soluble polymers, e.g. dialysis, ultrafiltration, Sephadex chromatography, have led us to investigate on the use of a support system soluble in pyridine as well as in water, thus retaining the advantages and circumventing the disadvantages of the carrier systems described above. Such a support system was developed on the basis of a copolymer of vinylacetate and N-vinylpyrrolidone. This communication describes the preparation of this support system together with some preliminary oligonucleotide studies aiming at the development of a rapid synthesis of short chain building blocks for fragment condensations of polynucleotides and primers for nucleotide polymerizing enzymes. Recently other soluble polymer systems of potential use for nucleic acid chemistry have been reported [5], however, they have not been shown to exhibit the solubility properties inherent to our system in the synthesis of oligonucleotides.

Results and discussion: The polymer support ((P) -OH, scheme 1) was prepared by bulk copolymerization of N-vinylpyrrolidone and vinylacetate with azobisisobutyronitrile as initiator for 10 h at 70° [6] and subsequent saponification of the ester groups in H_2SO_4/t butanol [7]. The molar ratios of monomers were chosen so as to give a product containing between 10 and 20 % vinylalcohol units. The polymer was purified by dialysis in water and lyophilized. Precipitation from chloroform solution into ether gave a number of fractions from which the one of average molecular weight about 42 000 (determined by viscosimetry) and an alcohol content of 1 mmol/g was chosen for further studies. For attachment of nucleosides 1 g of copolymer, dissolved in 5 ml dimethylformamide and 1 ml triethylamine, was reacted with 2 mmoles 3'-0-β-benzoylpropionyl deoxythymidine-5'-chloroformate [8] for 24 h at room temperature. The unreacted alcohol groups were then blocked with acetic anhydride in pyridine, each reaction step being followed by dialysis. 960 mg of support were obtained, the nucleoside content of which was determined from UV measurements ($\lambda_{max}^{H_2O}$ = 248 nm, shoulder at 272 nm, ϵ_{248} = 16 400 l/mol · cm) to be approximately 0,11 mmoles/g. After nucleoside deblocking by treatment with hydrazine hydrate [9] the carrier system (ⓟ -OCO-dT-OH, scheme 1) was ready for oligonucleotide synthesis.

In a test run for nucleotide polycondensations 800 mg ⓟ -OCO-dT-OH were reacted with 3 mmoles 5'-deoxythymidylic acid in 5 ml pyridine, using 6 mmoles triisopropylbenzene sulfonylchloride as condensing agent. After 72 h at room temperature the mixture was dialyzed in water. For complete removal of the non-bound oligonucleotides chromatography on Sephadex G 100 was necessary. The resulting polymer (ⓟ -OCO-dT(-T)$_n$-OH) was analyzed for its oligothymidylate content by cleavage of a sample of the support in concentrated ammonia at room temperature and chromatographic separation. The products obtained and their yields are given in table 1. It is clear from these results that the grafting of the oligonucleotide chains to a support favors the formation of short chain lengths. This parallels earlier findings reported by T. Kusama and H. Hayatsu for studies with a derivatized Merrifield resin [4]. With the aim of synthesizing short sequences in a most rational fashion the conversion of the nucleotidic material on the support was completed by addition of a suitably blocked nucleotide. In one case 320 mg ⓟ -OCO-dT(-T)$_n$-OH was reacted with 1 mmol 3'-0-p-methoxytrityl deoxythymidine-5'-phosphate (prepared analogously to a method of R.L. Letsinger et al. [10]) and 2 mmoles triisopropylbenzenesulfonylchloride in pyridine for 24 h at room temperature. After dialysis and Sephadex G 100 chromatography the elongated support (ⓟ -OCO-dT(-T)$_m$-O MMTr, scheme 1) was analyzed for its oligonucleotide content by treating it first with 80 % acetic acid, then with concentrated ammonia and separation of the oligomer chains by chromatography. The yields obtained after this reaction cycle are given in line 2 of table 1. In a similar reaction ⓟ -OCO-dT(-T)$_n$-OH was elongated by condensation with $0^{2'}$, $0^{3'}$-diacetyl-5'-ribouridylic acid to give, after cleavage, products of the type dT(-T)$_n$-rU (n = 0 - 2). It can be seen from table 1 that a drastic improvement of oligonucleotide yields has been

Oligonucleotide Synthesis on a Polymer Support Soluble in Water and Pyridine

scheme 1

abbreviations:
Ⓟ = vinylalcohol / N-vinylpyrrolidone copolymer
B.P = ß-benzoylpropionyl-
Ac = acetyl-
MMTr = p-methoxytrityl-
TPS = triisopropylbenzene-sulfonylchloride

Table 1: Yields of products obtained from oligonucleotide syntheses on a support based on a copolymer of N-vinylpyrrolidone and vinylacetate

Products [+]:	dT	d T-T	d T(-T)$_2$	d T(-T)$_3$
Yield I [++] (Mol %):	70	24	6	1
Yield II [++] (Mol %):	28,5	56	13	2

[+] Higher oligomers, present in very small amount (< 1%), were not characterized.

[++] Yield I = after polycondensation

Yield II = after stepwise elongation (for conditions see text)

attained which by far exceeds the yield improvement obtained by multiple polycondensations in the work cited above [4]. Thus, although the individual reaction steps have to be further worked out, the fact that a satisfactory conversion of the initially attached nucleoside (about 80 %) was shown in these orienting experiments and that affinity chromatography of the material released from the carrier will easiliy separate the chain extension products bearing trityl [11] or uridine [12] termini demonstrates the applicability of this support system for rapid syntheses of linear oligonucleotides of the general composition $A\,B_n\,C$, where A, B and C can be different deoxy- or ribonucleotides. Studies on the preparation of such oligomeric sequences and their use for the chemical and enzymic synthesis of longer oligonucleotide chains are in progress.

Acknowledgement: The authors wish to thank Frau Professor Dr. E. Husemann for her interest and encouragement. Financial support by the Deutsche Forschungsgemeinschaft is gratefully acknowledged.

References:

1) Communication no. 5 of a series on polymer support synthesis. No. 1 = ref. 2; no. 2 = ref. 9; no. 3: H. Seliger, R.L. Letsinger, in preparation; no. 4: H. Seliger, in preparation.

2) F. Cramer, R. Helbig, H. Hettler, K.H. Scheit, H. Seliger, Angew. Chem. 78, 640 (1966); Angew. Chem. Internat. Edit. 5, 601 (1966).

3) H. Hayatsu, H.G. Khorana, J. Amer. Chem. Soc. 88, 3182 (1966).

4) T. Kusama, H. Hayatsu, Chem. Pharm. Bull. (Tokyo) 18, 319 (1970).

5) H. Köster, Tetrahedron Lett. 1972, 1535; T. Seita, K. Yamauchi, M. Kinoshita, M. Imoto, Makromol. Chem. 164, 15 (1973); H. Schott, Angew. Chem. 85, 263 (1973).

6) J.F. Bork, L.E. Coleman, J. Poly. Sci. 43, 413 (1960).

7) A similar procedure is described for polyvinylacetate in : Houben-Weyl, Methoden der Organischen Chemie, Vol 14/II, E. Müller, editor, G. Thieme-Verlag, Stuttgart, 1963, p. 700.

8) H. Seliger, Tetrahedron Lett. 1972, 4043.

9) R.L. Letsinger, H. Seliger, Macromolecular Preprints, XXIIIrd Internat. Congress of Pure and Applied Chemistry, Boston, 1971, p. 1261

10) R.L. Letsinger, K.K. Ogilvie, J. Amer. Chem. Soc. 89, 4801 (1967).

11) K.L. Agarwal, A. Yamazaki, P.J. Cashion, H.G. Khorana, Angew. Chem. 84, 489 (1972); Angew. Chem. Internat. Edit. 11, 451 (1972).

12) H. Seliger, E. Rössner, M. Philipp, publication in preparation.

Oligonucleotide Synthesis on Polar Polymer Supports. The Use of a Polypeptide Support

By Toby M. Chapman and Dennis G. Kleid

(*Department of Chemistry, University of Pittsburgh, Pittsburgh, Pennsylvania* 15213)

Summary The solid-phase synthesis of a trinucleotide phosphate, pTpTpT, using a polar polypeptide support is described.

The use of polymer supports has proved to be of great importance in the synthesis of peptides;[1] their use in the synthesis of oligonucleotides has yet to reach this potential.[2] The major problem is the lack of a simple, high yield phosphorylation reaction in the construction of the internucleotide phosphate ester linkage. Another problem may be the incompatibility of the highly polar oligonucleotide with the non-polar polymers usually used as supports, mostly derivatized polystyrenes; indeed, the use of polymers containing polar backbone structures has hardly been investigated. Anfinsen[3] proposed the use of polypeptide supports for peptide synthesis and recently Köster[4] has investigated the use of polysaccharide and polyethylene glycol supports for the synthesis of oligonucleotides. We report the synthesis of a trinucleotide on a polypeptide support.

Poly-L-lysine hydrobromide (1) (mol. wt. 80,000) was modified so as to permit an aromatic phosphoramidate linkage between the support and the oligonucleotide. *p*-Trifluoroacetaminobenzamide linkages to polymer (1) were produced with *p*-trifluoroacetaminobenzoic isobutyl-carbonic anhydride (2) in dimethylformamide (DMF) and triethylamine; further reaction with phenyl isocyanate[5] to block unreacted aliphatic amine gave polymer (3). The amidation reaction went in 61% yield based on fluorine analysis. Anhydride (2) (m.p. 95—95.5°) was prepared from *p*-trifluoroacetaminobenzoic acid[6] and isobutylchloroformate. Treatment of (3) with saturated methanol-ammonia (12 h) liberated the aromatic amine; cross-linking with hexamethylenedi-isocyanate (2·4 mole %) gave polymer (4). The polymer swells in DMF and pyridine. The arylaminopolymer (4) was mixed with the dipyridinium salt of 5′-phosphoro-3′-acetylthymidine (5) and dicyclohexyl-carbodi-imide (DCC), in pyridine to give the nucleotide phosphoramidate (6) (15%). The method of Blackburn[7] was used to remove the 3′-acetyl group from (6) and the product was further elaborated by reaction with (5) and tri-isopropylbenzene sulphonyl chloride (TPS). After reaction with 1-naphthylisocyanate[5] to block unreacted mononucleotide, the 3′-acetate was removed and the product extended with (6) and TPS, followed by isocyanate treatment and removal of 3′-acetate.

Cleavage of the product from the polymer was effected using isoamyl nitrite in pyridine-acetic acid (1:1 v/v).[2] Paper chromatography of the product gave four u.v. absorbing spots, corresponding to pT-OH, pTpT-OH, pTpTpT-OH, and pTOCONHC$_{10}$H$_{7}$. The R_F values obtained were in good agreement with literature values[2,8] and were also checked against authentic samples kindly provided by Dr. K. L. Agarwal and Dr. H. G. Khorana. U.v. analysis showed 19 μmol of pT-OH, 4·3 μmol of pTpT-OH, and 3·3 μmol of trinucleotide pTpTpT-OH. This corresponds to a 14% yield of trinucleotide from phosphoramidate polymer (6), with the conversion of dinucleotide to desired product occurring in at least 43% yield.

We thank the Health Research and Services Foundation, Pittsburgh, Pennsylvania for support of this work and the University of Pittsburgh Department of Chemistry for providing an assistantship to D.G.K.

(*Received,* 12*th January* 1973; *Com.* 039.)

[1] R. B. Merrifield, *J. Amer. Chem. Soc.*, 1964, **86**, 304.
[2] E. Ohtsuka, S. Morioka, and M. Ikehara, *J. Amer. Chem. Soc.*, 1972, **94**, 3229 and references therein.
[3] C. B. Anfinsen, in 'Chemistry of Natural Products', I.U.P.A.C., Butterworths, London, 1968, p. 461.
[4] H. Köster, *Tetrahedron Letters*, 1972, 1531, 1535.
[5] K. L. Agarwal, A. Yamazaki, P. J. Cashion, and H. G. Khorana, *Angew. Chem. Internat. Edn.*, 1972, **11**, 451.
[6] F. Weygand and E. Leising, *Chem. Ber.*, 1954, **87**, 248.
[7] G. M. Blackburn, M. J. Brown, and M. R. Harris, *J. Chem. Soc.*, 1967, 2438.
[8] H. G. Khorana and J. P. Vizsoly, *J. Amer. Chem. Soc.*, 1961, **83**, 675; S. A. Narang, J. J. Michniewicz, and S. K. Dheer, *ibid.*, 1969, **91**, 936.

V

The Solid-Phase Synthesis of Saccharides

Editors' Comments on Papers 34 Through 39

34 **Fréchet and Schuerch:** *Solid-Phase Synthesis of Oligosaccharides: I. Preparation of the Solid Support. Poly[p-(1-propen-3-ol-1-yl)styrene]*

35 **Fréchet and Schuerch:** *Solid-Phase Synthesis of Oligosaccharides: II. Steric Control by C-6 Substituents in Glucoside Synthesis*

36 **Fréchet and Schuerch:** *Solid-Phase Synthesis of Oligosaccharides: III. Preparation of Some Derivatives of Di- and Tri-Saccharides via a Simple Alcoholysis Reaction*

37 **Guthrie, Jenkins, and Stehlicek:** *Synthesis of Oligosaccharides on Polymer Supports: I. 6-O-(p-Vinylbenzoyl) Derivatives of Glucopyranose and Their Copolymers with Styrene*

38 **Excoffier, Gagnaire, Utille, and Vignon:** *Solid-Phase Synthesis of Oligosaccharides: II. Synthesis of 2-Acetamido-6-O-(2-Acetamido-2-Deoxy-β-D-Glucopyranosyl)-2-Deoxy-D-glucose*

39 **Zehavi and Patchornik:** *Oligosaccharide Synthesis on a Light-Sensitive Solid Support: I. The Polymer and Synthesis of Isomaltose (6-O-α-D-Glucopyranosyl-D-glucose)*

The three papers (34, 35, 36) by Fréchet and Schuerch describe the use of a new type of polymer suitable for SPS in the saccharide field. The polymer is basically a polystyrene with the important modification of having alkyl alcohol groups attached. Thus the cleavage of a completed oligosaccharide could occur by an oxidative cleavage. A number of 6-substituted-1-halo or benzyl monomer units were studied preliminary to their use as the first saccharide unit to be attached to the polymer.

Application of the monosaccharides to the polymer proceeded in high yield. However, the anticipated stereochemical control of the alcoholysis reaction (adding the next monosaccharide) did appear to be highly selective. Perhaps the use of a resin limited the reaction conditions to the extent that the normal stereochemical effects of substituents, solvent, ion pairing, and so on, were lost.

The paper of Guthrie, Jenkins, and Stehlicek (Paper 37) describes the use of copolymers containing preformed monosaccharide units as esters of vinyl benzoate and styrene. Although Guthrie and coworkers were successful in obtaining good yields of the desired copolymers, some difficulty was encountered with the hydrolysis of the saccharides from the polymer. The Letsinger "popcorn" polymer (Paper 3) was used by Excoffier et al. (Paper 38) as a means of attachment of a monosaccharide. The polymer support apparently increases the selectivity of reactions at the 6-O position of the glucose unit over that at the 4-O position.

As mentioned in the early papers of saccharide synthesis on a solid support, one of the more difficult problems is to release the sugar moiety from the polymer at the conclusion of the synthetic steps. A different approach to this problem has been developed by Zehavi and Patchornik (Paper 39) in the use of a polymer that is light-sensitive (see Paper 18 for similar photolytic cleavage reaction). Cleavage is accomplished by a photochemical process that proceeds in fairly high yield.

Solid-Phase Synthesis of Oligosaccharides. I. Preparation of the Solid Support. Poly[p-(1-propen-3-ol-1-yl)styrene]

Jean M. Frechet and Conrad Schuerch

Contribution from the Department of Chemistry, State University College of Forestry at Syracuse University, Syracuse, New York 13210. Received July 13, 1970

Abstract: Solid-phase synthesis of oligosaccharides requires the use of a saccharide derivative with a reactive leaving group at C-1, one hydroxyl protected by a readily removable blocking group, the remaining hydroxyls protected by a stable blocking group, and a resin which can be separated from the formed oligosaccharide derivative without removing the alcoholic substituents. A polystyrene resin with allyl alcohol functional groups has been prepared and a procedure for the solid-phase synthesis of oligosaccharides has been tested with an appropriately substituted glycosyl halide.

The use of insoluble, functionalized polymer supports has been extensively studied for the synthesis of polypeptides[1-8] and polynucleotides.[9-11] The process of sequential synthesis on solid support has many attractive features and should be applicable to the preparation of oligosaccharides.

The main advantage of the solid-phase method is that once the growing molecule is firmly attached to a completely insoluble resin, purification is effected at each intermediate step merely by filtering and washing. Furthermore, reaction rates can be increased by using a large excess of reagent which, after the reaction has gone to completion, can be easily separated.

The requirements for application of this procedure to carbohydrates are the following.

1. The monomer must be a sugar derivative with a reactive leaving group at C-1, one hydroxyl group protected by an easily removable "temporary" blocking group, R_1, and the remaining hydroxyl groups protected by means of a "persistent" blocking group, R_2.

2. The solid substrate must contain an appropriate functional group to link to the glycosidic center. Separation of the solid substrate from the completed oligosaccharide must be possible under conditions which do not affect the persistent blocking groups, since their removal prior to or simultaneously with cleavage from the resin would cause formation of a hydrophilic molecule encaged in a network of hydrophobic insoluble resin.

Scheme I outlines our proposed preparation of a 1,6-linked oligomer of glucose. In the first step, the activated monomer unit (glycosyl bromide in which $R_1 =$

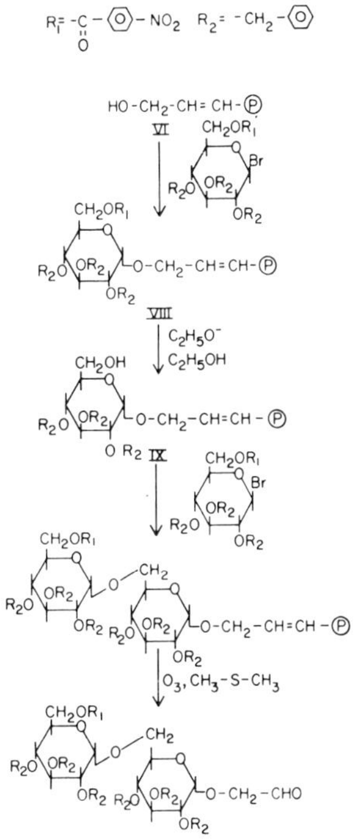

(1) R. B. Merrifield, *J. Amer. Chem. Soc.*, **85**, 2149 (1963); **86**, 304 (1964).
(2) R. B. Merrifield, *Biochemistry*, 3, 1385 (1964).
(3) R. B. Merrifield, *Science*, **150**, 178 (1965).
(4) B. Gutte and R. B. Merrifield, *J. Amer. Chem. Soc.*, **91**, 501 (1969).
(5) B. F. Gisin, R. B. Merrifield, and D. C. Tosteton, *ibid.*, **91**, 501 (1969).
(6) S. Wang and R. B. Merrifield, *ibid.*, **91**, 6488 (1969).
(7) M. Fridkin, A. Patchornik, and E. Katchalski, *ibid.*, **90**, 2953 (1968); **88**, 3164 (1966).
(8) E. Bayer, H. Eckstein, K. Hägele, W. A. König, W. Brüning, H. Hagenmaier, and W. Parr, *ibid.*, **92**, 1735 (1970).
(9) R. L. Letsinger and V. Mahadevan, *ibid.*, **87**, 3526 (1965); **88**, 5319 (1966).
(10) R. L. Letsinger, M. H. Caruthers, and D. M. Jerina, *Biochemistry*, **6**, 1379 (1967).
(11) H. Hayatsu and H. G. Khorana, *J. Amer. Chem. Soc.*, **88**, 3182 (1966).

Scheme I. Solid-Phase Synthesis of a Dissacharide

p-nitrobenzoate and $R_2 =$ benzyl) is coupled to a suitably functionalized allylic alcohol resin. In the next step, the temporary blocking group R_1 is removed under mild conditions which leave the persistent blocking groups and the glycosidic linkage to the resin untouched. The following step involves reaction of a new monomer unit with the reactive end of the unit previously attached to the resin via a simple alcoholysis reaction. Further steps consist of sequential deblocking and coupling until an oligomer of the desired length is obtained. Finally, the oligomer is cleaved from the resin support by oxidation and the persistent blocking groups are removed from the soluble derivative.

In the following discussion, Ⓟ-X refers to a styrene–divinylbenzene copolymer in which some of the aromatic rings contain substituent X. Thus, Ⓟ-CH_2Cl refers to a chloromethylated resin. It should be noted, however, that only a fraction (typically 10–15%) of the units carries the substituent X.

As is apparent above, the resin should contain a hydroxyl group for coupling with the glycosyl halide. A chloromethylated polystyrene–divinylbenzene copolymer[1] Ⓟ-CH_2Cl has been used extensively in solid-phase peptide synthesis; from this resin, Ⓟ-CH_2OH could easily be prepared. However, this resin would not be satisfactory, since cleavage of the oligomer would amount to a debenzylation requiring the use of sodium and liquid ammonia. These conditions would cause total collapse of the resin and removal of the persistent blocking group, making cleavage and recovery of the desired product difficult.

Since other solid supports which have been prepared[1,6,12] for use in solid-phase peptide synthesis are not satisfactory for our purposes, we have developed a new resin prepared from commercial cross-linked polystyrene. This resin, Ⓟ-$CH=CHCH_2OH$, contains approximately 10% allyl alcohol side chains. This type of side chain is very convenient, since it contains one site, the hydroxyl group, for coupling with the glycosyl halide and another site, the double bond, for oxidative cleavage of the finished oligomer under conditions to which the blocking groups and glycosidic linkages are resistant. The preparation of the resin Ⓟ-$CH=CH$-CH_2OH is outlined in Scheme II.

Scheme II. Preparation of the Solid Support

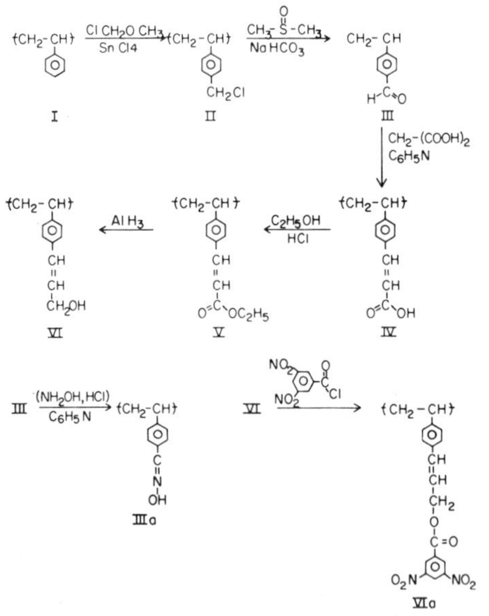

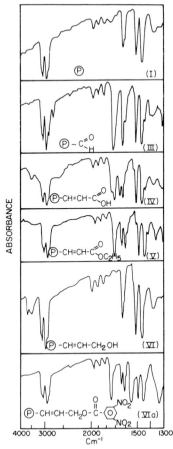

Figure 1. Infrared spectra of copolystyrene–2% divinylbenzene resin (I), Ⓟ—CHO resin (III), Ⓟ—CH=CHCOOH resin (IV), Ⓟ—CH=CHCOOC$_2$H$_5$ resin (V), Ⓟ—CH=CHCH$_2$OH resin (VI), and Ⓟ—CH=CHOCOC$_6$H$_3$(NO$_2$)$_2$ resin (VIa).

The starting material was a commercially available polystyrene cross-linked with 1 or 2% divinylbenzene (Bio-Rad Laboratories). This resin, in the form of 200–400-mesh beads, was readily swollen by a number of solvents such as benzene, pyridine, dioxane, and chloroform, in which no major accessibility problems were encountered, and the resin could easily be penetrated by various reagents.

The degree of chloromethylation was kept in the range 10–15% to avoid creation of reactive end groups at more difficultly accessible sites. Oxidation of the chloromethylated material, II, with dimethyl sulfoxide[13] gave an aldehydic resin, Ⓟ-CHO, III, which showed an intense carbonyl absorption at 1690 cm^{-1} and an aldehydic CH stretching absorption at 2700 cm^{-1} in the infrared spectrum (Figure 1). These values are in good agreement with those recorded by Ayres and Mann.[13] An estimation of the yield of the dimethyl sulfoxide oxidation was made by measuring the nitrogen content of the oxime IIIa derived from the formylstyrene polymer III. Conversion of III into its oxime

(12) R. L. Letsinger, M. J. Kornet, V. Mahadevan, and D. M. Jerina, *J. Amer. Chem. Soc.*, **86**, 5163 (1964).

(13) J. T. Ayres and C. K. Mann, *Polym. Lett.*, **3**, 505 (1965).

seemed to be quantitative, since no carbonyl absorption could be detected in the infrared spectrum of IIIa. By this method, it was determined that the dimethyl sulfoxide oxidation proceeded in 75% yield, which is higher than the recorded yield[14] (~60%) for the oxidation of benzyl chloride. Condensation of the Ⓟ-CHO resin with malonic acid in the presence of pyridine proceeded smoothly to yield 90% Ⓟ-CH=CHCOOH resin, IV. The infrared spectrum of the latter resin showed a very broad hydroxyl absorption in the 2500–2800-cm^{-1} region, an intense carbonyl absorption at 1695 cm^{-1} which corresponds to hydrogen-bonded carboxyl groups, and a weaker band at 1725 cm^{-1}, corresponding to free carboxyl groups. The main C=C olefin absorption was observed at 1625 cm^{-1}, while a weaker band at 975 cm^{-1} suggested a trans arrangement for the double bond. Other bands of interest included a CO stretching band at 1260 cm^{-1}. The 90% conversion was estimated by titration of the carboxylic acid with butyllithium in benzene. This yield is comparable to the yield of a similar reaction run on benzaldehyde.[15]

The influence of the solvent medium on the availability of the functional groups is illustrated by the fact that the resin could not be titrated satisfactorily by either sodium hydroxide in 80% ethanol, sodium ethoxide in ethanol, or potassium hydroxide in diethylene glycol without requiring long refluxing periods prior to back-titration by acid. Thus, in sodium ethoxide–ethanol, only 25% of the carboxyl groups could be neutralized at room temperature.

The reduction of the Ⓟ-CH=CHCOOH resin to the unsaturated alcohol Ⓟ-CH=CH₂OH was first studied on model compounds. Aluminum hydride reduction[16] of ethyl cinnamate and cinnamic acid in anhydrous ether gave, respectively, 95 and 83% yields of cinnamyl alcohol, while very little saturated alcohol was detected by vapor phase chromatography. Since a better yield of unsaturated alcohol was obtained by reduction of the ester, IV was esterified to V, which could be characterized by its infrared spectrum featuring a strong carbonyl absorption at 1705 cm^{-1}, a C=C stretch at 1630 cm^{-1}, and a broad CO stretch at 1155 cm^{-1}. Aluminum hydride reduction of V gave the desired resin, Ⓟ-CH=CHCH₂OH, VI.

The infrared spectrum of VI showed no carbonyl absorption, indicating that the reduction was complete. The presence of a double bond in VI was confirmed by the rapid consumption of bromine when VI was shaken with a solution of bromine in carbon tetrachloride and by the formation of a black osmate ester upon addition of osmium tetroxide in pyridine. Reaction of VI with 3,5-dinitrobenzoyl chloride produced the ester VIa. Nitrogen analysis on VIa gave results confirming that all the carboxyl groups of V had been reduced.

Two other approaches toward the synthesis of similar resins are presented in Scheme III; both methods were first studied on model compounds, then applied to the resin.

The first approach (path a) takes advantage of the availability of the chloromethylated resin. In the model reaction benzyltriphenylphosphonium chloride was prepared in quantitative yield from benzyl chloride.[17] A Wittig reaction carried out on the phosphonium salt with inverse addition of the phosphorane to chloroacetaldehyde, freshly prepared from its trimer,[18] yielded cinnamyl chloride in ~50% yield. When this reaction sequence was applied to the chloromethylated resin, it was found that formation of the phosphonium salt was slow but could be forced to completion as evidenced by phosphorus analysis. Reaction of Ⓟ-CH₂P(C₆H₅)₃⁺ + Cl⁻ with butyllithium produced the bright colored ylide which reacted rapidly with chloroacetaldehyde to yield the desired resin; 52% of the theoretical amount of triphenylphosphine oxide was collected, indicating that the yield of this reaction was comparable to that of the model reaction.

In the second approach (path b) benzaldehyde was used as a model for the formyl-substituted resin, Ⓟ-CHO. The reaction was carried out in one step by reaction of the alkoxide of hydroxyethylidenephosphorane with freshly distilled benzaldehyde. Unfortunately, the yields of cinnamyl alcohol were low (~30%) and when the reaction was applied to the resin, the product was found to contain some carbonyl groups.

The resin prepared above, Ⓟ-CH=CHCH₂OH, was allowed to react at room temperature with an excess of 2,3,4-tri-O-benzyl-6-O-p-nitrobenzoyl-β-D-glucopyranosyl bromide[19] in a dry solvent such as benzene or methylene chloride.

The reaction time was determined by model reaction of the glycosyl bromide with cinnamyl alcohol. The infrared spectrum of the resin-bound glycoside VIII, taken after rinsing and drying, exhibited a strong carbonyl absorption at 1720 cm^{-1} and a CO stretching band at 1280 cm^{-1}. Both bands are characteristic of the p-nitrobenzoyl ester substituent on carbon 6 of the sugar molecule. The yield of the coupling step, estimated from the gain in weight of the resin after coupling, varied from 70 to 91%, depending on the reaction conditions.

A solution of sodium ethoxide in ethanol–benzene was found to give excellent results for the removal of the temporary blocking group, as evidenced by the complete disappearance of the C=O absorption in the infrared spectrum of the resin (IX) after ester interchange.

After removal of the temporary blocking group, the resin containing a monomer unit with a free hydroxyl group on C-6 (IX) was condensed again with the protected glycosyl halide. The yield of this coupling step was found to be ~75% and the product (X) was

---

(14) H. R. Nace and J. J. Monagle, *J. Org. Chem.*, **24**, 1792 (1959).
(15) G. Lock and E. Bayer, *Chem. Ber.*, **72**, 1064 (1939).
(16) M. J. Jorgenson, *Tetrahedron Lett.*, 559 (1962).
(17) K. Friedrich and H. Henning, *Chem. Ber.*, **92**, 2756 (1959).
(18) H. Gross, *J. Prakt. Chem.*, **4**, 99 (1963).
(19) T. Ishikawa and H. G. Fletcher, *J. Org. Chem.*, **34**, 563 (1969).

again characterized by the appearance of a strong carbonyl absorption.

A resin-bound protected trisaccharide was synthesized from X in comparable yield by repetition of the ester interchange and coupling reaction sequence.

These experiments demonstrate the feasibility of solid-phase oligosaccharide synthesis. The preparation of a suitable resin was an important step in this direction. In preliminary experiments, the oligosaccharides prepared by this method have been cleaved from the solid support by ozonolysis. We are now proceeding with the characterization of these products, optimization of the various yields, and preparation of longer and varied sequences.

Experimental Section

Infrared spectra were taken on a Perkin-Elmer 137 infrared spectrophotometer in potassium bromide pellets and are shown in Figure 1. Elementary analyses were performed by Galbraith Laboratories and in this laboratory. The resin used was a copolymer of styrene-divinylbenzene in the form of 200–400 mesh resin beads containing 1 or 2% divinylbenzene and available commercially (Bio-Rad Laboratories: Bio-Beads S-X1 and S-X2). All the Wittig reactions were carried out in a drybox under nitrogen atmosphere.

Preparation of Ⓟ-CH₂Cl (II). This polymer was prepared as described by Merrifield.[1,2] The degree of chloromethylation could be controlled by changes in temperature and reaction time. In most cases resins with a capacity of 1–1.7 mequiv/g were prepared.

Preparation of Ⓟ-CHO (III). Chloromethylated Bio-Beads S-X2 (45 g, capacity 1.05 mequiv/g) were stirred in 300 ml of dimethyl sulfoxide with 19 g of sodium bicarbonate for 6 hr at 155°. The resin was then collected on a glass filter, washed with dimethyl sulfoxide, hot water, and a 2:1 mixture of dioxane and water, then rinsed with dioxane, acetone, ethanol, methylene chloride, and benzene. A yield of 44 g of cream colored Ⓟ-CHO was obtained after drying at 100° under vacuum. Microanalysis indicated that this resin contained no chlorine. III (0.5 g) was transformed into its oxime by reaction with excess hydroxylamine hydrochloride (0.3 g) in 5 ml of pyridine for 6 hr at 90–100°. After washing with pyridine and the solvents mentioned above, the oxime IIIa was dried in a vacuum oven at 50°. By microanalysis it was shown that this resin contained 1.11% nitrogen which corresponds to 0.79 mmol of functional group/g of resin, indicating a yield of 75% for the dimethyl sulfoxide oxidation.

Preparation of Ⓟ-CH=CHCOOH (IV). III (40 g) was suspended in 300 ml of pyridine. After addition of 7 g of malonic acid and 5 ml of piperidine, the mixture was stirred and heated to 80–90° for 1 hr, then brought to 110–115° for 3 hr. After cooling, the resin was collected on a glass filter and washed with benzene, dioxane, a 4:1:1 mixture of dioxane–HCl–water, a 2:1 mixture of dioxane–water, acetone, benzene, and methylene chloride, then dried at 100° under vacuum.

Titration was effected by treating 0.5 g of polymer with an excess of BuLi in benzene and back-titrating with 0.1 N H₂SO₄. This method was preferred over a titration in ethanolic medium which required a refluxing period prior to back-titration. IV was found to contain 0.71 mmol of functional groups/g of resin and was, therefore, obtained from III in 90% yield.

Preparation of Ⓟ-CH=CHCOOC₂H₅ (V). IV (40 g) was suspended in 160 ml of absolute ethanol. The mixture was heated to reflux and a current of dry HCl passed through the suspension for 2 hr. After cooling, the resin was collected on a glass filter, rinsed, and dried in a vacuum oven.

Preparation of Ⓟ-CH=CHCH₂OH (VI). The reduction of IV and V to the corresponding unsaturated alcohol VI was first studied on model compounds. To a cooled solution of 0.09 mol of lithium aluminum hydride in 50 ml of anhydrous ether was added 0.03 mol of aluminum chloride. After dissolution of the aluminum chloride, the solution was brought to room temperature and a solution of 0.1 mol of cinnamic acid in 250 ml of anhydrous ether was added slowly. Reaction was allowed to proceed at room temperature for 30 min. The reaction mixture was then treated with water and 3 N HCl, and after work-up cinnamyl alcohol was isolated in 83% yield. The same reaction was carried out on ethyl cinnamate with a 3:1 M ratio of lithium aluminum hydride to aluminum chloride. Cinnamyl alcohol was obtained in 95% yield.

The preceding reactions were also carried out as above with a 1:1 ratio of lithium aluminum hydride to aluminum chloride. Under these conditions, the reducing agent was the hydridoaluminum halide, AlH₂Cl,[20] which proved to be very satisfactory and gave cinnamyl alcohol from cinnamic acid and ethyl cinnamate in 80 and 92% yields, respectively.

The yields were measured by weight of crystalline product and by vapor phase chromatography on a 6-ft 10% Carbowax 20M column in a Hewlett-Packard 5750 research chromatograph. In the reduction of ethyl cinnamate, the only side product found in significant quantity (∼5%) was 3-phenylpropanol. In the case of the reduction of cinnamic acid, 3-phenylpropanol was also found (∼7%) together with another side product, possibly 1-phenyl-2-propen-1-ol (∼8–10%).

To a solution of 0.104 mol of aluminum hydride (made by addition of 0.078 mol of lithium aluminum hydride to 0.026 mol of aluminum chloride) in 50 ml of anhydrous ether was added a suspension of 39 g of resin V (Ⓟ-CH=CHCOOC₂H₅) in dry tetrahydrofuran at a rate sufficient to maintain gentle boiling. When the addition was complete, the mixture was stirred for an additional 30 min. The excess reducing agent was then destroyed by addition of an excess of 3 N HCl. The resin was then collected on a glass filter, washed, and dried.

The above reaction was also carried out on a sample of resin IV (Ⓟ-CH=CHCOOH). The product of the reaction had an infrared spectrum identical with that of VI.

Functional derivatives of VI were prepared as follows.

3,5-Dinitrobenzoate Ester (VIa). Resin VI (0.5 g) was swollen in 10 ml of pyridine and 0.5 g of dinitrobenzoyl chloride was added; the mixture was heated and stirred for 1 hr. The resin was then collected on a glass filter, rinsed, and dried. Microanalysis showed that VIa contained 1.76% nitrogen, indicating that at least 98% of the carboxyl groups of V had been reduced.

Dibromo Derivative. Resin VI (0.5 g) was suspended in 10 ml of a 9:1 mixture of CCl₄ and C₂H₅OH. The mixture was treated with an excess of bromine in CCl₄. Bromine consumption was rapid and after 20 min the resin was collected, washed, and dried. Microanalysis showed that the reaction product contained 8.8% bromine, indicating that at least 87% of the alcohol groups in resin VI were allylic.

Osmate Ester. When a suspension of VI (0.25 g) in 2 ml of pyridine was treated with a solution of 50 mg of osmium tetroxide in 2 ml of pyridine, a black osmate ester was formed immediately.

Preparation of Ⓟ-CH₂P(C₆H₅)₃⁺ + Cl⁻ (VII). Chloromethylated resin (32 g) containing 1.27 mmol of functional group/g was refluxed in 350 ml of dry dioxane containing 40 g of triphenylphosphine for 1 week. The resin particles were then collected on a glass filter, rinsed, and dried. Microanalysis of the resin showed that it contained 3% phosphorus, indicating that conversion of the chloromethylated resin to the phosphonium salt was quantitative.

Preparation of Chloroacetaldehyde. Monochloroethylene carbonate was prepared by chlorination of ethylene carbonate as described by Marder and Schuerch.[21] Pyrolysis of 20 g of monochloroethylene carbonate at 200° in the presence of 2 drops of triethylamine[8] yielded 10 g of crude chloroacetaldehyde which was collected as a green liquid. The crude material was redistilled and the fraction boiling at 82–88° was placed in the freezer compartment of a refrigerator and allowed to stand 2–3 days. The white solid (8 g) which was formed was dried under vacuum at room temperature and was identified as chloroacetaldehyde trimer (mp 85–87°). The trimer had no absorption in the 6-μ region of the infrared, attributable to a carbonyl function, and could be recrystallized from ethanol with little change in melting point. Distillation of this trimer immediately before use yielded chloroacetaldehyde, bp 85–87°.

Preparation of Ⓟ-CH=CHCH₂Cl. The Wittig reaction of the phosphonium salt VII with butyllithium and chloroacetaldehyde was first studied on a model compound. To 6 g (0.0155 mol) of finely powdered benzyltriphenylphosphonium chloride,[17] suspended in 30 ml of dry benzene, was added 9.3 ml (0.0149 mol) of a 1.6 M butyllithium solution in hexane. The wine-colored ylide was then added slowly, with stirring, to 1.22 g (0.0156 mol) of freshly distilled chloroacetaldehyde dissolved in 10 ml of methylene chloride. The red coloration of the ylide disappeared immediately and after 3 hr

---

(20) E. C. Ashby and J. Prother, *J. Amer. Chem. Soc.*, **88**, 729 (1966).

(21) H. L. Marder and C. Schuerch, *J. Polym. Sci.*, **44**, 129 (1960).

of reflux the solution was concentrated on a flash evaporator. The bulk of the triphenylphosphine oxide was precipitated by addition of petroleum ether and filtered. The remaining solution was concentrated and distilled to yield 50% of the theoretical amount of cinnamyl chloride as a fraction boiling at 107–109° (12–13 mm).

This reaction was carried out on resin VII using 0.98 molar equiv of butyllithium and 1.1 molar equiv of chloroacetaldehyde for each molar equivalent of resin. To a stirred suspension of the phosphonium salt VII (0.016 equiv) in 100 ml of dry benzene was added 9.8 ml (0.0157 mol) of a 1.6 M solution of butyllithium in hexane. A deep red coloration developed immediately, indicating formation of the phosphorane VIIa. After 20 min, the suspension of VIIa was added slowly, with stirring, to 1.4 g (0.0178 mol) of freshly prepared chloroacetaldehyde in 10 ml of methylene chloride. The red coloration of the resin vanished immediately and the reaction mixture was refluxed for 3 hr, then allowed to stand overnight. The cream-colored resin was then collected on a glass filter and rinsed several times with benzene and ethanol to remove the triphenylphosphine oxide. The yield of the reaction, estimated from the weight of triphenylphosphine oxide recovered from the filtrate, was about 50% and, therefore, comparable to that of the model reaction.

Preparation of Ⓟ-CH=CHCH$_2$OH by Wittig Reaction. The model reaction was performed on benzaldehyde as follows. To a stirred suspension of 10 g (0.029 mol) of finely powdered 2-hydroxyethyltriphenylphosphonium chloride[22] in 200 ml of tetrahydrofuran was added 40 ml (0.056 mol) of a 1.4 M butyllithium solution in hexane. After the solution became homogeneous, 3 ml of freshly distilled benzaldehyde was added. The deep red coloration changed to orange immediately, and the reaction mixture was refluxed for 2 hr, then stirred overnight at room temperature. After the usual work-up, the product was distilled under reduced pressure to yield ~30% cinnamyl alcohol.

An identical procedure was used for the reaction of hydroxyethylidenephosphorane with Ⓟ-CHO (II). However, the reaction product, after washing and drying, was found to contain some residual carbonyl groups.

---

(22) G. Aksnes, *Acta Chem. Scand.*, **15**, 438 (1961).

Preparation of 2,3,4-Tri-*O*-benzyl-6-*O*-*p*-nitrobenzoyl-β-D-glucopyranosyl Bromide. This monomer was obtained in 35% yield from 2,3,4-tri-*O*-benzyl-D-glucopyranose as described by Ishikawa and Fletcher.[19] The slightly colored material, as crystallized from ether, had mp 94–97° and was, therefore, used without further purification.

Formation of the Glycosidic Linkage to the Resin. To 1 g of resin VI was added a solution of 2 g of 2,3,4-tri-*O*-benzyl-6-*O*-*p*-nitrobenzoyl-β-D-glucopyranosyl bromide in 9 ml of dry benzene containing 0.07 ml of pyridine (molar ratio of 1:4:1 for resin–monomer–pyridine). The mixture was then stirred slowly at room temperature and the reaction allowed to proceed for 60 hr in the dark. The resin (VIII) was collected on a glass filter, washed, dried, and weighed. The weight gain of 370 mg indicated a conversion of 90% based on the number of allylic alcohol groups available for reaction on the resin. Nitrogen microanalysis gave results indicating that the actual yield might be somewhat higher than 90%. Lower yields were obtained when either lower monomer concentrations or shorter reaction times were used.

Preparation of a Disaccharide. To 1.2 g of resin VIII, suspended in 10 ml of benzene, was added 5 ml of an 0.21 N solution of sodium ethoxide in ethanol. After stirring the mixture for 30 min, resin IX was collected on a glass filter, rinsed, and dried.

IX (1 g) was allowed to react with a solution of 1.5 g of 2,3,4-tri-*O*-benzyl-6-*O*-*p*-nitrobenzoyl-β-D-glucopyranosyl bromide in 9 ml of dry benzene (molar ratio of 1:4 resin–monomer). The reaction was allowed to proceed in the dark with stirring for 60 hr. The resin X was then collected on a glass filter, washed, and dried. The yield of the coupling step, as estimated from the gain in weight of the resin (235 mg), was 75%. The presence of a new acyl group on the resin-bound disaccharide was again confirmed by the presence of a strong carbonyl absorption at 1720 cm^{-1}.

Acknowledgments. The present work has been supported by Research Grant GM06168 of the Division of General Medical Sciences, National Institutes of Health. The authors acknowledge the assistance of Dr. W. Bracke.

Solid-Phase Synthesis of Oligosaccharides. II. Steric Control by C-6 Substituents in Glucoside Syntheses

Jean M. Fréchet and Conrad Schuerch

Contribution from the Department of Chemistry, State University College of Forestry at Syracuse University, Syracuse, New York 13210. Received April 16, 1971

Abstract: A number of 2,3,4-tri-O-benzyl-α-D-glucopyranosyl bromides containing various acid residues on C-6 were prepared and tested in glycoside-forming reactions. The fraction of anomeric glucosides produced varied from over 90% α to over 90% β depending on the nature of the C-6 acyl group. In the para-substituted benzoic acid series the proportion of α-glucoside formed was shown to increase with increasing σ Hammett substituent constant value. This degree of steric control should be sufficient for the synthesis of lower 1→6 linked oligosaccharides and is probably due primarily to orbital overlap of the carbonyl function in the transition state. Reaction rates were measured polarimetrically and product composition was estimated by nmr spectroscopy with the help of deuterated samples. A few glucopyranosyl chlorides tested were shown to react by a different mechanism. The difference between the chlorides and bromides may be due to the relative tightness of the intermediate ion pairs.

The feasibility of synthesizing oligosaccharides in a solid-phase system was suggested by our previous work[1] in which a suitably functionalized solid support was prepared and a reaction sequence leading to an oligomer of glucose was tested in an exploratory fashion. Among the requirements cited for application of the solid-phase method to the synthesis of 1→6-linked oligosaccharides was the use of a monomer having a structure such as I (Chart I) in which X is a leaving

Chart I

IA, R = –C(=O)–⌬–OCH₃
IB, R = –C(=O)–⌬–CH₃
IC, R = –C(=O)–⌬
ID, R = –C(=O)–⌬(NO₂)(NO₂)
IE, R = –C(=O)–⌬–CN
IF, R = –C(=O)–C(CH₃)₃
IG, R = –C(=O)–CH₃
IH, R = –C(=O)–⌬–NO₂ (with ONO₂)
IJ, R = –C(=O)–⌬–NO₂

R' = –CH₂–⌬   X = Br

group, R is an easily removable group, and R' is a persistent blocking group.

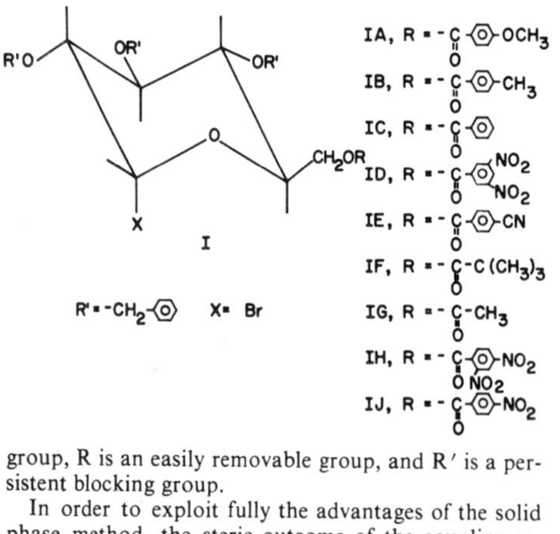

In order to exploit fully the advantages of the solid phase method, the steric outcome of the coupling reactions should be controlled and stereospecificity achieved. The addition of acid acceptors such as nucleophilic tertiary amines or catalysts such as metal ions[2,3]

should be avoided if possible, since their use induces side reactions and complicates the reaction sequence.

One method most frequently used for control of the configuration of C-1 during glycoside formation involves the use of a participating substituent at C-2. This method has been applied to produce low yields of oligosaccharides by means of some variations of the Koenigs–Knorr method[2] or Helferich's modifications.[3] The products isolated have, however, invariably contained a mixture of anomers and side products. Furthermore, the C-2 participating substituent is usually an ester group which is most conveniently used as a temporary blocking group rather than one which is retained through several steps of a solid-phase synthesis. The obvious solution would seemingly reside in the use of a monomer possessing a more stable (persistent) participating group at C-2. Such monomers are, however, not presently available though their synthesis is being actively investigated.[4] We have followed another approach involving glucosyl halides of type I, since monomers of this type have been effectively used in glucoside-forming reactions to yield products containing a high percentage of α or β configuration, although no general method was found to control the steric outcome of the reaction.

To this date few glycosyl halides of this type have been studied. Zemplén, Csürös, and Angyal[5] were the first to prepare 6-O-acetyl-2,3,4-tri-O-benzyl-D-glucopyranosyl bromide and Pravdić and Keglević[6] prepared the corresponding chloride in their synthesis of glucuronic acid esters. Both groups used a modified Koenigs–Knorr synthesis to prepare the corresponding β-glycosides (isolated in 15 and 63% yields, respectively).

(1) J. Frechet and C. Schuerch, *J. Amer. Chem. Soc.*, **93**, 492 (1971).
(2) W. Koenigs and E. Knorr, *Chem. Ber.*, **34**, 957 (1901).
(3) B. Helferich and J. Zinner, *ibid.*, **95**, 2604 (1962).
(4) R. Eby and C. Schuerch, unpublished results.
(5) G. Zemplén, Z. Csürös, and S. Angyal, *Chem. Ber.*, **70**, 1848 (1937).
(6) N. Pravdić and D. Keglević, *Tetrahedron*, **21**, 1897 (1965).

222

Table I. Preparation of 1,6-Diacyl-2,3,4-tri-O-benzyl-D-glucopyranose

Compd	Product composition β:α	Nmr spectrum[a] α	Nmr spectrum[a] β	Crystalln solvent[b]	Crystalline form	Yield of crystalline material, %	Mp, °C	$[\alpha]^{23}D,$[c] deg	Anal.[d] %C	%H	%N
A		e	6.02 (7.1)	1, 2	β	81	108–109	−10.4	71.85 71.86	5.89 5.77	
B	7:3	6.72 (3)	6.1 (7)	3, 2	β	65	109	−6.3	75.20 74.97	6.16 6.43	
C	2:1	6.68 (3)	6.06 (7)	3, 4	β	51	90–91	−6.2	74.76 74.99	5.81 5.86	
D	55:45	6.64 (3)	6.07 (7)	5, 6	β	50	118–119	−17	58.71 58.45	4.08 3.88	6.68 6.55
E	8:2	6.61 (3)	5.99 (7)	7, 8	β	67	127–128	−1.7	72.87 72.68	5.13 5.03	3.95 3.77
F	8:2	6.43 (3)	5.68 (7.3)	9, 10	β	70	76–77	+25	71.82 71.96	7.49 7.59	
G	3:8	6.41 (3.3)	5.70 (7.5)	3, 3	α	65[f]	63.5–65[f]	+68			
H	7:3	6.71 (3.5)	6.0 (6.5)						58.71 58.64	4.08 4.07	6.68 6.63
K[h]	7:9[g]	6.38 (3.3)	5.75 (7.3)								

[a] Upper line, chemical shift δ in parts per million; lower line, coupling constant in hertz. [b] First code number for crystallization solvent, second code number for recrystallization solvent: 1, methanol; 2, ethyl acetate–pentane; 3, ethanol; 4, ethanol–pentane; 5, ethyl acetate–petroleum ether; 6, methylene chloride–ethyl acetate; 7, carbon tetrachloride; 8, carbon tetrachloride–ethanol; 9, petroleum ether–pentane; 10, petroleum ether. [c] Measured in chloroform; c 1–2. [d] Upper line, calculated value; lower line, experimental value. [e] Anomeric proton covered by aromatic resonance of p-methoxybenzoyl group. [f] Pravdić and Keglević report 67–70% yield; mp 64–65.5°. [g] Pure α-K was obtained by reaction of 2,3,4-tri-O-benzyl-D-glucopyranose with trifluoroacetic anhydride in the presence of 2,4-dinitrobenzoic acid. [h] Acyl group, –COCF₃.

Ishikawa and Fletcher[7] in their investigation of the methanolysis of a number of α-D-glucopyranosyl bromides prepared 2,3,4-tri-O-benzyl-6-O-p-nitrobenzoyl-β-D-glucopyranosyl bromide which anomerized in solution to the corresponding α-D anomer. Methanolysis of this α-D-glucosyl bromide yielded almost exclusively (96%) the methyl α-D-glucopyranoside. This result is of considerable interest to our work since it should be applicable to the solid-phase synthesis of α-linked oligomers of glucose. Recently Anderson and his coworkers[8] prepared a disaccharide mixture by condensing 3-phenylpropyl 2,3,4-tri-O-benzyl-1-thio-β-D-glucopyranoside with 6-O-acetyl-2,3,4-tri-O-benzyl-α-D-glucopyranosyl chloride in the presence of a silver salt. The thioglucoside unit was used by these authors as a model for a planned solid-phase synthesis on a resin containing a thiol functional group. Anderson's approach differs, therefore, from ours in the method of coupling to resin and the use of metal ion as catalyst.

Experimental Section

Nuclear magnetic resonance spectra were measured on a Varian A-60 or Jeolco 100-MHz spectrometer in deuterated chloroform or deuterated acetone with tetramethylsilane as internal reference. Optical rotations were determined in a Perkin-Elmer Model 141 polarimeter using jacketed 1-dm cells kept at a constant temperature with a thermostated water bath. All melting points recorded are corrected.

Preparation of 1,6-Di-O-acyl-2,3,4-tri-O-benzyl-D-glucopyranose. The starting material for all the monomer syntheses was 2,3,4-tri-O-benzyl-D-glucopyranose prepared from 1,6-anhydro-2,3,4-tri-O-benzyl-β-D-glucopyranose as described by Zemplén, et al.[5] While 1,6-di-O-acetyl-2,3,4-tri-O-benzyl-α-D-glucopyranose was prepared as described by Zemplén, et al.[5] all the other monomers were prepared by the following general procedure.

(7) T. Ishikawa and H. G. Fletcher, *J. Org. Chem.*, **34**, 563 (1969).
(8) P. J. Pfaffli, S. H. Hixson, and L. Anderson, Abstracts, Chemical Institute of Canada–American Chemical Society Joint Conference, Toronto, Ont., May 24–29, 1970, Carbohydrate Division, paper No. 15.

To a cooled and stirred solution of 5 g (11.1 mmol) of 2,3,4-tri-O-benzyl-D-glucopyranose in 15 ml of dry pyridine was added an excess (23.6 mmol) of the desired acid chloride. A precipitate of pyridinium salt usually appeared after a few minutes and the colored reaction mixture was then stirred at room temperature overnight. The mixture was then poured into 300 ml of ice-water and the gummy precipitate which formed was extracted with chloroform. After extracting the aqueous phase with chloroform, the chloroform solutions were combined and washed successively with water, aqueous sodium bicarbonate solution, and water. The chloroform phase was dried on anhydrous magnesium sulfate, then evaporated to an oil. The nmr spectrum of the oil was recorded and the proportion of each anomer estimated from the area of the respective anomeric protons. On crystallization from the solvents indicated in Table I, the β isomer was obtained in every case. After recrystallizing from the proper solvent or solvent system (Table I), the melting point and specific rotation of the product were recorded with the results of the elemental analysis. These results are summarized in Table I.

Preparation of 6-O-Acyl-2,3,4-tri-O-benzyl-α-D-glucopyranosyl Bromides. 2,3,4-Tri-O-benzyl-6-O-p-nitrobenzoyl-α-D-glucopyranosyl bromide (IJ) was prepared from its β isomer by anomerization in dichloromethane as described by Ishikawa and Fletcher.[7] All the other glucopyranosyl bromides were prepared as follows. One gram of 1,6-di-O-acyl-2,3,4-tri-O-benzyl-D-glucopyranose was dissolved in a saturated solution of hydrogen bromide in dichloromethane (10–20 ml) and dry hydrogen bromide was bubbled through the solution for 10–30 min at room temperature. When p-methoxybenzoic acid, p-methylbenzoic acid, p-cyanobenzoic acid, 3,5-dinitrobenzoic acid, and 2,4-dinitrobenzoic acid had precipitated, they were collected on a fritted glass filter. The solution was then concentrated *in vacuo* and succesive portions of dichloromethane were evaporated *in vacuo* from the residual syrup to remove any trace of hydrogen bromide. In three instances the remainder of the acid produced by the reaction was removed by precipitation from other solvents: from ethyl acetate–pentane in the case of p-methoxybenzoic acid and from carbon tetrachloride in the cases of p-methylbenzoic acid and p-cyanobenzoic acid. The total yield of acid collected was in each case 95% or higher. In another instance, acetic acid was eliminated by azeotropic distillation with anhydrous benzene. Finally, in all other cases where less than 80% of the acid could be collected by filtration (3,5-dinitrobenzoic acid and 2,4-dinitrobenzoic acid) or no acid precipitated (benzoic acid, 2,2-dimethylpropanoic acid) the syrup was dissolved in methylene chloride. The solution was then washed twice with cold aqueous sodium bicarbonate solution, rinsed rapidly with water, and dried on anhydrous magnesium sulfate, and the solvent was evaporated *in vacuo*. After drying under high vacuum the nmr spectrum and optical rotation of each sample were recorded and the data can be found in Table II. In all cases the high optical rotation and nmr data indicated that the α anomer had been obtained.

Table II. Preparation of
6-O-Acyl-2,3,4-tri-O-benzyl-D-glucopyranosyl Bromides

Substrate	Reaction time, min	[α]^{23}D, deg in CHCl$_3$, c 1–2	Nmr spectrum δ, ppm	$J_{1,2}$, Hz
A	20	+123	6.43	3.7
B	45	+123	6.47	3.8
C	30	+124.5	6.46	3.6
D	10	+114	6.43	3.9
E	10	+129.5	6.46	3.9
F	30	+115	6.42	3.8
G	30	+123.5	6.41	3.9
H	15	+116	6.46	4

Preparation of 6-O-Acetyl-2,3,4-tri-O-benzyl-α-D-glucopyranosyl Chloride (IIG). This monomer was prepared from 1,6-di-O-acetyl-2,3,4-tri-O-benzyl-α-D-glucopyranose as described by Pravdić and Keglević.[6] The product had [α]^{23}D +88.5° (c 2, chloroform); Pravdić and Keglević report [α]D +90°. Its nmr spectrum included a doublet ($J_{1,2}$ = 3.6 Hz) centered at δ 6.07.

Preparation of 2,3,4-Tri-O-benzyl-6-O-p-nitrobenzoyl-α-D-glucopyranosyl Chloride (IIJ). One gram of 2,3,4-tri-O-benzyl-1,6-di-O-p-nitrobenzoyl-β-D-glucopyranose,[7] dissolved in 25 ml of dichloromethane, was added to a saturated solution of hydrogen chloride in ether (10 ml). Dry hydrogen chloride was bubbled through the solution for 6 hr. The solution containing some precipitated p-nitrobenzoic acid was then placed overnight in a refrigerator. The reaction mixture was then filtered through fritted glass filter to remove the p-nitrobenzoic acid (212 mg, 95%). The filtrate was then concentrated and the residual syrup evaporated three times from dichloromethane at room temperature. After drying on high vacuum the product had [α]^{23}D +102° (c 1.7, chloroform). Its nmr spectrum showed the expected proton ratio and a doublet ($J_{1,2}$ = 3.1 Hz) centered at δ 6.11 indicating that the product was the α anomer.

Preparation of 6-O-Benzoyl-2,3,4-tri-O-benzyl-α-D-glucopyranosyl Chloride (IIC). To a solution of 2.1 g of dry hydrogen chloride in 10 ml of absolute ether is added 2 g of 1,6-di-O-benzoyl-2,3,4-tri-O-benzyl-β-D-glucopyranose. After thorough mixing the solution was kept 4 days in a refrigerator. The solvent was then evacuated *in vacuo* and the remaining hydrogen chloride removed by repeated evaporation from absolute benzene. The benzoic acid was removed by washing a dichloromethane solution of the product successively with cold aqueous sodium bicarbonate and cold water. After drying over magnesium sulfate the solvent was evaporated *in vacuo*. After drying under high vacuum the product had [α]^{23}D +104° (c 1.7, chloroform). Its nmr spectrum showed the expected proton ratio and a doublet ($J_{1,2}$ = 3.4 Hz) centered at δ 6.09 indicating that the product was the α anomer.

Solvolysis of the D-Glucopyranosyl Halides. The sample of glucopyranosyl halide (~30 mg or 0.047 mmol) was placed in a 1-ml volumetric flask and dissolved in dry acetone (0.2 ml). Methanol (0.15–0.75 ml or 3.7–18.5 mmol) was added and the solution brought to 1 ml by addition of dry acetone. The optical rotation of the solution as a function of time was followed polarimetrically in a 1-dm tube at 23°. When the reaction was complete the solvents were removed *in vacuo* at room temperature and the resulting product was evaporated from several portions of dichloromethane. After drying on high vacuum the product was dissolved in deuterated chloroform for nmr analysis. The various samples were then saponified or transesterified and the nmr spectra of the resulting 2,3,4-tri-O-benzylmethyl-D-glucopyranosides were recorded. An alternate work-up procedure included washing the methanolysis product with aqueous bicarbonate to neutralize the solution before processing. The reaction was frequently performed on a larger scale (fivefold). In some cases deuterated methanol and deuterated acetone were used for the methanolysis. When tetrabutylammonium halides were used in the methanolyses two additional washings with distilled water were necessary to eliminate the tetrabutylammonium halide from the reaction product prior to nmr analysis.

When other alcohols were used the molar proportion alcohol–halide was kept constant. The nmr analysis was performed by comparison of the expanded spectra of the methyl D-glucopyranosides in the region δ 3–4 with the expanded spectra of corresponding deuterated samples in some cases prior to and in all cases after removal of the C-6 ester group. The anomeric composition determined by this method was most accurate for those samples having a high proportion of one isomer and the results were generally reproducible within 3–5%.

Results and Discussion

In an attempt to find a glucosyl halide susceptible of yielding a product of high stereospecificity when used in a glycoside forming reaction, a series of monomers of type I with benzyl substituents at C-2, C-3, and C-4 and esterified with various acid residues at C-6 was prepared by conventional methods from 2,3,4-tri-O-benzyl-D-glucopyranose.

Methanolysis of the various glucosyl halides was used as a model for the glucoside forming reaction since solvolyses using other alcohols showed that the reaction rate was practically independent of the nature of the primary alcohol chosen (Table III). Further-

Table III. Rates of Alcoholysis of
6-O-p-Methoxybenzoyl-2,3,4-tri-O-benzyl-α-D-glucopyranosyl Bromide (IA) and
6-O-Acetyl-2,3,4-tri-O-benzyl-α-D-glucopyranosyl Chloride (IIG)

Substrate	Alcohol	Half-life, hr	k ln	sec^{-1}
IA	Methanol[a]	2.14	9	10^{-5}
IA	1-Butanol[a]	2.77	7	10^{-5}
IA	2-Methylpropanol[a]	2.85	6.7	10^{-5}
IA	Methanol[a,b]	0.23	8.2	10^{-4}
IIG	Cinnamyl alcohol[c]	34	5.6	10^{-6}
IIG	Methanol[c]	32	6	10^{-6}

[a] Molar ratio, alcohol:monomer, 156:1. [b] 4 mol of Bu$_4$NBr added/mol of substrate. [c] Molar ratio, alcohol:monomer, 67:1.

more, the stereospecificity of the methanolyses could conveniently be estimated by nuclear magnetic resonance as the methoxyl resonances of methyl 2,3,4-tri-O-benzyl-D-glucopyranosides appear at different chemical shifts for the α and β isomers (δ 3.37 and 3.55 ppm, respectively). Fully deuterated methanol was also used in parallel experiments and the reaction products were examined by nmr to determine the exact area of the spectrum which could be assigned to the methoxyl resonance of each anomer. As can be seen in Table IV, drastic differences in product composition were observed for the methanolyses of the glucopyranosyl bromides containing different C-6 acyl groups. A product containing a high proportion (93%) of methyl β-D-glucopyranoside was obtained by methanolysis of 6-O-p-methoxybenzoyl-2,3,4-tri-O-benzyl-α-D-glucopyranosyl bromide in sharp contrast with the high hield (96%) of the corresponding methyl α-D-glucopyranoside obtained by Ishikawa and Fletcher[7] from the 6-O-p-nitrobenzoyl derivative. When the C-6 acyl groups were other substituted benzoates, intermediate results were obtained.

When the C-6 acyl group was an acetyl or 2,2-dimethylpropanoyl group, similar results were obtained with the higher β stereospecificity observed in the case of the 2,2-dimethylpropanoyl group (86 vs. 65% methyl β-D-glucopyranoside). In this series a trifluoroacetyl group would have been expected to cause formation of a

Table IV. Methanolysis of α-D-Glucopyranosyl Bromides[a]

Substrate	k ln	sec⁻¹	$t_{1/2}$, min	Methyl D-glucopyranosides formed, β:α
IA	8.7	10^{-4}	12.8	93:7
IA[b]	2.5	10^{-3}	4.5	
IB	8	10^{-4}	14.4	88:12
IB[b]	2.1	10^{-3}	5.4	40:60
IC	9.8	10^{-4}	11.8	84:16
IC[b]	2.1	10^{-3}	5.4	43:57
ID	5.4	10^{-4}	21.3	75:25
ID[b]				16:84
IE	6.2	10^{-4}	18.5	63:37
IE[b]	2.2	10^{-3}	5.1	30:70
IF	9	10^{-4}	13.5	86:14
IF[b]	2.2	10^{-3}	5	
IG	2	10^{-3}	6	65:35
IG[b]	3.9	10^{-3}	3.1	
IH	4.9	10^{-4}	23.7	55:45
IH[b]	1.5	10^{-3}	7.5	30:70
IJ	6.5	10^{-4}	17.9	8:92[c]
IJ[b]	2.6	10^{-3}	4.5	d

[a] Reaction temperature, 23°; molar ratio, methanol:monomer, ~390:1. [b] Indicates reaction in which 4 mol of tetrabutylammonium bromide was added for each mole of substrate. [c] Ishikawa and Fletcher[7] report 4:96. [d] Ishikawa and Fletcher[7] report 10:90.

higher percentage of the methyl α-D-glucopyranoside. However, the reaction of 2,3,4-tri-O-benzyl-1,6-di-O-trifluoroacetyl-D-glucopyranose with methanol was found to be difficult to control and 2,3,4-tri-O-benzyl-D-glucopyranose was the major product. The methanolysis of glucopyranosyl halides having a C-6 acetyl group was also complicated by loss of the C-6 acetyl group.

The solvolyses of all the halides were performed under identical conditions with the same alcohol:halide ratio and the reactions were followed polarimetrically. The polarimetric data from the various solvolyses were plotted against time and the first-order rate constants calculated from the classical polarimetric expression $k = 1/t \ln (\alpha_0 - \alpha_\infty)/(\alpha_t - \alpha_\infty)$. The rate constants for the methanolyses of the various glucopyranosyl halides are shown in Tables IV and V. In general the

Table V. Methanolysis of α-D-Glucopyranosyl Chlorides[a]

Substrate	k ln	sec⁻¹	$t_{1/2}$, hr	Methyl D-glucopyranosides formed β:α
IIC	4.8	10^{-5}	4	94:6
IIG	6.7	10^{-5}	2.85	94:6
IIJ	3.6	10^{-5}	5.2	84:16

[a] Reaction temperature, 23°; molar ratio, methanol:monomer, ~390:1.

initial rate was found to be slightly lower than the rate at half-life, probably due to the effect of the bromide ion liberated by the reaction. All the methanolyses were accompanied by levomutarotation and reaction rates were increased by the addition of halide ion. The largest rate increase was observed when the added ion was bromide or iodide rather than chloride and when the concentration of alcohol was low. As can be seen in Table IV, the addition of halide ion always caused also a change in product composition with a tendency to yield a product of lower stereospecificity.

As expected, changes in temperature were accompanied by changes in reaction rates as evidenced by the methanolysis of 6-O-p-methoxybenzoyl-2,3,4-tri-O-benzyl-α-D-glucopyranosyl bromide for which the relative reaction rates were the following: 1 at 22°, 0.085 at 0°, and 6.4 at 45°.

In some cases, it was possible to increase the stereospecificity of the reaction by decreasing the reaction temperature. The methanolysis of 6-O-benzoyl-2,3,4-tri-O-benzyl-α-D-glucopyranosyl bromide at −15° yielded a product containing 90% of the methyl β-D-glucopyranoside vs. 84% at 22° and 76% at 60°. Similarly, the corresponding 6-O-p-toluoyl derivative yielded 91% of the methyl β-D-glucopyranoside at −15° vs. 88% at 22°.

In all cases the glucopyranosyl chlorides were found to solvolyse approximately 20–30 times slower than the corresponding bromides under the reaction conditions chosen (Tables IV and V). Their methanolysis yielded products having a different anomeric composition, the most striking difference occurring in the case of 2,3,4-tri-O-benzyl-6-O-p-nitrobenzoyl-α-D-glucopyranosyl chloride which upon methanolysis yielded 84% of the methyl β-D-glucopyranoside, while the corresponding bromide yielded 96% of the methyl α-D-glucopyranoside. These results indicate that the two types of halides must react by different mechanisms.

Mechanism

The mechanism of solvolysis of glycosyl halides with nonparticipating groups at C-2 was first studied on 2,3,4,6-tetra-O-methyl-α-D-glucopyranosyl and mannopyranosyl chlorides by Rhind-Tutt and Vernon.[9] These authors found that the methanolysis of the glucopyranosyl chloride proceeded by an S$_N$1 mechanism with essentially complete inversion by backside attack on a specifically oriented ion pair to yield 94% of the methyl β-D-glucopyranoside. The formation of the small amount of α anomer was explained by a mechanism involving chloride ion exchange to yield the β ion and therefore some α product (Scheme I). The possi-

Scheme I. Possible Configuration Control in Glycoside Synthesis (Modified from Reference 9)

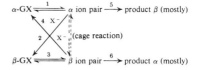

bility of inversion of β to α ion within a solvent cage was not discussed by the authors.

Ishikawa and Fletcher[7] in their investigation of the methanolysis of a number of α- and β-D-glucopyranosyl bromides having a benzyl group at C-2 used a somewhat similar mechanistic interpretation to explain the formation of their products which were mainly the methyl α-D-glucopyranosides. They suggest that these products were obtained from the β-bromides which were either present in small quantities in the starting material or formed rapidly in a concurrent anomerization presumably by reaction with bromide ion. Both groups

(9) A. J. Rhind-Tutt and C. A. Vernon, *J. Chem. Soc.*, 4637 (1960).

Synthesis of Oligosaccharides

Scheme II

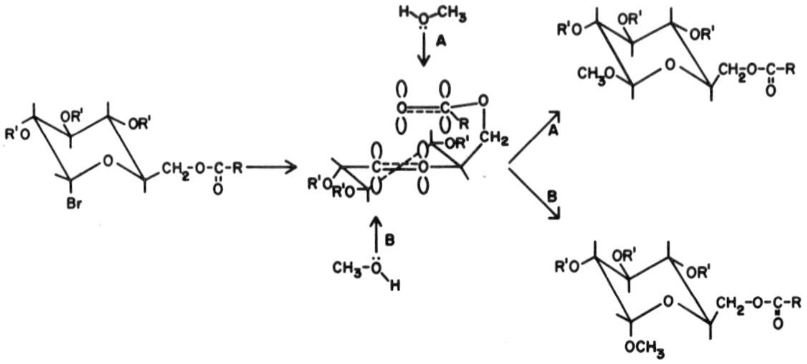

thus interpreted the steric course of the reaction in terms of the same equilibria although the products they obtained were opposite in C-1 configuration.

Although the mechanism used by these authors undoubtedly interprets successfully several features of these reactions, it appears not to be a complete explanation. Specifically, it is difficult to explain the wide differences in the configuration of products from quite similar starting materials, unless one considers the probable structure of the intermediate ions in more detail. Why, for example, should 2,3,4,6-tetra-O-methyl-α-D-glucopyranosyl chloride produce nearly pure methyl β-D-glucopyranoside whereas 2,3,4,6-tetra-O-benzyl-α-D-glucopyranosyl bromide yields essentially a 50:50 mixture of both methyl D-glucopyranosides? Similarly, why should the product derived from 2,3,4-tri-O-benzyl-6-O-p-nitrobenzoyl-α-D-glucopyranosyl chloride be nearly as sterically pure β-glucoside as that from the fully etherified glucopyranosyl chloride, while on the other hand the 6-O-acetylated glucopyranosyl bromides yield products varying from 95% α to 95% β under apparent electronic control? The mechanism shown in Scheme I suggests that the results are due to widely different relative rates of reaction of very similar ion pairs with two nucleophiles, halide ion and methanol, or perhaps are due to unrealistic differences in equilibria. Neither interpretation clarifies the results.

We suggest that the results are more interpretable if one postulates that the glucopyranosyl chlorides generate tight ion pairs which react as postulated by Rhind-Tutt and Vernon by backside approach either by halide ion or alcohol, but that the glucopyranosyl bromides generate loose or solvent-separated ion pairs. In that case, the carbonium ion derived from 2,3,4,6-tetra-O-benzyl-α-D-glucopyranosyl bromide is most probably in a half-chair conformation and approach from either the α or β direction is nearly equally probable. Since the substituents on asymmetric centers on the ion do not effectively block either mode of attack, the product mixture is nearly 50:50 $\alpha:\beta$.

As indicated above, however, the methanolyses of 6-O-p-methoxybenzoyl-2,3,4-tri-O-benzyl-α-D-glucopyranosyl bromide and of the corresponding 6-O-p-nitrobenzoyl derivative yielded products of high stereospecificity and opposite configuration. Since the starting materials (IA and IJ) differ only in the nature of the para substituent on the C-6 benzoyl group and the substituents are separated from the reaction center by a number of single (as well as some conjugated double) bonds, their effect cannot be ascribed to a simple inductive effect. We propose that the changes in product composition can be attributed to the polar effects of the substituents on the carbonyl group in the following manner. As the glucopyranosyl bromides are converted into loose ion pairs of half-chair conformation, overlap between the vacant orbital on the C-1 atom and the filled orbitals of ring oxygen develops (Scheme II). The Lewis acid character of the electron-deficient center serves to attract the carbonyl group of the C-6 substituent into a position in which there is substantial overlap with the p orbitals of the C-6 carbonyl. Although this position is not the most probable rotamer, it can be readily stabilized by Lewis acid–base coordination with no distortion of the ring conformation.

When the C-6 acyl group is the p-methoxybenzoyl group, the electron density on the carbonyl group is enhanced through resonance and the positive charge is concentrated more on the C-6 carbonyl group where it is best stabilized. Attack by the nucleophile will, therefore, proceed in the direction A shown in Scheme II and cause collapse of the carbonium ion with formation of the methyl β-D-glucopyranoside. (An alternate interpretation would involve hydrogen bonding of the alcohol with the more electron-rich oxygen followed by nucleophilic attack as in A.) The small amount of α isomer is presumably produced by direct attack of the nucleophile on C-1. If, however, the C-6 acyl group contains a relatively electron-deficient carbonyl (as in a p-nitrobenzoyl group), the positive charge will remain concentrated on C-1 and attack by the nucleophile will proceed from the least shielded side in direction B as shown in Scheme I to yield mainly the methyl α-D-glucopyranoside. In an intermediate case such as that of a benzoyl group on C-6 the reaction is less stereospecific as the electron density is balanced and does not cause localization of the positive charge on either site.

In all cases in which the C-6 acyl group was a para-substituted ester of benzoic acid, the steric outcome of the reaction could be predicted qualitatively from the value of the substituent constant as defined by σ in the classical Hammett equation. The proportion of methyl β-D-glucopyranoside produced decreased in the series: p-methoxybenzoyl > p-methylbenzoyl > benzoyl > p-cyanobenzoyl > p-nitrobenzoyl. Disubstitution on the benzene ring apparently introduces additional complexities for the product composition in these cases is not so readily interpretable. However, the same electronic influence was evident in the behavior of the bromides IG and IF in which the C-6 acyl group was an acetyl and a 2,2-dimethylpropanoyl group. The highest proportion of methyl β-D-glucopyranoside was obtained in the case of the second halide (IF) for which dispersion of the positive charge on the carbonyl group is facilitated by the inductive effect of the three methyl group.

The electronic interaction visualized in this mechanism differs from the participation expected of carbonyl functions at C-2. In the latter case, sp^3 hybridization of C-1 and tetrahedral geometry results. If sp^3 hybridization of C-1 occurred with participation of a C-6 substituent, it appears that the ring conformation would have to be a boat form or a chair with all substituents axial. The energetics of these conformations are too unfavorable to be considered.

When bromide ion is added to the reaction mixture the reaction mechanism becomes much more complex and presumably takes on the character of the reactions shown in Scheme I. The change in product composition observed in every case is probably due to both direct attack of bromide ion on the glucopyranosyl bromide and nucleophilic attack on the carbonium ion.

The degree of steric control achieved by using monomers such as 6-O-p-methoxybenzoyl-2,3,4-tri-O-benzyl-α-D-glucopyranosyl bromide and 6-O-p-nitrobenzoyl-2,3,4-tri-O-benzyl-α-D-glucopyranosyl bromide is probably sufficient for the synthesis of 1,6-linked oligosaccharides through classical or solid-phase synthesis. These monomers are now being tested in the synthesis of some otherwise relatively inaccessible α-glycosides and oligosaccharides of complex structure.

Acknowledgments. The present work has been supported by Research Grant No. GM06168 of the Division of General Medical Sciences, National Institutes of Health. The authors wish to thank Dr. L. Anderson of the University of Wisconsin for a stimulating discussion on the solid-phase synthesis of oligosaccharides and Mrs. H. Jennison of this laboratory for measuring most of the nmr spectra. We also wish to thank Drs. T. Ishikawa and H. G. Fletcher for sending us their manuscript[6] prior to publication.

SOLID-PHASE SYNTHESIS OF OLIGOSACCHARIDES
III. PREPARATION OF SOME DERIVATIVES OF DI- AND TRI-SACCHARIDES *VIA* A SIMPLE ALCOHOLYSIS REACTION

JEAN M. FRÉCHET AND CONRAD SCHUERCH

Department of Chemistry, State University College of Forestry at Syracuse University, Syracuse, New York 13210 (U. S. A.)

(Received October 6, 1971; accepted in revised form, December 20, 1971)

ABSTRACT

The solid phase method has been applied to the synthesis of di- and tri-saccharides using a simple alcoholysis of 6-*O*-*p*-nitrobenzoyl- and 6-*O*-*p*-methoxybenzoyl-2,3,4-tri-*O*-benzyl-α-D-glucopyranosyl bromides. High coupling yields were obtained with high monomer-to-resin ratios and in the presence of a small amount of 2,6-dimethylpyridine as acid acceptor. Cleavage of the completed oligomers by ozonolysis gave good yields of di- and tri-saccharides as glycosides of hydroxyethanol. All the cleaved disaccharides, irrespective of the monomer coupling sequence used in their preparation, had approximately the same optical rotation indicating that the reaction had not proceeded with the steric control characteristic of methanolysis of the same monomers.

INTRODUCTION

While several approaches are possible for the solid phase synthesis of oligosaccharides, we chose to test first a simple alcoholysis reaction in which the solid support carries the hydroxyl group and becomes linked to the anomeric center of the sugar unit. The rationale for this choice was that in the reaction of a glycosyl halide with an alcoholic hydroxyl group, side reactions are likely to occur at the anomeric center and not at the resin-bound hydroxyl group. The byproducts would therefore remain in the soluble phase and be eliminated by simple washing, while high yields of resin-bound glycosides should be obtained in the presence of a large excess of the glycosyl halide. In an alternate procedure of linking the resin to a hydroxyl group of the sugar unit, the glycosyl halide would be carried by the resin on subsequent steps and, therefore, each side reaction would cause chain termination.

The preparation[1] of a solid support with allylic alcohol functional groups and of suitable monomer units[2] carrying three types of functional groups has been described and rationalized in our previous work. Two features of a model glycoside-forming reaction have also been considered: the rate and the stereochemistry of the reaction of 6-*O*-acyl-2,3,4-tri-*O*-benzyl-α-D-glucopyranosyl bromides (**1** and **2**). It has been shown that the proportion of anomeric glycosides produced could be varied from over 90% α-D, in the case of the 6-O-p-nitrobenzoyl group[3] (**1**), to over 90% β-D, in the case of the 6-O-p-methoxybenzoyl group[2] (**2**). It was hoped that this type of steric control would also be applicable to the problem of solid phase oligosaccharide synthesis.

Recently, Flowers[4] prepared two α-D-glucopyranosyl disaccharides of D-galactose by the use of a glycosyl bromide having both a non-participating group at C-2 and a C-6 acyl group. By the use of a Koenigs–Knorr reaction with mercuric cyanide as condensing agent, Flowers treated 2-O-benzoyl-3,4,6-tri-O-p-nitrobenzoyl-β-D-glucopyranosyl bromide[3] with 1,2:3,4-di-O-isopropylidene-α-D-galactose or with 4,6-O-ethylidene-1,2-O-isopropylidene-α-D-galactose to produce the corresponding α-D-(1→6) or α-D-(1→3)-linked disaccharides in 63 and 47% yield, respectively. A similar treatment of the corresponding α-D-glucopyranosyl bromide was also shown to be stereospecific, although the reaction rate and yields obtained were lower. These results are consistent with those obtained by Ishikawa and Fletcher[3] in the methanolysis of these glucosyl halides, although the Koenigs–Knorr conditions are quite different from the conditions used in a methanolysis. In another recent publication, Dejter-Juszynski and Flowers[5] report that the Koenigs–Knorr condensation of 2,3,4-tri-O-benzyl-α-L-fucopyranosyl bromide with benzyl 2-acetamido-3,4-di-O-acetyl-2-deoxy-α-D-glucopyranoside yields a disaccharide mixture containing a 7:3 ratio of α-D to β-D anomer. The lack of stereospecificity in this system and in the methanolysis of 2,3,4,6-tetra-O-benzyl-α-D-glucopyranosyl bromide[3] is consistent with our suggestion that, in systems such as those used by Ishikawa and Fletcher[3] or by Flowers[4], the C-6 acyl group is responsible for the stereospecificity of the reaction.

However, the reaction conditions in a solid-phase system are very different from those used in the methanolysis or in Flowers' system. We chose, for example, to use a large excess of the glycosyl halide to ensure complete blocking of the reactive hydroxyl groups of the resin and to obtain reasonable rates. Furthermore, in a solid phase system, accessibility problems, steric hindrance, solvent differences, and the slower rate of glycosidation, may be important factors affecting steric control.

RESULTS AND DISCUSSION

Having selected the solid-support and monomers to be used in the synthesis of oligosaccharides, the problem remained to insure complete coupling of the sugar unit with the active hydroxyl groups of the resin, to find optimum reaction conditions for the cleavage of the finished product, and to test the stereospecificity of the reaction sequence.

The first problem was that of the reaction rate, as the relatively high rates of reaction obtained in the solvolysis of 6-O-acyl-2,3,4-tri-O-benzyl-α-D-glucopyranosyl bromides could not be achieved under the conditions of a solid-phase synthesis. Metal ions, such as Ag^+ and Hg^{2+}, have been used in Koenigs–Knorr glycoside syntheses to enhance reaction rates; however, we chose not to use these metal ions as their presence also enhances side reactions[6]. We, therefore, used a simple alcoholysis reaction in which the only reactants were the alcoholic hydroxyl on the resin support and the glycosyl bromide. Preliminary tests indicated that useful rates could not be obtained with a 1:1 ratio of resin to monomer, so higher concentrations of glycosyl halide were used.

TABLE I

INFLUENCE OF VARIOUS SOLVENTS ON THE EXTENT OF COUPLING

Solvent	Yield (%)	Relative Ratio[a]
2,6-Dimethylpyridine	59.2	1.67
Carbon tetrachloride	47.3	1.34
Chlorobenzene	35.7	1.01
Benzene	35.4	1.00
Nitrobenzene	28.9	0.84
Acetone	19.2	0.54
Pyridine	5.1	0.14

[a]Reaction in benzene taken as 1.00.

As can be seen in Table I, the extent of coupling increased in the series pyridine < acetone < nitrobenzene < benzene ≈ chlorobenzene < carbon tetrachloride < 2,6-dimethylpyridine. The fact that little or no reaction occurred in pyridine is probably an indication that pyridine itself acts as a nucleophile and reacts with the glycosyl bromide, as shown by Lemieux and Morgan[7]. The extent of coupling obtained in acetone, benzene or substituted benzenes, and carbon tetrachloride is presumably related to their ability to swell the resin. The reaction should proceed more readily in a good swelling agent, like carbon tetrachloride, than in acetone in which the degree of swelling is significantly lower. The higher yield of coupling obtained in 2,6-dimethylpyridine probably reflects the fact that 2,6-dimethylpyridine is a good swelling agent and also a relatively nonnucleophilic base which can accommodate the hydrogen bromide produced in the reaction. As shown in Table II, the highest coupling yields were obtained with solvents, such as benzene or carbon tetrachloride, used in combination with a small amount of base, such as 2,6-dimethylpyridine. The 2,3,4-tri-O-benzyl-6-O-p-nitrobenzoyl-α-D-glucopyranosyl bromide monomer (**1**) seemed to react more slowly with the solid support than the corresponding 6-O-p-methoxybenzoyl derivative (**2**), and with both monomers it was found that yields of 90% or over could only be obtained in one stage, at room temperature, if the monomer-to-resin molar ratio was ~8.1:1. In most cases where the molar ratio was of the order of 4 or 5 to 1 the reaction proceeded to an extent of 80–85%, and a second stage coupling was necessary to bring the reaction near completion. This difficulty was most often encountered in the coupling of the first unit to the resin but rarely in the coupling of subsequent units. It might have been preferable to block the less accessible hydroxyls after 80% of the reaction had been completed rather than to introduce a second coupling to complete the first reaction. A marked increase in

TABLE II
INFLUENCE OF ADDED BASE ON THE EXTENT OF COUPLING[a]

Monomer	Monomer, mmoles	Monomer concentration[b]	Molar ratio monomer to resin	Solvent[c]	Time (days)	Base[d]	Coupling yield, %
2	1.1	21	3.9:1	A	4	2,6-dimethylpyridine	82.5
2	1.1	21	3.9:1	A	4	1,8 BDAN[e]	72
2	1.1	21	3.9:1	B	4	2,6-dimethylpyridine	57
2	1.31	25	4.7:1	A	2	2,6-dimethylpyridine	77.8
2	1.31	25	4.7:1	A	2	pyridine	82.2
1	0.96	19.5	3.4:1	A	4.2	none	39.9
1	0.96	19.5	3.4:1	A	4.2	2,6-dimethylpyridine	65.2
1	0.96	19.5	3.4:1	A	4.2	pyridine	69.9
1	0.96	19.5	3.4:1	C	4.2	none	44.9
1	0.96	19.5	3.4:1	C	4.2	2,6-dimethylpyridine	87.9

[a]By use of a 1%-crosslinked resin (0.4 g or 0.28 mmole). [b]In g/100 ml of solution. [c](A) Benzene, (B) 2,6-dimethylpyridine, (C) carbon tetrachloride. [d]Concentration 0.43 mmole, molar ratio base to resin 1.5:1. [e]1,8-Bis(dimethylamino)naphthalene.

reaction rate was also observed when the reaction was performed at 60–65°, a 96% yield being obtained in 48 h with a monomer to resin ratio of ~6:1.

Coupling of the sugar units onto the resin could be monitored conveniently by gravimetry or by i.r. spectrometry. Weight increases of the order of 40–45% of the weight of the starting resin were recorded in the coupling of the first unit. Carbonyl absorptions appeared in the i.r. spectrum at about 1715 cm^{-1} for the coupling of both monomers. In addition, several bands corresponding to the 6-O-acyl groups could be identified. After each coupling reaction, excess monomer was reclaimed and, after elimination of the 2,6-dimethylpyridinium hydrobromide and 2,6-dimethylpyridine, the monomer was used again in another coupling reaction.

Removal of the temporary blocking group was achieved by ester interchange with methanol in tetrahydrofuran (70%) containing a catalytic amount of sodium methoxide. Once again the reaction was monitored both by gravimetry and i.r. spectrometry. Weight reductions amounting to 23.5–25.5% (depending on the nature of the temporary blocking group) of the weight gained in the coupling of the previous unit could be followed accurately. Disappearance of the carbonyl absorption and of the bands corresponding to the C-6 acyl substituent confirmed the successful removal of the temporary blocking group.

Coupling of subsequent sugar units was effected with good yields, in a single stage in most cases. The various coupling yields reported in Table III are computed with respect to the number of sugar units attached to the resin in the previous step, since unreacted hydroxyl groups were assumed to be located in unfavorable sites of the polymeric network. The preceding assumption is ascertained by the fact that while

TABLE III

PREPARATION OF RESIN-BOUND GLYCOSIDES

Exp. No.	Cross-linking of resin (%)	Monomer-coupling sequence[a]	Yield (%)		
			1st Unit	2nd Unit	3rd Unit
1	1	1*	100		
2	2	1*	91		
3	1	2*	99		
4	2	2*	91		
5	1	1*, 1*	97	98	
6	1	2*, 2	99	93	
7	2	1*, 2	90	94	
8[b]	2	1*, 2	94	100	
9	2	2*, 1	92	96	
10	2	1*, 1, 1	90	91	100
11	2	1*, 2, 1	90	94	98
12	2	2*, 2, 2	92	94	99
13	2	2*, 2, 1	92	94	92

[a]Asterisk indicates 2-stage coupling of unit preceding asterisk. [b]The 2nd stage of first-unit coupling and the 2nd-unit coupling were performed at 50–55°.

all the hydroxyl groups of the starting resin can be etherified with dihydropyran (phosphorus oxychloride catalyst) under mild conditions[8], the remaining hydroxyl groups of a 2%-crosslinked resin having 93% of its hydroxyl groups blocked by sugar units can only be etherified after long refluxing periods. As can be seen in Table III, better yields were obtained for the attachment of the first sugar unit to a 1%-crosslinked resin rather than to a 2%-crosslinked resin, probably due to the higher degree of swelling of the former resin.

Cleavage of the completed glycosides was most easily accomplished by ozonolysis followed by reduction with dimethyl sulfide. To minimize side reactions which may occur at the numerous benzylic positions of both the solid support and the substituted sugar, the reaction was run at low temperature ($-78°$) by the use of ozone in a nitrogen carrier which gave better results than the mixture of ozone and oxygen. Partial collapse of the polymer matrix was observed when ozone–oxygen was used. The yields reported in Table IV for the ozonolysis of the various glycosides are seemingly very variable. However, the low yields obtained in some experiments (e.g., products No. 10A, 11, Table IV) were most likely due to the poor design of one of the reaction vessels used, which did not allow for complete mixing of ozone–nitrogen with the polymer suspension, rather than to an intrinsic failure of the ozonolysis

TABLE IV

CLEAVAGE OF THE FINISHED GLYCOSIDES

Product No.[a]	Coupling sequence	Yield of ozonolysis (%)	Yield of reduction at C-1 and deblocking of C-6 (%)	$[\alpha]_D^{24}$ of product (degrees)
2	1	51[b]		+41.5
4	2	82	95	+40.8
5	1, 1	87	93	+60.5
8	1, 2	61	85[c]	+61.5
6A	2, 2	87	90[c]	+60
6B	2, 2	91	95	+60.2
9A	2, 1	63	94[c]	
9B	2, 1	71	90	+59
10A	1, 1, 1	51	95[c]	
10B	1, 1, 1	83	95	+62
11	1, 2, 1	57	79[c]	+68
12A	2, 2, 2	82	93	+70
12B	2, 2, 2	74	92	+78
13A	2, 2, 1	80	94[c]	
13B	2, 2, 1	72	89	+70

[a]Refers to experiment number in Table III; where more than one experiment was performed using the same starting material, the products are designated A and B. [b]Ozonolysis performed after removal of the C-6 acyl group in the solid phase. Ozonide reduction performed with sodium borohydride in methylene chloride–ethanol. [c]Lithium aluminum hydride used to reduce the aldehyde and C-6 acyl group. All other reductions were performed with sodium borohydride and followed by saponification with potassium hydroxide.

TABLE V

OPTICAL ROTATIONS OF GLYCOSIDES PREPARED BY SOLID-PHASE SYNTHESIS AND OF MODEL COMPOUNDS

Compound	Method of preparation[a]	$[\alpha]_D^b$ (degrees)	$[\phi]$ (degrees)
2-Hydroxyethyl 2,3,4-tri-O-benzyl-D-glucopyranoside	Prod. 4	+40.8	+202
2-Hydroxyethyl 2,3,4-tri-O-benzyl-D-glucopyranoside	Prod. 2	+41.5	+205
2-Hydroxyethyl 2,3,4-tri-O-benzyl-D-glucopyranoside	8[c]	+17.5	+119
Methyl 2,3,4-tri-O-benzyl-β-D-glucopyranoside	Ref. 10	+10	+46
Methyl 2,3,4-tri-O-benzyl-α-D-glucopyranoside	Ref. 9	+26.3	+122
Methyl 2,3,4,6-tetra-O-benzyl-β-D-glucopyranoside	Ref. 11	+12.6	+70
Methyl 2,3,4,6-tetra-O-benzyl-α-D-glucopyranoside	Ref. 12	+32.2	+179
2-Hydroxyethyl 2,3,4-tri-O-benzyl-6-O-(2,3,4-tri-O-benzyl-D-glucopyranosyl)-D-glucopyranoside	Prod. 5, 6, 8, 9	+59 to +61.5	+547 to +569
Methyl 2,3,4-tri-O-benzyl-6-O-(2,3,4,6-tetra-O-benzyl-β-D-glucopyranosyl)-α-D-glucopyranoside	Ref. 9	+17.4	+172
Methyl 2,3,4-tri-O-benzyl-6-O-(2,3,4,6-tetra-O-benzyl-α-D-glucopyranosyl)-α-D-glucopyranoside	Ref. 9	+59	+582

[a]Refers to product number in Table IV or to the literature. [b]In chloroform. [c]Model compound 8 is expected to contain a large proportion of the β-D-glucoside.

reaction. After reduction of the ozonides with dimethyl sulfide, the products, isolated as glycosides of hydroxyethanal, had n.m.r. spectra consistent with those expected for the mono-, di-, or tri-saccharides. Specifically, protons characteristic of the terminal C-6 blocking group were in proper proportion to other protons, within experimental error. In most cases, a small amount of impurity (usually amounting to 2% or less of the total number of protons) was observed at δ 1–1.4 p.p.m.

Although removal of the C-6 acyl group was not desired since it provided a convenient standard for n.m.r. analysis, it was usually removed at the same time as the aldehyde group was reduced to the corresponding alcohol group. Cleavage of the C-6 ester group was complete with lithium aluminum hydride as the reducing agent, and only partial with sodium borohydride. Sodium borohydride reduction, followed by saponification with potassium hydroxide, was preferred to lithium aluminum hydride reduction, as the separation of the final product was easier in the borohydride reaction (separation of the product from an acid salt vs. separation from an alcohol) and as lithium aluminum hydride is known to react with nitro groups.

As can be seen in Table IV, the optical rotations of the disaccharides prepared by the solid-phase method are in the range 59–61.5°, irrespective of the monomer-coupling sequence, thus indicating that the method, as applied, did not produce the expected stereochemical results. Since no anomeric protons can be observed in the n.m.r. spectra, it is difficult to draw any conclusion on the optical purity of the products from the data available. Conceivably, these products could be made up of a mixture of all possible isomers or of one major component with smaller amounts of the other isomers. Although no comparable glycol glycosides are described in the literature, a comparative study of the optical rotations of the products obtained in this work and of the optical rotations of various methyl glycosides can be attempted. The results of this comparative study, shown in Table V, suggest that the products obtained by the solid-phase method contain a high proportion of α-D linkages. This indication is supported further by the high positive optical rotation of compound **14** which was obtained by debenzylation of disaccharides prepared by the solid-phase method, followed by acetylation of the resulting product. A comparison of the optical rotation of **14** with those of a few disaccharide methyl glycosides found in the literature is presented in Table VI.

The solid-phase synthesis of oligosaccharides is probably a feasible proposition, since the present work is only a first approach to this new method of glycoside synthesis. In this system, however, reaction rates are too low and stereochemical results observed in related solvolyses are not obtained. Obviously, steric control by C-6-participating substituents[2–5] is markedly sensitive to reaction conditions. Until new glycoside-forming reactions are developed, efforts should be directed toward an increase in reaction rates, possibly by addition of metal ions in a system comparable to that used by Flowers[4,5]. This change may make lower monomer concentrations practical and may also tend to produce loose ion-pair intermediates. The latter have been postulated as the intermediates which favor steric control by C-6 acyl groups[2].

TABLE VI

OPTICAL ROTATION OF **14** AND OF MODEL COMPOUNDS

Derivatives of 6-O-(D-glucopyranosyl)-D-glucopyranoside	Method of Preparation[a]	$[\alpha]_D^b$ (degrees)	$[\phi]$ (degrees)
2-Acetoxyethyl (2,3,4,6-tetra-O-acetyl)-2,3,4-tri-O-acetyl-	**14**	+122.6	+886
2-Acetoxyethyl (2,3,4,6-tetra-O-acetyl-β)-2,3,4-tri-O-acetyl-α-	**11**	+49.5	+358
Methyl (2,3,4,6-tetra-O-acetyl-β)-2,3,4-tri-O-acetyl-β-	Ref. 13	−17	−111
Methyl (2,3,4,6-tetra-O-benzoyl-β)-2,3,4-tri-O-benzoyl-β-	Ref. 14	+8.3	+90
Methyl (2,3,4,6-tetra-O-benzoyl-α)-2,3,4-tri-O-benzoyl-β-	Ref. 14	+56.4	+611
Methyl (2,3,4,6-tetra-O-benzoyl-α)-2,3,4-tri-O-benzoyl-α-	Ref. 14	+108	+1170

[a]Refers to products described in Experimental or to the literature. [b]In chloroform.

EXPERIMENTAL

General methods. — I.r. spectra were measured with a Perkin–Elmer 137 infrared spectrophotometer on samples at 4–7% concentration in potassium bromide pellets. Since the pellets were usually opaque, a Perkin–Elmer reference beam attenuator was used during the recording of the spectra. N.m.r. spectra were measured with a Varian A-60 spectrometer in chloroform-*d* with tetramethylsilane as internal reference. Optical rotations were determined in a Perkin–Elmer model 141 polarimeter with jacketed 1-dm cells kept at 24° by circulation of water from a thermostatted bath. Microanalyses were performed by Galbraith Laboratories.

Preparation of the solid-support resin. — The resin was prepared as described previously[1]. A 1%-crosslinked resin having a capacity of 0.70 mmole/g and 2%-crosslinked resins having capacities of 0.78 and 0.70 mmole/g were prepared and used in this work.

Preparation of the monomers. — Crystalline 2,3,4-tri-O-benzyl-1,6-di-O-p-nitrobenzoyl-β-D-glucopyranose and 2,3,4-tri-O-benzyl-1,6-di-O-p-methoxybenzoyl-β-D-glucopyranose were prepared as described previously[2,3]. The corresponding α-D anomers were isolated from the mother liquor by column chromatography on Silica Gel with 200:1 benzene–ether for the 1,6-di-O-p-nitrobenzoyl derivative and 100:1 benzene–ether for the 1,6-di-O-p-methoxybenzoyl derivative. Both anomers were used in the preparation of the corresponding α-D glucopyranosyl bromides (**1,2**).

Glycoside-forming reaction: determination of optimum experimental conditions. — The influence of the nature of the solvent was studied as follows: to 0.195 mmole of 2%-crosslinked resin (capacity 0.78 mmole/g) was added 1.9 ml of a solution containing 0.453 g (0.7 mmole) of 2,3,4-tri-O-benzyl-6-O-p-methoxybenzoyl-α-D-glucopyranosyl bromide (**2**) in each of the following solvents: acetone, benzene, carbon tetrachloride, chlorobenzene, 2,6-dimethylpyridine, nitrobenzene, and pyridine. The reaction was allowed to proceed with stirring for 4.5 days at room temperature. The resin was then collected on a glass filter and extensively rinsed with various solvents. The i.r. spectra of the resin after reaction indicated that the reaction had proceeded to various extents in the different solvents, and the yields of the coupling step could be estimated from the gain in weight of the resin after reaction. The results of this study are shown in Table 1.

A procedure similar to the one just described was used to determine the influence of an added base to the extent of the coupling reaction. The experimental conditions and results are summarized in Table II.

The influence of the monomer concentration was studied in a similar way: to 0.4 g (0.28 mmole) of 1%-crosslinked resin was added 3.5 ml of a benzene solution containing either 1.6 g (2.41 mmole, 5 samples) or 0.4 g (0.6 mmole, 4 samples) of monomer **1**. After addition of 2,6-dimethylpyridine (0.05 ml, 0.43 mmole) each sample was stirred at room temperature. The samples were processed at appropriate intervals and the coupling yields estimated from the weight increase of each sample. The results of this study are shown in Fig. 1.

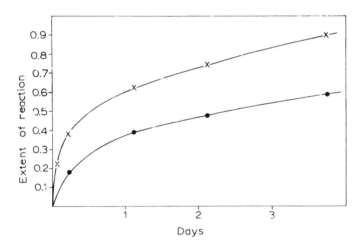

Fig. 1. Reaction of monomer **1** with resin. The molar ratio of monomer to resin was 2.1:1 (●) and 8.5:1 (×).

Preparation of a disaccharide. — A typical procedure for the preparation of a disaccharide from monomer **2** is given in the next paragraph. The same procedure was used for the coupling of monomer **1** with minor adjustments to take into account the difference in the molecular weights of the two monomers. Detailed information on the various yields obtained in the preparation of each oligomer can be found in Tables III and IV.

*Preparation of **3** by coupling of the first sugar unit to the resin.* — To 1.8276 g (1.28 mmole) of 1%-crosslinked resin was added a solution (15.5 ml) containing the monomer **2** (3.9 g, 6.02 mmoles) and 2,6-dimethylpyridine (0.20 ml, 1.72 mmole) in dry benzene. The molar ratio of monomer to 2,6-dimethylpyridine to resin was 4.7:1.34:1 and the monomer concentration 25 g/100 ml. The air in the flask containing the mixture was evacuated while the mixture was stirred to remove any air which might have remained trapped in the resin. After reaction for 72 h with stirring at room temperature, the resin was collected on a glass filter and rinsed with dry benzene, and the excess of monomer reclaimed. Removal of any remaining monomer or side product was effected by washing the resin successively with acetone, 4:1 pyridine–water, 1:1 pyridine–water, water, tetrahydrofuran, 4:1 tetrahydrofuran–water, 1:1 tetrahydrofuran–water, water, tetrahydrofuran, acetone, ethanol, acetone, benzene, and methylene chloride. After being dried overnight in a vacuum oven at 70°, the product (3) weighed 2.4249 g. The gain in weight of 597.3 mg indicated a coupling yield of 82% (calc.: 725.6 mg for 100% coupling). All the resin collected was then allowed to react again, as just described, for 4 additional days. After the usual processing, 2.5460 g of product 3 was obtained for a total weight gain of 718.4 mg (99%) in the two-stage coupling reaction.

Preparation of 4 by removal of the temporary blocking group of 3. — To a stirred suspension of resin 3 (2.4551 g) in tetrahydrofuran–methanol (20 ml) was added a catalytic amount of sodium methoxide. The mixture was heated for 90 min at reflux. After being cooled to room temperature, the resin was collected on a fritted-glass filter and rinsed with 7:3 tetrahydrofuran–methanol, 4:1 tetrahydrofuran–water, 1:1 tetrahydrofuran–water, water, ethanol, acetone, tetrahydrofuran, acetone, benzene, and methylene chloride. After being dried overnight in a vacuum oven, the resin 4 weighed 2.2893 g indicating that 165.8 mg had been lost in this step, and that removal of the C-6 acyl group was essentially complete (calc.: 24% of weight gained in first step or 166.2 mg). The absence of absorption in the carbonyl group region of the i.r. spectrum confirmed that the temporary blocking group had been cleaved.

Preparation of 5 by coupling of a second sugar unit to the resin. 4. — To 2.1486 g (~1.16 mmole) of the resin-bound monosaccharide 4 was added a solution (15 ml) containing 3.6 g (5.56 mmoles) of monomer 2 and 0.25 ml (2.15 mmoles) of 2,6-dimethylpyridine in dry benzene. The mixture was stirred for 5 days at room temperature. The resin was then collected on a fritted-glass filter and rinsed as described for the preparation of 3. After being dried overnight, 2.7508 g of resin-bound disaccharide 5 was obtained. The gain in weight of the resin was 0.6022 g indicating a coupling yield of 93% with respect to the resin-bound monosaccharide 3.

Cleavage of the disaccharide 6 from the resin 5 by ozonolysis. — In most cases the resin used in the decoupling step still contained the C-6 acyl group. The ozone used in this experiment was prepared in advance and stored under nitrogen in a glass coil cooled to $-78°$ and containing 30 g of previously ozonized and activated Silica Gel Davison grade 05 (6–16 mesh). Only a fraction of the coil was saturated with ozone. Nitrogen was used as carrier gas for the desorption of ozone and was passed through the cooled ($-78°$) coil at a rate of 150 ml/min.

A portion (1.3703 g containing 0.5463 g of bound sugar) of the resin-bound sugar 5 was suspended in dry dichloromethane (10 ml). After the suspension had been cooled to $-78°$, a stream of ozone in nitrogen carrier was bubbled through the mixture for 2 h. During the course of the reaction, the color of the resin changed from brown to light yellow. After flushing the excess of ozone with nitrogen, approximately 3–4 ml of the dichloromethane was evaporated at a temperature not exceeding −10°, and methanol (3 ml) pre-cooled to −78° was added and the mixture shaken. Dimethyl sulfide (2 ml) was added, and the mixture was kept for 1 h at −78° with intermittent shaking. The temperature was then allowed to rise to ambient, and the resin was filtered off on a fritted-glass filter and thoroughly rinsed with dichloromethane and chloroform. The resin was then extracted repeatedly with chloroform, 7:3 chloroform–acetone, and 7:3 chloroform–ethanol, heated at reflux. The combined filtrates were evaporated in a flash evaporator, leaving a yellow oil which was dissolved in chloroform and then washed with distilled water to remove the dimethyl sulfoxide produced in the reduction of the ozonide. After drying and evaporation of the solvent, the residual oil was dried under high vacuum to yield 0.4961 g of the desired disaccharide 6 obtained as a glycoside of hydroxyethanal. The n.m.r. spectrum of the product corresponded to that expected and showed the presence of a small amount (∼2% of the total number of protons) of impurity with high field resonance in the region δ 1–1.4 p.p.m.

Preparation of 7 by reduction of the hydroxyethanal glucoside 6 and removal of the C-6 blocking group. — The disaccharide 6 (0.496 g, 0.461 mmole) was dissolved in dry tetrahydrofuran (3 ml), and the solution was added dropwise through a dropping funnel to sodium borohydride (50 mg, 1.3 mmole) in ethanol (8 ml) at 0°. The solution was stirred for 90 min at room temperature, 0.5M potassium hydroxide (2 ml) was added, and the mixture was heated for 3 h at reflux. After evaporation of approximately half the solvent, the mixture was extracted with chloroform, the chloroform extract was rinsed with water, dried, and evaporated to yield 414 mg of the disaccharide glycol glycoside 7.

2-Hydroxyethyl 2,3,4-tri-O-benzyl-D-glucopyranosides (**8 and 9**). — To a stirred solution of 2,3,4-tri-O-benzyl-6-O-p-methoxybenzoyl-α-D-glucopyranosyl bromide (**2**, 1 g, 1.54 mmole) in acetone (4 ml) was added purified 1,2-ethanediol (10 ml, 145.3 mmoles) and 2,6-dimethylpyridine (0.2 ml, 1.72 mmole). After 4 days at room temperature, the acetone was evaporated in a flash evaporator and the mixture extracted with chloroform and water. The chloroform extract was washed with water, dried, and evaporated to a light yellow oil which was saponified in ethanol–water–potassium hydroxide. The resulting product was chromatographed on a column of silica gel using 100:1 benzene–methanol for elution. After evaporation of the solvent and drying under high vacuum, the product **8** (412 mg, 54%), isolated as a slightly yellow oil, had $[\alpha]_D^{24}$ +17.5° (c 1.5, chloroform).

A similar reaction was carried out on a 1-g sample of 2,3,4-tri-O-benzyl-6-O-p-nitrobenzoyl-α-D-glucopyranosyl bromide (**1**) and yielded product **9** with $[\alpha]_D^{24}$ +25.4° (c 3, chloroform). The two products prepared by this method had similar n.m.r. spectra which could not be assigned to either anomer, because of the lack of identifiable anomeric proton.

2-Hydroxyethyl 2,3,4-tri-O-acetyl-6-O-(2,3,4,6-tetra-O-acetyl-β-D-glucopyranosyl)-α-D-glucopyranoside (**10**). — β-Gentiobiose octaacetate (1 g, 1.47 mmole) was dissolved in dry dichloromethane (10 ml). A current of dry hydrogen bromide was bubbled through the solution for 10 min at room temperature. The solvent was evaporated in a flash evaporator, and successive portions of dichloromethane were evaporated *in vacuo* from the residual syrup at room temperature. Acetic acid produced in the reaction was eliminated by azeotropic distillation with dry benzene. The resulting hepta-*O*-acetylgentiobiosyl bromide was dissolved in acetone (3 ml) and treated with 1,2-ethanediol (10 ml, 145.3 mmoles) and 2,6-dimethylpyridine (0.17 ml, 1.47 mmole), and the mixture was stirred for 4.5 days at room temperature. After evaporation of the acetone, the mixture was extracted with chloroform and washed repeatedly with water. Evaporation of the chloroform yielded a clear syrup which was crystallized from ether (0.793 g, 79% yield). After recrystallization from ether–pentane, **10** had m.p. 144–45°, $[\alpha]_D^{24}$ +51° (*c* 1.3, chloroform), Its n.m.r. exhibited the expected proton ratio and included a low-field doublet (δ 6.21 p.p.m., $J_{1,2}$ 4 Hz) suggesting that the α-D anomer had been obtained.

Anal. Calc. for $C_{28}H_{40}O_{19}$: C, 49.41; H, 5.92. Found: C, 49.26; H, 5.88.

2-Acetoxyethyl 2,3,4-tri-O-acetyl-6-O-(2,3,4,6-tetra-O-acetyl-β-D-glucopyranosyl)-α-D-glucopyranoside (**11**). — Compound **10** (0.618 g, 0.91 mmole) was acetylated in the usual manner with dry pyridine (3 ml) and acetic anhydride (2 ml) to yield 0.592 g (90%) of **11** crystallized from ether. After recrystallization from chloroform–ether the product had m.p. 182–83°, $[\alpha]_D^{24}$ +49.5°; its n.m.r. spectrum included a doublet ($J_{1,2}$ 3.4 Hz) corresponding to 1 proton and centered at δ 6.39 p.p.m. assigned to the α-D anomer.

Anal. Calc. for $C_{30}H_{42}O_{20}$: C, 49.86; H, 5.86. Found: C, 49.49; H, 5.79.

2-Hydroxyethyl 2,3,4-tri-O-acetyl-6-O-(2,3,4,6-tetra-O-acetyl-β-D-glucopyranosyl)-D-glucopyranoside (**12**). — Hepta-*O*-acetylgentiobiosyl bromide, prepared as just described from octa-*O*-acetyl-β-gentiobiose (1 g, 1.47 mmole) and hydrogen bromide in dichloromethane, was dissolved in acetone (10 ml) and added dropwise to a mixture of 1,2-ethanediol (15 ml, 218 mmoles), silver carbonate (0.3 g), and powdered Drierite (2 g). The mixture was stirred for 24 h at room temperature, acetone was evaporated *in vacuo*, and the remaining syrup was extracted with chloroform and water. The chloroform extract was washed with water, dried, and evaporated *in vacuo*. Attempts to crystallize the resulting oil (0.81 g, 80%) failed, and the product was freeze-dried from benzene to a white solid having $[\alpha]_D^{24}$ +13.9° (*c* 1.9, chloroform).

2-Acetoxyethyl 2,3,4-tri-O-acetyl-6-O-(2,3,4,6-tetra-O-acetyl-β-D-glucopyranosyl)-D-glucopyranoside (**13**). — Acetylation of **12** (0.61 g, 0.9 mmole) in pyridine (2 ml) and acetic anhydride (1.5 ml) followed by the usual work-up of the product yielded 0.589 g (90%) of a slightly yellow oil. Its n.m.r. spectrum included two low-field doublets accounting for a total of one proton and centered at δ 5.79 p.p.m. ($J_{1,2}$ 5.9 Hz) and at δ 6.39 p.p.m. ($J_{1,2}$ 3.4 Hz) in a 7:3 ratio indicating that the product contained a mixture of **13** and **11** in 7:3 ratio. The product had $[\alpha]_D^{24}$ +22.5° (*c* 2.2, chloroform).

Anal. Calc. for $C_{30}H_{42}O_{20}$: C, 49.86; H, 5.86. Found: C, 49.69; H, 5.88.

*Preparation of **14** by debenzylation of **7** and acetylation.* — To a stirred solution of **7** (0.472 g, 0.5 mmole) in liquid ammonia (75 ml) and dry tetrahydrofuran

(10 ml) was added, portionwise, a total of 140 mg (6.1 mmoles) of sodium. After addition of each piece of sodium, a transient blue color developed around the metal and subsided rapidly. After addition of the last piece of sodium, a blue coloration developed throughout the solution but disappeared within 30 sec. Upon addition of another 40 mg (1.7 mmole) of sodium, the intense blue coloration reappeared and persisted until a small excess of ammonium chloride was added after 30 min of stirring. The solvents were evaporated under a stream of nitrogen, and the solid residue was extracted with chloroform, and then water. The water phase was washed several times with benzene, chloroform, and ether, and then evaporated to dryness. After being dried for 24 h in a vacuum oven at 85° the solid residue, which contained both the debenzylated product and appreciable quantities of inorganic byproducts, was acetylated with acetic anhydride (2 ml) in pyridine (3 ml) for 2 days at room temperature. The acetylation mixture was worked-up in the usual manner, and the desired product was isolated as an oil (**14**) which failed to crystallize. After elimination of a small amount of colored impurities on a small Silica Gel column, the product was freeze-dried to a slightly colored solid (191 mg, 52%) which had $[\alpha]_D^{24} +122.6°$ (c 1.3, chloroform). The n.m.r. spectrum of **14** was not identical with either that of **11** nor that of **13** but the proton amount of 18 in the region δ 3.5–3.9 and 24 in the acetyl region of the spectrum was similar to that of **11** and **13**.

Anal. Calc. for $C_{30}H_{42}O_{20}$: C, 49.86; H, 5.86. Found: C, 49.74; H, 5.95.

ACKNOWLEDGEMENTS

The present work has been supported by a research grant (GM 06168) from the Division of General Medical Sciences, National Institutes of Health, U. S. Public Health Service. A stimulating discussion with Dr. R. D. Guthrie is acknowledged.

REFERENCES

1 J. Fréchet and C. Schuerch, *J. Amer. Chem. Soc.*, 93 (1971) 492.
2 J. Fréchet and C. Schuerch, *J. Amer. Chem. Soc.*, 94 (1972) 604.
3 T. Ishikawa and H. G. Fletcher, Jr., *J. Org. Chem.*, 34 (1969) 563.
4 H. M. Flowers, *Carbohyd. Res.*, 18 (1971) 211.
5 M. Dejter-Juszynski and H. M. Flowers, *Carbohyd. Res.*, 18 (1971) 219.
6 H. R. Goldschmid and A. S. Perlin, *Can. J. Chem.*, 39 (1961) 2025.
7 R. U. Lemieux and A. R. Morgan, *J. Amer. Chem. Soc.*, 85 (1963) 1889.
8 F. Bohlmann and J. Ruhnke, *Chem. Ber.*, 93 (1960) 1945.
9 S. Hixon, *Ph. D. Thesis, University of Wisconsin*, University Microfilm No. 70, 8294, 1970, pp. 49, 50, 53, 54, and 56.
10 N. Pravdić and D. Keglević, *Tetrahedron*, 21 (1965) 1897.
11 F. Weygand and H. Ziemann, *Ann.*, 657 (1962) 179.
12 O. T. Schmidt, T. Auer, and H. Schadel, *Chem. Ber.*, 93 (1960) 556.
13 G. Zemplén and Z. Bruckner, *Chem. Ber.*, 64 (1931) 1852.
14 H. Bredereck, A. Wagner, D. Geissel, and H. Ott, *Chem. Ber.*, 95 (1962) 3064.

Synthesis of Oligosaccharides on Polymer Supports. Part I. 6-O-(p-Vinylbenzoyl) Derivatives of Glucopyranose and Their Copolymers with Styrene

By **R. D. Guthrie,** * **A. D. Jenkins,** and **J. Stehlícek,**† School of Molecular Sciences, University of Sussex, Brighton BN1 9QJ

In order to study the reactions leading to glucopyranose 1,2-O-orthoacetate units attached to a polymer support, methyl 2,3,4-tri-O-benzoyl-6-O-(p-vinylbenzoyl)-α-D-glucopyranoside, 1,2,3,4-tetra-O-acetyl-6-O-(p-vinylbenzoyl)-β-D-glucopyranose, 2,3,4-tri-O-acetyl-6-O-(p-vinylbenzoyl)-β-D-glucopyranosyl chloride, and 2,3,4-tri-O-acetyl-6-O-(p-vinylbenzoyl)-α-D-glucopyranosyl bromide were prepared, as well as their polymers and copolymers with styrene. The reactivity ratios of co-monomers were determined. Specific rotation measurements were used for the determination of the compositions of copolymers and to follow their chemical conversions.

OLIGOSACCHARIDES of strictly defined structure would be materials of great use to biochemists, especially in the study of enzyme mechanisms. The methods available at present, namely sequential synthesis in a homogenous phase, or fractionation of partly hydrolysed natural polysaccharides, are not versatile enough to provide tailor-made oligosaccharides. Synthesis on a polymer support might solve this problem. This method was pioneered by Merrifield,[1] originally for polypeptide synthesis, to which it has been extensively applied, and it has also been used for polynucleotides.[2-6] The polymer supports employed were generally styrene–divinylbenzene cross-linked copolymers, into which reactive groups had been introduced for bonding the first unit. The method of polymer support synthesis and the related method using polymer-supported reagents [7] require that the basic reaction for attachment of further units gives almost quantitative yields and that satisfactory protecting groups for the reactive sites that are not involved in the synthesis steps are available. This requirement creates difficulties in the formation of glycosidic linkages, where there is a limited choice of stereospecific reactions giving reasonably high yields.

There are two possible approaches to the oligomer synthesis. In Scheme A the first unit is attached by an alcoholic hydroxy-group to the polymer, leaving the C-1 group free for attack on a suitably blocked second unit having one free hydroxy-group; the C-1 group on the second unit is then converted into its reactive form and the process is repeated.

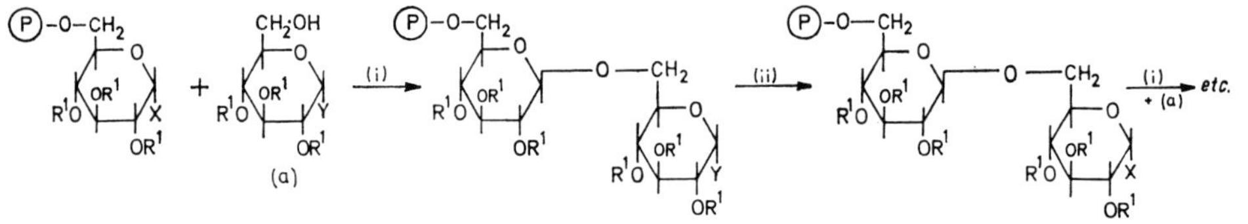

SCHEME A *

(i) glycoside formation with C-1 of (a) protected in unreactive form Y, (ii) conversion of C-1 group Y into reactive form X.

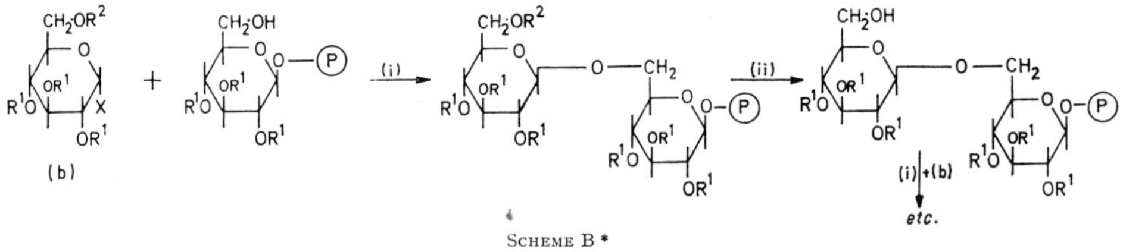

SCHEME B *

(i) glycoside formation, (ii) removal of blocking group R² in presence of blocking group R¹.

* The use of formation of a β 1 ⟶ 6 linkage between D-glucose units is purely illustrative.

† On leave from the Institute of Macromolecular Chemistry, Czechoslovak Academy of Sciences, Prague.

[1] R. B. Merrifield, *Biochemistry*, 1964, **3**, 1385; *J. Amer. Chem. Soc.*, 1963, **85**, 2149.
[2] R. L. Letsinger and V. Mahadevan, *J. Amer. Chem. Soc.*, 1965, **87**, 3526; 1966, **88**, 5319.
[3] R. L. Letsinger, M. H. Caruthers, and D. M. Jerina, *Biochemistry*, 1967, **6**, 1379.
[4] H. Hayatsu and H. G. Khorana, *J. Amer. Chem. Soc.*, 1966, **88**, 3182; 1967, **89**, 3880.
[5] F. Cramer, R. Helbig, H. Hettler, K. H. Sheit, and H. Seliger, *Angew. Chem. Internat. Edn.*, 1966, **5**, 601.
[6] L. R. Melby and D. R. Strobach, *J. Amer. Chem. Soc.*, 1967, **89**, 450; *J. Org. Chem.*, 1969, **34**, 421, 427.
[7] M. Fridkin, A. Patchornik, and E. Katchalski, *J. Amer. Chem. Soc.*, 1966, **88**, 3164; 1968, **90**, 2953.

The second approach (Scheme B) would be to attach the first unit to the polymer *via* the anomeric centre and let its one free hydroxy-group be attacked by a second unit with a reactive group on C-1. This second unit would also bear two types of blocking group on the alcoholic hydroxy-groups, one of which could be preferentially unblocked for a repetition of the process.

We felt that Scheme A would be the easier one, and chose to study the reaction of polymer-bound sugar 1,2-orthoesters with hydroxy-compounds. Orthoesters have been much used by Kochetkov and his group [8-10] for the synthesis of di- and tri-saccharides. The reaction gives β-glycosides and, especially when modified,[11] gives higher yields than other methods.

To simplify the procedures as much as possible in our initial studies we decided to use a soluble [12] non-cross-linked polymer. Conversions on insoluble polymer supports can usually be followed from changes of weight and from i.r. spectra, which are often not conclusive. It seemed desirable that the first monosaccharide unit should be attached by a specific hydroxy-group, to provide an identical steric environment for all units along the polymer. We therefore decided to construct our basic polymer in a way that we believe to be unique in solid-support synthesis work, in that we co-polymerised styrene with the required first sugar unit bearing a polymerisable functional group. 6-O-(p-Vinylbenzoyl) derivatives of D-glucose were therefore prepared and co-polymerised with styrene.

RESULTS AND DISCUSSION

Monomers.—All the monomers prepared have been 6-O-vinylbenzoylglucopyranose derivatives: methyl 2,3,4-tri-O-benzoyl-6-O-(p-vinylbenzoyl)-α-D-glucopyranoside (1), 1,2,3,4-tetra-O-acetyl-6-O-(p-vinylbenzoyl)-β-glucopyranose (2), and 2,3,4-tri-O-acetyl-6-O-(p-vinylbenzoyl)-α-D-glucopyranosyl chloride (3) and the corresponding bromide (4). They were made from the corresponding known D-glucose derivatives with a free C-6 hydroxy-group and p-vinylbenzoyl chloride, which was prepared from 2-phenylethanol *via* 2-phenylethyl bromide, p-(2-bromoethyl)acetophenone, p-(2-bromoethyl)benzoic acid, and p-vinylbenzoic acid by known methods [13-15] in an overall yield of 33%. Acylation was carried out in dry pyridine in the preparation of compounds (1) and (2), and in an inert solvent containing only an equivalent amount of base to avoid the formation of N-glucosides in the preparation of compounds (3) and (4). In spite of this, the preparation of the bromide (4) was accompanied by formation of side products, and its final purification was difficult.

All the monomers were crystalline compounds and were polymerised readily by free-radical catalysis. They also polymerised on heating above the m.p., especially the bromide (4), which when pure did not melt before polymerising.

Copolymerisation with Styrene.—The great difference between the molecular weights of the glucose monomers and that of the co-monomer styrene as well as the poor solubility of the rather polar glucopyranose esters in styrene made bulk copolymerization impossible with a reasonable starting ratio of monomers. Therefore the copolymers were prepared by solution copolymerization in toluene, and were isolated by precipitation into methanol. Conditions of copolymerization are given in Table 1, including those for the preparation of homopolymers. The copolymers were soluble in most solvents which dissolve polystyrene, such as benzene, toluene, acetone, chloroform, and dioxan, but not ether.

The kinetic results showed that copolymerization proceeded faster than the polymerization of styrene. Thus, under standard polymerization conditions (see Experimetal section) the yield of polystyrene after 6 h was 10.7%, while monomer (1)–styrene mixtures of the molecular ratios ($M_0 : S_0$) 0.05, 0.2, 1.0, and 2.0 gave 15.5, 28.3, 48.5, and 56.7% of copolymer, respectively. Also the fraction of the glucose monomer incorporated into the copolymer was higher than that in the starting mixture.

To obtain at least approximate values of reactivity ratios in copolymerization, we sought a sufficiently accurate method for analysis of the copolymer composition. In this case the accuracy of the most usual analytical methods was lowered by the difference between the molecular weights of the co-monomers, and by the necessity of handling small amounts of copolymers, as the glucose monomers were available only in limited amount.

The specific optical rotation was used in calculation

---

[8] N. K. Kochetkov, A. J. Khorlin, and A. F. Bochkov, *Tetrahedron*, 1967, **23**, 693.
[9] A. F. Bochkov, T. A. Sokolovskaya, and N. K. Kochetkov, *Izvest. Akad. Nauk S.S.S.R. Ser. khim.*, 1968, 1570.
[10] N. K. Kochetkov, A. F. Bochkov, and T. A. Sokolovskaya, *Doklady Akad. Nauk S.S.S.R.*, 1969, **187**, 96.
[11] A. F. Bochkov, V. I. Snyatkova, and N. K. Kochetkov, *Izvest. Akad. Nauk S.S.S.R. Ser. khim.*, 1967, 2684.
4 z

[12] M. M. Shemyakin, Yu. A. Ovchinnikov, A. A. Kiryushkin, and I. V. Kozhevnikova, *Tetrahedron Letters*, 1965, 2323.
[13] E. L. Foreman and S. M. McElevain, *J. Amer. Chem. Soc.*, 1940, **62**, 1435.
[14] E. Osawa, K. Wang, and O. Kurihara, *Makromol. Chem.*, 1965, **83**, 100.
[15] N. A. Adrova and K. K. Khomenkova, *Zhur. obshchei khim.*, 1962, **32**, 2267.

TABLE 1

Copolymers of *p*-vinylbenzoylglucose derivatives with styrene and their composition (copolymerized at 60°, concn. of monomers in toluene 1M, 1 mol % of azobisisobutyronitrile)

M_0/S_0 [a]	Time t/h	Yield (%)	$[\alpha]_D$	m/s [b] from $[\alpha]_D$	I.r.	Other methods
			Monomer (1)			
Homopolymer [e]	20	60·2	+81·7°			
Homopolymer [d]	46	78·9	(+81·2)			
2·0	1	14·0	+78·8	8·105	(1·455)	
2·0	6	56·7	+81·8	—	(1·511)	
1·0	6	48·5	+74·4	1·666	(0·776)	
0·983	1·2	10·5	+74·2	1·618 [m]	(0·686)	
0·2	97 [e]	79·1	+50·4	0·263	0·217	
0·199	6	28·3	+55·3	0·343 [m]	0·389	
0·2	2·25	9·5	+55·7	0·350 [m]	0·656	
0·1	97 [e]	81·7	+36·8	0·134	0·113	
0·1	4·5	15·8	+42·8	0·180 [m]	0·155	
0·05	6	15·5	+29·6	0·093 [m]	0·088	
			Monomer (2)			
Homopolymer [f]	68	85·1	+29·9			
1·0	2	17·9	+29·5	(16·023)	4·001	2·896 [g,m]
0·25	2	9·7	+21·6	0·566	0·548	0·665 [g,m]
0·2	300	87·7	+17·4	0·303	0·283	0·275,[g] 0·206,[h] 0·229,[h] 0·288,[h] 0·253 [i]
0·125	300	86·8	+14·0	0·192	0·203	0·154,[g] 0·161,[h] 0·153,[h] 0·147 [i]
0·1	3·5	11·0	+14·4	0·202	0·279	0·243 [g,m]
			Monomer (3)			
Homopolymer [j]	148	94·7	+152·2			
1·0	1·25	14·5	+137·7	2·173 [m]	3·153	
0·25	2	9·1	+110·7	0·611 [m]	0·526	
0·166	334	87·1	+75·4	0·225	0·214	
0·1	4	12·7	+81·0	0·260 [m]	0·248	
			Monomer (4)			
Homopolymer [k]	43	90·6	+168·0			
0·166 [l]	113	10·1	+100·4	0·310		

[a] Initial ratio of monomers. [b] Ratio of structure units in copolymer. [c] Concn. 0·8M. [d] Concn. 0·23M. [e] Polymerization temp. 80°; benzoyl peroxide instead of azobisisobutyronitrile. [f] Concn. 0·35M. [g] From saponification. [h] From $[\alpha]_D$ of the copolymer converted into C-1 bromo-derivative, based on $[\alpha]_D$ of homopolymer of (4). [i] From n.m.r. spectra. [j] Concn. 0·33M. [k] Concn. 0·5M. [l] Impure monomer. [m] Result was used for calculation of reactivity ratios.

of the weight fraction of units of optically active monomer in the copolymer, with the assumption of linear proportionality of both quantities. The polymers of the optically active monomers, having the asymmetric carbon atoms in the side chains at a distance more than that of the β-position to the main chain, should have practically the same rotation power as a low-molecular-weight model compound or the parent monomer.[16] Our case suits this condition, but however, the specific rotations of the homopolymers of compounds (1)—(4) were always slightly lower than those of the corresponding monomers. This may be because the monomer was not the best model for the polymer segment, or because the linkage to the polymer chain caused increased conformational rigidity in the sugar units.[17] The calculation of the composition of the copolymers was therefore based on $[\alpha]_D$ values of the corresponding homopolymers. The method is reasonably accurate only if the absolute value of $[\alpha]_D$ of the homopolymer is high enough, and this was the case with poly-[(1)-co-styrene] and poly-[(3)-co-styrene].

Poly-[(2)-co-styrene] was analysed by saponification of the ester groups. As the results in Table 2 show, the calibration of this method with homopolymer and the

TABLE 2

The composition of poly-[(2)-co-styrene] determined by saponification

	(2)			
	Determined		Corrected	
M_0/S_0	Wt. fraction	Mol. fraction	Wt. fraction	Mol. fraction
Monomer	1·124		1·004	
	1·127		1·006	
	1·119		0·999	
Homopolymer	1·117		0·997	
	1·122		1·002	
0·1	0·5904	0·2388	0·5272	0·1953
0·125	0·4632	0·1581	0·4136	0·1331
0·2	0·6255	0·2666	0·5585	0·2159
0·25	0·8437	0·5402	0·7533	0·3993
1·0	1·0417		0·9301	0·7434

monomer (2) gave values 12% higher than expected from stoicheiometry (five hydrolysable groups per molecule were assumed), with good reproducibility. The

[16] M. Goodman, A. Abe, and You-Ling Fan, *Macromol. Rev.*, 1967, **1**, 1.

[17] A. Abe and M. Goodman, *J. Polymer Sci.*, Part A-1, *Polymer Chem.*, 1963, 2193.

analyses of copolymers were corrected accordingly and they agreed better with the rotation data. The higher consumption of hydroxide was apparently caused by the degradation of the free glucose formed by strong base. This reaction prevented reproducible data being obtained from the saponification of poly-[(3)-co-styrene], where the cyclic glucal may be formed, besides the open-chain glucose, both differing in sensitivity to alkali. The benzoate groups in poly-[(1)-co-styrene] were resistant to the saponification conditions used and no consumption of hydroxide was found after the mixture had been refluxed for 5·5 h.

Comparisons of different analytical methods with the method of optical rotation are given in Table 1. The other values were obtained from the intensities of ester carbonyl i.r. absorptions, and from the ratio aromatic : backbone : acetate protons shown by the n.m.r. spectrum.

The reactivity ratios were determined by the method of intersecting slopes [18] for a copolymerization temperature of 60° (Figure). For these determinations the

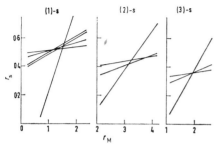

Determination of reactivity ratios by the method of intersecting slopes; r_S for styrene, r_M for glucose monomer

copolymers used were prepared with the starting monomer ratio $M_0 : S_0 < 1$, because at higher proportions of the glucose monomer the error of the analysis increased, owing to the difference between the molecular weights of the co-monomers. The pairs of reactivity ratios r_M (glucose monomer) and r_S (styrene) are only approximate as they were obtained with conversions to copolymer ranging from 9 to 18%, and are as follows:

Poly-[(1)-co-styrene] $r_M = 1·33 \pm 0·10$;
$r_S = 0·53 \pm 0·01$

Poly-[(2)-co-styrene] $r_M = 3·3 \pm 0·2$;
$r_S = 0·44 \pm 0·03$

Poly-[(3)-co-styrene] $r_M = 1·95$; $r_S = 0·36$

Monomer (4) of sufficient purity was prepared in a small amount only and was used for preparation of the homo-

[18] T. Alfrey, J. J. Bohrer, and H. Mark, 'Copolymerization,' High Polymer Series Vol. 8, Interscience, New York, 1952.
[19] L. J. Haynes and F. W. Newth, Adv. Carbohydrate Chem., 1955, **10**, 207.

polymer, which was required as a standard for reactions of poly-[(2)-co-styrene]. In the presence of a small proportion of impurities from its preparation, compound (4) partly decomposed during copolymerization with styrene. Because the copolymerization was not controlled only by the reactivity ratios, these were not determined.

Reactions of Copolymers.—As already mentioned, reactions on the polymer support were studied in solution, as an approximation to reactions on cross-linked copolymer. We used the measurement of specific rotation of the copolymers as a suitable and sensitive method for following the chemical changes undergone by the sugar residues in the copolymers.

We intended to use poly-[(1)-co-styrene] for the preparation of the glucopyranosyl bromide by acetobromolysis of the methyl glucosidic group, because Kochetkov [8] had obtained a higher yield in glycosidation of benzoyl derivatives. However, the copolymer was resistant to hydrogen bromide in acetic acid under the conditions commonly used in this reaction; [19] $[\alpha]_D^{22}$ values before and after treatment were $+36·8$ and $+32·8°$. Under similar conditions the C-1 acetate group in poly-[(2)-co-styrene] was replaced by bromide easily, giving quantitatively the α-D-glucopyranosyl bromide. This was proved by similar M/S values of the original and the converted copolymers, which were calculated from their $[\alpha]_D$ values and that of poly-(4) (Table 1).

1,2-Orthoesters of pyranoses have been prepared both from *cis*-halides [8,20] (*e.g.* from 2,3,4,6-tetra-O-acetyl-α-D-glucopyranosyl bromide) and from *trans*-halides [21] (*e.g.* 2,3,4,6-tetra-O-acetyl-β-D-glucopyranosyl bromide). However, Lemieux and Morgan [21] supposed that *cis*-halides of pyranoses must be anomerized to *trans*-halides prior to orthoesterification. The anomerization is catalysed by halide anions, supplied for example by tetrabutylammonium bromide or chloride.[21] In the preparation of orthoesters from *cis*-halides, the halide anions are formed in the beginning of the reaction by conversion of a small proportion of *trans*-halide, which is present, and catalyse the anomerization after that. With poly-[(3)-co-styrene] and poly-[(4)-co-styrene], the ethyl orthoacetate groups were not significantly formed in the absence of catalyst as judged by the small change of $[\alpha]_D$ (Table 3). In the presence of bromide anions, poly-[(4)-co-styrene] was converted in to the corresponding ethyl orthoacetate [Scheme C]. The presence of orthoacetate groups was proved by acid hydrolysis under mild conditions,[8] which changed dramatically the specific rotation of the copolymer (Table 3). The catalysed orthoesterification of poly-[(3)-co-styrene] was probably too slow owing to the lower reactivity of the C-1 chloride group.

The splitting of the sugar units from the polymer support was complete under conditions of methanolytic deacylation as commonly used in carbohydrate chemistry.

[20] B. Helferich and K. Weiss, Chem. Ber., 1959, **89**, 314.
[21] R. U. Lemieux and A. G. Morgan, Canad. J. Chem., 1965, **43**, 2199.

Table 3
Reactions of copolymers followed by optical rotation measurement ($[\alpha]_D$)

Substitution on C-1	A		B			C
Acetate	+14.0	→	+17.4	→	→	
Bromide	+73.3	+71.1	+83.4	+97.4	+88.0	
Chloride						+75.4
Orthoesterification without catalyst	+68.4					+74.5
Catalysed orthoesterification	+12.0	+15.0	+18.6	+19.8		+66.8
Acid hydrolysis of orthoester	+43.1		+59.6			+70.8
Methanolysis	−5.4	+0.6	−1.9	−2.3	−3.8	+0.9

A, Reactions of poly-[(2)-co-styrene], M_0/S_0 0.125. B, Reactions of poly-[(2)-co-styrene], M_0/S_0 0.2. C, Reactions of poly-[(3)-co-styrene], M_0/S_0 0.166.

The recovered polymer had entirely lost its optical activity (Table 3).

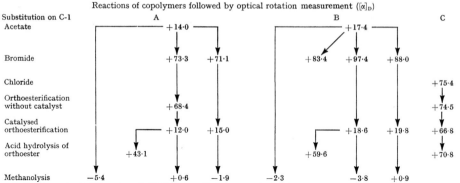

Scheme C

The results demonstrate that it is possible to carry out reactions leading to sugar orthoesters on a polymer support.

[22] J. F. Norris, M. Watt, and R. Thomas, *J. Amer. Chem. Soc.*, 1916, **38**, 1078.

EXPERIMENTAL

Rotations were measured for solutions in chloroform.

p-Vinylbenzoic Acid.—2-Phenylethyl bromide (92%; b.p. 105—107° at 16—18 mmHg) was obtained from 2-phenylethanol (B.D.H.) and 60% hydrobromic acid,[22] and was converted into *p*-(2-bromoethyl)acetophenone [13] (63%; b.p. 126.5—127.7° at 0.25—0.30 mmHg). *p*-(2-Bromoethyl)benzoic acid (96%), m.p. 208—210° (from benzene–light petroleum), was prepared by oxidation of the phenone with hypobromite,[13] and gave *p*-vinylbenzoic acid (79%), m.p. 142—144° [from water–ethanol (8 : 2)] on refluxing with water–ethanolic sodium hydroxide.[14]

p-Vinylbenzoyl Chloride.—*p*-Vinylbenzoic acid (7.0 g) was treated with thionyl chloride (20 ml) at room temperature for 40 h, in the dark. The excess of thionyl chloride was removed *in vacuo* and the residue was distilled through a short Vigreux column, giving *p*-vinylbenzoyl chloride (6.0 g, 76%), b.p. 54—58° at 0.2 mmHg (lit.,[15] 65—68° at 0.15 mmHg).

Methyl 2,3,4-Tri-O-benzoyl-α-D-glucopyranoside.—Methyl 2,3,4-tri-*O*-benzoyl-6-*O*-trityl-α-D-glucopyranoside was prepared by tritylation and subsequent benzoylation of methyl α-D-glucopyranoside,[23] and had m.p. 106.5—107.5° (from ethyl acetate–ethanol) $[\alpha]_D^{24}$ +130° (*c* 2). It was treated in dry chloroform with hydrogen chloride [23, 24] and gave, after separation of triphenylmethanol, methyl 2,3,4-tri-*O*-benzoyl-α-D-glucopyranoside (71%), m.p. 141—143° (from benzene–light petroleum), $[\alpha]_D^{20}$ +54.5° (*c* 2).

Methyl 2,3,4-Tri-O-benzoyl-6-O-p-vinylbenzoyl-α-D-glucopyranoside (1).—Methyl 2,3,4-tri-*O*-benzoyl-α-D-glucopyranoside (6.0 g) was treated with *p*-vinylbenzoyl chloride (2 g) in dry pyridine (25 ml) for 15 h at room temperature. The mixture was poured into ice–water (200 ml). The precipitate was separated and dissolved in benzene, the solution was washed with 0.5N-sulphuric acid and water, dried (MgSO₄), and evaporated *in vacuo*, giving the crude product (6.85 g, 91%), m.p. 134—136°. Two recrystallizations from ethanol gave *the glycoside* (1) (5.62 g), m.p. 137—138°, $[\alpha]_D^{20}$ +84.5° (*c* 2) (Found: C, 69.7; H, 5.1. $C_{37}H_{32}O_{10}$ requires C, 69.8; H, 5.1%).

[23] B. Helferich and J. Becker, *Annalen*, 1924, **440**, 1.
[24] H. Ohle and K. Tessmar, *Ber.*, 1938, **71**, 1843.

1,2,3,4-*Tetra*-O-*acetyl*-6-O-(*p*-*vinylbenzoyl*)-β-D-*glucopyranose* (2).—1,2,3,4-Tetra-O-acetyl-β-D-glucopyranose[25] (6·2 g) and *p*-vinylbenzoyl chloride (3·0 g) in dry pyridine (20 ml) were kept for 1 h at 0° and for 15 h at room temperature. The mixture was diluted with chloroform (100 ml) and poured into ice–water (200 ml). The chloroform layer was separated, washed with N-hydrochloric acid and water, dried (MgSO$_4$), and evaporated in vacuo to dryness giving the crude product (5·7 g, 67%). Recrystallisation from ethanol gave *compound* (2), m.p. 135—136°, $[\alpha]_D^{21}$ +35·2° (*c* 1) (Found: C, 57·5; H, 5·5. C$_{23}$H$_{26}$O$_{11}$ requires C, 57·7; H, 5·5%).

2,3,4-O-*acetyl*-α-D-*glucopyranosyl Chloride*.—1,6-Anhydro-β-D-glucopyranose, prepared by pyrolysis of wheat starch,[26] was acetylated and then treated with titanium tetrachloride in 0·5% ethanol–chloroform.[27] The yield of 2,3,4-tri-O-acetyl-α-D-glucopyranosyl chloride was 57%; m.p. 125·5—126°, $[\alpha]_D^{22}$ −155° (*c* 1).

2,3,4-*Tri*-O-*acetyl*-6-O-*p*-*vinylbenzoyl*-α-D-*glucopyranosyl Chloride* (3).—*p*-Vinylbenzoyl chloride (1·94 g) and 2,3,4-tri-O-acetyl-α-D-glucopyranosyl chloride (3·97 g) were agitated in dry benzene (100 ml) and pyridine (2 ml) for 69 h at room temperature. The solution was washed with water, dried (MgSO$_4$), and evaporated to dryness in vacuo, to give a crystalline mixture of compound (3) and unchanged starting compound. This was separated by preparative layer chromatography (p.l.c.) [Kieselgel GF$_{254}$; benzene–ether (1:1)] to give the *product* (3) (4·30 g, 81%), m.p. 86—87° (from benzene–light petroleum), $[\alpha]_D^{22}$ +166° (*c* 1) (Found: C, 55·5; H, 5·1. C$_{21}$H$_{23}$ClO$_9$ requires C, 55·45; H, 5·1%). Compound (3) polymerized rapidly above its m.p. to give a glassy polymer, m.p. 107—110°.

2,3,4-*Tri*-O-*acetyl*-α-D-*glucopyranosyl Bromide*.[28,29]—This was prepared from 2,3,4-tri-O-acetyl-1,6-anhydro-β-D-glucose and titanium tetrabromide, by a method analogous to that used for the chloride (yield 58%); the product had m.p. 112—112·5° (decomp.) (from ether–cyclohexane) (lit.,[28] 126—127°) (Titanium tetrabromide was prepared from the tetrachloride and anhydrous hydrogen bromide according to the method of Johannesen and Gordon.[30])

2,3,4-*Tri*-O-*acetyl*-6-O-*p*-*vinylbenzoyl*-α-D-*glucopyranosyl Bromide* (4).—A suspension of 2,3,4-tri-O-acetyl-α-D-glucopyranosyl bromide (3·01 g) in ether (100 ml) was treated with *p*-vinylbenzoyl chloride (1·87 g) and tributylamine (2·73 ml) for 24 h at room temperature. The resulting solution was washed with N-hydrochloric acid and water, dried (MgSO$_4$), and concentrated in vacuo to about 15 ml; addition of light petroleum and cooling gave crystalline material (1·26 g), m.p. 52—56°, which on p.l.c. [Kieselgel GF$_{254}$; toluene–ether (1:1)] and crystallized from cyclohexane to give the pure bromide (4) (0·49 g, 11·5%). This material polymerized before apparent melting into a glassy solid, m.p. 124—125°. The monomer had $[\alpha]_D^{22}$ +180° (*c* 1) (Found: C, 50·6; H, 4·7. C$_{21}$H$_{23}$BrO$_9$ requires C, 50·5; H, 4·6%). As the loss by decomposition during chromatography was high, other batches of compound (4) were crystallized from ether–cyclohexane, providing fractions of variable purity and m.p. 60—65° (yield 52%). All fractions solidified above the m.p., and the polymers formed melted at 100—117° (depending on the purity of monomer).

Polymerisation Experiments.—Copolymerization of the glucose monomers with styrene was carried out in toluene (AnalaR), generally at a monomer concentration of 1 mol l^{-1} with 1 mol % of azobisisobutyronitrile, at 60° in sealed evacuated ampoules, flushed several times with nitrogen. The copolymers were precipitated into methanol, filtered off, washed with methanol, and dried at room temperature and 10 mmHg for 15—35 h. For homopolymerization the concentrations of the monomers in toluene were chosen according to their solubility and are given in Table 1.

Determination of the Compositions of Copolymers.—Optical rotations were measured for chloroform solutions (1·0 g in 100 ml) in a 10 cm cell with a Perkin-Elmer Polarimeter 141.

Saponification of Ester Groups.—A solution of poly-[(2)-co-styrene] (0·01—0·03 g) in acetone (10 ml) and 0·1N-sodium hydroxide (5 ml) was shaken for 6 h at room temperature. The excess of hydroxide was titrated with 0·1N-sulphuric acid (phenolphtalein). The required time was estimated from the kinetics of saponification of the copolymer with M_0/S_0 0·125. The value of a blank determination with acetone alone was subtracted, and the results were corrected according to the data obtained with monomer (2) and its copolymer (see Table 2).

Spectra.—I.r. spectra were measured for solutions in chloroform (0·5—1·5 g l^{-1}) with a Perkin-Elmer 257 spectrophotometer (1 mm sodium chloride cell). The ester carbonyl absorption in the region 1750—1760 cm^{-1} was used for calculations based on the molar extinction coefficient of the related homopolymer.

N.m.r. spectra were recorded with a Varian HA100 (100 MHz) spectrometer for solutions in deuteriochloroform at room temperature. The areas of overlapping peaks of acetate and backbone protons and of aromatic nuclei protons were used for calculation of the composition.

Reactions of Copolymers.—*Conversion of poly-[(2)-co-styrene] into glucopyranosyl bromide derivative.* The copolymer (0·5 g) was dissolved in glacial acetic acid (45%; 2 ml). The solution was kept at 0° for 16 h, and then the copolymer was precipitated into ice-cold methanol (400 ml), collected on a sintered-glass filter, washed with methanol, water, dilute sodium hydrogen carbonate solution, water, and methanol, and then dried at room temperature in vacuo (10 mmHg), giving a light yellow solid. Poly-[(1)-co-styrene] was treated similarly without apparent effect.

Formation of ethyl 1,2-O-orthoacetate on the support. Poly-[1-C-bromo-(2)-co-styrene] (0·50 g) was dissolved in dry dioxan (6 ml), dry nitromethane (3 ml), 2,6-lutidine (0·3 ml), and absolute ethanol (0·4 ml). In the catalysed orthoesterification, tetrabutyl ammonium bromide (0·20 g) was added. The mixture was either heated at 50—58° for 20 h or kept at room temperature for 140 h, and then precipitated into methanol or water. The copolymer was collected on a sintered-glass filter, washed with water and methanol, and dried at room temperature in vacuo. The attempt to convert poly-[(3)-co-styrene] into orthoester was carried out similarly, either without catalyst or with tetraethylammonium chloride.

Acid hydrolysis as a proof of orthoester formation. The solution of the orthoesterified copolymer (0·02 g) in chloroform (2 ml) was thoroughly shaken with 2N-hydrochloric acid (1 drop) for about 3 min, then anhydrous magnesium

[25] E. A. Talley, *Methods Carbohydrate Chem.*, 1963, **2**, 337.
[26] R. B. Ward, *Methods Carbohydrate Chem.*, 1963, **2**, 222.
[27] Z. Czuros, G. Deak, and M. Haraszthy, *Acta Chim. Acad. Sci. Hung.*, 1959, **21**, 181, 193.
[28] G. Zemplen and A. Gerecz, *Ber.*, 1931, **64**, 1545.
[29] S. Haq and W. J. Whelan, *J. Chem. Soc.*, 1956, 4543.
[30] R. B. Johannesen and C. L. Gordon, *Inorg. Synth.*, 1967, **9**, 46.

sulphate (*ca.* 0·1 g) was added. After 30 min at room temperature the solution was filtered and used for measurement of $[\alpha]_D$.

Splitting of the glucose units from the copolymer. The copolymer (100 mg) was dissolved in dry dioxan (15 ml) and 0·1N-sodium methoxide in methanol (2 ml). The solution or fine suspension was agitated in a stoppered flask for 21 h at room temperature and then precipitated into 250 ml of methanol saturated with carbon dioxide. The polymer was collected, washed with water and methanol, and dried *in vacuo* at room temperature.

We thank Dr. G. A. F. Roberts for the preparation of 1,2,3,4-tetra-*O*-acetyl glucopyranose, Dr. M. G. Rayner for n.m.r. measurements, and the Gillette Company for an International Research Fellowship (to J. S.).

[1/175 *Received, February* 24*th,* **1971**]

SOLID-PHASE SYNTHESIS OF OLIGOSACCHARIDES.
II.[*] SYNTHESIS OF 2-ACETAMIDO-6-O-(2-ACETAMIDO-2-DEOXY-β-D-GLUCOPYRANOSYL)-2-DEOXY-D-GLUCOSE.

G. EXCOFFIER, D. GAGNAIRE, J.P. UTILLE and M. VIGNON

Centre de Recherche sur les Macromolécules Végétales, Centre National de la Recherche Scientifique, Domaine Universitaire BP 53, 38 041 Grenoble, France.

(Received in UK 23 October 1972; accepted for publication 7 November 1972)

The use of a solid support for the stepwise synthesis of oligosaccharides is, in principle, a promising technique, but it has only been attempted quite recently[1-4] and is subject to numerous difficulties.

As part of a general study[5] on the adaptation of the Merrifield type of synthesis on a polymer support[6] for use in sugar chemistry, we describe here the synthesis, by use of an insoluble, modified, polystyrene resin, of the known[7-9] disaccharide 2-acetamido-6-O-(2-acetamido-2-deoxy-β-D-glucopyranosyl)-2-deoxy-D-glucose (8).

Following the procedure of Letsinger[10], a 0.17% cross linked styrene-divinylbenzene "popcorn" polymer (1 g) having acid chloride functional groups (1.7 mmole/g of resin) was treated in a dry mixture of benzene (10 ml) and pyridine (2 ml) for 48 h at room temperature with benzyl-2-acetamido-2-deoxy-4,6-O-benzylidene-α-D-glucopyranoside[11] (1, 0.452 g) to attach the polymer to 1 through O-3 ; after washing the resin successively with benzene, chloroform, dioxane, 1:1 dioxane-water, acetone, methanol and ether, product 2 contained (according to the gain in weight) 1 mmole of 1 per g of resin. Residual acid chloride groups were esterified with methanol (1 ml) for 12 h. The benzylidene groups were split off by treating 2 with 9:1 trifluoroacetic acid-water[12] to give the polymer-fixed derivative 3 having O-4 and O-6 free. Glycosylation of 3 was effected either (a) with 2-methyl-4,5-(3,4,6-tri-O-acetyl-2-deoxy-α-D-glucopyrano)-2-oxazoline[13] (4, 3 moles per mole of 3) in a mixture of nitromethane (7 ml) and toluene (10 ml) containing a catalytic amount of p-toluenesulfonic acid (20 mg) for 1.5 h at 120° (conditions of Zurabyan[14]) or (b) with 2-acetamido-3,4,6-tri-O-acetyl-2-deoxy-α-D-glucopyranosyl chloride[15]

---

[*] Part I, see ref. 5.

(**5**, 2 moles per mole of **3**) in benzene containing mercuric cyanide (1 mole per mole of **5**) for 48 h at 60°. The resulting product was presumed to be the polymer-fixed, protected β-D-(1→6)-linked dissacharide **6**, on the basis of the favored stereochemistry of glycosylation reactions[16] of **4** and **5**, and the higher reactivity of primary hydroxyl groups (6-OH) over secondary ones (4-OH). At the elevated temperature of procedure (a) the polymer tended to become partially solubilized, decreasing the net yield, but three successive treatments led to a product whose gain in weight, allowing for loss of polymer by solubilization, indicated a yield of 80% in the coupling reaction.

Detachment of the protected disaccharide from the resin was effected by suspending **6** in 1:1 methanol-toluene saturated with ammonia or, more effectively, by use of 0.5 M sodium methoxide in methanol (2 ml) in 1:1 methanol-p-dioxane* (8 ml) for 1.5 h at 40°C ; sodium ions were subsequently removed with Amberlite IR-120 (H$^+$). The mixture was filtered, the filtrate evaporated, and the chloroform insoluble portion of the residue was acetylated (acetic anhydride-pyridine) to give benzyl 2-acetamido-6-O-(2-acetamido-3,4,6-tri-O-acetyl-2-deoxy-β-D-glucopyranosyl)-2-deoxy-α-D-glucopyranoside (**7**) yield 420 mg (58% based on **1**), m.p. 242° (from chloroform-ether), $[\alpha]_D^{25}$ +65° (c 1 in chloroform); m/e 364 (1.1% of base peak m/e 330) attributable[17] to ion A ; n.m.r. (CDCl$_3$: δ 5.65 and 5.99 bd, $J_{2,NH}$ 8.5 and 9.5 Hz, disappear slowly on deuteration (NHAc of disaccharide) ; 4.9 d, $J_{1,2}$ 3.5 Hz (H-1 equatorial of benzyl glycoside moiety) ; 4.5 d, $J_{1',2'}$ 8 Hz (H-1' axial of interglycosidic link) ; 4.93-5.4 (triplets, J 8-9 Hz, 4 protons diaxially disposed, H-3, H-4, H-3', H-4').

Anal. Calc. for C$_{33}$H$_{44}$N$_2$O$_{16}$: C 54.68, H 6.13, N 3.87, O 35.32.
Found : 54.70, 5.98 3.98 35.33.

These data in comparison with previous n.m.r. spectral analyses of amino sugar derivatives[18], indicate that the interglycosidic link in **7** is β-D through position O-6 (and not through O-4).

A second product formed in the coupling reaction by procedure (b) appeared to be a trisaccharide derivative, presumably the product glycosylated at O-6 and O-4.

Conventional O-deacetylation of **7** with methanolic ammonia, followed by hydrogenolytic cleavage of the benzyl glycoside group, gave the known disaccharide **8**[7-9] m.p. 200° (from ethanol), $[\alpha]_D^{25}$ +8° (equilibrium) (c 0.3 in water).

---

* As some carboxyl groups of the functionalized polymer were not transformed into acid chlorides, an excess of sodium over the usual quantities had to be used.

Although the yields here recorded are probably considerably lower than optimal yields achievable, the results do show that the solid-support technique does provide a valid procedure for stepwise oligosaccharide synthesis applicable with amino sugars.

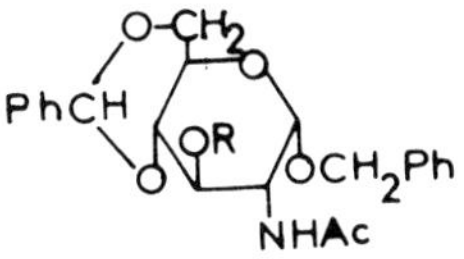

1 R = H
2 R = polymer-CO-

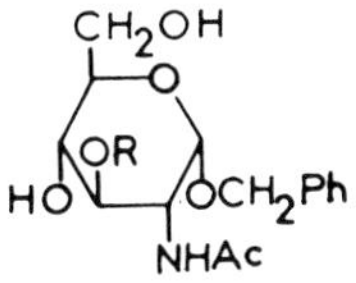

3 R = polymer-CO-

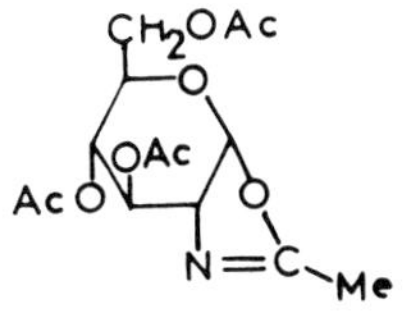

4

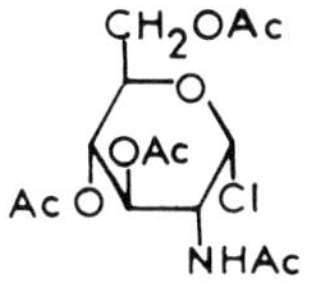

5

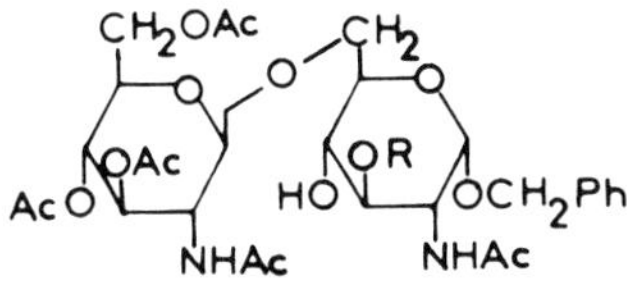

6 R = polymer-CO-

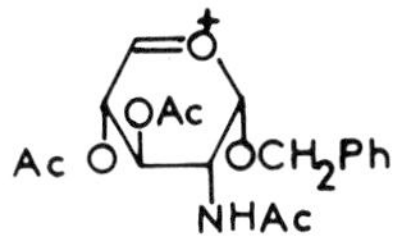

A: m/e 364

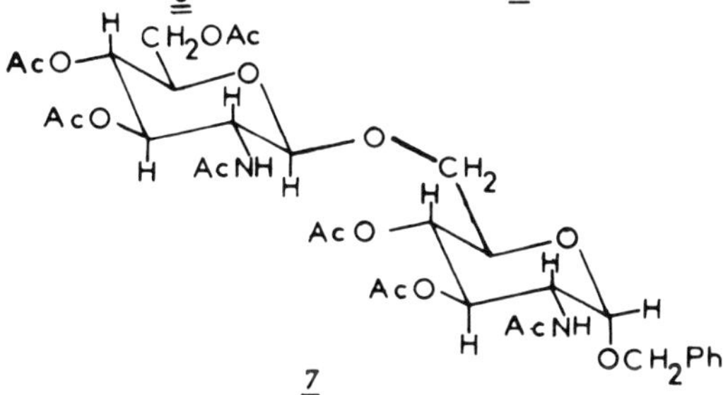

7

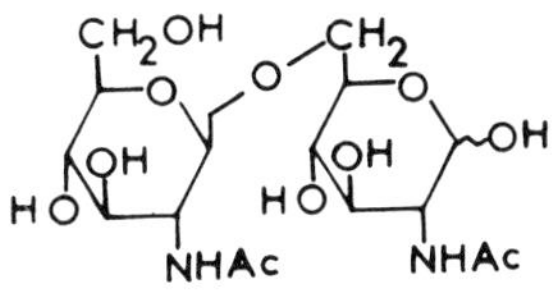

8

Acknowledgements.

We thank Professor Derek Horton for stimulating discussions and his help for the redaction of the manuscript.

REFERENCES

1. J.M. FRECHET and C. SCHUERCH, J. Amer. Chem. Soc., **93**, 492 (1971).

2. R.D. GUTHRIE, A.D. JENKINS and J. STEHLICEK, J. Chem. Soc. (C), 2690 (1971).

3. J.M. FRECHET and C. SCHUERCH, J. Amer. Chem. Soc., **94**, 604 (1972).

4. J.M. FRECHET and C. SCHUERCH, Carbohyd. Res., **22**, 399 (1972).

5. N. BELORIZKY, G. EXCOFFIER, D. GAGNAIRE, J.P. UTILLE, M. VIGNON and P. VOTTERO, Bull. Soc. chim. Fr., in press.

6. R.B. MERRIFIELD, J. Amer. Chem. Soc., **85**, 2149 (1963).

7. A.B. FOSTER and D. HORTON, J. Chem. Soc., 1890 (1958).

8. YU WANG and HSING-I TAI, Hua Hsueh Hsueh Pao, **25**, 50 (1959) ; Chem. Abstr., **54**, 6561a (1960).

9. D. BUNDLE and N. SHAW, Carbohyd. Res., **21**, 211 (1972).

10. R.L. LETSINGER, M.J. KORNET, V. MAHADEVAN and D.M. JERINA, J. Amer. Chem. Soc., **86**, 5163 (1964).

11. P.H. GROSS and R.W. JEANLOZ, J. Org. Chem., **32**, 2759 (1967).

12. J.E. CHRISTENSEN and L. GOODMAN, Carbohyd. Res., **7**, 510 (1968).

13. A.J. KHORLIN, M.L. SHUL'MAN, S.E. ZURABYAN, I.M. PRIVALOVA and Y.L. KOPAEVICH, Izv. Akad. Nauk. S.S.S.R. Ser. Khim., **227**, 2094 (1968).

14. S.E. ZURABYAN, T.P. VOLOSYUK and A.J. KHORLIN, Carbohyd. Res., **9**, 215 (1969).

15. D. HORTON, Organic Syntheses, **46**, 1 (1966).

16. D. HORTON, in R.W. JEANLOZ and E.A. BALAZS (Eds.), The Amino sugars, Volume IA, Academic Press Inc., New York, p. 97 ff (1969).

17. N.K. KOCHETKOV and O.S. CHIZHOV, Advan. Carbohyd. Chem., **21**, 39 (1966).

18. D. HORTON, J.B. HUGHES, J.S. JEWELL, K.D. PHILLIPS and W.N. TURNER, J. Org. Chem., **32**, 1073 (1967).

Copyright © 1973 by the American Chemical Society

Reprinted from *J. Amer. Chem. Soc.*, **95**(17), 5673–5677 (1973)

Oligosaccharide Synthesis on a Light-Sensitive Solid Support. I. The Polymer and Synthesis of Isomaltose (6-O-α-D-Glucopyranosyl-D-glucose)[1]

Uri Zehavi and Abraham Patchornik

Contribution from the Department of Organic Chemistry, The Weizmann Institute of Science, Rehovot, Israel. Received November 27, 1972

Abstract: A light-sensitive polymer was synthesized by attaching 6-nitrovanillin, through an ether linkage, to a chloromethylated styrene–divinylbenzene copolymer. The aldehyde functions on the polymer were reduced by sodium borohydride to alcohols, HO–[PNV], which were utilized for condensation. The polymer was filtered and analyzed after every synthetic step. 6-O-p-Nitrobenzoyl-2,3,4-tri-O-benzyl-β-D-glucopyranosyl bromide was allowed to react with the resin in benzene with pyridine as a base to give 6-O-p-nitrobenzoyl-2,3,4-tri-O-benzyl-D-glucopyranosyl-O-[PNV]. The p-nitrobenzoate was removed by sodium ethoxide in ethanol–dioxane, leaving the 6-OH free for a similar condensation with another molecule of the same bromide. Following this reaction the p-nitrobenzoyl group was removed, resulting in a resin to which substituted di- and monosaccharide units were attached. Irradiation (>320 nm) of the polymer, suspended in dioxane, released the saccharide, leaving a resin showing typical aldehyde absorption at 1720 cm^{-1}. The soluble products were reduced (Pd) to give isomaltose and glucose, which were separated chromatographically and identified.

The chemical synthesis of oligosaccharides is of fundamental importance for biological and chemical applications in many fields of contemporary interest including immunology, enzymology, and pharmacology. It is also of great importance in structure determination of oligo- and polysaccharides. The synthesis of oligosaccharides involves, however, complicated processes which make it difficult to tackle a variety of problems relating to these substances, including those of medicinal relevance. In many cases it is difficult, or impossible, to synthesize conventionally a desired structure. In many more cases the yields are low and the isolation of products becomes tedious.

The synthesis of oligosaccharides on a solid support, in general,[2] is a simple technique having the advantage that after every condensation or deblocking step the resin simply has to be washed to remove impurities before proceeding with the next step. There is also the possibility of using a great excess of soluble reagents in order to push the reaction closer to completion. In this case, the excess reagents can be recovered, if desired. One drawback to this method, however, is that mistakes in sequence occur due to incomplete reactions. Furthermore, oligosaccharide purification can only be carried out after detachment of the product from the polymeric carrier. In addition, one has to anticipate the possible formation of both α- and β-anomers.

Initial work in solid phase synthesis of oligosaccharides was carried out by two groups. Fréchet and Schuerch[3,4] have prepared substituted di- and trisaccharides attached to a polymer *via* a glycoside linkage. Following ozonolysis the saccharide was released from the resin as the 2-ethanal glycoside that could be reduced to the 2-hydroxyethyl glycoside. The method did not lead, however, to free oligosaccharides. Guthrie, Jenkins, and Stehlicek[5] have synthesized a soluble polymer to which a 1,2-ortho ester or 1-bromo derivative of glucose is bound. Such a polymeric reagent could, hopefully, serve in oligosaccharide synthesis.

In view of this situation, we were encouraged to try a new approach: the synthesis of oligosaccharides on a light-sensitive solid support. The synthetic route is similar to that of Fréchet and Schuerch,[3] except that we hoped to achieve a smooth release of the saccharide from the polymer as a reducing sugar by means of a photochemical reaction. The possibility of designing a resin for this purpose results from our study on 2-nitrobenzyl glycosides. These can be photolyzed in very high yields to the parent reducing sugars under conditions that do not affect O-benzyl groups.[6] Thus, quantitative release of D-glucose was observed following the irradiation of 2-nitrobenzyl β-D-glucopyranoside or 6-nitroveratryl β-D-glucopyranoside (Scheme I). After presenting this work,[1] we learned of another approach to oligosaccharide synthesis on a solid support.[7]

Results and Discussion

An aldehydo polymer (**1**) was synthesized by attaching 6-nitrovanillin[8] through an ether linkage to a chloromethylated styrene–divinylbenzene copolymer.[9] In addition to the binding of 6-nitrovanillin, some

---

(1) A preliminary account of this work has appeared: U. Zehavi and A. Patchornik, Abstracts, 164th National Meeting of the American Chemical Society, New York, N. Y., Aug 1972, No. CARB-27; Abstracts, 42nd Meeting of the Israel Chemical Society, Rehovot, Israel, Dec 1972, p 61.
(2) J. M. Stewart and J. D. Young, "Solid Phase Peptide Synthesis," W. H. Freeman, San Francisco, Calif., 1969.
(3) J. M. Fréchet and C. Schuerch, *J. Amer. Chem. Soc.*, **93**, 492 (1971).
(4) J. M. Fréchet and C. Schuerch, *Carbohyd. Res.*, **22**, 399 (1972).

(5) R. D. Guthrie, A. D. Jenkins, and J. Stehlicek, *J. Chem. Soc. C*, 2690 (1971).
(6) U. Zehavi, B. Amit, and A. Patchornik, *J. Org. Chem.*, **37**, 2281 (1972); U. Zehavi and A. Patchornik, *ibid.*, **37**, 2285 (1972).
(7) In a personal communication by Professor L. Anderson, he mentioned initial attempts towards an oligosaccharide synthesis on a solid support employing a similar sequence to ours, but using an S-glycoside as the temporary hook to the polymer. Recently, the scheme was described for a model reaction sequence in solution: P. J. Pfäffli, S. H. Hixson, and L. Anderson, *Carbohyd. Res.*, **23**, 195 (1972).
(8) M. Julia, P. Manoury, and C. Voillaume, *Bull. Soc. Chim. Fr.*, **47**, 1417 (1965).
(9) R. B. Merrifield, *J. Amer. Chem. Soc.*, **85**, 2149 (1963).

Scheme I

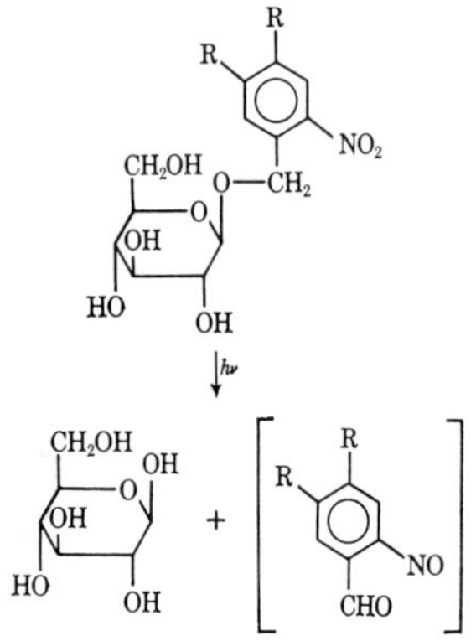

R = H, OCH₃

triethylamine, the base used in the reaction, was also bound as triethylammonium chloride. The aldehyde functions on the polymer were reduced to alcohols (2) by sodium borohydride which could be condensed with a sugar bromide. The attachment of 6-nitrovanillin to the polymer, the reduction step, and further synthetic steps were followed by ir spectroscopy and analysis. The data were in accord with the proposed structures. The substituted 6-nitroveratryl glycoside prepared thereby, as well as compound **2** itself, is light sensitive, as can be expected from the behavior of their low molecular weight analogs.[6]

Scheme II

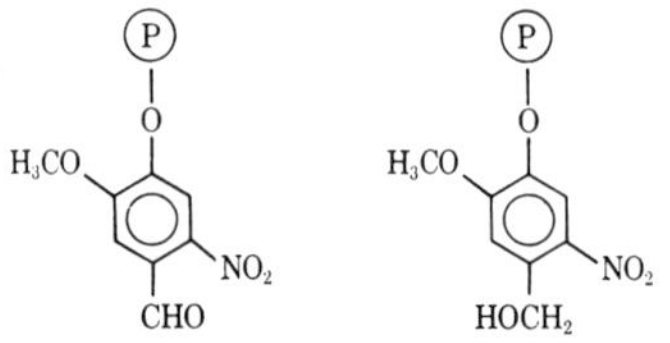

(P) = styrene-divinylbenzene copolymer (CH₂)

Thus, 6-O-p-nitrobenzoyl-2,3,4-tri-O-benzyl-β-D-glucopyranosyl bromide[10] reacts with the resin (**2**) in benzene, with pyridine as a base, to give a polymer to which 6-O-p-nitrobenzoyl-2,3,4-tri-O-benzyl-D-glucopyranosyl units are attached (**3**). The p-nitrobenzoate was removed by sodium ethoxide in ethanol–dioxane, leaving the 6-OH free (compound **4**) for a similar condensation with another molecule of the same bromide (elongation), yielding compound **5**. In principle, applying to compound **5** another cycle of deacylation and elongation would have led to a trisaccharide derivative. In the present work we describe only the deacylation of compound **5** to give the isomaltose derivative **6**.

Irradiation of polymer **6** suspended in dioxane at wavelengths longer than 320 nm released a mixture of benzylated saccharides and left an aldehydo polymer (**7**). The benzylated saccharides were reduced with palladium to give a mixture of D-glucose and isomaltose (6-O-α-D-glucopyranosyl-D-glucose). Hence, the synthesis of oligosaccharides on a light-sensitive solid support is feasible and the method has led to an α-glucoside (compare ref 10, 3, and 4).

The glucose and isomaltose obtained by synthesis were separated by preparative paper chromatography, and the former was estimated by its oxidation with glucose oxidase[11] and by the phenol–sulfuric acid test.[12] The synthetic isomaltose migrated on paper chromatography with an R_f value identical with that of an authentic sample of isomaltose, although the R_f is similar to that of gentibiose (6-O-β-D-glucopyranosyl-D-glucose). It could, however, be differentiated from gentibiose by glc of its trimethylsilyl derivative, by its susceptibility to α-glucosidase but not to β-glucosidase (emulsin),[13] and by its optical rotation (108°). The optical rotation reported for isomaltose is +119-+122° (ref 14) and for gentiobiose is +10.5° (ref 15).

The isolation of a mixture of glucose and isomaltose at the end of the synthesis, as well as the spectrophotometric determination of p-nitrobenzoate released upon deacylation of compounds **3** and **5**, indicated that the two condensation steps were incomplete. The photochemical cleavage of 2-nitrobenzyl glycosides in solution proceeds in very high yields.[6] The photolysis of resin-bound saccharide derivatives proceeds likewise, in analytical scale, in very high yields, although after much longer irradiation times. Irradiation of resin-bound saccharide derivatives in preparative scale is, apparently, not as effective and the next step of catalytic hydrogenolysis did not approach completion. Spots corresponding to partially benzylated saccharides can normally be observed on paper chromatograms.

A plausible mechanism of the photochemical reaction was already discussed by us to some extent.[6] It involves an intramolecular oxidation–reduction formation of a hemiacetal which splits to the reducing sugar and the nitroso aldehyde.

The main obstacle in the synthesis remains the incomplete condensation steps. In a paper that appeared after the completion of this work,[4] a few possibilities to eliminate this problem were described. These and others are presently being studied. Another obstacle, the partial hydrogenolysis of benzyl ethers, might be overcome by using different procedures for their removal or might be avoided altogether by using blocking groups other than benzyl ethers. One possibility might be 2-nitrobenzyl ethers, which should be cleaved during the irradiation of the polymer.

It is hoped that after improvements, refinement, and widening of the scope, oligosaccharide synthesis on a light-sensitive solid support will become a method of choice.

Experimental Section

Optical rotations were determined with a Bendix polarimeter. Ir spectra were measured with a Perkin-Elmer Model 237 spectro-

(10) T. Ishikawa and H. G. Fletcher, Jr., *J. Org. Chem.*, **34**, 563 (1969).

(11) E. Raabo and T. C. Terkildsen, *Scand. J. Clin. Lab. Invest.*, **12**, 402 (1966).
(12) M. Dubois, K. A. Gilles, J. K. Hamilton, P. A. Rebers, and F. Smith, *Anal. Chem.*, **28**, 350 (1956).
(13) U. Zehavi and A. Patchornik, to be published.
(14) R. W. Bailey, "Oligosaccharides," Pergamon, Oxford, 1965, p 67.
(15) A. Thompson and M. L. Wolfrom, *J. Amer. Chem. Soc.*, **75**, 3605 (1953).

Scheme III

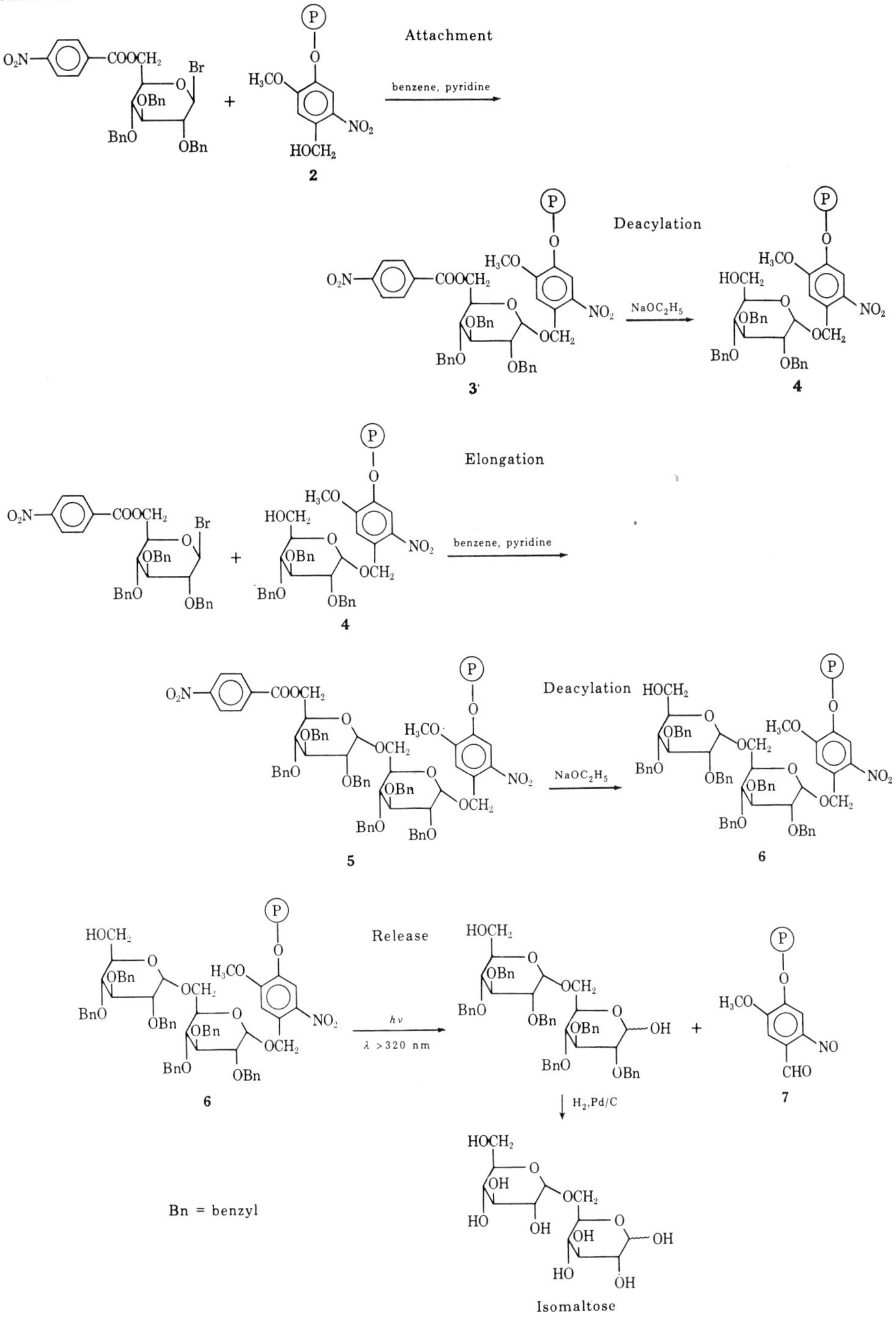

Bn = benzyl

Synthesis of Isomaltose

Scheme IV

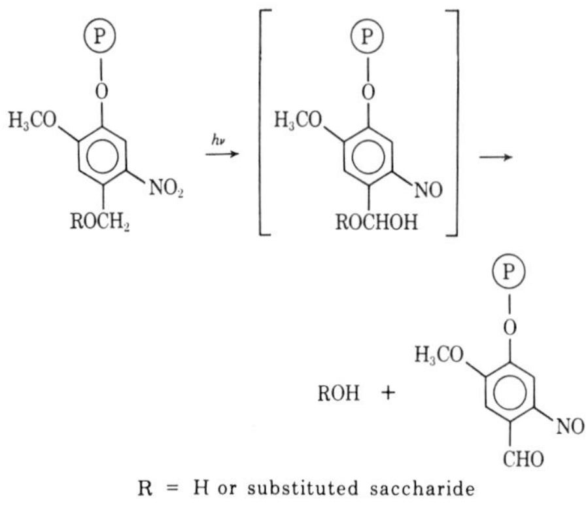

$$H_3CO-\underset{ROCH_2}{\underset{NO_2}{\bigcirc}}-P \xrightarrow{h\nu} \left[H_3CO-\underset{ROCHOH}{\underset{NO}{\bigcirc}}-P \right] \longrightarrow$$

$$ROH + H_3CO-\underset{CHO}{\underset{NO}{\bigcirc}}-P$$

R = H or substituted saccharide

photometer using KBr disks. Uv spectra were taken on a Cary Model 14 spectrophotometer. Colorimetric measurements were performed on a Klett–Summerson colorimeter equipped with filter no. 50. Descending paper chromatography was performed on Schleicher and Schull 2040a paper for qualitative work and on Whatman no. 3 MM paper, previously washed with water and methanol, for preparative work. The paper was developed with descending n-butyl alcohol–acetic acid–water (25:6:25, v/v/v, upper phase) (I) or with descending ethyl acetate–pyridine–water (2:1:2, v/v/v, upper phase) (II). The sugars were detected by silver nitrate.[16] The gas chromatograph Packard Model 7821 was used in conjunction with the Model 811 hydrogen flame ionization detector and 2 m × 4 mm silanized coiled glass column. The column packing used was 1% QS-1 on Gas Chrom Q of 80–100 or 100–120 mesh (Applied Science Laboratories; ref 17). Standards and samples were dissolved in "Sil-Prep" (Applied Science Laboratories) 30 min before injection. Photolysis was carried out in a RPR-100 apparatus (Rayonet, The Southern Co., Middletown, Conn.) with 320-nm lamps. α-Glucosidase (from yeast) and glucose oxidase (fungal) were purchased from Sigma Chemical Co. Peroxidase was a product of Boeringer and Sohne and β-glucosidase (emulsin) was obtained from Worthington Corp. Isomaltose (Sigma) and gentiobiose (Calbiochem) served as standards after being purified by preparative paper chromatography (System II).

6-Nitrovanillin. This compound was prepared as described in the literature[8] except that the benzyl ether was cleaved by hydrobromic acid. Thus, 4-O-benzyl-6-nitrovanillin (8.0 g) was dissolved in acetic acid (50 ml) at 85°. A 33% solution of hydrobromic acid in acetic acid (18 ml) was added and the solution was kept, with a condenser and a calcium chloride seal, for 10 min in a 85° bath. The reaction mixture was then cooled to room temperature and the product precipitated. It was collected by filtration, washed with boiling water, dissolved in boiling ethanol, treated with some active charcoal, and filtered, and the resulting solution was left for crystallization: yield, 3.0 g (55% pure by tlc), mp 208° (lit.[8] 208°).

The Resin. Chloromethylated (6.60% Cl) 2% cross-linked styrene–divinylbenzene copolymer[9] (3.3 g, 200–400 mesh) and 6-nitrovanillin (1.77 g) were suspended in dioxane (40 ml). Triethylamine (0.9 ml) was added and the reaction was stirred magnetically for 24 hr in a 85° bath under reflux and with a calcium chloride seal. Triethylamine (0.9 ml) was added once again and the reaction was continued, under the same conditions, for an additional 3 days. The brown resin (1) was collected on a sintered glass filter and washed with dioxane and methanol: yield, 4.14 g (Thr 4.32 g). The ir spectrum of the polymer (1) had absorption bands (CH) characteristic of the styrene–divinylbenzene copolymer moiety at 3025, 2915, 1495, 1455, and 1390 cm^{-1} and bands due to the nitrovanillin moiety at 1680 (C=O), 1570 and 1330 cm^{-1} (NO$_2$).

Anal. Found: N, 2.06; Cl, 2.03.

(16) N. Sharon and R. W. Jeanloz, *J. Biol. Chem.*, **235**, 1 (1960).
(17) E. C. Horning, W. J. A. Vanden Heuvel, and B. G. Creech, *Methods Biochem. Anal.*, **11**, 69 (1963).

The aldehydo polymer (1) (3.0 g) was suspended in dimethylformamide (200 ml). Sodium borohydride (1 g) was added and the mixture stirred for 90 min. The reaction was then stopped by the addition of 30% acetic acid (60 ml) and the resin was subsequently washed with dimethylformamide, water, dioxane, ethanol containing 1.5% of triethylamine, ethanol–dioxane, and methanol: yield, 3.0 g of brown resin (2, HO– [PNV]). The resin had lost the carbonyl absorption at 1680 cm^{-1} present in compound 1.

Anal. Found N, 1.96; Cl, 0.194; OMe, 2.48.

According to the methoxy analysis the resin had 0.8 mmol/g of bound nitroveratrol.

Irradiation for 18 hr of the resin (10 mg) suspended in dioxane (4 ml) did not release any significant amount of chromophore into the solution (measured at 258 nm), while the resulting yellow resin had acquired a new absorption at 1720 cm^{-1} (C=O).

Anal. Found: N, 2.17.

6-O-p-Nitrobenzoyl-2,3,4-tri-O-benzyl-β-D-glucopyranosyl Bromide.[10] The bromo derivative was either isolated in its crystalline form or the methylene chloride solution[10] was diluted and washed with water, saturated sodium bicarbonate solution, and water, dried over sodium sulfate, evaporated under vacuum, and dissolved in benzene.

6-O-p-Nitrobenzoyl-2,3,4-tri-O-benzyl-D-glucopyranosyl-O-[PNV] (3). Resin 2 (2.0 g) was suspended, with stirring, in benzene (30 ml) containing 6-O-p-nitrobenzoyl-2,3,4-tri-O-benzyl-β-D-glucopyranosyl bromide (prepared from 1,6-di-O-p-nitrobenzoyl-2,3,4-tri-O-benzyl-β-D-glucopyranoside,[10] 3 g, 2.5 molar equiv). Pyridine (0.36 ml) in benzene (5 ml) was added and the stirring was continued for 40 hr in a 80° bath and under calcium chloride seal. The product was collected by filtration on a sintered glass filter, and was washed with benzene, dioxane, and methanol: yield, 2.62 g (Thr 2.93 g). Compound 3 had a strong absorption at 1730 cm^{-1} (C=O). The amount of saccharide attached to the resin was 0.34 mmol/g (62%), as determined by transesterification, and 0.38 mmol/g (69%), determined by photolysis (Thr 0.55 mmol/g).

Anal. Found: N, 2.02.

Estimation of the Amount of Saccharide Attached to the Polymer (Compounds 3 and 5). a. Estimation as p-Nitrobenzoate Following Transesterification. The resin (2–3 mg) was suspended, by stirring, in a solution (1 ml) composed of dioxane and 0.2 M sodium ethoxide in ethanol (2:1). The stirring was continued overnight in a closed tube. The suspension was then filtered through a glass fiber filter (Whatman GFC), and diluted tenfold with the same solution and the absorbance at 272 nm was determined. p-Nitrobenzoic acid has λ_{max} 272 nm (ϵ 0.97 × 10^4) in this solution.

b. Estimation Following Photolysis. The resin (2–4 mg) was suspended, by stirring, in dioxane (1 ml) in a closed Pyrex test tube and irradiated for 12 hr. The suspension was then filtered, the solution was diluted 20-fold, and the absorbance at 258 nm was determined. 1,6-Di-O-p-nitrobenzoyl-2,3,4-tri-O-benzyl-β-D-glucopyranoside absorbs at 258 nm (ϵ 2.42 × 10^4 in dioxane). It is assumed that the released saccharide has half this extinction.

2,3,4-Tri-O-benzyl-D-glucopyranosyl-O-[PNV] (4). Compound 3 (1.0 g) was mixed in a solution (40 ml) composed of dioxane and 0.2 M sodium ethoxide in ethanol (2:1) for 18 hr in a closed erlenmeyer flask. The product (4) was then collected by filtration on a sintered glass filter, and washed with dioxane and methanol: yield, 0.84 g (Thr 0.94 g). The carbonyl absorption present in compound 3 (1730 cm^{-1}) is absent in compound 4. Assuming quantitative transesterification of compound 3 and taking into account the theoretical decrease in weight of resin, the calculated amount of saccharide attached to resin 4 is 0.38 mmol/g.

Anal. Found: N, 1.63.

6-O-(6-O-p-Nitrobenzoyl-2,3,4-tri-O-benzyl-α-D-glucopyranosyl)-2,3,4-tri-O-benzyl-D-glucopyranosyl-O-[PNV] (5). Compound 4 (0.5 g) was suspended, by stirring, in benzene (10 ml) containing 6-O-p-nitrobenzoyl-2,3,4-tri-O-benzyl-β-D-glucopyranosyl bromide that was prepared from 1,6-di-O-p-nitrobenzoyl-2,3,4-tri-O-benzyl-β-D-glucopyranoside (1.5 g, 5 molar equiv). Pyridine (0.2 ml) was added and the stirring was continued for 48 hr in a 65° bath and under calcium chloride seal. The product was collected by filtration on a sintered glass filter and washed with benzene, dioxane, and methanol: yield, 0.65 g (Thr 0.61 g). The ir spectrum showed a carbonyl absorption at 1730 cm^{-1}. The amount of saccharide attached to the resin was 0.29 mmol/g (93%) as determined by transesterification and 0.27 mmol/g (88%) as determined by photolysis (Thr 0.3 mmol/g, arbitrarily assuming the formation of the disaccharide derivative only).

Anal. Found: N, 2.16.

6-O-(2,3,4-Tri-O-benzyl-α-D-glucopyranosyl)-2,3,4-tri-O-benzyl-D-glucopyranosyl-O-[PNV] (6) was prepared from compound **5** (0.197 g) as described for the preparation of compound **4**: yield, 0.160 g (Thr, 0.187 g). The carbonyl absorption at 1730 cm^{-1} is not present in this compound. Assuming quantitative transesterification of compound **5** and taking in account the theoretical decrease in weight of resin, the calculated amount of saccharide attached to resin **6** is 0.32 mmol/g.

Anal. Found: N, 1.58; OCH$_3$, 2.36.

Release of Saccharide. Resin **6** (0.2 g) was suspended, by stirring, in dioxane (250 ml) and irradiated for 32 hr. The resin (**7**) was filtered off (it had an absorption at 1720 cm^{-1}) and the solution was evaporated *in vacuo* to a third of its original volume. Water (100 ml), acetic acid (1 drop), and a small amount of 10% palladium on charcoal were added and the solution was hydrogenated for 18 hr at room temperature and at 46 psi. The catalyst was then filtered off and the solution evaporated *in vacuo*. The residue was dissolved in water (10 ml) and was checked (40–100-μl samples) by paper chromatography (in the two solvent systems indicated) and was shown to contain a material migrating like isomaltose, a material migrating like glucose, and a few additional spots, probably corresponding to partially benzylated products. A portion of the aqueous solution (1.0 ml) was freeze dried and separated by preparative paper chromatography (system II). The products were eluted from the paper as follows. The bands containing glucose and isomaltose were cut into slices and placed into the outer part of a 5-ml disposable syringe that had been plugged with some washed cotton. The paper was then wet with water and the syringe centrifuged inside a 12-ml conical tube. The last step was repeated several times using a total of about 2.5 ml of water. Finally the yield of products was determined by the phenol–sulfuric acid test: glucose, 0.058 mmol/g (17.7%); isomaltose, 0.041 mmol/g (12.5%). Yields are based on the saccharide content of compound **6**. The yield of free disaccharide, isomaltose, remains similar (10.8%) also when calculated on the basis of the monosaccharide derivative **4**. Isomaltose was obtained on freeze drying. The trimethylsilyl derivative of the synthetic isomaltose was compared to trimethylsilylated isomaltose and gentiobiose by glc (22 psi, 178°) and was shown to contain the peaks corresponding to isomaltose at retention time 13.7 and 18.5 min. No evidence for gentiobiose (10% contamination should have been detected), retention time 16.8 and 18.5 min, was obtained. The synthetic isomaltose had $[\alpha]^{23}$D + 108° (c 0.14, water) and its purity was evaluated by a method developed for the determination of glucose, isomaltose, and gentiobiose in admixture.[13] A solution of the synthetic isomaltose (120 μg/ml, by phenol–sulfuric acid test minus a small amount of contaminating glucose determined with glucose oxidase) was digested by α-glucosidase and the glucose obtained was determined with glucose oxidase (97 μg/ml). When the same was repeated with β-glucosidase, no evidence for digestion was obtained.

Acknowledgments. The authors express their gratitude to Mrs. S. Ehrlich-Rogozinsky for the microanalysis and to Miss D. Levinson and Mr. I. Jacobson for the preparation of starting materials. This work was supported by Grant AM 05098 from the National Institutes of Health, Public Health Service.

VI

Synthetic Applications of the Solid-State Method

Editors' Comments on Papers 40 Through 53

40 **Camps, Castells, Ferrando, and Font:** *Organic Syntheses with Functionalized Polymers: I. Preparation of Polymeric Substrates and Alkylation of Esters*

41 **Camps, Castells, Font, and Vela:** *Organic Syntheses with Functionalized Polymers: II. Wittig Reaction with Polystyryl-p-Diphenylphosphoranes*

42 **McKinley and Rakshys:** *Wittig Resins: The Preparation and Application of Insoluble Polymeric Phosphoranes*

43 **Weinshenker and Shen:** *Polymeric Reagents: I. Synthesis of an Insoluble Polymeric Carbodiimide*

44 **Weinshenker and Shen:** *Polymeric Reagents: II. Preparation of Ketones and Aldehydes Utilizing an Insoluble Carbodiimide*

45 **Shambhu and Digenis:** *Insoluble Resins in Organic Synthesis: I. Preparation and Reactions of Polymeric Anhydrides*

46 **Yaroslavsky, Patchornik, and Katchalski:** *Unusual Brominations with N-Bromopolymaleimide*

47 **Grubbs and Kroll:** *Catalytic Reduction of Olefins with Polymer-Supported Rhodium(I) Catalyst*

48 **Grubbs, Gibbons, Kroll, Bonds, and Brubaker:** *Activation of Homogeneous Catalysts by Polymer Attachment*

49 **Collman, Hedegus, Cooke, Norton, Dolcetti, and Marquardt:** *Resin-Bound Transition Metal Complexes*

50 **Čapka, Svoboda, Černý, and Hetflejš:** *Hydrogenation, Hydrosilylation and Hydroformylation of Olefins Catalyzed by Polymer-Supported Rhodium Complexes*

51 **Neckers, Kooistra, and Green:** *Polymer-Protected Reagents: Polystyrene–Aluminum Chloride*

52 **Blossey, Turner, and Neckers:** *Polymer Protected Reagents: II. Esterifications with $\textcircled{P}$-$AlCl_3$*

53 **Blossey, Neckers, Thayer, and Schaap:** *Polymer-Based Sensitizers for Photooxidations*

Reagents Immobilized on Polymeric Substrates

The advantages of immobilizing a reactive intermediate or chemical reagents on a polymer accrue because (1) the products can be easily separated from the spent reactants, and (2) the starting materials may be regenerated on the polymer and the reagent recycled.

There have been several reports on chemical reagents attached to polymer supports and their use in organic synthesis. Reports from Camps and his group, as well as from McKinley and Rakshys, demonstrate that Wittig reagents attached to polymers are useful in synthesis. The idea is to attach the phosphonium salt, and hence an ylid, to the polystyrene–divinylbenzene copolymer bead. The reaction is carried out between the Wittig reagent on the polymer and the aldehyde or ketone in solution.

Both research groups indicate that the polymeric substrates can be prepared either from the available polystyrene–divinylbenzene copolymer following the original Merrifield report, or by copolymerization of the appropriate, respective monomers. Details of the synthetic methods are given in the first Camps paper which follows (Paper 40). That actual methodology involved in the use of the polymer-based reagent is given in both the Camps et al. report (Paper 41) and the paper by McKinley and Rakshys (Paper 42).

Other chemical reagents that have been prepared on a polystyrene–divinylbenzene copolymer include a carbodiimide, used for anhydride formation from carboxylic acids and to convert alcohols to aldehydes as reported by Weinshenker and Shen (Paper 44). This carbodiimide, prepared from chloromethylated polystyrene–divinylbenzene (see Paper 43), appears to give high yields of products in benzene–ether, 2:1. At least in the reported cases, yields using the polymer-based carbodiimide are similar to, or better than, those in solution-phase methods.

Polymer-based acylating agents have also been reported by Shambhu and Digenis in Paper 45.

Still another report, of Yaroslavsky, Patchornik, and Katchalski, describes a polymer-based allylic and benzylic brominating agent (Paper 46).

Immobilization of Catalysts

A second application of polymers in synthetic operations is the use of the polymer as a basis for a heavy metal catalyst. There have been several reports describing this kind of operation, many of which are in the patent literature. Two of the more interesting recent reports were those by Grubbs et al. (Papers 47 and 48), who employed immobilized rhodium complexes as catalysts for hydrogenation. The copolymer beads were prepared by the following sequence of reactions:

$$\text{\textcircled{P}}-CH_2Cl \xrightarrow{LiP(Ph)} 2\text{\textcircled{P}}-CH_2P(Ph)_2 \xrightarrow{RhClLn} \text{\textcircled{P}}-CH_2P(Ph)_2 + RhClLn$$

The complex, like many enzymes, selects less sterically hindered olefins in hydrogenation. Thus, the rate of hydrogenation of 1-hexene is decidedly faster than that of Δ^2-cholestene under identical conditions.

Collman and his coworkers in Paper 49 have also reported use of a polymer-based rhodium catalyst, as have Capka et al. (Paper 50). The catalyst favored by the Czech workers was prepared from highly cross-linked "macroporous" styrene–divinylbenzene copolymer, with the result that the polymer is much less easily penetrated by solvents and reagents than is the more normal, less highly cross linked polymer. This probably accounts for the larger reaction rates observed by Grubbs than by the Czech group when using the catalysts for hydrogenation.

Still another use of polymeric reagents is to protect anhydrous materials from the atmosphere and from moisture. The most thoroughly developed case is for the polymer-protected aluminum chloride reagent prepared from anhydrous aluminum chloride in several synthetic dehydration sequences (Papers 51 and 52).

There are two reports and several isolated inferences in the literature that photosensitizers attached to polymers can be quite useful. The report of Leermakers and James (1967) focuses on the use of a polymer-based carbonyl sensitizer. Although there is some reason to believe that the results of Leermakers and James may well be due to the presence of dissolved monomer, the idea still has a great deal of merit.

Another report, that of Blossey et al. (Paper 53), describes a polymer-based dye, rose bengal, and its use in the formation of singlet oxygen. Polymer-based photosensitizers have several advantages: they are easy to separate from products and reactants, they can be reused, they do not appear to fade or bleach, and they can be used in solvents not normally compatible with the dye sensitizers.

Reference

Leermakers, P. A., and F. C. James (1967). *J. Org. Chem.,* **32,** 2898.

ORGANIC SYNTHESES WITH FUNCTIONALIZED POLYMERS:

I. PREPARATION OF POLYMERIC SUBSTRATES AND ALKYLATION OF ESTERS.

F. Camps, J. Castells, M. J. Ferrando and J. Font

Instituto de Química. Centro de Investigación y Desarrollo. Patronato "Juan de la Cierva". Jorge Girona Salgado s/n. Barcelona-17. Spain.

(Received in UK 26 March 1971; accepted in UK for publication 1 April 1971)

SOME advantages of the use of polymeric substrates in general organic synthesis have been recently explained and exploited[1,2] in the acylation and Dieckmann cyclization of aliphatic esters. This fact has prompted us to report our results in this field.

We have employed two approaches for the preparation of the required functionalized polymer (that of p-(2-hydroxyethyl)polystyrene (I) is given as example):

a) Copolymerization of styrene, p-divinylbenzene (DVB) and p-functionalized styrene, in a ratio depending on the desired cross-linking and "dilution" of the functional group.

Thus, pearl copolymerization of a 1:7 p-vinylphenetyl alcohol[3]: styrene mixture with 2% DVB afforded white solid beads of I which showed free and associated OH stretching bands in the ir spectrum (KBr). I was also obtained by pearl copolymerization of a 1:7 p-vinylphenetyl acetate:styrene mixture with 2% DVB, followed by hydrolysis with aqueous NaOH-dioxane. (It was thought that by avoiding possible intermolecular hydrogen-bond associations of the monomer, a more regular dilution of the functionalization would be attained).

b) Direct functionalization of cross-linked polystyrene by the sequence: p-bromination, metalation and reaction with the appropiate functionalizing agent. A stoichiometric control of the p-bromination step was achieved by using $Tl(AcO)_3$ as catalyst and CCl_4 as solvent[4]. When conventional aromatic bromination catalysts were used unreproducible results were obtained[5].

Thus, 24 g of cross-linked polystyrene and 3.8 g of 84% $Tl(AcO)_3$ were suspended in 400 ml of CCl_4; to this mixture, 4.8 g of bromine were slowly added at room temperature, and after complete decoloration, the polymer was washed with aq.HCl-dioxane, water, methanol, dioxane and ether, and dried. Elementary analysis showed 9.7% of bromine. The brominated polymer was treated at -13°C with excess of n-butyl lithium in toluene, washed with the same solvent, reacted with ethylene oxide in toluene, and hydrolyzed with aq.HCl-dioxane. After washing thoroughly with water-dioxane, methanol and ether, the polymeric beads were dried, showing the same ir spectrum as the hydroxylic polymer obtained by method a).

As a first application of polymeric substrates in organic synthesis, we have studied the direct alkylation of p-(2-hydroxyethyl)polystyrene esters of type $\circledP$-CH$_2$CH$_2$OCOCHR$_1$R$_2$. We selected the particular polymeric alcohol I to avoid the benzylic nature of esters prepared from chloromethylated polystyrene[1,2].

The polymer ester of isobutyric acid (0.2 mmol of ester/g of polymer) was prepared by reaction of I with isobutyryl chloride. This polymer was swelled in benzene and an equivalent amount of sodium triphenylmethylide in ether was slowly added and the mixture stirred until disappearance of the red color of the base (2-3 hrs); then, two equivalents of benzyl chloride in benzene were added and the mixture stirred overnight at room temperature. The polymer was filtered, washed successively with benzene, ether, water-dioxane, dioxane and ether , and dried. Upon hydrolysis in refluxing aq.KOH-dioxane (60 hrs), a mixture of 80% unreacted isobutyric acid and 20% benzyldimethylacetic acid was obtained[6].

At the present we are studying the alkylation of esters having more than one α-hydrogen and the Dieckmann cyclization reaction of esters of type $\circledP$-CH$_2$CH$_2$OCO(CH$_2$)$_n$COOC$_2$H$_5$.

REFERENCES

1. A. Patchornik and M.A. Kraus, J. Amer. Chem. Soc., 1970, 92, 7587.
2. J.I. Crowley and H. Rapoport, J. Amer. Chem. Soc., 1970, 92, 6363.
3. S. Tanimoto and R. Oda, Kogyo Kagaku Zasshi, 1962, 64, 932; C. A., 1962, 57, 4854i.
4. A. McKillop, D. Bromley and E.C. Taylor, Tetrahedron Letters, 1969, 1623.
5. D. Braun and E. Seelig, Chem. Ber., 1964, 97, 3098.
6. B. E. Hudson and Ch. R. Hauser, J. Amer. Chem. Soc., 1940, 62, 2457.

ORGANIC SYNTHESES WITH FUNCTIONALIZED POLYMERS:

II. WITTIG REACTION WITH POLYSTYRYL-p-DIPHENYLPHOSPHORANES.

F. Camps, J. Castells, J. Font and F. Vela

Instituto de Química. Centro de Investigación y Desarrollo. Patronato "Juan de la Cierva". Jorge Girona Salgado s/n. Barcelona-17. Spain.

(Received in UK 15 March 1971; accepted in UK for publication 1 April 1971)

IN connection with a research program undertaken in this laboratory[1] to investigate the scope of the solid phase method in general organic synthesis, we describe herein the results obtained in the Wittig reaction with polymeric ylide III prepared from polystyryl-p-diphenylphosphine[2,3] (I) (Scheme 1).

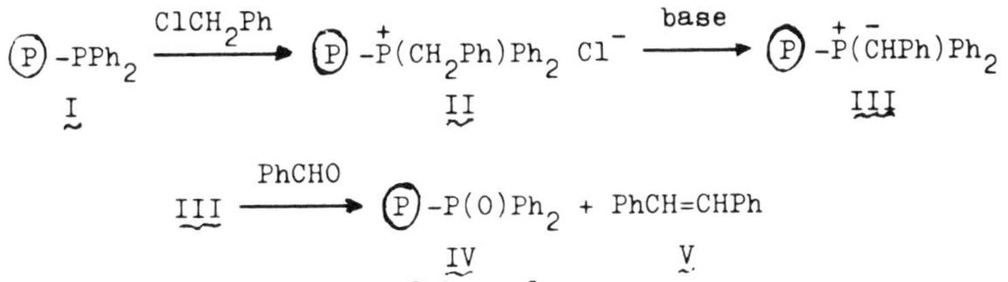

$$\text{(P)}-PPh_2 \xrightarrow{ClCH_2Ph} \text{(P)}-\overset{+}{P}(CH_2Ph)Ph_2\ Cl^- \xrightarrow{\text{base}} \text{(P)}-\overset{+}{P}(\overset{-}{C}HPh)Ph_2$$
$$\text{I} \qquad\qquad \text{II} \qquad\qquad\qquad \text{III}$$

$$\text{III} \xrightarrow{PhCHO} \text{(P)}-P(O)Ph_2 + PhCH=CHPh$$
$$\qquad\qquad\qquad \text{IV} \qquad\qquad \text{V}$$

Scheme 1

The substrate I here employed was a copolymer of 3:1 styrene:p-styryldiphenylphosphine[4], containing 2% divinylbenzene as cross-linking agent, prepared after the pearl technique using water as dispersant and azoisobutyronitrile as radical source. I, obtained in the form of white opaque beads was thoroughly washed with water, dioxane, ether and benzene, and dried over P_2O_5 under high vacuum. Its ir spectrum revealed the presence of aromatic p-substitution and its elementary analysis exhibited the expected percentage of phosphorus.

Treatment of I (60-140 mesh) with benzyl chloride without solvent afforded p-polystyryldiphenyl-benzyl-phosphonium chloride (II) in practically quantitative yield (P, Cl analysis).

A suspension of II in THF was reacted with a stoichiometric amount of base, then an equivalent amount of benzaldehyde was added, and the mixture was allowed to react overnight. The resulting polymeric phosphine oxide IV was removed by filtration and washed thoroughly with THF. The filtrate was evaporated at reduced pressure and the residue chromatographed on silica gel to give variable quantities of cis-, trans-stilbene (V), benzyl alcohol, benzoic acid and unreacted benzaldehyde. A 40% yield of $\underset{\sim}{V}$ was obtained when potassium $\underline{t}$-butoxide was used as base and 60% with sodium hydride.

No substantial yield improvement was observed when an excess of base or aldehyde was used. The removal of the excess of certain bases, like butyllithium, by filtration and repeated washings, previous to the addition of the aldehyde, was difficult, whereby in these cases an increase in the amount of side products, benzyl alcohol and benzoic acid, was observed. The presence of base can promote the formation of these products by a Cannizaro reaction that can compete with the Wittig reaction[5].

Further work is in progress to improve the results obtained and to make them comparable to those reported for the conventional Wittig procedure[6].

REFERENCES

1. F. Camps, J. Castells, M.J. Ferrando and J. Font, <u>Tetrahedron Letters</u>, preceeding paper. A first account of some results contained in Part I and II was given in the XIVth "Reunión Bienal de la Real Sociedad Española de Física y Química" held at Seville in September 1969.
2. D. Braun, <u>Angew. Chem.</u>, <u>1961</u>, 73, 197. R. Rabinowitz and R.W. Marcus, <u>C. A.</u>, <u>1963</u>, 58, P11401f.
3. When this investigation was under way, we learnt from the successful use of polystyryl dimethyl sulfonium methylide, described by S. Tanimoto, J. Horikawa and R. Oda, <u>Kogyo Kagaku Zasshi</u>, <u>1967</u>, 70, 1269; <u>C. A.</u>, <u>1968</u>, 68, 69406h.
4. R. Rabinowitz and R.W. Marcus, <u>J. Org. Chem.</u>, <u>1961</u>, 26, 4157.
5. A.K. Sen Gupta, <u>Tetrahedron Letters</u>, <u>1968</u>, 5205.
6. G. Wittig and U. Schollkopf, <u>Chem. Ber.</u>, <u>1954</u>, 87, 1318. G. Wittig and W. Haag ibid., <u>1955</u>, 88, 1654.

Wittig Resins: The Preparation and Application of Insoluble Polymeric Phosphoranes

By S. V. McKinley and J. W. Rakshys, Jun.

(*The Dow Chemical Company, Eastern Research Laboratory, Wayland, Massachusetts 01778*)

Summary A solid-phase method has been applied to Wittig olefin synthesis; insoluble poly-*p*-styryldiphenyl-alkylidenephosphorane resins (III) have been prepared and treated with carbonyl compounds to give olefins in good yield.

The application of solid-phase functionalized polymers as reaction substrates is of current interest, and has been explored extensively in the multi-stage polypeptide synthesis,[1,2] although few examples of general synthetic nature have been reported.[3,4] We now report the preparation of polymeric phosphorane derivatives (III; R^1 = H, Me, or Ph) and their treatment with aromatic and aliphatic ketones and aldehydes [equation (1)] to afford a wide variety of olefins in good yield. The use of the Wittig resin reagent (III) allows filtration of the olefinic product from the poly-phosphine oxide by-product.

Two approaches were used for the preparation of the ylide precursor, poly-*p*-styryldiphenylphosphine (I):

(a) *p*-Bromostyrene was suspension copolymerized with three parts of styrene and 2% divinylbenzene. The white solid beads of polymer were converted into the lithiopolymer with BuLi in benzene, and were then treated with chlorodiphenylphosphine to give partial incorporation of Ph_2P groups. The lithiation step was quantitative based on bromine analysis, but only *ca.* 60% of the theoretical number of Ph_2P groups were incorporated. This resulted in a resin containing *ca.* 1·0 mmol of functional group per gram.

(b) Alternatively, the monomer *p*-styryldiphenylphosphine was prepared[5] (70%) and then suspension copolymerized with three parts of styrene and 2% divinylbenzene. The final copolymer contained the expected proportion of phosphorus (*ca.* 1·5 mmol/g resin).

Polymer (I) was treated with various alkyl halides to give polymeric phosphonium salts (II). Me_2SO was a particularly effective solvent for alkylations with methyl and ethyl iodide, since it readily caused (II; R = H or Me; X = I) to swell. Polymer (II; R^1 = Ph) was obtained by alkylation with benzyl bromide, either neat or in solution in tetrahydrofuran (THF). In all cases the weight gain upon alkylation as well as phosphorus and halogen analyses confirmed the extent of alkylation to be *ca.* 85—100%.

$$\text{(I) } Ph_2P\text{-Ar} \xrightarrow{RCH_2X} \text{(II) } Ph_2P^+\text{-}CH_2R^1\ X^- \xrightarrow{Base} \text{(III) } Ph_2P=CHR^1$$

$$\xrightarrow{R^2R^3C=O} Ph_2P=O\text{-Ar} + R^2R^3C=CHR^1 \quad (1)$$

The Wittig reaction[6] was carried out by suspending the polymer beads (II) in a mixture of THF and Me_2SO and adding an excess of methylsulphinyl carbanion (sodium salt). After several hours, unchanged base and alkali halide were removed by repeated washings of the beads. Finally (III), suspended in THF, was treated with an approximately stoicheiometric amount of carbonyl compound and allowed to react for several hours. The olefin was recovered as a THF solution by combining the supernatant fluid with several washings of the beads. Product yields and conversions of starting carbonyl compounds were obtained by quantitative g.l.c. techniques using standard solutions in direct comparisons.

The results of the carbonyl-olefination reactions are listed in the Table. The reactions proceed in reasonable yield with polymeric ylides of widely varying reactivity, i.e. highly reactive methylene, ethylidene, and moderated benzylidene derivatives using aromatic and aliphatic ketones and aldehydes. It is worth noting that conversion of the carbonyl compound is generally incomplete, suggesting that a portion of the resin sites is inaccessible for reaction. Yields in the Table are based on recovered carbonyl reactant in order to account for unavailable reaction sites, and in general compare favourably with yields reported for the same monomeric reactions. Yields based on initial carbonyl reactant are shown in parentheses.†

For the last two reactions listed in the Table the formation of cis- and trans-olefins was possible and the isomer distribution was determined. 84% of cis-β-methylstyrene was formed while stilbene contained 58% of the trans-isomer. Nearly identical ratios of 87%,[7a] and 56%,[7b] respectively,

TABLE. *Olefin yields in Wittig reactions using poly-p-styryldiphenylalkylidene-phosphoranes* (III)[a] *with carbonyl compounds*

(III) R^1	R^2R^3CO	Product	Yield[b] (%)	Lit. yield (%)
H	Cyclohexanone	Methylenecyclohexane	90(63)	52[c]
H	Ph_2CO	$Ph_2C=CH_2$	94(72)	84[d]
H	PhCHO	$PhCH=CH_2$	82(50)	67[c]
H	PhCOMe	$PhC(Me)=CH_2$	76(14)	74[c]
H	$Me[CH_2]_4$·CHO	$Me[CH_2]_4$·CH=CH_2	58(56)	—
Me	Cyclohexanone	Ethylidenecyclohexane	90(50)	—
Me	Ph_2CO	$Ph_2C=CHMe$	93(24)	98[e]
Me	PhCHO	PhCH=CHMe	f	—
Ph	PhCHO	PhCH=CHPh	72(35)	82[c]

[a] Data were obtained using polymers prepared by method (b) (see text). [b] Yields are based on amount of carbonyl reactant consumed. Yields based on the initial amount of carbonyl are in parentheses. [c] G. Wittig and U. Schöllkopf, *Chem. Ber.*, 1954, **97**, 1318. [d] G. Wittig and G. Geissler, *Annalen*, 1953, **580**, 44. [e] E. J. Corey and M. Chaykovsky, *J. Amer. Chem. Soc.*, 1962, **84**, 866. [f] The geometric isomer ratio only was determined: 84% cis, 16% trans (see text).

have been reported for the analogous monomeric Wittig reactions in similar solvents and in the absence of alkali halide salts. Although limited to only two cases, this observation suggests that the reaction mechanism in the resin closely resembles that in solution.[6]‡

We thank Drs. J. Martin, H. Small, and R. Hansen for helpful discussions.

(*Received, November 2nd, 1971; Com. 1904.*)

† Side reactions of the carbonyl compounds independent of the resin are possible. However, in one case where benzyl alcohol was a possible product of the base-catalysed Cannizzaro reaction of benzaldehyde, we observed less than 1%, in contrast to the observation of Camps, *et al.*[4b]

‡ In other types of systems significant differences in the course of a reaction on a resin and in solution have been reported.[3b,c]

[1] R. B. Merrifield, *J. Amer. Chem. Soc.*, 1963, **85**, 2149.
[2] J. M. Stewart and J. D. Young, 'Solid Phase Peptide Synthesis,' Freeman, San Francisco, 1969.
[3] Some of the more recent examples are: (a) T. Takagi, *Polymer Letters*, 1967, **5**, 1031; (b) J. I. Crowley and H. Rapoport, *J. Amer. Chem. Soc.*, 1970, **92**, 6363; (c) A. Patchornik and M. A. Kraus, *ibid.*, 1970, **92**, 7587.
[4] (a) F. Camps, J. Castells, M. J. Ferrando, and J. Font, *Tetrahedron Letters*, 1971, 1713; (b) F. Camps, J. Castells, J. Font, and F. Vela, *ibid.*, p. 1715.
[5] R. Rabinowitz and R. Marcus, *J. Org. Chem.*, 1961, **26**, 4157.
[6] The Wittig reaction in solution is discussed in: A. W. Johnson, 'Ylid Chemistry,' Academic Press, New York, 1966.
[7] M. Schlosser in 'Topics in Stereochemistry,' vol. 5, ed. E. L. Eliel and N. L. Allinger, Wiley-Interscience, New York, 1970 (a) p. 15; (b) p. 10.

POLYMERIC REAGENTS I. SYNTHESIS OF AN INSOLUBLE POLYMERIC CARBODIIMIDE

Ned M. Weinshenker and C.-M. Shen[1]

ALZA Research, Palo Alto, California 94304

(Received in USA 10 May 1972; received in UK for publication 30 June 1972)

A major practical problem in organic synthesis is the isolation of the pure reaction product free from contamination due to reagents, solvents and catalysts. The rapid progress being made in new synthetic methods leads one to readily foresee a time when conditions will be found for virtually any type of stereospecific transformation. The isolation step, however, will remain as the most time consuming part of the procedure.

In spite of the success of the Merrifield Synthesis[2] of polypeptides, which has been in use for over a decade, the complementary reagents bound to polymers are only currently receiving attention. Several examples include linear soluble reagents (carbodiimide[3], N-chloroamide[4]) and crosslinked insoluble reagents (N-bromosuccinimide[5], phosphoranes[6] and ligands for metal catalysts[7]). Several carbodiimide derivatives of the usual polystyrene support were mentioned briefly in a report by Fridkin et al.[8]

In our work we have concentrated on crosslinked supports for the reagents. A major consideration is the regeneration of the reagent via a *single* chemical step after its use in a chemical transformation. This feature will lend versatility to the reagent and might eventually allow for their usage and regeneration on a column similar to ion-exchange resins[9]. The first reagent we chose to study was a carbodiimide[10]. This paper outlines its preparation and reaction with carboxylic acids.

Synthesis of an Insoluble Polymeric Carbodiimide

Crosslinked polystyrene beads (Bio-Rad Laboratories, S-X2, 200-400 mesh) were chloromethylated according to the general method of Merrifield[12] yielding a polymer 1 containing 3.1 mmoles Cl/gram (11.05% Cl). Reaction with potassium phthalimide (2.5 eq/Cl, DMF, 100°, 5 hr) followed by hydrazinolysis (hydrazine hydrate, 1 ml/gm polymer in EtOH, reflux 6 hr) produced the primary amine 2. Treatment with isopropylisocyanate (2 eq, THF, 25°, 15 hr) and dehydration of the resulting urea 3 (2 eq TsCl, Et$_3$N, CH$_2$Cl$_2$ reflux 15 hr) afforded the polymeric carbodiimide 4. The infrared spectrum (KBr) displayed strong absorption at 2110 cm^{-1} and elemental analysis (6.8% N) indicated a <u>maximum</u> of 2.42 mmoles of active carbodiimide/gram of polymer. The overall maximum yield for conversion of chloride to carbodiimide (4 steps) was 78 mole %.

The reaction of carboxylic acids with carbodiimide to form anhydrides was used as a measure of the minimum carbodiimide content of the polymer. Unfortunately the side reaction forming N-acyl ureas interferes, but a minimum value for carbodiimide content could be estimated. Conversion of acetic acid to acetic anhydride was followed by gas chromatography (7' x 1/4" Carbowax 20M, 160°) using triglyme as an internal standard. Treatment of polymer 4 (1.0 gm) with an excess of acetic acid in benzene:ether (2:1) yielded 1.4 mmoles of acetic anhydride. Subsequent experiments indicated that some N-acetyl urea was formed and it was assumed that the polymer contained between 1.4 and 2.4 mmoles of available carbodiimide/gram.

Both stearic and glutaric acid were converted to their anhydrides by use of this reagent. Glutaric acid (220 mg) was added to a suspension of 0.99 g of the polymer 4 in 20 ml of 2:1 benzene:ether. After 2 days at room temperature the beads were filtered and the filtrate concentrated to give 195 mg (quantitative) of glutaric anhydride mp 49-51° (commercial sample mp 50°C). Tlc and infrared analysis indicated no trace of glutaric acid. Stearic acid (567 mg) and polymer 4 (1 gm) in 10 ml of 2:1 benzene:ether after 19 hours yielded 360 mg (65%) of crystalline, pure (infrared and mp) stearic anhydride simply by filtration and evaporation of the filtrate.

In several of these experiments the recovered polymer could be easily recycled to the carbodiimide form; however, the activity was generally less than the starting material due to blockage of some active sites by the N-acyl urea rearrangement product.

In an accompanying paper the usefulness of this reagent in the Moffatt oxidation is explored. Work on peptide coupling is presently in progress.

References

1) ALZA Postdoctoral Fellow 1971-72.
2) G. R. Marshall and R. B. Merrifield in "Biochemical Aspects of Reactions on Solid Supports," (G. R. Stark, ed.), p. 111. Academic Press, New York, 1971.
3) Y. Wolman, S. Kivity and M. Frankel, Chem. Communications, 629 (1967).
4) H. Schuttenberg and R. C. Schulz, Angew. Chem. Internat. Edit., 10, 856 (1971).
5) C. Yaroslovsky, A. Patchornik and E. Katchalski, Tetrahedron Letters, 3629 (1970).
6) F. Camps, J. Castells, J. Font and F. Vela, Tetrahedron Letters, 1715 (1971); S. V. McKinley and J. W. Rakshys, jun., Chem. Communications, 134 (1972).
7) R. H. Grubbs and L. C. Kroll, J. Amer. Chem. Soc., 93, 3062 (1971); M. Capka, P. Svoboda, Mr. Cerny and J. Hetfleje, Tetrahedron Letters, 4787 (1971).
8) M. Fridkin, A. Patchornik and E. Katchalski, Peptides, Proc. Eur. Symp., 10th, 1969, p. 166.
9) Ion exchange resins are actually the simplest examples of polymeric reagents and should be so recognized.
10) We chose the carbodiimide moiety for several reasons: 1. ease of preparation; 2. versatility in various types of organic syntheses[11]; 3. the difficulty in many cases of removing the usual urea by-product.
11) See for examples: "Reagents for Organic Synthesis," Fieser and Fieser, John Wiley and Sons, Inc., 1967, pp. 231-235.
12) R. B. Merrifield, Biochem., 3, 1385 (1964).

POLYMERIC REAGENTS II. PREPARATION OF KETONES AND ALDEHYDES
UTILIZING AN INSOLUBLE CARBODIIMIDE

Ned M. Weinshenker and C.-M. Shen[1]

ALZA Research, Palo Alto, California 94304

(Received in USA 10 May 1972; received in UK for publication 30 June 1972)

The preceding paper[2] described the synthesis of a carbodiimide linked to a crosslinked polystyrene matrix (1). In this article the use of the reagent in the Moffatt[3] oxidation is reported.

The Moffatt oxidation appeared to provide a test of the usefulness and applicability of our polymeric carbodiimide to organic synthesis. This oxidation technique has proven to be an extremely versatile method that is particularly useful in dealing with highly sensitive compounds[4]. Oxidation of a range of alcohols to aldehydes and ketones is shown in the table. It should be noted that even the labile prostaglandin intermediate[5] 2 is converted readily to the desired aldehyde 3.

273

Oxidation of Alcohols Using Polymeric Carbodiimide 1

Alcohol[a]	Reaction Time (Hrs)	GC[c]	2,4DNP (Crude)	2,4DNP (Pure)	Isolated Carbonyl Product
Cyclohexanol	3-5	85	79	66[d]	---
Heptyl Aldehyde	2-3	97	80	68[e]	---
Benzyl Alcohol	2	95	88	82[f]	---
Geraniol	3	63	--	--	---
PG Alcohol 2	3	--	--	--	91[g]
PG Alcohol 2	7[b]	--	--	--	89[g]
4-Phenylcyclohexanol	5	--	--	--	67[g]
4-Phenylcyclohexanol	16[b]	--	--	--	67[g]

(a) Reactions were carried out using 1 mmole of alcohol, $\geq$ 2 mmole of carbodiimide, 16 µl of orthophosphoric acid as catalyst, and 2:1 φH-DMSO solvent at ambient temperature.

(b) Pure DMSO as solvent.

(c) Yields based on comparison with the internal standard (triglyme or DMF).

(d) mp 162°; lit[7a] mp 162°.

(e) mp 108°; lit[7b] mp 108°.

(f) mp 245-6°; lit[7b] mp 237°.

(g) Isolated as the solid.

Using this method the products are free from contamination by ureas, a major purification problem in the general synthetic method. Although the advent of water soluble carbodiimides[6] has partially solved that problem, the isolation of water soluble oxidation products still causes some difficulties. The present method should be universally applicable to substrates of varying physical properties.

The preceding table presents the scope of this method to date. Although some steps were taken to maximize yields, further refinements are possible. In general, orthophosphoric acid functioned as a superior catalyst. Use of trifluroracetic acid-pyridine resulted in competitive formation of trifluoroacetates of the starting alcohols, especially in the case of benzyl alcohol. It should also be noted that reaction times were generally longer in pure DMSO than in the mixed solvent containing benzene. Since the yields are the same in the two cases studied, this rate change merely reflects the capacity of the solvents to swell the polymer and the resultant changes in diffusion rate of the substrate to the active sites in the beads. In several cases the polymeric reagent was regenerated generally with some loss of activity[2]. The following example is illustrative of the method:

Oxidation of Alcohol 2 to Aldehyde 3. - The alcohol 2 (359 mg, 1.0 mmole) and 1.44 g (minimum of 2 mmoles) of carbodiimide were stirred in 12 ml of 2:1 benzene-dimethylsulfoxide for 20 minutes to cause swelling of the polymer beads. Then 16 µl of orthophosphoric acid was added and stirring at ambient temperature was continued for 3 hours at which time tlc analysis (silica gel; 2:1 benzene-ethyl acetate) indicated that the oxidation was complete. The beads were filtered and washed with benzene-ethyl acetate and the combined filtrates were concentrated (aspirator pressure) to 5 ml. Dilution with 50 ml of water and stirring precipitated the aldehyde 3 which, after vacuum drying, amounted to 327 mg (91.2%). This material was identical in all respects (nmr, ir, and tlc) to aldehyde prepared by existing methods[6].

References

1) ALZA Postdoctoral Fellow 1971-72.

2) N. M. Weinshenker and C.-M. Shen, preceding paper in this Journal.

3) For leading references see: "Reagents for Organic Synthesis," Fieser and Fieser, John Wiley and Sons, Inc., New York, 1967, Vol. I, pp. 303-307.

4) For example see: N. M. Weinshenker and F. D. Greene, J. Amer. Chem. Soc., **90**, 506 (1968).

5) E. J. Corey, S. M. Albonico, V. Koelliker, T. K. Shaaf, and R. K. Varma, ibid., **93**, 1491 (1971).

6) See "Reagents for Organic Synthesis," Vol. I, Fieser and Fieser, John Wiley and Sons, Inc., New York, 1967, p. 274.

7) "Semimicro Qualitative Organic Analysis," Second Edition, Cheronis and Entrikin, Interscience Publishers, Inc., New York, 1961. a) p. 633, b) p. 582.

INSOLUBLE RESINS IN ORGANIC SYNTHESES I.
PREPARATION AND REACTIONS OF POLYMERIC ANHYDRIDES

Manvendra B. Shambhu and George A. Digenis

College of Pharmacy, University of Kentucky
Lexington, Kentucky 40506

(Received in USA 22 February 1973; received in UK for publication 26 March 1973)

Recently many workers have pointed out the advantages of utilizing polymeric reagents in organic syntheses[1,2,3], a technique similar to the solid phase synthesis of peptides[4]. By employing reactive chemical functions on an insoluble polymer support, the usual purification procedures can be avoided. The excess reagent and other reaction products can be removed by simple filtration so that the desired reaction product alone is left in the filtrate in an ideal case. The simplicity of operation, quantitative yields (as an excess of the reagents may be employed) and mild reaction conditions are of great importance in the derivatization of complex molecules of biological importance and in the synthesis of radiolabelled compounds.

In our preliminary studies we have chosen the acylation of alcohols and amines as a model reaction to be performed by the polymeric reagents. As anhydrides are commonly used as acylating reagents, here we report the preparation of benzoic anhydrides (I and II) on polystyrene and their reactions with ethanol and aniline.

"Popcorn" polystyrene[5] was chosen as the insoluble polymeric support. Polymer <u>I</u> was prepared from the polystyrene bearing carboxylic acid functions obtained by the copolymerization of 40.0 g styrene, 0.15 g divinylbenzene and 5.0 g p-vinylbenzoic acid according to the procedure described by Letsinger, et al.[5] The carboxylic acid functions (0.73 meq per gram of the polymer) were converted to the acid chloride by treatment with excess oxalyl chloride in benzene[6]. After washing with hot benzene, the product (5.0 g) was treated with 1 ml pyridine in 50 ml benzene followed by a solution of benzoic acid in benzene (1.0 g in 20 ml). After maintaining at 50° for 2 hr, the polymer was washed repeatedly with hot benzene and the suspended particles of pyridine hydrochloride were removed[8]. The structure of the final product <u>I</u> (5.3 g) was confirmed by its ir spectrum (no bands at 3700-3350 cm^{-1}, carbonyl bands at 1750 and 1710 cm^{-1}). Polymer <u>II</u> was prepared from the succinylated polystyrene (polystyrene bearing -$COCH_2CH_2COOH$ groups). The latter was obtained by the reaction of succinic anhydride on "popcorn" polystyrene in the presence of aluminum chloride according to the procedure described by Yip and Tsou[9]. The carboxylic acid functions (0.9 meq per gram; 5.0 g polymer) were converted to the anhydride by the procedure described above (5.35 g product, ir - no bands at 3700—3350 cm^{-1}, additional bands at 1750 and 1715 cm^{-1}).

The reactions of the polymeric anhydrides were carried out by refluxing 5.0 g polymer with 1 ml aniline (or ethanol) in 50 ml toluene for 1 hr. After removal of the polymer beads by filtration, the solution was carefully distilled (8" Vigraux column) to remove most of the solvent. The residue was then dissolved in chloroform and extracted with sodium bicarbonate solution to remove any benzoic acid. The acid was subsequently precipitated by the addition of hydrochloric acid and isolated by the extraction with benzene. The chloroform solution was then extracted with dilute hydrochloric acid to remove unreacted aniline and benzanilide (ethyl benzoate when ethanol was employed) was obtained by the removal of chloroform. The results are summarized below:

Polymer[10]	Total benzoic[11] acid on 5.0 g polymer	Product with aniline	Product with ethanol	% Anhydride functions reacted in 1 hr
I	0.35	benzanilide	ethyl benzoate	90%
II	0.4	benzoic acid	benzoic acid	85%

Two products can be formed by the reaction of $\underline{I}$ or $\underline{II}$ with the nucleophile (aniline or ethanol). An attack at carbonyl $\underline{1}$ would give rise to benzoic acid, an undesirable product, while an attack at carbonyl $\underline{2}$ would result in the formation of the desired product (amide or ester). We were able to isolate and identify only one product in each case. Anhydride I was effective in carrying out benzoylation while anhydride II gave only benzoic acid.

The selectivity exhibited by the anhydride II is not totally unexpected. Benton and Perry[12] have shown that the hydrolysis of acetic benzoic anhydride by water enriched in ^{18}O results in the incorporation of ^{18}O in acetic acid. The authors have suggested an A-2 mechanism for the reaction, the acetyl-oxygen bond being broken in both acid and neutral hydrolysis. As the anhydride II is a similar mixed anhydride, an attack at the aliphatic carbonyl is to be expected.

As the anhydride I is structurally similar to benzoic anhydride, the electronic effects alone cannot explain the selectivity of the attack at the carbonyl $\underline{2}$. In our opinion, the proximity of carbonyl $\underline{1}$ to the polystyrene backbone makes the approach of the large nucleophiles more difficult.

At the present we are exploring the possibilities of carrying out the acylations under milder conditions. The potential of these resins in the preparation of mixed anhydrides is also under investigation.

References

1. F. Camps, J. Castells, M. J. Ferrando and J. Font, Tetrahedron Lett., 20, 1713 (1971).
2. N. M. Weinshenker and C. M. Shen, ibid., 32, 3281 (1972).
3. D. C. Neckers, D. A. Kooistra and G. W. Green, J. Amer. Chem. Soc., 94, 9284 (1972).
4. R. B. Merrifield, ibid., 85, 2149 (1963).
5. R. L. Letsinger, M. J. Kornet, V. Mahadevan and M. J. Jerina, ibid., 86, 5163 (1964).
6. Oxalyl chloride was preferred over the more conventional thionyl chloride as the use of the latter is known to incorporate considerable amount of sulfur in the final product (ref. 7).
7. G. L. Southard, G. S. Brooke and J. M. Pettee, Tetrahedron, 27, 2701 (1971).
8. No attempts were made to remove all traces of pyridine hydrochloride.
9. K. F. Yip and K. C. Tsou, J. Amer. Chem. Soc., 93, 3272 (1971).
10. Polymer beads of 40-100 mesh size were used in the reactions.
11. This is the amount isolated from the polymer after refluxing the beads with excess potassium hydroxide in dioxane-water (5:1) for 4 hours.
12. C. A. Bunton and S. G. Perry, J. Chem. Soc., 3070 (1960).

Copyright © 1970 by Microforms International Marketing Corp.

Reprinted from *Tetrahedron Letters*, No. 42, 3629–3632 (1970), with permission of Microforms International Marketing Corporation as exclusive copyright licensee of Pergamon Press journal back files

UNUSUAL BROMINATIONS WITH N-BROMOPOLYMALEIMIDE
Carmela Yaroslavsky, Abraham Patchornik and
Ephraim Katchalski
Department of Biophysics, The Weizmann Institute of Science,
Rehovot, Israel
(Received in UK 13 July 1970; accepted for publication 5 August 1970)

We wish to report the formation of several unexpected products from the treatment of some benzylic and olefinic compounds with N-bromopolymaleimide (PNBS).

Polymaleimide[1] was prepared by free radical polymerization of maleimide, in the presence of 2.5 - 5% divinylbenzene. It was brominated by addition of bromine to a suspension of the polymer in aqueous sodium hydroxide solution. The average yield of bromination amounted to 60% of the theoretical.

Reaction of 1.75 mmole of N-bromosuccinimide residues of PNBS (I) with 0.75 mmole of cumene (II) in 5 ml boiling carbon tetrachloride in the presence of 3.15 mg benzoyl-peroxide for 1.5 hours resulted in a mixture containing α,β,β'-tribromocumene (III, 48%), β-bromo-α-methylstyrene[2] (IV, 15%) and α-bromomethylstyrene[3] (V, 13%), and unchanged cumene (II, 24%), as shown by analysis of the NMR spectra.[4] When the molar ratio of I to II was 3.7:1, and the reaction proceeded for 3 hours, the yield of III increased to 85%.

Compound III was isolated from the reaction mixture by thick layer chromatography (Rf = 0.3) on Kieselgel GF 254, using petroleum ether as eluent. Anal. Calcd. for $C_9H_9Br_3$: C, 30.25; H, 2.52; Br, 67.23. Found: C, 30.73; H, 2.50; Br, 66.1. Compound III was characterized by its NMR spectrum, which consisted of an AB quartet(4H), δ 4.39, 4.17 ppm (J = 11 Hz), and a multiplet, centered at 7.38 ppm (5H) and its mass spectra contained a 3-bromine quarted, M-1 at m/e = 353. III was also obtained by bromination of cumene with excess bromine or by addition of bromine to α-bromomethylstyrene (V) in carbontetrachloride solution.

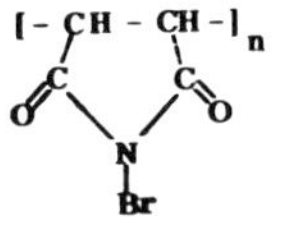

I

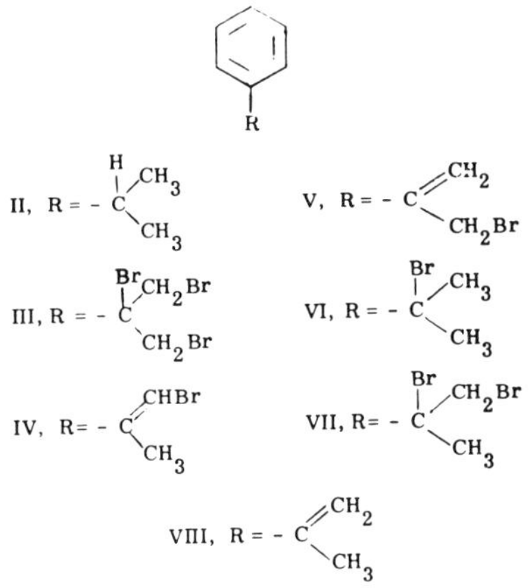

It is of interest to note that neither α-bromocumene (VI)[5] nor α,β-di-bromocumene (VII)[6] were formed upon reacting polymer I with cumene (II) at any of the molar residue ratios of I to II investigated (1:1, 1.2:1, 2.3:1, 3.7:1, respectively). Compounds VI and VII, however, are the only bromination products obtained when cumene is reacted with N-bromosuccinimide (NBS) in boiling carbontetrachloride. For example, when the molar ratio of NBS to cumene was 1:1, the NMR spectrum of the mixture of products indicated the presence of VI (80%), VII (10%) and starting material (10%). Even when the molar ratio of NBS to cumene was increased to 10:1, no α,β,β'-tribromocumene could be detected in the reaction mixture which consisted mainly of VII (95%). A similar behaviour was exhibited by p-cymene and β-bromocumene, which yielded on treatment with PNBS, the corresponding tribromoderivatives. The expected α-bromination products were obtained on treatment with NBS.

The formation of compound III from cumene on treatment with PNBS most likely proceeds by a set of consecutive reactions, the first of which consists of benzylic bromination to yield VI. VI undergoes dehydrobromination to form α-methylstyrene (VIII). The hydrobromic acid formed reacts with the N-Br moiety of the polymer to liberate a bromine molecule, which adds to the double bond of VIII to form VII. Compound VII undergoes further dehydrobromination to form IV and V. Compound III is finally formed by addition of bromine to V. Compound III is stable under the experimental conditions employed and thus accumulates in the reaction mixture.

The dehydrobromination capacity of the maleimide residues of PNBS was checked by refluxing carbon tetrachloride solutions of III, VI and VII in the presence of polymaleimide. When

0.07 ml of a mixture containing III (43%) and VII (50%) was treated with 0.344 g of polymaleimide in 3 ml boiling carbon tetrachloride for 2 hours, analysis of the reaction mixture showed that while the amount of III was essentially unchanged, VII underwent extensive dehydrobromination to yield IV (7% in the reaction mixture) and V (26% in the reaction mixture). No such changes were observed when the same mixture of III and VII was treated with succinimide or kept for 2 hours in boiling carbon tetrachloride. Treatment of 0.1 ml of a mixture containing VI (50%) and VII (26%) with 0.537 gr of polymaleimide in 6 ml of boiling carbon tetrachloride for 2 hours yielded a mixture containing IV (8%), V (15%), VI (6%), VII (3%) and VIII (44%).

α-Methylstyrene (VIII) does not react with PNBS or NBS at the temperature of boiling carbon tetrachloride. At higher temperatures (160-170°C) both PNBS and NBS[3] react with VIII to yield V as the major product and IV as the minor product. This leads to the conclusion that if VIII appears as an intermediate in the reaction leading to the formation of III, it adds bromine rather than undergoing allylic bromination. Bromine is thus available only after the dehydrobromination of VI; benzylic bromination is thus a prerequisite step in the formation of III.

Different reaction products were obtained also when ethylbenzene was reacted with PNBS and NBS in boiling carbon tetrachloride and in the presence of free radical initiators. When the molar ratio of NBS residues in I to ethylbenzene was 1.13:1, the NMR spectrum of the mixture of products indicated 40% of α-bromoethylbenzene (compared to 92% using NBS) and 31% of α,β-dibromoethylbenzene (0% with NBS). A molar ratio of PNBS to ethylbenzene of 3.3:1, resulted in 71% of α,β-dibromoethylbenzene. Reactions of NBS with ethylbenzene, in molar ratios of 1:1, 1.5:1 and 2.5:1, respectively did not yield any α,β-dibromoethylbenzene. Cyclohexene, when reacted with PNBS, yielded exclusively 1,2-dibromocyclohexane, while, with NBS, 3-bromocyclohexene was obtained.[8] Toluene and diphenylmethane, when reacted with PNBS or NBS, yielded benzylbromide and α-bromodiphenylmethane.[9]

It is of interest to note that reaction of NBS in polar medium with the above mentioned benzylic and olefinic compounds, gave similar products to those obtained with PNBS in carbon tetrachloride. Thus, when NBS (3.2 mmole) reacted with cumene (1 mmole) in 3 ml boiling acetonitrile in the presence of 7 mg azobisisobutyronitrile, for 75 min, α,β,β'-tribromocumene (III) was obtained in a yield of 70%. Reaction of NBS with ethylbenzene in acetonitrile yielded α-bromoethylbenzene and also α,β-dibromoethylbenzene. Treatment of VII with NBS in acetonitrile caused dehydrobromination of VII to IV and V.

The unusual reactions of PNBS may be due to its polymeric structure, by which neighbouring groups effect the chemical properties of the active functional groups of the polymer. The N-bromosuccinimide residues on the polymer are located in a polar medium provided by the two succinimide residues adjacent to them. In an alternative copolymer of styrene-maleimide, the N-bromosuccinimide moiety in the polymer will be situated between two styrene residues and should thus not undergo the same polar interactions. Such a copolymer[9] was prepared and brominated. This brominated copolymer did not yield on reaction with cumene α,β,β'-tribromo-cumene, nor any α,β-dibromoethylbenzene on reaction with ethylbenzene. The unbrominated copolymer did not dehydrogenate α,β-dibromocumene in boiling carbon tetrachloride.

PNBS can be recovered quantitatively from the reaction mixture, rebrominated and used again either as brominating or oxidating agent.

Acknowledgement. This work was supported by a grant from the Israel National Council for Research and Development.

References

1. P.O. Tawney, R.H. Snyder, R.P. Conger, K.A. Leibbrand, C.H. Stiteler and A.R. Williams, J. Org. Chem. 26, 15 (1961).
2. S.W. Tobey, J. Org. Chem. 34, 1281 (1969).
3. H. Pines, H. Alul and M. Kolobielski, J. Org. Chem. 22, 1113 (1957); S.F. Reed, Jr., J. Org. Chem. 30, 3258 (1965).
4. All NMR spectra were recorded in Varian A-60 instrument, in carbon tetrachloride with tetramethyl silane as internal reference.
5. Chemische Werke Huels A.-G. Belg. Patent 645,807, July 16, 1964. Chem. Abstr. 63, 9863 d.
6. D.R. Davies and J.D. Roberts, J. Am. Chem. Soc. 84, 2252 (1962).
7. High Resolution NMR Spectra Catalog, Varian Associates, 1962-1963, Spectrum No. 503.
8. K. Ziegler, A. Späth, E. Schaaf, W. Schumann and E. Winkelmann, Ann. Chem. Liebigs, 551, 80 (1942).
9. Y. Yanagisawa, M. Akiyama and M. Okawara, Kogyo Kagaku Zasshi, 72, 1399 (1969). [C.A. 71, 113410t(1969)].

Catalytic Reduction of Olefins with a Polymer-Supported Rhodium(I) Catalyst

Sir:

Many biological systems are capable of selecting substrates for reaction from solution on the basis of bulk molecular properties. These same systems, which are generally catalytic, also allow a large turnover of substrates without loss of the catalytic site. This is accomplished in most cases by attaching the reagent to the inside of large, semiordered, insoluble polymer. Solvent channels in the polymer allow soluble substrates to enter and leave the stationary reagent site. In this way, the substrates are easily separated from reagents and the overall geometric and polar properties of the solvent channels determine the substrates that are able to enter into the catalytic center.

We now wish to report that this rather simple principle has been used to develop a new class of olefin hydrogenation catalysts which demonstrates many of the best properties of homogeneous and heterogeneous catalysts.[1] In addition they are capable of selecting olefins from solution on the basis of overall molecular size.

Polystyrene beads (200–400 mesh) with 1.8% cross-linking of divinylbenzene were chloromethylated on 10% of the aromatic rings by the procedure of Pepper, Paisley, and Young.[2] The chloromethylated polymer was treated with a 1 M tetrahydrofuran solution of lithiodiphenylphosphine[3] for 1 day to replace 80% of the chlorines with diphenylphosphine groups.[4] These beads were then equilibrated with a twofold excess of tris(triphenylphosphine)chlororhodium(I)[5] for 2–4 weeks.

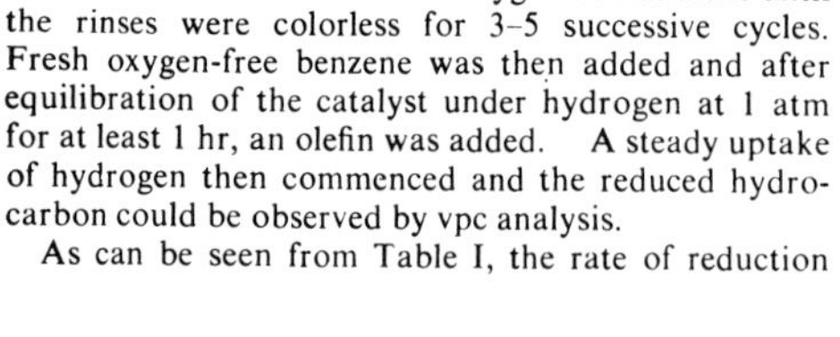

At the conclusion of the equilibration period the deep red beads were washed with deoxygenated benzene until the rinses were colorless for 3–5 successive cycles. Fresh oxygen-free benzene was then added and after equilibration of the catalyst under hydrogen at 1 atm for at least 1 hr, an olefin was added. A steady uptake of hydrogen then commenced and the reduced hydrocarbon could be observed by vpc analysis.

As can be seen from Table I, the rate of reduction

Table I

Olefin[a,d]	Beads rel rate[b]	RhCl(Ph₃P)₃ rel rate[b,c]
Cyclohexene	1	1.0
1-Hexene	2.55	1.4
Δ²-Cholestene	1/32	1/1.4 (1/2.3)[e]
Octadecene (isom mix)	1/2.06	1/1.4
Cyclooctene	1/2.54	1.0
Cyclododecene (cis and trans)	1/4.45	1/1.5

[a] Measured by rate of hydrogen uptake at 1 atm. [b] 1 M olefin concentrations. [c] (Ph₃P)₃RhCl 2.5 mM, in benzene at 25°. [d] The olefins were purified by distillation from sodium under nitrogen. [e] Ratio of H₂ uptake with 10% palladium on carbon.

depended on the molecular size of the olefin. Increasing the ring size of a cyclic olefin or going from an acyclic to a cyclic olefin decreased the rate of reduction. Large rigid olefins such as Δ²-cholestene showed a dramatic decrease in reduction rate. We attribute this decrease in reduction rate to the restriction of the size of the solvent channels by the random cross-links in the polymer.[6] These observations also demonstrated that the major portion of the reductions was taking place inside of the polymer bead.[7] A surface reduction reaction would have shown a much lower specificity for the larger olefins (see Table I).

The catalyst was easily recovered from the reagents by filtration and could be reused many times. The catalyst activity increased slowly to a maximum activity with use[8] and varied by ±5% over ten runs after the activity had maximized. The data given in Table I were obtained with one batch[9] of catalyst which reduced 1 M cyclohexene in benzene at a rate of 0.97 ml/min at 1 atm. The relative rates were determined by setting the rate of reduction of cyclohexene before and after the run on each olefin equal to 1.0. Different batches of the catalyst vary in activity.

Work is now in progress to further define the substrate specificity of these catalysts on the basis of substrate size, per cent of cross-linking in the polymer, and substrate polarity.

Acknowledgments. The authors wish to thank Mr. Terence K. Brunck for aid with the experimental work and The Dow Chemical Company, Midland, Mich., for a gift of the polymer beads used in this work.

(1) P. Legzdins, G. L. Rempel, and G. Wilkinson, *Chem. Commun.*, 825 (1969); and Mobil Oil, Patent, U. S. No. 1,800,371 (1969).
(2) K. W. Pepper, H. M. Paisley, and M. A. Young, *J. Chem. Soc.*, 4097 (1953).
(3) C. Tamborski, *et al.*, *J. Org. Chem.*, 27, 619 (1962).
(4) Determined by microanalysis.
(5) C. O'Conner and G. Wilkinson, *J. Chem. Soc. A*, 2665 (1968).
(6) A plot of average molecular size, as determined from molecular models, *vs.* corrected relative rates of reduction shows a reasonably linear relationship.
(7) R. B. Merrifield, *Science*, 150, 178 (1965).
(8) R. L. Augustine and J. Van Pepper, *Chem. Commun.*, 571 (1970).
(9) Obtained by equilibrating 3.2 g of beads with 3.22 g of rhodium complex in 40 ml of benzene for 4 weeks under argon. Analysis showed 0.13 mmol of rhodium/g of beads.

Robert H. Grubbs, LeRoy C. Kroll
Department of Chemistry, Michigan State University
East Lansing, Michigan 48823
Received March 9, 1971

Activation of Homogeneous Catalysts by Polymer Attachment

ROBERT H. GRUGGS, CARL GIBBONS,
LEROY C. KROLL, WESLEY D. BONDS, JR.,
C. H. BRUBAKER, JR.

Sir:

The attachment of homogeneous catalysts to polystyrene–divinylbenzene copolymer produces a new class of catalysts with many of the best properties of both heterogeneous and homogeneous catalysts.[1] These catalysts also showed selectivity toward substrates of different molecular bulk[2] and polarities.[3]

We now wish to report that catalysts can be significantly activated by such attachment. A basic requirement for homogeneous catalysis by a transition metal complex is the presence of an open coordination site. In most cases, attempts to open a coordination site on a metal result in the polymerization of the complex, and the production of bridged species in which the required site of unsaturation is blocked.[4] Attachment of a saturated complex that is a potential catalyst to a rigid support, followed by the reductive elimination of a ligand, should produce higher concentrations of monomeric coordinatively unsaturated species than is obtained in solution.[5]

Collman and his coworkers[5] have found that 2% divinylbenzene–styrene copolymers are mobile enough to allow ligands attached to the polymer beads to act as chelates. Consequently, this polymer is not rigid enough to prevent dimerization of attached unstable species.

We have found, however, that treatment of the chlorobis(cyclooctadiene)rhodium(I) dimer with a 3-equiv excess of polystyryl–diphenylphosphine prepared from 20% divinylbenzene–styrene macroreticular copolymer (600 Å av pore size)[6] releases only 1.4 mol of cyclooctadiene per mole of complex absorbed. A similar reaction performed with phosphenated 2% divinylbenzene–styrene copolymer produces 2.0 mol of cyclooctadiene per mole of absorbed metal. The 2% cross-linked material gives the same results as the

(1) R. H. Grubbs and L. C. Kroll, *J. Amer. Chem. Soc.*, **93**, 3062 (1971).
(2) M. Capka, P. Suobda, M. Cerny, and J. Hetfleje, *Tetrahedron Lett.*, 4787 (1971), and references therein.
(3) R. H. Grubbs, L. C. Kroll, and E. M. Sweet, *J. Macromol. Sci., Chem.*, in press.
(4) J. P. Collman, *Accounts Chem. Res.*, **1**, 137 (1968).
(5) J. P. Collman, *et al.*, *J. Amer. Chem. Soc.*, **94**, 1789 (1972).
(6) Obtained from The Dow Chemical Co., Midland, Mich.

treatment of 1 mol of the cyclooctadiene dimer with 3 equiv of triphenylphosphine. The 20% cross-linked material yields the same results as the treatment of 1 mol of the cyclooctadiene rhodium dimer with 1 mol of triphenylphosphine. It is apparent that there is much less chelation, *i.e.*, less mobility of the polymer structure, in the 20% cross-linked than in the 2% cross-linked copolymer.

Titanocene has been suggested as a reactive intermediate in the reduction of olefins,[7] acetylenes, and nitrogen.[8] Brintzinger[8] has recently demonstrated that titanocene is rapidly converted into an inactive polymeric compound. We have therefore used the principles outlined above to produce a highly activated titanocene catalyst. Titanocene dichloride ($TiCp_2Cl_2$) was attached to 20% cross-linked macroreticular copolymer by the procedure shown in Scheme I.

Starting with 13.6% (1.13 mequiv/g) substitution of chloromethyl groups,[9] a polymer containing 0.79 mequiv of $TiCp_2Cl_2$ per gram of beads (20% by wt) was obtained. Visible and far-infrared spectra obtained from mulled samples of the metallocene-substituted polymer were comparable with spectra recorded for the benzyl–titanocene monomer (benzyl-Cp)CpTiCl$_2$. On treatment with excess butyllithium or sodium naphthalide, the salmon pink titanocene dichloride polymer was converted to a highly reactive gray polymer which catalyzed the reduction of olefins and acetylenes (see Table I).[7]

Treatment of titanocene dichloride or benzyltitanocene dichloride with 2 equiv of butyllithium produced a gray catalyst that was only 0.15 times as active per milliequivalent of metal in the reduction of cyclohexene as was the polymer-attached metallocene (Table II). This contrasts with the observation that the activity of the rhodium-attached catalysts are only 0.06 times[3] as reactive as an equivalent amount of homogeneous reagent. We attribute the faster reduction rate of the polymer-attached catalyst to the presence of a higher

(7) Y. Tajima and E. Kunioka, *J. Org. Chem.*, **33**, 1689 (1968).
(8) J. E. Bercaw, R. H. Marvich, L. G. Bell, and H. H. Brintzinger, *J. Amer. Chem. Soc.*, **94**, 1219 (1972), and references therein.
(9) K. W. Pepper, H. M. Paisley, and M. A. Young, *J. Chem. Soc.*, 4097 (1953).

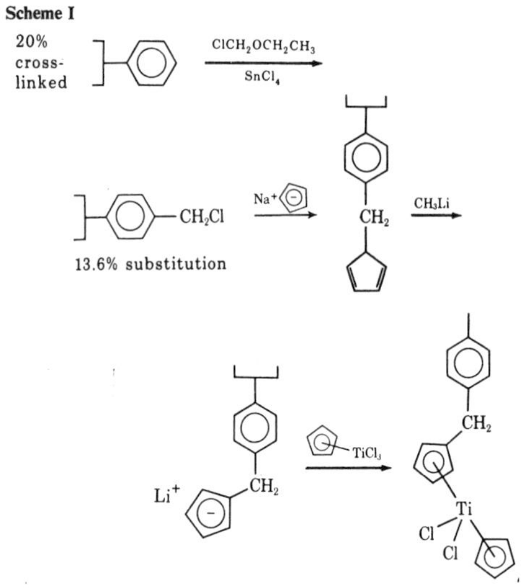

Scheme I

[Scheme I shows: 20% cross-linked polystyrene bead + ClCH₂OCH₂CH₃/SnCl₄ → chloromethylated polystyrene (13.6% substitution); then reaction with Na⁺ cyclopentadienide to give polymer-CH₂-cyclopentadienyl; then CH₃Li; then with Li⁺ cyclopentadienide and TiCl₃ to give polymer-attached titanocene dichloride]

thalide to polymer-bound titanocene dichloride resulted in the formation of species with two distinctly different esr signals of comparable intensity near $g = 2$.[10] These data suggest that both stages of reduction yield paramagnetic species. Treatment of monomeric benzyltitanocene with sodium naphthalide under identical conditions yields solutions containing a rapidly disappearing paramagnetic species and mediocre catalytic ability. These observations support the suggestion that dimerization of the reduced titanocene complexes has been avoided by polymer attachment.

Acknowledgments. The authors wish to thank the Dow Chemical Co. for the gift of the polystyrene beads used in this work. Carl Gibbons thanks Dow Chemical Co. for support while completing this research. This work was also supported in part under NSF Grant GP 17422X.

(10) H. H. Brintzinger, *J. Amer. Chem. Soc.*, **89**, 6871 (1967).
(11) On leave from The Dow Chemical Co., Midland, Mich.

Department of Chemistry, Michigan State University
East Lansing, Michigan 48823
Received November 29, 1972

Table I. Rates of Reduction with Polymer Attached Titanocene

Olefin (~0.5 M in hexane)	Rate of reduction, ml of H_2/min (1 atm of H_2)
40 mg of Catalyst (in 10 ml of hexane)	
1,3-Cyclooctadiene	1.81
1,5-Cyclooctadiene	1.54
Styrene	2.05
3-Hexyne	1.26 (hexane)
1-Hexene	1.80
1-Hexyne	Polymer
Cholestenone	0
Vinyl acetate	0
420 mg of Catalyst	
Diphenylacetylene	3.62 (eq)
Cyclohexene	8
1-Methylcyclohexene	0.92
1,2-Dimethylcyclohexene	0

Table II. The Reduction Rate of Various Olefins with Attached and Homogeneous Titanocene Species

mequiv of cat.	Olefin	Solvent	Rate of reduction
Benzyltitanocene Dichloride			
0.2	Cyclohexene	Hexane	1.9 ml of H_2/min
0.2	1-Methylcyclohexene	Hexane	No uptake in 115 min less than 9×10^{-3} ml/min
Polymer–Titanocene Dichloride			
0.032	Cyclohexene	Hexane	1.8 ml of H_2/min
0.32	1-Methylcyclohexene	Hexane	0.09 ml/min
0.032	1-Hexene	Hexane	1.7 ml of H_2/min
0.032	1-Hexene	THF	0.92 ml of H_2/min
Titanocene Dichloride			
0.2	Cyclohexene	Hexane	2.1 ml of H_2/min
0.2	Cyclohexene	THF	0

concentration of monomeric species. This conclusion is supported by esr studies.

Stepwise addition of 1 and 2 equiv of sodium naph-

Resin-Bound Transition Metal Complexes

JAMES P. COLLMAN, LOUIS S. HEGEDUS, MANNING P. COOKE, JACK R. NORTON, GIULIANO DOLCETTI, and DONALD N. MARQUARDT

Sir:

Organic reactions in which one component is fastened to a porous solid support offer strategic synthetic advantages in the dilution[1] or isolation[2] of that component and the removal of byproducts.[3] Homogeneous transition metal catalysts bound to a resin may also exhibit properties different from those in solution. For example, a rhodium hydrogenation catalyst attached to polystyrene exhibits selectivity toward smaller olefins.[4,4a] Since the active forms of homogeneous catalysts are unsaturated,[5] immobilization on a solid support might prevent self-aggregation affording high concentrations of unsaturated complexes. As is shown below, this effect will be difficult to achieve with phosphine-substituted polystyrene because of the pronounced tendency of such polymeric ligands to chelate.

Resin-substituted triphenylphosphine **1** was prepared from cross-linked polystyrene (Biobeads SX-2, 2% crosslinking) by sequential bromination (Br$_2$, FeBr$_3$), lithiation (*n*-BuLi, THF), and treatment with (C$_6$H$_5$)$_2$PCl. The following experiments with **1** employed

polymer—⟨◯⟩—P(C$_6$H$_5$)$_2$

1

MCl(CO)(P(C$_6$H$_5$)$_3$)$_2$ MCl(CO)L$_2$
2a, M = Rh   **3a**, L = **1**; M = Rh
b, M = Ir   **b**, L = **1**; M = Ir

1.2 mmol of P/g corresponding to 10% ring substitution. Treatment of **1** with the rhodium(I) or iridium(I) complexes, **2**, afforded the resin-bound complexes, **3**, releasing two P(C$_6$H$_5$)$_3$ per metal atom introduced into the polymer.[6] Similarly two P(C$_6$H$_5$)$_3$ and one CH$_2$=CH$_2$ were displaced from **4** affording the red polymeric complex **5**, an analog of the Wilkinson hydrogenation catalyst.[7] In these and similar cases, intermediate levels of coordination per metal atom in the polymer by resin-bound phosphine could not be achieved either by varying the ratio of starting complex to resin or the level of phosphine incorporated[8] in the resin. That nonstatistical functionalization

---

(1) R. L. Letsinger, M. J. Kornet, V. Mahadevan, and D. M. Jermia, *J. Amer. Chem. Soc.*, **86**, 5163 (1964).
(2) (a) I. T. Harrison and S. Harrison, *ibid.*, **89**, 5724 (1967); (b) J. I. Crowley and H. Rapoport, *ibid.*, **92**, 6363 (1970).
(3) R. B. Merrifield, *Fed. Proc.*, **21**, 412 (1962); *J. Amer. Chem. Soc.*, **85**, 2149 (1963); **86**, 304 (1964).
(4) R. H. Grubbs and L. C. Kroll, *ibid.*, **93**, 3062 (1971).
(4a) NOTE ADDED IN PROOF. Other resin-bonded catalysts have been described: Mobil Oil, U. S. Patent 1,800,371 (1969); British Petroleum Co., Belgian Patents 739,607 and 739,609.
(5) J. P. Collman, *Accounts Chem. Res.*, **1**, 136 (1968).

(6) These quantities were determined by analysis of the filtrates from reactions using measured quantities of **1** and the homogeneous complex. A small portion of resin phosphine sites were too sterically hindered to act as ligands. Resin-bound complexes exhibited the same ν_{CO} and colors as their homogeneous counterparts.
(7) J. A. Osborn, F. H. Jardine, J. F. Young, and G. Wilkinson, *J. Chem. Soc. A*, 1711 (1966).
(8) Phosphine substitution levels down to 2% of the benzene rings in the resin were employed with similar results. Analytical errors become correspondingly larger at these low levels.

leading to pockets of highly substituted resin was not the cause of this chelation was demonstrated by carrying out the above experiments with the same result using a phosphine-substituted "popcorn" copolymer[1] prepared from styrene (0.9 mmol), *p*-bromostyrene (0.1 mmol), and divinylbenzene (0.002 mmol). The most plausible explanation is that the polymer chain is sufficiently mobile, especially when the resin is solvent swelled, to bring nonadjacent sites together.[9]

In cases where tertiary phosphines are not displaced and monosubstitution is kinetically favored resin chelation does not result. For example, treatment of [Rh(COD)Cl]$_2$ with **1** yielded a 1:1 resin complex by splitting the chloro bridge. Similarly 1 equiv of **6** reacted with 2 equiv of **1** under mild conditions (C$_6$H$_6$, 25°) to form the 1:1 complex **7** (ν_{CO} 2022, 1965; ν_{NO} 1750). However, further heating (70°, 24 hr, diglyme) converted **7** into the 1:2 complex **8** (ν_{CO} 1940; ν_{NO} 1700 cm^{-1}).[10]

RhCl(P(C$_6$H$_5$)$_3$)$_2$(C$_2$H$_4$)	MClL$_3$	Co(NO)(CO)$_3$
4	**5a**, L = **1**; M = Rh **b**, L = **1**; M = Ir	**6**
Co(NO)(CO)$_2$L	Co(NO)(CO)L$_2$	Rh$_4$(CO)$_{10.5}$L$_{1.5}$
7, L = **1**	**8**, L = **1**	**9**, L = **1**

The resin-bonded complexes exhibit normal reactions. For example, **3b** reacts with *p*-nitrobenzoyl azide to afford a resin-bound N$_2$ complex, ν_{N_2} 2095. Treatment of **3b**, **5a**, or **5b** with hydrogen afforded the spectral changes expected for oxidative addition. The resin-bound mononuclear hydrogenation catalysts **5a,b** were superior to their homogeneous analogs only in catalyst lifetime and ease of product separation.

Novel polynuclear catalysts were prepared by treating **1** with the readily substituted Rh$_4$(CO)$_{12}$ and Rh$_6$(CO)$_{16}$.[11] Aerial oxidation (accelerated by light) led to the disappearance of ν_{CO}; H$_2$ (1 atm, THF) afforded active metal particles of extremely small size, presumably 4 and 6 Rh atoms, respectively. The substance derived from Rh$_6$(CO)$_{16}$ catalyzed the hydrogenation of arenes at 25° and 1 atm of H$_2$, exhibiting reactivity and substrate selectivity similar to Engelhard's 5% Rh/Al$_2$O$_3$,[12] whereas that derived from Rh$_4$(CO)$_{12}$ exhibited diminished activity which we attribute to Rh$_6$(CO)$_{16}$ formed spontaneously in solutions of Rh$_4$(CO)$_{12}$. Evidence that the polymer prevents aggregation is afforded by comparing reactions of these catalysts with CO: from the Rh$_4$ catalyst over 50% of the Rh was removed as Rh$_4$(CO)$_{12}$; from the Rh$_6$ catalyst a small amount of Rh$_4$(CO)$_{12}$ and much Rh$_6$(CO)$_{16}$ were obtained at <40 psi; and from 5% Rh/Al$_2$O$_3$[13] only a barely detectable amount of Rh$_4$(CO)$_{12}$ was obtained after 24 hr at >1000 psi. Polymer-free Rh$_4$(CO)$_{11}$-(P(C$_6$H$_5$)$_3$) similarly oxidized and reduced also catalyzed arene reduction but failed to give Rh$_4$(CO)$_{12}$ upon CO treatment suggesting that irreversible aggregation to larger metal units occurred without the use of the polymeric ligand.

Acknowledgments. We are indebted to Rohm and Haas Company for experimental polymer samples and to Professor M. Boudart for experimental aid and R. Bacskai, Chevron Research Company, for advice. This work was supported by the National Science Foundation Grant No. GP 20273X.

(9) More highly cross-linked polystyrene resins were employed: Biobeads SX-12 and Rohm and Haas XAD-2 and XE-305. Substantial amounts of phosphine could only be introduced into XE-305 but 2 mequiv of triphenylphosphine was liberated for each milliequivalent of **2b** taken up.
(10) (a) R. F. Heck, *J. Amer. Chem. Soc.*, **85**, 657 (1963); (b) E. M. Thorsteinson and F. Basolo, *Inorg. Chem.*, **5**, 1691 (1966).
(11) (a) R. Whyman, *Chem. Commun.*, 230 (1970); (b) B. L. Booth, M. J. Else, R. Fields, and R. N. Hazeldine, *J. Organometal. Chem.*, **27**, 119 (1971).
(12) P. N. Rylander and L. Hasbrouck, "Technical Bulletin," Vol. X, No. 2, Engelhard Industries, Murray Hill, N. J., 1969, p 50.
(13) Hydrogen adsorption studies of this catalyst by H. Uchida showed 43% dispersion. Assuming a cubic shape, this corresponds to an average particle size of 40 or 80 Å corresponding to each surface Rh taking up one or two H atoms.

Department of Chemistry, Stanford University
Stanford, California 94305
Received January 7, 1972

HYDROGENATION, HYDROSILYLATION AND HYDROFORMYLATION OF OLEFINS CATALYSED BY POLYMER-SUPPORTED RHODIUM COMPLEXES[1]

M. Čapka, P. Svoboda, M. Černý and J. Hetflejš

Institute of Chemical Process Fundamentals, Czechoslovak

Academy of Science, Prague-Suchdol, Czechoslovakia

(Received in UK 1 November 1971; accepted for publication 10 November 1971)

In recent years several attempts have been made to combine the advantages of homogeneous and heterogeneous catalysis[2,3]. A recent work of Grubbs and Kroll on the use of a polymer-supported rhodium(I) complex as hydrogenation catalyst[4] prompt us to report our preliminary results in this very promising field. In conjuction with our studies of hydrosilylation catalysed by soluble rhodium(I) complexes[5,6] we were also interested in catalytic activity of rhodium complexes coordinatively bonded to a polymer containing phosphine groups. In the present communication we wish to report briefly the preparation of such complexes and their catalytic activity in hydrogenation, hydrosilylation and hydroformylation of some olefins.

Preparation of Catalysts. A reaction of easily accessible macroreticular chloromethylated styrene-divinylbenzene copolymer(I) (20% crosslinking, 14,4%Cl particle size 1-2 mm) with benzene solution of lithiodiphenylphosphine[7] (reaction 1) yielded polymeric ligand II (2.4 % Cl, 9.8 % P). Treatment of ethanolic solution of $RhCl_3 \cdot 3H_2O$ with polymer II (reaction 2) afforded substance III which contained rhodium fixed via Rh-P coordination to the polymer. In an attempt to prepare the analogue of Wilkinson[8] catalyst (compound VII) by reaction 3 we found that the action of triphenylphosphine on compound III in ethanol results in extensive displacement of ligand II from the coordination sphere of rhodium, the main product being Wilkinson catalyst, tris(triphenylphosphine)chlororhodium (I) (compound IVa), as proved by comparison of its IR spectrum with the spectrum of authentic sample[8].

SCHEME

On the other hand, due to lower coordination ability of diphenylphosphine, reaction 4 is accompanied by the above displacement to much lesser extent. By passing ethylene through a mixture of compound III, benzene and ethanol we prepared compound VI (reaction 5) which contains coordinated ethylene. The other compounds were prepared by exchange of phosphino ligand in the corresponding soluble complexes for polymeric ligand II. Compounds VII and VIII were obtained by shaking a mixture of polymer II and methylene chloride solution of triphenyl- or diphenylphosphine for 24 h (reaction 6 and 7, respectively). All the compounds so prepared were washed several times with the solvents and dried under vacuum. All the procedures were carried out under argon.

Hydrosilylation was carried out in the following way. A mixture of catalyst (1 mg), olefin (0.5 ml) and organosilicon hydride (0.5 ml) was heated in a sealed ampoule to 80° for 2 h. Yields of hydrosilylation products were determined by g.l.c.

We have found that the addition of triethoxysilane to 1-hexene in the presence of compounds III, IV, VI and VII yields 91, 22, 99 and 19% of n-hexyltriethoxysilane, respectively. The addition of the same hydride to ethylvinyl ether acrylonitrile and trans-2-heptene catalysed by complex III (VII) afforded 88 (50), 18(14), 5(5)% of the corresponding n-alkyltriethoxysilane. Compound III turned out to be very efficient hydrosilylation catalyst. Its catalytic activity did not change after the catalyst was exposed to air for 3 days. The hydrosilylation of 1-hexene by triethoxysilane proceeded readily already at 25°C and the catalyst could be reused several times.

Hydrogenation was carried out in an autoclave. 1-Heptene was hydrogenated

Table: Hydrogenation of 1-Heptene[a]

| Catalyst | Heptane | 1-Heptene | 2-Heptene | | 3-Heptene | |
			cis-	trans-	cis-	trans-
III	98(63.6)	0(1)	0(7.3)	1(16.3)	0(2.7)	1(9.1)
IV	68.2(43.6)	0(0.5)	4.9(10.2)	17.1(20.4)	2.4(2)	7.3(6.1)
V	74.5(49.7)	0(13.2)	9.5(15)	12.7(17)	1(1.9)	2(3.8)
VI	54.2(43.6)	3.9(9)	15(19.5)	20.5(24)	1.8(1.5)	4.7(3)
VII	19.7(4.3)	70(78.2)	5.5(6.5)	4.4(10.8)	0(0)	0(0)
VIII	6(2)	91.8(96)	1(2)	1(2)	0(0)	0(0)

[a] Composition of reaction mixtures was determined by g.l.c.

in the presence of 1%(w.) of catalysts (hydrogen pressure 24 atm., temperature 65°, reaction time 10 h). The reaction mixture was then allowed to stand in air for 30 h, the catalyst was recovered by filtration and reused under the same experimental conditions (see Table, the data in parentheses).

Some of the complexes were catalytically active at room temperature. So, for instance, in the presence of complex III 1-heptene was converted into n--heptane with 60% yield (25°, 35 atm. H_2, 16 h), crotonaldehyde was selectively hydrogenated to butyraldehyde (35%), vinyl acetate gave ethyl acetate (50%) and vinylethyl ether afforded diethyl ether (55%) (110°, 40 atm H_2, 6 h).

Hydroformylation. Compound VI was found to be good hydroformylation catalyst. Hydroformylation of 1-heptene in the presence of 5% (w.) of complex VI (H_2:CO=3:4, total pressure 40 atm, 20 h) gave a mixture of n-octanal (56% yield) and iso-octanal (24% yield). The reaction was accompanied by concurrent isomerisation of the unreacted olefin. Somewhat worse results were obtained with complexes IV, V and VII.

References

1. Catalysis by Metal Complexes, Part V; Part IV: P. Svoboda, M. Čapka, J. Hetflejš, V. Chvalovský, V. Bažant, H. Jahr and H. Pracejus, Z. Chem. in press.
2. J. Rony, J. Catalysis 14, 142 (1969).
3. J. Manassen, Israel J. Chem.-Supl. Proc. Israel Chem. Soc. 8, 5 (1970).
4. R. H. Grubbs and LeRoy C. Kroll, J. Am. Chem. Soc. 93, 3062 (1971).
5. P. Svoboda, M. Čapka, V. Bažant and V. Chvalovský, Coll. Czech. Chem. Commun. 36, 2785 (1971).
6. P. Svoboda, M. Čapka, J. Hetflejš and V. Chvalovský, Coll. Czech. Chem. Commun. in press.
7. K. Issleib and A. Tzschach, Chem. Ber. 92, 1118 (1959).
8. J. A. Osborn, F. H. Jardine, J. F. Young and G. Wilkinson, J. Chem. Soc. A 1966, 1711.

Polymer-Protected Reagents: Polystyrene–Aluminum Chloride

D. C. NECKERS, D. A. KOOISTRA, and G. W. GREEN

Sir:

Insoluble resin techniques offer several advantages in preparative procedures. Resins can be used as diluents to hold reagents from other reagents during a reaction,[1] polymers can be used to hold catalysts during a chemical transformation,[2] and polymers can be used as centers upon which large molecules can be grown.[3-5]

Some time ago, we initiated studies of applications of solid state techniques in organic preparations. In this communication, we report the first use of a tightly bound complex of styrene–divinylbenzene copolymer and anhydrous aluminum chloride, as a mild Lewis acid catalyst for certain organic preparations. The complex of the polymer and $AlCl_3$ provides a shelf-stable acidic material, the active ingredient of which can be called out by an appropriate polymer swelling solvent at the time it is desired.

Polystyrene–divinylbenzene (1.8%) copolymer beads form a water-stable aluminum chloride complex hereafter denoted as Ⓟ–$AlCl_3$. In a typical preparation, 31.0 g (0.46 mol, phenyl residues) of polystyrene–divinylbenzene copolymer beads (1.8%, 50–100 mesh) was added to a 1-l. flask equipped with a stirrer, condenser, and dropping funnel. Carbon disulfide (450 ml) was added followed by 7.5 g (0.06 mol) of anhydrous $AlCl_3$ powder. The mixture was stirred at reflux for 40 min and cooled and 400 ml of cold water was cautiously added to hydrolyze the excess $AlCl_3$. The mixture was stirred until the deep orange color disappeared and the polymer became light yellow. The polymer beads were then filtered and washed with 1 l. of water and successively with 150 ml of ether, acetone, hot isopropyl alcohol, and ether. After these washings, the polymer was dried for 18 hr in a vacuum oven.

Complex formation is demonstrated by the increased color of polymers prepared using higher concentrations of $AlCl_3$, by new bands at 1650 cm^{-1} in the infrared spectrum of the polymer, and by the high general stability of the $AlCl_3$ polymer preparations.[6]

For synthetic purposes, the polymer-bound anhydrous $AlCl_3$ can be released from the polymer by swelling the polymer with certain kinds of solvents. Thus, the $AlCl_3$ can be used to catalyze ether formation and other acid-catalyzed reactions when the polymer holding the reagent is swollen by solvents such as benzene. This is the unique feature of Ⓟ–$AlCl_3$. The polymer protects the easily hydrolyzed Lewis acid until placed in an appropriate solvent where it can be spent in a chemical reaction.

Typical of the reactions of Ⓟ–$AlCl_3$ are its reactions with certain carbinols. For example, dicyclopropylcarbinol, when treated with Ⓟ–$AlCl_3$, produces di-(dicyclopropylcarbinyl) ether in yields as high as 81%,[7-9] (eq 1).

$$2 \text{ H}-\underset{\triangle}{\overset{\triangle}{\text{C}}}-\text{OH} \xrightarrow{\text{Ⓟ}-AlCl_3} \text{H}-\underset{\triangle}{\overset{\triangle}{\text{C}}}-\text{O}-\underset{\triangle}{\overset{\triangle}{\text{C}}}-\text{H} \quad (1)$$

The yield of ether was dependent on the nature of the Ⓟ–$AlCl_3$. The presence of the polymer mediates the effect of the strong Lewis acid catalyst producing higher yields of the desired ether and lower yields of the competing, higher molecular weight side products. In addition, sensitive carbinols react more cleanly with Ⓟ–$AlCl_3$ than they do with $AlCl_3$ directly. Data at 50° for 1:1 mixtures of alcohol and copolymer of different Al content are given in Table I.

The scope of the procedure is demonstrated by the preparation of the mixed ethers, Table II.

Finally, the data in Table III demonstrate the susceptibility of the reaction to the solvent in which the reaction is carried out. Thus, for the reaction of dicyclopropylcarbinol with isopropyl alcohol, much higher product yields are obtained with solvents capable of swelling the polymer. They serve to swell the polymer and make the aluminum chloride more accessible.

After washing with 150 ml each of water, ether, acetone, hot isopropyl alcohol, and ether, the polymer retained 3.67% Al as the chloride. The 1650-cm^{-1} band was also retained in the polymer after the above wash procedures. In addition, certain Ⓟ–$AlCl_3$ preparations have been left open to the atmosphere for over 1 year without losing either their catalytic activity or the characteristic infrared bands.

(1) P. Patchornik and M. A. Kraus, *J. Amer. Chem. Soc.*, **92**, 7587 (1970).
(2) A. I. Vogel, "Practical Organic Chemistry," Longmans, London, 1961, p 387.
(3) R. B. Merrifield, Abstracts of Papers, 163rd National Meeting of the American Chemical Society, Boston, Mass., spring 1972.
(4) R. B. Merrifield, *Science*, **150**, 178 (1965).
(5) R. L. Letsinger, M. J. Kornet, V. Mahedevon, and D. M. Jerina, *J. Amer. Chem. Soc.*, **86**, 5163 (1964).
(6) For example, 5 g of Ⓟ–$AlCl_3$ prepared as above but not washed with solvents other than water retained 5.28% Al as the chloride.
(7) H. Hart and J. M. Sandri, *J. Amer. Chem. Soc.*, **78**, 320 (1956).
(8) R. H. Mazur, W. N. White, D. A. Semanov, C. C. Lee, M. S. Silver, and J. D. Roberts, *ibid.*, **81**, 4390 (1959).
(9) Di(dicyclopropylcarbinyl) ether was identified on the basis of its spectral properties and confirmed by independent synthesis: nmr δ 2.78 (2 H, t), 1.10–0.20 (20 H, m); ir 9.02 μ; mass spectrum, m/e 95 (100%), 111 (13%), 165 (0.1%), 178 (0.1%).

Our conclusion is that polymers may provide extremely useful ways for protecting anhydrous and pyrophoric reagents and attempts are underway to further explore the concept using other reactive reagents. The complete scope of ⓟ–AlCl₃ reactions will be described in a full paper to be published on the work in the near future.

Acknowledgment. The authors acknowledge the support of the Research Corportion and the National Science Foundation, Grant No. GP-33566, for this work. Helpful discussions with Professor Erich Blossey are also acknowledged.

Table I. Product Ratios from the Reaction of Dicyclopropylcarbinol and ⓟ–AlCl₃

% Al in copolymer[a]	Yield of ether,[b] %	% conversion[c]
0	0	0
0.57	74.5	33.1
1.83	64.5	56.8
2.20	58.5	66.5
Pure AlCl₃	48.0	75.0

[a] Aluminum analyses by atomic absorption. [b] Analyses after 30 min at 50°. [c] The remainder of the products under these conditions were dimeric and of higher molecular weight.

Table II. ⓟ–AlCl₃ Reactions of Carbinols[a]

Carbinol	Solvent alcohol	Polymer (% Al)	% ether
Dicyclopropyl-carbinol	MeOH	0.57	90.4
	EtOH	0.57	65.0
	i-PrOH	0.57	42.5
	t-BuOH	0.57	19.0
Triphenylcarbinol	MeOH	2.20	81.5
	EtOH	2.20	83.0
Diphenylmethyl-carbinol	MeOH	2.20	75.8[b]
	EtOH	2.20	59.0[b]
	MeOH	0.57	93.0[b]
Phenyldimethyl-carbinol	MeOH	0.57	50.0
tert-Butyl alcohol	MeOH	0.57	0

[a] All reactions were carried out at 60 ± 5° for 90 min. [b] 1,1-Diphenylethylene is the other product.

Table III. Yield of Dicyclopropylcarbinyl Isopropyl Ether as a Function of Solvent Composition[a]

Solvent	Concn of solvent to isopropyl alcohol, wt:wt	Yield of dicyclopropyl-carbinyl isopropyl ether, %	Carbinol remaining, %
Hexane	0	57.0	43.0
	1:3	50.1[b]	2.0
	1:1	76.5[b]	2.0
	3:1	80.6[b]	2.0
Benzene	0	57.0	43.0
	1:3	65.0[b]	2.1
	1:1	69.6[b]	1.8
	3:1	81.0[b]	2.0
Carbon disulfide	0	57.0[b]	43.0
	1:3	48.7[b]	2.0
	1:1	77.7[b]	1.7
	3:1	81.4[b]	2.0

[a] All experiments were conducted with polymer containing 0.57% Al. [b] The remainder of the product in these cases was the ring-opened ether.

Department of Chemistry, The University of New Mexico
Albuquerque, New Mexico 87106
Received August 11, 1972

POLYMER PROTECTED REAGENTS: (II) ESTERIFICATIONS WITH ⓟ-AlCl₃

E.C. Blossey[1], L.M. Turner and D.C. Neckers[2]
Department of Chemistry, University of New Mexico
Albuquerque, New Mexico 87106

(Received in USA 26 February 1973; received in UK for publication 3 April 1973)

In a recent communication from our laboratories, the advantageous use of a polymer protected reagent, polymer protected anhydrous aluminum chloride (ⓟ-AlCl₃) was reported.[3,4] This polymer protected reagent, a tightly bound complex between anhydrous aluminum chloride and polystyrene-divinylbenzene copolymer beads, was described as a polymer protected dehydrating agent which could be released from the polymer by placing it in appropriate swelling solvents.

As part of a general program designed to investigate reactions of polymer protected anhydrous reagents, we report studies of ⓟ-AlCl₃[5] as a mild catalyst for esterification. ⓟ-AlCl₃ combines the advantages of being both a Lewis acid and a dehydrating agent and therefore is ideal for condensation reactions in which very mild conditions are required.

As an example, n-butyl propionate was prepared from propionic acid (0.02 mole) and n-butanol (0.04 mole) by stirring the reagents at room temperature for 46 hrs. with 0.5 g of ⓟ-AlCl₃ in 2.0 ml benzene. After filtering the polymer and removal of the benzene and n-butyl alcohol, n-butyl propionate[6] was isolated in 96.8% yield. A similar experiment carried out with all reagents but ⓟ-AlCl₃ produced only 5.6% of n-butyl propionate under identical conditions.

Similar reactions with various acids and alcohols using ⓟ-AlCl₃ as a catalyst are compared with uncatalyzed reactions in Table I. From these results it is clear that the more sterically hindered alcohols react slowly and electron withdrawing groups at the α-position of carboxylic acids enhance esterification, Table II.

Table I

Product Yields - Esterification Reactions

Acid	Alcohol	Mole Ratio (Acid/alcohol)	T(C°)	Time (hr.)	Yield(%)[a] No Catalyst	ⓟ-AlCl$_3$
Propionic	1-butanol	1:2	95	1	5.5	56.8
" "	" "	1:2	95	2	11.1	70.0
" "	" "	1:2	25	46	5.6	96.8
Propionic	2-butanol	1:2	95	2	0	36.0
" "	" "	1:2	25	70	0	16.2
Propionic	2-methyl-2-propanol	1:2	95	41	0	0
Acetic	trans-4-t-butylcyclohexanol	1:2	95	26	17.0	35.1
"	" " "	1:2	95	53	38.7	52.5
"	cholesterol	1:1	95	68	3.8	58.0
Cyclopropane carboxylic	1-butanol	1:1	95	18	9.5	56.1
Benzoic	" "	1:1.1	95	5	0.5	22.9
" "	" "	1:1.1	95	46	6.4	44.3
p-Nitrobenzoic	" "	1:1.1	95	42	0	5.5

a Yields determined by g.l.c. with an added internal standard

Table II[a]

Acid	Yield, no catalyst[b]	ⓟ-AlCl$_3$
Acetic	0.0	30%
Adipic	0.0	67%[c]

a All reactions at 25° for 24 hours with 1-butanol in benzene
b Ratio of product (ester) to a specific internal standard by g.l.c.
c Combined total of mono- and diester at 25°

The stereochemistry of the ⓟ-AlCl$_3$ reactions was shown by use of trans-4-t-butylcyclohexanol as well as with cholesterol to occur with retention of configuration. Thus, only trans-4-t-butylcyclohexyl acetate was produced from reaction of trans-4-t-butylcyclohexanol with acetic acid using ⓟ-AlCl$_3$ and cholesteryl acetate was the only observed product from the reaction of cholesterol with acetic acid catalyzed by (P)-AlCl$_3$. Other products could not be detected either by gas-liquid chromatography or nuclear magnetic resonance experiments. This information suggests that the carbon oxygen bond in the alcohol is not broken during the esterification process and that the acid catalyst complexes with the carboxylic acid only. The transition state below seems likely from these experiments.

$$R-C\underset{OH}{\overset{\overset{\displaystyle (P)}{\underset{\delta-}{\overset{|}{AlCl_3}}}}{\overset{\delta-}{\cdots}}}O\overset{H}{\underset{\delta+}{\diagdown}}_{R'}$$

All substituted benzoic acids which were examined in competitive esterification rate studies reacted less rapidly than did benzoic acid itself under identical conditions. We believe this results from complexation between the substituent and the (P)-AlCl$_3$ and are currently investigating substituted benzoic acids with substituent groups which are not able to complex. Relative esterification rates for substituted benzoic acids are also given in Table III.

Table III
Relative Rates - Substituted Benzoic Acids[a]

Substituent	Rates, k_x/k_H
H	1.00
p-Cl	0.57
m-Cl	0.65
p-Br	0.52
p-OEt	0.33

a. At 65° with 1-butanol, 20:1 excess over the total acids.

Greater reactivity was noted in reactions run in benzene solutions than in carbon disulfide. This effect is due to the swelling of the polymer in benzene which is greater than the comparable swelling possible in carbon disulfide as well as to the higher maximum temperature possible in the former solvent.

The polymer protected aluminum chloride was also shown to be an effective catalyst for transesterification. Treatment of n-butyl propionate with 1-hexanol and (P)-AlCl$_3$ for 43 hours at 95° gave n-hexyl propionate in 57% yield. The non-catalyzed reaction under identical conditions produced only 8% yield of the n-hexyl ester.

Investigations are in progress concerning other applications of (P)-AlCl$_3$ and its derivatives in synthetic transformations.

Acknowledgements: Support from the National Science Foundation (Grant No. GP 33566) and the Research Corporation is gratefully acknowledged. In addition, we thank the National Institute of Health-General Medical Sciences for a special post-doctoral fellowship for one of us (ECB).

REFERENCES

1. On leave (1972-1973) from Rollins College, Winter Park, Florida.

2. Fellow of the Alfred P. Sloan Foundation, 1971-73; to whom correspondance concerning this work should be addressed.

3. D.C. Neckers, D.A. Kooistra, and G.W. Green, J. Am. Chem. Soc., $\underline{94}$, 9284 (1972).

4. Not all ways in which polymer protected reagents can be used are advantageous. For example, throughout the literature, various workers have used Friedel Crafts reactions on polymer supports. Recent examples were reported by Hayatsu and Khorana, J.Am.Chem.Soc., $\underline{88}$, 3182 (1966); ibid., $\underline{94}$, 4855 (1972). In the latter case, an asymmetric synthesis on a polymer support was reported. We urge precaution in the use of AlCl$_3$ in polystyrene reactions. The bound complexes which it forms with polystyrene-divinylbenzene copolymer beads are very hard to destroy so that AlCl$_3$ surely remains after typical and even not so typical wash procedures. The presence of AlCl$_3$ can be detected by the new band at 1650 cm^{-1}. If AlCl$_3$ is held on the polymer it may be partially released, perhaps not advantageously, later in the reaction sequence.

5. The water soluble complex is formed[3] from styrene-divinylbenzene (1.8%) copolymer beads (100-200 mesh). It contains 0.57% Al.

6. Products were identified by comparison of spectral characteristics with those of known samples.

53

Reprinted from *J. Amer. Chem. Soc.*, **95**(17), 5820–5822 (1973)

Polymer-Based Sensitizers for Photooxidations

ERICH C. BLOSSEY, DOUGLAS C. NECKERS, ARTHUR L. THAYER, and A. PAUL SCHAAP

Sir:

Insoluble polymer supports were introduced several years ago by Merrifield[1] and by Letsinger[2] to facilitate polypeptide synthesis. The technique involves the use of an insoluble styrene–divinylbenzene copolymer bead to provide a foundation upon which successive chemical transformations can be carried out.

For some time we have been interested in the use of insoluble polymer supports in photochemical reactions. In this report, we describe the preparation and use of the first example of a synthetically applicable, polymer-based photosensitizer. The reagent, polymer-based Rose Bengal (Ⓟ-Rose Bengal), is utilized to sensitize the generation of singlet molecular oxygen. Rose Bengal[3] is attached to a chloromethylated polystyrene support *via* the following procedure: Rose Bengal, 2.0 g (2.1 mmol), was stirred at reflux in 60 ml of reagent grade dimethylformamide with 2.0 g of chloromethylated styrene–divinylbenzene copolymer beads (1.38 mequiv of CH_2Cl, 50–100 mesh). After 20 hr, the polymer (now dark red) was filtered and washed successively with 150-ml portions of benzene, ethanol, ethanol–water (1:1), water, methanol–water (1:1), and methanol. After these washings, the final filtrate was colorless. The polymer beads[4] were dried in a vacuum oven to a final weight of 2.17 g.

Singlet molecular oxygen exhibits three modes of reaction with alkenes:[5] 1,4-cycloaddition with conjugated dienes to yield cyclic peroxides, an "ene" type reaction to form allylic hydroperoxides, and 1,2 cycloaddition[6] to give 1,2-dioxetanes which cleave thermally to carbonyl-containing products. Examples of all of these three reaction types have been carried out utilizing Ⓟ-Rose Bengal as a sensitizer (see Table I).

Table I. Photoxidations with Ⓟ-Rose Bengal

Singlet oxygen acceptor	Product	% yield (isolated)
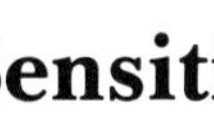 1	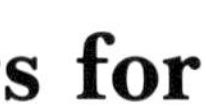 2	95
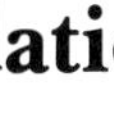 3	4	69
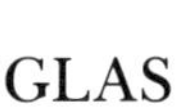 5	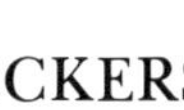 6	82

To a solution of 140 mg (0.6 mmol) of 1,2-diphenyl-*p*-dioxene (**1**) in 6 ml of CH_2Cl_2 was added 200 mg of sensitizer beads. The resultant mixture contained in a Pyrex vessel was vigorously stirred at 10° under O_2 and irradiated with a 500-W tungsten–halogen lamp through a uv-cutoff filter. Gas chromatography indicated complete oxidation of **1** after 6 hr. Removal of the sensitizer beads by filtration of the reaction mixture through a sintered glass disk[7] and removal of the solvent under vacuum gave colorless crystals of **2** in 95% yield. The photooxidation product was compared with an authentic sample of **2**.[8] Absorption spectra of the reaction solution before and after photolysis indicated that no Rose Bengal or other sensitizer is leached into the reaction solution.

The following control experiments indicate that the conversion of **1** to **2** is a singlet oxygen-mediated reaction. The reaction is inhibited by the addition of 10 mol % (based on **1**) of 1,4-diazabicyclo[2.2.2]octane

(1) R. B. Merrifield, *Science*, **150**, 178 (1965).
(2) R. L. Letsinger, M. J. Kornet, V. Mahedevon, and D. M. Jerina, *J. Amer. Chem. Soc.*, **86**, 5163 (1964).
(3) Rose Bengal

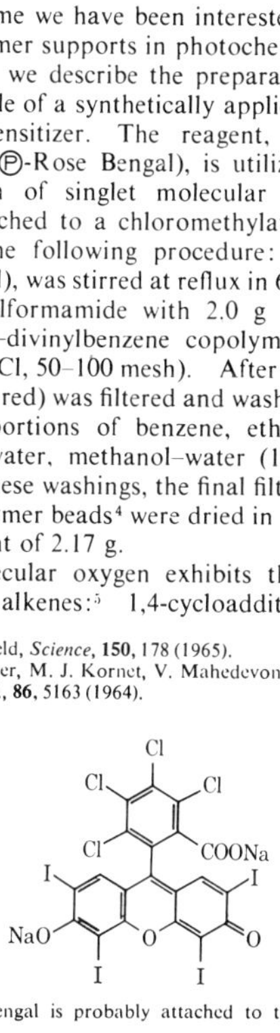

(4) The Rose Bengal is probably attached to the polymer as the carboxylate ester.
(5) (a) C. S. Foote, *Accounts Chem. Res.*, **1**, 104 (1968); (b) D. R. Kearns, *Chem. Rev.*, **71**, 395 (1971); (c) K. Gollnick and G. O. Schenck in "1,4-Cycloaddition Reactions," J. Hammer, Ed., Academic Press, New York, N. Y., 1967, p 255; (d) K. Gollnick, *Advan. Photochem.*, **6**, 1 (1968); (e) W. R. Adams in "Oxidation," Vol. 2, R. L. Augustine and D. J. Trecker, Ed., Marcel Dekker, New York, N. Y., 1971, p 65; (f) J. T. Hastings and T. Wilson, *Photophysiology*, **5**, 49 (1970).
(6) (a) A. P. Schaap and G. R. Faler, *J. Amer. Chem. Soc.*, **95**, 3381 (1973); (b) N. M. Hasty and D. R. Kearns, *ibid.*, **95**, 3380 (1973).
(7) The dried beads can be reused with no detectable decrease in efficiency.
(8) E. J. Bourne, M. Stacey, J. C. Tatlow, and J. M. Tedder, *J. Chem. Soc.*, 2976 (1949).

(DABCO), a singlet oxygen quencher.[9] The photooxidation of **1** can be carried out in the presence of 10 mol % of 2,6-di-*tert*-butylcresol, a free radical inhibitor. The conversion of **1** to **2** can also be effected by photooxidation with 562-nm radiation using a Bausch and Lomb grating monochromator and SP-200 mercury light source.[10] It should also be noted that a suspension of solid Rose Bengal in CH_2Cl_2 is relatively ineffective in photosensitizing the generation of singlet oxygen.

1,3-Cyclohexadiene (**3**) and tetramethylethylene (**5**) undergo the 1,4-cycloaddition and ene reactions, respectively, with singlet oxygen produced by Ⓟ-Rose Bengal sensitization. The reactions were carried out as described for the photooxidation of **1**. Products **4**[11] and **6**[12] were isolated by distillation under vacuum and compared with authentic samples.

One criterion for the generation of free singlet oxygen from various sources has been the product distribution obtained from 1,2-dimethylcyclohexene (**7**). Photooxidation of **7** with polymer-based Rose Bengal yields a similar distribution of the two possible ene products **8** and **9** (see Table II).

Table II. Oxidation of 1,2-Dimethylcyclohexene (**7**) Using Various Singlet Oxygen Sources

Sources	8	9	10
Ⓟ-Rose Bengal[a]	87	13	0
Photooxidation (soluble sens.)[b]	89	11	0
$OCl^- - H_2O_2$[b]	91	9	0
$(C_6H_5O)_3PO_3$[c]	96	4	0
K_3CrO_8[d]	82	18	0
Radical autoxidation[b]	6	39	54

[a] Products from this reaction were analyzed by gas chromatography as the alcohols obtained by triphenylphosphine reduction of **8** and **9**. [b] See ref 5a. [c] R. W. Murray and J. W.-P. Lin, *Ann. N. Y. Acad. Sci.*, **171**, 121 (1970). [d] J. W. Peters, J. N. Pitts, Jr., I. Rosenthal, and H. Fuhr, *J. Amer. Chem. Soc.*, **94**, 4348 (1972).

Dihydropyran (**11**) is a singlet oxygen acceptor that also yields two products: **13** is obtained by thermal cleavage of the 1,2-dioxetane **12** and **15** is formed upon dehydration under the reaction conditions of the ene product **14**. Photooxidation of **11** in CH_2Cl_2 with tetraphenylporphine gives 73% **13** and 27% **15**.[13]

With the Ⓟ-Rose Bengal sensitizer in CH_2Cl_2, the photooxidation of **11** gives an identical product distribution.

We conclude, on the basis of the experiments described in this report, that free singlet oxygen is efficiently formed by energy transfer from Ⓟ-Rose Bengal to oxygen. The possible uses for an insoluble, easily recovered sensitizer in preparative photochemical reactions are obvious. Insoluble polymer-based sensitizers may also be useful in mechanistic investigations in which the particular sensitizer is itself insoluble in the solvent of choice. Experiments with other types of Ⓟ sensitizers are in progress.

Acknowledgment. Financial support to D. C. N. from the Research Corporation (Cottrell Research Grant) and the National Science Foundation (GP-33566), to E. C. B. from the National Institutes of Health for a special post-doctoral fellowship, and to A. P. S. from the Research Corporation (Cottrell Research Grant) and the U. S. Army Research Office—Durham is gratefully acknowledged. The authors wish to thank Dow Chemical Co. for a gift of styrene–divinylbenzene copolymer beads.

(14) On leave from Rollins College, 1972–1973.
(15) Fellow of the Alfred P. Sloan Foundation, 1971–1973.

Erich C. Blossey,[14] *Douglas C. Neckers*[15]
Department of Chemistry, The University of New Mexico
Albuquerque, New Mexico 87106

Arthur L. Thayer, A. Paul Schaap
Department of Chemistry, Wayne State University
Detroit, Michigan 48202
Received June 15, 1973

(9) C. Ovannès and T. Wilson, *J. Amer. Chem. Soc.*, **90**, 6527 (1968).
(10) Rose Bengal: $\lambda_{max}^{(CH_3)_2CO}$ 562 nm.
(11) G. O. Schenck and D. E. Dunlap, *Angew. Chem.*, **68**, 248 (1956).
(12) C. S. Foote and S. Wexler, *J. Amer. Chem. Soc.*, **86**, 2879 (1964).
(13) The product distribution from **11** is independent of the sensitizer used but a function of the solvent employed for the reaction: P. D. Bartlett, G. D. Mendenhall, and A. P. Schaap, *Ann. N. Y. Acad. Sci.*, **171**, 79 (1970).

VII

Synthetic Applications of Polymer Supports in Specific Reactions

Editors' Comments on Papers 54 Through 62

54 **Patchornik and Kraus:** *Reactive Species Mutually Isolated on Insoluble Polymeric Carriers: I. The Directed Monoacylation of Esters*

55 **Kraus and Patchornik:** *The Directed Mixed Ester Condensation of Two Acids Bound to a Common Polymer Backbone*

56 **Crowley and Rapoport:** *Cyclization via Solid Phase Synthesis: Unidirectional Dieckmann Products from Solid Phase and Benzyl Triethylcarbinyl Pimelates*

57 **Crowley, Harvey, and Rapoport:** *Solid Phase Synthesis: Evidence for and Quantification of Intraresin Reactions*

58 **Harrison:** *The Effect of Ring Size on Threading Reactions of Macrocycles*

59 **Letsinger and Saverside:** *Selectivity in Solvolyses Catalyzed by Poly-(4-vinylpyridine)*

60 **Letsinger and Klaus:** *Investigation of a Synthetic Catalytic System Exhibiting Substrate Selectivity and Competitive Inhibition*

61 **Letsinger and Klaus:** *Selective Catalysis Involving Reversible Association of a Synthetic Polymeric Catalyst and Substrate*

62 **Letsinger and Wagner:** *Regulation of Rate of Reaction of Polyuridylic Acid Derivative by Use of Suppressor and Antisuppressor Molecules*

Monoacylation of Esters

The monoacylation of esters is often a difficult reaction to control in solution. Thus, in attempts to acylate esters that have more than one *alpha*-hydrogen, self-condensation of the ester, as well as diacylation, competes with the desired monoacylation process. An illustration of the problem, given with ethyl phenyl acetate, follows.

Desired reaction:

$$C_6H_5CH_2COOEt \xrightarrow{B^-} C_6H_5\bar{C}HCOOEt \xrightarrow{RCOX} C_6H_5\underset{\underset{R}{\underset{\|}{C=O}}}{C}HOOEt$$

Competing reactions:
Diacylation:

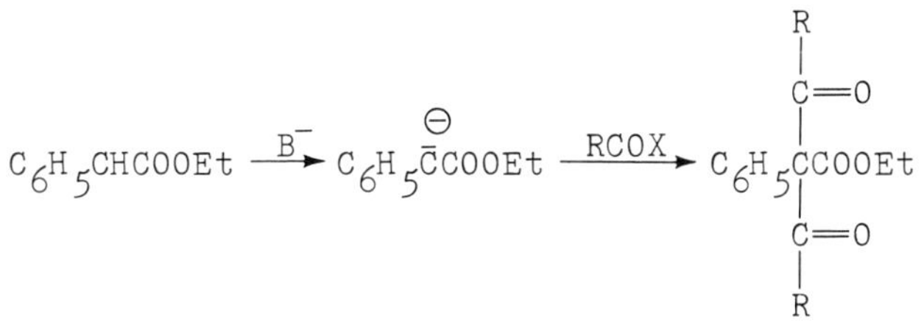

$$C_6H_5CHCOOEt \xrightarrow{B^-} C_6H_5\bar{C}COOEt \xrightarrow{RCOX} C_6H_5\underset{\underset{R}{\overset{\|}{C=O}}}{\overset{\overset{R}{\overset{\|}{C=O}}}{C}}COOEt$$

Self-condensation:

$$C_6H_5\bar{C}HCOOEt \xrightarrow{C_6H_5CH_2COOEt} C_6H_5CH_2\underset{OEt}{\overset{\overset{\ominus}{O}}{C}}-\underset{C_6H_5}{\overset{C_6H_5}{C}}HOOEt$$

$$\downarrow$$

$$C_6H_5CH_2\overset{O}{\overset{\|}{C}}\underset{C_6H_5}{C}HCOOEt$$

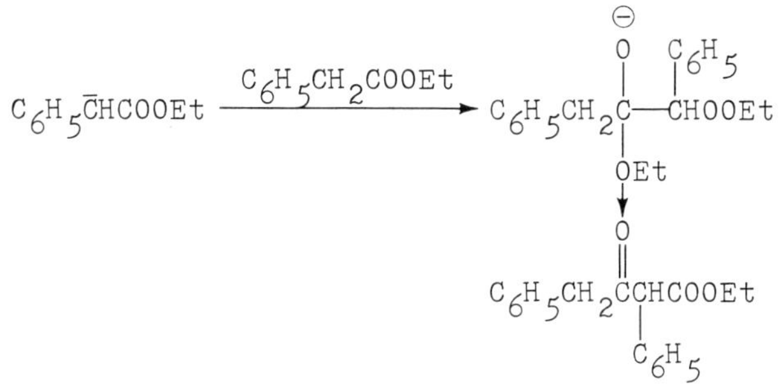

Self-condensation is particularly important in the case of ethyl phenyl acetate, and, in fact, dibenzoylketone is made synthetically using this reaction.

Patchornik's idea was that if the ester residue was attached to a polymer and if the residue was held there in very low concentration, it could essentially be isolated on a polymer support. The two communications of Patchornik and Kraus (Papers 54 and 55) illustrate this point.

Dieckmann Condensation

A related problem arises when one attempts to carry out a Dieckmann condensation of a mixed ester. Crowley and Rapoport (Paper 56) have used a polymer support to immobilize one end of a diester so that the only products isolated in a mixed Dieckmann condensation are the desired products. Their scheme is shown next.

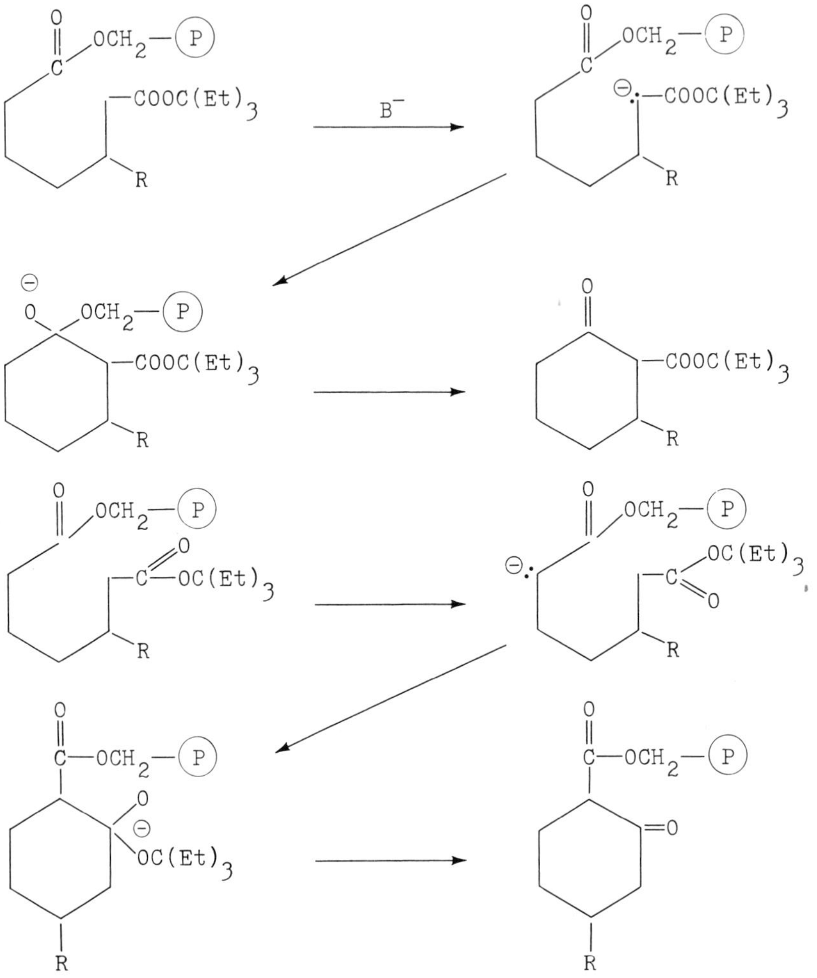
OCH2—P
COOC(Et)3
R
B⁻
still on the polymer
and not isolated

The Rapoport method, although interesting, appears limited to isotopically labeled compounds. In its present state, more experiments are needed to determine the generality of the method (Paper 57).

The synthesis of hooplane demonstrates that chemists, too, can have a sense of humor. The ingenuity and cleverness of the synthesis speak for themselves in Paper 58 by Harrison.

One of the many other interesting ways in which polymer supports can be used in chemical reactions is as selective catalysts. Letsinger's many reports describe the use of poly(4-vinylpyridine), poly(N-vinylimidazole), and polyuridylic acid for this purpose (Papers 59–62):

Reactive Species Mutually Isolated on Insoluble Polymeric Carriers: I. The Directed Monoacylation of Esters

ABRAHAM PATCHORNIK and MENAHEM A. KRAUS

Sir:

When molecules of a substance, A, are bound to an insoluble, cross-linked polymer, their free motion will be restricted due to the relative rigidity of the polymeric lattice. If the molecules are bound at the appropriate mutual distances, a situation approaching infinite dilution can thus be obtained, while actual "concentration" may still be relatively high. In this manner intermolecular reactions between the molecules of A (or reactive intermediates derived therefrom) are minimized, and, depending on the case, these molecules can be made to react either with a soluble reagent, B, or intramolecularly to give cyclic compounds.

This concept of "immobilization-on-polymer" has recently been used in the synthesis of cyclic peptides.[1-4] In this communication we wish to illustrate the general applicability of this novel approach by reporting on the directed monoacylation of polymer-bound ester enolates.

When attempting to acylate esters having more than one α-hydrogen by treating the corresponding enolate with an acyl halide, two competing reactions are reported to occur: self-condensation of the ester to be acylated and diacylation due to proton transfer from the monoacylated ester to still unreacted enolate.[5,6]

In the following it is shown that these side reactions can be avoided by "immobilizing" the ester enolates to be acylated on insoluble polymeric carriers, thus separating them from each other and from un-ionized ester molecules.

Polymer esters of phenylacetic and acetic acids (3; 0.1–0.3 mmol of ester/g of polymer[7]) were prepared by reaction of either acid (2) with chloromethylated polystyrene–2% divinylbenzene (1).[9] The remaining chloromethyl groups were treated with excess ethyl mercaptan.

The polymer ester was swelled in toluene–20% 1,2-dimethoxyethane and then converted into the enolate by the action of an equivalent amount of trityllithium in tetrahydrofuran[10] at 0° under dry argon.

$$\text{ⓟ–CH}_2\text{Cl} + \text{R}_1\text{CH}_2\text{CO}_2\text{H} \longrightarrow \text{ⓟ–CH}_2\text{OCOCH}_2\text{R}_1 \xrightarrow{(\text{Ph})_3\text{CLi}}$$
$$\quad\quad 1 \quad\quad\quad\quad 2 \quad\quad\quad\quad\quad\quad 3$$

$$\text{ⓟ–CH}_2\text{OCOCHR}_1 \xrightarrow{\text{R}_2\text{COX}} \text{ⓟ–CH}_2\text{OCOCH}(\text{R}_1)\text{COR}_2 \xrightarrow[-\text{CO}_2]{\text{HBr–CF}_3\text{CO}_2\text{H}}$$
$$\quad\quad\quad 4 \quad\quad\quad\quad\quad\quad\quad\quad 5$$

$$\text{R}_2\text{COCH}_2\text{R}_1$$
$$6$$

ⓟ = insoluble polymer
X = Cl or OCOR$_2$

(1) M. Fridkin, A. Patchornik, and E. Katchalski, *J. Amer. Chem. Soc.*, **87**, 4646 (1965).
(2) E. Bondi, M. Fridkin, and A. Patchornik, *Israel J. Chem.*, **6**, 22p (1968).
(3) L. Yu. Sklyarov and I. V. Shashkova, *Zh. Obshch. Khim.*, **39**, 2778 (1969).
(4) F. Flanigan and G. R. Marshall, *Tetrahedron Lett.*, 2403 (1970).
(5) B. E. Hudson, Jr., and C. R. Hauser, *J. Amer. Chem. Soc.*, **63**, 3156 (1941).
(6) D. F. Thomson, P. L. Bayless, and C. R. Hauser, *J. Org. Chem.*, **19**, 1490 (1954).
(7) Determined in the case of phenylacetic acid by gas chromatography of methyl phenylacetate, obtained by hydrolysis of polymer phenylacetate and esterification of the released acid. In the case of acetic acid esters, acetyl residues on the polymer were determined by Wiesenberger's method[8] as modified in our laboratory by S. Rogozinski.
(8) G. Ingram in "Comprehensive Analytical Chemistry," Vol. IB, C. L. Wilson and D. W. Wilson, Ed., Elsevier, Amsterdam, 1960, p 642.
(9) R. B. Merrifield, *J. Amer. Chem. Soc.*, **85**, 2149 (1963).
(10) P. Tomboulian and K. Stehower, *J. Org. Chem.*, **33**, 1509 (1968).

Table I. The Acylation of Polymer Esters

Ester	Acylating agent	Product	Mp, °C (lit.)	Yield,[a] %	Yield of unreacted acid, %
ⓟ–CH$_2$OCOCH$_2$C$_6$H$_5$	p-NO$_2$C$_6$H$_4$COCl	p-NO$_2$C$_6$H$_4$COCH$_2$C$_6$H$_5$ **7**	159–160 (160–160.5)[b]	43	40
ⓟ–CH$_2$OCOCH$_2$C$_6$H$_5$	p-BrC$_6$H$_4$COCl	p-BrC$_6$H$_4$COCH$_2$C$_6$H$_5$ **8**	104–105 (103)[c]	40	45
ⓟ–CH$_2$OCOCH$_2$C$_6$H$_5$	(α-C$_{10}$H$_7$CH$_2$CO)$_2$O	α-C$_{10}$H$_7$CH$_2$COCH$_2$C$_6$H$_5$ **9**	57–58 (58.5–59)[d]	40	55[e]
ⓟ–CH$_2$OCOCH$_3$	p-NO$_2$C$_6$H$_4$COCl	p-NO$_2$C$_6$H$_4$COCH$_3$ **10**	80–81 (78.5–80)[f]	20	45

[a] Based on the amount of purified product relative to that of starting ester. [b] H. Zimmer and J. P. Bercz, *Justus Liebigs Ann. Chem.*, **686**, 107 (1965). [c] G. I. Chervenyuk and V. P. Kravets, *Ukr. Khim. Zh.*, **30**, 1335 (1964); *Chem. Abstr.*, **62**, 9051e (1965). [d] Ch. Ivanov and L. Mladenova-Orlinova, *Angew. Chem.*, **76**, 301(1964). [e] In this case polystyrene-25% divinylbenzene was used as the carrier. [f] H. G. Walker and C. R. Hauser, *J. Amer. Chem. Soc.*, **68**, 1386 (1946).

After the disappearance of the red color of the base (1–5 min), 1.5 equiv of an acid chloride or anhydride was added and the mixture stirred for 1 hr at room temperature. The polymer was then filtered, washed thoroughly with benzene, water, and methanol, and dried. Upon cleavage with dry HBr in trifluoroacetic acid at room temperature a single ketone **6** and unreacted acid were obtained in every case [thin layer chromatography (tlc)]. The ketones were identified by their melting points and by nmr, ir, and mass spectra. Results are summarized in Table I.

In an analogous reaction in solution, performed under identical conditions, several ketonic products were formed. Ethyl phenylacetate was added to a tritylithium solution until the red color disappeared (final concentration of the ester was 0.2 M). Treatment with p-nitrobenzoyl chloride (as described above) and workup yielded a mixture containing five 2,4-dinitrophenylhydrazine (DNP)-positive compounds (tlc). After hydrolysis with H$_2$SO$_4$–HOAc–H$_2$O still five DNP-positive compounds were detected.[11] The yield of 4′-nitro-2-phenylacetophenone (**7**) was 22%.

In order for the "immobilization" method to succeed, concentrations of the species bound to the polymer must not be too high. Thus when polymer phenylacetate, 1.5 mmol/g, was treated with sodium *tert*-amylate for 45 min at 80°, a product of ester condensation, dibenzyl ketone, was obtained upon cleavage (mp 35°; 4% yield). Also, when polymer acetate, 2 mmol/g, was treated with p-nitrobenzoyl chloride (tritylithium as base, room temperature) two ketones were obtained upon cleavage: the desired 4′-nitroacetophenone (**10**, 4% yield) and p-nitrobenzoylacetone (10%, mp 110°;

(11) No attempt was made to identify all ketones either before or after hydrolysis.

lit.[12] 111–113°; molecular peak m/e 207). The latter is apparently formed by nitrobenzoylation of the ester condensation product, ⓟ–OCOCH$_2$COCH$_3$.

Spectroscopic evidence for a polymeric dilution effect was observed in the ir spectra of compounds of the type ⓟ–O(CH$_2$)$_n$OH. These alcohols were obtained by treating chloromethylated polystyrene-2% divinylbenzene with an excess of the monosodium salts of α,ω-diols. The compound derived from ethylene glycol, ⓟ–OCH$_2$CH$_2$OH (1 mmol/g), showed mainly a free OH absorption (3580 cm^{-1}, sharp; KBr), while the corresponding 1,4-butanediol derivative, ⓟ–O(CH$_2$)$_4$OH (1 mmol/g), showed both a weak free OH band and a much stronger hydrogen-bonded OH absorption (3300 cm^{-1} broad).[13]

At present we are studying other directed carbanionic reactions of polymer-bound species, including various cyclization reactions. In preliminary experiments, we have obtained by this method good yields of pure carbocyclic compounds having simple (five- and six-membered) rings. We used the reaction of α,ω-dibromoalkanes with polymer-bound malonate, ⓟ–OCOCH$_2$COOC$_2$H$_5$, and the Dieckmann cyclization of diesters of the type ⓟ–OCO(CH$_2$)$_n$COOR. When n was 8 and 14 in the Dieckmann cyclization, there were indications of the formation of 9- and 15-membered rings, respectively.

(12) A. Sieglitz and O. Horn, *Chem. Ber.*, **84**, 607 (1951).
(13) L. J. Bellamy, "The Infra-red Spectra of Complex Molecules," 2nd ed, Methuen, London, 1958, pp 96–99.

Department of Biophysics, The Weizmann Institute of Science
Rehovot, Israel
Received July 20, 1970

The Directed Mixed Ester Condensation of Two Acids Bound to a Common Polymer Backbone

MENAHEM A. KRAUS and ABRAHAM PATCHORNIK

Sir:

In previous reports from this laboratory, it has been demonstrated that a cross-linked polymer exerts a strong immobilizing effect on molecules covalently bound to it. The effect was illustrated by the directed monoacylation[1] and monoalkylation[2] of polymer esters. It was shown that at sufficiently low concentration and temperature, polymer-bound molecules behave virtually as in an infinitely dilute solution.

In this communication we wish to report on the interaction of *two* compounds bound to the same polymer bead occurring at the other extreme of *high* concentrations of bound species. Apparently under these conditions a close mutual proximity is imposed on some of the bound molecules by the rigid polymeric lattice. The selectivity of such "intrapolymeric" reactions and the high yields obtainable will be illustrated by the mixed ester condensation of two carboxylic acids bound to a common polymer backbone.[3]

A polymer, **3**, containing a low concentration of moieties of an enolizable acid[4] was obtained by treating chloromethylated polystyrene–2% divinylbenzene (**1**)[5] with a limited amount of acid **2**[6] (see Scheme I). After thorough

Scheme I

$$\text{P}\begin{array}{c}\text{CH}_2\text{Cl}\\-\text{CH}_2\text{Cl}\\\text{CH}_2\text{Cl}\end{array} + \text{R}_1\text{CH}_2\text{CO}_2\text{H} \longrightarrow \text{P}\begin{array}{c}\text{CH}_2\text{Cl}\\-\text{CH}_2\text{OCOCH}_2\text{R}_1\\\text{CH}_2\text{Cl}\end{array}$$

1 **2** **3**

$$\mathbf{3} + \text{R}_2\text{CO}_2\text{H} \longrightarrow \text{P}\begin{array}{c}\text{CH}_2\text{OCOR}_2\\-\text{CH}_2\text{OCOCH}_2\text{R}_1\\\text{CH}_2\text{OCOR}_2\end{array}$$

4 **5**, Ⓟ = insoluble polymer

washing and drying, polymer **3** was treated with an excess of a second, nonenolizable acid, **4**, to yield a polymer, **5**, which contained a relatively high concentration of ester moieties of **4**[4] (see Table I).

(1) A. Patchornik and M. A. Kraus, *J. Amer. Chem. Soc.*, **92**, 7587 (1970).
(2) M. A. Kraus and A. Patchornik, *Israel J. Chem.*, **9**, 269 (1971).
(3) The *self*-condensation of a polymer ester was mentioned previously.[1]
(4) Determined as described before.[1]
(5) Bio-Rad, SX-2; chlorine content, 5–10%.
(6) R. B. Merrifield, *J. Amer. Chem. Soc.*, **85**, 2149 (1963). The yield of this step is only about 30% but unreacted acid is easily recovered.

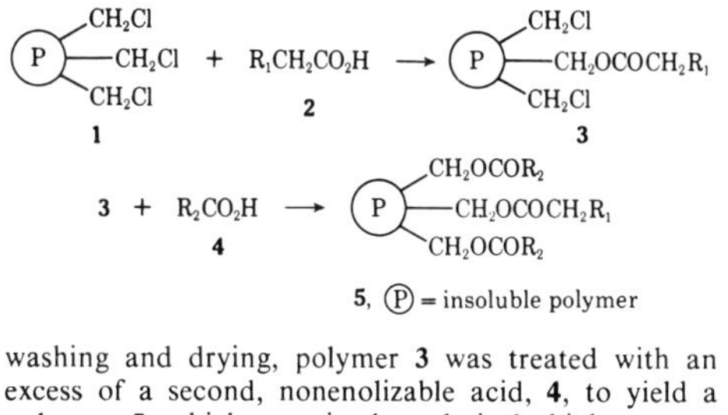

The ester condensation (Scheme II) was carried out by suspending polymer **5** in dry toluene–20% 1,2-dimethoxyethane under argon, and adding a solution of trityllithium (1 equiv, corresponding to moieties of **2**) in tetrahydrofuran[7] at room temperature. The red color of the base disappeared within 2 min. After a further 5-min stirring the reaction mixture was neutralized and the polymer filtered and washed thoroughly. Cleavage of the benzyl ester groups in polymer **7** was effected by HBr in trifluoroacetic acid (TFA) (2 hr at room temperature). The polymer was filtered and washed and the filtrate evaporated. The residue was dissolved in toluene and refluxed for 30 min to effect decarboxylation of the β-keto acid, **8**. In every case a single ketone, **9**, and unreacted starting acids were the only products obtained (thin layer chromatography [tlc]).

Scheme II

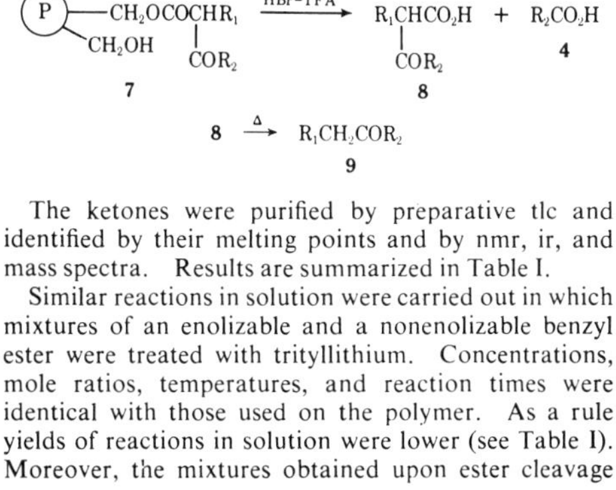

$$\text{P}\begin{array}{c}\text{CH}_2\text{OCOR}_2\\-\text{CH}_2\text{OCOCH}_2\text{R}_1\\\text{CH}_2\text{OCOR}_2\end{array} \xrightarrow{(\text{Ph})_3\text{CLi}}$$

5

$$\text{P}\begin{array}{c}\text{CH}_2\text{OCOR}_2\\-\text{CH}_2\text{OCO}\overline{\text{C}}\text{HR}_1\\\text{CH}_2\text{OCOR}_2\end{array} \xrightarrow{(\text{H}^+)}$$

6

$$\text{P}\begin{array}{c}\text{CH}_2\text{OCOR}_2\\-\text{CH}_2\text{OCOCHR}_1\\\text{CH}_2\text{OH}|\\\text{COR}_2\end{array} \xrightarrow{\text{HBr-TFA}} \text{R}_1\text{CHCO}_2\text{H} + \text{R}_2\text{CO}_2\text{H}$$
$$|$$
$$\text{COR}_2\mathbf{4}$$

7 **8**

$$\mathbf{8} \xrightarrow{\Delta} \text{R}_1\text{CH}_2\text{COR}_2$$

9

The ketones were purified by preparative tlc and identified by their melting points and by nmr, ir, and mass spectra. Results are summarized in Table I.

Similar reactions in solution were carried out in which mixtures of an enolizable and a nonenolizable benzyl ester were treated with trityllithium. Concentrations, mole ratios, temperatures, and reaction times were identical with those used on the polymer. As a rule yields of reactions in solution were lower (see Table I). Moreover, the mixtures obtained upon ester cleavage

(7) P. Tomboulian and K. Stehower, *J. Org. Chem.*, **33**, 1509 (1968).

Table I. The Condensation of Two Acids Bound to a Common Polymer Backbone

Enolizable acid (2)	Concn of 2 in 5, mmol/g	Nonenolizable acid 4	Concn of 4 in 5, mmol/g	Product, mp, °C (lit.)	Yield,[f] %	Yield of analogous reaction in solution, %
$CH_3(CH_2)_6CO_2H$	0.20	$p\text{-}ClC_6H_4CO_2H$	0.60	$p\text{-}ClC_6H_4CO(CH_2)_6CH_3$,[a] 58–59 (57.5–58)	35	30
$C_6H_5CH_2CO_2H$	0.11	$C_6H_5CO_2H$	0.52	$C_6H_5COCH_2C_6H_5$,[b] 56 (55–56)	45	
$C_6H_5(CH_2)_2CO_2H$	0.10	$C_6H_5CO_2H$	1.04	$C_6H_5CO(CH_2)_2C_6H_5$,[c] 72 (72–73)	85	42
$CH_3(CH_2)_4CO_2H$	0.07	$C_6H_5CO_2H$	1.71	$C_6H_5CO(CH_2)_4CH_3$[d] 23–24 (24.7°)	95	
$C_6H_5(CH_2)_2CO_2H$	0.04	$p\text{-}ClC_6H_4CO_2H$	1.70	$p\text{-}ClC_6H_4CO(CH_2)_2C_6H_5$,[e] 76–77 (78)	85	20

[a] L. N. Nikolenko and K. K. Babievskii, *Zh. Obshch. Khim.*, **25**, 2231 (1955). [b] C. F. H. Allen and W. E. Barker in "Organic Syntheses," Collect. Vol. II, Wiley, New York, N. Y., 1959, p 156. [c] D. Bar and Mme. Erb-Debruyne, *Ann. Pharm. Fr.*, **16**, 235 (1958). [d] F. L. Breusch and M. Oguzer, *Chem. Ber.*, **87**, 1225 (1954). [e] H. Burton and C. K. Ingold, *J. Chem. Soc.*, 904 (1928). [f] Based on the amount of the pure ketone relative to that of moieties of **2** in **5**.

(HBr in TFA) and decarboxylation were of much greater complexity, consisting of at least six major components (tlc). It will be noted that in reactions on the polymer the yield of ketone **9** increases with increasing ratio of nonenolizable to enolizable ester. A similar increase in ratio in solution did not improve the yield beyond 42%.[8]

Apparently in reactions on the polymer, at sufficiently high ratios the majority of enolizable ester groups (separated from each other by the polymer lattice) have some nonenolizable ester moieties in their close vicinity.

In order to confirm the mechanism proposed in Scheme II, namely the interaction of ester groups within the same polymer bead, equal amounts of two different batches of polymer, each containing a different ester, were mixed and treated with trityllithium for 10 min. One batch contained *p*-chlorobenzoate groups (1 mmol/g), the other 3-phenylpropionate (0.1 mmol/g). Upon cleavage and work-up as described above, no ketones whatsoever could be detected (tlc), the only products being unreacted starting acids. This result indicates that the condensations described are truly intrapolymeric and that no mechanism such as cleavage followed by condensation is involved.[11]

Acknowledgment. Stimulating discussions with C. Yaroslavsky and B. Amit are gratefully acknowledged.

(8) The mixed condensation of aliphatic with aromatic esters is well documented in the literature.[9] Though in some cases good yields of the condensation product are reported, they do not usually exceed 60%.[10] Self-condensation of the aliphatic ester is often difficult to avoid.

(9) C. R. Hauser and B. E. Hudson, Jr., *Org. React.*, **1**, 266 (1942).

(10) E. E. Royals, J. C. Hoppe, A. D. Jordan, Jr., and A. G. Robinson III, *J. Amer. Chem. Soc.*, **73**, 5857 (1951); E. E. Royals and D. G. Turpin, *ibid.*, **76**, 5452 (1954).

(11) Preliminary attempts to react polymer-bound acid **2** with an excess of a *soluble* ester of **4** have as yet been unsuccessful.

Department of Biophysics
The Weizmann Institute of Science
Rehovot, Israel

Received August 18, 1971

56

Reprinted from *J. Amer. Chem. Soc.*, **92**(21), 6363–6365 (1970)

Cyclization via Solid Phase Synthesis. Unidirectional Dieckmann Products from Solid Phase and Benzyl Triethylcarbinyl Pimelates

JOHN I. CROWLEY and HENRY RAPOPORT

Sir:

The Dieckmann cyclization of mixed esters has not been previously reported. In a study of the application of solid phase synthesis[1] to cyclization, we have isolated the products from the essentially unidirectional closure of unsubstituted and 3-ethyl-substituted 1-triethylcarbinyl-7-aralkyl pimelates-*7*-^{14}*C* (**1**) to triethylcarbinyl 2-oxocyclohexanecarboxylates-*2*-^{14}*C* (**2**). The preparation of **2b** is particularly significant since the closure of dialkyl 3-alkyladipates, pimelates, and suberates in the opposite direction is well documented.[2-4] We also have observed that replacement of the triethylcarbinyl by ethyl in **1a** and **1b** yields β-keto ester with the label extensively scrambled. This scrambling results from intervention of kinetically competitive transesterification of the original diester prior to cyclization. Use of the triethylcarbinyl moiety as one of the ester groups and as the base minimizes this transesterification, and the extent of transesterification can be measured by the distribution of label in the keto ester (Table I).

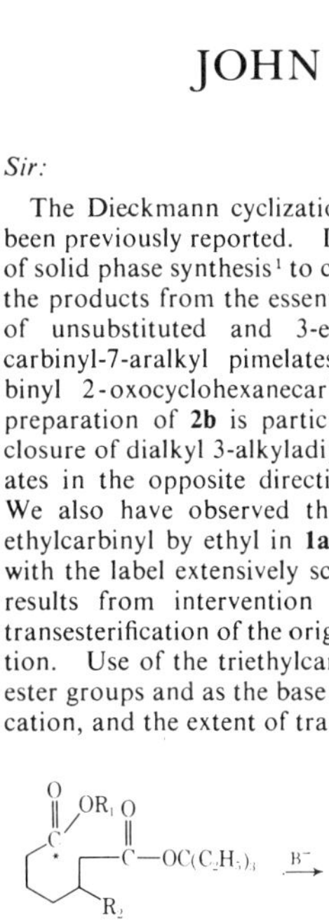

1a, $R_1 = CH_2$Ⓟ; $R_2 = H$
b, $R_1 = CH_2C_6H_5$; $R_2 = H$
c, $R_1 = CH_2$Ⓟ; $R_2 = C_2H_5$
d, $R_1 = CH_2C_6H_5$; $R_2 = C_2H_5$
Ⓟ = polystyrene

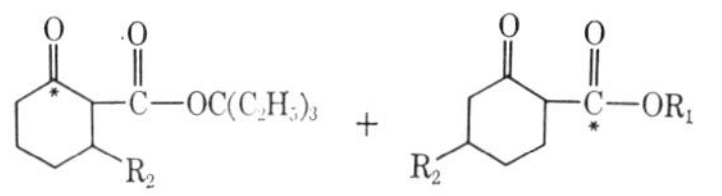

2a, $R_2 = H$ — **3a**, $R_1 = CH_2C_6H_5$; $R_2 = H$
b, $R_2 = C_2H_5$ — **b**, $R_1 = CH_2C_6H_5$; $R_2 = C_2H_5$

Table I. Yield and Distribution of Label in Dieckmann Products

Ester	% yield[a]	Mol % of benzyl product[b]	Specific activity, dpm/mmol × 10^{-4}: Keto ester	BaCO$_3$	Ketone	% activity in ester carbonyl
1a	46		7.8	0.025	7.8	0.31
1b	34	10	7.8	0.13	7.7	1.70
1c	15		53	1.5	52	2.75
1d	37	21	53	1.7	51	3.20

[a] Radiochemical yield, *i.e.*, total disintegrations per minute of chromatographically pure keto ester/total disintegrations per minute of pure diester or resin ester. [b] Benzyl-*tert*-alkyl ratio determined from ratio of aromatic H, δ 7.30, to triethylcarbinyl CH_3, δ 0.77.

Monoethyl pimelic acid-*1*-^{14}*C* was prepared and the unique location of its label established in the following way. Dibenzyl sodiomalonate-*1*-^{14}*C* was alkylated with ethyl 5-bromovalerate in DMF. Crude triester was converted to diacid mono ester by hydrogenolysis after which rapid distillation (170–190° (5 mm)) and fractionation yielded monoethyl pimelic acid-*1*-^{14}*C*. That the label was exclusively in the carboxylic acid carbon was established by decarboxylation[5,6] to radioinactive ethyl 6-bromohexanoate. Monotriethylcarbinyl pimelic acid-*1*-^{14}*C* was obtained from the monoethyl ester by transesterification.

To prepare the polystyrene beads, SX-2,[7] as support for the pimelate ester, the benzoate groups resulting from the benzoyl peroxide used to catalyze polymerization of the styrene were removed by refluxing with potassium *tert*-butoxide in toluene for 1 hr. These debenzoylated beads were chloromethylated by a modified procedure permitting greater control of the extent of chloromethylation. Using one-fifth as much stannic chloride as originally[1] in chloromethyl methyl ether at 0° we obtained resin with 1.5% chlorine in 8 min.

The pimelic half-ester was converted to its potassium salt by heating with a 10% excess of potassium carbonate in DMF for 30 min at 150°. Chloromethylated beads equivalent to 67 mol % of the carboxylate groups were added and the mixture held at 150° for 10 hr.

(1) R. B. Merrifield, *Biochemistry*, **3**, 1385 (1964).
(2) J. Schaeffer and J. J. Bloomfield, *Org. React.*, **15**, 1 (1967).
(3) H. L. Lochte and A. G. Pittman, *J. Org. Chem.*, **25**, 1462 (1960).
(4) O. S. Bhanot, *Indian J. Chem.*, **5**, 127 (1967).
(5) K. B. Wiberg and G. M. Lampman, *J. Amer. Chem. Soc.*, **88**, 4429 (1966).
(6) S. J. Cristol and W. C. Firth, *J. Org. Chem.*, **26**, 280 (1961).
(7) Bio Beads SX-2, 200–400 Mesh, Bio-Rad Laboratories, Richmond, Calif.

The resin ester **1a** was removed, washed serially with dioxane, water, methanol, methylene chloride, methanol, methylene chloride, methanol, and ether, and dried (12 hr at 55° (5 mm)). Esterifications employing triethylammonium salts[1] resulted in the competitive production of quaternary ammonium sites[8] on the resin while the use of hydroxylic solvents[9,10] led to the solvolytic formation of hydroxy- and alkoxybenzylic moieties as well as resin ester.

Changes in the resin functionality were followed by direct[11] scintillation counting of the resin and by infrared difference spectra employing pellets containing 10 mg of resin and 190 mg of potassium bromide. Reference beam pellets contained debenzoylated polystyrene. The resulting difference spectra were much more easily interpreted than were the direct spectra of functionalized beads and a satisfactory Beer's law plot could be obtained for chloromethylation based upon the H–C–Cl bending vibration at 1250 cm^{-1}. In a number of other cases, the difference spectra were essentially superimposable upon the spectra of p-cym-7-yl model compounds.

Cyclization was accomplished by addition of resin ester to a refluxing toluene solution of potassium triethylcarbinylate, 4.5 M equiv of base in 15 ml of toluene per gram of resin ester. After 2 min, the reaction was cooled, quenched with glacial acetic acid, and filtered, and the resin was washed twice with toluene. Concentration of the filtrate and washings, after aqueous bicarbonate extraction, yielded keto ester which was chromatographed on silica gel,[12] eluting with hexane followed by hexane–ether, 50:1, to yield keto ester: tlc R_F 0.35 (hexane–ether, 10:1); nmr (CCl$_4$) δ 0.77 (t, 9 H, CH$_3$), 1.67 (m, 6 H, CH$_2$), 1.75 (q, 6 H, CH$_2$), 2.13 (t, 2 H, CH$_2$), 12.3 (s, 1 H, enol).

The alkyl-substituted pimeloyl resin ester **1c** was prepared[13] and cyclized in an analogous manner, except that the triethylcarbinyl half-ester of 3-ethylpimelic acid could not be obtained by transesterification. The *tert*-butyl half-ester was prepared by transesterification, converted to the 1-*tert*-butyl 7-methyl diester with methyl iodide, cleaved with anhydrous TFA to 7-methyl half-ester, and converted through the ester acid chloride to 1-triethylcarbinyl 7-methyl diester with lithium triethyl carbinylate.[14] The methyl ester of this diester was selectively hydrolyzed with 1 equiv of potassium hydroxide in diglyme–water–THF, then lyophilized, and the lyophilizate in DMF was attached to the resin and cyclized as above to give keto ester: nmr (CCl$_4$) δ 0.77 (overlapping triplets, 12 H, CH$_3$); 1.05–2.40 (m, br, 13 H, CH$_2$, CH), includes 1.80 (q, 0.7 H, CO$_2$-CHC=O); 12.5 (s, 0.3 H, enol).

We obtained the unsubstituted benzyl triethylcarbinyl diester **1b** from benzyl bromide and potassium monotriethylcarbinyl pimelate. Transesterification of the 1-triethylcarbinyl-7-methyl 3-ethylpimelate at 25° (25 mm) yielded 7-benzyl 1-triethylcarbinyl 3-ethylpimelate (**1d**). Cyclization by the procedure already described yielded mixtures of triethylcarbinyl and benzyl β-keto esters, which could not be separated by chromatography. The mixed keto esters were hydrogenolyzed using 10% Pd–C as catalyst in ether, washed with aqueous bicarbonate, and concentrated to yield triethylcarbinyl keto esters; nmr established the presence of less than 1% of benzyl esters in both cases.

The distribution of label in the keto esters was determined by decarboxylation in refluxing 25% aqueous ethanolic 3 N hydrochloric acid; the evolved CO$_2$ was counted as barium carbonate.[15] The cyclohexanone and 3-ethylcyclohexanone were extracted into benzene and total activity was determined; the solutions then were analyzed by glpc to obtain ketone specific activities. Cyclization yields and distribution of label in the β-keto esters are shown in Table I.

The unidirectional closure of 1-triethylcarbinyl-7-aralkyl pimelates described here provides a route to unambiguously labeled or substituted triethylcarbinyl cyclohexanone-2-carboxylates which is superior to the existing multistep method.[16–18] Our method requires fewer steps, can be used with compounds containing acid-sensitive substituents, and is capable of extension to the synthesis of other than six-carbon rings.

The use of esters in which one carboxyl provides attachment to a resin support affords a clear benefit over the benzyl case in specificity and greatly simplified isolation and purification of the mixed diester and β-keto ester. We are presently investigating the generalization of this method, including the utilization of any hyperentropic effects available from solid phase mixed esters.

(8) R. B. Merrifield, *Advan. Enzymol.*, **32**, 221 (1969).
(9) M. Bodansky and J. T. Sheehan, *Chem. Ind. (London)*, 1597 (1966).
(10) J. M. Stewart and J. D. Young, "Solid Phase Peptide Synthesis," W. H. Freeman, San Francisco, Calif., 1969, p 27.
(11) C. L. Krumdieck and G. M. Baugh, *Biochemistry*, **8**, 1568 (1969).
(12) Silica Gel 0.05–0.2 mm extra pure for column chromatography, E. Merck (AG), Darmstadt, Germany.
(13) We are grateful to Dr. James Cason for a generous gift of 3-ethylglutaric acid, which was used in the preparation of methyl 3-ethyl-5-bromovalerate.
(14) C. R. Hauser and W. J. Chambers, *J. Amer. Chem. Soc.*, **78**, 3837 (1956).

(15) F. H. Woeller, *Anal. Biochem.*, **2**, 508 (1961).
(16) D. Vorlander, *Ann. Chem.*, **294**, 300 (1877).
(17) R. von Schilling and D. Vorlander, *ibid.*, **308**, 195 (1899).
(18) S. Mukherjee, *J. Indian Chem. Soc.*, **39**, 347 (1962).
(19) Supported in part by the U. S. Army Research Office, Durham, N. C.

Department of Chemistry, University of California
Berkeley, California 94720
Received July 18, 1970

Copyright © 1973 by Marcel-Dekker, Inc.

Reprinted from *J. Macromol. Sci. Chem.*, **A7**(5), 1117–1126 (1973)

Solid Phase Synthesis. Evidence for and Quantification of Intraresin Reactions

J. I. CROWLEY,* T. B. HARVEY, III, and H. RAPOPORT

Department of Chemistry
University of California
Berkeley, California 94720

ABSTRACT

Intrareaction occurs between moieties attached to copolystyrene-2% divinylbenzene resin as used in solid phase synthesis even when only 0.5% of the phenyl residues are functionalized. Evidence for this interaction has been obtained from the dimeric products resulting from Dieckmann cyclization of resin bound sebacates and ω-cyanopelargonyl thiol resin esters, from kinetic and product data on radio-activity scrambling during the Dieckmann cyclization of uniquely singly labeled tertiary alkyl pimeloyl resin esters, and from anhydride formation with carboxymethyl resin. The extent to which site-site interactions can occur as a function of the percentage functionalization has been measured quantitatively by radiotracer studies on intraresin anhydride formation from carboxymethyl substituted resin. The synthesis and characterization of the resin bound reactants is described, and the significance of these observations is discussed.

---

*Present address, IBM Research Laboratory, San Jose, California.

INTRODUCTION

The view is prevalent in solid phase synthesis that copolystyrene-2% divinylbenzene is a rigid polymer in which specific loci maintain their separation during reaction [1-6]. In connection with our investigation of the hyperentropic efficacy of solid phase cyclization [7], we have developed several lines of evidence which establish that intraresin or site-site reactions can occur to the extent of 50-80% of the total functionality with resin functionalized in the range normally employed in solid phase synthesis. Other reports of intraresin reactions have been made. Kraus and Patchornik [8] have observed such reactions using resin with percent substitution considerably higher than that usually employed in solid phase synthesis, and that fact is explicitly emphasized. In two recent reports, Collman et al. [9] and Beyerman et al. [10], in widely different reactions of resin bound moieties, also have reported the intrusion of undesirable intraresin reactions.

RESULTS AND DISCUSSION

We originally shared the belief that individual sites on the resin would remain separated during reaction. It appeared possible to take advantage of this separation to enhance the rate of cyclization relative to polymerization for moieties covalently bond to the copolystyrene-divinylbenzene resin, provided that the pendant reagent molecules were far enough from each other so that intramolecular reactions could not occur.

The Dieckmann cyclization of alkyl sebacyl resin esters was selected as a useful test of solid phase cyclization because the ineffectiveness of solution Dieckmann conditions, even at high dilution, has been very well documented [11]. For this purpose, t-butyl sebacyl resin ester (3) was prepared from potassium mono-t-butyl sebacate (2) and chloromethylated resin as previously described [7].

$$\text{\textcircled{P}}-CH_2Cl + (CH_3)_3COC(CH_2)_8CO_2^-K^+ \xrightarrow{DMF} \text{\textcircled{P}}-CH_2OC(CH_2)_8COC(CH_3)_3$$
$$\qquad\qquad\qquad\quad \parallel \qquad\qquad\qquad\qquad\qquad\qquad\qquad \parallel \quad\;\; \parallel$$
$$\qquad\qquad\qquad\quad O \qquad\qquad\qquad\qquad\qquad\qquad\qquad O \quad\;\; O$$
$$\quad 1 \qquad\qquad\qquad 2 \qquad\qquad\qquad\qquad\qquad\qquad\qquad\qquad 3$$

Dieckmann cyclization with potassium tert-butoxide in refluxing xylene yielded chiefly transesterification product 4.

$(CH_3)_3COC(CH_2)_8CO(CH_3)_3$
         ∥   ∥
         O   O

         4

However, small yields of cyclic dimeric diketoesters were obtained as both autocleaved and resin retained products:

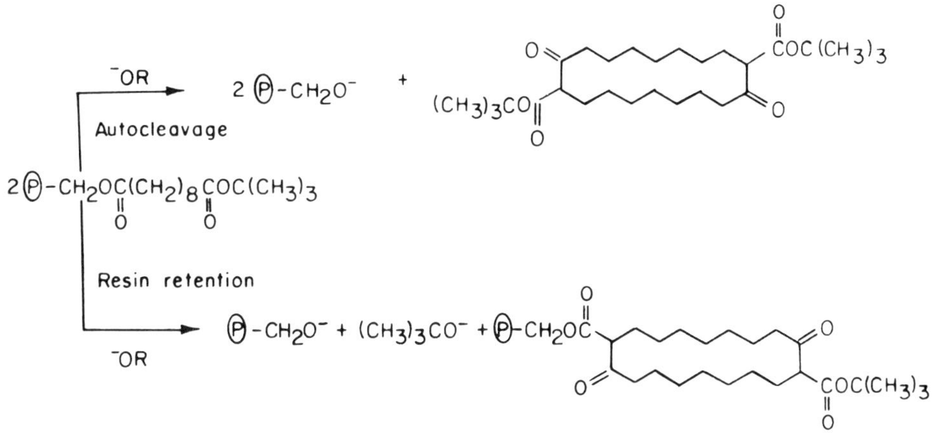

$2 \; \textcircled{P}-CH_2OC(CH_2)_8COC(CH_3)_3$

[Reaction scheme showing autocleavage pathway yielding $2\;\textcircled{P}-CH_2O^-$ + cyclic dimer with $(CH_3)_3CO$ groups, and resin retention pathway yielding $\textcircled{P}-CH_2O^-$ + $(CH_3)_3CO^-$ + resin-bound cyclic dimer]

The cleavage product was isolated as diketodiacid 5, while the resin retained product was identified as diketomonoacid 6, after HBr cleavage in methylene chloride.

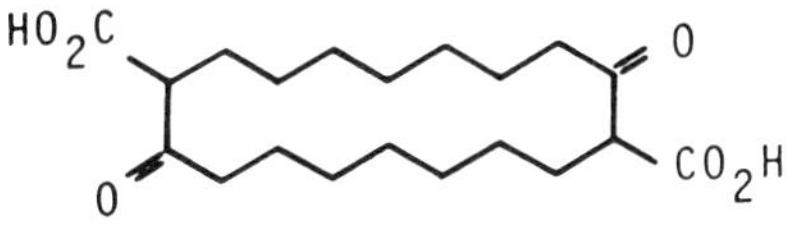

[Structures 5 and 6]

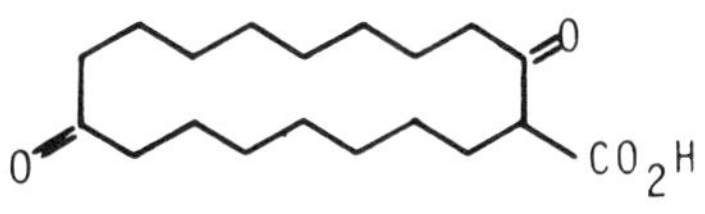

We selected ω-cyanopelargonyl thiol resin ester 9 as a sebacyl derivative which would be more reactive toward cyclization and less labile to cleavage under Dieckmann conditions. The thiol resin ester was prepared from ω-cyanopelargonyl chloride 7 [12, 13] and the previously unreported thiolmethyl resin 8, obtained from chloromethyl resin 1 (0.4 meq Cl/g resin) and KHS in DMF (160, 30 min).

$NC(CH_2)_8CCl \; + \; \textcircled{P}-CH_2SH \longrightarrow \textcircled{P}-CH_2SC(CH_2)_8CN$
           ∥   ∥
           O   O

    7   8   9

The authenticity of our synthesis was established by IR difference spectra [7] and cleavage of the cyanothiol ester from the resin as N,N-diethyl ω-cyanopelargonamide with silver acetate-diethylamine in dioxane at 120° for 16 hr. The cyanoamide was identified by IR, NMR, mass spectroscopy (MS), and elemental analysis.

$$\text{Ag}[\text{NH}(C_2H_5)_2]_2\text{OAc} + 9 \longrightarrow (C_2H_5)_2\underset{\underset{O}{\|}}{N}C(CH_2)_8CN$$

          10   11

Treatment of this resin thiol ester with lithium diethylamide yielded 40% cyanoamide, 10% dimeric diketodinitrile, and about 5% of 2-cyanocyclononanone, established by MS and by comparison (UV, TLC, GC) of the 2-cyanocyclononanone with authentic material [14].

Our yield of 2-cyanocyclononanone was seriously impaired by rapid conversion (several hours at room temperature) to azelaic acid when stored in ethyl ether, a decomposition which was confirmed with the independently synthesized cyanoketone. The azelaic acid was identified by IR, NMR and MS comparison with authentic material.

      12   13   14

When larger bases, e.g., lithium bistriethyl-disilazide [15], were employed, displacement to amide was prevented and the isolated product, 19% yield, was exclusively diketodinitrile; uncyclized cyanothiol ester was cleaved from product resin in 75% yield.

                            15

This predominant dimerization undoubtedly occurs because the individual moieties are not effectively separated from each other during reaction.

Further evidence against site isolation during reaction was provided by the label distribution in resin retained products from the Dieckmann cyclization of uniquely singly labeled alkyl pimeloyl resin esters, 16a*-c*.

$$R'O\overset{*}{\underset{\|}{C}}(CH_2)_5\overset{O}{\underset{\|}{C}}-OR$$

16

a, $R=C_2H_5$, $R'=$ⓟ-CH_2
b, $R=tert-C_4H_9$, $R'=$ⓟ-CH_2
c, $R=(C_2H_5)_3C$, $R'=$ⓟ-CH_2
d, $R=$ⓟ-CH_2, $R'=$ⓟ-CH_2
e, $R,R'=(C_2H_5)_3C$

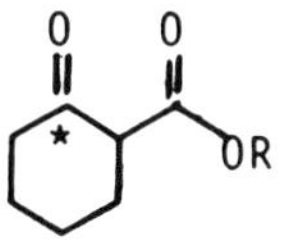

17

a, $R=C_2H_5$
b, $R=tert-C_4H_9$
c, $R=(C_2H_5)_3C$
d, $R=$ⓟ-CH_2, labeled carboxy, not keto

ⓟ = Polystyrene; 16c**, 16d**, and 16e** signify transesterification reaction products containing pimeloyl residues with the labeled carbon located at the original resin ester carbonyl together with structures in which the label has been scrambled into the nonresin ester carbonyl, or effectively scrambled because both ester groups are the same.

The preparation and analysis of label purity of starting materials has been described [7] together with some data on label distribution in autocleaved products 17a*-c*.

The kinetic relationships required to explain the observed label distribution of ketoester retained on the resin (17d**) prior to HBr/CH_2Cl_2 cleavage support our conclusion that pendant moieties on this type resin do not maintain the separation statistically implied by percent substitution and the well-defined bond lengths in polystyrene.

The pattern of label distribution in Dieckmann products from triethyl carbinyl pimeloyl resin ester are shown in Table 1. To obtain the label distribution in the resin retained ketoester 17d, the ketoacid was cleaved from the support with HBr in methylene chloride and, after concentration, the crude product was decarboxylated in refluxing alcoholic HCl. The CO_2 produced in this reaction was captured as $BaCO_3$ and counted [16].

We interpret the data shown in Table 1 as indicating rapid resin alkoxide participation in the transesterification process through Reactions (1) or (2a) and (2b).

TABLE 1. Mode and Relative Rate of Formation of 2-Carboxycyclohexanone Esters from Triethylcarbinyl Resin Pimelate

Cyclization[a]	% Yield[b]	Distribution[c]	Product[d]	Relative rate[e]
1. Attack at α-ester	46			
Direct		99.4	17c*	163
Transester		0.6	17c**	1.0
2. Attack at ω-ester	10			
Direct		4.0	17d*	1.4
Transester		96.0	17d**	35

[a]Cyclization direction is described by α for attack on the carbonyl nearest the resin, accompanied by autocleavage, or ω for the alternative direction. The process is further described as direct for cyclizations which do not result in scrambled label, and transester for those following a transesterification.

[b]Percent yield was obtained from the sum of activities of previously reported purified autocleaved ketoester and purified resin-retained activity prior to HBr/CH_2Cl_2 cleavage.

[c]Percent label distributions were obtained by computing twice the % activity observed in the scrambled location as transester and 100% minus transester as direct.

[d]Doubly labeled formulas indicate equimolar mixtures of singly labeled ketoesters.

[e]Relative rates of formation obtained from relative values of % yield times % label distribution.

$$\text{(P)}-CH_2O^- + \text{(P)}-CH_2O\overset{O}{\overset{\|}{C}}(CH_2)_5\overset{O}{\overset{\|}{C}}OC(C_2H_5)_3 \rightarrow \text{(P)}-CH_2O\overset{O}{\overset{\|}{C}}(CH_2)_5\overset{O}{\overset{\|}{C}}OCH_2-\text{(P)} \quad (1)$$

16c* 16d**

$$(C_2H_5)_3CO^- + \text{(P)}-CH_2O\overset{O}{\overset{\|}{C}}(CH_2)_5\overset{O}{\overset{\|}{C}}OC(C_2H_5)_3 \rightarrow (C_2H_5)_3CO\overset{O}{\overset{\|}{C}}(CH_2)_5\overset{O}{\overset{\|}{C}}OC(C_2H_5)_3 \quad (2a)$$

16c* 16e**

$$(C_2H_5)_3CO\overset{O}{\overset{\|}{C}}(CH_2)_5\overset{O}{\overset{\|}{C}}OC(C_2H_5)_3 + \text{(P)}-CH_2O^- \rightarrow \text{(P)}-CH_2O\overset{O}{\overset{\|}{C}}(CH_2)_5\overset{O}{\overset{\|}{C}}OC(C_2H_5)_3 \quad (2b)$$

16e** 16c**

The relative importance of the two reaction sequences can be assessed for the case of triethylcarbinyloxide catalyzed cyclization of triethylcarbinyl resin pimelate (16). The autocleaved ketoester has 99.7% of its activity in the keto group and 0.3% in the carboxyl [7], while the ketoester retained on the resin has 48% in the keto and 52% in the carboxyl. The data establish that the pattern of uniquely located labeling observed for triethylcarbinyl ketoester cleaved from the resin is not observed for ester retained on the resin. If the major source of 17d** is 16c**, then the product ratio 17c**/17d** must equal 17c*/17d*, or 163/1.4. In fact, this ratio is 1/35, requiring the major pathway to scrambled resin-retained ketoester to be through 16d** and not through 16c**.

We rule out independent routes to the components of the scrambled resin retained ketoester 17d** by Reactions (3) and (4).

$$\text{P}-CH_2O^- + \underset{17c^*}{\overset{*}{\bigcirc}}\overset{O}{\overset{\|}{C}}\overset{O}{\overset{\|}{C}}OC(C_2H_5)_3 \longrightarrow \underset{17d \text{ (labelled keto)}}{\overset{*}{\bigcirc}}\overset{O}{\overset{\|}{C}}\overset{O}{\overset{\|}{C}}OCH_2-\text{P} \quad (3)$$

$$\text{P}-CH_2O\overset{O}{\overset{\|}{\underset{*}{C}}}(CH_2)_5\overset{O}{\overset{\|}{C}}OC(C_2H_5)_3 + RO^- \longrightarrow \underset{17d \text{ (labelled carboxyl)}}{\text{P}-CH_2O\overset{O}{\overset{\|}{\underset{*}{C}}}\bigcirc\overset{O}{\overset{\|}{C}}} \quad (4)$$

16c*

If Reaction (3) were occurring to a significant extent, we could expect the observed mixture of keto and carboxyl labeled 17d** to yield scrambled 17c** by Reaction (5):

$$(C_2H_5)_3CO^- + \underset{17d^{**}}{\overset{*}{\bigcirc}\overset{*}{}}\overset{O}{\overset{\|}{C}}\overset{O}{\overset{\|}{C}}OCH_2-\text{P} \longrightarrow \underset{17c^{**}}{\overset{*}{\bigcirc}\overset{*}{}}\overset{O}{\overset{\|}{C}}\overset{O}{\overset{\|}{C}}OC(C_2H_5)_3 \quad (5)$$

Reactions (3) and (5) should have very similar transition states, and Reaction (5) should be faster in view of the severalfold higher concentration of triethyl carbinyl oxide relative to resin alkoxide. Furthermore, reaction between anionic alkoxide and anionic enolate is not a promising competitor for reaction between alkoxide and neutral ester. We can rule out Reaction (5), and therefore Reaction (3), on the basis of the very specific labeling pattern observed in the autocleaved 17c*.

For quantitative evaluation of intraresin reactions, carboxymethyl resin was prepared, and its ability to form anhydride was measured. Chloromethyl resin treated with sodium cyanide in DMF (100°, 24 hr) followed by hydrolysis in ethanolic potassium hydroxide [17] gave carboxymethyl resin. Carboxyl groups were converted to either intraresin or mixed trifluoroacetic anhydrides by treating this resin in methylene chloride with trifluoroacetic anhydride and silver trifluoroacetate. Repeated codistillation with methylene chloride then removed trifluoroacetic acid and anhydride, optimizing intraresin anhydride formation [18-20]. Washing this resin with 2% aqueous dioxane hydrolyzed mixed anhydride (IR 1860, 1792 cm^{-1}) without destroying intraresin anhydride (1821, 1748 cm^{-1}). The resin product also showed absorption at 1704 cm^{-1} due to resin acid formed by hydrolysis of the mixed anhydride.

The extent of intraresin anhydride formation was measured by reaction with 1-aminobutane-1-^{14}C (1.26 × 10^6 dpm/mmole) in methylene chloride, and the resin was counted in dioxane scintillation solution [21]. Ionically bound butylamine was removed by a 10% TFA/CH_2Cl_2 wash; control experiments established complete removal of amine from carboxymethyl resin which had not been treated with anhydride and from anhydrized resin after complete anhydride hydrolysis. The initial resin carboxylic acid content was determined by formation of radioactive butylamide through resin 2,4,5-trichlorophenyl ester. Percent intraresin anhydride formation vs percent carboxymethyl substitution of resin were 78 vs 9.8, 64 vs 4.7, and 53 vs 0.5, respectively. These data are plotted in Fig. 1.

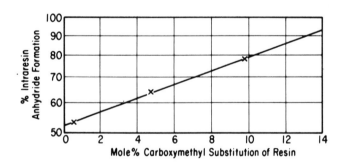

FIG. 1. Intraresin anhydride formation as a function of degree of resin carboxymethyl substitution.

CONCLUSIONS

These results demonstrate that the carboxylic acid groups are not isolated during reaction, and that intraresin reactions occur readily even at 0.5% substitution and even though the functionalized sites are randomly distributed. This inference of random site location was reached [4] on the basis of autoradiography of microtomed sections of peptide-labeled resin peptide ester. We found radioactive resin ester beads sized by screening and by CH_2Cl_2 flotation [22] displayed constant specific activity even though differing in diameter by a factor of 5. Our evidence for resin site interaction suggests that the observed random functionalization may be achieved by translocation of strands of solvent swollen beads and that neither line of evidence may bear on the question of diffuse vs surface reactivity, which was originally inferred from the previous observation [4]. We present this evidence, in agreement with that presented earlier [6], but independent of it, to show that statistically impressive uniformity of the functional distribution in the resin beads exists. This fact can be compatible with all the observations of interactions at high and low levels of functionalization only if the observable distribution and interactins are not related. We find that resin flexibility during chloromethylation and during further reactions is the best explanation for both kinds of data.

With longer molecules attached to the resin, this reactivity should increase due to added flexibility afforded by the increased chain length. Thus molecules attached to copolystyrene-2% divinylbenzene are not isolated to a significant extent during reaction.

Our evidence is directed at establishing that divinyl benzene-copolystyrene is quite flexible, and that those results reported to the present time which imply site separation must be explained on the basis of favorable kinetic relationships between the desired reactions and competitive reactions, the very real purification advantages offered by the solid phase method, and the introduction of steric influences around the substituted site, which cannot be involved in parallel reactions in solution. A further implication of these conclusions is that, as less kinetically favorable reactions are attempted with polymer bound reagents, the extent of site-site interactions will increase.

ACKNOWLEDGMENT

Supported in part by the U.S. Army Research Office, Durham, North Carolina.

REFERENCES

[1] M. Fridkin, A. Patchornik, and E. Katchalski, J. Amer. Chem. Soc., 87, 4646 (1965).
[2] E. Bondi, M. Fridkin, and A. Patchornik, Israel J. Chem., 6, 22p (1968).
[3] R. B. Merrifield, Prog. Hormone Res., 23, 469 (1967).
[4] R. B. Merrifield, Advan. Enzymol., 32, 265 (1969).
[5] E. Flanigan and G. R. Marshall, Tetrahedron Lett., 1970, 2403.
[6] A. Patchornik and M. A. Kraus, J. Amer. Chem. Soc., 92, 7587 (1970).
[7] J. I. Crowley and H. Rapoport, Ibid., 92, 6363 (1970), and other observations to be published.
[8] M. A. Kraus and A. Patchornik, Ibid., 93, 7387 (1971).
[9] J. P. Collman, L. S. Hegedus, M. P. Cooke, J. R. Norton, G. Dolcetti, and D. N. Marquardt, Ibid., 94, 1789 (1972).
[10] H. C. Beyerman, E. W. B. deLeer, and W. van Vossen, J. Chem. Soc., Chem. Commun., 1972, 929.
[11] J. P. Schaefer and J. J. Bloomfield, Org. Reactions, 15, 1 (1967).
[12] H. Rapoport and H. D. Baldridge, Jr., J. Amer. Chem. Soc., 73, 343 (1951).
[13] B. S. Biggs and W. S. Bishop, Org. Syntheses, Coll. Vol. III, 768 (1955).
[14] M. E. Kuehne, J. Amer. Chem. Soc., 81, 5400 (1959).
[15] C. R. Kruger and E. G. Rochow, J. Organometal. Chem., 1, 476 (1963).
[16] F. H. Woeller, Anal. Biochem., 2, 508 (1961).
[17] IR absorptions for cyanomethyl and carboxymethyl resins: T. Kusama and H. Hayatsu, Chem. Pharm. Bull., 18, 319 (1970).
[18] W. D. Emmons, K. S. McCallum, and A. F. Ferris, J. Amer. Chem. Soc., 75, 6047 (1953).
[19] E. J. Bourne, M. Stacey, J. C. Tatlow, and R. Worrall, J. Chem. Soc., 1954, 2006.
[20] A. F. Ferris and W. D. Emmons, J. Amer. Chem. Soc., 75, 232 (1953).
[21] C. L. Krumdieck and C. M. Baugh, Biochemistry, 8, 1568 (1969).
[22] J. W. M. Baxter, M. Manning, and W. H. Sawyer, Ibid., 8, 3592 (1969).

Received for publication January 29, 1973

The Effect of Ring Size on Threading Reactions of Macrocycles

By I. T. Harrison

(*Institute of Organic Chemistry, Syntex Research, Stanford Industrial Park, Palo Alto, California* 94304)

Summary 1,10-Bis(triphenylmethoxy)decane forms stable threaded compounds with macrocyclic hydrocarbons in the range C_{25}–C_{29} only.

Statistical and directed syntheses of compounds consisting of macrocycles threaded by methylene chains bearing large end groups, such as (**9**), have been described[1] previously. Described here is the application of the statistical method to the determination of the effect of macrocycle ring size on the formation and stability of threaded compounds.

A mixture of cyclic hydrocarbons (**7**), containing all homologues from cyclotetradecane to cyclodotetracontane, was first prepared as follows. A mixture of the five dicarboxylic acid monoesters (**1**) was subjected to a crossed Kolbe electrolysis in methanol–benzene containing sodium methoxide, forming the ester mixture (**2**) which was hydrolysed[2] to the monoester mixture (**3**). A mixture of (**1**) and (**3**) was further electrolysed forming the ester mixture (**4**) containing all homologues from dimethyl dodecane-1,12-dicarboxylate to dimethyl tetracontane-1,40-dicarboxylate. Cyclisation of (**4**) with sodium in refluxing xylene in the presence of chlorotrimethylsilane[3] gave, after acid hydrolysis, the acyloin mixture (**5**). Hydrogen iodide in acetic acid converted (**5**) into the ketone mixture (**6**) which was purified through the Girard derivative. Borohydride reduction[4] of the tosylhydrazone of (**6**) yielded the hydrocarbon mixture (**7**) which was freed from unsaturated impurities by treatment with alkaline sodium permanganate solution and purified by filtration of a solution in hexane through silica gel.

The macrocycle mixture (**7**) and 1,10-bis(triphenylmethoxy)decane (**8**) were heated together at 120° rapidly forming a small equilibrium concentration of the threaded compound (**9**), formed from the C_{29} macrocycle only (t.l.c.). Separation by column chromatography or by t.l.c. gave the compound (**9**) as well as the unchanged starting materials (**7**) and (**8**). Heating (**9**) to 120° caused the expected extrusion of the threading piece (**8**), $t_{1/2}$ *ca.* 10 min, releasing the C_{29} macrocycle (**12**) (g.l.c.). Since only the C_{29} macrocycle forms a stable threaded compound it appears that the C_{29} and smaller macrocycles do not allow passage of the blocking group at 120° while the C_{30} and higher macrocycles form only transient compounds (10) which separate into their components even at room temperature.†

In contrast to the above thermal synthesis of threaded compounds, the reaction of (7) with (8), catalysed by small quantities of trichloroacetic acid at 120° which reversibly cleaves the triphenylmethyl group allowing insertion‡ by the unblocked chain, formed a small amount of the macrocycle mixture (11), containing all rings from C_{25} to C_{29}. Acid hydrolysis of this product gave a mixture (13) of the five macrocycles cyclopentacosane to cyclononacosane (relative weight yields 1, 9, 23, 43, and 57) (g.l.c.–m.s.).§ Very small peaks in the gas chromatogram possibly correspond to the C_{23} and C_{24} macrocycles. However, no traces of the C_{30} and higher macrocycles were observed.

Molecular models¶ are consistent with the above results; a ring of 22 methylenes and a threaded methylene chain are in close contact, while the limiting ring size for passage of a triphenylmethyl group is about C_{29}. Models also show that the triphenylmethyl group does not pass through the C_{29} macrocycle by a symmetrical transition state but rather that the group must tilt relative to the ring allowing one or a group of two phenyls to pass first.

(*Received,* 10*th December* 1971; *Com.* 2102.)

† Previous work (ref. 1) has shown that stable threaded compounds can be prepared using the triphenylmethyl blocking group and C_{30} macrocyclic acyloins.

‡ The inserting species is presumably the monotriphenylmethyl ether of decane-1,10-diol.

§ We thank Dr. L. Tőkés and Mr. B. Amos for this analysis.

¶ Constructed from Corey-Pauling (CPK) atomic models.

[1] I. T. Harrison and S. Harrison, *J. Amer. Chem. Soc.*, 1967, **89**, 5723; G. Schill and H. Zollenkopf, *Annalen*, 1969, **721**, 53.
[2] L. J. Durham, D. J. McLeod, and J. Cason in 'Organic Syntheses', Interscience, New York, 1963, Coll Vol 4, p. 635.
[3] U. Schräpler and K. Rühlmann, *Chem. Ber.*, 1964, **97**, 1383.
[4] L. Caglioti and P. Grasselli, *Chem. and Ind.*, 1964, 153.

Erratum

On p. 231, the fourth line from the bottom of the second column should read: "the C_{29} macromolecule . . . ."

[CONTRIBUTION FROM THE CHEMISTRY DEPARTMENT OF NORTHWESTERN UNIVERSITY, EVANSTON, ILL.]

Selectivity in Solvolyses Catalyzed by Poly-(4-vinylpyridine)[1]

BY ROBERT L. LETSINGER AND THOMAS J. SAVEREIDE[2]

RECEIVED FEBRUARY 12, 1962

Partially protonated poly-(4-vinylpyridine) in an ethanol water solution serves as a particularly effective catalyst, relative to 4-picoline or to either non-protonated or highly protonated poly-(4-vinylpyridine), in solvolyses of nitrophenyl acetates which bear a negative electric charge. It is a poorer catalyst than 4-picoline for the solvolysis of 2,4-dinitrophenyl acetate, an electrically neutral substance. The selectivity of the partially protonated polymer with respect to charged substrates is attributed to polymer–counter ion electrostatic interaction, which increases the local concentration of an anionic substrate in the region of the polymer coil.

A catalyst is here termed "selective" if it distinguishes between substrates that differ only at positions far removed from the functional group undergoing transformation. Although highly selective catalysts, enzymes, are elaborated in abundance in living systems, little progress has been made in synthesizing agents with comparable properties. The action of an enzyme may be represented schematically by the sequence: $E + S \rightarrow E\text{-}S \rightarrow E +$ products, where E, S and E-S represent an enzyme, the substrate or substrates, and an enzyme-substrate complex, respectively. Selectivity in these reactions may be achieved by association of the substrate with the catalyst prior to the major covalent change in the substrate.

We report in this paper a study of the catalytic properties of poly-(4-vinylpyridine) in aqueous ethanol solutions. This polymer appeared promising as a selective catalyst since in mildly acidic solution it would possess both cationic sites and basic nitrogen sites. The former should serve to bind anionic substrates[3] and the latter should act as catalytic centers for the hydrolysis of nitrophenyl esters.[4] In addition, the relative number of basic and cationic sites would be subject to control, and, as a result of the flexibility of the polymer chain, both functions would coexist within a given molecule in a great variety of spatial relationships.

2,4-Dinitrophenyl acetate (DNPA), potassium 3-nitro-4-acetoxybenzenesulfonate (NABS), 5-nitro-4-acetoxysalicylic acid (NAS) and 3-nitro-4-acetoxybenzenearsonic acid (NABA) were selected as substrates. The dinitrophenyl acetate was chosen to illustrate the solvolytic behavior of an uncharged ester; the remaining esters, to reveal the effect of charge interaction involving polymer and substrate on the course of a catalyzed solvolysis.

DNPA: 2,4-dinitrophenyl acetate (O_2N–C$_6$H$_3$(NO$_2$)–OAc)

NABS: potassium 3-nitro-4-acetoxybenzenesulfonate (O_2N–C$_6$H$_3$(SO$_3^-$K$^+$)–OAc)

NAS: 5-nitro-4-acetoxysalicylic acid (O_2N–C$_6$H$_2$(OH)(COOH)–OAc)

NABA: 3-nitro-4-acetoxybenzenearsonic acid (O_2N–C$_6$H$_3$(AsO$_3$H$_2$)–OAc)

Of the previous attempts to achieve selectivity with synthetic catalysts, the most favorable results were obtained with a cross-linked sulfonated polystyrene resin, Dowex-50.[5] Whitaker and Deathe-

(1) This research was supported in part by a grant from the National Science Foundation, G7414. For a preliminary report, see R. L. Letsinger and T. J. Savereide, J. Am. Chem. Soc., **84**, 114 (1962).
(2) Hercules Powder Co. Fellow, 1958; Public Health Service Research Fellow, 1960.
(3) Numerous studies of counter ion binding by polyelectrolytes have been reported. See, for example: F. T. Wall and W. B. Hill, J. Am. Chem. Soc., **82**, 5599 (1960); F. T. Wall and M. J. Eitel, ibid., **79**, 1550, 1556 (1957); A. M. Liquori, F. Ascoli, C. Botre, V. Cresenzi and A. Mele, J. Polymer Sci., **40**, 169 (1959); M. Nagasawa and I. Kagawa, ibid., **25**, 61 (1957); I. Kagawa and K. Katsuura, ibid., **17**, 365 (1955); H. P. Gregor and D. H. Gold, J. Phys. Chem., **61**, 1347 (1956); P. Doty and G. Ehrlich, Ann. Rev. Chem., (1952); H. Morawetz, A. M. Kotliar and H. Mark, J. Phys. Chem., **58**, 19 (1954). It is noteworthy that H. Ladenheim, E. M. Loebl and H. Morawetz, J. Am. Chem. Soc., **81**, 20 (1959), found that poly-(4-vinylpyridine) functioned selectively in a non-catalytic reaction, quaternization with α-bromoacetamide and bromoacetate ion. See also H. Ladenheim and H. Morawetz, ibid., **81**, 4860 (1959).
(4) The subject of nucleophilic catalysis in the hydrolysis of nitrophenyl esters was reviewed recently by M. L. Bender, Chem. Rev., **60**, 53 (1960).
(5) J. R. Whitaker and F. E. Deatherage, J. Am. Chem. Soc., **77**, 3360, 5298 (1955).

rage found that glycylglycine hydrolyzed 35 times faster than acetylglycine with the resin as catalyst, whereas glycylglycine hydrolyzed at only $1/6$th the rate of acetylglycine in hydrochloric acid. The resin system was heterogeneous and it may be assumed that the basic peptide was preferentially held in the acid medium within the resin particles. Comparable selectivity has not been reported for non-enzymatic hydrolyses in homogeneous media; however, good evidence for "electrostatic catalysis" in the hydrolysis of o-nitrophenyl hydrogen oxalate catalyzed by 2-aminopyridine was obtained by Bender and Chow,[6] and polystyrenesulfonic and polyethylenesulfonic acid were shown by Kern and Sherhag[7] to be somewhat more effective than hydrochloric acid in hydrolyzing dipeptides. An appreciable effect of certain anions on the course of hydrolysis of proteins in dilute acid solutions was noted by Steinhardt and Fugitt.[8] Thus, in the presence of dodecyl sulfate ions, amide bonds in egg albumin were cleaved in preference to peptide bonds.

Experimental Section

Carbon, hydrogen and nitrogen analyses were performed by Miss Hilda Beck. Melting points were taken on a Fisher-Johns melting point block. The infrared spectra were determined with a Baird double beam recording spectrophotometer with the sample in potassium bromide unless otherwise stated.

5-Nitro-4-acetoxysalicylic acid was best prepared by partial acetylation of 2,4-dihydroxy-5-nitrobenzoic acid by a modification of the method of Hemmelmayr.[9] In our hands the original procedure yielded a product, m.p. 147-149°, which appeared to be the diacetate; λ^{KBr} 5.62, 5.88, 12.1 (absent in the monoacetate) μ.

Anal. Calcd. for $C_{11}H_9NO_8$: C, 46.7; H, 3.20; N, 4.95. Found: C, 47.3; H, 3.12; N, 5.16.

However, by using a molar ratio of sodium acetate to dihydroxynitrobenzoic acid of 1:10 rather than 1:1, we obtained a 30% yield of the monoacetate, m.p. 152-154°, lit.[9] m.p. 150°; λ^{KBr} 5.7, 6.05 7.02 (absent in the diacetate) μ.

Anal. Calcd. for $C_9H_7NO_7$: C, 44.8; H, 2.93; N, 5.81. Found: C, 44.6; H, 2.73; N, 6.12.

5-Nitro-4-acetoxysalicylic acid also was prepared independently by nitration of 4-acetoxysalicylic acid. The acetoxysalicylic acid (9.8 g.)[10] was added to a well stirred mixture of concd. nitric acid (13.5 g.) and acetic anhydride (30 ml.) at $-20°$. After 90 minutes of stirring at $-20°$ the mixture was poured into water and extracted with ether.[11] Distillation at reduced pressure of the volatile matter from the ether extract left 10.5 g. of yellow powder. After several recrystallizations of a portion (9.5 g.) of this powder, 0.35 g. of pale yellow plates of 5-nitro-4-acetoxysalicylic acid was obtained. This product was identical with that produced by acetylation of 2,4-dihydroxy-5-nitrobenzoic acid. Another

(6) M. L. Bender and Y. L. Chow, *J. Am. Chem. Soc.*, **81**, 3929 (1959).
(7) W. Kern and B. Sherhag, *Makromol. Chem.*, **28**, 209 (1958); W. Kern, W. Herold and B. Sherhag, *ibid.*, **17**, 231 (1955).
(8) J. Steinhardt, *J. Biol. Chem.*, **141**, 995 (1941); J. Steinhardt and C. H. Fugitt, *J. Res. Natl. Bur. Stand.*, **29**, 315 (1942). For related cases see G. Schramm and J. Primosigh, *Z. physiol. Chem.*, **283**, 34 (1948); E. Waldschmitt-Leitz and Fr. Zinnert, *Makromol. Chem.*, **6**, 272 (1951).
(9) F. V. Hemmelmayr, *Monatsh.*, **25**, 21 (1904).
(10) This acid was prepared by the procedure of R. Lesser and G. Gad, *Ber.*, **59**, 234 (1926).
(11) It is important that the ether extraction be carried out promptly. In cases where the mixture was first allowed to stand an hour, an isomeric acetate, probably 4-hydroxy-5-nitro-2-acetoxybenzoic acid, was isolated as well as the desired compound. The isomeric acetate melted at 188-189°; analysis: Found: C, 45.0; H, 2.90; N, 5.94. The infrared spectrum was very similar to that of 5-nitro-4-acetoxysalicylic acid but showed a peak at 10.7 μ absent in the spectrum of the lower melting isomer.

portion (0.234 g.) of the crude reaction mixture was treated with diazomethane and the products separated by chromatography on silica gel (10% ether in hexane solvent). **Methyl 2-methoxy-5-nitro-4-acetoxybenzoate** (0.11 g., m.p. 100-101°; λ^{CHCl_3} 5.65, 5.82 μ, no O-H absorption) was isolated. The same ester was also obtained by the action of diazomethane on a pure sample of 5-nitro-4-acetoxysalicylic acid.

Anal. (methyl ester). Calcd. for $C_{11}H_{11}NO_7$: C, 49.3; H, 3.76; N, 5.22. Found: C, 48.9; H, 3.97; N, 5.48.

Alkaline hydrolysis of methyl 2-methoxy-5-nitro-4-acetoxybenzoate afforded a good yield of 2-methoxy-4-hydroxy-5-nitrobenzoic acid, m.p. 191-192°,[12] 192°; λ^{KBr} 2.90, 5.86 μ; and the alkaline hydrolysis of 4-acetoxy-5-nitrosalicylic acid yielded 2,4-dihydroxy-5-nitrobenzoic acid, m.p. 215°, λ^{KBr} 5.97μ.

The midpoint in the titration curve of 5-nitro-4-acetoxysalicylic acid (titrated in 50% aqueous ethanol 0.04 M in potassium chloride at 37.6°) occurred at pH 2.69.

Potassium 3-Nitro-4-acetoxybenzenesulfonate.—Potassium 3-nitro-4-hydroxybenzenesulfonate (1.0 g.), prepared by the method of Kolbe and Gauhe,[13] was heated 1 hour at reflux in a solution of acetic acid (6 ml.) and acetic anhydride (4 ml.). After addition of 20 ml. of methanol the solution was refluxed for 2 hours, cooled, filtered, and evaporated to dryness. Recrystallization of the residue from acetic acid-benzene (~10% benzene) gave 0.47 g. (40%) of potassium 3-nitro-4-acetoxybenzenesulfonate as a pale tan powder which did not melt below 300°; λ^{KBr} 5.65 μ.

Anal. Calcd. for $C_8H_6NO_7SK$: C, 32.0; H, 2.02; N, 4.68. Found: C, 31.8; H, 2.20; N, 5.19.

Hydrolysis with a dilute potassium hydroxide solution, followed by neutralization with hydrochloric acid, yielded potassium 2-nitro-4-hydroxybenzenesulfonate, as evidenced by the infrared spectrum.

3-Nitro-4-acetoxybenzenearsonic acid was obtained in 53% yield by addition of acetic anhydride to an alkaline solution of 3-nitro-4-hydroxybenzenearsonic acid (Eastman Kodak Co. white label) according to the procedure used by Chattaway[14] for acetylation of phenols. It was washed well with ice-water, dried, and recrystallized from acetone (alcohol free) to yield a white, crystalline acid melting above 300°; λ^{KBr} 3.6, 4.4, 5.66, 6.2 μ.

Anal. Calcd. for $C_8H_8AsNO_7$: C, 31.5; H, 2.64; N, 4.59. Found: C, 31.6; H, 2.71; N, 4.41.

The apparent pK_a of 3-nitro-4-acetoxybenzenearsonic acid (determined by titration in 50% aqueous ethanol 0.04 M in potassium chloride at 36.8°) was 3.85.

2,4-Dinitrophenyl acetate was kindly provided by B Zerner.

Poly-(4-vinylpyridine) was obtained as a transparent, solid rod by bulk polymerization of freshly distilled, deoxygenated 4-vinylpyridine (Reilly Tar and Chemical) by the procedure of Fitzgerald and Fuoss.[15] It was fractionally precipitated from solution in *t*-butyl alcohol by successive additions of benzene. The first portion precipitated was redissolved in *t*-butyl alcohol and recovered as a white, porous cake by sublimation of the butanol at low temperature. This sample, which was used for the subsequent studies, was soluble in alcohol and in aqueous acid and had a softening point about 175°. The intrinsic viscosity in 95% ethanol was 0.377. On the basis of the equation of Boyes and Strauss[16] (for 92% ethanol) the average molecular weight was about 60,000.

Anal. Calcd. for C_7H_8N: N, 13.33. Found: N, 13.37.

Poly-(N-vinylimidazole) was generously donated by the Badische Anilin und Soda Fabrik AG., Ludwigshafen A. Rhein, Germany. It came as a white powder and was used without further treatment.

Anal. Calcd. for $(C_5H_6N_2)_n$: N, 29.78. Found: N, 30.17.

Spectrophotometric Titration of Pyridine Bases.—For analysis of the data from the catalyzed solvolyses it was desirable to know the state of ionization of the catalysts as a function of the pH of the solution. This information was obtained by spectrophotometric titrations.

(12) H. Goldstein and A. Jaquet, *Helv. Chim. Acta*, **24**, 30 (1941).
(13) H. Kolbe and F. Gauhe, *Ann.*, **147**, 71 (1868).
(14) F. D. Chattaway, *J. Chem. Soc.*, 2495 (1931).
(15) E. G. Fitzgerald and R. M. Fuoss, *Ind. Eng. Chem.*, **42**, 1603 (1950).
(16) A. G. Boyes and U. P. Strauss, *J. Polymer Sci.*, **22**, 463 (1956).

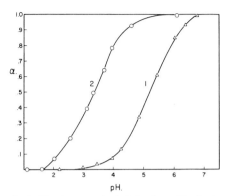

Fig. 1.—Titration curves for 4-picoline (curve 1) and poly-(4-vinylpyridine) (curve 2) in 50% ethanol, 0.04 M in KCl.

An aqueous solution of poly-(4-vinylpyridine) (75 ml. of 50% v.v. ethanol–water 5.0×10^{-4} M in pyridine units and 0.04 M in potassium chloride) was warmed to 37.8° in a water-jacketed beaker. Small portions of 1.2 M hydrochloric acid in 50% ethanol–water were added from a syringe microburet, model SB2, Micro-Metric Instrument Co. After each addition, the pH of the solution was determined with a Photovolt pH meter; a 3-ml. sample then was withdrawn with prewarmed pipets and the absorbance determined at 254 mμ with a Cary model 11 spectrophotometer (temperature in cell compartment, 36.6°). The fraction of non-protonated nitrogen (α) was determined by interpolation from the absorbancies of the polymer in strong alkaline solution (pH 11.2, A 0.743) and in strong acid solution (pH 1.1, A 1.750). The titration of 4-picoline was carried out in the same way, with the spectrophotometer set at 255 mμ (at pH 2.2, A was 1.650; at pH 11.0, A was 0.813). The data are presented in Fig. 1.

Plots of $\log((1-\alpha)/\alpha)$ versus pH were linear for both 4-picoline and poly-(4-vinylpyridine) over the range α 0.06–0.94, in accordance with the modified Henderson–Hasselbach equation[17]: $pK = 5.20 = pH + 1.10 \log((1-\alpha)/\alpha)$ for 4-picoline and $3.34 = pH + 1.14 \log((1-\alpha)/\alpha)$ for poly-(4-vinylpyridine).[18]

Kinetic Determinations.—The solvent consisted of equal volumes of ethanol and water, mixed at room temperature. In this system poly-(4-vinylpyridine) remained in solution throughout the pH range of interest (1 to 7). Except where noted, the solutions were 0.04 M in potassium chloride. The salt was added to minimize salt effects arising from differences in the amounts of hydrochloric acid used to adjust the pH of the solutions.

Solvolysis of 5-Nitro-4-acetoxysalicylic Acid.—To an ethanol water solution of poly-(4-vinylpyridine), 0.02 base M and 0.04 M in potassium chloride, was added concd. hydrochloric acid (vol. less than 1/200 vol. of the polymer solution) to bring the pH to 3.50. A. 3.00-ml. portion of this solution was warmed to 39° and added to a quartz cell (1-cm. path) in a thermostatically controlled cell compartment of a Cary spectrophotometer. The polymer solution was allowed to stand 30 minutes in order to attain temperature equilibrium, after which the temperature in the cell remained at 36.8 ± 0.1°. At zero time 0.050 ml. of a freshly prepared ethanol solution of 5-nitro-4-acetoxysalicylic acid (1.2 × 10^{-2} M) was injected into the cell by a microburet and the solution was shaken briefly. The spectrum between 310 and 360 mμ was then scanned repeatedly and the final reading taken after 5 hours (16 half-lives). As the reaction proceeded a broad maximum in the range of 320 mμ (max. for ester) decreased (A 1.48 → 1.14) and a new maximum (for 2,4-dihydroxybenzoic acid) developed at 356 mμ (A 0.97 → 1.77). A well defined isosbestic point appeared at 331.5 mμ, indicating the absence of any medium-lived intermediates. The changes at both wave lengths corresponded to first-order kinetics through 90% reaction. The rate constant determined from the decrease in absorption at 318 mμ was 0.037 min.$^{-1}$; that determined from the increase in absorption at 356 mμ was 0.038 min.$^{-1}$.

General Procedure A.—The procedure just described was followed except that the spectrophotometer was set to record absorbance as a function of time at a particular wave length. Variations in the quantities of reagents and the pH are indicated in the tables and figures. Checks on the initial and final pH values showed that the pH did not change detectably in the course of the reaction. The solvolyses were followed at 356 mμ for NAS and 360 mμ for the remaining substrates. For the very slow reactions the infinity values of the absorbancies were obtained from solutions containing the appropriate concentrations of the aromatic solvolysis products; otherwise, infinity values were obtained by carrying out the reactions until the spectra no longer changed. In general, good first-order kinetics were obtained; however, at the higher pH values (>4) some deviation was noted for the solvolyses of the charged substrates (NAS, NABS, NABA) catalyzed by the polymer. The rate constants tended to decrease as the reaction proceeded. In these cases the first-order rate constants corresponded to about the first 50% of the reaction.

The products of the reactions were identified by the similarity of their ultraviolet spectra with the spectra of the corresponding non-acetylated materials. Furthermore, 2,4-dihydroxy-5-nitrobenzoic acid (68% yield after recrystallization, identified by m.p., mixture m.p. and infrared spectrum) was isolated from a polymer-catalyzed solvolysis carried out with 0.120 g. of NAS. Likewise, 3-nitro-4-hydroxybenzenearsonic acid was isolated from a reaction with NABA. In this case a cellophane dialysis bag was submerged in the reaction solution; it was subsequently removed and the contents evaporated under vacuum to yield 3-nitro-4-hydroxybenzenearsonic acid.

The applicability of Beer's law to polyvinylpyridine solutions of the solvolysis products was tested with standard solutions; Beer's law was obeyed throughout the range investigated (10^{-4} to 4×10^{-4} M 3-nitro-4-hydroxybenzenesulfonate and 3-nitro-4-hydroxybenzenearsonic acid, and 5×10^{-5} to 19×10^{-5} M 2,4-dihydroxy-5-nitrobenzoic acid, in 0.01 base molar polymer at pH 3.6 in 50% ethanol–water, 0.04 M in potassium chloride).

Procedure B.—A different technique was used in following the relatively slow reactions of NAS in which catalytic quantities of poly-(4-vinylpyridine) were used. The reaction solvent was 50% (v. v.) aqueous ethanol made 0.04 M in potassium acid phthalate and adjusted to the appropriate pH by addition of concd. hydrochloric acid. To 22 ml. of this solution was added 1 ml. of the catalyst solution (0.005 M poly-(4-vinylpyridine) in ethanol); the mixture then was warmed to 38.3° in a constant temperature bath. One ml. of a standard substrate solution (0.05 M 5-nitro-4-acetoxysalicylic acid) was added, the volumetric flask was filled to the 25-ml. mark with the phthalate buffer solution, and the mixture was shaken vigorously for 30 seconds. At periodic intervals 2-ml. aliquots were removed and diluted to 100 ml. with 50% ethanol–water buffered at pH 4.9 with 0.02 M sodium acetate–acetic acid. The absorbancies were determined at 317 and 357 mμ with a Beckman DU spectrophotometer. The reactions followed first-order kinetics with respect to substrate disappearance and product appearance.

Equilibrium Dialysis Studies.—Since rapid solvolysis precluded direct binding experiments involving the ester substrates and polymer, a substrate model, potassium 4-hydroxy-3-nitrobenzenesulfonate, was used in the binding experiments. This hydroxy derivative should be bound at least as extensively as the acetoxy compound (NABS).

A 10-ml. portion of a poly-(4-vinylpyridine) solution (0.01 M in base units, 50% ethanol–water solvent), for which the pH had been adjusted to the appropriate value by addition of concd. hydrochloric acid, was placed in a thoroughly washed and soaked 23/100 sausage casing (Visking Corp.). After the ends had been tied, the bag was immersed in 10

---

(17) (a) W. Kern, *Z. physik. Chem.*, **181**, 249 (1938); (b) A. Katchalsky and P. Spitnik, *J. Polymer Sci.*, **2**, 432 (1947); (c) A. Katchalsky, N. Shavit and H. Eisenberg, *ibid.*, **13**, 69 (1954).

(18) One would expect the coefficient of the log term for 4-picoline to be 1.00. The deviation from unity may therefore reflect an experimental error. The relatively low value of the coefficient (1.14) for the polymer is probably a consequence of the high percentage of alcohol in the solvent. For example, the corresponding coefficient for polymethacrylic acid in aqueous solution is 2; however, the value is markedly reduced in dioxane–water solutions; see ref. 17b.

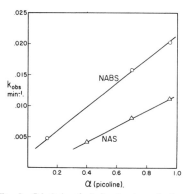

Fig. 2.—Solvolysis of 5-nitro-4-acetoxysalicylic acid (NAS) (2×10^{-4} M) and sodium 3-nitro-4-acetoxybenzenesulfonate (4×10^{-4} M) in 50% ethanol, 0.0157 M in 4-picoline and 0.04 M in KCl at 36.8°.

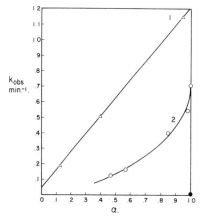

Fig. 3.—Solvolysis of dinitrophenyl acetate (DNPA) (2×10^{-4} M) in 50% ethanol, 0.04 M in KCl at 36.8°: curve 1, 0.0157 M 4-picoline present; curve 2, 0.01 base molar poly-(4-vinylpyridine) present. The pH of the points at $\alpha = 1$ was 6.5; the ● point indicates a reaction carried out in absence of nitrogen bases at pH 6.5.

ml. of ethanol–water (50% v.v.) which was 0.08 M in potassium chloride and 8×10^{-4} M in potassium 3-nitro-4-hydroxybenzenesulfonate. The tube was tightly stoppered and agitated in a wrist action shaker for 24 hours at 25°, whereupon a sample of the solution outside the bag was analyzed

TABLE I

BINDING OF POTASSIUM 4-HYDROXY-3-NITROBENZENE-SULFONATE TO POLY-(4-VINYLPYRIDINE)

pH	Absorbance "outside" soln.	Sulfonate bound by PVPy, %
5.20[a]	0.900	..
5.00	.860	4
4.47	.830	8
3.52	.805	10
2.93	.820	9

[a] A blank run; no poly-(4-vinylpyridine) in the bag.

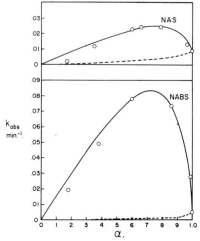

Fig. 4.—Solvolysis of 5-nitro-4-acetoxysalicylic acid (NAS) (2×10^{-4} M) and potassium 3-nitro-4-acetoxybenzenesulfonate (NABS) (2×10^{-4} M) in 50% ethanol, 0.01 base molar in poly-(4-vinylpyridine) and 0.04 M in KCl at 36.8°: experimental points -O-; the solid and dashed lines represent calculated curves (see text).

by its absorption at 343 mμ in a Cary spectrophotometer. At this wave length the absorption did not vary with pH. The absorbance of the "outside solution," which is a measure of the concentration of the sulfonate, varied from 0.900 for a blank run in which no polymer was in the bag to 0.805 for a run with polymer at pH 3.52. In these experiments the polymer nitrogen/substrate mole ratio was 12.5:1; so in the acidic solution sufficient polymer was present to bind all the substrate. The data are presented in Table I as percentage of the substrate which was bound by the polymer.

Results

As controls for the polymer work, the rates of solvolysis of three substrates, 2,4-dinitrophenylacetate (DNPA), 5-nitro-4-acetoxysalicylic acid (NAS) and potassium 3-nitro-4-acetoxybenzenesulfonate (NABS), in aqueous ethanol solutions containing 4-picoline were investigated. The pH of the picoline buffers was varied from 4.3 to 6.5. In Figs. 2 and 3, the pseudo-first-order rate constants for these reactions are plotted as a function of α, the fraction of the picoline present as the free base. The spontaneous solvolysis rates in 50% ethanol–water were negligible compared to the rates for the catalyzed reactions. It was found that the rate constants, k_{obs}, increased linearly with α. This result is in accord with a considerable body of data on catalyzed solvolyses of nitrophenyl esters conducted in aqueous solution.[4]

The shape of the k_{obs}–α plot was quite different when poly-(4-vinylpyridine) was used in place of 4-picoline (Fig. 3). With DNPA, an uncharged substrate, poly-(4-vinylpyridine) was less effective as a catalyst than an equivalent amount of 4-picoline and the k–α plot curved upward with increasing α. This behavior of poly-(4-vinylpyridine) may be attributed to a decrease in the nucleo-

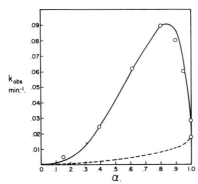

Fig. 5.—Solvolysis of 3-nitro-4-acetoxybenzenearsonic acid (NABA) (4×10^{-4} M) in 50% ethanol, 0.01 base molar in poly-(4-vinylpyridine) and 0.04 M in KCl at 36.8°; the solid and dashed lines represent calculated curves (see text).

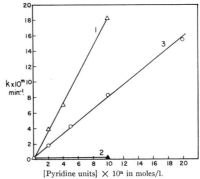

Fig. 6.—Solvolysis of 5-nitro-4-acetoxysalicylic acid (2×10^{-3} M) at pH 3.6 by experimental method B; $m = n = 4$: curve 1, in presence of poly-(4-vinylpyridine); curve 2, in presence of 4-picoline; curve 3: solvolysis of potassium 3-nitro-4-acetoxybenzenesulfonate (2×10^{-4} M) in 50% ethanol in presence of poly-(4-vinylpyridine) at constant salt concentration (adjusted to 0.04 M Cl⁻ by addition of KCl) by exp. method A; $m = 2$, $n = 3$.

philicity of the free basic groups attendant on an increase in positive charges on the polymer chain.

A striking change in the pattern of solvolysis was observed in the poly-(4-vinylpyridine)-catalyzed reactions of NAS, NABS and NABA, substrates which exist completely or extensively as anions in the pH range studied (see Figs. 4 and 5). The reaction rates *increased* as the pH was *lowered* below 6.5. A maximum was attained in the range pH 3.8–4.0 (α 0.7–0.8), with further decreases in pH the values of k_{obs} decreased. For α 0.6 (pH 3.6 and 5.45 for the polymer and the picoline solutions, respectively) poly-(4-vinylpyridine) was 9.3 times *more* effective per mole of free amine than picoline in catalyzing the solvolysis of NABS. The corresponding factor was 5.1 for the reaction of NAS.[19] By contrast, under the same conditions the polymer was only two-fifths as active as picoline in the solvolysis of DNPA.

The relation between rate and polymer concentration was explored for NABS at pH 3.6. As shown in Fig. 6, the rate of solvolysis was proportional to the polymer concentration in the range 2×10^{-3} to 20×10^{-3} M pyridine unit.[20] In these experiments nitrogen base was present in large excess relative to substrate. Significant effects also were produced by relatively small or "catalytic" quantities of polyvinylpyridine. Data are presented in Fig. 6 for experiments in solutions buffered with potassium acid phthalate for which the NAS-polymer nitrogen ratio ranged from 2 to 10. The rates of these reactions were linearly related to the poly-(4-vinylpyridine) concentration. Control reactions in which the polymer was omitted or replaced by 4-picoline were extremely slow.

Potassium chloride had a marked effect upon the rates of the catalyzed solvolyses of the anionic substrates but not upon the pH optimum. A comparison of the data in Table II and Fig. 4 reveals that the rates for NABS were approximately twice as great in the salt-free solutions (*i.e.*, those in which potassium chloride was omitted) as in solutions 0.04 M in potassium chloride.

TABLE II
CATALYSIS BY POLY-(4-VINYLPYRIDINE) AND POLY-(N-VINYLIMIDAZOLE); NO SALT ADDED (METHOD A)

Substrate	Vol. % ethanol in solvent	Base (0.01 M)	pH	$k_{obs.}$, min.⁻¹
NABS	50	PVPy	3.50	0.190
			3.60	.220
			4.55	.176
			6.20	.036
			6.80	.010
NABS	50	PVIm	3.10	0.244
			3.85	.438
			4.25	.596
			5.75	.428
			7.70	.133
NABA	10	PVIm	2.50	0.064
			3.20	.553
			3.80	1.087
			4.60	1.553
			5.60	0.649
			7.50	0.553

Poly-(4-vinylpyridine) is not unique in its catalytic properties. In Table II are also summarized data on solvolyses of NABS and NABA catalyzed by poly-(N-vinylimidazole). A distinct maximum was found in the k_{obs}–pH profile in both cases. As expected, the imidazole unit was a more effective catalyst than the pyridine unit. However, for purposes of the present study polyvinylimidazole was less suitable than poly-(4-vinylpyridine) since the change in the ultraviolet spectrum accompanying protonation was so small that an accurate determination of α could not be made.

Discussion

The existence of maxima in the k_{obs}–α profiles for the solvolyses of NAS, NABS and NABA

(19) The relative superiority of the polymer would, of course, appear much greater if solutions at the same pH were compared.

(20) For this relationship to hold it is necessary to keep the salt concentration constant.

catalyzed by poly-(4-vinylpyridine) show at least two different factors to be involved in determining the reaction rates; one is favored by a decrease and the other by an increase in the acidity of the reaction medium. It is reasonable that these factors are the amounts of basic nitrogen and protonated nitrogen in the polymer. The solvolytic data may be rationalized qualitatively on the basis that (a) the basic nitrogen catalyzes the actual solvolysis and (b) the protonated nitrogen sites, being positively charged, serve to hold negatively charged substrates within the region of the polymer coils. In the sense that the polymer contains both binding sites and catalytic sites it may be considered to be a model for an enzyme. Substrates such as DNPA which do not carry a negative charge are not subject to the rate enhancement derived from ionic binding.

An estimate of the contribution of binding to the solvolysis rates can be made in the following way. The values of k_{obs} at $\alpha = 1$ should indicate the extent of conventional nucleophilic catalysis by non-protonated polymer. On the assumption that the curve for the DNPA solvolysis (curve 2, Fig. 3) provides a measure of the nucleophilicity of poly-(4-vinylpyridine) in acid solutions, one can derive the dashed curves shown in Figs. 4 and 5, which represent the values of k_{obs} that would be expected if poly-(4-vinylpyridine) functioned only in the conventional sense. The difference between the dashed curves and the experimental points indicates the extent of reaction associated with ionic interaction between the substrate and catalyst. In the most favorable case (NABS) it is estimated that the solvolytic rate at $\alpha = 0.6$ is enhanced by a factor of eighty as a consequence of the electrostatic association.

As a first approximation it could be assumed that the rates of the catalyzed solvolyses, at constant polymer concentration, would be proportional to the product of α, the fraction of catalytic sites, and $1 - \alpha$, the fraction of binding sites. In this case a symmetrical k–α plot with a maximum at α 0.5 would be observed. Actually, the k_{obs}–α curves for the charged substrates do not show this symmetry; so some other factor, or factors, must be operative. One possibility is that the equilibrium constant for binding substrate to polymer is sufficiently large that most of the substrate is in the bound state in the moderately acidic solutions (pH 3.6). This characteristic would account for the shape of the k_{obs}–α profile; however, it would also lead to the prediction that the rates of the catalyzed solvolyses would not be proportional to the polymer concentration, a consequence contrary to fact. Furthermore, equilibrium dialysis studies with poly-(4-vinylpyridine) and potassium 3-nitro-4-hydroxybenzenesulfonate, a model for NABS, indicate that the binding constant would be small (Table I). A second possibility is that asymmetry in the k_{obs}–α curve is produced by the reduction in nucleophilicity of the basic sites which accompanies an increase of positive charges on the polymer chain. However, calculations based on the curvature of the k_{obs}–α profile for an uncharged substrate, DNPA (curve 1, Fig. 3), indicate that this property alone would not account for the bias in the k_{obs}–α plots for the charged substrates. We therefore favor a third possibility, that the polymer undergoes an unfavorable conformational change when the acidity of the medium increases. The change might result either in less binding of substrate or in unfavorable geometrical relationships between the nucleophilic and binding sites. It has been established in other systems that the conformation of a polymer depends upon the charge state.[21]

With substrates which are weak acids or salts of weak acids an additional factor, the state of ionization of the substrate, may influence the k_{obs}–α curve. At high pH the substrate would be ionized and subject to electrostatic binding whereas at low pH it would exist as a neutral molecule with little affinity for the polymer. This factor is probably not significant in the reactions of NABS; however, it may be partially responsible for the decrease in rate observed with NAS at high acidities and is surely involved in the solvolysis of NABA, which is a relatively weak acid. The k_{obs} values for NABA reached a maximum at a somewhat higher α-value and decreased more rapidly on the acid side of the maximum than the k_{obs} values for NAS and NABS. These results support the view that the unusually rapid solvolyses catalyzed by poly-(4-vinylpyridine) in the region of pH 4 involve the substrate in the anionic rather than the acid form.

On the basis of these considerations we suggest that the catalytic rate constant, k_c, for poly-(4-

$$k_c = k(\alpha_p)(1 - \alpha_p)f(\alpha_p)\alpha_s + k_N$$

vinylpyridine) may to a good approximation be represented by the equation where k_N is the rate constant for conventional nucleophilic catalysis, α_s is the fraction of substrate in the anion form, α_p is the fraction of the polymeric base in the free base form, and $f(\alpha_p)$ is the function relating the change in the spacing and intrinsic effectiveness of the binding and catalytic sites with the change in acidity of the medium. This equation cannot be tested directly since $f(\alpha_p)$ is unknown; however, it is interesting and perhaps significant that, as shown by the continuous curves in Figs. 4 and 5, a reasonably good fit to the experimental data may be obtained by using this equation and assuming a single expression for $f(\alpha_p)$ for all solvolyses catalyzed by poly-(4-vinylpyridine). The continuous curves in Figs. 3 and 4 were calculated on the basis that $f(\alpha_p) = 1/(1 - 0.86\alpha)$ that $k = 0.04, 0.156$ and 0.221 min.$^{-1}$ for NAS, NABS and NABA, respectively, and that $\alpha_s = 1$ for NAS and NABS. The values for α_s for NABA were calculated on the assumption that the effective pK_a of this acid in the reaction medium was 3.60, and the k_N values were determined from the k_{obs}'s at $\alpha = 1$ and curve 2, Fig. 3, as previously indicated.

Acknowledgment.—Helpful discussions with Professors H. Morawetz and M. Bender are gratefully acknowledged.

(21) For example, A. Oth and F. Doty, *J. Phys. Chem.*, **56**, 45 (1952), found that molecules of polymethacrylic acid in solution extend about sevenfold as the extent of ionization increases.

Investigation of a Synthetic Catalytic System Exhibiting Substrate Selectivity and Competitive Inhibition[1,2]

Robert L. Letsinger and Irvin S. Klaus[3]

Contribution from the Department of Chemistry, Northwestern University, Evanston, Illinois 60201. Received March 20, 1965

Catalysis by poly(N-vinylimidazole) of the solvolysis of mono- and dinitrophenyl ester groups in polymeric substrates bearing carboxyl groups exhibits several features characteristic of enzymatic reactions, namely, appreciable catalytic activity at low catalyst concentration, kinetics indicative of saturation of the catalyst at high substrate concentration, competitive inhibition by a substance (polyacrylic acid) possessing the same binding groups as the substrate, and catalysis of the reaction of the polymeric substrates (e.g., acrylic acid–2,4-dinitrophenyl p-vinylbenzoate copolymer) possessing binding sites in preference to structurally similar substrates (e.g., 2,4-dinitrophenyl p-isopropylbenzoate), which lack binding sites for the catalyst. These properties are attributed to the formation of a catalyst–substrate complex which increases the probability of encounter between nucleophilic sites on the catalyst and the labile ester groups on the substrate.

In a previous paper we reported that partially protonated poly(4-vinylpyridine) and poly(N-vinylimidazole) are unusually effective catalysts for the solvolysis of nitrophenyl esters, such as potassium 4-acetoxy-3-nitrobenzenesulfonate, which bear a negative charge in the nitrophenyl portion of the molecule.[4] Recently Overberger, et al., studying 4-acetoxy-3-nitrobenzoic acid and poly-4(5)-vinylimidazole, provided another example of the high activity of a cationic, polymeric catalyst toward an anionic substrate.[5] The enhanced activity of the partially protonated basic polymers was attributed to electrostatic attraction between the substrates and the polymer, which tends to concentrate the substrate in the region of the active nucleophilic centers.[4]

These systems clearly demonstrate the significant role that electrostatic charge may play in nucleophilic catalysis in polymeric systems. As examples for "selective catalysis"[4] and as models for enzymatic reactions, however, they suffer several deficiencies. (1) The reaction rates increased linearly with the catalyst concentration; no saturation phenomena characteristic of strong association between catalyst and substrate could be observed. (2) While negatively charged substrates solvolyzed unusually rapidly, no cases were found in which the relative order of solvolysis of two closely related substrates could be reversed by changing from a non-selective to a selective catalyst. (3) Competitive inhibition was not demonstrated.

For development of selective, synthetic catalytic systems, which in kinetic behavior and specificity of action more closely resemble the enzymatic ones, it appeared that substrates should be sought which would be much more strongly bound to the catalyst than in the cases heretofore examined. With the expectation that a polyanionic substrate would satisfy this condition we undertook a study of polymeric substrates which contained multiple binding sites as well as nitrophenyl ester groups (reaction sites). The substances selected for investigation were copolymers of acrylic acid and p-nitrophenyl p-vinylbenzoate (I) or 2,4-dinitrophenyl p-vinylbenzoate (II). In the range of pH 7 most of the carboxyl groups should be ionized and therefore serve as binding sites. Moreover, the carboxyl groups would be sufficiently far removed from the neighboring ester groups in these substances that intramolecular catalysis of solvolysis by the carboxylate anions should

I, X = H
II, X = NO$_2$
(R = H or C$_2$H$_5$)

III

IV, X = H
V, X = NO$_2$

---

(1) Paper IV in the series on Selective Catalysis. For a preliminary account see R. L. Letsinger and I. Klaus, *J. Am. Chem. Soc.*, **86**, 3884 (1964). Papers I–III: *ibid.*, **84**, 3122 (1962); **85**, 2230 (1963); **85**, 2223 (1963).
(2) This research was supported by a grant from the National Science Foundation (G25069).
(3) Toni-Gillett Fellow, 1961; Lubrizol Foundation Fellow, 1962.
(4) R. L. Letsinger and T. J. Savereide, *J. Am. Chem. Soc.*, **84**, 114 (1962); **84**, 3122 (1962).
(5) C. G. Overberger, T. St. Pierre, N. Vorchheimer, and S. Yaroslavsky, *ibid.*, **85**, 3513 (1963); **87**, 296 (1965).

be negligible.[6] The present paper reports the results obtained with the copolymers and also with three low molecular weight substrates, p-nitrophenyl hydrogen terephthalate (III), p-nitrophenyl o-isopropylbenzoate (IV), and 2,4-dinitrophenyl p-isopropylbenzoate (V), which served as models for evaluating the reactivity of the copolymers. The stoichiometry is typified by the following equation which represents the solvolysis of one of the ester groups in the polymer chain.

Experimental

p-Nitrophenyl p-Vinylbenzoate. To 2.5 g. (0.0165 mole) of p-vinylbenzoic acid[7] and an excess of p-nitrophenol (5.3 g., 0.0381 mole) in 50 ml. of dry pyridine at room temperature was added dropwise with stirring 3.5 g. (0.017 mole) of N,N'-dicyclohexylcarbodiimide dissolved in 25 ml. of pyridine. Stirring was continued for 24 hr., whereupon the solid which had precipitated was removed by filtration. The filtrate was evaporated and the residue was dissolved in ether and washed successively with several portions of dilute acid, aqueous sodium carbonate, and water. Evaporation of the ether layer afforded p-nitrophenyl p-vinylbenzoate, which after two recrystallizations from ethanol melted sharply at 103° and weighed 1.4 g. (31.5%). Strong bands in the infrared spectrum (5–9-μ region) were found at 5.85, 6.20, 6.28, 6.59, 6.70, 7.12, 7.40, and 7.86 μ.

Anal.[8] Calcd. for $C_{15}H_{11}NO_4$: C, 66.91; H, 4.12; N, 5.20. Found: C, 66.95; H, 4.27; N, 5.37.

2,4-Dinitrophenyl p-Vinylbenzoate. A solution of 3.76 g. (0.0185 mole) of N,N'-dicyclohexylcarbodiimide in 40 ml. of pyridine was added to 100 ml. of a solution of 2.75 g. (0.0185 mole) of p-vinylbenzoic acid and 3.41 g. (0.0185 mole) of 2,4-dinitrophenol in pyridine. After the mixture had stood overnight at room temperature with stirring, it was filtered and the solvent was removed at reduced pressure. Two recrystallizations of the residue from chloroform–hexane afforded 3.95 g. (66%) of 2,4-dinitrophenyl p-vinylbenzoate, m.p. 143–144°. The infrared spectrum contained strong bands at 5.72, 6.20, 6.60, 7.45, and 7.86 μ. The analytical sample, m.p. 145°, was further recrystallized from benzene–hexane. A mixture melting point taken with p-vinylbenzoic acid was 123–129°.

Anal. Calcd. for $C_{15}H_{10}N_2O_6$: C, 57.33; H, 3.21; N, 8.92. Found: C, 57.36; H, 3.28; N, 9.00.

p-Nitrophenyl p-Isopropylbenzoate. p-Isopropylbenzoic acid (9.0 g., 0.055 mole, m.p. 116.5–117.5°) was mixed with excess thionyl chloride and allowed to stand at room temperature for 3 hr.; thereafter excess thionyl chloride was removed under reduced pressure and 9.0 g. (0.065 mole) of p-nitrophenol in 60 ml. of dry pyridine was added. After the mixture was warmed on the steam bath for several hours, the pyridine was evaporated under reduced pressure and the residue was dissolved in chloroform. Washing the chloroform solution several times with dilute acid, water, aqueous sodium carbonate, and water, followed by evaporation

(6) The copolymer of acrylic acid and p-nitrophenyl acrylate, though more readily available, was unattractive since it solvolyzes very rapidly in aqueous solutions as a consequence of intramolecular catalysis. See E. Gaetjens and H. Morawetz, *J. Am. Chem. Soc.*, **83**, 1738 (1961).
(7) J. R. Leebrick and H. E. Ramsden, *J. Org. Chem.*, **23**, 935 (1958).
(8) The analyses were made by Miss H. Beck and by the Micro-Tech Laboratories, Skokie, Ill.

of the chloroform in a rotatory evaporator, yielded p-nitrophenyl p-isopropylbenzoate, which after recrystallization several times from ethanol weighed 9.3 g. (60%) and melted at 104.5–105°, λ_{max} 5.80 μ.

Anal. Calcd. for $C_{16}H_{15}NO_4$: C, 67.36; H, 5.30; N, 4.91. Found: C, 67.46; H, 5.44; N, 4.97.

2,4-Dinitrophenyl p-Isopropylbenzoate. A solution containing p-isopropylbenzoyl chloride, prepared from 1.64 g. (0.01 mole) of the acid as in the previous experiment, 1.84 g. (0.01 mole) of 2,4-dinitrophenol, and 1 ml. of pyridine in 50 ml. of benzene was heated at reflux for 3 hr. After the mixture had cooled, it was filtered and the filtrate was evaporated to dryness. The residue was extracted with boiling ether. On cooling the extract, 1.22 g. (37%) of 2,4-dinitrophenyl p-isopropylbenzoate crystallized, m.p. 116.5–117.5°, λ_{max} 5.71 μ. A mixture with p-isopropylbenzoic acid melted at 95–105°.

Anal. Calcd. for $C_{16}H_{14}N_2O_6$: C, 58.18; H, 4.27; N, 8.48. Found: C, 58.41; H, 4.19; N, 8.70.

p-Nitrophenyl Hydrogen Terephthalate. p-Nitrophenol (6.58 g., 0.047 mole) in 100 ml. of pyridine was added dropwise to 10.1 g. (0.050 mole) of terephthalyl chloride in 30 ml. of pyridine at about 100°. After standing overnight the mixture was filtered to remove a small amount of di-p-nitrophenyl terephalate, m.p. 245°.

Anal. Calcd. for $C_{20}H_{12}N_2O_8$: C, 58.82; H, 2.96; N, 6.91. Found: C, 58.36; H, 3.28; N, 6.80.

The filtrate was then evaporated and the residue, after washing with ether and dilute hydrochloric acid, was recrystallized from benzene–hexane to give 3.15 g. (22%) of p-nitrophenyl hydrogen terephthalate, m.p. 210–211.5°. The analysis correspond to the half ester plus $1/_6$ molecule of benzene.

Anal. Calcd. for $C_{14}H_9NO_6 \cdot 1/_6C_6H_6$: C, 60.00; H, 3.36; N, 4.67. Found: C, 60.31; H, 3.32; N, 4.82.

The infrared spectrum of p-nitrophenyl hydrogen terephthalate had bands characteristic for the 2,4-dinitrophenyl ester carbonyl (5.72 μ), carboxyl carbonyl (5.90 μ), carboxyl O–H (3.5, 3.8 and 4.0 μ), and the nitro group (6.59 and 7.40 μ). The compound was completely soluble in aqueous sodium bicarbonate solution and it reprecipitated from the solution on addition of acid.

Acrylic Acid–p-Nitrophenyl p-Vinylbenzoate Copolymer (I). A mixture of 0.40 g. (5.7 mmoles) of acrylic acid, 0.60 g. (2.23 mmoles) of p-nitrophenyl p-vinylbenzoate, and 10 mg. of benzoyl peroxide in 20 ml. of dry benzene was warmed at 60° for 72 hr. under a nitrogen atmosphere. Benzene was evaporated, and the solid residue was ground under ether and extracted several times with ether to give 0.7 g. of copolymer. The milliequivalent of ester groups per gram of polymer was determined by hydrolyzing 0.0144 g. of polymer in 10 ml. of 0.1 M aqueous sodium hydroxide. From the absorbance (0.598 at 404 mμ) of a solution obtained by diluting 1.0 ml. of this solution to 100 ml. with water, and the extinction coefficient of p-nitrophenol in alkali (1.84 $\times$ 10^4), the milliequivalents of ester per gram of polymer was found to be 2.3.

Acrylic Acid–2,4-Dinitrophenyl p-Vinylbenzoate Copolymer (II). A vial containing 1.95 g. (6.2 mmoles) of 2,4-dinitrophenyl *p*-vinylbenzoate, 0.81 g. (11.2 mmoles) of acrylic acid, 10 mg. of benzoyl peroxide, and 30 ml. of benzene was flushed with nitrogen, sealed, and suspended in an oven at 65° for 96 hr. The vial was then opened and the liquid was decanted from the precipitate and evaporated at reduced pressure. The combined solids were ground under ether and washed with ether to give 1.0 g. of polymer which softened at approximately 160°. A portion (0.6 g.) of this substance was dissolved in dioxane and partially reprecipitated by addition of a mixture of ether and hexane (3:1). Two-tenths of a gram of white polymer (softening point approximately 255°) was obtained as a precipitate. From the absorbance of a solution obtained by hydrolyzing 33.9 mg. of this copolymer with aqueous sodium hydroxide and from the extinction coefficient of 2,4-dinitrophenol in alkaline solution (1.13×10^4) it was found that the polymer contained 1.9 mequiv. of ester groups per gram.

Poly(acrylic Acid). Freshly distilled acrylic acid was warmed at 120° until polymerization appeared to be complete. The resulting material was stirred with water and the mixture was centrifuged to remove the insoluble portion. Addition of concentrated hydrochloric acid to the solution afforded a precipitate, which was recovered by centrifugation and redissolved in water. On freezing the solution and subliming the water at reduced pressure a porous cake of poly(acrylic acid) was obtained.

Viscosity Determinations. Relative viscosities were determined in a modified Oswald viscometer Serial dilutions were made and the intrinsic viscosity, $[\eta] = \lim_{c \to 0} 2.3/C \log t/t_0$, was obtained from the plot of reduced viscosity, $2.3/C \log t/t_0$, vs. C, where C is the concentration of the polymer in grams per deciliter, t is the average flow time for the polymer solution, and t_0 is the average flow time for the solvent. The intrinsic viscosities obtained were: poly(N-vinylimidazole), 0.97; acrylic acid–*p*-nitrophenyl *p*-vinylbenzoate copolymer (I), 1.26; and acrylic acid–2,4-dinitrophenyl *p*-vinylbenzoate copolymer (II), 0.087.

Reagents. 2,4,6-Trimethylpyridine (collidine), b.p. 171–172°, was distilled at atmospheric pressure. The sample of N-methylimidazole was a fraction which distilled at 93° (3 mm.). Reagent grade imidazole, m.p. 90.5–91°, was used directly. Dioxane was refluxed over sodium for 60 hr. and distilled (b.p. 99.9 ± 0.1°) prior to use. The N,N-dimethylformamide was first shaken with potassium hydroxide and distilled from calcium oxide. It was then warmed with phthalic anhydride and redistilled at reduced pressure. The fraction collected boiled at 52.5° (22 mm.). Solutions of nitrophenyl esters in N,N-dimethylformamide thus purified were stable and did not turn yellow.

Poly(N-vinylimidazole) was kindly provided by the Badische Anilin und Soda Fabrik, Ludwigshafen, Germany. Prior to use the polymer was washed successively with boiling benzene, hexane, and ethyl ether, and then freed of adhering solvents by evacuation at 1 mm. for 24 hr.

Spectrophotometric Titration. A titration was carried out by adding portions of 10% hydrochloric acid in 50% (weight) aqueous ethanol from a syringe microburet (0.2 μl. per division) to 80 ml. of 50% aqueous alcohol which was 8×10^{-4} M in poly(N-vinylimidazole) and 0.01 M in potassium chloride. The temperature was 25 ± 0.2°. After addition of each increment a sample of the polymer was withdrawn and the absorbance determined at 225 mμ. The absorbance varied from 1.199 at pH 3.70 to 1.325 at pH 7.90. Addition of aqueous sodium hydroxide brought the absorbance to 1.326 at pH 9.30. The results are given in Figure 1.

Kinetics. The reactions were carried out in 50% (weight) ethanol–water solution 1×10^{-2} M in potassium chloride and 2×10^{-2} M in heterocyclic base units. When the base unit concentration of the catalyst (polyvinylimidazole,[9] imidazole, or N-methylimidazole) was less than 2×10^{-2} M, sufficient collidine, which does not function as a nucleophilic catalyst, was added to bring the total concentration to 2×10^{-2} M. The pH was adjusted by addition of hydrochloric acid. For the reactions of the monomeric esters 0.25 ml. of the substrate solution (6.14×10^{-3} M ester in dioxane) was added to 25 ml. of the catalyst solution, which had previously been brought to the reaction temperature (30 ± 0.2° or 25 ± 0.2°). With the dinitrophenyl esters, a portion of the reaction mixture was transferred to a cuvette and the absorbance change at 412 mμ followed in a Beckman DU spectrophotometer. A_∞ was obtained after sufficient time had elapsed that no further change in absorbance could be detected. With the *p*-nitrophenyl esters, which hydrolyzed very slowly in 50% aqueous ethanol, the reaction mixture was kept in a thermostated bath. At suitable intervals 2-ml. aliquots were removed and added to 1 ml. of 10% aqueous hydrochloric acid to quench the reaction. The absorbance was read at 344 mμ, where the absorbances of acidic and alkaline solutions of *p*-nitrophenol are the same. A_∞ values were obtained from aliquots of the reaction mixture which were added to aqueous sodium hydroxide to complete the solvolysis.

With the polymeric substrates it was necessary to exercise considerable care in mixing the substrate solution with the solution of poly(N-vinylimidazole) in order to avoid formation of a precipitate or development of turbidity. In the case of copolymer II equal volumes (12.5 ml.) of solvent and of catalyst solution, each 0.01 M in potassium chloride, were brought to 25°. To the solvent portion was added with vigorous stirring 0.25 ml. of a dioxane solution of copolymer II 6.14×10^{-3} M in ester units. The catalyst solution was then added, again with vigorous stirring, to the substrate solution thus prepared. A sample of the mixture was transferred quickly to a cuvette for the absorbance reading. The mixing procedure for copolymer I was similar except that N,N-dimethylformamide was used as a solvent for the polymer in place of dioxane. Clear solutions were obtained in all cases except for the experiment involving poly(N-vinylimidazole) and copolymer II at the highest ester concentration (4.9×10^{-4} M, see Figure 6), for which a bluish cast, indicative of some colloidal suspension, was apparent. For the reactions involving polyacrylic acid as an inhibitor, experiments were carried out in which polyacrylic acid was added first to the solution

---

(9) Throughout this paper the concentration of poly(N-vinylimidazole) is expressed as moles/l. of imidazole groups in the polymer. Similarly, the concentration of the polymeric substrate is expressed as moles/l. of ester groups.

3382

Investigation of a Synthetic Catalytic System

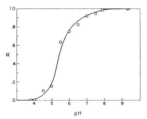

Figure 1. Titration of poly(N-vinylimidazole) in 50% aqueous ethanol at 25°.

of poly(N-vinylimidazole) and also in which the polyacrylic acid was added first to the substrate solution to see if the order of mixing influenced the inhibitor's action. No effect was found, indicating that equilibrium was achieved before appreciable solvolysis of the substrate.

To see if the spectral changes observed actually reflected the conversion of ester groups to the nitrophenol, a reaction of copolymer II (6.14×10^{-5} M) catalyzed by poly(N-vinylimidazole) (1×10^{-4} M) was followed by repeatedly scanning the spectrum between 290 and 400 mμ. The final spectrum corresponded to that of 2,4-dinitrophenol and an isosbestic point occurred at 313 mμ. This experiment shows that the absorbance increase was due to formation of 2,4-dinitrophenol, not to development of turbidity that was visually undetectable.

Treatment of Kinetic Data. The rate data for individual experiments with the monomeric substrates (*p*-nitrophenyl hydrogen terephthalate, *p*-nitrophenyl *p*-isopropylbenzoate, and 2,4-dinitrophenyl *p*-isopropylbenzoate) and for the copolymer of acrylic acid and *p*-nitrophenyl *p*-vinylbenzoate were satisfactorily accommodated by a first-order rate equation. Pseudo-first-order rate constants are indicated in the tables as k_{obsd}. For the reactions of acrylic acid–2,4-dinitrophenyl *p*-vinylbenzoate copolymer, curvature in the first-order plots was observed. This feature was a characteristic of the substrate and did not depend upon the catalyst employed. It is not clear why copolymers I and II differ in their kinetic pattern; however, it may be noted that complex kinetics are often encountered in reactions of polymeric substrates.[6,10] The reactions of copolymer II may be compared by means of initial velocities. The results are essentially the same as for the treatment that follows, which we prefer since it shows that the effect of changing the catalyst or the initial concentrations is a characteristic of the system over a wide range of substrate conversion. It was found that reasonably good straight lines were obtained when $(A - A_0)/(A_\infty - A)$ was plotted against time for reactions of copolymer II catalyzed by imidazole, N-methylimidazole, and poly(N-vinylimidazole) as well as for a reaction conducted in the buffer solution in absence of an imidazole-type catalyst. Typical results are given in Figure 2. A_0, A_∞, and A represent the absorbancies at initial time, at completion of the reaction, and at the time of a measurement. As shown

(10) See, for example, J. Moens and G. Smets, *J. Polymer Sci.*, **23**, 931 (1957).

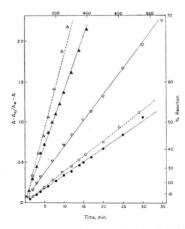

Figure 2. Solvolysis of copolymer II (6.14×10^{-5} M in ester groups): ○, without catalyst; ●, with 0.02 M N-methylimidazole; △, with 2.5×10^{-5} M poly(N-vinylimidazole); ▽, with 5×10^{-4} M poly(N-vinylimidazole); ▲, with 0.02 M poly(N-vinylimidazole). Solid lines, read bottom time scale; broken lines, read top time scale.

in Figure 3, the fit was somewhat less satisfactory for the experiments in which the initial substrate concentration was varied, but it was good enough to permit a meaningful comparison of the data. These plots correspond to rate eq. 1. The relatively slow rate of

$$(A - A_0)/(A_\infty - A) = k't \qquad (1)$$

solvolysis at high conversions may be ascribed to the low intrinsic reactivity of the ester groups in the partially hydrolyzed polymer. Equation 1 represents the case in which the intrinsic reactivity of the ester groups is proportional to C/C_0 and the instantaneous rate is given by eq. 2, where C and C_0 are the concentration of the ester groups at a given time and at zero time.

$$\frac{-dC}{dt} = k'C(C/C_0) \qquad (2)$$

Results

Previous studies[4,5] of selective catalysis of nitrophenyl esters, RCOOR′, involved substrates with a negative charge in the nitrophenyl portion of the molecule, R′. In contrast, the negative charges in copolymers I and II are located in the R portion of the ester. To ascertain whether compounds of this general type also exhibit enhanced catalytic rates in the presence of positively charged catalysts, we first investigated the reaction of a simple monomeric substrate, *p*-nitrophenyl hydrogen terephthalate. Data for solvolysis of this ester in the presence of poly(N-vinylimidazole) are plotted in Figure 4. Throughout most of the pH range investigated the substrate exists largely as a monoanion and the catalyst possesses positive sites as well as nucleophilic centers. A pronounced maximum is apparent in the rate–pH profile, indicative of a rate enhancement stemming from the negative charge on the substrate. It may be noted that *p*-nitrophenyl benzoate, an electrically neutral substrate used as a control,

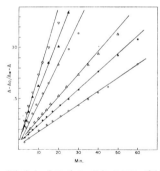

Figure 3. Solvolysis of copolymer II in presence of 5×10^{-4} M poly(N-vinylimidazole) at 25°. Concentration of ester groups in substrate $\times$ 10^4, for curves left to right, 0.246, 0.430, 1.23, 3.46, 3.68, and 4.91, respectively.

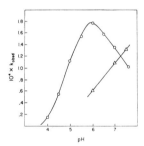

Figure 4. Solvolysis of p-nitrophenyl hydrogen terephthalate (○) and p-nitrophenyl benzoate (△), each 2×10^{-4} M, in presence of 0.02 M poly(N-vinylimidazole) at 30°.

behaved in a normal manner in the solvolytic reaction (see Figure 4).

Supported by this evidence, we continued with the preparation and study of the polymeric substrates. Results obtained with the acrylic acid–p-nitrophenyl p-vinylbenzoate copolymer (I) and with p-nitrophenyl p-isopropylbenzoate, a model for the ester portion of the polymer, are summarized in Table I. These substrates constitute a good pair for study of selective solvolysis since they differ only at positions eight carbon atoms or more removed from the ester function. Several features merit comment. First, the rate of solvolysis of copolymer I in the presence of poly(N-vinylimidazole) is faster at pH 7.9 than at either 4.7 or 8.5 (experiments 7–9), in accord with the findings for p-nitrophenyl hydrogen terephthalate. Second, whereas imidazole is more effective than poly(N-vinylimidazole) in catalyzing the solvolysis of p-nitrophenyl p-isopropylbenzoate (experiments 11 and 12 and 14 and 15), poly-(N-vinylimidazole) is considerably more effective than imidazole in catalyzing the solvolysis of copolymer I (experiments 2–4). Indeed, the experiments provide a good example of selectivity in catalysis; p-nitrophenyl p-isopropylbenzoate solvolyzes three times faster than copolymer I in the presence of 5×10^{-4} M imidazole (experiments 2 and 11) whereas copolymer I solvolyzes eight times faster than p-nitrophenyl isopropylbenzoate in the presence of 5×10^{-4} M poly(N-vinylimidazole) (experiments 4 and 12). The rate constants, k_{obsd}, represent the actual solvolysis of the substrates and include the contribution from the spontaneous reaction characteristic of the catalyst-free solutions. On the basis of rate constants for the catalyzed portion of the reactions alone, $k_{obsd} - k_{solvent}$, the selectivity would appear considerably greater. Finally, it is interesting that $k_{obsd} - k_{solvent}$ for copolymer I increases less than fivefold when the poly(N-vinylimidazole) concentration increases 40-fold. This result suggests that the substrate may be saturated with catalyst at the higher catalyst concentration and that catalyst in excess of that needed to complex with the substrate has little effect upon the solvolytic reaction.

The reactions of the p-nitrophenyl esters in 50% aqueous alcohol are very slow, requiring several days for completion, even in favorable cases. For a detailed study of the effect of substrate and catalyst concentrations upon the rate of solvolysis it was desirable to work with more reactive esters. Accordingly, the copolymer of acrylic acid and 2,4-dinitrophenyl p-vinylbenzoate (II) was prepared and examined. This substrate proved to be relatively labile, particularly in the presence of poly(N-vinylimidazole). Rate data for typical reactions of copolymer II are given in Table II. At pH 7.5 and an imidazole unit concentration of

Table I. Solvolysis of p-Nitrophenyl p-Vinylbenzoate–Acrylic Acid Copolymer (I)[a] and p-Nitrophenyl p-Isopropylbenzoate (IV)[b] at 30 ± 0.2°

Expt.	Substrate	Catalyst	Catalyst concn., mole/l.	pH	k_{obsd} × 10^5, min.$^{-1}$
1	I	0	...	7.5	1.4
2	I	Im	5×10^{-4}	7.5	1.8
3	I	Im	2×10^{-2}	7.5	13.0
4	I	PVI	5×10^{-4}	7.5	15.0
5	I	0	0	7.9	2.7
6	I	PVI	5×10^{-4}	7.9	12.0
7	I	PVI	2×10^{-2}	7.9	47.0
8	I	PVI	2×10^{-2}	8.5	14.0
9	I	PVI	2×10^{-2}	4.7	3.3
10	IV	0	...	7.5	1.6
11	IV	Im	5×10^{-4}	7.5	5.3
12	IV	PVI	5×10^{-4}	7.5	1.8
13	IV	0	...	8.5	6.4
14	IV	Im	5×10^{-4}	8.5	14.0
15	IV	PVI	5×10^{-4}	8.5	8.1

[a] 1.68×10^{-4} M ester groups. [b] 2.0×10^{-4} M; PVI = poly(N-vinylimidazole); Im = imidazole.

Table II. Solvolysis of Acrylic Acid–2,4-Dinitrophenyl p-Vinylbenzoate Copolymer (6.14×10^{-5} M in Ester)

Catalyst	Catalyst concn., mole/l.	$10^3 \times k'_{obsd}$, min.$^{-1}$
None	...	1.54
N-Methylimidazole	5×10^{-4}	2.0
Poly(N-vinylimidazole)	5×10^{-4}	65
Poly(N-vinylimidazole)	0.25×10^{-4}	7.0

5×10^{-4} M, poly(N-vinylimidazole) is 140 times as effective as N-methylimidazole in catalyzing the solvolysis of copolymer II (measured by $k'_{obsd} - k'_{solvent}$).

A striking result was obtained when the reaction of copolymer II was studied as a function of the concentration of poly(N-vinylimidazole). As shown in Figure 5, $k'_{obsd} - k'_{solvent}$ increases rapidly with catalyst concentration in the range of 2.5×10^{-5} to 5×10^{-4} M imidazole units. At high concentrations, however, the curve levels off and $k'_{obsd} - k'_{solvent}$ is independent of the catalyst concentration. The kinetic picture is markedly different when N-methylimidazole is used as the catalyst. This substance, which is comparable to poly(N-vinylimidazole) with respect to nucleophilic properties but lacks the binding sites of the polymeric catalyst, is ineffective at low concentrations. As the concentration is increased the rate of solvolysis of the substrate increases throughout the range of the experiments (5×10^{-4} M to 1×10^{-2} M).

A similar curve was obtained when the amount of poly(N-vinylimidazole) was held constant (5×10^{-4} M) and the concentration of copolymer II varied over a 20-fold range (2.46×10^{-5} to 4.91×10^{-4} M in initial ester groups). The results are presented in Figure 6 in the form of initial velocity, $v_i = (k'_{obsd} - k'_{solvent}) \cdot C_0$, plotted against C_0, the initial ester concentration. As in many enzymatic reactions the initial velocity is proportional to the substrate concentration at low values of the substrate concentration but it levels off at high substrate concentrations.

If the high activity of poly(N-vinylimidazole) at pH 7.5 in catalyzing solvolysis of copolymer II is due to formation of a substrate–catalyst complex, it would be expected that a substance possessing binding groups similar to the substrate should act as a competitive inhibitor. This expectation was realized in experiments with poly(acrylic acid) as the inhibitor. The data are given in Table III.

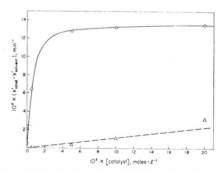

Figure 5. Reaction of copolymer II (6.14×10^{-5} M in ester) catalyzed by poly(N-vinylimidazole) (O) and by N-methylimidazole (△) at 25°. The solid line is calculated from eq. 6.

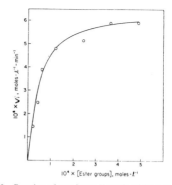

Figure 6. Reaction of copolymer II catalyzed by 5×10^{-4} M poly(N-vinylimidazole) at 25°. The solid line is calculated from eq. 4.

Table III. Inhibition by Poly(acrylic Acid)[a]

Poly(acrylic acid), mole/l.[b]	$10^3 k'_{obsd}$, min.$^{-1}$
0	65
5×10^{-4}	28[c]
10×10^{-4}	14[c]

[a] Copolymer I, 6.14×10^{-5} M in ester groups; poly(N-vinylimidazole), 5×10^{-4} M in imidazole units. [b] Of acid groups. [c] k' decreased somewhat as the reaction progressed. These values correspond to the first 30% of the reaction. They are close to values for k'_{obsd} (25×10^{-3} and 16×10^{-3} min.$^{-1}$, respectively), calculated from the substrate and inhibitor concentrations and from binding and rate data derived from Figure 6. In the calculations it was assumed that the binding constants for carboxylate in copolymer I and in polyacrylic acid are the same and that the ratio of carboxylate to ester in copolymer I is 3:1, as indicated by ultraviolet data.

Finally, it was demonstrated that copolymer II solvolyzes selectively in preference to 2,4-dinitrophenyl p-isopropyl benzoate when poly(N-vinylimidazole) is employed as the catalyst. In noncatalyzed solvolyses and with N-methylimidazole as a catalyst, the monomeric substrate reacts faster than copolymer II. Pertinent data are summarized in Table IV. The result is similar to that for copolymer I with poly(N-vinylimidazole) and imidazole as catalysts, and it provides another good example of discrimination by a catalyst on the basis of substrate structural variations distant from the functional groups that undergo covalent change. The reversal in reactivity stems from the unusually high reactivity of a substrate in the presence of a catalyst which possesses complementary binding sites.

Table IV. Selective Catalysis

Catalyst	$10^3 k$, min.$^{-1a}$	$10^3 k'$, min.$^{-1b}$
None	2.2	1.54
N-Methylimidazole (2×10^{-2} M)	56.5	33
Polyvinylimidazole (5×10^{-4} M in imidazole units)	3.1	65

[a] Dinitrophenyl isopropylbenzoate. [b] Copolymer I.

Discussion

It is clear that poly(N-vinylimidazole) in low concentration at pH 7.5 is a highly effective catalyst for the solvolysis of the copolymers of acrylic acid with p-nitrophenyl p-vinylbenzoate and 2,4-dinitrophenyl p-vinylbenzoate. Neither imidazole nor N-methylimidazole show comparable activity toward these substrates. Furthermore, poly(N-vinylimidazole) does *not* show unusual activity toward p-nitrophenyl p-isopropyl benzoate or 2,4-dinitrophenyl p-isopropyl benzoate, substrates which serve as models for the ester portion of copolymers I and II. This fact permits a demonstration of selective catalysis in which the relative solvolytic rates of two similar substrates may be reversed by changing the catalyst (see Results).

The striking reactivity of copolymers I and II in the presence of poly(N-vinylimidazole) seems best explained on the basis of the reversible formation of a complex between the substrate and catalyst, within which nucleophilic nitrogen sites on the catalyst act on the ester groups of the substrate. The complex is held together by attraction between the carboxylate ions in the substrate and protonated nitrogen atoms in the catalyst. Since the binding sites in both the substrate and the catalyst are insulated electronically from the active nucleophilic and reaction sites by a chain of two or more saturated carbon atoms, rate enhancements resulting from inductive or resonance effects of the binding groups may be eliminated from consideration. At pH 7.5 most of the carboxyl groups in the polymeric substrates should be ionized. A spectrophotometric titration of the poly(N-vinylimidazole) revealed that very few, ~5%, of the basic groups in the catalyst are protonated at this pH. That only a relatively low percentage of the nitrogen sites need be protonated to obtain strong binding is reasonable in view of the fact that both components in the complex are polymeric. When p-nitrophenyl hydrogen terephthalate served as the substrate, a higher fraction of protonated sites was necessary for effective binding. In this case the maximum effective catalytic activity of poly(N-vinylimidazole) was obtained at pH 6.0, at which one-fourth of the nitrogen sites are protonated.

Strong support for the concept that poly(N-vinylimidazole) and copolymer II form a complex is provided by the competitive inhibition exhibited by poly(acrylic acid) and by the shape of the curves shown in Figures 5 and 6. The solid lines in Figures 5 and 6 are drawn from equations based on the assumption that substrate II and the polymeric catalyst form a complex reversibly (eq. 3), and that the catalytic effect of poly(N-vinylimidazole) outside the complex is negligible compared to that within the complex. The equilibrium constant, K, is defined by eq. 5, and the rate constant characteristic of the catalyzed reaction, $k'_{\text{obsd}} - k'_{\text{solvent}}$, is given by eq. 6. In these equations $[C_i]$ is

$$\text{II} + \text{PVI} \rightleftharpoons C_i \xrightarrow{\text{ROH}} \text{products} \quad (3)$$

$$v_i = k[C_i] \quad (4)$$

$$\frac{[C_i]}{[E_0 - C_i]\left[\dfrac{P_0}{n} - C_i\right]} = K \quad (5)$$

$$k'_{\text{obsd}} - k'_{\text{solvent}} = \frac{v_i}{E_0} \quad (6)$$

the initial concentration of the complex expressed as mole/l. of the initial ester groups, v_i is the initial reaction velocity, E_0 and P_0 are the concentrations in mole/l. of the total substrate and catalyst introduced into the solution, and n is the number of imidazole units per ester group in the initial complex. The lines in Figures 5 and 6 were constructed on the assumption that K is 4.2×10^4 l./mole, n is 11, and k is 0.137 min.$^{-1}$. The agreement between the experimental points and the calculated curves supports the general picture developed for the catalytic reaction and the choice of values for K, n, and k. Since a single set of parameters suffices to correlate the data in both Figures 5 and 6, it appears that the composition of the complex is essentially the same in these solutions whether the polyester or the poly(N-vinylimidazole) is in excess. The value of n seems quite reasonable in view of the composition of the polymers and the low extent of protonation of the poly(N-vinylimidazole) at pH 7.5.

As catalysts, the enzymes are distinguished by their high efficiency and by their selectivity toward structurally similar substrates. The former characteristic appears to depend upon the concerted action of two or more groups in the enzyme on the labile functional group of the substrate. Numerous studies on synthetic model compounds have clearly demonstrated that high reaction rates may indeed result from the concerted action of catalytically active groups. The latter feature apperas to depend, in part at least, upon the preferential binding of certain substrates to the enzyme. The present study demonstrates that selective catalysis may also be achieved in synthetic systems in which reversible substrate–catalyst binding occurs.

Copyright © 1964 by the American Chemical Society

Reprinted from *J. Amer. Chem. Soc.*, **86**, 3884–3885 (Sept. 20, 1964)

Selective Catalysis Involving Reversible Association of a Synthetic Polymeric Catalyst and Substrate[1]

ROBERT L. LETSINGER and IRVIN KLAUS

Sir:

The high degree of substrate specificity exhibited by many enzymes is attributable, in part at least, to a specific binding of the substrate to the enzyme. Attempts to develop homogeneous systems of synthetic substances which exhibit selective catalytic features similar to those characterizing the enzymatic systems have met with only limited success. The best examples appear to be the solvolysis of substituted nitrophenyl acetates catalyzed by partially protonated poly(4-vinylpyridine),[2] poly(N-vinylimidazole),[2] and poly(4(5)-vinylimidazole).[3] In these cases substrates bearing a negatively charged substituent group exhibited unusually high reactivity, presumably as a consequence of electrostatic attraction between the substrate and catalyst.[4]

We now report results of an investigation of the solvolysis of copoly(acrylic acid-2,4-dinitrophenyl *p*-vinylbenzoate) (I) catalyzed by partially protonated poly(N-vinylimidazole).[5] This polymer system exhibits a number of features of enzymatic reactions not found with the synthetic substances previously studied[2,3] and provides kinetic evidence for association of the catalyst with the substrate.

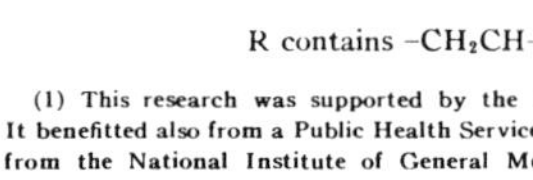

```
      CH2    O
       |    ||
R—CH——⟨ ⟩—C—O—⟨ ⟩—NO2
                  NO2
           I

      CH3    O
       |    ||
CH3—CH——⟨ ⟩—C—O—⟨ ⟩—NO2
                   NO2
            II
```

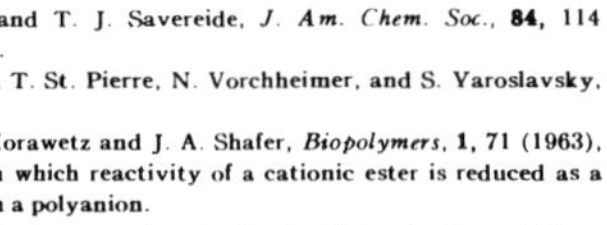

R contains –CH2CH–COOH

(1) This research was supported by the National Science Foundation. It benefitted also from a Public Health Service training grant (5T1 GM 626) from the National Institute of General Medical Science, Public Health Service.
(2) R. L. Letsinger and T. J. Savereide, *J. Am. Chem. Soc.*, **84**, 114 (1962); **84**, 3122 (1962).
(3) C. G. Overberger, T. St. Pierre, N. Vorchheimer, and S. Yaroslavsky, *ibid.*, **85**, 3413 (1963).
(4) In addition, H. Morawetz and J. A. Shafer, *Biopolymers*, **1**, 71 (1963), have described a case in which reactivity of a cationic ester is reduced as a result of association with a polyanion.
(5) Polyvinylimidazole was supplied by the Badische Anilin und Soda Fabrik, Ludwigshafen A. Rhein, Germany.

Copolymer I (η 0.09; 62% by weight dinitrophenyl vinylbenzoate) was prepared by polymerizing acrylic acid and 2,4-dinitrophenyl *p*-vinylbenzoate in benzene. Solvolyses were conducted in 50% (w./w.) ethanol–water at pH 7.5[6] and 25.0 ± 0.1° in a cuvette in a Beckman DU spectrophotometer, formation of 2,4-dinitrophenol being followed by the increase in absorbance at 412 mµ.

Satisfactory pseudo-first-order kinetics were obtained for 2,4-dinitrophenyl *p*-isopropylbenzoate; however, curvature in the plots of log $(A_\infty - A)$ vs. time for the reaction of copolymer I was found.[7] In treating the data for I it was convenient to use the empirical relation[8]

$$(A - A_0)/(A_\infty - A) = k't$$

As shown in Fig. 1, a reasonably good fit to this equation was obtained for solvolyses conducted in the absence of active nucleophilic catalysts as well as in those catalyzed by N-methylimidazole and poly(N-vinylimidazole).

Rate data for a series of reactions in which the catalyst concentration was varied are presented in Fig. 2. With polyvinylimidazole good activity was found when the imidazole unit concentration was low (5 × 10⁻⁴ *M*). Moreover, the rate attained a limiting value at high catalyst concentrations. For experiments in which the concentration of copolymer I was varied, a plot of the rate, $(k' - k'_{\text{solvent}})$(initial concentration of ester groups), as a function of the initial ester concentration afforded a similarly shaped curve (Fig. 3). Both results are explicable on the basis that copolymer I and poly(N-vinylimidazole) associate in solution and that catalysis is highly efficient in the complex that is pro-

(6) This value was selected since clear solutions, free of turbidity, were obtained on mixing the polymeric substrate and catalyst (turbidity developed on mixing at pH 7 or lower) and the pH rate profile exhibited a maximum at this value. All solutions were 0.01 *M* in potassium chloride and 0.02 *M* in 2,4,6-trimethylpyridine. That the spectral change was due to solvolysis was shown by repeatedly scanning the spectrum from 290 to 400 mµ in a Cary spectrophotometer as the reaction proceeded. The final spectrum was that of 2,4-dinitrophenol, and an isosbestic point occurred at 313 mµ.
(7) Observation of complex kinetics in hydrolysis of functional groups on polymer chains is not uncommon. See, for example, J. Moens and G. Smets, *J. Polymer Sci.*, **23**, 931 (1957), and E. Gaetjens and H. Morawetz, *J. Am. Chem. Soc.*, **83**, 1738 (1961).
(8) "*k*" may be considered to be a pseudo-first-order rate constant defined by the expression $-dC/dt = k'C(C/C_0)$, where C/C_0 represents the relative reactivity of the ester groups on a polymer chain as a function of extent of reaction.

336

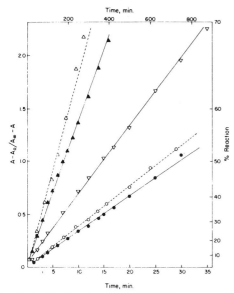

Fig. 1.—Solvolysis of copolymer I (6.14 M in ester groups): $-\bigcirc-$, without catalyst; $-\bullet-$, with 0.02 M N-methylimidazole; $-\triangle-$, with 2.5×10^{-5} M (concentration expressed in imidazole units) poly(N-vinylimidazole); $-\nabla-$, with 5×10^{-4} M poly(N-vinylimidazole); $-\blacktriangle-$, with 0.02 M poly(N-vinylimidazole). Solid lines, read bottom time scale; broken lines, read top time scale.

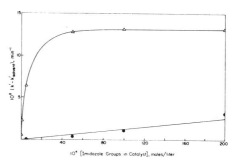

Fig. 2.—Solvolysis of copolymer I (6.14 × 10^{-4} M in ester groups) catalyzed by poly(N-vinylimidazole) ($-\triangle-$), 2.5×10^{-5}—2×10^{-2} M in imidazole groups, and by N-methylimidazole ($-\bullet-$), 5×10^{-4}—2×10^{-2} M.

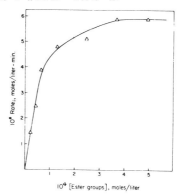

Fig. 3.—Solvolysis of copolymer I, catalyzed by poly(N-vinylimidazole), 5×10^{-4} M.

duced. The polymeric catalyst–substrate system is particularly favorable for demonstrating these phenomena since the individual substrate molecules possess numerous anionic sites that are attracted by the multiple cationic sites on the catalyst.

In support of the concept that reversible association between catalyst and substrate is involved in these reactions it was found that poly(acrylic acid), which competes with the copolymer for binding sites on poly(N-vinylimidazole), functioned as an inhibitor in the catalyzed solvolyses (Table I).

TABLE I[a]

Poly(acrylic acid) (moles/l. of ester groups)	$10^3 k'$, min.$^{-1}$
0	65
5×10^{-4}	28[b]
10×10^{-4}	14[b]

[a] Copolymer I, 6.4×10^{-5} M in ester groups; poly(N-vinylimidazole), 5×10^{-4} M in imidazole units. [b] k' decreased somewhat as the reaction progressed. These values correspond to the first 30% of the reaction.

TABLE II

Catalyst	$10^3 k$, min.$^{-1}$ (dinitrophenyl isopropylbenzoate)	$10^3 k$, min.$^{-1}$ (copolymer I)
None	2.2	1.54
N-Methylimidazole (2×10^{-2} M)	56.5	33
Polyvinylimidazole (5×10^{-4} M in imidazole units)	3.1	65

Data demonstrating selective solvolysis of a pair of substituted dinitrophenyl benzoates are given in Table II. With N-methylimidazole as catalyst 2,4-dinitrophenyl p-isopropylbenzoate solvolyzed somewhat faster than copolymer I. On changing the catalyst to poly(N-vinylimidazole) it was found that copolymer I solvolyzed markedly faster than dinitrophenyl p-isopropylbenzoate. Selectivity results from *acceleration* in the rate of solvolysis of the substrate capable of binding to the catalyst.

(9) Toni-Gillett Fellow, 1961; Lubrizol Foundation Fellow, 1962.

DEPARTMENT OF CHEMISTRY
NORTHWESTERN UNIVERSITY ROBERT L. LETSINGER[9]
EVANSTON, ILLINOIS IRVIN KLAUS

RECEIVED JUNE 25, 1964

Regulation of Rate of Reaction of a Polyuridylic Acid Derivative by Use of Suppressor and Antisuppressor Molecules [1,2]

ROBERT L. LETSINGER and THOMAS E. WAGNER

Sir:

It has been demonstrated that the rate of a reaction involving transformations of functional groups X and Y in RX and R′Y may be unusually high when R and R′ bear ionic charges of opposite sign.[1,3] This phenomenon may be attributed to the molecular organization effected by the attractive forces between R and R′. In the present communication it is shown that the rate of reaction in such systems may be controlled, that is, successively decreased and increased at will, by additions to the reaction medium of substances which are chemically inert in the usual sense yet in extremely low concentrations serve as rate regulators. The principle of controlling reaction rates by means of chemical regulators may prove useful in developing a chemistry of complex molecules which react sequentially in a highly selective manner, such as found in biological systems.

A nitrophenyl ester derivative of polyuridylic acid was selected as RX. In solution at pH 8.5 this polymer is negatively charged, and it possesses ester groups which on reaction with imidazole derivatives afford nitrophenoxide, easily detected by the ultraviolet spectrum. It was prepared by warming polyuridylic acid ammonium salt[4] with excess succinic anhydride and pyridine in formamide for 6 hr at 65°, isolating the succinoylated derivative by chromatography in water on Sephadex G-25 followed by lyophilization, and esterifying the succinate derivative with nitrophenol and dicyclohexylcarbodiimide in formamide and pyridine. The resulting polymer (I), purified by chromatography on Sephadex, contained 0.44 nitrophenyl group per uridine unit as indicated by the ultraviolet spectra before and after alkaline hydrolysis. From the method of preparation and the properties of the substance it is probable that the nitrophenyl succinate

---

(1) Part V in the series on Selective Catalysis. Part IV: R. L. Letsinger and I. S. Klaus, *J. Am. Chem. Soc.*, **87**, 3380 (1965).
(2) This research was supported by the Division of General Medical Sciences, National Institutes of Health, Grant 10265, and by a Predoctoral Public Health Science Fellowship awarded to T. E. W. (1-Fl-GM-25, 197-01).
(3) R. L. Letsinger and T. J. Savereide, *J. Am. Chem. Soc.*, **84**, 114 (1962); **84**, 3122 (1962); C. G. Overberger, T. St. Pierre, N. Vorchheimer, and S. Yaroslavsky, *ibid.*, **85**, 3513 (1963); **87**, 296 (1965).
(4) Miles Chemical Co., Clifton, N. J.

[Structure I shown: nucleotide with phosphate, sugar, and -O-C(O)-CH₂-CH₂-C(O)-O-C₆H₄-NO₂ group, labeled I]

moiety is joined primarily at the 2' position. Poly-(N-vinylimidazole)[5] was chosen for the R'Y component. The net reaction with I in water may be represented by eq 1. The experiments were carried out in 3 ml of 0.067 M aqueous sodium bicarbonate solution,

$$R-C(O)-O-C_6H_4-NO_2 + \text{imidazole-R'} \longrightarrow$$

$$R-C(O)-N^+\text{(imidazole)}-R' + ^-O-C_6H_4-NO_2$$

$$\xrightarrow{H_2O} R-COOH + \text{N-R imidazole} + H^+ \quad (1)$$

formation of nitrophenol being followed by the increase in absorbance at 400 mμ. Rate data are expressed in Table I as "initial first-order rate constants,"

Table I. Reaction of Imidazole Derivatives with p-Nitrophenyl Poly U Succinate (I) at 25.0°, pH 8.5

Imidazole derivative	Concn × 10⁵, moles/l. of imidazole groups	$10^5 \times k^i_{obsd}$ sec⁻¹	$(k^i_{obsd} - k^i_{solvent})/$ [imidazole groups], l. mole⁻¹ sec⁻¹
None	0	1.23	
Histidine	32	1.4	(0.006)
Imidazole	32	3.9	0.08
Imidazole	333	32	0.09
Polyvinylimidazole	3.3	103	30
	10	162	16
	33	270	8
Poly-L-histidine	33	515	...

k^i_{obsd}.[6] A striking feature of the reactions is that polyvinylimidazole at low concentrations (3.3 × 10⁻⁵ M in imidazole units) is about 300 times as effective as an equivalent amount of imidazole in liberating nitrophenol from p-nitrophenyl poly U succinate (I). It is interesting that poly-L-histidine, which forms a gelatinous precipitate in water at pH 8.5, reacts even more rapidly than poly(N-vinylimidazole) with substrate I. This system provides an interesting model for chemical reactions in protein–nucleic acid complexes.

As a suppressor for the reaction of I with poly(N-vinylimidazole), an agent was sought which at low con-

(5) Kindly provided by the Badische Anilin und Soda Fabrik, Ludwigshafen, Germany.
(6) As in the polyvinylimidazole-catalyzed hydrolyses of 2,4-dinitrophenyl p-vinylbenzoate–acrylic acid copolymer,¹ plots of log $(A_\infty - A_0)/(A_\infty - A)$ were curved. k^i_{obsd} represents the initial slope and satisfactorily accommodates the data through the first 30% of reaction. By 62% conversion the slope, $\Delta \log (A_\infty - A_0/A_\infty - A)/\Delta t$ was one-fourth the initial value. No precipitation occurred on mixing or during the reaction of polyvinylimidazole.

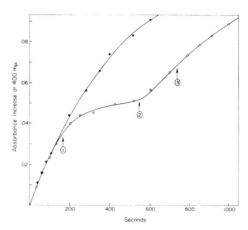

Figure 1. Regulation of rate of reaction of p-nitrophenyl poly U succinate (I) with poly(N-vinylimidazole): —●—control; —O— regulated reaction. (1) 0.2 cc of 2 × 10⁻⁴ M hexadecyltrimethylammonium bromide added to 2.8 cc of reaction mixture; (2) 0.2 cc of 2 × 10⁻⁴ M sodium dodecyl sulfate added; (3) 0.2 cc of 2 × 10⁻⁴ M sodium dodecyl sulfate added.

centration would interact with the polymer–polymer complex but would not precipitate it or cause a change in the covalent bonding. Hexadecyltrimethylammonium bromide proved to be effective. Curves for two reactions are shown in Figure 1. One, the control, was allowed to proceed to completion under the standard conditions. To the other was added at time = 160 sec sufficient hexadecyltrimethylammonium bromide to give a 1.3 × 10⁻⁵ M solution of the detergent.[7] Within 100 sec the rate of formation of nitrophenol fell to 0.12 that of the control. Sodium dodecyl sulfate was found to function as an "antisuppressor" in this reaction. Thus when the solution was subsequently made 1.2 × 10⁻⁵ M in sodium dodecyl sulfate (time = 545 sec), the rate increased to 0.65 that of the control and on further addition of the sulfate (to give a total concentration of 2.3 × 10⁻⁵ M) it was restored to that of the control. Sodium dodecyl sulfate did not significantly influence the reaction of polyvinylimidazole with I in the absence of the ammonium detergent. In another experiment it was found that the cycle could be repeated, that is, the reaction could be retarded, accelerated, retarded, and again accelerated by successive additions of the cationic and anionic detergents (6.8 × 10⁻⁶ M increments). Additional reactions of p-nitrophenyl poly U succinate and the role of the detergent molecules in these reactions will be discussed in the full paper.

(7) This value is below the critical micelle concentration for this detergent in water or salt solutions: see T. Nash, *J. Appl. Chem.* (London), **8**, 440 (1958). In another test it was found that hexadecyltrimethylammonium bromide caused polyuridylic acid to precipitate (the solution developed turbidity) in the absence of poly(N-vinylimidazole); however, no precipitation occurred when poly(N-vinylimidazole) was present in the solution with polyuridylic acid and the detergent.

Department of Chemistry, Northwestern University
Evanston, Illinois
Received February 28, 1966

Bibliography

Polypeptides

Barton, M. A., R. U. Lemieux, and J. Y. Savoie, "Solid-Phase Synthesis of Selectively Protected Peptides for Use as Building Units in the Solid-Phase Synthesis of Large Molecules," *J. Amer. Chem. Soc.* **95,** 4501 (1973).

Bilibin, A. U., and N. U. Kozhevnikova, "Solid-Phase Synthesis of Brady-Kinin Using Sephadex LH 20 as a Support," *Zh. Obshch. Khim.* **43,** 2046 (1973).

Garden, J., Jr., and A. M. Tometsko, "Fluorometric Method for Quantitative Determination of Free Amine Groups in Peptide-Containing Merrifield Resins," *Anal. Biochem.* **52,** 377 (1972).

Lösse, A., "Use of Macroporous Resins in Solid-Phase Peptide Synthesis," *Tetrahedron* **29,** 1203 (1973).

Monahan, M. W., and C. Gilon, "A Rapid Method for the Preparation of Amino Acid Resin Esters for Merrifield Solid-Phase Peptide Synthesis," *Biopolymers* **12,** 2513 (1973).

Mutter, M., H. Hagenmaier, and E. Bayer, "New Method of Polypetide Synthesis," *Angew. Chem. Intern. Ed.* **10,** 811 (1971).

Rivaille, P., and G. Milhaud, "Solid-Phase Synthesis of Peptides and Preparation of Luteinizing Hormone-Releasing Hormone Decapeptides," *Compt. Rend. Acad. Sci.* **C227,** 343 (1973).

Tometsko, A. M., "Determining the Availability of Activated Amino Acids for Solid Phase Peptide Synthesis," *Biochem. Biophys. Res. Comm.* **50,** 886 (1973).

Wünsch, E., "Synthesis of Naturally Occurring Polypeptides, Problems of Current Research," *Angew. Chem. Intern. Ed.* **10,** 786 (1971).

Yamashiro, D., and C. H. Li, "Adrenocorticotropins. 44. Total Synthesis of the Human Hormone by the Solid-Phase Method," *J. Amer. Chem. Soc.* **95,** 1310 (1973).

Oligonucleotides

Köster, H., "Polymer Support Oligonucleotide Synthesis. VI. Use of Inorganic Carriers," *Tetrahedron Letters,* 1527 (1972).

Köster, H., "Polymer Support Oligonucleotide Synthesis. VIII. Use of Polyethyleneglycol," *Tetrahedron Letters,* 1535 (1972).

Köster, H., and K. Heyens, "Polymer Support Oligonucleotide Synthesis. VII. Use of Sephadex LH 20," *Tetrahedron Letters,* 1531 (1972).

Oligosaccharides

Kawana, M., and S. Emoto, "Asymmetric Synthesis with Sugar Derivatives. V. The Synthesis of α-Hydroxy Acids on Insoluble Polymer Supports," *Bull. Chem. Soc. Japan* **46,** 160 (1974).

Bibliography

Polymer-Bound Reagents and Catalysts

Immobilized Reagents

Gorecki, M., and A. Patchornik, "Polymer-Bound Dihydrolipoic Acid: A New Insoluble Reducing Agent for Disulfides," *Biochim. Biophys. Acta* **303**, 36 (1973).

Heitz, W., and R. Michels, "Polymeric Wittig Reagents," *Angew. Chem. Intern. Ed.* **11,**298(1972).

Leznoff, C.C., "The Use of Insoluble Polymer Supports in Organic Chemical Synthesis," *Chem. Soc. Rev.,* 65 (1974).

Leznoff, C. C., and J. Y. Wong, "The Use of Polymer Supports in Organic Synthesis. The Synthesis of Monotrityl Ethers of Symmetrical Diols," *Canad. J. Chem.* **50**, 2892 (1972); **51**, 3756 (1973).

Polymer Protected Chemicals

Beasley, M. L., and R. L. Collins, "Water-Degradable Polymers for Controlled Release of Herbicides and Other Agents," *Science,* **169,** 769 (1970).

Catalysts

Collman, J. P., and C. A. Reed, "Synthesis of Ferrous Porphyrin Complexes. A Hypothetical Model of Deoxymyoglobin." *J. Amer. Chem. Soc.* **95,** 2048 (1973).

Moffat, A., "The Use of Solid Polymeric Ligands in a New Oxo Catalysts Recovery and Recycle System," *J. Catal.,* **19**, 322 (1970).

Overburger, C. G., and K. N. Sameness, "Polymer Supports and Organic Synthesis," *Angew. Chem. Intern. Ed.* **13,** 99 (1974).

Pittman, C. U., and G. O. Evans, "Polymer Bound Catalysts and Reagents," *Chem. Tech.,* 560 (1973).

Rony, P. R., "Diffusion Kinetics and Supported Liquid-Phase Catalysts,"*J. Catal.,* **19,** 142 (1968).

Enzymes

Many, many enzymes have been bound to polymer supports. No examples are given in this volume, but the interested reader is referred to the following general reviews:

Mosbach, K., "Enzymes Bound to Artifical Matrixes," *Sci. Amer.* **224**, 26 (1971).

Starks, G., ed., *Biochemical Aspects of Reactions on Solid Supports,* Academic Press, Inc., New York, 1971.

Wingard, L. B., ed., *Enzyme Engineering,* John Wiley & Sons, Inc. (Interscience Division), New York, 1972.

Zaborsky, O., *Immobilized Enzymes,* CRC Press, Cleveland, 1973.

Author Citation Index

Abdulaev, N. D., 80
Abe, A., 244
Abel, E. W., 113
Abita, J. P., 140
Adams, W. R., 299
Adrova, N. A., 243
Agarwal, K. L., 213, 214
Agtarap, A., 80
Ailhaud, G. P., 140
Akabori, S., 101
Akiyama, M., 283
Aksnes, G., 221
Alberts, A. W., 140
Albertson, N. F., 130
Albonico, S. M., 276
Aldanova, N. A., 80
Alfrey, T., 245
Allen, C. F. H., 308
Allende, J. E., 43
Allewell, N. M., 43, 58
Alul, H., 283
Amit, B., 117, 253
Amos, J. L., 17
Anderson, G. W., 101, 130
Anderson, L., 223, 253
Andreoli, T. E., 80
Anfinsen, C. B., 42, 43, 58, 86, 90, 96, 140, 147, 158, 165, 214
Angyal, S., 222
Antonov, V. K., 80
Arakawa, K., 65
Arkhipova, S. F., 80
Arnon, R., 67
Ascoli, F., 323
Ashby, E. C., 220
Audrieth, L. F., 158
Auer, T., 241
Augustine, R. L., 284
Avitabile, G., 80
Ayres, J. T., 218

Baas, J. M. A., 147
Babievskii, K. K., 308
Bailey, J. L., 42, 68
Bailey, R. W., 254
Baldridge, H. D., Jr., 320
Bar, D., 308
Bardakos, V., 43
Barker, W. E., 308
Barltrop, J. A., 117
Barnard, E. A., 42
Barrollier, J., 42
Barry, G. T., 13
Barstow, L. E., 170
Baugh, G. M., 310
Bartlett, P. D., 300
Barton, M. A., 341
Basolo, F., 288
Bath, R. J., 163
Bator, B., 158
Battersby, A., 65
Baxter, J. W. M., 320
Bayer, E., 113, 116, 118, 120, 140, 144, 145, 147, 150, 153, 160, 217, 219, 341
Bayless, P. L., 305
Bažant, V., 113, 292
Beasley, M. L., 342
Beck, J., 81
Becker, B., 156
Becker, J., 246
Begg, R. W., 80
Bell, L. G., 285
Bell, R. M., 140
Bellamy, L. J., 306
Bello, J., 42, 58
Belorizky, N., 252
Bender, M. L., 157, 323, 324
Benesek, W. F., 42
Ben-Ishai, D., 12, 67
Benjamini, E., 130
Bercaw, J. E., 285
Berger, A., 12, 42, 67
Bergmann, M., 102
Bernardi, L., 93
Bevan, K., 80
Beyerman, H. C., 96, 140, 147, 150, 158, 320

Author Citation Index

Bhanot, O. S., 309
Bhatia, P. L., 65
Biggs, B. S., 320
Bilibin, A. U., 341
Birge, C. H., 140
Birr, C., 91, 109, 147
Birrell, P., 119
Bishop, W. S., 320
Blackburn, G. M., 214
Blaha, K., 82
Blair, J., 194
Bloomfield, J. J., 309, 320
Blotny, G., 158
Blout, E. R., 79
Blumenfeld, O. O., 65
Boaz, H. E., 81
Bochkov, A. F., 243
Bodanszky, M., 12, 42, 65, 79, 82, 96, 109, 130, 155, 163, 310
Bohlmann, F., 241
Bohrer, J. J., 245
Boissonnas, R. A., 11, 71, 72, 97
Bondi, E., 305, 320
Booth, B. L., 288
Borin, G., 43
Bork, J. F., 213
Bosisio, G., 93
Bossinger, C. D., 164
Botre, C., 323
Bourne, E. J., 299, 320
Bowman, R. L., 154
Boyes, A. G., 324
Bradshaw, R. A., 161
Brandenburg, D., 140
Braun, D., 202, 264, 266
Braun-Menendez, E., 65
Bray, G. A., 140
Bredereck, H., 241
Breitmeier, E., 113
Bremer, H., 140
Brenner, M., 158
Breusch, F. L., 308
Bircas, E., 65
Brinkhoff, O., 140
Brintzinger, H. H., 285, 286
Brockmann, H., 80
Bromley, D., 264
Brooke, G. S., 91, 279
Brown, J. F., Jr., 157
Brown, M. J., 214
Bruckner, Z., 241
Brunfeldt, K., 140, 148
Brüning, W., 113, 116, 120, 147, 217
Buchanan, D. L., 171

Büchi, H., 194
Buka, M., 109
Bumpus, F. M., 11, 65, 96
Bundle, D., 252
Bunton, C. A., 279
Burgermeister, W., 80
Burton, H., 308
Bush, M. A., 81
Butterworth, P. H. W., 140
Buyle, R., 158
Buzhinsky, E. P., 80
Bystrov, V. F., 80

Caglioti, L., 322
Callahan, F. M., 101
Calvin, M., 101
Camble, F., 96
Cammarata, P. S., 13, 68
Campbell, J. R., 79
Camps, F., 266, 268, 272, 279
Canfield, R. E., 161
Čapka, M., 285, 292
Caro, G. de, 93
Carpino, L. A., 130
Carroll, W. R., 42, 58
Caruthers, M. H., 217, 242
Cary, P., 171
Cashion, P. J., 213, 214
Cason, J., 322
Cass, A., 80
Castells, J., 266, 268, 272, 279
Castriglione, R. de, 93
Caviezel, M., 81
Cerny, M., 272, 285
Chamberlin, J. W., 80
Chambers, W. J., 310
Chang, M. M., 153
Chang, W. C., 43
Chapman, T. M., 79
Charles-Sigler, R., 118
Chattaway, F. D., 324
Chawla, R. K., 147
Chen, C. C., 43
Chen, P. S., 181
Cheng, L. L., 43
Cherkasov, I. A., 162
Cheronis, N. D., 276
Chew, L., 96
Chi, A. H., 43
Chibnall, A. C., 65
Chillemi, F., 93
Chizhov, O. S., 252
Choate, W. L., 42, 58
Chou, F. C. H., 147

Chow, Y. L., 324
Christensen, J. E., 252
Christensen, J. J., 80
Chu, S. Q., 43
Chung, D., 11, 65
Chvalovský, V., 113, 292
Ciani, S. M., 80
Cleland, W. W., 161
Close, V., 96
Cockrell, R. S., 80
Cohen, S. G., 180, 193
Cole, R. D., 42
Coleman, J., 42
Coleman, L. E., 213
Colescott, R. L., 164
Collins, R. L., 342
Collman, J. P., 285, 287, 320, 342
Colonge, J., 86
Conger, R. P., 283
Conn, J. B., 43
Cook, D. I., 164
Cook, P., 80, 81
Cooke, J., 42
Cooke, M. P., 320
Corcoran, B. J., 171
Corey, E. J., 276
Corigliano, M. A., 89, 94
Corley, M., 86
Cornelius, D., 170
Cotlove, E., 154
Coulter, K. E., 17
Craig, L. C., 13
Cramer, F., 192, 195, 202, 213, 242
Creech, B. G., 256
Cresenzi, V., 323
Crestfield, A. M., 42, 58
Criddle, R. S., 140
Cristol, S. J., 309
Crowley, J. I., 109, 264, 268, 287, 320
Csürös, Z., 222
Curtius, T., 96
Czuros, Z., 247

Dale, J., 158
Danho, W., 43
Davies, D. R., 283
Davies, J. S., 80
Davis, W., Jr., 43
Dawkins, P., 80
Deak, G., 247
Deatberage, F. E., 323
Deér, A., 96
Dejter-Juszynski, M., 241
Denkewalter, R. G., 42, 43, 140

Destefano, J. J., 113
Dewald, H. A., 71
Dheer, S. K., 214
Diebler, H., 80
Dietrich, B., 80
Dietrich, H., 97
Dobler, M., 81
Dohner, R. E., 80
Dolcetti, G., 320
Dominguez, J., 81
Dorer, F. E., 171
Dorman, L. C., 109, 147, 148
Dorsey, J., 140
Doty, P., 323, 328
Drees, F., 97
Du, Y. C., 43
Dubois, M., 254
Dubos, R. J., 43, 58
Dunitz, J. D., 81
Dunlap, D. E., 300
Durham, L. J., 322

Eastlake, A., 86
Eby, R., 222
Eckstein, H., 113, 116, 120, 140, 147, 217
Edelhoch, H., 42
Edery, H., 72
Egami, F., 42
Ehrlich, G., 323
Eigen, M., 80
Eisenberg, H., 325
Eisenman, G., 80
Eitel, M. J., 323
Elkins, E., 15
Ellenbroek, B. W. J., 96
Elliott, D. F., 69
Ellis, B. W., 13, 68
Elovson, J., 140
Else, M. J., 288
Emmons, W. D., 320
Emoto, S., 341
Entrikin, J. B., 276
Epstein, C. J., 42, 58, 161
Erb-Debruyne, M., 308
Erspamer, V., 93
Evans, G. O., 342
Evstratov, A. V., 80
Excoffier, G., 252

Faesel, J., 81
Faler, G. R., 299
Fan, Y. L., 244
Farr, A. L., 140
Fasciolo, J. C., 65

Faulstich, H., 80
Feibush, B., 118
Feigina, M. Y., 80
Ferrando, M. J., 266, 268, 279
Ferris, A. F., 320
Fessel, H. H., 144
Feurer, M., 158
Fields, R., 288
Fieser, L. F., 161, 192, 272, 276
Fieser, M., 161, 272, 276
Finkelstein, A., 80
Finlayson, J. S., 162
Finn, F. M., 43
Firth, W. C., 309
Fischer, G., 118
Fischer, W., 151
Fisher, W., 109
Fitzgerald, E. G., 324
Flanigan, E., 91, 109, 305, 320
Fleckenstein, P., 109
Fletcher, H. G., Jr., 219, 223, 241, 254
Flowers, H. M., 241
Föhles, J., 96
Font, J., 266, 268, 272, 279
Fontaine, J., 176, 179
Foote, C. S., 299, 300
Foreman, E. L., 243
Foster, A. B., 252
Frankel, M., 272
Frant, M. S., 80
Fréchet, J. M., 222, 241, 252, 253
Fridkin, M., 69, 71, 73, 79, 209, 217, 242, 272, 305, 320
Friedrich, K., 219
Fruchter, R. G., 42, 58
Fruton, J. S., 102, 158
Fuchs, S., 67
Fugitt, C. H., 324
Fukuda, K., 43, 99
Fuoss, R. M., 324

Gabor, G., 158
Gad, G., 324
Gaetjens, E., 330, 336
Gagnaire, D., 252
Gallop, P. M., 65
Ganis, P., 80
Garden, J., Jr., 341
Garner, R., 96
Garson, L. R., 96, 109
Gauhe, F., 324
Gavilov, N. I., 68
Geiger, R., 147
Geissel, D., 241

Gerecz, A., 247
Gerlach, H., 67, 81
Gil-Av, E., 118, 119
Gilham, P. T., 178, 179, 192, 204
Gilles, K. A., 254
Gilon, C., 341
Gish, D. T., 65, 130
Gisin, B. F., 81, 82, 84, 109, 155, 158, 217
Glaeser, A., 93
Glasoe, P. K., 158
Goffredo, O., 93
Gold, D. H., 323
Goldberg, I. H., 198
Goldberger, R. F., 42, 58, 161
Goldman, P., 140
Goldschmid, H. R., 241
Goldstein, H., 324
Gollnick, K., 299
Goodman, J. W., 79
Goodman, L., 252
Goodman, M., 65, 244
Gordon, C. L., 247
Gordon, S., 11, 57
Gorecki, M., 342
Gorman, M., 81
Graff, B. van de, 147
Grahl-Nielsen, O., 42
Grasselli, P., 322
Green, B., 96, 109
Green, G. W., 279, 298
Greene, F. D., 276
Greenstein, J. P., 11, 101
Gregor, H. P., 323
Gregory, J. D., 13
Greig, C. G., 13
Grohmann, K., 153
Gross, E., 42
Gross, H., 219
Gross, P. H., 252
Grubbs, R. H., 272, 285, 287, 292
Guber, J. F. K., 145
Gurd, F., 162
Gut, U., 164
Guthrie, R. D., 252, 253
Gutte, B., 42, 43, 58, 79, 82, 90, 140, 147, 157, 160, 217
Guttmann, S., 11, 71, 72, 97

Haber, E., 42, 43
Hägele, K., 113, 116, 120, 144, 147, 217
Hagenmaier, H., 113, 116, 118, 120, 140, 144, 145, 147, 150, 160, 217, 341
Halász, I., 113, 116, 153
Haley, E. E., 171

Hall, M. J., 80
Halpern, B., 96, 118
Halsman, J. A. R., 145
Halstrøm, H., 96
Halstrøm, J., 140, 148
Hamilton, J. K., 254
Hamilton, S. B., 17
Hammill, R. L., 81
Hancock, W. S., 140, 147, 171
Hanson, A. W., 42, 58
Haq, S., 247
Haraszthy, M., 247
Hardman, K. D., 43, 58
Hargitay, B., 158
Harker, D., 42, 58
Harris, E. J., 80
Harris, M. R., 214
Harrison, I. T., 287, 322
Harrison, S., 287, 322
Hart, H., 293
Hartley, B. S., 81
Hasbrouck, L., 288
Hassall, C. H., 80, 81
Hastings, J. T., 299
Hasty, N. M., 299
Haugland, R. P., 96, 158
Hauser, C. R., 264, 305, 308, 310
Hausmann, W., 15
Hayatsu, H., 178, 192, 194, 202, 213, 217, 242, 298, 320
Haymore, B. L., 80
Haynes, L. J., 245
Hazeldine, R. N., 288
Heck, R. F., 288
Hegedus, L. S., 320
Heitz, W., 342
Helbig, R., 192, 195, 202, 213, 242
Helferich, B., 222, 245, 246
Helmer, O. H., 65
Hemmelmayr, F. V., 324
Henning, H., 219
Herlmann, J., 42
Herold, W., 324
Hess, G. P., 12, 42, 43, 65, 69, 73, 82, 87, 130, 155
Hetfleje, J., 272, 285, 292
Hettler, H., 192, 195, 202, 213, 242
Heyens, K., 341
Higgens, C. E., 81
Hill, R. L., 140, 161
Hill, W. B., 323
Hindriks, H., 96, 140
Hirs, C. H. W., 42, 58, 162
Hirschmann, R., 23, 42, 43, 140

Hiskey, R. G., 43
Hixon, S., 241
Hixson, S. H., 223, 253
Hofer, W., 158
Hofle, G., 94
Hofmann, A., 156
Hofmann, K., 11, 15, 43, 65
Hogenkamp, P. C., 192
Holley, R. W., 66
Hollinden, C. S., 96
Holly, F. W., 42, 140
Honzl, J., 96
Hoppe, J. C., 308
Horikawa, J., 266
Horn, O., 306
Horning, E. C., 256
Horton, D., 252
Horvath, C. G., 145
Houben, J., 213
Hruby, V. J., 170
Hsing, C. Y., 43
Hsu, J. Z., 43
Hu, S. C., 43
Huang, H. T., 68, 158
Huang, W. T., 43
Hubert, A. J., 158
Hudson, B. E., Jr., 264, 305, 308
Hughes, J. B., 252
Hunger, K., 91
Hunzicker, P., 113, 116, 140, 147
Hutton, J. J., 158

Ikehara, M., 214
Ilgenfritz, G., 80
Imai, K., 161
Immergut, E. H., 17
Imoto, M., 213
Inagami, T., 43, 58
Ingold, C. K., 308
Ingram, G., 305
Inoue, K., 120
Inukai, N., 71, 96
Iselin, B., 42, 65, 68, 86, 91, 96, 158
Isemura, T., 161
Ishikawa, T., 219, 223, 241, 254
Issleib, K., 292
Ivanov, V. T., 80
Izatt, R. M., 80
Izumiya, N., 140, 147

Jacob, E. J., 140
Jacob, T. A., 42
Jacob, T. M., 176, 177, 178, 192, 194, 204
Jacobs, P. M., 79

Author Citation Index

Jahr, H., 292
Jakes, R., 81
James, F. C., 262
Jaquenoud, P. A., 11, 71, 97
Jaquet, A., 324
Jarabak, R. R., 65
Jardine, F. H., 287, 292
Jarry, R. L., 43
Jauregui-Adell, J., 161
Jeanloz, R. W., 252, 256
Jencks, W. P., 158
Jenkins, A. D., 252, 253
Jentsch, J., 97
Jerina, D. M., 96, 176, 179, 217, 218, 242, 252, 279, 287, 293, 299
Jernberg, N., 42, 58, 150
Jewell, J. S., 252
Jiang, R. Q., 43
Johannesen, R. B., 247
Johnson, A. W., 268
Johnson, B. J., 79
Johnson, J. J., 140
Johnson, L. N., 43, 58
Johnston, H. W., 192
Jollès, J., 161
Jollès, P., 161
Jones, F. D., 130
Jones, W., 58
Jones, W. C., Jr., 43
Jordan, A. D., Jr., 308
Jordan, P., 80
Jorgensen, E. C., 93, 140
Jorgenson, M. J., 219
Joshua, H., 43
Julia, M., 253
Jung, G., 113, 116, 120, 140, 144, 145, 147, 150, 153, 160

Kabat, B. A., 140
Kagawa, I., 323
Kahn, J. R., 65, 66
Kaiser, E., 164
Kamen, M. D., 153, 160, 164, 170
Kamp, P. R. M. van der, 150
Kanarek, L., 161
Kappeler, H., 11, 65, 99, 130
Karlsson, S., 147, 163, 170
Kartha, G., 42, 58
Kashelikar, D. V., 65, 130
Katchalski, E., 69, 71, 73, 79, 194, 209, 217, 242, 272, 305, 320
Katchalsky, A., 325
Kato, I., 43
Kato, T., 140, 147

Katsoyannis, P. G., 11, 42, 43, 57, 65, 99, 130
Katsuura, K., 323
Kaufman, S., 15
Kawana, M., 341
Kawatani, H., 109
Ke, L. T., 43
Kearns, D. R., 299
Keglević, D., 222, 241
Keller-Schierlein, W., 80, 81
Kende, A. S., 65, 130
Kenner, G. W., 91
Kern, W., 324, 325
Khairallah, P. A., 65
Khalilulina, K. K., 80
Khomenkova, K. K., 243
Khorana, H. G., 164, 176, 177, 178, 179, 180, 192, 193, 194, 195, 198, 202, 203, 204, 205, 209, 213, 214, 217, 242, 298
Khorlin, A. J., 243, 252
Khosla, M. C., 96
Kibler, R. F., 147
Kilbourn, B. T., 81
Kinoshita, M., 213
Kiryushkin, A. A., 96, 194, 243
Kirkland, J. J., 113
Kishida, Y., 42, 72, 88
Kivity, S., 272
Klaus, I. S., 329, 338
Kleinberg, J., 158
Klostermeyer, H., 96, 140
Knorr, E., 222
Knox, J. R., 42, 58
Kochetkov, N. K., 243, 252
Koelliker, V., 276
Koenigs, W., 222
Kolbe, H., 324
Kolobielski, M., 283
König, W. A., 113, 116, 120, 144, 147, 217
Konz, W., 81
Kooistra, D. A., 279, 298
Kopaevich, Y. L., 252
Kopple, K. D., 65
Kornet, M. J., 96, 176, 179, 192, 218, 252, 279, 287, 293, 299
Kössel, H., 194
Köster, H., 213, 214, 341
Kotliar, A. M., 323
Kovács, K., 140, 148
Kozhevnikova, I. V., 96, 194, 243
Kozhevnikova, N. U., 341
Krajewski, J., 81
Kraus, M. A., 264, 268, 293, 307, 320
Kravchenko, N. A., 162
Kroll, L. C., 272, 285, 287, 292

Kruger, C. R., 320
Krumdieck, C. L., 310, 320
Kuehne, M. E., 320
Kun, K. A., 153
Kung, Y. T., 43
Kunioka, E., 285
Kunitz, F. W., 69
Kunitz, M., 42, 43, 58
Kun Sun, K. Z., 144
Kurihara, M., 42, 140, 160
Kurihara, O., 243
Kurtz, J., 70
Kusama, T., 213, 320
Kusch, P., 96

Ladenheim, H., 323
Laemmli, U. K., 140
Laine, I. A., 80
Lampman, G. M., 309
Lande, S., 11
Lansbury, P. T., 65
Lanz, P., 67
Lardy, H., 80
Larrabee, A. R., 140
Laufer, D. A., 79
Laüger, P., 80
Laurell, C. B., 162
Laurell, S., 162
Lawson, W. B., 161
Lazdunski, M., 140
Leaback, D. H., 13
Lee, B., 42, 58
Lee, C. C., 293
Lee, T. C., 93
Leebrick, J. R., 330
Leeman, S. E., 153
Leer, E. W. B. de, 96, 140, 147, 150, 320
Leermakers, P. A., 262
Legzdins, P., 284
Lehn, J. M., 80
Leibbrand, K. A., 283
Leising, E., 214
Leloir, L. F., 65
Lemieux, R. U., 241, 245, 341
Lenard, J., 42, 43, 86, 96
Lentz, K. E., 66
Leone, R. E., 176, 179
Lerch, B., 178, 192, 195
Lergier, W., 67
Lesser, R., 324
Letsinger, R. L., 17, 96, 176, 179, 192, 194, 202, 209, 213, 217, 218, 242, 252, 279, 287, 293, 299, 323, 329, 336, 338
Leung, C. Y., 130

Lev, A. A., 80
Lewalter, J., 91
Lewis, G. P., 72
Leznoff, C. C., 342
Li, C. H., 11, 43, 65, 140, 160, 341
Li, H. S., 43
Li, L., 43
Liener, I. E., 79, 91, 109
Lin, M. C., 58
Lindeberg, G., 147, 163, 170
Lipsky, S. R., 145
Liquori, A. M., 323
Littau, V., 42, 57, 113
Liu, A. K., 161
Liu, T., 11
Lo, T., 11
Lochte, H. L., 309
Lock, G., 219
Loebl, E. M., 323
Loffet, A., 96, 109
Loh, T. P., 43
Lohrmann, R., 192, 194, 209
Lösse, A., 341
Losse, G., 91, 96
Lowry, O. H., 140
Lozier, R., 83, 109, 140, 147, 158, 161
Lüben, G., 81
Lübke, K., 57, 82, 147, 158
Lunkenheimer, W., 96
Lynen, F., 140

Maas, G., 80
McCallum, K. S., 320
McDermott, J. R., 91
McElevain, S. M., 243
McGregor, A. C., 130
McGregor, D. N., 158
McKay, F. C., 130
McKerrow, J. H., 168, 171
McKillop, A., 264
McKinley, S. V., 272
McLeod, D. J., 322
McMullen, A. I., 80
Mahadevan, V., 96, 176, 179, 192, 194, 202, 217, 218, 242, 252, 279, 287, 293, 299
Maier, C. A., 80
Majerus, P. W., 140
Malenkov, G. G., 80
Manassen, J., 292
Mangan, J. L., 65
Mann, C. K., 218
Manning, J. M., 82, 92, 120
Manning, M., 320
Manoury, P., 253

Author Citation Index

Marchiori, F., 43
Marcus, R. W., 266, 268
Marder, H. L., 220
Marglin, A., 42, 109, 140, 147
Mark, H., 245, 323
Markham, R., 180
Markley, L. D., 109, 147
Marlborough, D. I., 79
Marquardt, D. N., 320
Marsh, W. H., 66
Marshall, D. L., 79, 91, 109
Marshall, G. R., 57, 65, 70, 79, 89, 90, 91, 109, 130, 140, 147, 158, 165, 171, 272, 305, 320
Marvel, D. S., 192
Marvich, R. H., 285
Massen van den Brink, W., 150
Masson, G. M. C., 65
Matsueda, R., 109
Mauger, A. B., 155
Mayer, H., 82
Mayer, M. M., 140
Mayers, D. F., 80
Mazur, R. H., 13, 68, 158, 293
Mechlinski, W., 80
Mehlis, B., 109, 151
Meienhofer, J., 11, 82, 90, 140, 158
Melby, L. R., 192, 195, 242
Mele, A., 323
Melnik, E. I., 80
Mendenhall, G. D., 300
Merrifield, R. B., 11, 17, 42, 43, 57, 58, 65, 70, 71, 73, 79, 81, 82, 83, 84, 86, 89, 90, 91, 94, 96, 98, 99, 109, 113, 116, 117, 118, 130, 140, 143, 147, 148, 150, 151, 154, 155, 157, 158, 160, 164, 194, 202, 214, 217, 242, 252, 253, 268, 272, 279, 284, 287, 293, 299, 305, 307, 309, 310, 320
Metz, B., 81
Meyer, C. E., 80
Michels, R., 342
Michelson, A. M., 176, 180
Michniewicz, J. J., 214
Mikhaleva, I. I., 80
Milhaud, G., 341
Milkowski, J. D., 43
Mitsuyasu, N., 140, 147
Miura, Y., 140
Mizoguchi, T., 91
Moens, J., 332, 336
Moffat, A., 342
Moffatt, J. G., 164
Möhle, W., 80
Monagle, J. J., 219

Monahan, M. W., 164, 341
Moon, M. W., 178, 194, 203
Moore, C., 80
Moore, C. E., 155
Moore, S., 13, 17, 42, 43, 58, 82, 84, 87, 92, 120, 140
Moras, D., 81
Morawetz, H., 323, 330, 336
Morioka, S., 214
Morgan, A. R., 241, 245
Moroder, L., 43
Morton, R. B., 80, 81
Mosbach, K., 342
Moser, P., 81
Mueller, P., 80
Mukaiyama, T., 109
Mukherjee, S., 310
Munoz, J. M., 65
Murakami, M., 96
Murray, W. C., 80
Mutter, M., 341

Nace, H. R., 219
Nagamatsu, A., 158
Nagasawa, M., 323
Najjar, V. A., 83, 96, 109
Nakamizo, N., 158
Narang, S. A., 194
Nakaparksin, S., 119
Narang, S. A., 178, 214
Nash, T., 339
Neckers, D. C., 279, 298
Nelson, D. P., 80
Neubert, K., 91
Neurath, H., 15
Newth, F. W., 245
Niall, H. D., 153
Nickless, G., 113
Nicolaides, E. D., 71
Nicot-Gutton, C., 65
Niedrich, H., 151
Niemann, C., 68, 158
Nikolenko, L. N., 308
Nitechi, D. E., 79
Niu, C. I., 43
Nobuhara, Y., 101
Noda, K., 140, 147
Nokano, K., 96
Nool, W., 113
Norris, J. F., 246
Norton, J. R., 320
Noubert, K., 96
Nulty, W. L., 140, 147

Nurok, D., 118
Nutt, R. F., 42

O'Conner, C., 284
Oda, R., 264, 266
Ogihara, Y., 81
Ogilvie, K. K., 192, 209, 213
Oguzer, M., 308
Ohle, H., 246
Ohly, K. W., 67
Ohnishi, M., 80
Ohno, M., 86, 96, 165
Ohtsuka, E., 178, 194, 203, 214
Oikawa, T. G., 192
Okada, M., 42, 72, 88, 140
Okawara, M., 283
Okuda, T., 140
Olofson, R. A., 82
Omenn, G. S., 158
Ondetti, M. A., 79
Ontjes, D. A., 86, 140, 147
Orlova, T. I., 68
Oró, J., 119
Osawa, E., 243
Osborn, J. A., 287, 292
Oth, A., 328
Ott, H., 241
Ottenheim, H., 81
Otting, W., 80
Ouchterlony, O., 140
Ovannès, C., 300
Ovchinnikov, Y. A., 80, 96, 194, 243
Overath, P., 140
Overberger, C. G., 329, 336, 338, 342
Owtschinnikow, J. A., 67

Pache, W., 80
Packham, D. I., 67, 73
Page, I. H., 11, 65
Page, J., 42, 58
Paisley, H. M., 12, 82, 109, 155, 284, 285
Parmentier, J. H., 150
Parr, W., 113, 116, 120, 140, 147, 153, 217
Pataki, G., 84
Patchornik, A., 69, 70, 71, 73, 79, 117, 161, 209, 217, 242, 253, 254, 264, 268, 272, 293, 305, 307, 320, 342
Patel, R. P., 158
Patton, W., 96
Paul, I. C., 80
Paul, R., 65, 130
Pauschmann, H., 144
Payne, J. W., 81
Peck, R., 155

Pedersen, C. J., 81
Penke, B., 109
Pepper, K. W., 12, 82, 109, 155, 284, 285
Perlin, A. S., 241
Perry, S. G., 279
Pettee, J. M., 91, 279
Pettit, G. R., 117
Peyton, M. P., 162
Pfäffli, P. J., 223, 253
Philipp, M., 213
Phillips, D. A. S., 80, 81
Phillips, K. D., 252
Phocas, I., 43
Photaki, I., 43
Pichat, L., 86
Pierre, T. St., 329, 336, 338
Pietta, P. G., 171
Pines, H., 283
Pinkerton, M., 80
Pioda, L. A. R., 80, 81
Pisano, J. J., 162
Pittman, A. G., 309
Pittman, C. U., 342
Plant, P. J., 117
Plattner, P. A., 81
Pless, J., 71
Pol, E. H., 180
Pollard, Z. H., 113
Popov, E. M., 80
Porath, J., 147, 170
Porter, J. W., 140
Potts, J. T., Jr., 42, 43, 153
Pourchot, L. M., 140
Pracejus, H., 292
Pravdić, N., 222, 241
Preiss, B. A., 145
Prelog, V., 67, 81
Prescott, D. J., 140, 147, 171
Pressman, B. C., 80, 82
Primosigh, J., 324
Privalova, I. M., 252
Prother, J., 220
Prox, A., 81, 144
Pugh, E. L., 140

Quitt, P., 81

Raabo, E., 254
Rabinowitz, R., 266, 268
Ragnarsson, U., 96, 147, 163, 170
Rakshys, J. W., 272
Ralph, R. K., 176
Ramachandran, G. N., 158
Ramachandran, J., 11

Author Citation Index

Rammler, D. H., 198
Ramsden, H. E., 330
Randall, R. J., 140
Rapoport, H., 109, 264, 268, 287, 320
Razzell, W. E., 192, 204
Rebers, P. A., 254
Redfield, R. R., 42, 58
Reed, C. A., 342
Reed, S. F., Jr., 283
Rees, M. W., 65
Reiher, M., 101
Rempel, G. L., 284
Ressler, C., 11, 57, 65, 130
Reusser, F., 80
Rexroth, E., 73
Richards, F. M., 42, 43, 58
Riehm, J. P., 43
Riniker, B., 65
Rittel, W., 65, 68
Ritter, E., 140
Rivaille, P., 341
Roberts, C. W., 11, 57
Roberts, J. D., 283, 293
Robinson, A. B., 42, 83, 86, 96, 109, 140, 147, 153, 158, 160, 161, 164, 166, 170, 171
Robinson, A. G., III, 308
Robinson, J. C., 65
Rocchi, R., 43
Rochow, E. G., 66, 130, 320
Roepstorff, P., 148
Roeske, R. W., 99, 101
Roncari, D. A. K., 140
Rony, J., 292
Rony, P. R., 342
Rosebrough, N. J., 140
Ross, J. W., Jr., 80
Rössner, E., 213
Rothe, M., 69, 96, 158
Royals, E. E., 308
Rudin, D. O., 80
Rudinger, J., 82, 96, 109, 164
Ruettinger, T. A., 79
Rühlmann, K., 322
Ruhnke, J., 241
Rupley, J. A., 42, 170
Russell, D. W., 69
Rhind-Tutt, A. J., 225
Ryabova, I. D., 80
Rylander, P. N., 288
Rytting, J. H., 80

Sakakibara, S., 42, 71, 72, 88, 93, 101, 140, 165
Sameness, K. N., 342

Sandri, J. M., 293
Sano, S., 42, 140, 153, 160
Sano, Y., 158
Sasisekharan, V., 158
Sato, S., 140
Sauer, F., 140
Sauvage, J. P., 80
Savereide, T. J., 323, 329, 336, 338
Savoie, J. Y., 341
Sawyer, W. H., 320
Saxena, V. P., 161
Schaaf, E., 283
Schaap, A. P., 299, 300
Schadel, H., 241
Schaefer, J. P., 309, 320
Schaffner, C. P., 80
Schallenberg, E. E., 101
Schaller, H., 178, 192, 195, 203
Schally, A. V., 43
Scheit, K. H., 192, 195, 202, 213
Schenck, G. O., 299, 300
Scheraga, H. A., 42, 43
Schexnayder, D. A., 176, 179
Schill, G., 322
Schilling, R. von, 310
Schlatter, J. M., 158
Schlosser, M., 268
Schlossman, F., 42
Schmid, J., 81
Schmidt, O. T., 241
Schmidt-Kastner, G., 80
Schnabel, E., 11, 65, 83, 120, 140, 145
Schneider, H., 96
Schofield, P., 117
Scholer, R. P., 80
Schollkopf, U., 266
Schöniger, W., 74
Schott, H., 213
Schramm, G., 324
Schräpler, U., 322
Schreiber, J., 109
Schröder, E., 57, 82, 158
Schuerch, C., 220, 222, 241, 252, 253
Schulz, R. C., 272
Schumann, W., 283
Schuttenberg, H., 272
Schwarz, H., 11, 65
Schwyzer, R., 11, 65, 67, 68, 81, 83, 97, 99, 130, 158
Scoffone, E., 43
Scotchler, J., 83, 109, 140, 147, 158, 161
Sebastian, I., 113, 116, 153
Seelig, E., 202, 264
Seita, T., 213

Sela, M., 67
Seliger, H., 192, 195, 202, 213, 242
Semanov, D. A., 293
Sen Gupta, A. K., 266
Senyavina, L. B., 80
Shaaf, T. K., 276
Shafer, J. A., 336
Shapira, R., 147
Sharon, N., 256
Sharp, J. J., 153, 160, 161, 170
Shashkova, I. V., 305
Shavit, N., 325
Shaw, N., 252
Sheehan, J. C., 12, 42, 65, 69, 73, 82, 87, 155, 158
Sheehan, J. T., 82, 96, 109, 130, 155, 310
Sheit, K. H., 242
Shemyakin, M. M., 80, 96, 194, 243
Shen, C.-M., 276, 279
Sheppard, R. C., 91, 147
Sherhag, B., 324
Shi, P. T., 43
Shigezane, K., 91
Shimizu, M., 130
Shimoda, T., 170
Shimonishi, Y., 42, 72, 88, 101, 140
Shkrob, A. M., 80
Shul'man, M. L., 252
Shumway, N. P., 65, 66
Sieber, P., 42, 65, 67, 68, 86, 91, 96, 130
Sieglitz, A., 306
Sievers, R. E., 140, 147
Silman, I. H., 194
Silver, M. S., 293
Simon, W., 80, 81
Simoni, R. D., 140
Singh, A., 117
Skeggs, L. T., Jr., 65, 66
Sklyarov, L. Y., 305
Skoog, N., 162
Smart, N. A., 66, 96
Smeby, R. R., 65, 96
Smets, G., 332, 336
Smith, F., 254
Smith, H., 164
Smith, M., 198
Smith, R. L., 43
Smyth, D. G., 42, 58, 158
Snyatkova, V. I., 243
Snyder, R. H., 283
Sokolovskaya, T. A., 243
Sokolovsky, M., 70
Söll, D., 194
Sondheimer, E., 66

Southard, G. L., 91, 279
Spackman, D. H., 13, 17, 84, 87, 140
Späth, A., 283
Spitnik, P., 325
Sroka, W., 140
Staab, H. A., 82, 155
Stacey, M., 299, 320
Stadtman, E. R., 140
Stark, G., 80
Stark, G. R., 140
Starks, G., 342
Stedman, R. J., 65, 99, 130
Stefanac, Z., 80
Steffen, K. D., 158
Steglich, W., 94
Stehlicek, J., 252, 253
Stehower, K., 305, 307
Stein, W. H., 13, 17, 42, 43, 58, 84, 87, 140
Steinhardt, J., 324
Steinrauf, L. K., 80
Stewart, F. M. C., 99
Stewart, J. M., 23, 42, 58, 66, 82, 87, 100, 109, 117, 130, 140, 150, 154, 155, 160, 253, 268, 310
Stiteler, C. H., 283
Stoll, A., 156
Strauss, U. P., 324
Strobach, D. R., 192, 195, 242
Stryer, L., 96, 158
Studer, R. O., 67, 81
Stumpf, P. K., 140
Sugihara, H., 42, 72, 140
Suobda, P., 285
Surbeck-Wegmann, E., 97
Svoboda, P., 272, 292
Swain, C. G., 157
Swan, J. M., 11, 57
Sweet, E. M., 285
Szabo, G., 80

Tai, H. I., 252
Tajima, Y., 285
Takagi, T., 140, 161, 268
Takahashi, K., 42
Takamura, N., 91
Talley, E. A., 247
Tamborski, C., 284
Tanford, C., 140
Tang, K. L., 43
Tanimoto, S., 264, 266
Taniuchi, H., 90
Taschner, E., 158
Tatlow, J. C., 299, 320
Tawney, P. O., 283

Author Citation Index

Taylor, E. C., 264
Taylor, J., 43
Tedder, J. M., 299
Tener, G. M., 176, 179, 180, 193
Tennant, F. M., 17
Terada, J. S., 140, 147
Terkildsen, T. C., 254
Tesser, G. I., 96
Tessmar, K., 246
Theodoropoulos, D., 66
Theysohn, R., 158
Thomas, A. B., 66, 130
Thomas, A. M., 43
Thompson, A., 254
Thompson, R. H. S., 43, 58
Thomsen, J., 148
Thomson, D. F., 305
Thornton, E. R., 117
Thorsteinson, E. M., 288
Tieffenberg, M., 80, 81
Tilak, M. A., 96
Titlestad, K., 158
Tobey, S. W., 283
Todd, A. R., 176, 180
Tokunaga, R., 153
Tomboulian, P., 305, 307
Tometsko, A. M., 43, 99, 341
Toribara, T. Y., 181
Tosteson, D. C., 80, 81, 84, 109. 158, 217
Trantham, H. V., 154
Tregear, G. W., 153
Tritsch, G. L., 42
Truter, M. R., 81
Tsernoglou, D., 42, 58
Tsou, K. C., 279
Tun-Kyi, A., 81
Turner, A. F., 179
Turner, W. N., 252
Turpin, D. G., 308
Turrian, H., 65
Tzschach, A., 292

Uchida, T., 42
Uder, P. C., 113
Ueki, M., 109
Urry, D. W., 80
Utille, J. P., 252
Uvarova, N. N., 80
Uziel, M., 43

Vagelos, P. R., 140, 147, 171
Vaidya, V. M., 79
Vanaman, T. C., 140
Vanden Heuvel, W. J. A., 256

Van Pepper, J., 284
Varma, R. K., 276
Varga, S. L., 42, 43
Veber, D. F., 42, 43, 140
Vela, F., 268, 272
Vernon, C. A., 225
Vigneaud, V. du, 11, 57, 65, 99, 130
Vignon, M., 252
Vinogradova, E. I., 80
Visser, J. P., 43
Vitali, R. A., 42
Vithayathil, P. J., 43
Vizsolyi, J. P., 176, 203, 214
Vogel, A. I., 293
Vogler, K., 67, 81
Voillaume, C., 253
Volosyuk, T. P., 252
Vorchheimer, N., 329, 336, 338
Vorlander, D., 310
Voser, W., 81
Vossen, W. van, 320
Vottero, P., 252
Vries, J. X. de, 81

Wachter, H. A., 80
Wagner, A., 241
Waki, M., 140, 147
Wakil, S. J., 140
Waldschmitt-Leitz, E., 324
Wall, F. T., 323
Wang, C. H., 193
Wang, K., 243
Wang, S. S., 42, 58, 86, 91, 117, 217
Wang, Y., 43, 252
Ward, R. B., 247
Warner, H., 181
Wasielewski, C., 158
Watanabe, H., 109
Watt, M., 246
Watzke, E., 42
Weber, H., 196
Weimann, G., 178, 192, 193, 195, 205
Weinshenker, N. M., 276, 279
Weintraub, J., 140, 147
Weiss, K., 245
Weiss, R., 81
Wells, R. D., 194
Westall, F. C., 140, 147, 166
Westerling, J., 150
Westheimer, F. M., 117
Westley, J. W., 118
Wetlaufer, D. B., 161
Weygand, F., 86, 91, 96, 101, 144, 214, 241
Weyl, T., 213

Whelan, W. J., 247
Whitaker, J. R., 323
White, F. H., Jr., 42, 58
White, W. N., 293
Whitson, J., 192
Whyman, R., 288
Wiberg, K. B., 309
Wieland, T., 67, 80, 81, 91, 109, 147
Wilchek, M., 70
Wilkinson, G., 284, 287, 292
Willecke, K., 140
Williams, A. R., 283
Williams, M. W., 66, 96
Wilson, T., 299, 300
Windridge, G. C., 93, 140
Wingard, L. B., 342
Winitz, M. W., 11, 101
Winkelmann, E., 283
Winkler, R., 80
Wipf, H. K., 80
Wissenbach, H., 147
Wissmann, H., 147
Witkop, B., 42, 161
Wittig, G., 266
Woeller, F. H., 310, 320
Wolfrom, M. L., 254
Wolman, Y., 272
Wong, J. Y., 342
Woodward, R. B., 82, 117
Woolley, D. W., 11, 42, 66, 97, 99, 130
Worrall, R., 320
Wünsch, E., 90, 97, 341
Wyckoff, H. W., 42, 43, 58

Yajima, H., 11, 15, 65, 109
Yamashiro, D., 140, 160, 341

Yamauchi, K., 213
Yamazaki, A., 213, 214
Yanagisawa, Y., 283
Yanaihara, N., 11
Yaron, A., 42
Yaroslavsky, S., 329, 336, 338
Yaroslovsky, C., 272
Yieh, Y. H., 43
Yip, K. F., 279
Yound, D. M., 58
Young, D. M., 42, 43
Young, G. T., 66, 96
Young, J. D., 23, 42, 82, 87, 100, 109, 117, 130, 140, 154, 155, 160, 253, 268, 310
Young, J. F., 287, 292
Young, M. A., 12, 82, 109, 155, 284, 285
Yovanidis, C., 43

Zabel, R., 140
Zaborsky, O., 342
Zagats, R., 109
Zahn, H., 43, 73, 140
Zähner, H., 80
Zalut, C., 99
Zehavi, U., 117, 253, 254
Zemplén, G., 222, 241, 247
Zervas, L., 43, 102
Ziegler, K., 283
Ziemann, H., 241
Zinner, J., 222
Zinnert, F., 324
Zollenkopf, H., 322
Zuber, H., 65
Zurabyan, S. E., 252

Subject Index

2-Acetamido-6-*O*-(2-acetamido-2-deoxy-β-D-glucopyranosyl)-2-deoxy-D-glucose, 249–251
Acetic anhydride, from acetic acid and polymer carbodiimide, 270–271
Alkenes, oxidation of, 300
Alkylation, of (2-hydroxyethyl)-polystyrene-isobutyrate with benzyl chloride, 264
Amino acids, and cesium salts of BOC, 108
Analysis, by direct scintillation counting, 310, 318
Anhydride formation, intraresin, 318
Antisuppressor molecules, 338, 339
Asparagine-(L), and *p*-nitrophenyl ester of *N*-*t*-BOC, 63
L-aspartyl-L-arginyl-L-valyl-L-tyrosyl-L-isoleucyl-L-histidyl-L-prolyl-L-phenylalanine (isoleucine-angiotensin II), 59, 60, 61, 64
L-aspartyl-nitro-L-arginyl-L-valine -*O*-benzyl-L-tryosyl-L-isoleucyl-imbensyl-L-histidyl-L-prolyl-L-phenylalanyl resin, 63, 64
Association, reversible, between synthetic polymer catalysis and substrate, 336, 337
Automated synthesis, 7–8, 25–26, 48, 131–140
  apparatus, 121–130

Carbonyl complexes, with polymer-bound transition metal ions, 287, 288
Catalytic reduction of alkenes, on polymer-supported rhodium catalyst, 284
Chemical regulators, polymeric, 338, 339
Condensation reactions
  fragment, 102
  stepwise, 102
Cyclic peptides, 67–68, 76–78
Cyclization reactions, 309–320

Deoxycytidylyl-(3′–5′)-thymidine, synthesis of, 176–177

Dieckmann cyclization, 309–310, 312–313, 315–317
Disaccharides, synthesis of, 228–241

Enzyme models, 329–337
Ester condensations, mixed, 307–308
Esterification
  of chloromethylated polystyrene with acids, 309–310
  with polymer-bound aluminum chloride, 295–298
Esters, monoacylation of, 305–306
Ether formation, with polymer-bound aluminum chloride, 293–294

Glucose, synthesis of, 254, 257
Glutaric anhydride, synthesis of, from glutaric acid with polymer carbodiimide, 271

Halogenation, allylic and benzylic, 280–283
Histidine, N^{α}-*t*-butyloxycarbonyl, 25
Hooplane, 321–322
Hydroformylation of alkenes, on polymer-supported rhodium complexes, 289, 292
Hydrogenation
  of alkenes, on polymer-supported rhodium complexes, 289–291
  of arenes, with polymer-bound rhodium carbonyl complexes, 288
Hydrosilylation of alkenes, on polymer-supported rhodium complexes, 290–292
Hyperentropic effects, 310–312

Immobilization, on a polymer, 67, 76, 305, 307
Intrapolymeric reactions, 307–308, 311, 318
Iridium–polymer complexes, 287–288
Isomaltose, synthesis of, 254, 257

Subject Index

Kinetic studies, of polystyrene-amino acid azides, 99–100
Kolbe electrolysis, crossed, 321

L-leucyl-L-alanylglycyl-L-valine, 102
L-leucylglycine, 101
Leucyl-leucyl-valyl-phenylalanine, methyl ester of, 94
Lysyl-phenylalanyl-phenylalanyl-glycyl-leucyl-methionine, amide of, 95

Macrocycles, 321–322
Merrifield, R. Bruce (biography), 4–8
Moffat oxidation
  of aldehydes and ketones, 273–275
  with a polymer carbodiimide, 273–275

Oligonucleotides
  attachment to resins
    by amide bonds, 176, 179
    by ester bonds, 185, 188, 189
  cleavage reactions from polymers, 177, 179, 204, 205, 212, 213, 214
  internucleotide bond formation
    kinetic study of, 199
    selectivity in, 177, 196
  phosphodiester activation, 177
  phosphorylation with β-cyanoethyl phosphate, 177, 179, 189
  removal of O-acetyl group, 197, 214
  by sulfonate ester bonds, 207, 208
  synthesis
    coupling reaction, 185, 188, 190, 211, 214
    removal of N-protecting groups, 196
    use of polymer-bound p-methoxytrityl groups, 194–201, 202–205
Oligosaccharides
  attachment to polymer, 228, 229, 232, 237, 240
  synthesis
    cleavage reactions, 223, 233, 235, 238, 245, 247, 248, 250, 254, 256
    coupling reactions, 231–233, 237, 238–239, 247, 249, 254, 256
    effect of solvents on coupling, 230
    photochemical cleavage from resin, 254, 256
    polymer disaccharide, 221
    steric control by C-6 substituents in glucoside synthesis, 222–227, 234–235

Phosphorylation, of polymer-bound nucleosides, 176
Photooxidations, with polymer-based sensitizers, 299–300
Pimeloyl resin esters, 309–310, 315–316
Polymer-based sensitizers, 299–300
Polymer catalysis
  poly-L-histidine, 339
  polyuridylic acid, 338–339
Polymer-protected reagents, 293, 295–298
Polymer reactions, solvent effects of, 297
Polymer substrates, poly(acrylic acid-p-nitrophenyl p-vinylbenzoate) as, 329–335
Polymer supports, 153
  p-alkoxybenzyl alcohol polystyrene, 90–95
  t-alkyloxycarbonylhydrazide polystyrene, 86–89
  p-aminomethyl polystyrene, 220
  N-bromopolymaleimide, 280–283
  ω-cyanopoclargonyl thio-polystyrene, 313–314
  p-(2-hydroxyethyl)polystyrene, 263–264
  p-iodopolystyrene, 204
  light-sensitive, 253–257
  p-lithiopolystyrene, 204
  macroreticular polystyrene versus 2% poly(styrene-2% divinylbenzene), 285
  methoxytrityl alcohol polystyrene, 204
  methylchloroformylated polystyrene, 96, 100
  modified polypeptide support for oligonucleotide synthesis, 214
  monomethoxytritylchloride-polystyrene, 198, 202–205
  poly(4-hydroxy-3-nitrostyrene-4% divinylbenzene), 67, 69, 71, 73
  poly-L-lysine hydrobromide, 214
  poly(DL-lysine + 3-nitro-L-tyrosine), acetylated, 67
  polymaleimide, 280
  polymer benzoic anhydrides, 277–279
  polymer-supported iridium complexes, 287–288
  polymer-supported rhodium complexes, 284, 287–290
  polypeptides on glass beads, 110–113, 114–116
  poly [p-(1-propen-3-ol-yl)styrene], 217–221, 228, 236
polystyrene, 11
  amino acid azides, 96–101
  bromo, 12

Subject Index

carboxy chloride, 17
carboxylic acid, 17
chloromethylated, 12
diphenylcarboxamido, 17
glass beads (pellicular), 145–146
hydroxymethyl, 17
nitro, 12
"popcorn," 17, 176, 179
polystyrene-aluminum chloride, 293–298
polystyrene-(benzyl-2-acetamido-2-deoxy-4,6-O-benzylidene-α-D-glucopyranoside), 249
poly(styrene-N-bromomaleimide), 283
polystyrene-carbodiimide, 270
polystyrene-cinnamic acid, 220
polystyrene-cinnamyl alcohol, 220–221
polystyrene-4-(6-nitrovanillin), 253, 256
polystyrene-4-(6-nitrovanillol), 256
polystyryl-p-diohenylphosphine, 265, 267
p-polystyryldiphenyl-benzyl-phosphonium chloride, 265
poly(N-vinylimidazole), 324, 327, 329–337, 339
poly(4-vinylpyridine), 323–324, 327–328
poly(N-vinylpyrrolidone-vinyl alcohol), 210–213
thiomethyl polystyrene, 313
titanocene polymer, 285–286
use of poly(3,5-diethylstyrene)-sulfonyl chloride for internucleotide bonds, 206–209
6-O-(p-vinylbenzoyl) derivatives of glucopyranose and copolymers, with styrene, 242–248
Polypeptides
  attachment groups
    p-bromomethyl phenyl silolyl glass beads, 110–113
    N,N'-dicyclohexylcarbodiimide on glass beads, 115
    p-(methylthio) phenyl, 76–79
  attempted synthesis of lysozyme, 160–171
  blocking groups, c-nitrophenylsulfenyl as, 71, 74
  cleavage reactions from resin, 12, 15, 26–27, 50, 61, 76–79, 133, 138–139, 154
  coupling reactions, 25–26, 49–50, 77, 86–89, 91–95, 97–98, 100–101, 103–109, 154
    in automated synthesis, 132–133
    total esterification procedure, 108–109
  failure sequences, 143–154

hydrophobic potassium binding peptide, 80–85
inverse coupling reactions, 69–73, 82–83, 86–89
synthesis
  acyl carrier protein (ACP), 131–140
  (alanyl-phenylalanyl)$_6$, 145
  bradykinnin, 7, 71–75
  cleavage reactions, 162, 164–165, 169, 171
  in columns, 114–116
  coupling reactions, 158–161, 164, 170
  deletion sequence, 147–154
  desamidosecretin, 119–120
  diketopiperazine formation, 157–158
  intramolecular aminolysis reactions, 155–159
  (leucyl-alanine)$_6$, 145
  L-leucyl-L-alanylglycyl-L-valine, 11–16
  leucylglycine, 17
  leucyl-leucyl-glutaconyl-glycine, 116
  lysine, ε-amino group protection, 165
  optical purity, 13, 15, 94, 118–120
  optimum chain length, 144, 145
  D-prolyl-D-valyl-L-proline, 155, 158–159
  protection of sulhydryl groups, 165–166
  quantitative analysis by Dorman procedure, 148, 151, 154
  quantitative analysis by picric acid, 155, 159
  reaction vessel, 14, 122–123
  reversed coupling reaction, 159
  ribonuclease, 7–8
  ribonuclease A, 24–58
  solvents, 12, 166
  tryptophan protection, 165
  truncation, 143–154
  use of cesium salts, 103–107
  valinomycin, 80, 81
Poly(styrene-1% divinylbenzene), chloromethylation of, 25, 82, 83, 100
Polystyrene-Rose Bengal, 299–300
Poly-p-styryldiphenylalkylidene-phosphoranes, 268
Potassium ion complexes, 80–85

Reactions of polymer carbodiimide, with carboxylic acids, 270–271
Reduction of alkenes, with titanocene polymer, 286
Relative reaction rates, of polymer reagents vs. free reagent, 285–286
Resin, N-t-BOC phenylalanine, 63

Subject Index

Resin polypeptides, stability of blocking group of, 89
Rhodium-polymer complexes, 284, 287–290

Saccharide derivatives, copolymerization of, with styrene, 243–248
Selective acyl transfer reactions, 279
Singlet-oxygen reactions, 299, 300
Solvolyses, selective, with polymer catalysis, 323–328
Solvolysis, of mono- and dinitrophenyl esters, 329–335
Stearic anhydride, from stearic acid and polymer carbodiimide, 271
Suppressor molecules, 338, 339
Synthetic catalytic system, 329–335

Thymidylyl-(3′–5′)-thymidine, 208, 209, 212
Thymidylyl-(3′–5′)-thymidylyl-(3′–5′)-thymidine, 191–192, 201, 204, 205, 212
Transesterification, with polymer-bound aluminum chloride, 298
Trityldeoxycytidine, and polystyrene, 176–177
  phosphorylation of, 176
5′-O-trityldeoxycitidylyl-(3′–5′)-thymidylyl-(3′–5′)-thymidylyl-(3′–5′)-thymidine, 182
Trityl protecting groups, for oligionucleotide synthesis on resins, 185, 203, 204

Wittig reactions, 265–268